Recent Progress in

# HORMONE RESEARCH

Proceedings of the Laurentian Hormone Conference

VOLUME 48

# RECENT PROGRESS IN
# HORMONE RESEARCH

*Proceedings of the*
*1991 Laurentian Hormone Conference*

### Edited by
### C. WAYNE BARDIN

### VOLUME 48

PROGRAM COMMITTEE

| | |
|---|---|
| C. W. Bardin | A. R. Means |
| J. Clark | M. New |
| D. K. Granner | D. Orth |
| P. A. Kelly | G. Ringold |
| I. A. Kourides | N. B. Schwartz |
| S. McKnight | W. Vale |

WITHDRAWN
FAIRFIELD UNIVERSITY
LIBRARY

ACADEMIC PRESS, INC.
**Harcourt Brace Jovanovich, Publishers**
San Diego   New York   Boston
London   Sydney   Tokyo   Toronto

This book is printed on acid-free paper. ∞

Copyright © 1993 by ACADEMIC PRESS, INC.

All Rights Reserved.

No part of this publication may be reproduced or transmitted in any form or by any means, electronic or mechanical, including photocopy, recording, or any information storage and retrieval system, without permission in writing from the publisher.

Academic Press, Inc.
1250 Sixth Avenue, San Diego, California 92101-4311

*United Kingdom Edition published by*
Academic Press Limited
24–28 Oval Road, London NW1 7DX

Library of Congress Catalog Number: Med. 47-38
International Standard Book Number: 0-12-571148-4

PRINTED IN THE UNITED STATES OF AMERICA

93 94 95 96 97 98   QW   9 8 7 6 5 4 3 2 1

# CONTENTS

List of Contributors  ........................................................................  ix

Preface  ........................................................................................  xi

1. Anti-Müllerian Hormone: The Jost Factor
   Nathalie Josso, Richard L. Cate, Jean-Yves Picard, Bernard Vigier,
   Nathalie di Clemente, Cheryl Wilson, Sandrine Imbeaud, R. Blake
   Pepinsky, Daniel Guerrier, Laurence Boussin, Laurence Legeai, and
   Danièle Carré-Eusèbe  ..........................................................................  1

2. Expression of the FSH Receptor in the Testis
   Leslie Heckert and Michael D. Griswold  ..............................................  61

3. Differential Gene Expression from a Single Transcription Unit during
   Spermatogenesis
   Anthony R. Means and Francisco Cruzalegui  .........................................  79

4. Retinoid Receptors
   David J. Mangelsdorf, Steven A. Kliewer, Akira Kakizuka, Kazuhiko
   Umesono, and Ronald M. Evans  ........................................................  99

5. The Growth Hormone/Prolactin Receptor Family
   P. A. Kelly, S. Ali, M. Rozakis, L. Goujon, M. Nagano, I. Pellegrini,
   D. Gould, J. Djiane, M. Edery, J. Finidori, and M. C. Postel-Vinay  .......  123

6. Molecular Genetics of Laron-Type GH Insensitivity Syndrome
   Michel Goossens, Serge Amselem, Philippe Duquesnoy, and Marie-Laure
   Sobrier  ..........................................................................................  165

7. Conventional and Nonconventional Uses of Growth Hormone
   Barbara M. Lippe and Jon M. Nakamoto  ............................................  179

8. Site-Directed Mutagenesis in the Mouse
   Allan Bradley  ................................................................................  237

270987

vi                                    CONTENTS

9.  The Molecular Basis for Growth Hormone–Receptor Interactions
    James A. Wells, Brian C. Cunningham, Germaine Fuh, Henry B. Lowman,
    Steven H. Bass, Michael G. Mulkerrin, Mark Ultsch, and Abraham M.
    deVos ........................................................... 253

10. Catecholamine Receptors: Structure, Function, and Regulation
    Marc G. Caron and Robert J. Lefkowitz .............................. 277

11. The Insulin Receptor and Its Substrate: Molecular Determinants of Early
    Events in Insulin Action
    C. Ronald Kahn, Morris F. White, Steven E. Shoelson, Jonathan M.
    Backer, Eiichi Araki, Bentley Cheatham, Peter Csermely, Franco Folli,
    Barry J. Goldstein, Pedro Huertas, Paul L. Rothenberg, Mario J. A. Saad,
    Kenneth Siddle, Xiao-Jian Sun, Peter A. Wilden, Kazunori Yamada, and
    Stacy A. Kahn ..................................................... 291

12. Thyrotropin-Releasing Hormone Receptor: Cloning and Regulation of Its
    Expression
    Marvin C. Gershengorn .............................................. 341

13. Bombesin-like Peptides: Of Ligands and Receptors
    Eliot R. Spindel, Eliezer Giladi, Thomas P. Segerson, and Srinivasa
    Nagalla ........................................................... 365

14. Thyroid Hormone Regulation of Thyrotropin Gene Expression
    William W. Chin, Frances E. Carr, Joan Burnside, and Douglas S.
    Darling ........................................................... 393

15. Prohormone Structure Governs Proteolytic Processing and Sorting in the Golgi
    Complex
    Linda J. Jung, Thane Kreiner, and Richard H. Scheller ............... 415

16. Endocrinology Alfresco: Psychoendocrine Studies of Wild Baboons
    Robert M. Sapolsky ................................................. 437


                            **Short Communications**


    Heterogeneous Secretory Response of Parathyroid Cells
    Lorraine A. Fitzpatrick ............................................ 471

## CONTENTS

Progesterone Inhibits Estrogen-Induced Increases in c-*fos* mRNA Levels in the Uterus
John L. Kirkland, Lata Murthy, George M. Stancel ..................... 477

Genotoxic Damage and Aberrant Proliferation in the Mouse Mammary Epithelial Cells
N. T. Telang, A. Suto, H. Leon Bradlow, G. Y. Wong, and Michael P. Osborne ......................................................... 481

Development of Hypophysiotropic Neuron Abnormalities in GH- and PRL-Deficient Dwarf Mice
Carol J. Phelps and David L. Hurley .................................. 489

Germ Cell Factor(s) Regulates Opioid Gene Expression in Sertoli Cells
Masato Fujisawa, C. W. Bardin, and Patricia L. Morris ................. 497

LHRH- and (Hydroxyproline$^9$)LHRH-Stimulated hCG Secretion from Perifused First-Trimester Placental Cells
W. David Currie, Gillian L. Steele, Basil Ho Yuen, Claude Kordon, J.-Pierre Gautron, and Peter C. K. Leung ............................. 505

Tyrphostins Inhibit Sertoli Cell-Secreted Growth Factor Stimulation of A431 Cell Growth
Dolores J. Lamb and Sankararaman Shubhada ......................... 511

Induction of Calcium Transport into Cultured Rat Sertoli Cells and Liposomes by Follicle-Stimulating Hormone
Patricia Grasso and Leo E. Reichert, Jr. .............................. 517

Pituitary *in Vitro* FSH and LH Secretion after Administration of the Antiprogesterone RU486 *in Vivo*
K. L. Knox and N. B. Schwartz ...................................... 523

Release of Immunoreactive Inhibin from Perifused Rat Ovaries: Effects of Forskolin and Gonadotropins during the Estrous Cycle
Jacqueline F. Ackland, Brigitte G. Mann, and Neena B. Schwartz ........ 531

Identification and Partial Purification of a Germ Cell Factor That Stimulates Transferrin Secretion by Sertoli Cells
C. Pineau, V. Syed, C. W. Bardin, B. Jégou, and C. Y. Cheng ........... 539

Index ................................................................ 543

# LIST OF CONTRIBUTORS

J. F. Ackland
S. Ali
S. Amselem
E. Araki
J. M. Backer
C. W. Bardin
S. H. Bass
L. Boussin
A. Bradley
H. L. Bradlow
J. Burnside
M. G. Caron
F. E. Carr
D. Carré-Eusèbe
R. L. Cate
B. Cheatham
C. Y. Cheng
W. W. Chin
N. di Clemente
F. Cruzalegui
P. Csermely
B. C. Cunningham
W. D. Currie
D. S. Darling
A. M. deVos
J. Djiane
P. Duquesnoy
M. Edery
R. M. Evans
J. Finidori
L. A. Fitzpatrick
F. Folli
G. Fuh
M. Fujisawa
J.-P. Gautron
M. C. Gershengorn
E. Giladi

B. J. Goldstein
M. Goossens
L. Goujon
D. Gould
P. Grasso
M. D. Griswold
D. Guerrier
L. Heckert
P. Huertas
D. L. Hurley
S. Imbeaud
B. Jégou
N. Josso
L. J. Jung
C. R. Kahn
S. A. Kahn
A. Kakizuka
P. A. Kelly
J. L. Kirkland
S. A. Kliewer
K. L. Knox
C. Kordon
T. Kreiner
D. J. Lamb
R. J. Lefkowitz
L. Legeai
P. C. K. Leung
B. M. Lippe
H. B. Lowman
D. J. Mangelsdorf
B. G. Mann
A. R. Means
P. L. Morris
M. G. Mulkerrin
L. Murthy
S. Nagalla
M. Nagano

J. M. Nakamoto
M. P. Osborne
I. Pellegrini
R. B. Pepinsky
C. J. Phelps
J.-Y. Picard
C. Pineau
M. C. Postel-Vinay
L. E. Reichert, Jr.
P. L. Rothenberg
M. Rozakis
M. J. A. Saad
R. M. Sapolsky
R. H. Scheller
N. B. Schwartz
T. P. Segerson
S. E. Shoelson
S. Shubhada
K. Siddle
M.-L. Sobrier
E. R. Spindel
G. M. Stancel
G. L. Steele
X.-J. Sun
A. Suto
V. Syed
N. T. Telang
M. Ultsch
K. Umesono
B. Vigier
J. A. Wells
M. F. White
P. A. Wilden
C. Wilson
G. Y. Wong
K. Yamada
B. H. Yuen

# PREFACE

Endocrine research dealing with all aspects of biological organization, including differentiation of the embryo, growth and development at puberty, maintenance of adult well-being, and aging, is the subject of this volume of *Recent Progress in Hormone Research*. The tools of laboratory and clinical science have led to an enhanced understanding of modern endocrinology, and the approaches used in the work described in each chapter of this book range from molecular regulation of genes to current topics in clinical endocrinology. In the chapters concerning growth, the role of oncogenes in fetal development is explored using gene knock-out experiments. The molecular descriptions of the genes for growth hormone and prolactin receptors are compared to receptors for a variety of other peptides, hormones, and growth factors. This work complements studies of gene defects that cause short stature. The three-dimensional structure of the growth hormone receptor is shown for the first time and it should be noted that this information has recently led to proposed inhibitors for growth hormone. The conventional and nonconventional uses of human growth hormone are reviewed emphasizing that the wide availability of this protein is possible only through molecular biology. Development of the male reproductive tract and germ cells is emphasized with studies of gene transcription during spermatogenesis and the work on anti-Müllerian hormone. Hormone signaling is reviewed, with emphasis on the catecholamine, insulin, TRH, and FSH receptors. New insights into the role of hormones in the modulation of CNS function and the ability of the CNS to regulate hormone secretion are demonstrated in reports on thyrotropin gene expression as well as on neuropeptide processing and packaging. Finally, the effect of hormones on primate behavior is reported in baboons, providing insights into the basis of good and bad manners.

Discussions of each of the major presentations have been omitted. In their place, we have included Short Communications prepared by investigators who submitted posters at the Laurentian Hormone Conference. Many of these communications amplify and expand some of the ideas presented in the major chapters.

The 1991 meeting of the Laurentian Hormone Conference is the basis for this volume of *Recent Progress in Hormone Research*. This meeting was sponsored in part by generous gifts from Genentech; Merck, Sharp and Dohme; and the Population Council.

C. Wayne Bardin

# Anti-Müllerian Hormone: The Jost Factor

NATHALIE JOSSO,[*] RICHARD L. CATE,[†] JEAN-YVES PICARD,[*] BERNARD
VIGIER,[*] NATHALIE DI CLEMENTE,[*] CHERYL WILSON,[†] SANDRINE
IMBEAUD,[*] R. BLAKE PEPINSKY,[†] DANIEL GUERRIER,[*] LAURENCE
BOUSSIN,[*] LAURENCE LEGEAI,[*] AND DANIÈLE CARRÉ-EUSÈBE[*]

[*] Unité de Recherches sur l'Endocrinologie du Développement (INSERM), Ecole
Normale Supérieure, Montronge, France, and [†] Biogen Inc.,
Cambridge, Massachusetts

## I. Introduction

Alfred Jost, the founder of fetal endocrinology, died February 3, 1991.
Death, however, cannot wipe out the impact of great personalities on
their surroundings. Among the many discoveries that earned the French
physiologist international fame, the identification of the mechanisms of
sex differentiation ranks highest. Here is the story, told in Jost's own
words[1]:

> From a biological point of view, it was a challenge that androgens . . . produced only
> a very partial masculinization of the fetus. They did not duplicate the freemartin condition
> known in cattle. It was specially intriguing that one did not obtain the disappearance of
> the Müllerian (female) ducts in treated animals. I decided to try the unorthodox way, and
> to directly investigate the endocrine activity—if any—of the fetal gonads, using surgery
> on the intrauterine fetus. This had not been done previously and the first success was not
> obtained before many hard months, actually one and a half years. It took so long not only
> because it was difficult to obtain enough rabbits for experimental research at that time,[2]
> but also because I had not the faintest experience in refined surgery. . . . It was found
> that whatever the genetic sex of the fetus, the genital tract becomes feminine in the absence
> of gonads. The fetal testis is the sex differentiator: it imposes masculinity against a female
> inherent program of the body. The testis produces a factor which inhibits the Müllerian
> (female) ducts and androgens that masculinize the genital tract. The explanatory scheme
> for human sexual abnormalities, first published in 1950 (Jost, 1950) was later enriched by
> important discoveries made by others, but it remains the basis of present concepts.

In Jost's laboratory, in which two of the present authors (NJ and BV)
received their training, the factor responsible for Müllerian regression was
called "hormone inhibitrice." In his English publications, Jost referred to

[1] Alfred Jost, outline of the research career, 1988.
[2] World War II.

it as the "Müllerian inhibitor," but, in the United States, the factor is usually called Müllerian inhibiting substance (MIS) or factor (MIF). The term "anti-Müllerian hormone" (AMH) was coined by two of the authors (NJ and JYP) with the aim of devising a name that would be similar in both French and English. For consistency's sake, it will be used throughout this article, regardless of the name originally used by the authors. Previous review articles (Josso *et al.*, 1977; Donahoe *et al.*, 1982; Josso and Picard, 1986; Donahoe *et al.*, 1987; Cate *et al.*, 1990) described the purification, cloning, and putative biological effects of AMH. Here, we will focus essentially on our joint efforts to dissect the molecular mechanisms of gene expression and protein bioactivity and on the clinical applications of AMH research.

## II. The AMH Molecule

### A. BIOCHEMICAL STRUCTURE

*1. AMH Purification*

*a. Bovine AMH.* As late as 1972, the only indication concerning the biochemical nature of AMH was a negative one: the testicular factor responsible for Müllerian duct regression was different from testosterone (Jost, 1953). The work of Picon (1969), providing an *in vitro* test for anti-Müllerian activity, was decisive for further progress. The target organ, the sexually undifferentiated rat fetal Müllerian duct, responds to mammalian AMH regardless of species; therefore, bovine testicular tissue was chosen as starting material for AMH purification, because fetal and newborn bovine testes of relatively large size can be conveniently obtained at slaughterhouses. Bovine AMH was shown to be a nondialyzable macromolecule (Josso, 1972; Josso *et al.*, 1975), and evidence was produced in favor of its glycoprotein dimeric nature (Picard *et al.*, 1978). Bovine AMH was finally purified to homogeneity by immunochromatography on a monoclonal antibody (Picard and Josso, 1984). Purified bovine AMH contains 13.5% carbohydrate, with both *O*- and *N*-oligosaccharide linkages, and a high proportion of hydrophobic amino acids (Picard *et al.*, 1986b). Incomplete purification can be achieved using dye affinity chromatography; the fraction containing anti-Müllerian activity has been named fraction Green-III (Budzik *et al.*, 1983).

*b. Human Recombinant AMH.* Recombinant DNA techniques are now preferentially used for AMH production. Chinese hamster ovary (CHO) cells are cotransfected with a pSV2 vector carrying the human AMH gene under the control of a viral promoter and with dihydrofolate reductase recombinant DNA contained in plasmid pAdD26. The transfection efficiency of pAdD26 is enhanced by the SV40 enhancer carried by the pSV2 vector, allowing coamplification of both the dihydrofolate reductase and the AMH genes (Kaufman et al., 1985). Mature AMH is purified from the culture medium by a combination of ion-exchange, lentil–lectin, and immunoaffinity chromatography on a monoclonal antibody (Wallen et al., 1989). Its electrophoretic pattern on polyacrylamide gels in the presence of sodium dodecyl sulfate (SDS–PAGE) is shown in Fig. 1. In addition to the 140K dimer, other higher molecular weight polymers are present. These larger forms were also obtained after pulse labeling with radioactive cysteine followed by direct analysis by SDS–PAGE, indicating that they were not generated by the purification procedure (Pepinsky et al., 1988). Following cleavage of disulfide bonds by a reducing agent, the 140K recombinant dimer dissociates into 70K

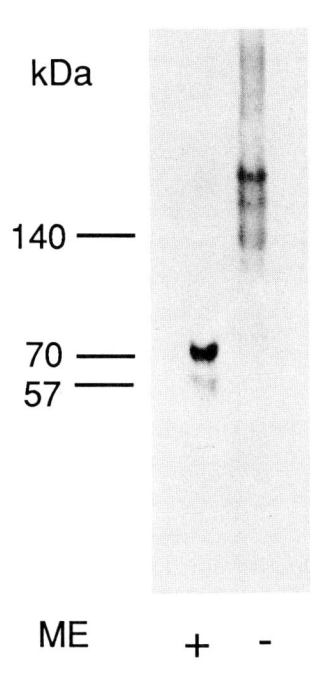

FIG. 1. Western blot (Mab 10.6) of recombinant human AMH, electrophoresed on a 4–30% gradient polyacrylamide gel in the presence of SDS. Under nonreducing conditions (no mercaptoethanol) AMH migrates as a 140K dimer and aggregates of higher molecular weight are visible. Under reducing conditions (mercaptoethanol present), the hormone dissociates into 70K subunits. The 57K band below it represents the N-terminal proteolytic fragment; the small C-terminus has migrated off the gel. ME, Mercaptoethanol.

```
                                                                              (*)
H  Mrdlptsla Lvlsalgall gtealraeep av------  -------gts GLIFEDlcW PpgiipqePLC LValggdsWg SsspLRVVga LsaYEqAFLg
B  MpGpsLS-La LvLsaMGAlL rpgtpqeevf stsalpreqa tqsgalifqa awdmplssW LpgsPLgLC LvtLngsGNg sRApLRVVGv LsSYEqAFLE
M  MqGpPhLSpLv LLLatMGAvL qpea------ -venlatnIr GLIFEDElW PpssPpEPLC LVetrgsGNg SrAsLRVVGg LnsYEpAFLE
R  MqGpPhLSlLl LLLatMGAvL qadtvee---  ---------- ----ltnIr GLIFLEDgvW PpssPpEPLC LVavrGeGdt ShsYEhAFLE

H  AVGraRKGPr DLATFGVCnt GDrdaaLPsL rRLGAWLrdp GgQrLVLHL eEVtWePtps LrFGEPPPGG AgppEIALLV LYPGPGPeVT VTAGLpGaQ
B  AVrrtHmGls DLCTFaVCpa gnGGpvLPhL QRLGAWLGEP GgrwLVVLHL eEVtWGEPtpL LrFGEPPPGG AsppEIALLV VYPGPGleVT VTGAGLpGTq
M  AVQesRKGPq DLATFGV-ss tDSqattPaL QRLGAWLGET GeGqLVLHL aEVIWEPall LKFGEPPPGG ASrwEqALLV LYSGPGPqVT VTGtGLrGTQ
R  AVQesRKGPq DLATFGVC-s tDSqttLPaL QRLGAWLGEt GeGqLVLHL aEVIWEqall LKFGEPPPGG ASrwEqALLV LYSGPGPqVT VTGAGLqGTQ

H  SLCPsRDTRY LVLaVDrPAG AWrgsGLALT LqPrgEdsrL StArLGALLF GqDDRCFTRM TPa-LLLLPr LPLLLLPT AATtGdPapL HdPtSAPWAt aLARRVAaeL
B  SLCITaDsdf LaLvVDhPeG AWrrpGLALT LrrrgnGALL GaDsRCFTRk TPALLLLP- arssaPmPAH QgPKSpLWAA QGARRVAAeL
M  nLCPTRDTRY LVLtVDfPAG AWsGfGLILT LQPsrEGATL GqDsRCFTRM TP-tLLvLPp ae-PsPqPAH GqLDTmFPPq PgLsIEPEaL
R  SLCPTRDTRY LVLtVhfPAG AWsGsGLALT tIAQLGAFLF GsDsRCFTRM TP-tLvLLPp -tgPTPqPAH GOLDTVFPPq PgLsIEPEdL

H  PpsADPFLET LTRLVRALrv PpaRASapRL ALDPdLLAgF PqGLVNLSDP AALeRLLDGE EPLLLLLrPT AATtGdPapL HdPtSAPWAt aLARRVAaeL
B  PpsADPFLET LTRLVRALaG PpaRASppRL ALDPGALAgF PqGGVNLSDP AALeRLLDGE EPLLLLLPPT AATtGvPatp QGPKSpLWAA GLARRVAAeL
M  PhSADPFLET LTRLVRALrG PtcqAsntqL ALDPGALAsF PqGLVNLSDP AALgRLLDME EPLLLLLSpT AAtLerePIrL HGPaSAPWAA GLqRRVAVeL
R  PhSADPFLET LTRLVRALrG PtcqAsntrL ALDPGALAsF PqGLVNLSDP VaLgRLLDGE EPLLLLLSPa AATvGePmrL HsPtSAPWAA GLARRVAVEL

                                               N-term. ↓ C-term.  * domain
H  QAAAeELrsL PGLPPataPL LARLLALCPg gpPsglGDPLR ALLLLKALqG LRvEwRGRcdp rGpgRAORsa GataeDGPCA LRELSVDLRA ERsVLiPETy
B  QAVAaEELRaL PGLPPaAPPL LARLLALCPg npdSPGpGPlR ALLLLKALqG LRAEWGRGeEN rGrtRAORSa sGSeaRAORSa GeaaaDGPCA LRELSVDLRA ERsVLiPETY
M  QAAAsEELRdL PGLPPtAPPL LARLLALCPn dSrSsGDPLR ALLLLKALqG LRAEWGREG rGrtRAOR-- GdkgqGpGPCA LRELSVDLRA ERsVLiPETY
R  QAAAaEELRdL PGLPPtAPPL LSRLLALCPn dsrSaGDPLR LRAEWGRE- GrgRAgRSK G-tgtDGICA LRELSVDLRA ERsVLiPETY

        *                                                              *                          * *
H  QANNCqGvCg WPqSDRNPRY GNhVLLLKm QARGAALCPg PCCVPTAYaG KLLISLSEER ISAhHVPNWV ATECGCR
B  QANNCqGACg WPQSDRNPRY GNhVLLLKM QARGAtLaRp PCCVPTAYG KLLISLSEER ISAhHVPNWV ATECGCR
M  QANNCqGACr WPQSDRNPRY GNhVVLLLKM QARGATLaRp PCCVPTAYG KLLISLSEER ISAdhHVPNWV ATECGCR
R  QANNCqGACa WPQSDRNPRY GNhVVLLLKM QARGAALqRl PCCVPTAYG KLLISLSEER ISAhHVPNWV ATECGCR
```

ANTI-MÜLLERIAN HORMONE 5

subunits. A minor species of 57K, also observed in native bovine AMH (Picard and Josso, 1984), is generated by posttranslational processing, as discussed below (Section II,B,2). Chicken AMH has also been isolated from 8-week-old chick testes; it resembles mammalian AMH with a subunit molecular weight of 74K (Teng *et al.*, 1987). Its amino acid composition is not known at the present time.

## 2. Protein Structure

a. *AMH Precursors.* The recent cloning of mouse (Münsterberg and Lovell-Badge, 1991) and rat (Haqq *et al.*, 1992) genes now allows comparison of the AMH protein sequence in four different species. Figure 2 shows the alignment of human, bovine, mouse, and rat AMH proteins, according to the multiple alignment program of Levin (1989). The bovine precursor is the longest, with 575 amino acids, followed by the human one, which contains 560 residues; mouse and rat AMH precursors have 555 and 553 amino acids, respectively. Table I shows the homology within the various species, computed according to Kanehisa (1984) separately for the N and C termini, defined by the monobasic RS cleavage site located in the fifth exon of human, bovine, and rat AMH. The C-terminal homology between different species is 90–96%; paradoxically, most mismatches in the C-terminal region occur near the cleavage site. All cysteines are conserved in the C-terminal domain. The N-terminal homology is much lower, approximately 68%, except for rat and mouse AMHs, which show 89% homology. One cysteine in the first exon is not conserved in mouse AMH, which has a serine at position 121 instead. Likewise, one of the two N-terminal potential N-glycosylation sites is not conserved in the rat.

b. *The Mature Proteins and Structure Predictions.* The structures of mature human AMH (hAMH) and bovine AMH molecules have been determined by sequencing their N termini (Cate *et al.*, 1990); the probable N termini of mature rat and mouse AMHs can be predicted by alignment (Fig. 2). A 24-amino acid leader is removed from the molecule prior to secretion, and probably corresponds to the signal sequence, although the presence of a proline residue at the putative bovine cleavage site is unusual according to Von Heijne (1986). For this reason, it has been postulated

---

FIG. 2. Alignment of the amino acid sequence of human (H), bovine (B), mouse (M), and rat (R) AMH, according to a multiple alignment program (Levin, 1989). The N-terminal residues of mature human and bovine proteins are underlined; the N termini of the mature mouse and rat proteins has not been determined. Potential glycosylation sites are marked with brackets and cysteines are denoted with asterisks (the cysteine denoted with an asterisk between parentheses is not conserved in the mouse protein). The locations of the four introns within the human protein are indicated by arrowheads. An arrow indicates the cleavage site between the N- and C-terminal fragments of human AMH.

## TABLE I
### Homology between the N- and C-Terminal Proteolytic Fragments of AMH in Different Species

| C-TER / N-TER | HUMAN | BOVINE | MOUSE | RAT |
|---|---|---|---|---|
| HUMAN | | 96.3% 105/109 | 91.6% 98/107 | 90.6% 96/106 |
| BOVINE | 69.2% 314/447 | | 90.7% 97/107 | 90.6% 96/106 |
| MOUSE | 67.2% 303/451 | 68.2% 279/409 | | 92.5% 98/106 |
| RAT | 68.4% 307/448 | 68.2% 277/406 | 89.0% 395/443 | |

that signal sequence cleavage could occur further upstream (Cate *et al.*, 1986), but the putative intermediate proprotein has not been detected.

A prediction of the secondary structure of the AMH molecule has been performed using hydrophobic cluster analysis (Lemesle-Varloot *et al.*, 1990). This method defines the shape, size, orientation, and distribution of two-dimensional hydrophobic clusters transcribed on a classic $\alpha$ helix and is particularly useful for detecting homology in sequences with low amino acid identity. Results for hAMH are shown in Fig. 3. A distinct hinge region, between residues 440 and 460, separates the C- and N-terminal regions; two other putative hinges located close to residues 155 and 270 may define three globular domains in the N terminus.

## B.  THE AMH C TERMINUS

### 1.  Relationship to the Transforming Growth Factor-β Family

The 109 carboxy-terminal residues of AMH exhibit distinct structural homology with transforming growth factor-$\beta$ (TGF-$\beta$) and its "family" of disulfide-linked factors (Cate *et al.*, 1986). The relationship of these proteins to one another has been revealed through systematic data base searches rather than predicted by common physiological effects, and is limited to the C-terminal domains of the precursors. TGF-$\beta$ is a ubiquitous disulfide-linked homodimer acting as a bifunctional regulator of cell growth (Roberts *et al.*, 1988). Under most conditions, TGF-$\beta$ inhibits the growth of normal or malignant epithelial cells while stimulating the proliferation of connective tissue. For instance, TGF-$\beta$ arrests mammary duct elongation and produces a fibrotic response at the bud tip (Daniel *et al.*, 1989).

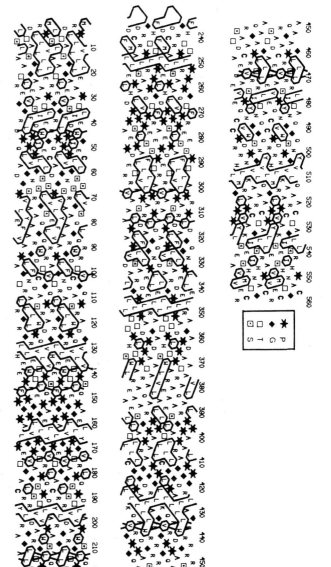

FIG. 3. Hydrophobic cluster analysis of hAMH, performed with software MANSEK, according to Lemesle-Verloot et al. (1990). The amino acid sequence, in one-letter code, is drawn on a classical α-helical net (3.6 residues per turn) and duplicated. The following amino acids are considered hydrophobic: V, I, L, F, W, M, and Y. In a hydrophobic environment, alanines and cysteines may sometimes also be hydrophobic. Clusters composed of adjacent hydrophobic residues not separated by prolines are circled. The clusters correspond to regular secondary structure elements (α helices or β sheets), which can be recognized through their shape and orientation. A distinct hinge region between residues 440 and 460, separates the C- and N-terminal regions of hAMH; two other putative hinges located close to residues 155 and 270 may define three globular domains in the N terminus.

In addition to TGF-$\beta$, the family includes inhibins (Mason *et al.*, 1985), activins (Ling *et al.*, 1986; Vale *et al.*, 1986), bone morphogenetic proteins (Wozney *et al.*, 1988), the predicted product of the decapentaplegic gene complex (DPP-C) of *Drosophila* (Padgett *et al.*, 1987), the *Xenopus* Vg-1 protein (Weeks and Melton, 1987), the related mammalian Vgr-1 protein (Lyons *et al.*, 1989), and a newly isolated growth factor GDF-1 (Lee, 1990).

Conservation between members of the family is usually limited to the C terminus and clusters around six of the seven cysteines present in each polypeptide. The bone morphogenetic proteins, the DPP-C gene product, Vg-1, and Vgr-1 exhibit at least 45% homology to each other and could represent the dipteran and mammalian derivatives of the natural mesoderm-inducing factor. The homology between AMH and TGF-$\beta$ is only 28% at the amino acid level. AMH and inhibin-$\alpha$ show the lowest conservation to the other polypeptides and to each other.

As reviewed by Gelbart (1989), TGF-$\beta$ family members are active as 110- to 130-amino acid disulfide-linked dimers, which derive from longer precursors. Processing occurs at dibasic sites to release the mature secreted product (Fig. 4). For instance, TGF-$\beta$ is generally synthesized and secreted in a biologically inactive high-molecular-weight complex in which the homodimeric 25K carboxy terminus is noncovalently associated with a dimer of the remainder of its precursor "pro" region, the so-called TGF-$\beta$ latency-associated peptide (Wakefield *et al.*, 1988). The C-terminal dimer must be released to activate the molecule. Latency-associated peptides regulate bioactivity by forming latent complexes with the C terminus (Gentry and Nash, 1990) and also aid in folding, disulfide bond formation, and export of their respective homodimers (Gray and Mason, 1990).

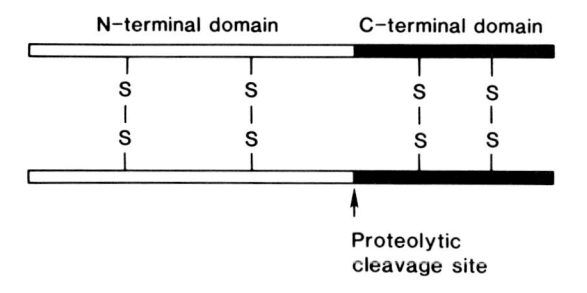

FIG. 4. Generic structure of the TGF-$\beta$ family proteins. From Cate *et al.* (1990) with permission.

## ANTI-MÜLLERIAN HORMONE

### 2. AMH Processing and Preparation of AMH Proteolytic Fragments

In contrast to other TGF-β family members, which must be cleaved from larger precursors to release an active polypeptide, AMH lacks a dibasic cleavage sequence at the appropriate site and the full-length AMH molecule is biologically active in organ culture. However, 5–20% of hAMH secreted by CHO cells is cleaved at a monobasic site located 109 amino acids away from the C terminus, between arginine-427 and serine-428, generating a TGF-β-like fragment (Pepinsky *et al.*, 1988). Processing can be driven to completion by incubation with plasmin for 1 hour at room temperature. The 110- and 25-kDa dimeric fragments remain associated as a noncovalent complex from which they can be released by acid treatment or by boiling; however, these methods abolish the bioactivity of the full-length AMH molecule and cannot be used to map the bioactive site to one of its proteolytic fragments.

This problem can be overcome by using deoxycholate as a dissociation agent (di Clemente *et al.*, 1992b). Alternatively, the C- and N-terminal fragments can be separated on a gel filtration column equilibrated in 1 $M$ acetic acid with 0.3 mg/ml polyethylene glycol 3350 at 4°C. Acid-generated samples were lyophilized and resuspended in 1 m$M$ HCl before use (Wilson *et al.*, submitted), thus preventing the aggregation that destroyed the bioactivity of our previous acid-generated C-terminal preparations (Pepinsky *et al.*, 1988). Mapping of the AMH bioactive site to the AMH C terminus is described in Section V,D.

### III. The AMH Gene

### A. GENE CLONING

#### 1. Bovine AMH cDNA

Bovine AMH cDNA was cloned first because, at the time, only the bovine protein had been purified. The cDNA was cloned independently by two laboratories using different techniques.

*a. The Immunological Approach.* Picard *et al.* (1986a), in Paris, took advantage of the fact that, having purified bovine AMH to homogeneity (Picard and Josso, 1984), a specific polyclonal antibody against the native protein was available to them. They used this polyclonal antibody to screen a DNA expression library complementary to mRNA extracted from testicular tissue of a 2-week-old calf, constructed in the phage expression vector λgt11. A library of 1.5 million primary clones with 94% recombinants was obtained, from which three positive clones were identified. The specificity of the isolated inserts was tested by antiepitope technology, i.e., by confirming that the antibodies eluted from the fusion proteins synthesized by the clones were capable of recognizing AMH on Western blots and of precipitating radioiodinated bovine AMH in a double-antibody precipitation test. In both tests, only two cross-hybridizing inserts gave positive signals; the larger one, 1.2 kb, was found to bind a 2-kb mRNA species present only in immature testicular tissue and in ovarian follicles. This mRNA moiety was not detected in other fetal bovine tissues and its size was in the range expected for an AMH-specific mRNA. The probe cross-hybridizes with human and, to a lesser degree, rat AMHs.

b. *The Degenerate Oligonucleotide Approach.* Cate *et al.* (1986), working in collaboration with P. K. Donahoe's group, electroeluted and sequenced the denatured AMH subunit from partially purified bovine AMH run on a polyacrylamide gel. The two oligonucleotide probes they constructed hybridized to 52 cDNA clones from a bovine testis cDNA library, all of which turned out to be false positives, due to the high degree of degeneracy of the probes. Reduction of probe degeneracy was obtained by hybridizing subpools to Northern blots of bovine testicular RNA. Subpools recognizing RNA only in testicular tissue were split further and subjected to the same analysis. Reduction of probe degeneracy by this method allowed the isolation of complementary DNA coding for bovine AMH.

### 2. *Cloning the AMH Gene in Other Species*

Probing of human genomic libraries with bovine cDNA probes easily yielded AMH human genomic clones due to the high homology between the two species. Bovine cDNA was used to isolate a genomic clone from a human cosmid library, which, placed under the control of a viral promoter and transfected into COS cells, produced conditioned medium exhibiting anti-Müllerian activity in the Müllerian duct assay (Cate *et al.*, 1986). A bovine cDNA probe was also successful in isolating the rat gene (Haqq *et al.*, 1992).

Isolation of the mouse gene proved more difficult: screens of genomic libraries using partial bovine and human cDNAs as probes yielded only false positives (B. Skene and R. Lovell-Badge, personal communication), suggesting that the mouse gene was fairly divergent from the probes used and that other sequences in the genome cross-hybridize to the probes. Because the AMH C terminus shows the greatest interspecies homology (Cate *et al.*, 1986), the probes were chosen within the fifth exon, which happens to exhibit an exceptionally high proportion of guanines and cytosines (GC), allowing cross-hybridization to noncoding regions of the genome or to ribosomal RNA genes. To overcome this difficulty, Mün-

ANTI-MÜLLERIAN HORMONE 11

sterberg and Lovell-Badge (1991) prepared a cDNA library from 14.5 postcoitum (pc) fetal mouse testes and screened it with bovine, human, and rat cDNA probes independently on triplicate screens. Eight cDNA probes that hybridized to all three probes were isolated, the longest of which contained a 1-kilobase (kb) *Eco*RI insert, starting within the third exon and missing the 5' end. This insert was used to clone the mouse gene from a genomic library. To the best of our knowledge, the chick AMH gene has not yet been cloned.

## B. GENE STRUCTURE

### 1. The Coding Region

The AMH gene is a relatively short one, wherein 2.75 kb are divided into five exons. Overall, the gene is characterized by a surprisingly high GC content: 72% in the exons and 68% in the introns and flanking regions. The fifth exon, the most interesting because it exhibits the greatest interspecies homology and is the only part of the AMH gene to show homology to the TGF-$\beta$ family (Cate *et al.*, 1986), contains up to 80% GC bases.

### 2. The AMH Promoter

*a. Transcription Initiation Sites.* Promoters are required for the accurate and efficient initiation of transcription. Typically, a TATA (or Hogness) box directs transcription to begin between 19 and 27 base pairs (bp) downstream. Bovine AMH contains a canonical TATA box and a single transcription initiation site, located 10 bp upstream of the ATG codon (Cate *et al.*, 1986); the mouse and rat promoters also contain an almost perfect TATA box. In contrast, the human gene has only a degenerate (TTAA) Hogness box (Cate *et al.*, 1986) and exhibits several transcription initiation sites, as demonstrated by S1 nuclease protection analysis (Fig. 5) and primer extension assays (Guerrier *et al.*, 1990): a major transcription initiation site, representing 88% of the transcripts, 10 bp upstream of the ATG codon, and three minor sites, 24, 34, and 44 bp upstream of the major one. Genes lacking a TATA box, such as many so-called housekeeping genes, often have multiple start sites for transcription; however, in the case of AMH, the predominance of the proximal start site over the others suggests that the degenerate TATA box has functional activity.

*b. A Search for Enhancers.* Most genes displaying a tightly regulated expression pattern contain enhancers, which interact with transcription factors, DNA-binding proteins produced by specific cells at specific times.

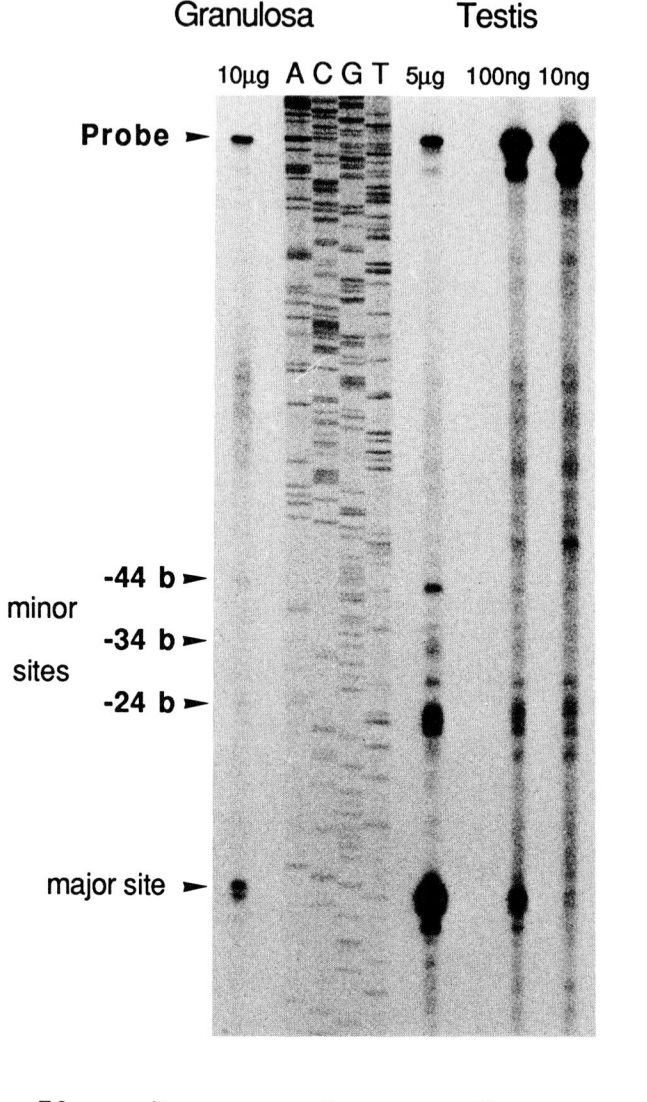

Consensus sequences for cognate transcription factors can be detected by computer analysis of flanking sequences.

Consensus nucleotide sequences have been described in the promoters of genes regulated by second messengers of polypeptide hormones or by steroids. Because gonadotropins, which use cyclic AMP (cAMP) for message transduction, have been implicated in the regulation of the AMH gene (Voutilainen and Miller, 1987; Kuroda *et al.,* 1990), it was of interest to see whether the cAMP response element (CRE) was present in the vicinity of the AMH promoter. The consensus CRE is (G/T)(A/T)CGTCA (Israel and Kaufman, 1989) and is not present in any of the AMH genes sequenced so far; a sequence with one mismatch is present at −56 in the bovine promoter, but its lack of conservation argues against any physiological significance.

c. *An Estrogen Response Element.* A 13-base palindrome, nearly identical to the estrogen response element (ERE) that activates the vitellogenin genes of *Xenopus,* is located further upstream of the human promoter, at −1772 (Guerrier *et al.,* 1990). Various fragments of the human AMH promoter encompassing this palindrome were inserted in a thymidine kinase, chloramphenicol acetyltransferase (TK–CAT) plasmid and cotransfected into HeLa cells together with a plasmid expressing the human estrogen receptor. Expression of this plasmid was measured in the presence or absence of estrogen to determine whether the ERE-like palindrome was capable of conferring estrogen responsiveness to the plasmid (Table II). This proved to be the case only if several copies of a synthetic oligonucleotide containing the putative ERE were present in the construct. No estrogen stimulation of CAT expression was seen when sequences flanking the ERE in the AMH promoter were present. These experiments do not tell us whether the AMH ERE confers estrogen inducibility to its natural promoter. The issue is confused by the fact that estrogen blocks anti-Müllerian activity of testicular tissue by abolishing

---

FIG. 5. S1 nuclease mapping of the start sites of transcription of the AMH gene. Total RNA, in the amounts indicated, was hybridized with a labeled probe (Guerrier *et al.,* 1990) and monocatenar strands were digested by nuclease S1. The size of the protected fragments in human granulosa cells and human fetal testicular tissue is determined through a sequencing reaction run on the same gel. In both testis and ovary, a major transcription initiation site is seen 10 bases upstream of the ATG codon. Three additional start sites, confirmed by primer extension experiments, are located −24, −34, and −44 bases from the major site. Other bands seen on the gel are artifactual, because they were not confirmed by primer extension. The results show that transcription initiation sites are similar in fetal Sertoli and adult granulosa cells. The amount of specific RNA is approximately 300 times less in the ovary than in the testis.

TABLE II

*Effect of Estradiol Stimulation on Chloramphenicol Acetyltransferase Expression of ERE–TK–CAT Constructs[a]*

| Plasmid | HE0 + E2 ($10^{-7}M$ E2) | HE0 only | No HE0, no E2 |
|---|---|---|---|
| KS–SV2–CAT | 100 | 100 | 100 |
| Vit–ERE–TK–CAT | 40 | 4 | 4 |
| 182-bp TK–CAT | 1.4 | <1 | <1 |
| 182-bp x2–TK–CAT | <1 | <1 | <1 |
| 35-bp x2–TK–CAT | 14 | 2 | 2 |
| 35-bp x4–TK–CAT | 98 | 1.2 | 7 |
| pBL–CAT8+ | nd | nd | 5 |

[a] Estradiol (E2)-stimulated chloramphenicol acetyltransferase (CAT) expression was evaluated using various ERE–TK–CAT constructs cotransfected into HeLa cells together with a plasmid (HE0) expressing the human estrogen receptor. Acetylated spots were cut out from the chromatograph and radioactivity was measured in a scintillation counter. Results are expressed as percentage of CAT activity in cells transfected by plasmid KS–SV2–CAT, in which CAT expression is constitutive, and are corrected for variations in transfection efficiency. pBL–CAT8+ is a plasmid carrying the CAT gene under the control of the TK promoter, with no estrogen response element (ERE) inserted; Vit–ERE–TK–CAT is the same plasmid, with insertion of the ERE of the vitellogenin gene; 182-bp TK–CAT is the same plasmid, with insertion of a 182-bp fragment of the AMH promoter, containing the ERE; 182-bp x2–TK–CAT is the same plasmid, with two copies of the promoter fragment; 35-bp x2–TK–CAT is the same plasmid, with insertion of two copies of a 35-bp DNA fragment containing the AMH ERE; 35-bp x4–TK–CAT is the same plasmid, with four copies of the DNA fragment. The AMH ERE is functional when contained in the 35-bp but not the 182-bp fragment of the AMH promoter. nd, Not determined. From Guerrier *et al.* (1990) with permission.

AMH responsiveness of the target organ (Newbold *et al.*, 1984); no quantitative measurement of the AMH output of estrogen-treated testicular tissue has been performed.

Computer analysis of the flanking sequences of the AMH gene does not detect enhancers specific for the AMH promoter, likely to be located in conserved regions. A 9-bp repeat, GGAGATAGG, is found twice at similar locations in the bovine and human promoter (Cate *et al.*, 1986) but not in the murine promoters. Before the significance of such regions can be assessed, it will be necessary to obtain cells in which the AMH promoter can be stably expressed. As is discussed later (Section IV,A,2), Sertoli cells in culture rapidly cease to produce the hormone (Vigier *et al.*, 1985) and the lack of a suitable cell model hampers studies of AMH gene regulation.

## C. GENE MAPPING

### 1. Mapping the Human Gene

Although involved in male sex differentiation, the AMH gene is not located on a sex chromosome, in keeping with the tenet that only the genetic trigger for gonadal sex determination needs to be sex specific. Preliminary evidence for the autosomal location of the human AMH gene was presented by Cate et al. (1986), who showed by Southern analysis that male and female DNAs have a similar hybridization pattern. Using a panel of human–rodent somatic hybrid cell lines segregating human chromosomes and probes derived from bovine cDNA, Cohen-Haguenauer et al. (1987) localized the human gene to chromosome 19. Because the human gene for TGF-$\beta$ also maps to chromosome 19 (Fujii et al., 1986), and to exclude the possibility of cross-hybridization, in situ hybridization was performed. The Amh gene maps to the extremity of the short arm of chromosome 19, band 19p13.3, whereas the gene for TGF-$\beta$ is located on the long arm, subband q13.1 → q13.3. Inhibin subunits, which also belong to the TGF-$\beta$ family, map to chromosomes 2 and 7 (Barton et al., 1987).

### 2. Other Species

Recent data on the location of the AMH gene in other species are now available. No association to genes located on human chromosome 19 has been detected, indicating complex rearrangements in chromosomal structure during evolution. For mapping in the domestic cow, DNA probes from bovine AMH and osteonectin were hybridized to cow–hamster and cow–mouse hybrid cell lines, segregating bovine chromosomes. Bovine osteonectin is homologous to secreted protein, cysteine rich (SPARC), in humans, and maps to human chromosome 5. Bovine AMH and osteonectin loci were fully concordant with each other and discordant with all other bovine syntenic groups described to date (Rogers et al., 1991). In the mouse, the Amh gene has been mapped to the distal region of chromosome 10, between the phenylhydroxylase and the zinc finger autosomal genes (King et al., 1991). Several polymorphic Amh-related sequences were mapped to other chromosomes and could represent pseudogenes.

## IV. AMH Ontogeny

AMH is produced exclusively by gonadal somatic cells in both sexes; however, there are major differences: immature Sertoli cells synthesize high amounts of AMH that accumulate in the rough endoplasmic reticulum (Tran and Josso, 1982; Hayashi et al., 1984), and the granulosa cells

16 NATHALIE JOSSO ET AL.

synthesize the hormone only after birth and in low quantities (Vigier *et al.*, 1984a).

## A. TESTICULAR PRODUCTION

In the testis, AMH is produced only by Sertoli cells, in contrast to inhibin and activin, which are also synthesized by interstitial cells (Lee *et al.*, 1989; Roberts *et al.*, 1989; Shaha *et al.*, 1989). Testicular AMH ontogeny in different animal species is shown in Table III.

### 1. In Vivo

*a. In the Fetus.* AMH is the earliest cognate protein expressed by developing Sertoli cells. Testicular anti-Müllerian activity is expressed at the time of formation of seminiferous tubules (Picon, 1970; Tran *et al.*, 1977; Vigier *et al.*, 1983). Peak values were found at the time of regression of Müllerian ducts in bovine fetuses, but relatively high levels were sustained up to birth. No AMH expression was found in the germ cells, nor outside the testis cords, nor in any other fetal tissue. In mice expressing the *W* locus, in which the germ cells fail to reach and colonize the gonadal ridge, AMH expression is normal, indicating that it is not dependent on interaction with germ cells (Münsterberg and Lovell-Badge, 1991). AMH levels have recently been measured by enzyme-linked immunoassay in

TABLE III
*Ontogeny of AMH Production in Different Species*

| Species | Detection technique | Beginning[a] (postcoitum) | End[b] (postpartum) | Ref. |
|---------|---------------------|---------------------------|---------------------|------|
| Human | Bioassay | 7 weeks | | Josso (1971) |
| | Bioassay | | 2 years | Donahoe *et al.* (1977a) |
| | Immunocytochemistry | | 7 years | Tran *et al.* (1987) |
| | ELISA | | Puberty | Hudson *et al.* (1990) |
| | ELISA | | Puberty | Josso *et al.* (1990b) |
| Rat | Bioassay | 13.5 days | 4 days | Picon (1969) |
| | Immunocytochemistry | 13.5 days | 9 days | Tran *et al.* (1977) |
| | Immunocytochemistry | | 20 days | Kuroda *et al.* (1990) |
| Mouse | *In situ* hybridization | 12.5 days | 2–3 weeks | Münsterberg and Lovell-Badge (1991) |
| Bovine | RIA | 43 days | 18 months | Vigier *et al.* (1983) |
| Pig | Bioassay | 25 days | 60 days | Tran *et al.* (1977) |

[a] dpc, Days postcoitum.
[b] dpp, Days postpuberty.

# ANTI-MÜLLERIAN HORMONE

human fetal sera in collaboration with Dr. Norman Davis (Ultrasound Diagnostic Services, London, England). Fetal sex, not known at the time of the assay, was determined by cytogenetic examination. Results are shown in Fig. 6. High levels were found in the sera of all male fetuses studied, at 19–22 weeks and at 40 weeks. No AMH was detected in female sera nor in unconcentrated amniotic fluid of either sex. These results are potentially interesting for assessment of testicular function in late fetal life, at a time when testosterone levels are more or less similar in both sexes (Winter *et al.*, 1977).

   *b.   After Birth.*   In the perinatal period, AMH is expressed at high levels, then progressively decreases and drops sharply at the beginning of pubertal maturation. AMH transcripts are undetectable by *in situ* hybridization in the testes of adult mice (Münsterberg and Lovell-Badge, 1991), but by immunoassay, low AMH concentrations are found in human adult serum (Hudson *et al.*, 1990) and in bovine rete testis fluid (Vigier *et al.*, 1983). Serum AMH is increased in children with delayed puberty, compared to age-matched controls (Josso *et al.*, 1990a) (see Section VI,A).

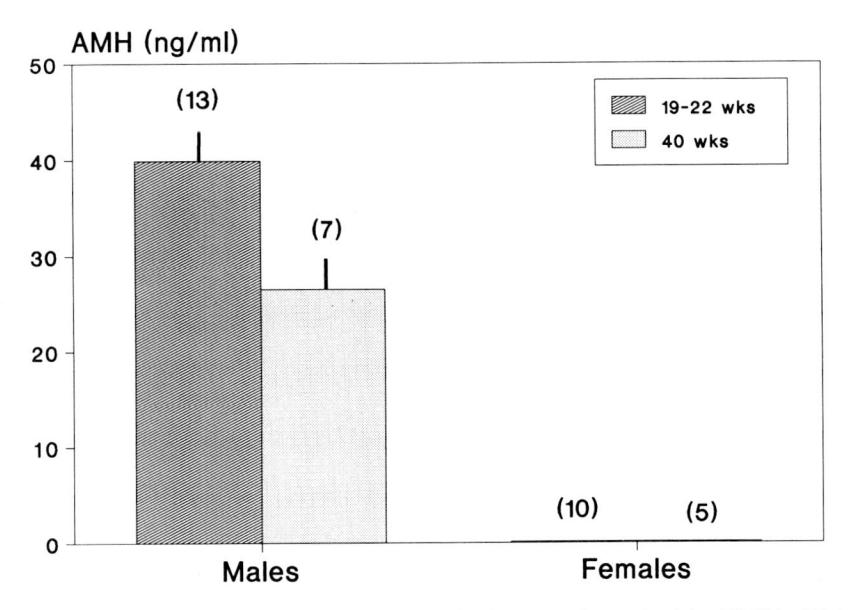

FIG. 6.   Concentration of AMH in human fetal serum, determined by ELISA. High concentrations were found in male fetuses, both at midpregnancy and near term; AMH was not detectable in the serum of female fetuses. The fetal serum samples were obtained through Dr. Norman Davies, London, England.

## 2. In Vitro

Bovine Sertoli cells rapidly cease to produce AMH when placed in culture (Vigier et al., 1985), even when they are obtained from immature testicular tissue, which would normally continue to produce AMH for several months in vivo. Extinction of AMH expression is not due to cell loss or damage, because after 3 days in culture the amount of DNA remains stable and the Sertoli cells increase their cyclic AMP production 10-fold in response to follicle-stimulating hormone (FSH) stimulation. AMH expression can be reactivated by pharmacological doses of cAMP [0.5 mM (Bu)$_2$cAMP for 2 days], as monitored by Northern blots of mRNA extracted from human fetal testicular cells cultured 4 days in the presence of serum (Voutilainen and Miller, 1987). The cause of extinction of AMH expression in cultured Sertoli cells is not clear at the present time. Hormonal deprivation does not seem responsible, because addition of FSH, testosterone (Vigier et al., 1985), estrogen, progesterone, gonadotropin-releasing hormone (GnRH), luteinizing hormone (LH) (LaQuaglia et al., 1986), or human chorionic gonadotropin (hCG) (Voutilainen and Miller, 1987) fails to restore AMH production.

## B. OVARIAN PRODUCTION

Granulosa cells of the ovary and testicular Sertoli cells share many structural and functional characteristics. The first indication for ovarian production of AMH was the demonstration by Hutson et al. (1981) of anti-Müllerian activity of ovarian tissue of hens. Follicular fluid extracted from small cow follicles contains significant concentrations of AMH, comparable to the levels found in the circulation of fetal males (Vigier et al., 1984a; Necklaws et al., 1986). Production of AMH in the mammalian ovary was localized to granulosa cells by Vigier et al. (1984a). Using avidin–biotin histochemistry with either a monoclonal (Takahashi et al., 1986a; Bézard et al., 1987) or a polyclonal (Ueno et al., 1989a,b) antibody, AMH was localized to the cytoplasm of granulosa cells of developing follicles. AMH is not detected in fetal ovaries, neither in the rat (Ueno et al., 1989b), where follicles form only after birth, nor in the ewe, where growing follicles with two- to six-cell-deep granulosa cell layers are already present at 120 days of gestation (Bézard et al., 1987). In the mouse ovary, AMH transcripts, studied by in situ hybridization, are first detectable 6 days after birth (Münsterberg and Lovell-Badge, 1991). No immunoreactive AMH can be detected in fetal female blood (Vigier et al., 1983; N. Davies and N. Josso, unpublished data) (Fig. 6). Both AMH expression and immunoreactivity are dependent on the degree of follicular maturation,

ANTI-MÜLLERIAN HORMONE                    19

rather than on the age of the animal. As the follicle grows, the immunoreactivity of granulosa cells close to the basal lamina progressively fades, while that of those lining the antrum and occupying the cumulus oophorus increases (Bézard *et al.*, 1987; Ueno *et al.*, 1989a). AMH production appears confined to the stem cell population of granulosa cells, which is incapable of undergoing terminal differentiation under FSH stimulation (Erickson, 1983). Just before ovulation, when the oocyte completes the first meiotic division, AMH immunoreactivity (Ueno *et al.*, 1989a) and transcripts (Münsterberg and Lovell-Badge, 1991) become undetectable in cumulus cells.

Whatever the degree of follicular maturation or the age of the animal, the production of AMH by granulosa cells is low compared to that of immature Sertoli cells, as shown by Northern blots of messenger RNA extracted from gonads of both sexes (Picard *et al.*, 1986a) and by differences in anti-Müllerian activity (Ueno *et al.*, 1989b). In contrast, when tested at similar concentrations, AMH purified from either fetal testis or adult ovarian follicles has similar anti-Müllerian activity (Vigier *et al.*, 1984a). Transcription initiation sites are similar in both tissues (Fig. 5), suggesting that the same promoter drives the gene in the testis and in the ovary.

## C.  REGULATION OF THE AMH GENE:
## CURRENT SPECULATIONS

The chronological pattern of gonadal AMH expression is of greatest importance in sex differentiation. In the male, AMH *must* be expressed early, before Müllerian ducts lose their responsiveness, i.e., before 15 days pc in the rat (Picon, 1969) and before 8 weeks in the human fetus (Josso *et al.*, 1977). On the other hand, in the female, it is essential that AMH *not* be expressed during fetal life, otherwise female reproductive organs would be destroyed (Behringer *et al.*, 1990). Therefore, the AMH gene is likely to be under strict transcriptional control. Paradoxically, very little is known concerning the regulation of gonadal production of AMH, mainly because transcription of AMH is rapidly extinguished in cultured Sertoli cells (Vigier *et al.*, 1985) and no other cell model characterized by stable AMH production under the control of the natural promoter has yet been developed. At this time, only speculations concerning the physiological regulation of the AMH gene can be presented.

*1.  The Testis-Determining Factor*

One way for nature to ensure that AMH is expressed only in the male during fetal life would be to place the AMH gene under the control of the epithelial–mesenchymal interactions (Dyche, 1979; Trelstad *et al.*, 1982), perhaps triggered by the dissolution of the basement membrane (Trelstad *et al.*, 1982). Epithelial cells become reoriented and extend into the mesenchyme, which condenses into a characteristic periductal ring. The effect of AMH on the cell components of Müllerian ducts is rather similar to that of TGF-$\beta$, known to stimulate the growth of fibroblastic cells and to promote the production of fibronectin while acting as a potent growth inhibitor of epithelial cells (see Wakefield, 1990, for review). It is, however, unlikely that TGF-$\beta$ could mediate the effect of AMH on the Müllerian duct, because it is inactive in this system (Tran, unpublished).

Müllerian ducts become insensitive to AMH very early in fetal development, after 14 days postcoitum in the rat (Picon, 1969) and after 8 fetal weeks in the human (Josso *et al.*, 1977). Persistence of Müllerian duct derivatives has been observed in male rabbit fetuses, passively immunized against AMH during fetal life (Tran *et al.*, 1986), and in normally virilized human males suffering from a mutation of the AMH gene (Knebelmann *et al.*, 1991).

The first bioassay described for AMH (Picon, 1969) was based on the capacity of the hormone, produced by testicular tissue or added to culture medium, to induce regression of 14.5-day-old Müllerian ducts after 3 days in organ culture. Rat Müllerian ducts respond to avian and mammalian AMH, but chick embryonic Müllerian ducts are insensitive to mammalian AMH (Tran *et al.*, 1977). Various grading methods have been proposed (see Josso and Picard, 1986, for review); in all cases, the degree of Müllerian duct regression is assessed by qualitative criteria on histological sections, and this should be borne in mind when choosing a method of statistical analysis. The Student's *t* test and analysis of variance may be used only after appropriate data transformation (Ben-David *et al.*, 1969).

## 2. Antiidiotypic Antibodies and the Elusive AMH Receptor

Polypeptides exert their physiological effects by binding to specific receptors on target cells, triggering signal transduction pathways leading to the activation of specific genetic programs. Several superfamilies of hormone receptors sharing a common structure and mechanism of action have been described recently, but not the faintest information is available on the AMH receptor.

The antiidiotype method for receptor visualization is based on the hypothesis that, regardless of functional differences, macromolecules of the same specificity will show structural homologies in their binding sites. Therefore, an antibody (antiidiotypic antibody) directed against the variable region of an antihormone antibody should resemble the hormone epitope against which the antihormone antibody is directed (Fig. 7). Pro-

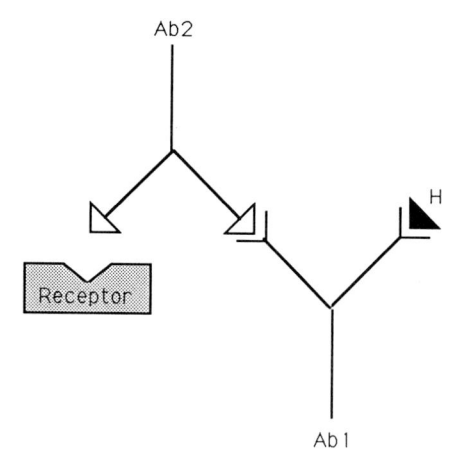

FIG. 7. Antiidiotype interactions. A monoclonal antibody (Ab1) is directed against the receptor-binding site of a hormone molecule (H). A second antibody (Ab2), raised against the Ab1 idiotype, is also able to interact with the hormone receptor. Adapted from Strosberg *et al.* (1985) with permission.

vided this hormone epitope is related to the receptor-binding site, the antiidiotypic antibody might interact with the receptor, and has indeed been shown to do so (reviewed in Strosberg *et al.*, 1985).

We generated antiidiotypic antibodies to an IgM monoclonal antibody to bovine AMH, Mab 168 (Legeai *et al.*, 1986). Mab 168 is directed against conserved epitopes located on the C terminus, which has been shown to carry the bioactive epitope(s) (see Section V,D). The antiidiotypic antibody to Mab 168 purified by immunochromatography (Lefèvre *et al.*, 1989), in contrast to antiidiotypes generated against N-terminal specific monoclonal antibodies, exhibited an AMH-like effect in the Müllerian duct bioassay, suggesting interaction with the AMH receptor. Yet it did not allow receptor visualization by immunocytochemistry, perhaps because of low receptor concentration.

*3. Morphological Virilization of the Fetal Ovary: The Freemartin Model*

Recently, another effect of AMH has been unequivocally demonstrated, namely a growth-inhibiting and masculinizing effect on the fetal ovary (Vigier *et al.*, 1987; Behringer *et al.*, 1990). Jost and associates should be credited for being the first to suspect that AMH could be responsible for the ovarian stunting and masculinization observed in freemartins, bovine heterosexual twins united by placental anastomoses. Ovarian growth is arrested at the time of Müllerian duct regression in the freemartin and her male twin (Jost *et al.*, 1972); AMH levels are correlated in both fetuses (Vigier *et al.*, 1984b). To test the effect of AMH on the fetal ovary, Vigier *et al.* (1987) cultured 14.5-day-old rat fetal ovaries in the presence of purified bovine AMH.

*a. Effect on Germ Cell Number.* The evolution of germ cell number in developing rat ovaries in the presence or absence of AMH is shown in Fig. 8. *In vivo*, the number of germ cells increases sharply between 14 and 17 days pc and then steadily decreases. *In vitro*, in the absence of AMH, the trend is similar, although at each time point, fewer germ cells survive. In the presence of AMH, a steep fall in germ cell population occurs between 14 and 17 days, indicating that AMH acts essentially by inhibiting germ cell proliferation. The effect on germ cell number is time and sex dependent: fetal ovaries explanted at 20 days pc are impervious to AMH, indicating a critical window of sensitivity, such as has been demonstrated for Müllerian ducts (Picon, 1969). AMH treatment has no effect on XY germ cells.

*b. Effect on Sertoli Cell Differentiation and Gonadal Structure.* AMH treatment of 14.5-day-old fetal rat ovaries consistently in-

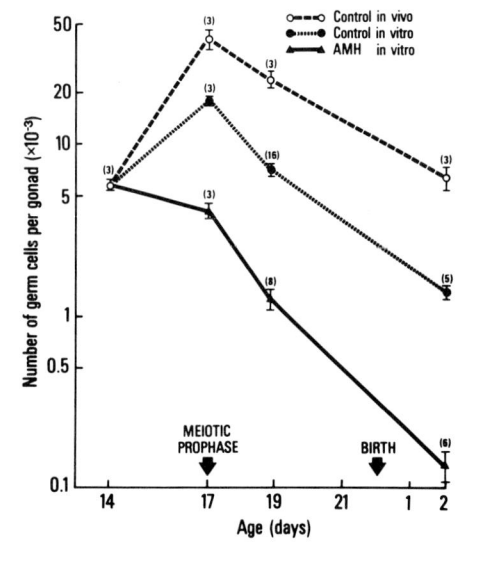

FIG. 8. Number of cells in developing rat ovaries, *in vivo* and *in vitro*. The number of germ cells increases markedly between 14 and 17 days postcoitum *in vivo*, and then decreases when the germ cells enter meiotic prophase. Control cultured ovaries show roughly the same pattern, but the number of germ cells is lower at each time point. In contrast, in cultured ovaries exposed to bAMH (7.5 μg/ml), germ cells do not proliferate and their number diminishes at 17 days. From Vigier *et al.* (1987) with permission.

duces the formation in the gonadal blastema of cordlike structures that resemble seminiferous tubules. Somatic cells contained within the tubules assume an epithelioid appearance and become polarized against a discontinuous basal membrane; under the electron microscope, they resemble differentiating Sertoli cells with an enlarged cytoplasm that is less electron dense than that of other cell types. At the surface of the gonad, a layer of flattened connective cells forms a tunica albuginea (Fig. 9). Control cultures, performed in the absence of AMH, are characterized by a homogeneous blastema containing numerous germ cells that are separated by somatic cells with a sparse, electron-dense cytoplasm. Seminiferous tubulelike structures developing in AMH-treated fetal rat ovaries are also functionally virilized and produce immunoreactive AMH (Vigier et al., 1988).

### 4. Transgenic Mice with Chronic, Unregulated Production of AMH

An *in vivo* simulation of the freemartin condition was generated by Behringer et al. (1990), who constructed a line of transgenic mice expressing the human AMH gene under the control of the metallothionein promoter. The metallothionein promoter can direct expression of heterologous genes to a variety of fetal and adult tissues in transgenic mice (Palmiter et al., 1983). Nine founder transgenic mice were generated—two sterile females, lacking a uterus, oviducts, and ovaries, and seven males that transmitted the transgene to their progeny. Levels of serum AMH in these animals varied from 32 to 6000 ng/ml, with wide individual variations, reflected in different phenotypes. Females usually lacked a uterus but retained ovaries. These ovaries, normal at birth, progressively lost germ cells and either disappeared or developed cordlike structures resembling testicular seminiferous tubules (Fig. 10). Transgenic males were usually normal, but testicular dysgenesis and Leydig cell dysfunction were observed in those with the highest level of expression of AMH.

### 5. Role of Germ Cell Deprivation in AMH-Induced Ovarian Lesions

In the normal ovary, oocytes induce the supporting cell lineage to develop as follicle cells, and it is the continued presence of oocytes that maintains follicles (McLaren, 1990). Ovaries without germ cells, for instance those of human 45X females, degenerate. It follows that the ovarian aplasia observed in freemartins and in female AMH transgenic mice is probably a consequence of the deleterious effect of AMH on XX germ cells.

Is ovarian virilization also a consequence of germ cell degeneration due to prolonged culture? Arguments in favor of this view have been presented by Taketo et al. (1985), who obtained development of seminiferous tubules

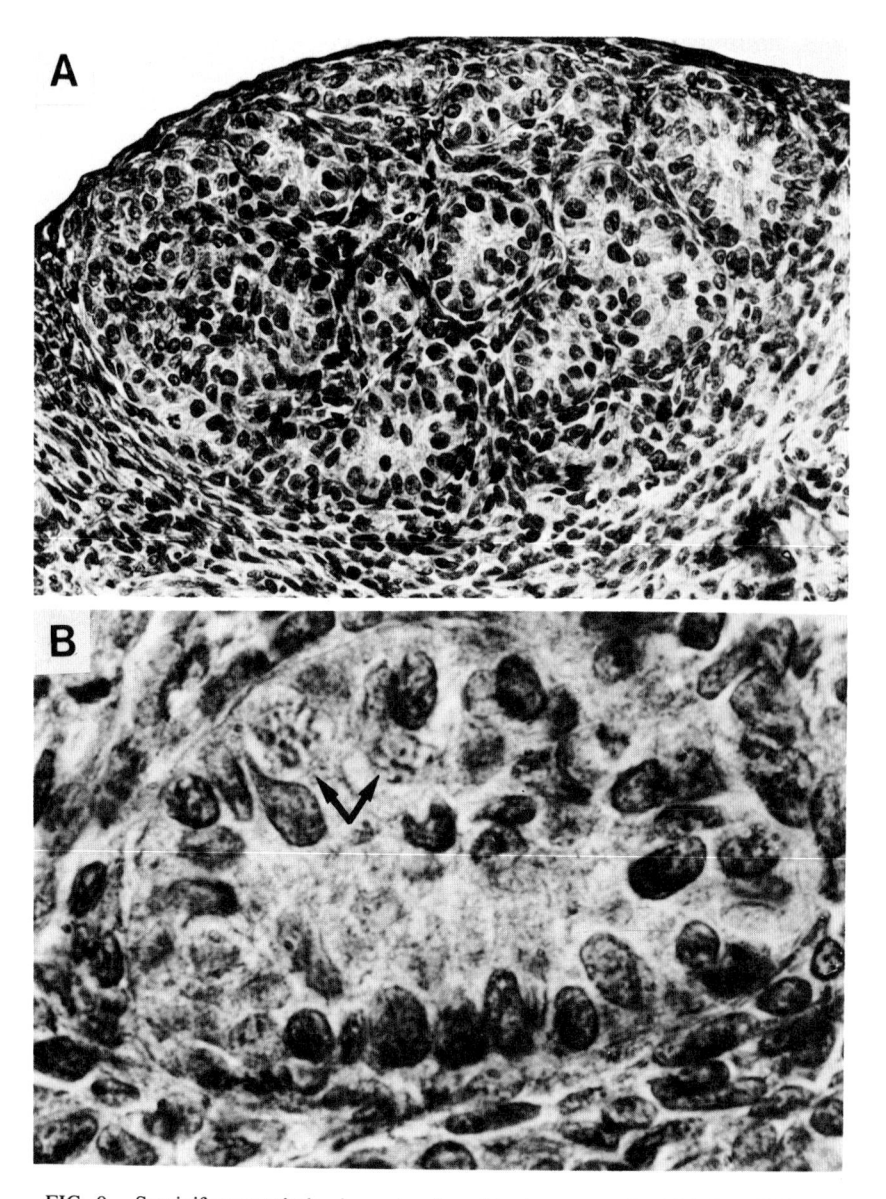

FIG. 9. Seminiferous tubules have developed in the ovary of a 14.5-day-old fetal rat, exposed in organ culture to bAMH, 5 μg/ml. (A) Seminiferous tubules have developed in the ovarian parenchyma (×350). (B) At a higher magnification (×1200), two surviving germ cells are seen entering meiotic prophase (arrows).

ANTI-MÜLLERIAN HORMONE 27

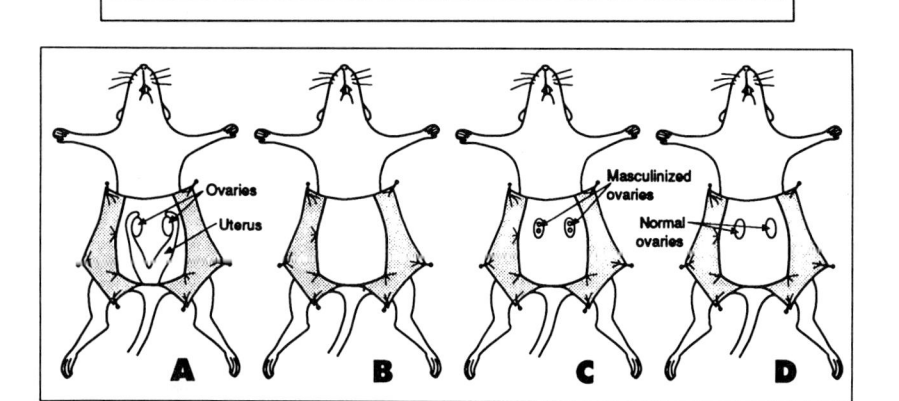

FIG. 10. Transgenic mice generated by placing the human AMH gene under the control of the metallothionein promoter (Behringer *et al.*, 1990). All transgenic mice lack a uterus and tubes and most adults (B) also lack ovaries. Seminiferous tubules may develop after birth (C); more rarely, ovaries are normally developed (D). From Josso *et al.* (1990b) with permission.

by transplantation of fetal ovaries under the kidney capsule of male mice and concluded that fetal ovaries at early stages of development can form testicular structures, even in the absence of any male-specific factors. Prépin and Hida (1989) reported that fetal rat ovaries maintained 16 days in organ culture lose their germ cell population and develop cordlike structures with anti-Müllerian activity. However, at the developmental period tested, normal ovaries already produce AMH (Ueno *et al.*, 1989b).

Several reasons prompt us to believe that germ cell depletion is not the cause of AMH-induced virilization (Josso *et al.*, 1991b). AMH-induced cords may still contain germ cells, some of them in meiotic prophase (Vigier *et al.*, 1987) (Fig. 9). In mammals, the well-known incapacity of XX spermatogonia to survive testicular differentiation has been attributed to a "spermatogenesis gene" on the Y chromosome (Burgoyne *et al.*, 1986), and therefore one would expect germ cell degeneration to accompany virilization of XX gonads, regardless of its cause. Apparently, this is not the case in birds. Complete spermatogenesis has been obtained in genetically female domestic fowls sex reversed by an extraembryonic testis graft and it has been suggested that the sex reversal is due to AMH

28        NATHALIE JOSSO ET AL.

secretion by the graft (Rashedi *et al.*, 1983). This awaits confirmation, however.

## B.  EFFECTS ON THE OVARIAN STEROIDOGENESIS PATHWAY

### 1.  AMH-Induced Sex Reversal of the Ovarian Steroidogenic Pathway

Endocrine differentiation of fetal gonads is detectable very early in development (George *et al.*, 1979). Fetal ovaries acquire the capacity to form estrogens from C19 steroids, whereas the fetal testis is characterized by testosterone synthesis. Sex reversal of the steroidogenesis pathway has been demonstrated in the young bovine freemartin ovary (Shore and Shemesh, 1981) and in fetal ovaries exposed to AMH *in vitro* (Vigier *et al.*, 1989). The effect of AMH on ovine fetal ovaries, characterized by high rates of estradiol production in early fetal life (Mauléon *et al.*, 1977), is shown in Fig. 11. Normally, fetal ovaries produce estradiol and fetal

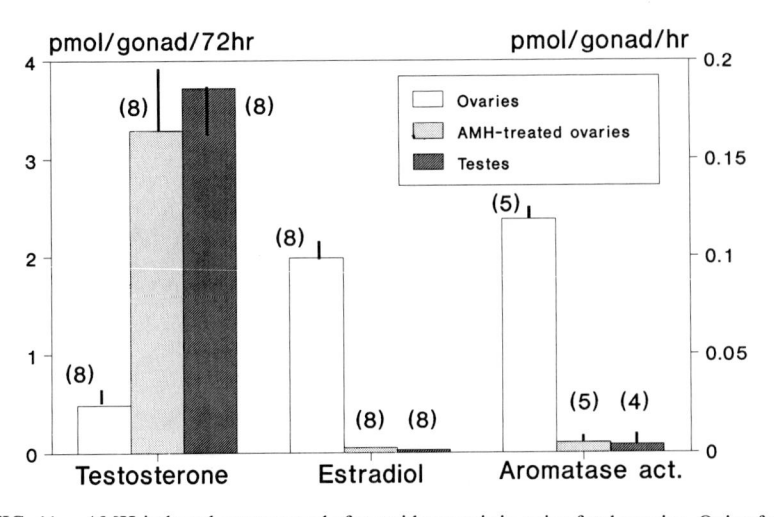

FIG. 11.   AMH-induced sex reversal of steroidogenesis in ovine fetal ovaries. Ovine fetal gonads, explanted at 29 days postcoitum, are maintained in organ culture, either in control medium or in medium containing bAMH (7.5 µg/ml). Estradiol and testosterone were assayed by specific radioimmunoassay in the medium between 6 and 9 days after explantation. Aromatase activity of the explants was assayed 9 days after explantation by measuring the amount of tritiated estradiol metabolized from a tritiated testosterone precursor. [For details, see Vigier *et al.* (1989).] AMH-treated fetal ovaries acquire a male pattern of steroidogenesis and their aromatase activity is dramatically decreased. This effect requires a relatively prolonged exposure to AMH and is not due to competitive aromatase inhibition.

ANTI-MÜLLERIAN HORMONE 29

testes secrete testosterone, a pattern reversed by AMH treatment. The effect of AMH is mediated by a decrease of ovarian aromatase activity. In rat fetal ovaries, the low spontaneous aromatase activity can be stimulated by cAMP (Picon *et al.*, 1985), an effect blocked by AMH (Vigier *et al.*, 1989).

## 2. The Fetal Ovary/Aromatase Assay: A Quantitative Test for AMH Bioactivity

Until recently, the only test for anti-Müllerian activity was based on the work of Picon (1969) showing that fetal rat Müllerian ducts in organ culture regress in the presence of AMH. This labor-intensive bioassay is at best semiquantitative and requires high amounts of AMH (7–35 n$M$) to obtain unequivocal histological evidence of Müllerian regression. The decrease in aromatase activity of fetal ovaries exposed to AMH is readily measurable and lends itself to quantitation of AMH biological activity. The ontogeny of cAMP-stimulated aromatase activity in fetal rat ovaries (Fig. 12) showed a peak at 16 days postcoitum and this developmental stage was chosen for the assay.

*a. Technique of the Fetal Ovary Aromatase Assay.* As previously described (Vigier *et al.*, 1987), 16-day-old rat fetal ovaries are explanted 3 days in organ culture in culture medium CMRL 1066 containing 0.25 mg/ml bovine serum albumin, 0.1 m$M$ 3-isobutyl-1-methylxanthine, and 1 m$M$ $N^6,O$-2'-dibutyryladenosine 3',5'-cyclic monophosphate (Bt$_2$cAMP). Aromatase activity is measured at the end of the culture period by the tritiated water technique (Ackerman *et al.*, 1981). Results of triplicate experiments are expressed as the percentage of decrease of aromatase activity in explants exposed to both Bt$_2$cAMP and AMH, compared to those receiving only Bt$_2$cAMP. Because a decrease in aromatase activity can also result from nonspecific tissue injury, it is essential that suitable controls be designed.

*b. Results.* A linear log/dose response to AMH treatment was demonstrated between 1.5 and 30 n$M$ of bovine AMH ($r = 0.964$, $p > 0.01$) and between 1 and 10 n$M$ for hAMH ($r = 0.918$, $p < 0.01$). Intraassay and interassay variations were, respectively, 12.3% ($n = 3$) and 14.6% ($n = 3$) for bAMH and 12.2% ($n = 3$) and 19.2% ($n = 4$) for hAMH. Recombinant hAMH was more active than bovine AMH in this system, with an ED$_{50}$ equal to 3 n$M$ compared to 8 n$M$ for bAMH. Monoclonal antibodies to bAMH and hAMH, added to the culture medium at a molar concentration equal to five times that of the hormone, decreased the bioactivity of their antigens by 90 and 70%, respectively (Fig. 13) (di Clemente *et al.*, 1992a).

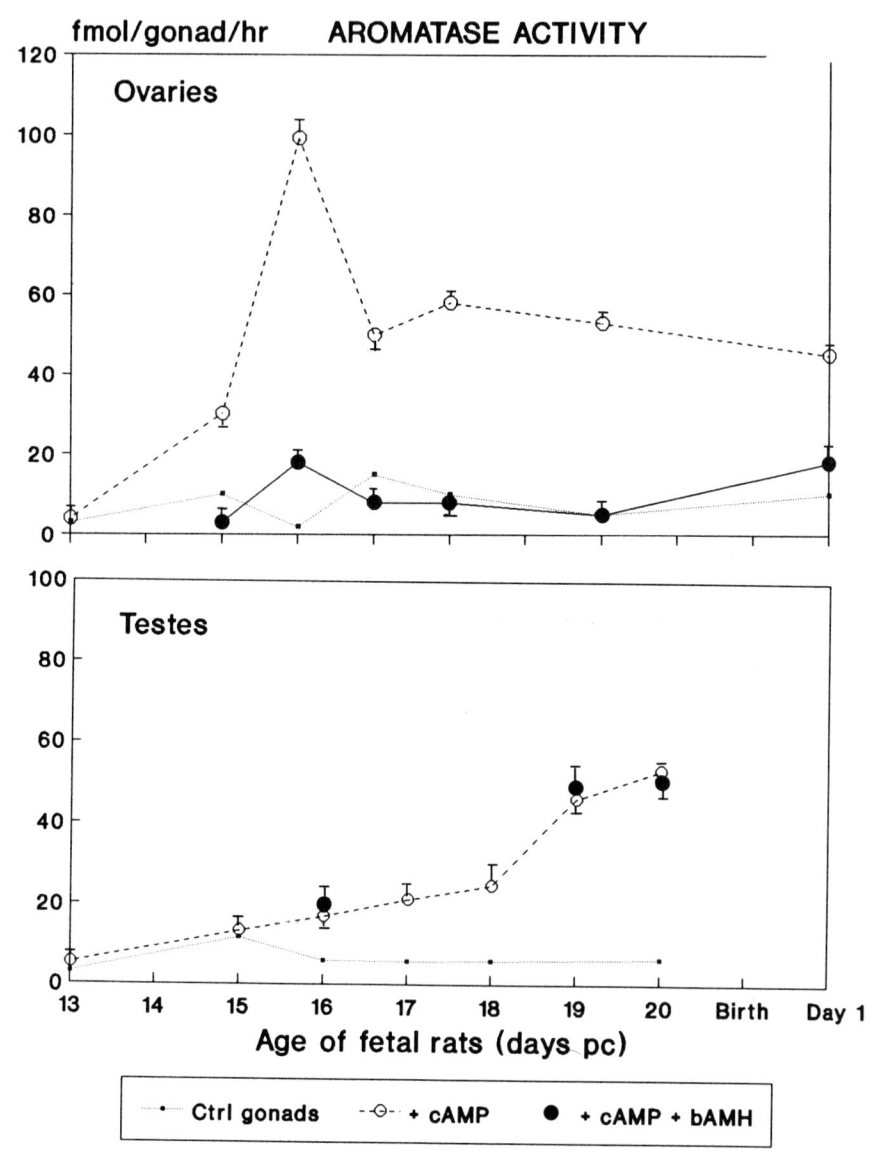

FIG. 12. Ontogeny of rat ovarian aromatase response to AMH in rat fetal gonads maintained in organ culture from 13 days postcoitum (pc) to 1 day after birth. In rat fetal ovaries, basal aromatase activity is low at all stages studied. Stimulation by 0.1 mM Bt₂cAMP is maximal at 16 days and can be inhibited by 10 μg/ml bAMH from 15 days p.c. to 1 day after birth. In fetal testes, stimulation of aromatase activity by 1 mM Bt₂cAMP increased with fetal age regardless of whether 10 μg/ml bAMH was added to the culture medium. Values shown represent the mean and SEM of experiments usually performed in triplicate. At 16 days pc, $n = 6$ for controls and Bt₂cAMP + AMH-treated ovaries, and $n = 15$ for Bt₂cAMP-treated ovaries.

ANTI-MÜLLERIAN HORMONE 31

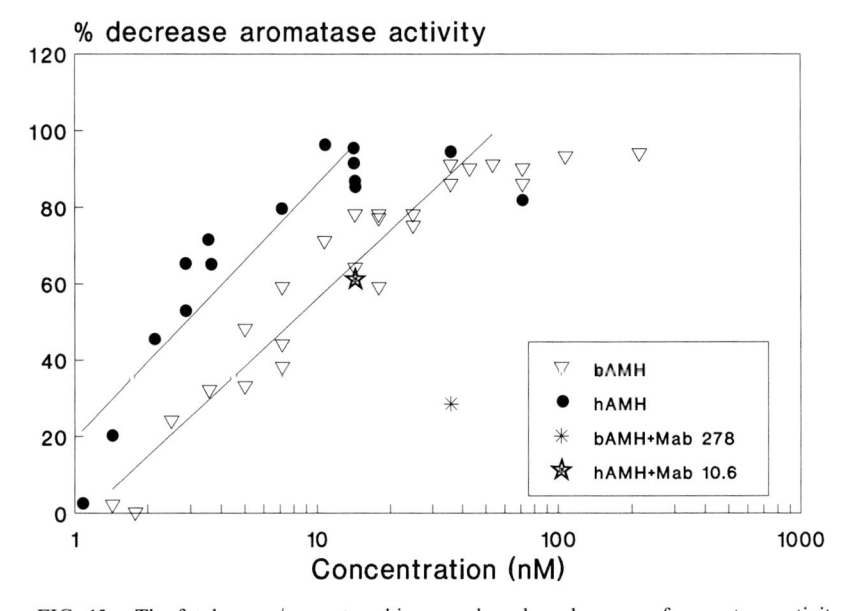

FIG. 13. The fetal ovary/aromatase bioassay, based on decrease of aromatase activity of Bt$_2$cAMP-stimulated 16-day-old rat fetal ovaries. Decrease of aromatase activity of AMH-treated ovaries was determined by comparison with the aromatase activity of controlateral ovaries cultured without AMH. Each point represents the mean of triplicate experiments. Decrease of aromatase activity was proportional to the logarithm of the concentration of either bAMH or hAMH in the culture medium. Half-maximal inhibition (ED$_{50}$) was obtained with 8 n$M$ and 3 n$M$ hAMH. Monoclonal antibodies to either bAMH or hAMH decreased hormone bioactivity at a hormone/Mab ratio of 1:5. The concentration of untreated bAMH (0.068 n$M$) and hAMH (0.084 n$M$) required to elicit the same biological effect as Mab-treated AMH of the same species (0.68 and 0.28 n$M$, respectively) was reduced by 90 and 70%, respectively. (From di Clemente *et al.*, 1992a.)

## C. OTHER PROPOSED EFFECTS

### 1. *Effect on the Testis*

*a. Effect on Testicular Differentiation.* Given the dramatic ovarian virilization produced by AMH, several investigators (Vigier *et al.*, 1989; Behringer *et al.*, 1990) have proposed that AMH might play a role in testicular differentiation. Repression of constitutive estrogen production is believed to be crucial for testicular differentiation to occur (Dorizzi *et al.*, 1991). AMH, however, does not affect aromatase in the fetal testis (Fig. 12) and testicular development is not impaired in patients with the persistent Müllerian duct syndrome due to AMH gene mutations (see

Section VII). At this time, in spite of its virilizing impact on fetal ovaries, AMH has not proved to be a testicular differentiating factor.

*b. Effect on Testicular Descent.* Hutson and Donahoe (1986) have championed the hypothesis that AMH plays a role in testicular descent. Their contention is based on the following arguments: testicular factors other than testosterone are involved in the first phase of testicular descent, patients with retained Müllerian derivatives often suffer from cryptorchidism, and cryptorchid testes have lower anti-Müllerian activity than do scrotal ones (Donahoe *et al.*, 1977b). None of these arguments is final. Mechanical restraint can explain failure of testicular descent in males with retained Müllerian ducts (Guerrier *et al.*, 1989), cryptorchid testes are more likely to exhibit impaired Sertoli cell function than are scrotal ones, and, finally, a testicular factor thought to play a role in the first phase of testicular descent has been identified, but it is different from AMH (Fentener van Vlissingen *et al.*, 1988).

### 2. Oocyte Meiosis-Retarding Effect

Although female germ cells enter the first stages of meiosis during fetal life, meiotic maturation does not progress beyond the dichtyate phase until the germ cell is released from the follicle (Edwards, 1965). Spermatogonia surrounded by Sertoli cells in seminiferous tubules do not enter meiosis, but those in extragonadal locations do (Upadhyay and Zamboni, 1982), suggesting that products of somatic gonadal cells may be involved: activin (Itoh *et al.*, 1990) stimulates and inhibin inhibits (Itoh *et al.*, 1990) oocyte maturation. Other growth factors, such as TGF-$\beta$, act through cumulus cells to accelerate meiosis (Feng *et al.*, 1988), Because AMH is produced by ovarian granulosa cells, particularly those surrounding the oocyte, a group of investigators believe that AMH also regulates meiotic maturation. Rat oocytes exposed to an incompletely purified preparation of bovine AMH exhibited a decreased rate of germinal vesicle breakdown, reflecting a partial inhibition of oocyte maturation (Takahashi *et al.*, 1986b). This effect was not obtained with purified human recombinant AMH unless detergent was added, and detergent, on its own, is a powerful inhibitor of oocyte maturation (Ueno *et al.*, 1988). Thus, AMH potentiates the effect of detergent on oocyte maturation, but has no specific effect.

### 3. Anticancer Effects

Because the growth of female genital primordia is repressed by AMH, and because malignant cells often revert to a fetal phenotype, Donahoe *et al.* (1979) postulated that AMH might impede the growth of cancers of female genitalia. This attractive hypothesis has not been confirmed (Rosenwaks *et al.*, 1984). Partially purified preparations of bovine AMH

ANTI-MÜLLERIAN HORMONE 33

(bAMH) have been shown to delay the growth of malignant cell lines originating from cancers of the human female genital tract (Fuller *et al.*, 1982). When, however, purified recombinant human AMH was tested for its anticancer activity by four independent assays, none of the 19 cell lines tested responded (Wallen *et al.*, 1989). A limited effect on the growth of 5 out of 43 primary tumor explants was the only positive result obtained.

### 4. Other Putative Effects

a. *Protein Phosphorylation.* Initial claims that AMH inhibits autophosphorylation of EGF receptors were based on a possible modulation of Müllerian regression by factors involved in intracellular phosphorylation, such as EGF (Hutson *et al.*, 1984). In order to demonstrate AMH/EGF antagonism, experiments were performed in A431 cells, which contain many EGF receptor molecules on their membranes. Incompletely purified bovine AMH decreased cell proliferation and blocked EGF-stimulated autophosphorylation in both intact cells and isolated membranes (Coughlin *et al.*, 1987) but, when purified to homogeneity, the human recombinant hormone was inactive in both respects (Wallen *et al.*, 1989; Cate *et al.*, 1990).

b. *Lung Maturation and Immune System.* Catlin *et al.* (1988) have reported that female lung fragments accumulate less phospholipid when cultured with testis or partially purified bovine or human recombinant AMH than do controls exposed to vehicle buffer. AMH has been reported to enhance the expression of major histocompatibility complex genes (Donahoe *et al.*, 1989).

In conclusion, although only anti-Müllerian and antiovarian effects have been proved unequivocally at the time of writing, one cannot exclude that AMH may play a wider role, in keeping with its relationship with the TGF-$\beta$ family, but clear-cut experiments, using AMH purified to homogeneity, are required to prove this point.

### D. MAPPING THE AMH BIOACTIVE SITE

As discussed in Section II,B,2, processing of the AMH molecule at a monobasic site generates a C-terminal fragment with homology to the TGF-$\beta$ active molecule. To map the AMH bioactive site, acid-generated C and N termini of recombinant human AMH were tested in the fetal ovary aromatase and in the Müllerian duct bioassays. The effect of AMH proteolytic fragments in the fetal ovary aromatase assay is shown in Fig. 14: The C terminus decreased the aromatase activity of rat fetal ovaries in a dose-dependent fashion, but exhibited only 3% of the activity of the full-length, plasmin-cleaved molecule, unless the N terminus was added

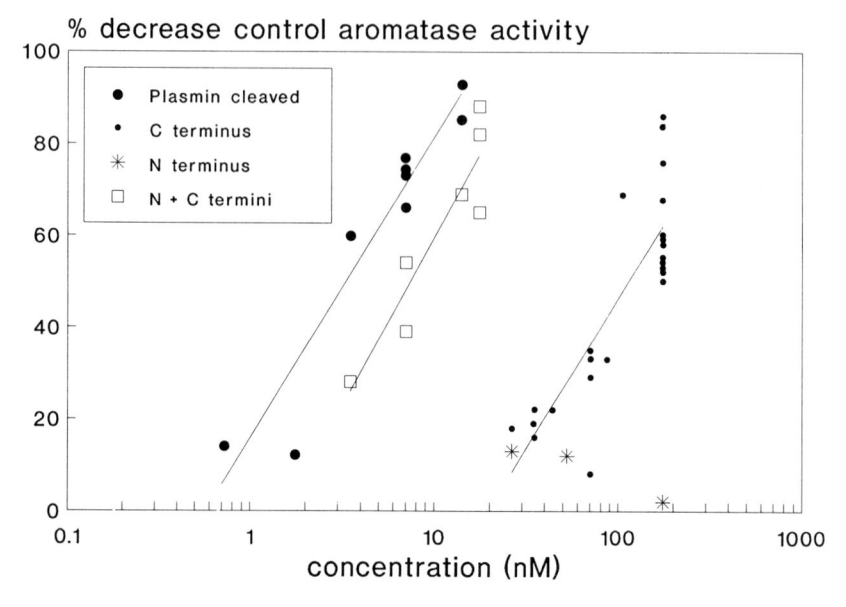

FIG. 14. Effect of acid-generated proteolytic AMH fragments in the fetal ovary/ aromatase assay. Alone, the C terminus exhibits only 3% of the activity of the plasmin-cleaved ($ED_{50}$ 120 and 3.4 n$M$, respectively). Addition of equimolar concentrations of the N terminus restores the bioactivity of the C terminus to 47% of that of the full-length molecule ($ED_{50}$ 7.2 n$M$). Alone, the N terminus has no significant activity. Reproduced, with permission, from Wilson et al. (1993).

back. Alone, the N terminus had no significant activity. Similar results were obtained using the Müllerian duct bioassay (Picon, 1969) (Fig. 15): although exhibiting no bioactivity per se, the N terminus dramatically enhanced the ability of the C terminus to inhibit Müllerian duct development. These data demonstrate that, like in other members of the TGF-$\beta$ family, the AMH C terminus alone is endowed with bioactivity. The N terminus, however, enhances the bioactivity of the AMH C-terminus, whereas it reimposes latency on the TGF-$\beta$ C terminus (Gentry and Nash, 1990). Deoxycholate (di Clemente et al., 1992b) and fetal calf serum (Wilson et al., submitted) also enhance the bioactivity of the AMH C terminus, suggesting that these reagents may act by helping to maintain a properly folded conformation.

## VI. AMH in Clinical Practice

Clinical applications of AMH research have been delayed by the lack of suitable reagents. Most anti-bAMH monoclonal antibodies are zoospecific and recognize only AMH from ruminants, such as bovine, ovine, and

ANTI-MÜLLERIAN HORMONE 35

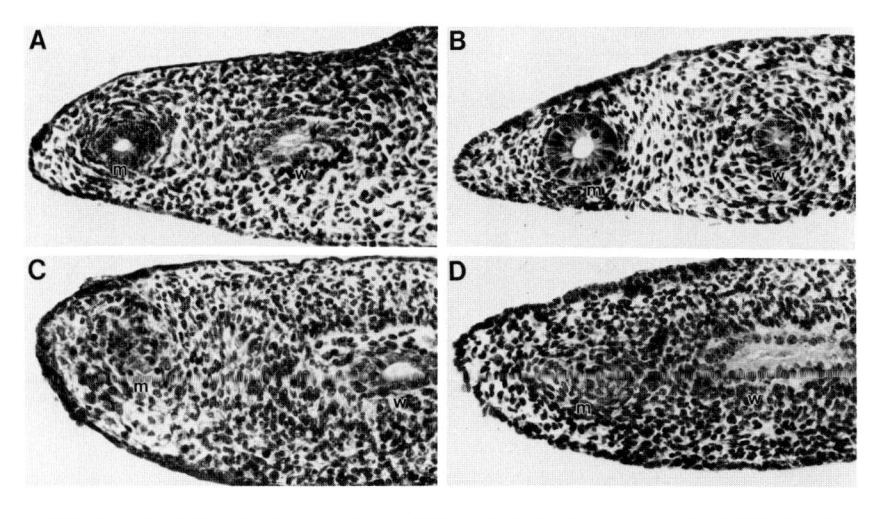

FIG. 15. Activity of acid-generated AMH proteolytic fragments in the Müllerian duct (M) assay (W, Wolffian duct). Undifferentiated (14.5-day-old) fetal rat reproductive tracts were maintained 3 days in organ culture in the presence of the following AMH acid-generated proteolytic fragments: (A) C terminus, 35 nM; the Müllerian duct shows signs of incipient regression and is surrounded by a ring of connective tissue. (B) N terminus, 35 nM; the Müllerian duct has developed normally. (C) C terminus, 35 nM, + N terminus, 35 nM; complete Müllerian regression is observed, similar to the one obtained in D using 140 nM C terminus (×235). Reproduced, with permission, from Wilson et al. (1993).

caprine species. Those recognizing conserved epitopes are IgMs, unsuited to immunochemical applications. The cloning of human AMH and the subsequent expression of the protein in heterologous cells (Cate *et al.*, 1986) opened a new area for the study of AMH in human subjects.

## A. AN IMMUNOENZYME-LINKED ASSAY FOR HUMAN AMH

In the January 1990 issue of the *Journal of Clinical Endocrinology and Metabolism* appeared three independent reports of the development of an enzyme-linked immunosorbant assay (ELISA) for hAMH. Two studies employed antibodies raised against bAMH and cross-reacting with hAMH (Baker *et al.*, 1990; Josso *et al.*, 1990a), and one (Hudson *et al.*, 1990) employed monoclonal and polyclonal antibodies raised against recombinant hAMH. Results obtained by all three groups in normal subjects were relatively similar except that values in the various age groups tended to be lower using the hAMH-specific antibodies, the only ones detecting AMH in female and adult male sera. Normal values obtained using polyclonal antibodies raised against bAMH are shown in Fig. 16. AMH levels

36                        NATHALIE JOSSO ET AL.

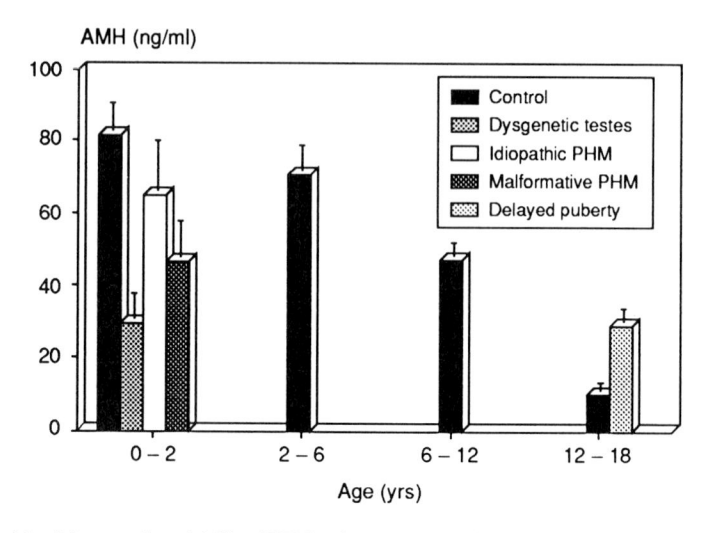

FIG. 16.   Mean and variability (SEM) of serum AMH in 117 normal boys according to age (in years): under 2 ($n = 33$), from 2 to 6 ($n = 30$), from 6 to 12 ($n = 41$), and from 12 to 18 ($n = 13$). Also depicted are values found in children with various developmental disorders: testicular dysgenesis ($n = 7$), malformative male pseudohermaphroditism ($n = 6$), idiopathic male pseudohermaphroditism ($n = 10$), and boys with delayed puberty. [See Josso *et al.* (1990a) for details.]

are high in infant boys up to 2 years of age, decrease slowly up to puberty, and fall sharply thereafter.

Figure 17 shows AMH levels in 81 boys, aged 12 to 18, with either normal or delayed pubertal maturation, measured using antibodies raised against human recombinant AMH (Carré-Eusèbe *et al.*, 1992). In the Fig. 17A, AMH serum concentration is plotted according to chronological age and exhibits great variability. In Fig. 17B, plotting according to stages of pubertal maturation shows that AMH concentration is similar in boys at either P1 [$n = 32$, mean concentration = $27.2 \pm 2.66$ (SEM) ng/ml] or P2 ($n = 23$, mean AMH = $28.5 \pm 5.7$ ng/ml) stage of pubertal maturation and falls when maturation reaches stage P3 ($n = 12$, mean AMH = $13.8 \pm 2.58$ ng/ml) or P4–P5 ($n = 14$, mean AMH = $6.4 \pm 1.2$). In Fig. 17C, AMH concentration is plotted against serum testosterone, which was measured in 28 subjects. AMH concentration is extremely variable in boys with testosterone levels below 1 ng/ml and is consistently low only after serum testosterone rises above 2 ng/ml.

Prior to puberty, serum AMH levels reflect the number and functional value of Sertoli cells. In this respect, AMH is more informative than inhibin, for which the serum level declines soon after birth (Burger *et al.*,

ANTI-MÜLLERIAN HORMONE 37

1991). Children with congenital gonadotropin deficiency due to hypopituitarism tend to have decreased serum AMH concentration (Swaenepoel *et al.*, 1991). Possibly a lack of gonadotropin stimulation impairs Sertoli cell function, as shown in newborn rats passively immunized against LHRH (Vogel *et al.*, 1983).

## B. CLINICAL APPLICATIONS OF SERUM AMH MEASUREMENTS

*A Test for the Presence of Testicular Tissue*

*a. Bilateral Cryptorchidism.* In clinical practice it is not unusual to question the existence of testicular tissue. The problem arises in normally virilized children in whom testes cannot be palpated (Fig. 18). To date, assay of plasma testosterone was the only available method of exploration, with the drawback that, after the newborn period, prolonged hCG stimulation is required to induce testosterone production by the testes. Assay of serum AMH requires no prior testicular stimulation; indeed, no means of increasing AMH production by the testis is known at the present time and, because only testes produce a measurable amount of AMH in children, no false positives due to adrenal hyperfunction, covert testosterone administration, etc., can arise. False negatives are, however, possible, if severe testicular dysgenesis is present or the patient is affected with the persistent Müllerian duct syndrome (see below).

*b. AMH in Intersex States.* In a recent study (Josso *et al.*, 1990a) we showed that serum AMH is significantly decreased in male pseudohermaphrodites with testicular dysgenesis compared to other forms of male pseudohermaphroditism. Long-term follow-up will show whether AMH values have a prognostic value, for instance, in the prediction of fertility in those patients raised as males. AMH values are normal in patients with androgen insensitivity (Fig. 18). AMH measurement is also useful in sexually ambiguous XX infants to discriminate between idiopathic female pseudohermaphroditism—where, by definition, no testicular tissue is present—and true hermaphroditism (Fig. 18). In true hermaphrodites, AMH levels are usually detectable but are low. In our series (Josso *et al.*, 1991a), only two true hermaphrodites out of 10 patients had AMH serum values within normal limits. Anatomical evidence for Müllerian regression and serum AMH levels are not always correlated, because Müllerian regression reflects testicular anti-Müllerian activity at the initiation of sex differentiation at 8 weeks of fetal life, whereas serum AMH provides information on testicular function at the time of study.

38

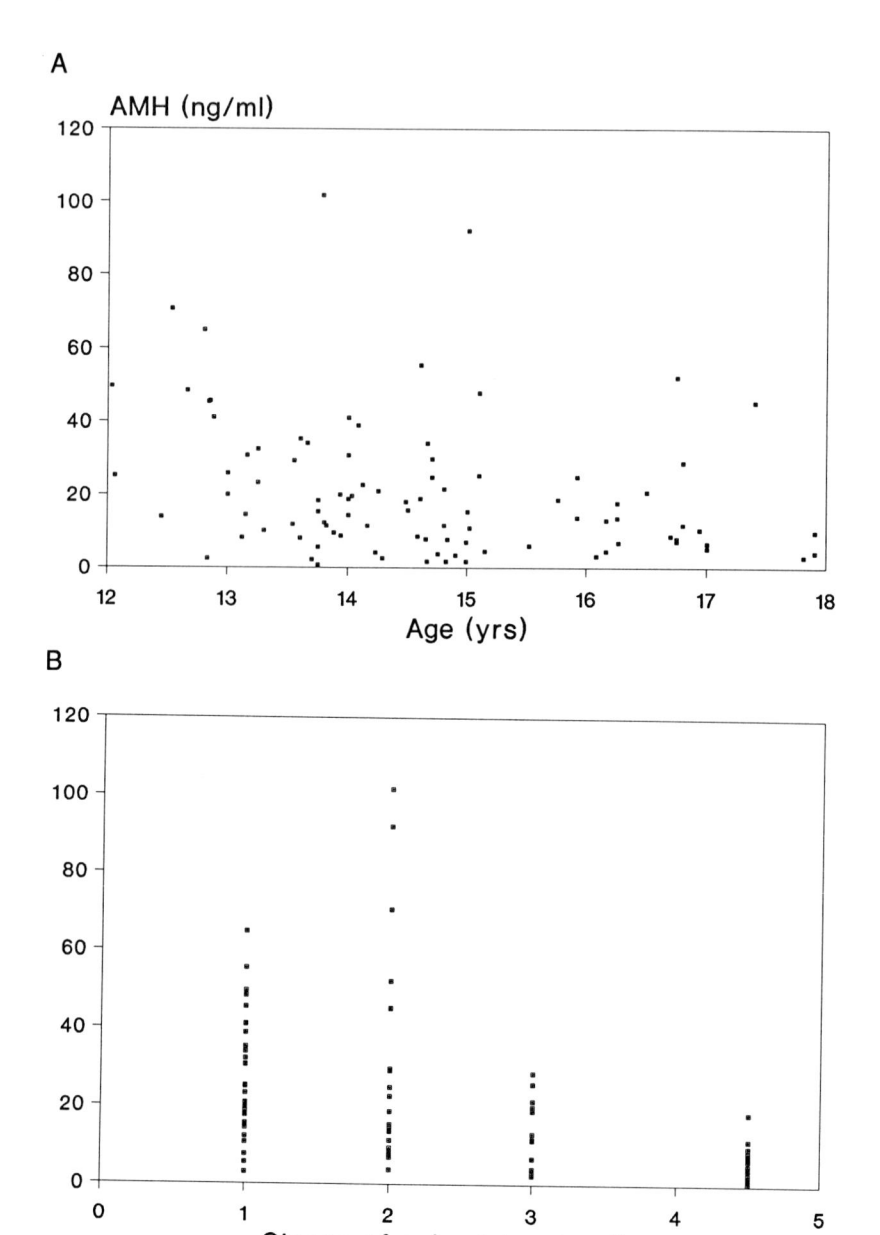

FIG. 17. Influence of various parameters of pubertal maturation on AMH serum concentration in boys aged 12–18 with normal or delayed puberty. AMH serum concentration is plotted against (A) chronological age ($n = 81$); (B) stage of pubertal maturation according to Tanner (P4 and P5 are grouped together, $n = 81$); (C) serum testosterone concentration ($n = 28$).

# ANTI-MÜLLERIAN HORMONE 39

C

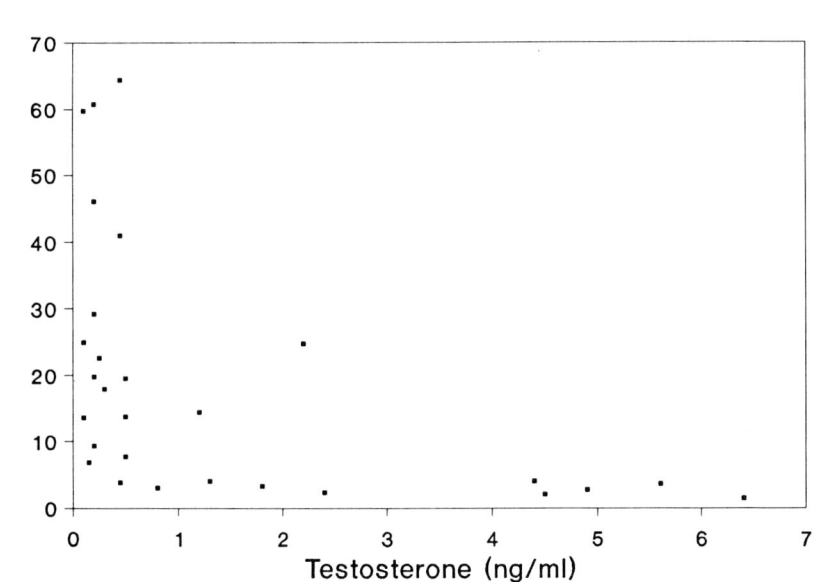

Testosterone (ng/ml)

FIG. 17.—Continued

## VII. The Persistent Müllerian Duct Syndrome

### A. CLINICAL PRESENTATION AND TREATMENT

*1. Anatomy of the Persistent Müllerian Duct Syndrome*

The persistent Müllerian duct syndrome (PMDS) is a rare form of inherited male pseudohermaphroditism characterized by the presence of uterus and tubes in otherwise normally virilized males. The condition has been reported in approximately 150 patients to date. Some years ago, it was usually discovered in adult or adolescent patients undergoing surgical repair of cryptorchidism or inguinal hernia. Because early treatment of undescended testes is now advocated, the incidence in children has increased in recent years.

The testes are tightly linked to the fallopian tubes (Fig. 19), and the clinical picture depends on the degree of mobility of the Müllerian derivatives. Usually, the Müllerian derivatives are not fixed in the pelvis and are dragged into the inguinal canal by the descending testis. The inguinal hernia, called "hernia uteri inguinalis" in the literature, contains the uterus and tubes. It may appear incarcerated, although there are no signs of

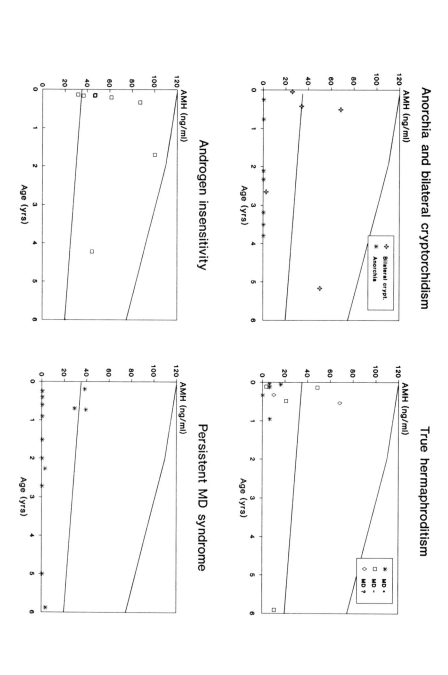

# ANTI-MÜLLERIAN HORMONE 41

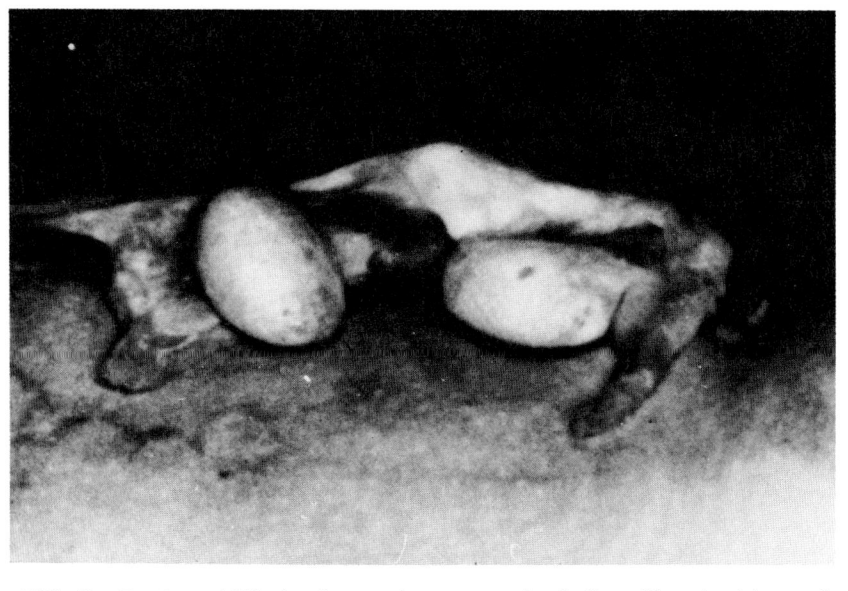

FIG. 19. Persistent Müllerian duct syndrome: operative findings. Note the tight attachment of the testes to the Fallopian tubes; the uterus is in the middle.

intestinal obstruction. Clinically, there is a negative correlation between cryptorchidism and inguinal hernia on the same side; that is, the testis and hernia are always palpable together, because of the tight attachment of the testis to the fallopian tube. Gentle traction on the uterus draws the controlateral tube and testis in the inguinal canal. Traction is not always necessary, because sometimes both testes are already in the same inguinal canal. This is known as "transverse testicular ectopia," almost never encountered when cases of PMDS are excluded. More rarely, the uterus and tube are maintained in the pelvis by the round ligament, and the testes are in an "ovarian" position: clinically, the patients are bilaterally cryptorchid and no inguinal hernia is present.

---

FIG. 18. Serum AMH values in sex ambiguity. Top left: AMH assay is usually able to distinguish between anorchia and bilateral cryptorchidism, except in one 2.5-year-old patient with severe gonadal dysgenesis. Top right: AMH serum concentration in true hermaphrodites, AMH concentration is usually decreased; only 2 patients out of 10 had values within normal limits. MD+: A uterus and tube were present, at least unilaterally. MD−: Müllerian ducts were normally regressed. MD?: Müllerian status unknown. Bottom left: AMH levels are normal in androgen insensitivity. Bottom right: AMH serum concentration is either undetectable or within normal limits in patients with the persistent Müllerian duct syndrome, indicating that not all cases are due to lack of AMH production.

In the absence of an index case in the family, PMDS is usually not suspected prior to surgery that has been scheduled for cryptorchidism or inguinal hernia. This creates practical difficulties, because the surgeon is often unacquainted with the condition and unprepared to perform the difficult separation of vas deferens from the Müllerian derivatives. When, for any reason, the diagnosis is suspected prior to surgery, inguinal or pelvic sonography can be helpful.

### 2. Treatment Aim: Preservation of Fertility

Testicular tissue differentiates normally in PMDS patients. If the testes are in a scrotal position or if early correction of cryptorchidism has been carried out, the germ cell population is normal and would be expected to undergo normal spermatogenesis. Fertility then depends on the preservation of the male excretory duct system. Sloan and Walsh (1976) have stressed that total hysterectomy is dangerous because of possible injury to the vasa deferentia embedded in the uterine wall and to the epididymis contained in the mesosalpynx. However, it is usually impossible to bring the testes into the scrotum if Müllerian derivatives are left intact, because the free segment of the spermatic cord is too short. This can be corrected by careful dissection of the cord, followed by salpingectomy and corporeal hysterectomy, leaving the fimbriae and cervix intact. In patients with transverse testicular ectopia, crossed orchidopexy gives good results.

## B. ETIOLOGY AND HEREDITARY TRANSMISSION

### 1. AMH-Positive and -Negative Forms of PMDS

PMDS is a heterogeneous disorder involving either AMH or the Müllerian duct. Elucidation of PMDS etiology (Fig. 20) is feasible only during childhood, when the testis normally expresses high amounts of AMH. A lack or severe decrease of testicular AMH production can be assessed by measuring the level of circulating AMH (Fig. 18), and is probably due to an AMH gene mutation.

When levels of circulating AMH are normal, a test of the anti-Müllerian activity of the testicular biopsy is required to distinguish between Müllerian duct insensitivity and an AMH gene mutation affecting the bioactivity but not the immunoreactivity of the molecule. This test gives valid results in children below the age of 2 years (Donahoe et al., 1977a). Figure 21 shows results obtained in two such cases, with normal AMH serum concentrations. In Fig. 21A, the testicular biopsy is able to induce regression of the rat Müllerian duct, and Müllerian duct insensitivity must be blamed for the occurrence of PMDS. Peripheral AMH resistance could

ANTI-MÜLLERIAN HORMONE 43

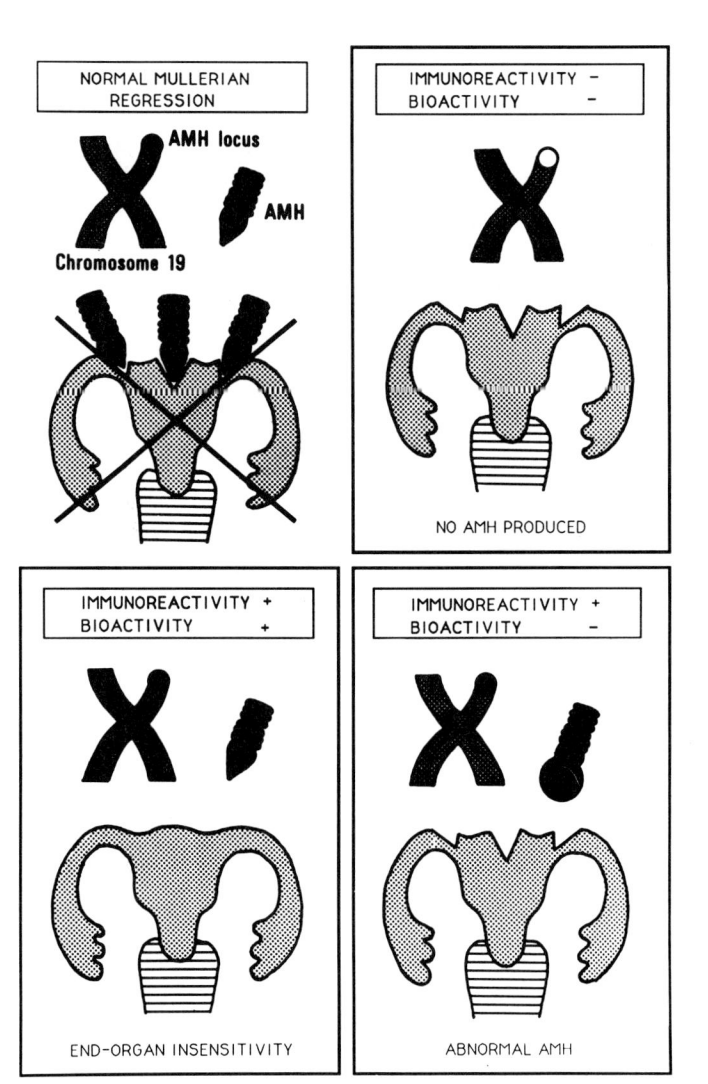

FIG. 20. Heterogeneity of the persistent Müllerian duct syndrome. Top left: Normal situation, wherein a normal gene on chromosome 19 directs the output of AMH, which interacts with putative receptors on the Müllerian ducts to induce their regression. Top right: Neither immunoreactive nor bioactive AMH can be detected; an AMH gene mutation affecting the synthesis or the stability of the AMH molecule is likely. Bottom left: Immunoreactive AMH is present in the serum and the testicular biopsy induces regression of a rat fetal Müllerian duct; Müllerian duct insensitivity is probably the cause of the persistent Müllerian duct syndrome. Bottom right: Immunoreactive AMH is present, but the testicular biopsy has no anti-Müllerian activity; an AMH mutation affecting the bioactivity but not the stability of the molecule is probably involved.

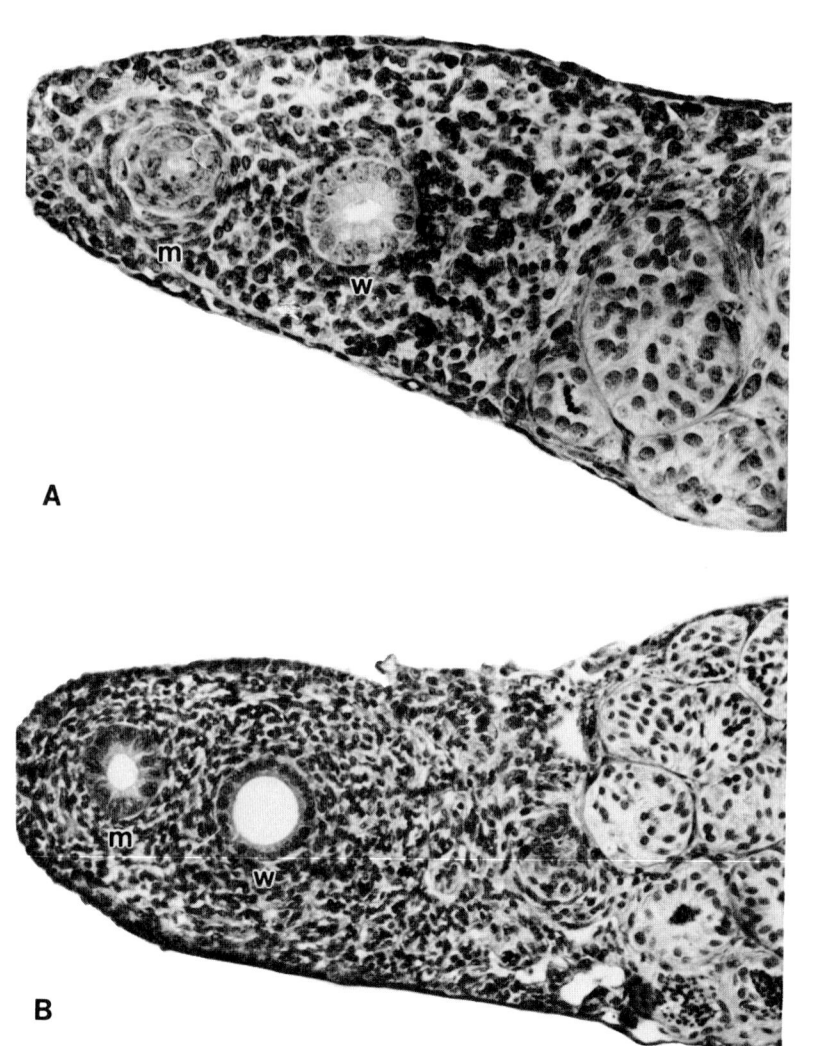

FIG. 21. Anti-Müllerian activity of testicular biopsies from two patients with AMH-positive persistent Müllerian duct syndrome. Aspect of a 14-day-old rat fetal Müllerian duct after 3 days of association in organ culture with a fragment of testicular biopsy. (A) Normal bioactivity, suggesting end-organ insensitivity. (B) No bioactivity, suggesting a mutation of the AMH gene (×175). M, Müllerian duct; W, Wolffian duct.

be due to a receptor abnormality or, alternatively, to desynchronization between AMH production by Sertoli cells and the window of Müllerian duct sensitivity to AMH. In the human fetus, Müllerian ducts respond to AMH only before 8 weeks of gestation (Josso *et al.*, 1977), and any delay in the initiation of AMH production by the fetal testis would result in

ANTI-MÜLLERIAN HORMONE 45

preservation of the Müllerian duct system. On the other hand, if the testicular biopsy does not induce rat Müllerian duct regression (Fig. 21B), a mutation leading to the production of an immunoreactive, but not bioactive, protein is the most likely possibility.

## 2. Genetic Transmission

PMDS is a genetic disorder; 12 sets of siblings, including one pair of identical twins (Weiss et al., 1978), have been described in the literature. The anatomical position of the uterus, fallopian tubes, and testes is not genetically determined (Guerrier et al., 1989). The mode of transmission of the disorder has been somewhat difficult to elucidate because, until recently, AMH-positive and AMH-negative cases could not be distinguished. Most family trees are compatible with the autosomal recessive, male-restricted transmission expected in patients with a defect in the autosomal AMH gene (Cohen-Haguenauer et al., 1987), but an apparently sex-linked PMDS transmission has been reported in two families (Sloan and Walsh, 1976; Naguib et al., 1989). The AMH status of these patients is unknown. The X-linked transmission could be explained in the AMH-positive form if the gene for the AMH receptor, which may possibly be involved, was located on the X chromosome. Against this hypothesis is the fact that, in dogs, AMH-positive PMDS is transmitted as an autosomal recessive trait (Meyers-Wallen et al., 1989) and one would expect X linkage to be conserved between species. The problem will be resolved when X-linked cases of PMDS become available for testing and the AMH receptor is cloned.

## C. MUTATIONS OF THE AMH GENE ASSOCIATED WITH PMDS

At the time of writing, several mutations of the AMH gene have been identified in AMH-negative forms of PMDS.

## 1. AMH$_{Bruxelles}$: A Nonsense Mutation

The index case was a 5-year-old Moroccan boy operated on for unilateral cryptorchidism and controlateral inguinal hernia; two brothers were also affected. Neither immunoreactive nor bioactive AMH was detectable in the testicular biopsies (Guerrier et al., 1989).

a. Detection of a Stop Codon on Both Alleles. DNA of the youngest sibling was amplified by the polymerase chain reaction (PCR) (Saiki et al., 1988). PCR products were cloned into an M13 vector and sequenced (Knebelmann et al., 1991). A guanine-to-thymidine transversion, at position 2096 in the fifth exon, changes a GAA triplet coding for Glu-382 (counted from the origin of translation) into a TAA ochre termination codon (Fig. 22).

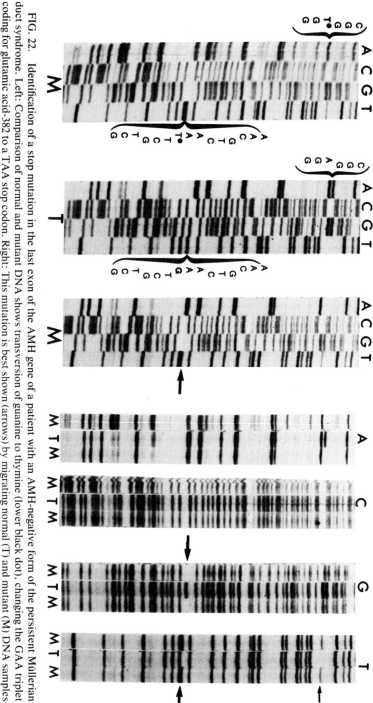

FIG. 22. Identification of a stop mutation in the last exon of the AMH gene of a patient with an AMH-negative form of the persistent Müllerian duct syndrome. Left: Comparison of normal and mutant DNA shows transversion of guanine to thymine (lower black dot), changing the GAA triplet coding for glutamic acid-382 to a TAA stop codon. Right: This mutation is best shown (arrows) by migrating normal (T) and mutant (M) DNA samples side by side for each base. From Knebelmann *et al.* (1991) with permission.

ANTI-MÜLLERIAN HORMONE 47

Homozygosity of the mutated allele was established by the results of allele-specific hybridization, which showed that the patient's DNA hybridized only with the mutant oligonucleotide probe and not with the wild-type one. Identical results were obtained with DNA from the two other affected siblings. When identical amounts of DNA were electrophoresed and hybridized with both a full-length AMH probe and an X-linked one, the ratio of the intensity of the AMH versus the X-linked gene was twice as high in control and PMDS males compared to females, thus ruling out the possibility that the second AMH allele was deleted instead of mutated.

b. *Illegitimate Transcription.* The mutation can also be shown using "illegitimate" transcripts in immortalized lymphoblasts. Chelly *et al.* (1989) have recently suggested that any gene is transcribed in any cell type, albeit in very low abundance. We took advantage of this "illegitimate transcription" to test whether it was possible to identify a gene mutation by this technique. Two rounds of amplification of the DNA complementary to RNA extracted from lymphoblastoid cells, which do not normally express AMH, generated a fragment whose sequence was that expected for AMH cDNA and carried the nonsense mutation characteristic of AMH$_{Bruxelles}$. This opens up new possibilities for the detection by PCR of mutations occurring in huge genes with numerous exons, or for the study of splice mutations in genes expressed in tissues difficult to obtain for laboratory study.

c. *Expression Analysis of the Mutant AMH Gene.* To obtain recombinant AMH$_{Bruxelles}$, the fragment of the mutant gene located between restriction sites *Xho*I and *Sac*I was amplified by PCR and ligated to a pSV2AMH (Knebelmann *et al.*, 1991) vector digested by the same enzymes. The pSV2 vectors carrying either the wild-type or the mutant AMH gene were transfected into CHO cells together with dihydrofolate reductase recombinant DNA carried in plasmid pAdD26 (Kaufman *et al.*, 1985). The transfection efficiency of pAdD26 is enhanced by the SV40 enhancer carried by the pSV2 vector, allowing coamplification of both the dihydrofolate reductase and the AMH genes. Transformants were selected by culture in Dulbecco's $\alpha$ minimal essential medium, without nucleosides, enriched with 10% dialyzed fetal calf serum. AMH production by the recombinant clones was measured by ELISA using a monoclonal antibody directed against the hAMH N terminus to exclude reaction with bovine AMH contained in fetal calf serum.

The normal and mutant AMH gene products were partially purified from culture medium, electrophoresed, and blotted onto a nitrocellulose membrane. All samples were analyzed under reducing and nonreducing conditions. The product of cells transfected with the mutant gene was much shorter than the product of the wild-type gene (Fig. 23). Visualiza-

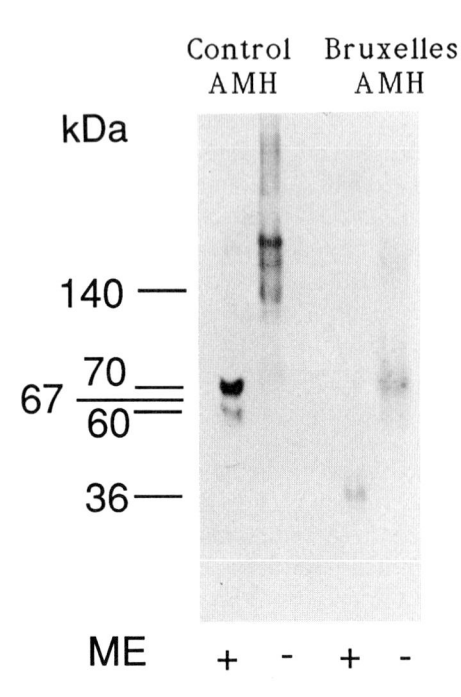

FIG. 23. Western blot of partially purified recombinant AMH secreted by CHO cells transfected with either wild-type (control) or mutant (Bruxelles) AMH genes. All samples were electrophoresed with (+) and without (−) prior reduction of disulfide bonds by β-mercaptoethanol (ME). Molecular mass of markers is shown on the left; the AMH dimer (140 kDa) and monomer (70 kDa) were included as markers. Computed molecular masses of the AMH$_{Bruxelles}$ dimer and monomer are 74 and 35 kDa, respectively. From Knebelmann *et al.* (1991) with permission.

tion of the truncated product of the AMH$_{Bruxelles}$ gene secreted by Chinese hamster ovary cells is surprising, given the AMH-negative phenotype of the mutation. Although immunocytochemically undetectable in testicular tissue, AMH$_{Bruxelles}$ was readily secreted by transfected CHO cells. Truncated gene products are expected to exhibit severe alterations in tertiary structure, making them more susceptible to posttranslational degradation by intracellular proteolytic enzymes (Adams *et al.*, 1990). The stability of the mutant AMH protein produced by CHO cells may be due to a difference in specificity in human testicular and hamster ovarian proteolytic enzymes.

### 2. AMH$_{Chicago}$: A Missense Mutation Leading to Protein Instability

The propositus, a patient of I. Rosenthal and D. Loeff (from the University of Illinois and the University of Chicago, respectively), was a 2-month-

ANTI-MÜLLERIAN HORMONE 49

old black male infant with right inguinal hernia. Penis and scrotum had a normal appearance. The parents were not related and there were no siblings. The patient was taken to surgery for inguinal herniorraphy. During the procedure, fallopian tubes and uterus were unexpectedly encountered when the hernia sac was incised. Traction on the hernia sac delivered the right testis, the Müllerian duct structures, and finally the left testis, which apparently was not fixed in the scrotum. No immunoreactive AMH was detected in the circulation by ELISA, or in testicular tissue by immunocytochemistry. Southern blots of genomic DNA failed to show any abnormality, suggesting that a point mutation, or minideletion or insertion, was responsible for PMDS in this AMH-negative patient.

*a. Screening for Point Mutations by Single-Strand Conformation Polymorphism.* To avoid sequencing of the entire mutant gene, screening methods have been developed to indicate potentially abnormal regions of the gene that can be later analyzed for sequence abnormalities. The screening method we chose, single-strand conformation polymorphism, is based on the assumption that the conformation of single-stranded nucleic acid is determined by the balance between thermal fluctuation and weak local stabilizing forces such as short intrastrand base pairings and base stacking (Orita *et al.*, 1989). Therefore, a single-stranded mutated DNA fragment is likely to exhibit a change in conformation, which can lead to mobility shifts. The effect of a given sequence change on electrophoretic mobility is unpredictable, and therefore this method may not detect all types of base changes. Nevertheless, it has been very useful in our hands, as shown in Fig. 24: some restriction fragments migrated differently from similar fragments of the control gene, suggesting that two separate point mutations or microinsertions or deletions were present. The fragments exhibiting abnormal migration were cloned in M13 and sequenced. Similar results were found in 1 sense and in 6 antisense strands. The first mutation, at nucleotide position 156, is conservative: the thymine in codon ATC coding for Ile-49 is changed to a guanine, leading to an AGC serine codon; bovine AMH also carries serine at that position (Cate *et al.*, 1986).

*b. The Chicago Mutation.* At nucleotide position 377, cytosine is replaced by a thymine, changing a CGG arginine codon to a TGG tryptophan. Direct sequencing of the product of polymerase chain reaction (Dodé *et al.*, 1990) yielded identical results (Imbeaud *et al.*, 1991). Hydrophobic cluster analysis (Lemesle-Varloot *et al.*, 1990) of the mutated portion of the molecule is shown in Fig. 25. The mutated amino acid, Arg-123, is located on the outside of a predicted amphiphilic $\alpha$ helix. In bAMH, another polar residue, serine, is found in this position. The substitution of tryptophan, a bulky aromatic residue normally within the hydrophobic

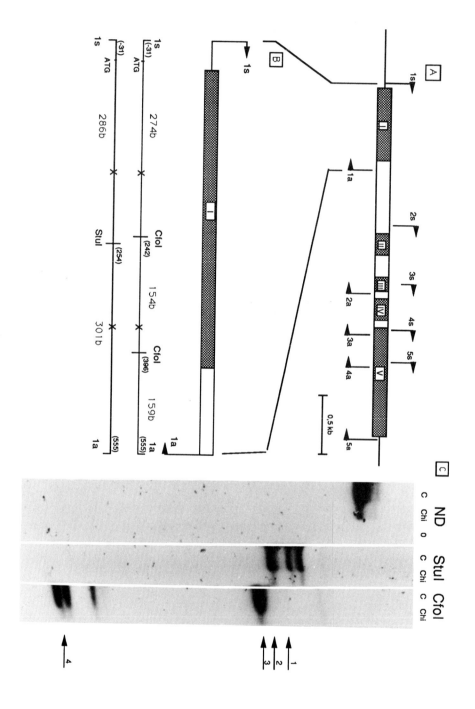

# ANTI-MÜLLERIAN HORMONE

cores (Richardson and Richardson, 1989), is expected to constrain the secondary structure, thus affecting protein folding and stability. The AMH-negative phenotype indicates that this mutation leads to instability of the AMH molecule. We cannot rule out that the replacement of Ile-49 by a serine residue amplifies the effect of the mutation at position 123 on protein stability (McPhaul et al., 1991).

## 3. AMH$_{New York}$: A Compound Heterozygote

The propositus, a baby of Italian descent, was born with normal male genitalia. At 1 month of age, an inguinal hernia became palpable on the right side, and surgery was performed, which revealed both testes to be within the right hernial sac. Attached to each testes was a vas deferens and a fallopian tube; an infantile uterus was present in the middle. No AMH was detected by ELISA (Harbison et al., 1991).

When restriction fragments of the DNA of the patient and his family, amplified by the polymerase chain reaction, were analyzed on a polyacrylamide gel, two retarded bands were visualized above the normal one in the propositus and his mother, but not in his father (Fig. 26). Sequencing revealed that the patient and his mother were both heterozygous for a 14-bp deletion in exon 2, leading to a frameshift mutation and a stop codon 10 amino acid residues downstream. The deletion is located in a region with repeated sequences, which probably favored defective recombination. The abnormal bands represent heteroduplexes between a deleted and nondeleted allele (Nagamine et al., 1989; Rommens et al., 1990). The

---

FIG. 24. Single-strand conformation polymorphism in AMH$_{Chicago}$ mutation. (A) Schematic representation of the AMH gene showing the five PCR primer pairs used to amplify the exons and splicing sites. (B) Blowup of the first exon, showing the restriction sites used to prepare fragments for single-strand conformation polymorphism. (C) Autoradiography of restriction fragments of the mutant (Chi) and normal (C) gene fragments under nondenaturing conditions. No difference in migration of the nondigested amplified 587-base fragment between patient (Chi) and control (C) can be detected; the PCR negative blank (0) shows the absence of contamination. Center: The 301-base StuI-digested fragment from both the control and the patient migrates as two separate single strands. One mutant base strand (arrow 1) exhibits abnormal migration. The strands of the 286-base fragment are not clearly individualized on the photograph; the mutant fragment (arrow 2) is retarded compared to the control fragment. Right: Similarly, the two strands of the 274-base CfoI-digested fragment migrate close to one another; the mutant fragment (arrow 3) is retarded. Migration of the 159-base fragment is normal, but migration of one of the mutant 154-base strands (arrow 4) is retarded. ND, Nondigested; StuI, CfoI, restriction enzyme-digested fragments. The location of the mutations, identified after sequencing, is indicated by Xs.

52        NATHALIE JOSSO ET AL.

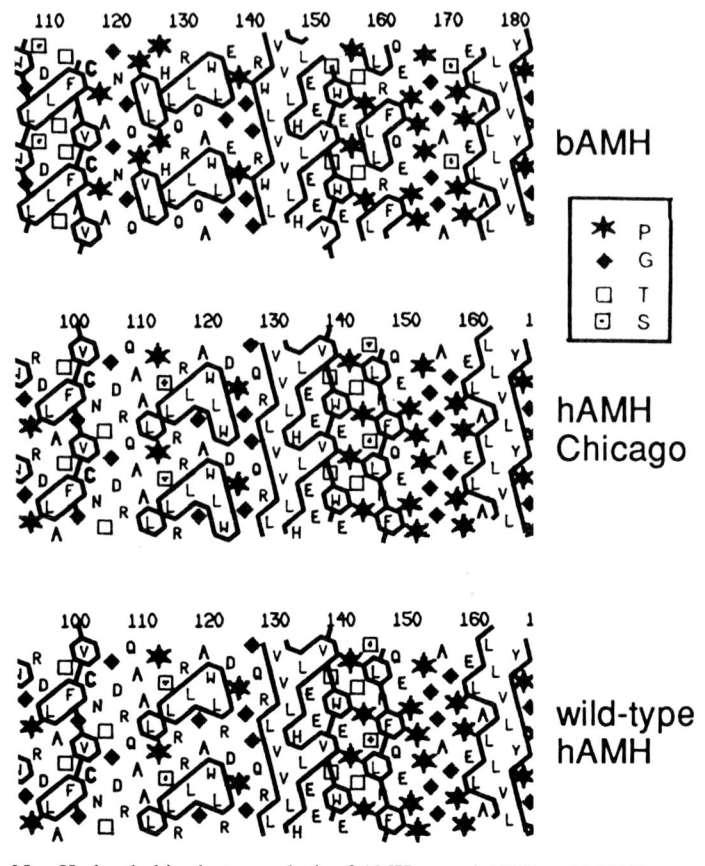

FIG. 25. Hydrophobic cluster analysis of AMH$_{Chicago}$, bAMH and hAMH, up to residue 200. The mutated amino acid, Arg-123, is located on the outside of a predicted amphiphilic α helix. In bAMH, another polar residue, serine is found in this position.

paternal allele carries a point mutation at position 1345 in the third exon. The transition from a cytosine to a thymine creates a stop codon, interrupting translation after residue 190. The Arg-191 stop mutation destroys an AvaI restriction site (Fig. 27) (Carré-Eusèbe et al., 1992).

### 4. Heterogeneity of the AMH Mutations

To date, allele-specific hybridization, performed with DNA from approximately 20 PMDS AMH-negative patients, has not revealed any mutational hot spots. This emphasizes the need for effective screening techniques to pinpoint the defective DNA region. Understanding relationships

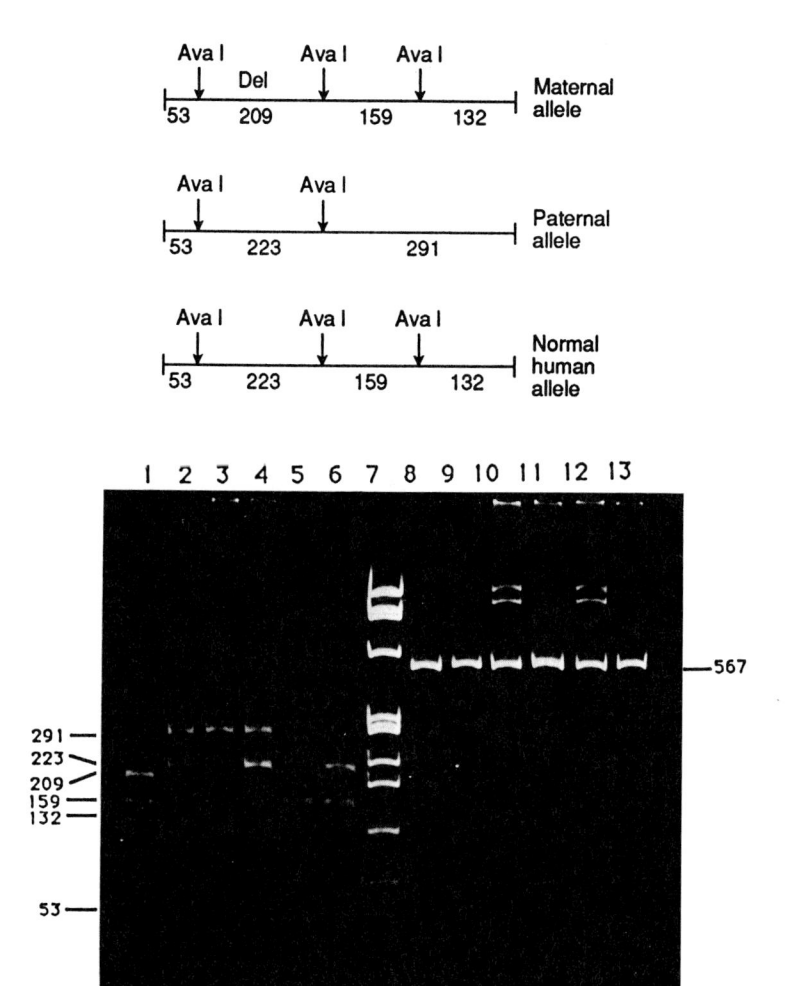

FIG. 26. Polyacrylamide gel electrophoresis of the AMH gene from a patient and his family with compound heterozygote mutations of the AMH gene, amplified by the polymerase chain reaction between bases 916 and 1472. Samples in lanes 1 to 6 were electrophoresed after digestion with restriction enzyme AvaI. The size of the expected restriction fragments in the normal gene and mutated alleles is indicated in base pairs (bp) on the left part of the figure. Lane 6, normal control; lane 7, size markers (φ×174, HaeIII cut). The maternal allele cloned in M13 (lane 1) exhibits one truncated (209 instead of 223 bp) fragment, due to the 14-bp deletion. The stop mutation in the paternal allele (lane 2) destroys a restriction site, therefore only three fragments are visible. Digestion of DNA from the father (lane 4) shows that he is heterozygous for the stop mutation and does not have the 14-base deletion. Digestion of DNA from the mother (lane 5) shows that she is heterozygous for the deletion and does not have the stop mutation. Digestion of DNA from the patient (lane 3) shows heterozygosity for both the deletion and the mutation. With the undigested samples (lanes 8–13), in addition to the normal 567-bp fragment, two retarded bands are observed in the propositus (lane 10) and his mother (lane 12), corresponding to heteroduplexes between nondeleted and deleted alleles. These fragments are not seen in maternal (lane 8) or paternal (lane 9) alleles cloned in M13, nor in DNA from the father (lane 11) or a normal control (lane 13).

MATERNAL ALLELES

Wild type
```
|G G A|G G A|G C T|G G C|C C C|C C A|G A G|C T G|G C G|C T G|C T G|G T G|
 154  Gly   Gly   Ala   Gly   Pro   Pro   Glu   Leu   Ala   Leu   Leu   Val
```

14-bp deletion
```
|G G A|G G A|G C T|G G C| .........................................  |G C T|G C T|G G T|
 154  Gly   Gly   Ala   Gly                                            Ala   Ala   Gly
```

PATERNAL ALLELES

Wild type
```
|T G C|C C C|T C C|C G A|G A C|A C C|C G C|
 188  Cys   Pro   Ser   Arg   Asp   Try   Arg
```

Stop mutation
```
|T G C|C C C|T C C|T G A|G A C|A C C|C G C|
 188  Cys   Pro   Ser   Stop
```

FIG. 27. Schematic representation of the mutations in the AMH alleles in the parents of a compound heterozygote with the persistent Müllerian duct syndrome. One of the maternal alleles exhibits a 14-bp deletion, facilitated by repeats (stippled) on both sides. The reading frame is disrupted by the deletion. One of the paternal alleles exhibits a transition from a cytosine to a thymine, leading to a TGA stop codon. The numbering refers to the amino acid residues, counted from the initiation of translation.

between structure and function should help to elucidate the basic requirements for biological activity, particularly in the AMH-positive form of the disorder, which affects the bioactivity, but not the stability of the molecule. The fetal ovary/aromatase assay should be particularly useful for testing the bioactivity of the mutant gene expressed in heterologous cells, emphasizing the benefits of active collaboration between medical geneticists, molecular biologists, and developmental biologists.

## ACKNOWLEDGMENTS

The authors are indebted to Dr. Jonathan Levin for performing the alignment shown in Fig. 2, to Dr. J. P. Mornon for interpretation of the hydrophobic analysis plots (Figs. 3 and 26), to Dr. Norman Davies for supplying samples of human fetal serum, and to Dr. Andrea Münsterberg for making unpublished data available to us. Dr. Dien Tran, Dr. Gilbert Lefèvre, Dr. Françoise Watrin, Dr. Bertrand Knebelmann, Dr. Isabelle Richard, Chris Ehrenfelds, Franck Louis, and Isabelle Lamarre participated in studies reported in this article.

## REFERENCES

Ackerman, G. E., Smith, M. E., Mendelson, C. R., MacDonald, P. C., and Simpson, E. R. (1981). *J. Clin. Endocrinol. Metab.* **53**, 412–417.

Adams, J. G., Steinberg, M. H., and Kazazian, H. H. (1990). *Br. J. Haematol.* **75**, 561–567.

Baker, M. L., Metcalfe, S. A., and Hutson, J. M. (1990). *J. Clin. Endocrinol. Metab.* **70**, 11–15.

Barton, D. E., Yang-Feng, T. L., Mason, A. J., Seeburg, P. H., and Francke, U. (1987). *Cytogenet. Cell. Genet.* **46**, 578.

Behringer, R. R., Cate, R. L., Froelick, G. J., Palmiter, R. D., and Brinster, R. L. (1990). *Nature (London)* **345**, 167–170.

Ben-David, M., Heston, W. E., and Rodbard, D. (1969). *J. Natl. Cancer Inst.* **42**, 207–218.

## ANTI-MÜLLERIAN HORMONE 55

Bercu, B. B., Morikawa, Y., Jackson, I. M. D., and Donahoe, P. K. (1978). *Pediatr. Res.* **12,** 139–142.

Bézard, J., Vigier, B., Tran, D., Mauléon, P., and Josso, N. (1987). *J. Reprod. Fertil.* **80,** 509–516.

Budzik, G. P., Powell, S. M., Kamagata, S., and Donahoe, P. K. (1983). *Cell (Cambridge, Mass.)* **34,** 307–314.

Burger, H. G., Yamada, Y., Bangah, M. L., McCloud, P. I., and Warne, G. L. (1991). *J. Clin. Endocrinol. Metab.* **72,** 682–686.

Burgoyne, P. S., Levy, E. R., and McLaren, A. (1986). *Nature (London)* **320,** 170–172.

Carré-Eusèbe, D., Imbeaud, S., Harbison, M., New, M. I., Josso, N., and Picard, J. Y. (1992). *Hum. Genet.,* In press.

Cate, R. L., Donahoe, P. K., and Mac Laughlin, D. T. (1990). Müllerian-inhibiting substance. *In* "Handbook of Experimental Pharmacology" (M. B. Sporn and A. B. Roberts, eds.), Vol. 95/II, pp 179–210. Springer-Verlag, Berlin.

Cate, R. L., Mattaliano, R. J., Hession, C., Tizard, R., Farber, N. M., Cheung, A., Ninfa, E. G., Frey, A. Z., Gash, D. J., Chow, E. P., Fisher, R. A., Bertonis, J. M., Torres, G., Wallner, B. P., Ramachandran, K. L., Ragin, R. C., Manganaro, T. F., MacLaughlin, D. T., and Donahoe, P. K. (1986). *Cell (Cambridge, Mass.)* **45,** 685–698.

Catlin, E. A., Manganaro, T. F., and Donahoe, P. K. (1988). *Am. J. Obstet. Gynecol.* **159,** 1299–1303.

Chelly, J., Concordet, J. P., Kaplan, J. C., and Kahn, A. (1989). *Proc. Natl. Acad. Sci. U.S.A.* **86,** 2617–2621.

Cohen-Haguenauer, O., Picard, J. Y., Mattei, M. G., Serero, S., Nguyen, V. C., de Tand, M. F., Guerrier, D., Hors-Cayla, M. C., Josso, N., and Frézal, J. (1987). *Cytogenet. Cell. Genet.* **44,** 2–6.

Coughlin, J. P., Donahoe, P. K., Budzik, G. P., and MacLaughlin, D. T. (1987). *Mol. Cell. Endocrinol.* **49,** 75–86.

Daniel, C. W., Silberstein, G. B., Van Horn, K., Strickland, P., and Robinson, S. (1989). *Dev. Biol.* **135,** 20–30.

di Clemente, N., Ghaffari, S., Pepinsky, R. B., Pieau, C., Josso, N., Cate, R. L., and Vigier, B. (1992a). *Development* **114,** 721–727.

di Clemente, N., Pepinsky, R. B., Saez, J., Josso, N., Cate, R. L., and Vigier, B. (1992b). (Abstract 15.04.095 presented at the Ninth International Congress of Endocrinology, Nice 1992).

Dodé, C., Rochette, J., and Krishnamoorthy, R. (1990). *Br. J. Haematol.* **76,** 275–281.

Donahoe, P. K., Cate, R. L., MacLaughlin, D. T., Epstein, J., Fuller, A. F., Takahashi, M., Coughlin, J. P., Ninfa, E. G., and Taylor, L. A. (1987). *Rec. Prog. Horm. Res.* **43,** 431–467.

Donahoe, P. K., Budzik, G. P., Trelstad, R., Mudgett-Hunter, M., Fuller, A., Jr., Hutson, J. M., Ikawa, H., Hayashi, A., and MacLaughlin, D. T. (1982). *Rec. Progr. Horm. Res.* **38,** 279–326.

Donahoe, P. K., Catlin, E., Kuroda, T., Barksdale, E. M., Epstein, J., and MacLaughlin, D. T. (1989). *In* "Development and Function of the Reproductive Organs" (N. Josso, ed.), Vol. 3, pp. 99–110. Raven Press, New York.

Donahoe, P. K., Ito, Y., Marfatia, S., and Hendren, W. H., III (1976). *Biol. Reprod.* **15,** 329–334.

Donahoe, P. K., Ito, Y., Morikawa, Y., and Hendren, W. H., III (1977a). *J. Pediatr. Surg.* **12,** 323–330.

Donahoe, P. K., Ito, Y., Price, J. M., and Hendren, W. H., III (1977b). *Biol. Reprod.* **16,** 238–243.

56 NATHALIE JOSSO ET AL.

Donahoe, P. K., Swann, D. A., Hayashi, A., and Sullivan, M. D. (1979). *Science* **205**, 913–914.

Dorizzi, M., Mignot, T. M., Guichard, A., Desvages, G., and Pieau, C. (1991). *Differentiation* **47**, 9–17.

Dyche, W. J. (1979). *J. Morphol.* **162**, 175–210.

Edwards, R. G. (1965). *Nature (London)* **208**, 349.

Erickson, G. F. (1983). *Mol. Cell. Endocrinol.* **29**, 21–49.

Feng, P., Catt, J., and Knecht, M. (1988). *Endocrinology (Baltimore)* **122**, 181–186.

Fentener van Vlissingen, F. M., Van Zoelen, E. J. J., Usem, P. J. F., and Wensing, C. J. G. (1988). *Endocrinology (Baltimore)* **123**, 2868–2877.

Fujii, D. M., Brissenden, J. E., Derynck, R., and Francke, U. (1986). *Somat. Cell. Mol. Genet.* **12**, 281–288.

Fuller, A. F., Guy, S., Budzik, G. P., and Donahoe, P. K. (1982). *J. Clin. Endocrinol. Metab.* **54**, 1051–1055.

Gelbart, W. M. (1989). *Development* **107**, 65–74.

Gentry, L. E., and Nash, B. W. (1990). *Biochemistry* **29**, 6851–6857.

George, F. W., Simpson, E. R., Milewich, L., and Wilson, J. D. (1979). *Endocrinology (Baltimore)* **105**, 1100–1106.

Gray, A. M., and Mason, A. J. (1990). *Science* **247**, 1328–1330.

Gubbay, J., Collignon, J., Koopman, P., Capel, B., Economou, A., Münsterberg, A., Vivian, N., Goodfellow, P., and Lovell-Badge, R. (1990). *Nature (London)* **346**, 245–250.

Guerrier, D., Boussin, L., Mader, S., Josso, N., Kahn, A., and Picard, J. Y. (1990). *J. Reprod. Fertil.* **88**, 695–706.

Guerrier, D., Tran, D., VanderWinden, J. M., Hideux, S., Van Outryve, L., Legeai, L., Bouchard, M., VanVliet, G., DeLaet, M. H., Picard, J. Y., Kahn, A., and Josso, N. (1989). *J. Clin. Endocrinol. Metab.* **68**, 46–52.

Haqq, C., Lee, M. M., Tizard, R., Wysk, M., DeMarinis, J., Donahoe, P. K., and Cate, R. L. (1992). *Genomics* **12**, 665–669.

Harbison, M. D., Magid, M. L. S., Josso, N., Mininberg, D. T., and New, M. I., (1991). *Ann. Génét.* **34**, 226–232.

Hayashi, H., Shima, H., Hayashi, K., Trelstad, R. L., and Donahoe, P. K. (1984). *J. Histochem. Cytochem.* **32**, 649–654.

Hudson, P. L., Dougas, I., Donahoe, P. K., Cate, R. L., Epstein, J., Pepinsky, R. B., and MacLaughlin, D. T. (1990). *J. Clin. Endocrinol. Metab.* **70**, 16–22.

Hutson, J. M., and Donahoe, P. K. (1986). *Endocrine Rev.* **7**, 270–283.

Hutson, J. M., Fallat, M. E., Kamagata, S., Donahoe, P. K., and Budzik, G. P. (1984). *Science* **233**, 586–589.

Hutson, J. M., Ikawa, H., and Donahoe, P. K. (1981). *J. Pediatr. Surg.* **16**, 822–827.

Imbeaud, S., Carré-Eusèbe, D., Boussin, L., Knebelmann, B., Guerrier, D., Josso, N., and Picard, J. Y. (1991). *Ann. Endocrinol.* **52**, 415–419.

Israel, D. I., and Kaufman, R. J. (1989). *Nucleic Acids Res.* **17**, 4589–4604.

Itoh, M., Igarashi, M., Yamada, K., Hasegawa, Y., Seki, M., Eto, Y., and Shibai, H. (1990). *Biochem. Biophys. Res. Commun.* **166**, 1479–1484.

Jean, C. (1968). *Arch. Anat. Microsc. Morphol. Exp.* **57**, 121–166.

Josso, N. (1971). *J. Clin. Endocrinol. Metab.* **32**, 404–409.

Josso, N. (1972). *J. Clin. Endocrinol. Metab.* **34**, 265–270.

Josso, N. (1974). *Pediatr. Res.* **8**, 755–758.

Josso, N., and Picard, J. Y. (1986). *Physiol. Rev.* **66**, 1038–1090.

Josso, N., Forest, M. G., and Picard, J. Y. (1975). *Biol. Reprod.* **13**, 163–167.

Josso, N., Picard, J. Y., and Tran, D. (1977). *Rec. Prog. Horm. Res.* **33**, 117–160.

Josso, N., Legeai, L., Forest, M. G., Chaussain, J. L., and Brauner, R. (1990a). *J. Clin. Endocrinol. Metab.* **70**, 23–27.

## ANTI-MÜLLERIAN HORMONE 57

Josso, N., Picard, J. Y., and Vigier, B. (1990b). *Med. Sci. (Paris)* **6**, 694–695.
Josso, N., Boussin, L., Knebelmann, B., Nihoul-Fékété, C., and Picard, J. Y. (1991a). *Trends Endocrinol. Metab.* **2**, 227–233.
Josso, N., Vigier, B., Magre, S., and Picard, J. Y. (1991b). *Seminars Dev. Biol.* **2**, 285–291.
Jost, A. (1950). *Gynecol. Obstet. (Paris)* **49**, 44–60.
Jost, A. (1953). *Rec. Prog. Horm. Res.* **8**, 379–418.
Jost, A., Vigier, B., and Prépin, J. (1972). *J. Reprod. Fertil.* **29**, 349–379.
Kanehisa, M. I. (1984). *Nucleic Acids Res.* **12**, 203–213.
Kaufman, R. J., Wasley, L. C., Spiliotes, A. J., Gossels, S. D., Latt, S. A., Larsen, G. R., and Kay, R. M. (1985). *Mol. Cell. Biol.* **5**, 1750–1759.
King, T. R., Lee, B. K., Behringer, R. R., and Eicher, E. M. (1991). *Genomics* **11**, 273–283.
Knebelmann, B., Boussin, L., Guerrier, D., Legeai, L., Kahn, A., Josso, N., and Picard, J. Y. (1991). *Proc. Natl. Acad. Sci. U.S.A.* **88**, 3767–3771.
Koopman, P., Münsterberg, A., Capel, B., Vivian, N., and Lovell-Badge, R. (1990). *Nature (London)* **348**, 450–452.
Kuroda, T., Lee, M. M., Haqq, C. M., Powell, D. M., Manganaro, T. F., and Donahoe, P. K. (1990). *Endocrinology (Baltimore)* **127**, 1825–1832.
LaQuaglia, M., Shima, H., Hudson, P., Takahashi, M., and Donahoe, P. K. (1986). *J. Urol.* **136**, 219–224.
Lee, S. J. (1990). *Mol. Endocrinol.* **4**, 1034–1040.
Lee, W., Mason, A. J., Schwall, R., Szonyi, E., and Mather, J. P. (1989). *Science* **243**, 396–397.
Lefèvre, G., Tran, D., Hoebeke, J., and Josso, N. (1989). *Mol. Cell. Endocrinol.* **62**, 125–133.
Legeai, L., Vigier, B., Tran, D., Picard, J. Y., and Josso, N. (1986). *Biol. Reprod.* **35**, 1217–1225.
Lemesle-Varloot, L., Henrissat, B., Gaboriaud, C., Bissery, V., Morgat, A., and Mornon, J. P. (1990). *Biochimie (Paris)* **72**, 555–574.
Levin, J. M. (1989). "Prédiction de la structure des protéines par homologie: Structures secondaires et modélisation de la structure tertiaire" Ph.D. dissertation, Paris XI University.
Ling, N., Ying, S. Y., Ueno, N., Shimazaki, S., Esch, F., Hotta, M., and Guillemin, R. (1986). *Nature (London)* **321**, 779–782.
Lyons, K., Graycar, J. L., Lee, A., Hashmi, S., Lindquist, P. B., Chen, E. Y., Hogan, B. L. M., and Derynck, R. (1989). *Proc. Natl. Acad. Sci. U.S.A.* **86**, 4554–4558.
Mason, A. J., Hayflick, J. S., Ling, N., Esch, F., Ueno, N., Ying, S. Y., Guillemin, R., Niall, H., and Seeburg, P. H. (1985). *Nature (London)* **318**, 659–663.
Mauléon, P., Bézard, J., and Terqui, M. (1977). *Ann. Biol. Anim. Biochem. Biophys.* **17**, 399–401.
McLaren, A. (1990). *Nature (London)* **345**, 111–112.
McPhaul, M. J., Marcelli, M., Tilley, W. D., Griffin, J. E., Isidro-Gutierrez, R. F., and Wilson, J. D. (1991). *J. Clin. Invest.* **87**, 1413–1421.
Meyers-Wallen, V. N., Donahoe, P. K., Ueno, S., Manganaro, T. F., and Patterson, D. F. (1989). *Biol. Reprod.* **41**, 881–888.
Münsterberg, A., and Lovell-Badge, R. (1991). *Development* **113**, 613–624.
Nagamine, C. M., Chan, K., and Lau, Y. F. C. (1989). *Am. J. Hum. Genet.* **45**, 337–339.
Naguib, K. K., Teebi, A. S., Al-Awadi, S. A., El-Khalifa, M. Y., and Mahfouz, E. S. (1989). *Am. J. Hum. Genet.* **33**, 180–181.
Necklaws, E. C., LaQuaglia, M. P., MacLaughlin, D. T., Hudson, P., Mudgett-Hunger, M., and Donahoe, P. K. (1986). *Endocrinology (Baltimore)* **118**, 791–796.
Newbold, R. R., Suzuki, Y., and MacLachlan, J. A. (1984). *Endocrinology (Baltimore)* **115**, 1863–1868.
Orita, M., Suzuki, Y., Sekiya, T., and Hayashi, K. (1989). *Genomics* **5**, 874–879.

58 NATHALIE JOSSO ET AL.

Padgett, R. W., St-Johnston, R. D., and Gelbart, W. M. (1987). *Nature (London)* **325,** 81–84.
Palmiter, R., Norstedt, G., Gelinas, R. E., Hammer, R. E., and Brinster, R. L. (1983). *Science* **222,** 809–814.
Pepinsky, R. B., Sinclair, L. K., Chow, E. P., Mattaliano, R. J., Manganaro, T. F., Donahoe, P. K., and Cate, R. L. (1988). *J. Biol. Chem.* **263,** 18961–18965.
Picard, J. Y., Benarous, R., Guerrier, D., Josso, N., and Kahn, A. (1986a). *Proc. Natl. Acad. Sci. U.S.A.* **83,** 5465–5468.
Picard, J. Y., Goulut, C., Bourrillon, R., and Josso, N. (1986b). *FEBS Lett.* **195,** 73–76.
Picard, J. Y., and Josso, N. (1984). *Mol. Cell. Endocrinol.* **34,** 23–29.
Picard, J. Y., Tran, D., and Josso, N. (1978). *Mol. Cell. Endocrinol.* **12,** 17–30.
Picon, R. (1969). *Arch. Anat. Microsc. Morphol. Exp.* **58,** 1–19.
Picon, R. (1970). *C.R. Acad. Sci. Sér. D. Paris* **271,** 2370–2372.
Picon, R., Pelloux, M. C., Benhaim, A., and Gloaguen, F. (1985). *J. Ster. Biochem.* **23,** 995–1000.
Prépin, J., and Hida, N. (1989). *J. Reprod. Fertil.* **87,** 375–382.
Rashedi, M., Maraud, R., and Stoll, R. (1983). *Biol. Reprod.* **29,** 1221–1228.
Richardson, J. S., and Richardson, D. C. (1989). *In* "Prediction of Protein Structure and the Principles of Protein Conformation" (G. D. Fasman, ed.), pp. 1–99. Plenum, New York.
Roberts, A. B., Flanders, K. C., Kondaiah, N. L., Thompson, N. L., Van Obberghen-Schilling, E., Wakefield, L. M., Rossi, P., de Crombrugghe, B., Heine, U., and Sporn, M. B. (1988). *Rec. Prog. Horm. Res.* **44,** 157–197.
Roberts, V., Meunier, H., Sawchenko, P. E., and Vale, W. (1989). *Endocrinology (Baltimore)* **125,** 2350–2359.
Rogers, D. S., Gallagher, D. S., and Womack, J. E. (1991). *Genomics* **9,** 298–300.
Rommens, J., Kerem, B. S., Greer, W., Chang, P., Tsui, L. C., and Ray, P. (1990). *Am. J. Hum. Genet.* **46,** 395–396.
Rosenwaks, Z., Liu, H. C., Picard, J. Y., and Josso, N. (1984). *J. Clin. Endocrinol. Metab.* **59,** 166–169.
Saiki, R. K., Gelfand, D. H., Stoffels, S., Scharf, S. J., Higuchi, R., Horn, G. T., Mullis, K. B., and Erlich, H. A. (1988). *Science* **239,** 487–491.
Shaha, C., Morris, P. L., Chen, C. L. C., Vale, W., and Bardin, C. W. (1989). *Endocrinology (Baltimore)* **125,** 1941–1950.
Shore, L., and Shemesh, M. (1981). *J. Reprod. Fertil.* **63,** 309–314.
Sinclair, A. H., Berta, P., Palmer, M. S., Hawkins, J. R., Griffiths, B. L., Smith, M. J., Foster, J. W., Frischauf, A. M., Lovell-Badge, R., and Goodfellow, P. (1990). *Nature (London)* **346,** 240–244.
Skinner, M. K., McLachlan, R. I., and Bremner, W. J. (1989). *Mol. Cell. Endocrinol.* **66,** 239–249.
Sloan, W. R., and Walsh, P. C. (1976). *J. Urol.* **115,** 459–461.
Strosberg, A. D., Guillet, J. G., Chamat, S., and Hoebeke, J. (1985). *In* "Current Topics in Microbiology and Immunology," Vol. 119, pp. 91–110. Edward Arnold, London.
Swaenepoel, C., Adamsbaum, C., Brauner, R., Chaussain, J. L., and Josso, N. (1991). *Pediatr. Res.* **35**(Suppl. S2), 50–51. [Abstr.]
Takahashi, M., Hayashi, M., Manganaro, T. F., and Donahoe, P. K. (1986a). *Biol. Reprod.* **35,** 447–454.
Takahashi, M., Koide, S. S., and Donahoe, P. K. (1986b). *Mol. Cell. Endocrinol.* **47,** 225–234.
Taketo, T., Koide, S. S., and Merchant-Larios, H. (1985). *In* "Origin and Evolution of Sex" (H. O. Halvorson and A. Monroy, Eds.), pp. 271–288. A. R. Liss, New York.
Teng, C. S., Wang, J. J., and Teng, J. I. N. (1987). *Dev. Biol.* **123,** 245–254.
Tran, D., and Josso, N. (1982). *Endocrinology (Baltimore)* **111,** 1562–1567.

## ANTI-MÜLLERIAN HORMONE

Tran, D., Meusy-Dessole, N., and Josso, N. (1977). *Nature (London)* **269**, 411–412.

Tran, D., Picard, J. Y., Vigier, B., Berger, R., and Josso, N. (1986). *Dev. Biol.* **116**, 160–167.

Tran, D., Picard, J. Y., Camparque, J., and Josso, N. (1987). *J. Histochem. Cytochem.* **35**, 733–743.

Trelstad, R. L., Hayashi, A., Hayashi, K., and Donahoe, P. K. (1982). *Dev. Biol.* **92**, 27–40.

Ueno, S., Manganaro, T. F., and Donahoe, P. K. (1988). *Endocrinology (Baltimore)* **123**, 1652–1659.

Ueno, S., Kuroda, T., MacLaughlin, D. T., Ragin, R. C., Manganaro, T. F., and Donahoe, P. K. (1989a). *Endocrinology (Baltimore)* **125**, 1060–1066.

Ueno, S., Takahashi, M., Manganaro, T. F., Ragin, R. C., and Donahoe, P. K. (1989b). *Endocrinology (Baltimore)* **124**, 1000–1006.

Upadhyay, S., and Zamboni, L. (1982). *Proc. Natl. Acad. Sci. U.S.A.* **79**, 6584–6588.

Vale, W., Rivier, J., Ying, S. Y., Vaughan, J., McClintock, R., Corrigan, A., Woo, W., Karr, D., and Spiess, J. (1986). *Nature (London)* **321**, 776–778.

Vigier, B., Tran, D., Du Mesnil du Buisson, F., Heyman, Y., and Josso, N. (1983). *J. Reprod. Fertil.* **69**, 207–214.

Vigier, B., Picard, J. Y., Tran, D., Legeai, L., and Josso, N. (1984a). *Endocrinology (Baltimore)* **114**, 1315–1320.

Vigier, B., Tran, D., Legeai, L., Bézard, J., and Josso, N. (1984b). *J. Reprod. Fertil.* **70**, 473–479.

Vigier, B., Picard, J. Y., Campargue, J., Forest, M. G., Heyman, Y., and Josso, N. (1985). *Mol. Cell. Endocrinol.* **43**, 141–150.

Vigier, B., Watrin, F., Magre, S., Tran, D., and Josso, N. (1987). *Development* **100**, 43–55.

Vigier, B., Charpentier, G., and Josso, N. (1988). *In* "Progress in Endocrinology Excerpta Medica" (H. Imura, K. Shizume, and S. Yoshida, eds.), pp. 679–684. Elsevier, Amsterdam.

Vigier, B., Forest, M. G., Eychenne, B., Bézard, J., Garrigou, O., Robel, P., and Josso, N. (1989). *Proc. Natl. Acad. Sci. U.S.A.* **86**, 3684–3688.

Vogel, D. I., Gunsalus, G. L., Bercu, B. B., Musto, N. A., and Bardin, C. W. (1983). *Endocrinology (Baltimore)* **112**, 1115–1121.

Von Heijne, G. (1986). *Nucleic Acids Res.* **14**, 4683–4690.

Voutilainen, R., and Miller, W. L. (1987). *Mol. Endocrinol.* **1**, 604–608.

Wakefield, L. (1990). *In* "Serono Symposia Publications," 70, "Hormonal Communicating Events in the Testis" (A. Isidori, A. Fabbri, and M. L. Dufau, eds.), pp. 181–190. Raven Press, New York.

Wakefield, L. M., Smith, D. M., Flanders, K. C., and Sporn, M. B. (1988). *J. Biol. Chem.* **263**, 7646–7654.

Wallen, J., Cate, R. L., Kiefer, D. M., Riemen, M. W., Martinez, D., Hoffman, R. M., Donahoe, P. K., Von Hoff, D. D., Pepinsky, B., and Oliff, A. (1989). *Cancer Res.* **49**, 2005–2011.

Weeks, D. L., and Melton, D. A. (1987). *Cell (Cambridge, Mass.)* **51**, 861–867.

Weiss, E. B., Kiefer, J. H., Rowlatt, U. F., and Rosenthal, I. M. (1978). *Pediatrics* **61**, 797–800.

Winter, J. S. D., Faiman, C., and Reyes, F. I. (1977). *In* "Morphogenesis and Malformation of the Genital System" (D. B. Bergsma and R. J. Blandau, Eds.), pp. 41–58. A. R. Liss, New York.

Wozney, J. M., Rosen, V., Celeste, A. J., Mitsock, L. M., Whitters, M. J., Kriz, R. W., Hewick, R. M., and Wang, E. A. (1988). *Science* **242**, 1528–1534.

# Expression of the FSH Receptor in the Testis

LESLIE HECKERT AND MICHAEL D. GRISWOLD

*Department of Biochemistry and Biophysics, Washington State University, Pullman, Washington 99164*

## I. Introduction

Sertoli cells are the target cells for the action of follicle-stimulating hormone (FSH) in the testes of mammals. FSH receptors are present on the Sertoli cells, and the gene for the FSH receptor is likely expressed only in these cells (Fritz, 1978; Means, 1975, 1977; Means et al., 1978; Means and Huckins, 1974). Sertoli cell function includes physical and biochemical support for germ cell development into spermatozoa. Sertoli cells create an environment in which germ cells are provided with metabolites, nutrients, and physical support and are influenced by paracrine factors. According to this scenario, FSH indirectly influences spermatogenesis by exerting influence on the Sertoli cells and on their functions (Fritz, 1978; Griswold, 1988; Griswold et al., 1988; Means et al., 1978; Means et al., 1976). In the Sertoli cells the primary action of FSH is mediated by increased concentrations of intracellular cAMP (Means et al., 1980; Means et al., 1976). There is also evidence that FSH can alter intracellular calcium levels in cultured Sertoli cells through mechanisms that are independent of both the protein kinase C pathway and the adenylate cyclase activity (Grasso et al., 1991; Grasso and Reichert, 1989). Therefore, any influence on Sertoli cell functions by FSH appears to involve cAMP and/or possibly calcium as second messengers.

## II. Physiological Response of Testis to FSH

The response of the testis to FSH is entirely dependent on the developmental status of the animal. In the rat, FSH produced by the fetal and early postnatal rat pituitary acts as a mitogen for Sertoli cells. The proliferation of Sertoli cells is maximal in 20- and 21-day-old fetuses and declines steadily until the second week after birth, when further cell division is rare (Orth, 1984). The adult population of Sertoli cells is established prior to the onset of meiosis in germ cells. Blocking the action of FSH during the early postnatal period results in testes with reduced numbers of Sertoli cells, which diminishes the ultimate spermatogenic capability of the testis. There is good experimental evidence to suggest that the size of the Sertoli cell population is limiting to overall sperm production (Orth *et al.*, 1988).

At 2 weeks after birth the proliferation of rat Sertoli cells *in vivo* and in culture becomes progressively less responsive to FSH or dibutyryl-cAMP. Most of the studies examining the actions of FSH have been done on cultured Sertoli cells from 10- to 30-day-old rats. The Sertoli cells from rats of this age are easily placed in culture and respond to FSH with increased levels of cAMP, increased protein synthesis, and increased estradiol production (Fritz *et al.*, 1976). Whereas the mitotic response is decreased and nearly finished at this age, it appears that FSH may be required for maturation of the Sertoli cells. FSH appears to be essential for the formation of tight junctional complexes between adjacent Sertoli cells and for the initiation of the first wave of spermatogenesis. It has been proposed that the action of FSH on spermatogenesis results from the stimulation of Sertoli cells to produce proteins such as androgen-binding protein (ABP), plasminogen activator, transferrin, sulfated glycoproteins 1 and 2, and a number of mitogens or growth factors (Fritz *et al.*, 1978; Griswold *et al.*, 1988). A number of morphological maturation-like features of Sertoli cells, such as the pattern of chromatin condensation, the development of large nucleoli and nuclear infoldings, and the accumulation of smooth endoplasmic reticulum, also appear to require the action of FSH (Solari and Fritz, 1978).

As the age of the rat increases to 40 days or more, the response of Sertoli cells both in culture and *in vivo* changes again. There is a large increase in the phosphodiesterase activity in the cells and in the accumulation of cAMP, and subsequent stimulation of specific protein synthesis is curtailed (Hugly *et al.*, 1988; Means *et al.*, 1980; Means *et al.*, 1976). Responses to FSH in Sertoli cells from the adult rat can usually only be measured in the presence of a phosphodiesterase inhibitor. When adult rats are hypophysectomized, spermatogenesis can be maintained by testosterone in the absence of added FSH. However, if hypophysectomized rats are allowed to regress for 20 days before hormone treatment, both FSH and testosterone are necessary to reinitiate spermatogenesis (Steinberger *et al.*, 1975; Steinberger *et al.*, 1978; Steinberger, 1971). Long-term passive immunization of adult rats with antiserum to FSH had little or no effect on spermatogenesis (Dym *et al.*, 1979). These experiments led to the suggestion that, in the rat, FSH is not required for adult spermatogenic function. The hormonal requirements of spermatogenesis in the adult rat appear to be satisfied by testosterone alone.

When rats were hypophysectomized at 40 or at 60 days of age and were maintained for 20 days with no treatment or daily injections of FSH or testosterone, the reduced action of FSH in the adult was easily illustrated (Hugly *et al.*, 1988). Daily treatment with FSH increased the testis weight, total RNA per testis, and transferrin mRNA in rats that were 40 days old at the time of hypophysectomy. FSH had no effect on these parameters in rats that were 60 days old at the time of hypophysectomy. In turn, daily injections of testosterone had no effect on these parameters in the younger group but increased all three in older rats. Figure 1 shows the increase in testis weight of the hormone-treated rats compared to the untreated hypophysectomized rats. Because the number of Sertoli cells is the same in 40- and 60-day-old rats, the testis weight is indicative of the extent to which spermatogenesis is recovered. These data do not take into account any possible cooperative interactions between FSH and testosterone.

The information presented so far suggests that FSH may not play a role in normal function of the adult rat testis. However, several other factors

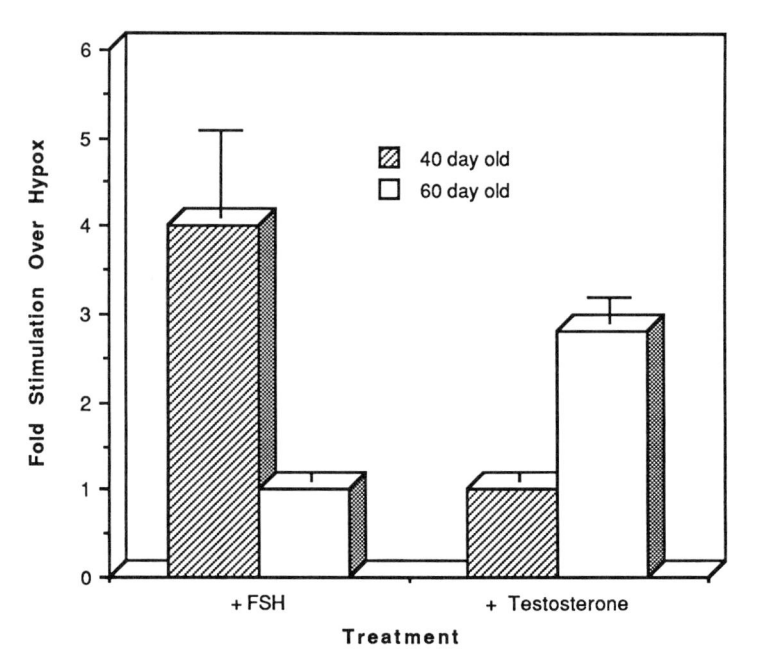

FIG. 1. Age dependence of FSH or testosterone actions on hypophysectomized rats. Rats were hypophysectomized at 40 or 60 days of age and maintained for 20 days with no treatment or daily injections of FSH (0.2 U NIH S13) or testosterone proprionate (2 mg). Testes were removed and weighed. Data are expressed as the fold stimulation in testis weight over the untreated hypophysectomized (Hypox) rats; $N = 6$.

need to be considered. First, spermatogenesis can be maintained in hypophysectomized rats in the absence of FSH, but this maintenance is only qualitatively normal and the total sperm output is compromised (Sharpe, 1987, 1989). Second, FSH receptors measured by binding studies appear to be present in the adult testis (Abou-Issa and Reichert, 1976; Thanki and Steinberger, 1978; Yoon et al., 1990). Third, a recent report shows that FSH in conjunction with low levels of testosterone is more effective at maintenance of quantitatively normal spermatogenesis than is testosterone alone (Bartlett et al., 1989). Finally, the situation is much clearer in primates, in which the action of FSH on spermatogenesis in the adult is more easily demonstrated. In primates there is a much clearer relationship between sperm output and FSH (Wickings et al., 1980).

The importance of the FSH receptor in spermatogenesis and the changing nature of the response of Sertoli cells to FSH (summarized in Table I) during development stimulated our interest in the structure of the FSH receptor (FSHR) gene and its expression in Sertoli cells.

## III. FSH Receptor

FSH receptor (FSHR), like receptors for luteinizing hormone/chorionic gonadotropin (LH/CG) and thyrotropin-stimulating hormone (TSH), is a member of a family of receptors that act through interactions with G proteins (Loosfelt et al., 1989; McFarland et al., 1989; Parmentiar, 1989; Sprengel et al., 1990). Other members of this receptor family include the adrenergic, muscarinic cholinergic, dopamine, and substance K receptors, as well as the visual pigment rhodopsin and the putative G-21 protein (Dal Toso et al., 1989; Kobilka et al., 1987a; O'Dowd et al., 1989; Sunahara, 1990). The predicted membrane topography of all of these receptors is such that they traverse the membrane with seven $\alpha$ helices oriented with an extracellular amino terminus and an intracellular carboxy terminus (Johnson and Dhanasedaran, 1989; O'Dowd et al., 1989). In this receptor

TABLE I

*Summary of the Actions of FSH in Spermatogenesis*

| Age of rat | Observed action of FSH |
| --- | --- |
| Prenatal and newborn | Sertoli cell mitogen, establishment of spermatogenic potential |
| Immature (10–45 days) | Maturation of Sertoli cells, formation of tight junctional complexes, initiation of spermatogenesis |
| Adult | Maintenance of spermatogenic output |
| Adult (hypophysectomized) | Needed for quantitatively normal spermatogenesis; if testes are allowed to regress, it is required for reinitiation of spermatogenesis |

family the primary amino acid sequence in the membrane-spanning regions is highly conserved (McFarland et al., 1989; O'Dowd et al., 1989). Cloned cDNAs for many receptors in this family have been produced and the deduced amino acid sequences reveal major differences in the size of the extracellular domain (Bunzow et al., 1988; Dal Toso et al., 1989; Gocayne et al., 1987; Heckert and Griswold, 1991; Kobilka et al., 1987a; Loosfelt et al., 1989; O'Dowd et al., 1989; Parmentiar, 1989; Sprengel et al., 1990; Sunahara, 1990). The cDNA for the FSH receptor was first cloned by Sprengel et al. (1990) and the derived amino acid sequence encodes a 675-amino acid, 75,000-Da protein (Sprengel et al., 1990). While many of the receptors in the G-protein-coupled family have very short amino termini (approximately 30–40 amino acids), the receptor for the FSH as well as LH/CG and TSH all have extensive amino-terminal domains (348, 333, and 398 amino acids for the FSH, LH/CG, and TSH receptors, respectively). The extracellular domains of these hormones share approximately 40–45% sequence similarity and are organized into a repeated series of conserved 25-residue leucine-rich motifs (for review see Vassart et al., 1991). It has been postulated that these leucine repeats are important in the protein–protein interactions leading to hormone receptor binding. The transmembrane domains of the FSH, LH/CG, and TSH receptors share about 70% sequence homology and have considerable homology to this region of other members of the G-protein-coupled family (Vassart et al., 1991).

## IV. Gene for FSH Receptor

The gene structures of several members of the G-protein-coupled family of receptors have been characterized. These include the $\alpha$- and $\beta$-adrenergic receptors (Kobilka et al., 1987b,c), D1 and D2 dopamine receptors (Dal Toso et al., 1989; Sunahara, 1990), several muscarinic receptors (Bonner et al., 1987), the putative G-21 protein (Kobilka et al., 1987a), and rhodopsin (Nathans and Hogness 1983). The genes for the adrenergic receptors, the D1 dopamine receptor, and the G-21 protein contain no introns and the muscarinic receptor genes lack introns within the coding region for the protein. In contrast, rhodopsin and the dopamine D2 receptor genes have several introns within the coding region of the protein. The structures of the LH receptor (LHR), TSH receptor (TSHR), and FSH receptor (FSHR) genes have been recently characterized and are more complex than the genes of other members in this receptor family (Gross et al., 1991; Heckert et al., 1992; Koo et al., 1991; Tsai-Morris et al., 1991).

We have characterized the FSHR gene and its promoter (Heckert et al., 1992). This gene encompasses at least 85 kb of DNA, which is divided into

10 exons, often separated by very large introns. The relationship of the 10-exon regions to the protein domains is described in Fig. 2. The first 9 exons are relatively small (68 to 185 base pairs) and code for the amino-terminal extracellular domain that contains the leucine repeat regions. The region of the molecule that encodes the seven transmembrane-spanning regions is contained within a single large exon (exon 10). The genes coding for FSHR, TSHR, and LHR have a number of similarities: (1) they are all very large genes with 10 (TSHR, FSHR) or 11 (LH) exons (summarized in Table II); (2) the last exon in each gene codes for the entire membrane-spanning region; (3) exons 2 through 9 in each receptor code for the amino-terminal repeats. This structure suggests that an ancestral gene arose from an exon coding for the leucine repeats and was duplicated several times, subsequently combining with an exon for the transmembrane region that has been conserved throughout evolution for the G-protein-coupled receptors. It has been proposed that these receptors (FSHR, LHR, and TSHR) constitute a closely related subfamily of the G-protein-coupled receptors, which have a relatively recent evolutionary origin (Vassart *et al.*, 1991).

We have also characterized the 5' upstream region of the FSHR gene (Heckert *et al.*, 1992). Primer extension and S1 nuclease experiments revealed the presence of two transcriptional start sites at positions $-80$ and $-98$ relative to the translational start site. The promoter region immediately upstream from the transcriptional start site did not contain TATA or CAAT elements. The promoter did contain a consensus AP-1-binding site at position $-214$. AP-1 is a transcriptional factor that interacts with the promoters of phorbol ester-inducible genes. There is evidence that treatment of Sertoli cells with phorbol esters in culture results in a decreased response of the cells to FSH (Monaco and Conti, 1987).

This promoter is of considerable interest because it should contain cell-specific enhancer elements that allow expression of transgenes in Sertoli cells. We constructed a fusion gene containing 830 bp of DNA 5' to the translational start site. This part of the FSHR promoter was linked to the chloramphenicol acetyltransferase reporter gene. When this gene construct was transfected into cultured Sertoli cells it was shown that this portion of the gene was capable of acting as a transcriptional promoter in Sertoli cells (Table III).

## V. Expression of FSH Receptor in the Testis

We were interested in relating the known physiological actions of FSH in the rat testis to the expression of FSH receptors in the testis. We have utilized cloned genomic DNA corresponding to part of the FSHR gene encoding only the exonic sequence as a probe on Northern blots to quantify

# FSH RECEPTOR IN THE TESTIS 67

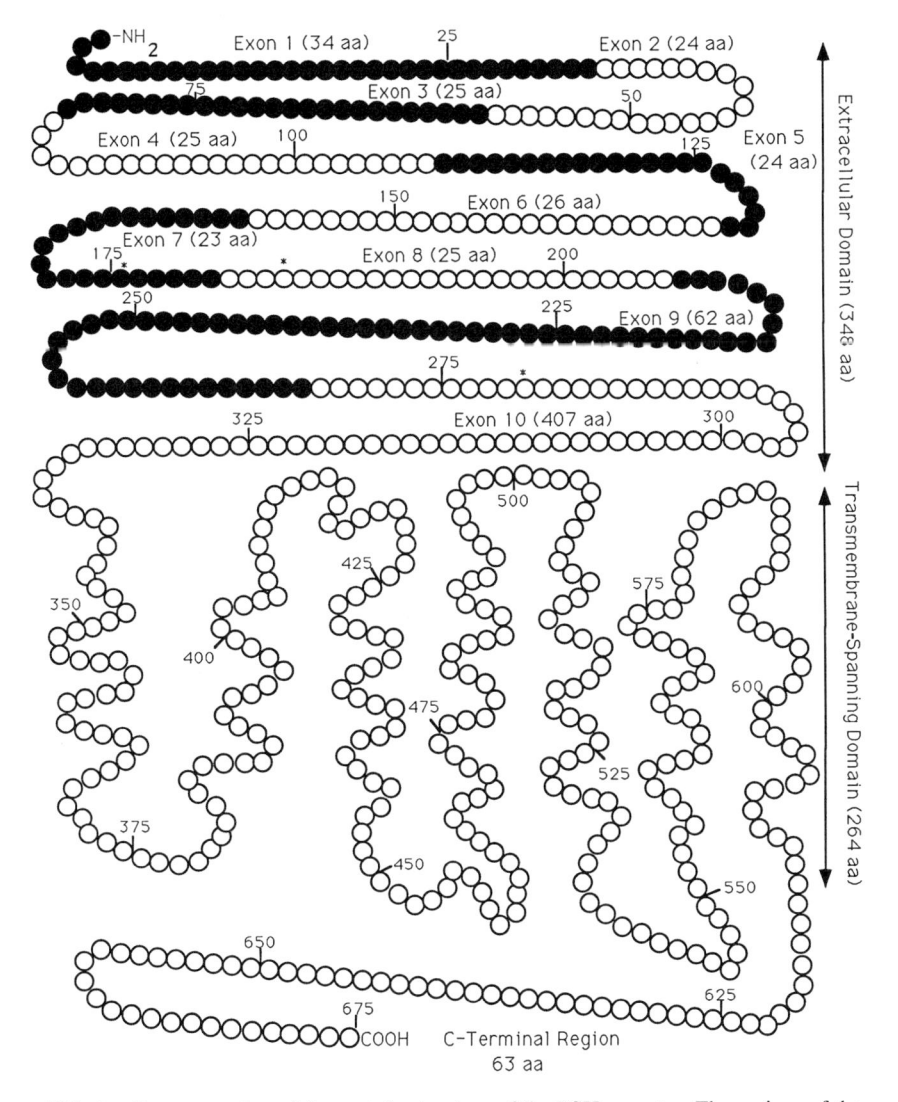

FIG. 2. Representation of the protein structure of the FSH receptor. The regions of the protein coded for by various exons were determined by Heckert *et al.* (1992). The amino acids coded for by the consecutive exons are denoted by alternating black and white circles. The three domains of the protein–extracellular, transmembrane, and C terminal—are denoted and the helical regions represent the seven transmembrane helices. The asterisks show the positions of three consensus glycosylation sites in the extracellular domain; there is a total of 675 amino acids. Redrawn from Sprengel *et al.* (1990).

## TABLE II
### Comparison of Gene Structures for the Glycoprotein Hormones

| Hormone | Species | Size (kb) | Number of exons | Ref. |
|---------|---------|-----------|-----------------|------|
| FSH | Rat | >85 | 10 | Heckert et al. (1992) |
| LH | Rat | 60 | 11 | Tsai-Morris et al. (1991) |
| | | | | Koo et al. (1991) |
| TSH | Human | >60 | 10 | Gross et al. (1991) |

the mRNA for the FSH receptor. Sertoli cells and testes were found to contain a major transcript of 2.6 kb and a minor transcript of 4.5 kb that hybridized to the FSH receptor DNA sequence (Fig. 3) (Heckert and Griswold, 1991). Our preliminary analysis of these two transcripts indicates that the 4.5-kb transcript may have an extended 3' terminus, suggesting a possible alternative polyadenylation site. If this proves to be the case, the protein product from each transcript would be identical. When a variety of tissues were screened, we detected hybridizing bands only in

## TABLE III
### Rat FSHR Promoter Activity in Transfected Sertoli Cells[a]

| Sample | Description | Butyrylated chloramphenicol (cpm) |
|--------|-------------|-----------------------------------|
| Mock transfection | No DNA | 197 |
| pCAT-basic | SV40 lacking promoter and enhancer | 350 |
| pCAT | SV40 promoter only | 589 |
| FSHR-basic | FSHR promoter only | 1,317 |
| pCAT-enhancer | SV40 enhancer only | 1,596 |
| pCAT-complete | SV40 promoter and enhancer | 10,628 |
| FSHR-enhancer | FSHR promoter plus SV40 enhancer | 13,260 |

[a] FSHR promoter DNA (830 bp) coding from region −30 to −860 relative to the translational start site was fused to the chloramphenicol acetyltransferase (CAT) reporter gene. This construct, FSHR-basic, and another construct that was essentially the same but contained the SV40 enhancer element, FSHR-enhancer, were used in transient transfection studies of rat Sertoli cells. Primary cultures of rat Sertoli cells from 20-day-old animals were transfected with 10 $\mu$g of DNA on day 4 of culture. DNA was introduced by $CaPO_4$ precipitation and after 4 hours the cells were subjected to a hyperosmotic shock (10% glycerol in culture medium) for 3 minutes. Cells were harvested 60 hours after transfection and the CAT activity was determined as described elsewhere using butyryl-CoA as a substrate. Assays were performed using 30 $\mu$l of extract (from a total of 100 $\mu$l) for 16–20 hours at 37°C. Butyrylated chloramphenicol was determined by scintillation counting after extractions with xylene. The experiments were done a minimum of three times using two different plasmid preparations to control for possible differences in transfection efficiency. The data presented are from one representative experiment.

FSH RECEPTOR IN THE TESTIS   69

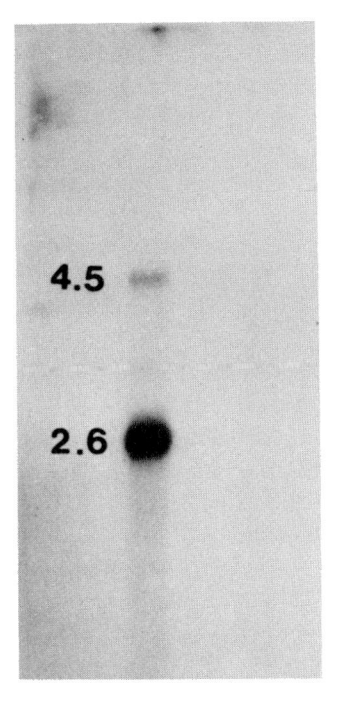

FIG. 3. Northern blot of Sertoli cell mRNA probed for the FSH receptor. The mRNA [4 μg poly(A$^+$)] was isolated from primary cultures of Sertoli cells. The Sertoli cells, obtained from 20-day-old rats, were placed in culture for 3 days prior to the isolation of the RNA. The isolation of the RNA and the Northern analysis were as described in Heckert and Griswold (1991). The probe was a randomly primed DNA from the exon that codes for the transmembrane region of the protein.

tissue from the testis or from the ovary. We found that both the 2.6- and the 4.5-kb transcripts were present in testes of rats from 10 to 60 days of age. In the adult rats the relative amount of FSHR mRNA was decreased but at least a portion of this decrease was due to the increase in testicular mRNA coming from the increased number of germinal cells. However, it was not clear from these studies if the adult rat testis contained reduced amounts of FSHR mRNA.

Sertoli cells from adult and 20-day-old rats were placed in culture in order to enrich significantly the Sertoli cell population. When the mRNA from these cells was analyzed on Northern blots, the relative amount of FSHR mRNA was found to be similar in Sertoli cells from adult and 20-day-old rats. From these results we conclude that the adult rat testis has the capability to express FSH receptors at a level that is at least equal to that of the 20-day-old rats (Heckert and Griswold, 1991).

Spermatogenesis in the adult rat is organized into a series of 14 stages that constitute the cycle of the seminiferous epithelium and are defined by the germinal cell composition of the tubules (Leblond and Clermont, 1952). In the normal rat the 14 stages are present simultaneously at different regions along the length of the tubule, and the release of spermatozoa (spermiation), which occurs at stage VIII, is asynchronous. There is good evidence that the functions of the Sertoli cells vary with the different stages of the cycle. Two methods have been employed to examine these functional changes. In the first method, segments of seminiferous tubules in defined stages of the cycle are manually dissected from the whole testis (Parvinen, 1982). The second method utilizes retinol deprivation and repletion, which results in the synchronization of the testis to three or four related stages of the cycle (Morales and Griswold, 1987a,b). Spermatogenesis appears to proceed normally, but because the entire testis is in roughly the same stage, spermiation occurs only every 12 to 13 days. We have utilized the retinol synchronization method to obtain synchronized testes, which represent all parts of the cycle. This method allows the collection of sufficient amounts of stage-synchronized testis to enable the analysis of low-abundance cell products (Morales *et al.,* 1989).

We have speculated that the function of FSH in the adult rat may be confined to very discrete regions of the cycle of the seminiferous epithelium. Utilizing the FSH receptor probe generated from the exonic regions of the gene and mRNA isolated from synchronized testes, we were able to measure the steady-state levels of FSHR mRNA during the different stages of the cycle of the seminiferous epithelium (Heckert and Griswold, 1991). The mRNA was isolated from the testes of stage-synchronized rats and analyzed by Northern blots. We found that the relative levels of FSHR mRNA varied in a cyclic manner with low levels in stages V–IX and fivefold higher levels in stages XIII, XIV, and I. If these data are plotted over two cycles of the seminiferous epithelium, the cyclic nature of the amounts of FSHR mRNA is apparent (Fig. 4). The binding of labeled FSH and the ability of FSH to stimulate cAMP production were determined in dissected tubules of defined stages and the results correlate very well with our analysis of mRNA levels (Kangasniemi *et al.,* 1990a,b). The tubules in stages XIII–II bound more FSH and responded to exogenous FSH with increased cAMP levels. The tubules in stages VII–VIII were essentially refractory to FSH stimulation of cAMP levels. Altogether, these studies suggest that the primary action of FSH in the adult rat is cyclic in nature and may be confined to stages XII–IV. These are the stages that encompass a number of the spermatogonial divisions, and an effect of FSH on these mitotic divisions would explain the decrease in the efficiency of spermatogenesis in FSH-deprived animals (Sharpe, 1989). Another factor

# FSH RECEPTOR IN THE TESTIS

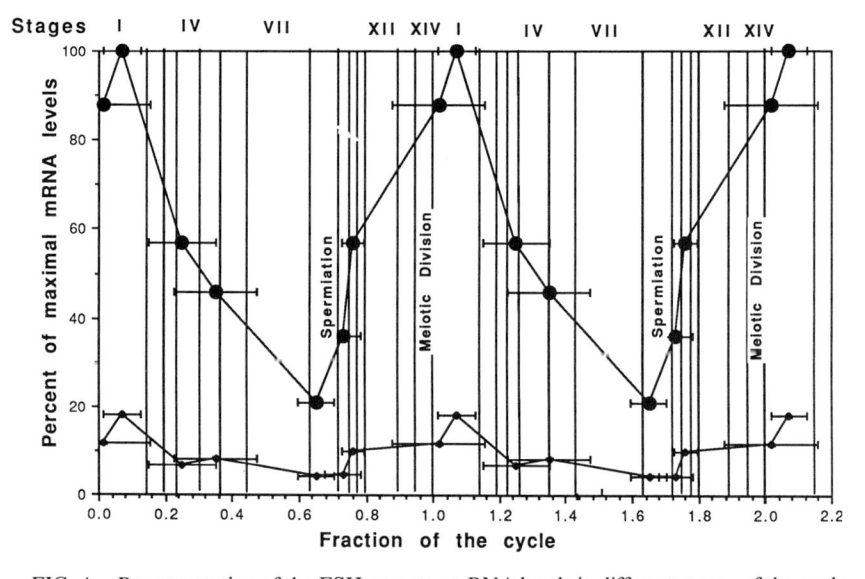

FIG. 4. Representation of the FSH receptor mRNA levels in different stages of the cycle of the seminiferous epithelium as determined in stage-synchronized rats. The data are plotted in duplicate such that two complete cycles are represented. The stages, designated at the top of the graph and by the vertical lines, have different durations and represent different fractions of a cycle. Each point represents the calculated midpoint of synchrony and the horizontal bars represent ±1 standard deviation of the distribution of the stages in any sample (Van Beek and Meistrich, 1990). Stage-synchronized testes were obtained from rats using the Vitamin A depletion–replacement procedure (Morales *et al.*, 1987). The distribution of stages (200 tubules) in each testis was determined by light microscopy. The total RNA was isolated from each sample and the mRNA was isolated by chromatography over oligo-dT columns. The mRNA (10 $\mu$g) from each testis was run on a denaturing agarose gel and blotted to a nylon filter. The probe was generated by random, primed replication of a plasmid containing the exonic DNA coding for the FSH receptor. The band densities were determined by densitometric scanning of the autoradiographs. The amount of RNA loaded was determined by hybridization to random, primed cDNA for the SGP-1 sequence. Parts of the cycle in which spermiation (stage VIII) and the meiotic divisions (stage XIV) take place are noted. The large black circles are the data for the 2.6-kb FSH receptor transcript and the small black diamonds are the data for the 4.5-kb FSH receptor transcript.

in the action of FSH in the adult testis is phosphodiesterase. Because the known actions of FSH are modulated by cAMP, then the presence of a cAMP phosphodiesterase could be inhibitory to this action. The phosphodiesterase activity of Sertoli cells in the adult rat appears to be maximal during stages VII–VIII of the cycle (Parvinen, 1982). This result would be consistent with a lack of FSH action during these stages.

## VI. Control of FSHR Expression

We wished to examine the possible regulation of FSHR mRNA levels by hormones and growth factors that play important roles in spermatogenesis. In particular, we wanted to determine whether FSH could downregulate the FSH receptor at the mRNA level and whether phorbol esters, potentially acting through the AP-1 site on the FSHR promoter, decreased the steady-state levels of FSHR mRNA. Sertoli cells from 20-day-old rats were placed in culture and treated with different combinations of hormones, phorbol esters, and vitamin A. The amount of FSHR mRNA in these cells was then assayed by Northern blots. The data were normalized to the mRNA levels of actin and SGP-2, which have been shown be unchanged in cell culture in the presence of different hormones (Fig. 5). It was found that after 5 days in culture, FSH, insulin, retinol, or testosterone added individually had no effect on the FSHR levels [relative to total poly($A^+$) mRNA]. Addition of all four reagents resulted in a 1.5- to 2-fold increase in the relative FSHR levels. Another study has shown that FSH added to cultured Sertoli cells down-regulated the level of FSHR mRNA within 4 hours of addition of hormone to cultured Sertoli cells (Themmen et al., 1991). In this study, transcriptional run-on experiments revealed that FSH did not inhibit the initiation of transcription of the FSHR gene and that by 16 hours after addition of hormone the FSHR mRNA levels had returned to normal. This study suggested that the FSH-induced downregulation of FSHR mRNA occurred through a posttranscriptional mechanism.

The presence of the AP-1 site in the FSHR promoter sequence and a report that showed that phorbol esters desensitize Sertoli cells to FSH prompted the examination of the effect of phorbol esters on the steady-state levels of FSHR mRNA (Heckert et al., 1992; Monaco and Conti, 1987). The addition of phorbol esters to cultured Sertoli cells results in a rapid and dramatic decrease in FSHR mRNA followed by a return to steady-state levels. However, in these experiments there is also a decrease in the levels of other specific mRNAs, such as SGP-1 mRNA, actin mRNA, and ribosomal protein mRNA. This suggests that phorbol esters may have a general effect on the transcription of many mRNAs in the Sertoli cells. The inhibitory action of phorbol esters on FSHR mRNA is greater than the inhibition of these other mRNAs. As shown in Fig. 6, the levels of FSHR mRNA are normalized to the levels of ribosomal protein S8 mRNA and there is a repeatable decrease in the FSHR mRNA relative to S8 mRNA. The mechanism of this inhibitory action is unknown (transcriptional or posttranscriptional), but the results clearly show that phorbol esters can alter the FSH response of Sertoli cells by lowering the steady-state levels of FSHR mRNA.

FSH RECEPTOR IN THE TESTIS 73

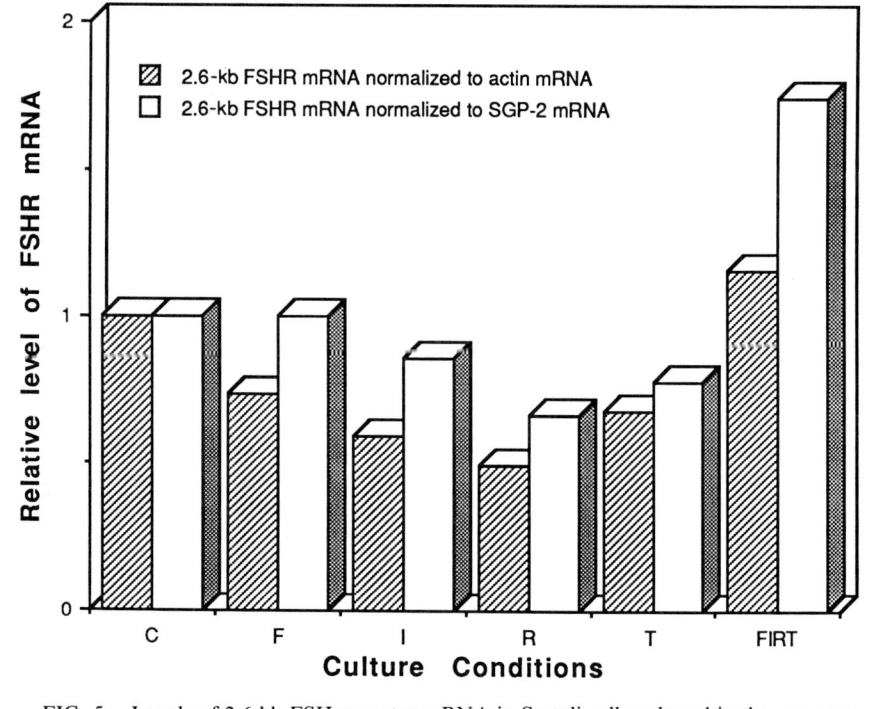

FIG. 5. Levels of 2.6-kb FSH receptor mRNA in Sertoli cells cultured in the presence of different hormones. Sertoli cells were cultured for 5 days in the presence of no added hormones, 25 ng/ml FSH (F), 5 $\mu$g/ml insulin (I), 0.3 $\mu M$ retinol (R), 0.7 $\mu M$ testosterone (T), or a mixture all four (FIRT). Poly(A$^+$) mRNA was isolated from the Sertoli cells and 2 $\mu$g was analyzed by Northern blots probed with random, primed exonic DNA derived from the FSH receptor gene. The density of the bands on the autoradiogram of the Northern blot was determined on a BioImage gel scanner. The same Northern blots were also probed with labeled actin and SGP-2 cDNA. These transcripts are not expected to change as a result of hormone treatment and are used as a control for gel loading. The FSHR data for the 2.6-kb transcript were normalized to the levels of actin mRNA and SGP-2 mRNA. This experiment represents the data from one Northern blot, which was representative of the total results.

## VII. Summary

FSH has multiple and changing roles in the regulation of spermatogenesis. The first function of FSH is to increase the number of Sertoli cells by stimulation of their mitotic activity. During the prepubertal phase of development, FSH is important for the maturation of the Sertoli cells. Hormonal stimulation of tight junction formation and specific protein secretion are essential. In the adult rat, some of the functions carried out by FSH in prepubertal animals are assumed by testosterone. However, there

74    LESLIE HECKERT AND MICHAEL D. GRISWOLD

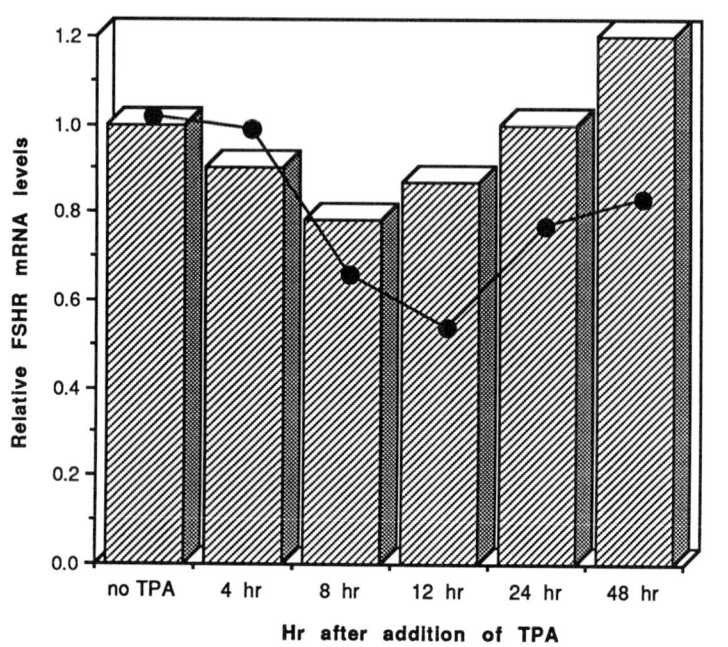

FIG. 6.    Response of FSHR mRNA levels to the phorbol ester 13-tetradecanoyl-O-phor-
bol acetate (TPA). Sertoli cells from 20-day-old rats were cultured in the presence of F-12
media, and on the third day of culture, TPA was added at a concentration of 100 ng/ml. Total
RNA was isolated from the cultured cells at the times indicated after TPA addition and
20 $\mu$g was used to analyze mRNA levels by Northern blot. The band intensities on Northern
blots were quantified by densitometry. In the bar graph, the levels of FSHR mRNA were
first normalized within each treatment to the level of mRNA for the ribosomal protein S8,
and these relative values were then normalized to the relative band intensity for the control
sample (no added TPA). The bars show that FSHR mRNA changes in relation to other
specific mRNAs (in this case the S8 mRNA). The line graph represents the average of three
separate experiments, showing the FSHR mRNA levels relative to total RNA.

is evidence that even in the adult rat, FSH is important for quantitatively
normal spermatogenesis. The gene for the FSH receptor is large (greater than
85 kb) and complex (10 introns) and is structurally similar to the genes for
the LH and TSH receptor. The promoter region of the FSHR gene has
been identified and is active in the expression of transgenes in transfected
Sertoli cells. We have shown that the FSH receptor mRNA is present in
the testes of the adult rat and that the levels of this mRNA are changing
during the cycle of the seminiferous epithelium. The presence of relatively
high levels of FSHR mRNA in stages XIV–II of the cycle and the relatively
low levels in stages VII–VIII suggest that the FSH receptor is carefully

## FSH RECEPTOR IN THE TESTIS 75

regulated in adult rats and presumably has an important function in spermatogenesis. The levels of FSHR mRNA in cultured Sertoli cells are immediately reduced in the presence of FSH or phorbol esters, but the levels soon return to normal.

## ACKNOWLEDGMENT

This work was supported by Grants HD 10808 and 25846 to MDG from NIH.

## REFERENCES

Abou-Issa, H., and Reichert, L. E. (1976). *J. Biol. Chem.* **251,** 3326–3337.
Bartlett, J. M. S., Weinbauer, G. F., and Nieschlag, E. (1989). *J. Endocrinol.* **121,** 49–58.
Bonner, T. I., Buckley, N. J., Young, A. C., and Brann, M. R. (1987). *Science* **237,** 527–532.
Bunzow, J. R., Van Tol, H. H. M., Grandy, D. K., Albert, P., Salon, J., Christie, M., Machida, C. A., Neve, K. A., and Civilli, O. (1988). *Nature (London)* **336,** 783–787.
Dal Toso, R., Sommer, B., Ewert, M., Herb, A., Pritchett, D. B., Bach, A., Shivers, B. D., and Seeburg, P. H. (1989). *EMBO J.* **8,** 4025–4034.
Dym, M., Raj, H. G., Lin, M. Y. C., Chemes, H. E., Kotite, N. J., Nayfeh, S. N., and French, F. S. (1979). *J. Reprod. Fert.* **26**(Suppl.), 175–181.
Fritz, I. (1978). "Biochemical Actions of Hormones." New York, Academic Press.
Fritz, I. B., Griswold, M. D., Louis, B. G., and Dorrington, J. H. (1978). *Can. J. Biochem.* **56**(9), 875–879.
Fritz, K. B., Rommerts, F. G., Louis, B. G., and Dorrington, J. H. (1976). *J. Reprod. Fert.* **46,** 17–24.
Gocayne, J., Robinson, D. A., Fitzgerald, M. G., Chiung, F. Z., Kerlavage, A. R., Lentes, K. U., Lai, J., Wang, C. D., Fraser, C. M., and Venter, J. C. (1987). *Proc. Natl. Acad. Sci. U.S.A.* **84,** 8296–8300.
Grasso, P., Joseph, M. P., and Reichert, L. E. (1991). *Endocrinology (Baltimore)* **128**(1), 158–164.
Grasso, P., and Reichert, L. J. (1989). *Endocrinology (Baltimore)* **125**(6), 3029–3036.
Griswold, M. D. (1988). *Int. Rev. Cytol.* **110**(133), 133–156.
Griswold, M. D., Morales, C., and Sylvester, S. R. (1988). *Oxford Rev. Reprod. Biol.* **10**(124), 124–161.
Gross, B., Misrahi, M., Sar, S., and Milgrom, E. (1991). *Biochem. Biophys. Res. Commun.* **177,** 679–687.
Heckert, L. L., Daley, I., and Griswold, M. D. (1992). *Mol. Endocrinol.* **6,** 70–80.
Heckert, L. L., and Griswold, M. D. (1991). *Mol. Endocrinol.* **5**(5), 670–677.
Hugly, S., Roberts, K., and Griswold, M. D. (1988). *Endocrinology (Baltimore)* **122**(4), 1390–1396.
Johnson, G. L., and Dhanasedaran, N. (1989). *Endocr. Rev.* **10,** 317–331.
Kangasniemi, M., Kaipia, A., Mali, P., Toppari, J., Huhtaniemi, I., and Parvinen, M. (1990a). *Anat. Rec.* **227**(1), 62–76.
Kangasniemi, M., Kaipia, A., Toppari, J., Perheentupa, A., Huhtaniemi, I., and Parvinen, M. (1990b). *J. Androl.* **11**(4), 336–343.
Kobilka, B. K., Frielle, T., Collins, S., Yang-Feng, T., Kobilka, T. S., Francke, U., Lefkowitz, R. J., and Caron, M. G. (1987a). *Nature (London)* **329,** 75–79.

# 76 LESLIE HECKERT AND MICHAEL D. GRISWOLD

Kobilka, B. K., Frielle, T., Dohlman, H. G., Bolanski, M. A., Dixon, R. A. F., Keller, P., Caron, M. G., and Lefkowitz, R. J. (1987b). *J. Biol. Chem.* **262,** 7321–7327.

Kobilka, B. K., Matsui, H., Kobilka, T., Yang-Feng, T., Francke, U., Caron, M., Lefkowitz, R., and Regen, J. (1987c). *Science* **238,** 650–656.

Koo, Y. B., Ji, I., Slaughter, R. G., and Ji., T. H. (1991). *Endocrinology (Baltimore)* **128,** 2297–2650.

Leblond, C. P., and Clermont, Y. (1952). *Ann. N.Y. Acad. Sci.* **55,** 548–573.

Loosfelt, H., Misrahi, M., Atger, M., Salesse, R., Thi, M. T. V. H.-L., Jolivet, A., Guiochon-Mantel, A., Sar, S., Jallal, B., Garnier, J., and Milgrom, E. (1989). *Science* **245,** 525–528.

McFarland, K. C., Sprengel, R., Phillips, H. S., Köhler, M., Rosemblit, N., Nikolics, N. K., Segaloff, D. L., and Seeburg, P. H. (1989). *Science* **245,** 494–499.

Means, A. R. (1975). "Handbook of Physiology," V, "Male Reproductive System," Section 7, pp. 203–218. American Physiological Society, Washington, D.C.

Means, A. R. (1977). "The Testis." Academic Press, New York.

Means, A. R., Dedman, J. R., Tash, J. S., Tindall, D. J., van Sickle, M., and Welsh, M. J. (1980). *Annu. Rev. Physiol.* **42**(59), 59–70.

Means, A. R., Dedman, J. R., Tindall, D. J., and Welsh, M. J. (1978). *In* "Endocrine Approach to Male Contraception," pp. 403–423. Scriptor, Copenhagen.

Means, A. R., Fakunding, J. L., Huckins, C., Tindall, D. J., and Vitale, R. (1976). *Recent Prog. Horm. Res.* **32**(477), 477–527.

Means, A. R., and Huckins, C. (1974). *In* "Hormone Binding and Target Cell Activation in the Testis" (M. L. Dufau, and A. R. Means, eds.) pp. 145–165. New York, Plenum.

Monaco, L., and Conti, M. (1987). *Mol. Cell. Endocrinol.* **49**(2,3), 277–236.

Morales, C., and Griswold, M. D. (1987a). *Endocrinology (Baltimore)* **121**(1), 432–434.

Morales, C. R., and Griswold, M. D. (1987b). *Ann. N.Y. Acad. Sci.* **513,** 292–293.

Morales, C., Hugly, S., and Griswold, M. D. (1987). *Biol. Reprod.* **36**(4), 1035–1046.

Morales, C. R., Alcivar, A. A., Hecht, N. B., and Griswold, M. D. (1989). *Mol. Endocrinol.* **3**(4), 725–733.

Nathans, J., and Hogness, D. S. (1983). *Cell (Cambridge, Mass.)* **34,** 807–814.

O'Dowd, B. F., Lefkowitz, R. J., and Caron, M. G. (1989). *Annu. Rev. Neurosci.* **12,** 67–83.

Orth, J. M. (1984). *Endocrinology (Baltimore)* **115**(4), 1248–1255.

Orth, J. M., Gunsalus, G. L., and Lamperti, A. A. (1988). *Endocrinology (Baltimore)* **122**(3), 787–794.

Parmentiar, M. (1989). *Science* **246,** 1620–1622.

Parvinen, M. (1982). *Endocr. Rev.* **3**(4), 404–417.

Sharpe, R. M. (1987). *J. Endocrinol.* **113,** 1–2.

Sharpe, R. M. (1989). *J. Endocrinol.* **121,** 405–407.

Solari, A. J., and Fritz, I. B. (1978). *Biol. Reprod.* **18**(3), 329–345.

Sprengel, R., Braun, T., Nikolics, K., Segaloff, D. L., and Seeburg, P. H. (1990). *Mol. Endocrinol.* **4**(4), 525–530.

Steinberger, A., Elkington, J. S., Sanborn, B. M., and Steinberger, E. (1975). *Curr. Top. Mol. Endocrinol.* **2**(399), 399–411.

Steinberger, A., Hintz, M., and Heindel, J. J. (1978). *Biol. Reprod.* **19**(3), 566–572.

Steinberger, E. (1971). *Physiol. Rev.* **51,** 1–22.

Sunahara, R. K. (1990). *Nature (London),* **347,** 80–83.

Thanki, K. H., and Steinberger, A. (1978). *Andrologia* **10**(3), 195–202.

Themmen, A., Blok, L., Post, M., Baarends, W., Hoogerbrugge, J., Parmentier, M., Vassart, G., and Grootegood, A. (1991). *Mol. Cell. Endocrinol.* **78,** R7–R13.

Tsai-Morris, C. H., Buczko, E., Wei, W., Xie, X. Z., and Dufau, M. L. (1991). *J. Biol. Chem.* **266,** 11355–11358.

Van Beek, M. E. A. B., and Meistrich, M. L. (1990). *Biol. Reprod.* **42,** 424–431.

Vassart, G., Parmentier, M., Libert, F., and Dumont, J. (1991). *Trends Endocrinol. Metab.* **2**(4), 151–156.

Wickings, E. J., Usadel, K. H., Dathe, G., and Nieschlag, E. (1980). *Acta Endocrinol.* **95,** 117–128.

Yoon, D. J., Reggiardo, D., and David, R. (1990). *J. Endocrinol.* **125**(2), 293–299.

# Differential Gene Expression from a Single Transcription Unit during Spermatogenesis

ANTHONY R. MEANS AND FRANCISCO CRUZALEGUI

*Department of Pharmacology, Duke University Medical Center, Durham, North Carolina 27710*

## I. Introduction

Spermatogenesis is a unique and continual process in mammals whereby stem cells divide mitotically to become cells that either remain quiescent or give rise to those that undergo a series of mitotic and meiotic divisions, culminating in the onset of a remarkably complicated differentiation pathway (Clermont, 1972). It is this differentiation process, called spermiogenesis, that results in formation of immature spermatozoa. The frequency of stem cell division is tightly controlled in rodents. Only each 13 days do unknown environmental and/or genetic cues result in mitosis, and whereas one daughter cell begins its tortuous path to produce many genetically identical spermatogonia, the other cell becomes arrested before it divides and apparently retains stem cell potential (Huckins, 1972). Because of the incredibly intractable nature of the process, virtually nothing is known about the molecular mechanisms that control spermatogenic stem cell renewal.

Once a cell is committed to the developmental pathway, a bit more information is available (Leblond and Clermont, 1952). This type A spermatogonium will divide mitotically several times, with each cell cycle resulting in morphological and presumably functional differentiation. Eventually type B spermatogonia are produced and divide to become premeiotic primary spermatocytes. The duration of the first meiotic division is amazingly protracted and requires more than 12 days to complete (Perey *et al.*, 1961). However, during this time many changes occur as a continuum and eventually are sufficient to trigger the second meiotic division. The second division requires only hours and gives rise to the round spermatid. This haploid cell can never again divide but serves as the most immature cell in the spermiogenic differentiation pathway.

Several events that occur during the life of a spermatogonium are quite unusual and are undoubtedly programmed to ensure survival of several genetically identical germ cells. First, all of the offspring of a single cell develop as a clone or cohort (Huckins, 1972). The cells within each cohort remain attached by cytoplasmic bridges that allow transfer of molecules, including proteins. Division within a cohort is synchronous and a certain number of such clones eventually die and degenerate at an identical stage within the cell cycle (Huckins, 1972). The molecular principles that govern this unique form of programmed cell death are unknown. Administration of exogenous follicle-stimulating hormone (FSH) to immature rats decreases this degenerative process, presumably due to stimulation of production of as yet uncharacterized factors from Sertoli cells (Means et al., 1976). Because Sertoli cells have been reported to produce a mitogenic factor in response to FSH (Feig et al., 1980), it is tempting to speculate that this molecule might be involved in germ cell proliferation. Second, the first meiotic cell becomes arrested in the cell cycle at metaphase II but continues to undergo remarkable morphological and biochemical changes. Chromatin condenses and the nucleus becomes quite large and distinct. Accompanying these morphological changes are an amazing array of transcriptional events. Nuclear RNA synthesis in the primary pachytene spermatocyte may be as active as in any other mammalian cell in situ (Monesi, 1964). Third, prior to completion of the first meiotic division, the cells pass from a basal to an adluminal compartment within the seminiferous tubule (Dym and Fawcett, 1970). The developing germ cells that can no longer divide mitotically thus become protected from the environment due to the "blood–testis barrier" that separates the two compartments. This barrier is formed by tight junctions between the adjacent Sertoli cells, within which the developing germ cells are imbedded. Such membrane specializations are remarkably impervious to environmental insult, as even lanthanides delivered into the circulatory system cannot permeate the barrier (Dym and Fawcett, 1970; Tindall et al., 1975). Finally, the entire process has evolved to function effectively at a temperature nearly 5°C lower than that of the body (Waites, 1970).

This introduction omits discussion of many other events that are unique to testicular germ cell development. Nevertheless, the few areas we have chosen to highlight should give an appreciation of this complicated and highly intricate process. It should not be surprising to learn, therefore, that many genes and gene products are uniquely expressed in the developing germ cell. Such molecules include enzymes involved in carbohydrate metabolism, isoforms of cytoskeletal proteins, DNA-binding proteins, intracellular signaling molecules, and protooncogenes [see Table I; a comprehensive list of such genes in the mouse has been published by Wolgemuth and Watrin (1991)]. The first such molecules to be investigated in molecular detail were the protamines (Dixon, 1972; Bellvé, 1979). These

## GENE EXPRESSION DURING SPERMATOGENESIS

TABLE I

*Examples of Genes Expressed Exclusively in Male Germ Cells*

| Function | Gene | Ref. |
|---|---|---|
| Carbohydrate metabolism | Phosphoglycerate kinase | Boer *et al.* (1987) |
| | Lactic dehydrogenase X | Tanaka and Fujimoto (1986) |
| Cytoskeletal structure | $\alpha$-Tubulin | Distel *et al.* (1984) |
| | $\beta_3$-Tubulin | Sullivan *et al.* (1986) |
| | $\gamma$-Actin | Kim *et al.* (1989) |
| DNA binding | Protamine | Dixon (1972) |
| | *Zfp35* | Cunliffe *et al.* (1990) |
| | *Mok-Z* | Ernoult-Lange *et al.* (1990) |
| Intracellular signaling | Phosphorylase kinase | Hanks (1989) |
| | Tyrosine kinase | Fischman *et al.* (1990) |
| | Ser/Thr protein kinase | Matsushime *et al.* (1990) |
| | cAMP kinase catalytic subunit | Beebe *et al.* (1990) |
| Protooncogene | Review | Propst *et al.* (1988) |
| | c-*abl* | Ponzetto and Wolgemuth |
| | *fps/fes; eph* | (1985) |
| | | Letwin *et al.* (1988) |

extremely basic proteins are produced only in postmeiotic male germ cells, replace histones as the primary DNA-binding proteins, and function to compact and protect the haploid nucleus of the developing sperm. It is now known that germ cell-specific expression is controlled by a germ cell-specific promoter/enhancer (Johnson *et al.*, 1991), that this regulatory DNA is sufficient to target germ cell-specific expression of a heterologous gene in transgenic mice (Braun *et al.*, 1989), and that the mRNA is stored in the cells that express it to be translated at a later stage of development (Kleene *et al.*, 1984). Even with this advanced state of knowledge the proteins and regulatory processes that orchestrate these events remain to be elucidated. However, work on the protamines has given rise to a rapidly expanding field that deals with mechanisms that govern the expression and function of proteins expressed exclusively in the haploid cell (see Means *et al.*, 1990, for review).

## II. Calspermin

The interest in haploid-specific gene products developed in this laboratory arose as a logical extension of our examination of $Ca^{2+}$ as an intracellular regulatory molecule (Means, 1981, 1988). Calcium is involved in the cellular actions of LH and FSH, participates in the regulation of cell proliferation, is intimately involved in the acquisition of sperm motility, and plays regulatory roles in specialized responses of the sperm required

82    ANTHONY R. MEANS AND FRANCISCO CRUZALEGUI

for fertilization as well as the actual fertilization process. The great majority of these regulatory roles of $Ca^{2+}$ require interaction with the ubiquitous intracellular $Ca^{2+}$ receptor calmodulin as an obligatory intermediate. One approach to the understanding of how calmodulin controls a specific cellular response has been to identify and characterize targets of the $Ca^{2+}/$ calmodulin complex that might be cell type specific.

While a graduate student in Japan, Tomio Ono was evaluating calmodulin and calmodulin-binding proteins in the testis. He discovered a soluble protein that was a potent competitive inhibitor of calmodulin-dependent cyclic nucleotide phosphodiesterase. Ono *et al.* (1984, 1985) purified the protein; it was found to be highly acidic (p$I$ = 3.9) and to migrate on SDS-containing polyacrylamide gels at a $M_r$ of 39,000. This was one of the most abundant calmodulin-binding proteins in adult pig and rat testes. Subsequent studies revealed that the protein was undetectable in testes of very young rats but increased during development coincident with the onset of complete spermatogenesis (Koide *et al.*, 1986). Antibodies were developed and used to examine tissue distribution of this calmodulin-binding protein (Ono *et al.*, 1987). Because of its abundance in adult testis and its apparent absence from all other tissues, the protein was named calspermin (Koide *et al.*, 1985).

Calmodulin is also a very acidic protein, whereas many calmodulin-binding proteins are basic. Calspermin was unique because of its acidic nature and due to its apparent tissue specificity. We therefore prepared a rat testis cDNA expression library and utilized synthetic oligonucleotides corresponding to the amino acid sequences of a few proteolytically derived peptides to clone and sequence the cDNA (Ono *et al.*, 1989). The deduced amino acid sequence corresponding to the open reading frame of the cDNA indicated a $M_r$ of 18,735. The amino acid sequence is shown in the portion of Fig. 1 between the brackets; the initiating M residue is amino acid 306. Analysis revealed a possible calmodulin-binding region (boxed in Fig. 1) close to the amino terminus and a very glutamic acid-rich area in the carboxyl half of the molecule that would contribute to the aberrant migration on polyacrylamide gels. No other functional domains were obvious from examination of the sequence. Whereas the amino-terminal half of the putative calmodulin-binding domain was identical to that present in the multifunctional calmodulin-dependent protein kinase (FNARRKLK), the COOH-terminal half was unique (Hanley *et al.*, 1987). The cDNA was cloned into a bacterial expression vector. The purified protein bound calmodulin with high affinity only in the presence of $Ca^{2+}$ and migrated with an apparent $M_r$ of 32,000 on SDS–PAGE. These features were characteristic of calspermin purified from rat testis.

# GENE EXPRESSION DURING SPERMATOGENESIS

FIG. 1. Amino acid sequence of CaM kinase IV. A box encloses the putative calmodulin-binding domain, based on the structure of CaM kinase II. Brackets enclose the amino acid sequence of calspermin. The last residue of the conserved kinase catalytic domain (aa 284) is underlined. The over- and underlined ATP-binding site is at residue 49. Arrows: location of exon–intron junctions identified in the genomic DNA sequence. Asterisk: point of divergence between the amino acid sequence of CaM kinase IV and the 5' untranslated sequence of the calspermin cDNA.

## III.  Calspermin and Calmodulin Kinase IV

The mRNA encoding calspermin is 1.1 kb and is restricted to adult rat testis (Ono *et al.*, 1989). Testis RNA also revealed a 2.0-kb species that cross-hybridized with the calspermin cDNA. Similar 2.0-kb RNAs were present in brain and spleen but were absent from eight other tissues examined by Northern blots (Ono *et al.*, 1989). Cerebellar RNA also contained a 3.5-kb cross-hybridizing RNA. Because the larger RNAs were detected using stringent hybridization conditions, it seemed likely that additional proteins could exist that would share sequence similarities with calspermin. A computer-assisted search of available protein data bases yielded only one additional RNA with apparent homology. This sequence, termed λ–ICM-1, had been obtained by Sikela and Hahn (1987) by screening a mouse brain cDNA expression library with $^{125}$I-labeled calmodulin. The calmodulin-binding regions of mouse λ–ICM-1 and rat calspermin were identical whereas the remaining amino acid sequence showed considerable diversity. In a note added in proof, Sikela and Hahn (1987) reported that they had sequenced an additional 200 nt upstream of the sequence reported in the paper and that this sequence contained consensus residues common to protein kinases. This additional sequence was subsequently published and Sikela et al. (1989) mapped the gene to mouse chromosome 5 within the region of bands q21 to q23. Thus it was suggested that λ–ICM-1 might encode a calmodulin-dependent protein kinase and this putative enzyme was christened $Ca^{2+}$/calmodulin-dependent protein kinase IV.

For the size of the open reading frame (510 nt), the calspermin mRNA contained an unusually long 5′ nontranslated region of 232 nt (Ono et al., 1989). Translation of this region in the same frame as that which encoded calspermin revealed an additional 33 amino acids before a stop codon was encountered (Means et al., 1990). These 33 amino acids contained residues indicative of protein kinases (Hanks et al., 1988) and were 58% identical to the α subunit of the rat brain multifunctional calmodulin-dependent protein kinase (CaM kinase IIα) between residues 248 and 281 (Hanley et al., 1987). Because the $NH_2$-terminal halves of the calmodulin-binding regions of the enzyme and calspermin were identical, we predicted that the larger cross-hybridizing mRNA might encode a new type of $Ca^{2+}$/calmodulin-dependent protein kinase. Polymerase chain reaction (PCR) was employed to extend the rat calspermin cDNA in the 5′ direction and DNA sequencing revealed further similarity to protein kinases (Means et al., 1991). Therefore, the PCR product was used to obtain a full-length cDNA from a rat brain cDNA library. The open reading frame of this clone shown in Fig. 1 would be predicted to encode a 474-amino acid protein kinase with a calculated $M_r$ of 53,159. Over the entire length, this molecule was only 32% identical to CaM kinase IIα. However, over the 70 amino acids that could be compared, it was more than 90% identical to mouse λ–ICM-1, now renamed $Ca^{2+}$/calmodulin-dependent protein kinase IV (Sikela et al., 1989), although it has yet to be established that this latter cDNA encodes a functional protein kinase. In addition, the sequence of our putative kinase was identical to the sequence of a partial rat brain cDNA predicted to encode a $Ca^{2+}$/calmodulin-dependent protein kinase specific to the granule cells of the cerebellum (CaM kinase $G_r$) (Ohmstede et al., 1989). We had decided to call our new enzyme calspermin kinase. However, it seemed likely that an identical functional protein was destined

# GENE EXPRESSION DURING SPERMATOGENESIS

to be referred to by three very different names—calspermin kinase, CaM kinase IV, and CaM kinase $G_r$. Because the rat and mouse proteins are probably homologues and the kinase is not granule cell specific, we capitulated to Sikela *et al.* (1989) and called our cDNA CaM kinase IV. Ohmstede *et al.* (1991) have now reported the sequence of their cDNA (CaM kinase $G_r$) beginning with amino acid 47 of CaM kinase IV. Throughout the remaining sequence the two cDNAs are identical except that the nucleic acid cloned by Ohmstede *et al.* (1991) contains a much longer 3' nontranslated region and apparently utilizes a distinct polyadenylation site from the molecule cloned by Means *et al.* (1991).

Hanks *et al.* (1988) compared the amino acid sequences of all protein kinases cloned at that time within the catalytic homology region. This region was defined as beginning with the ATP-binding site (generally G-X-G-X-X-G) and ending some 240 amino acids later in an invariant R residue. The location of these residues in rat CaM kinase IV are indicated in Fig. 1 by over- and underlines (amino acids 49–284). As shown in Table II, CaM kinase IV is most similar in this region to rat and *Drosophila* CaM kinase II homologues. Percent identities are as great between rat CaM kinase IV and *Saccharomyces cerevisiae* CaM kinase (both CMK1 and CMK2) as they are between our enzyme and other mammalian calmodulin-dependent protein kinases, except for the catalytic subunit of phosphorylase kinase (Table II). Computer algorithms that predict evolutionary relationships suggest CaM kinase IV to be much more closely related to the CaM kinase II isoforms than to any other protein kinases.

TABLE II

*Comparison of the Catalytic Domains of CaM Kinase IV and Other Protein Kinases*[a]

| Kinase | Source | Identity | Ref. |
|---|---|---|---|
| CaM kinase II$\alpha$ | Rat | 42 | Lin *et al.* (1987) |
| CaM kinase | *Drosophila* | 43 | Cho *et al.* (1991) |
| Phosphorylase kinase | Rabbit | 41 | da Cruz e Silva and Cohen (1987) |
| S6 kinase | *Xenopus* | 36 | Jones *et al.* (1988) |
| CaM kinase (CMK1) | *Saccharomyces cerevisiae* | 36 | Pausch *et al.* (1991) |
| Smooth muscle MLCK | Chicken | 35 | Guerriero *et al.* (1986) |
| Skeletal muscle MLCK | Rabbit | 35 | Herring *et al.* (1990) |
| PKC $\beta$ | Rat | 30 | Housey *et al.* (1988) |
| PKA $\beta$ | Mouse | 28 | Uhler *et al.* (1986) |

[a] Amino acid sequences of conserved kinase domains were aligned with residues 1–337 of CaM kinase IV (including the putative CaM-binding domain) using Bestfit (GCG package, Genetics Computer, Inc.).

To confirm that the CaM kinase IV cDNA did encode a functional protein kinase, it was cloned into an expression system and produced in a rabbit reticulocyte lysate (Means *et al.*, 1991). The protein migrated at 61,000 Da on SDS–PAGE, was recognized by an affinity-purified antibody to rat calspermin, and bound $^{125}I$-labeled calmodulin only in the presence of $Ca^{2+}$. The protein also demonstrated $Ca^{2+}$/calmodulin-dependent protein kinase activity and could phosphorylate synapsin I, myosin light chains, and a synthetic peptide analog based on the sequence of glycogen synthase called GS-10. These three molecules are known substrates of CaM kinase II$\alpha$ and in parallel experiments were also phosphorylated by this enzyme, as produced in reticulocyte lysates from the coding strand of the gene. Because of the low abundance of CaM kinase IV produced in the coupled transcription–translation system and the presence of other $Ca^{2+}$/calmodulin-dependent protein kinases in the lysate, it has not been possible to further characterize the properties of this enzyme. We have now expressed the cDNA in baculovirus using the expression vector pVLCaMkIV shown in Fig. 2. The protein can be visualized by Coomassie blue staining of total soluble protein and is very active as a

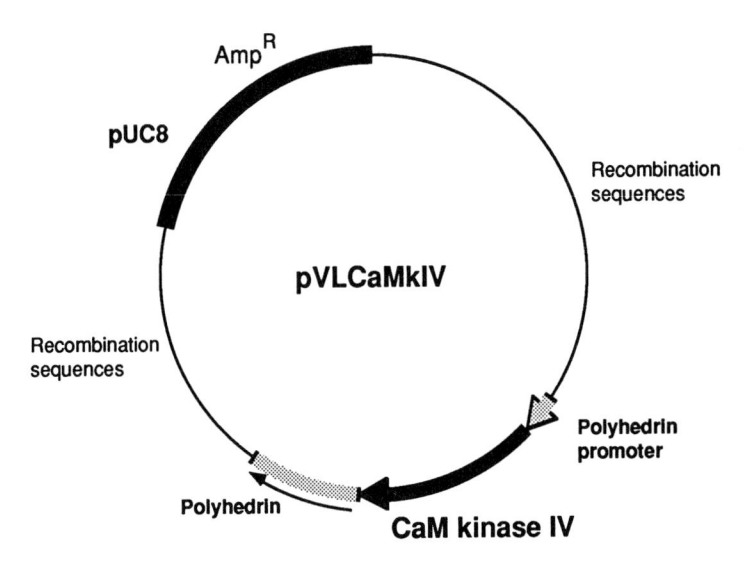

FIG. 2. Transfer vector for expression of CaM kinase IV in Sf9 cells. The full-length CaM kinase IV cDNA (1.4 kb) was inserted in the *Sma*I site of pVL1393 (kindly provided by Dr. Ming-Jer Tsai, Baylor College of Medicine). The resulting vector was used to cotransfect *S. frugiperda* cells with wild-type baculovirus DNA.

GENE EXPRESSION DURING SPERMATOGENESIS                87

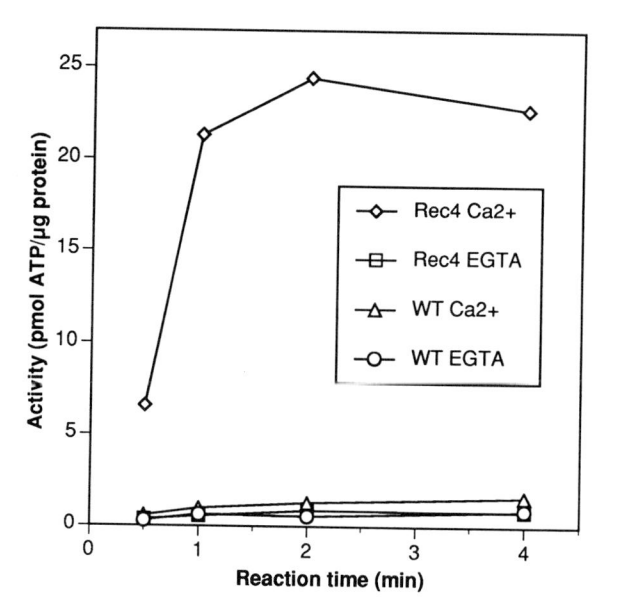

FIG. 3.   Activity of baculovirus-expressed CaM kinase IV. Sf9 cells were infected with recombinant baculovirus expressing CaM kinase IV (Rec4, recombinant clone 4). Soluble extracts (40 mg of protein) were used in standard kinase assays in the presence of calmodulin and either $Ca^{2+}$ or EGTA. GS-10 (synthetic peptide derived from the CaM kinase II phosphorylation site on glycogen synthase) was used as substrate. Control reactions were carried out using extracts of cells infected with wild-type (WT) baculovirus.

$Ca^{2+}$/calmodulin-dependent GS-10 kinase. As shown in Fig. 3, all of the enzyme activity is $Ca^{2+}$/calmodulin-dependent and 10 $\mu$l of the extract has a specific activity of 24 pmol ATP incorporated per microgram of protein. We are in the process of purifying this enzyme from insect Sf9 cell extracts. Availability of large quantities of CaM kinase IV will enable us to pursue studies to determine how it is regulated, its kinetic properties, and its substrate specificity in ways analogous to experiments with CaM kinase II. Based on sequence differences between these two enzymes, one would predict some distinct properties.

Calmodulin kinase II is inactive in the absence of $Ca^{2+}$/calmodulin and, in cells, consists of a heteromultimer composed of 10–12 subunits (Schulman, 1988). Binding of $Ca^{2+}$/calmodulin results in a rapid autophosphorylation on Thr-286 that renders the enzyme independent of $Ca^{2+}$/calmodulin (Miller et al., 1988; Thiel et al., 1988; Schworer et al., 1988). When phosphorylated on Thr-286, $Ca^{2+}$/calmodulin can be removed from the enzyme and it remains active. When Thr-286 is altered to Ala or Leu by site-specific mutagenesis, the enzyme does not rapidly autophosphorylate on activation by $Ca^{2+}$/calmodulin and requires the continued presence of the regulatory molecule for activity (Fong *et al.*, 1989; Hanson *et al.*, 1989; Waxham *et al.*, 1990). Such studies have led to the hypothesis that the inactive state is maintained by an intramolecular interaction that effectively blocks the catalytic site (Payne *et al.*, 1988). $Ca^{2+}$/calmodulin relieves this inhibition and the presence of a phosphate group on Thr-286 prevents restoration of the autoinhibited state even when $Ca^{2+}$/calmodulin is removed. Calmodulin kinase IV does not contain a phosphorylatable residue in a position similar to Thr-286 of CaM kinase II (see Figs. 1 and 4). However, Frangakis *et al.* (1991a) have isolated what appears to be CaM kinase IV from rat cerebellum and maintain that it does undergo an autophosphorylation reaction that releases it from dependence on $Ca^{2+}$/calmodulin for catalytic activity. The kinetics of this reaction is very different from that of CaM kinase II and the phosphorylated residues have not yet been identified. Sahyoun *et al.* (1991) have also reported that CaM kinase IV isolated from rat cerebellum will phosphorylate a low-$M_r$ GTP-binding protein called Rap-1b, whereas CaM kinase II will not. Finally, it has been suggested that CaM kinase IV may be monomeric or, in the most complicated scenario, a heterodimer (Frangakis *et al.*, 1991a). Therefore, provocative preliminary data argue that CaM kinases II and IV will be structurally and functionally different as well as be regulated by distinct molecular mechanisms. Because baculovirus expression systems for both CaM kinases II$\alpha$ and IV now exist, it will be possible to rapidly pursue answers to these questions.

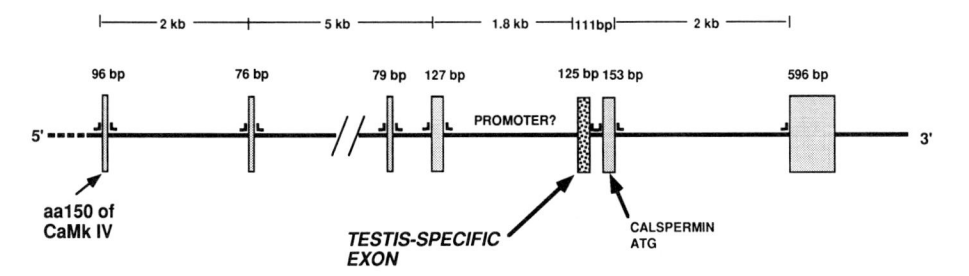

FIG. 4. Schematic representation of the CaM kinase IV/calspermin gene structure. Exons are shown as rectangles. Acceptor/donor splice sites are represented by right angle symbols. The broken line at the left indicates that the gene is incomplete. Arrows in Fig. 1 represent where the introns interrupt the amino acid sequence of the kinase.

## IV. CaM Kinase IV and Calspermin mRNAs Generated from Contiguous DNA

Primer extension of rat testis mRNA was used to determine the transcriptional start site of calspermin mRNA (Means *et al.*, 1991). Dideoxynucleotide sequencing of the extended product using oligonucleotide probes, shown in Fig. 5, revealed that the CAP site was only 9 nt upstream of the calspermin cDNA sequence reported by Ono *et al.* (1989) at the CAG residues highlighted by asterisks in Fig. 5. The first 130 nt downstream of the calspermin CAP site are unique. However, nucleotide 131 is a G residue that corresponds to nucleotide 816 of the CaM kinase IV cDNA. From this nucleotide to the poly(A) addition site, the kinase and calspermin mRNA sequences are identical. Such results are compatible with a mechanism whereby alternative transcriptional initiation and/or exon usage could generate both mRNAs. To distinguish between these possibilities, several probes derived from the kinase or calspermin cDNA were used to screen a rat genomic library. To date we have identified six exons and five introns of the kinase gene. The first exon identified begins with nucleotide 449 of the cDNA (amino acid 150; see Fig. 1) and the final exon contains the distal portion of the amino acid coding region as well as all of the 3′ nontranslated sequence. Figure 4 presents a schematic representation of

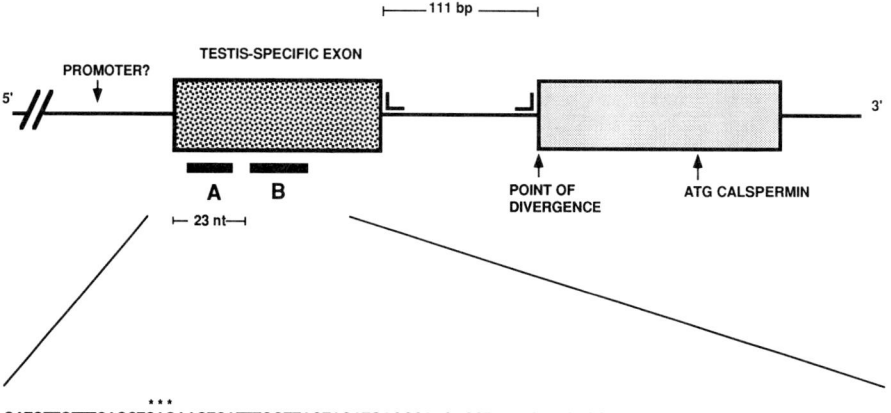

FIG. 5. Transcriptional start site of the calspermin mRNA. The testis-specific exon has no acceptor/donor splice site present at its 5′ end. Underlined sequences are oligonucleotides used in primer-extension experiments (Means *et al.*, 1991). Asterisks denote nucleotides identified as transcriptional start sites.

90     ANTHONY R. MEANS AND FRANCISCO CRUZALEGUI

this genomic organization, and arrows in Fig. 1 indicate where the amino acid sequence of the enzyme is interrupted by introns. The next to the last exon begins with nucleotide 816 of the kinase cDNA, which is the first residue of the identical sequence between calspermin and CaM kinase IV (the asterisk above the arrow in Fig. 1). Thus all but the 5' 130 nt of the calspermin mRNA are encoded by two exons that are common to CaM kinase IV. The next upstream exon of the kinase is about 1.8 kb from the 5' boundary of the first common exon (Fig. 4). The entire 130 nt of sequence unique to calspermin mRNA, including the transcription initiation sites, is found as a contiguous piece of DNA within this intron of the kinase gene (Means et al., 1991). The location of this sequence is shown diagrammatically in Fig. 4. A consensus AG/GT exon/intron boundary exists at the 3' end of the unique calspermin mRNA 5' nontranslated sequence. These data imply that calspermin mRNA transcription could be regulated by a testis-specific enhancer/promoter that is nested within an intron of the larger kinase gene. Differential transcriptional initiation controlled by the putative promoter upstream of the testis-specific exon shown in Fig. 5 would result in an obligate unique slicing event that would connect the testis-specific sequence to the first nucleotide of the first common exon (at the point of divergence; Fig. 5). The reading frame so generated would result in a 33-amino acid coding region of the kinase (amino acids 273–305; Fig. 1) being used as 5' nontranslated sequence in calspermin. In support of this suggestion, a probe consisting of the 130 nt of DNA unique to calspermin mRNA (i.e., the entire sequence of the testis-specific exon; Fig. 5) hybridized exclusively to the 1.1-kb calspermin mRNA present only in testis (Means et al., 1991). We have now ligated the fragment of DNA predicted to contain a testis-specific transcriptional element to a gene encoding β-galactosidase. This DNA construct is being used to produce transgenic mice in which to evaluate whether and in what tissues the transgene is expressed. A genomic fragment of DNA encoding the CaM kinase IV has also been isolated by Ohmstede et al. (1991). Whereas the diagram presented in Fig. 3 of that paper is consistent with our results, these authors suggest alternative exon usage as the primary mechanism involved in creating the calspermin transcript. Because no DNA sequence is presented and the transcriptional start site of calspermin was not mapped, it is impossible to determine the validity of their claim. However, we believe that the case for differential transcriptional initiation is compelling.

We have analyzed the 1.8-kb intron immediately upstream of the testis-specific exon for the presence of potential transcription factor binding sites. The results of this computer-assisted analysis are represented in Fig. 6. A possible TATA box exists at about −35, which is preceded by two

# GENE EXPRESSION DURING SPERMATOGENESIS

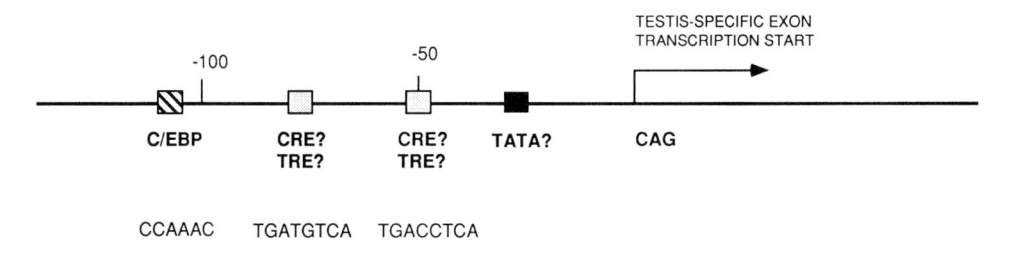

FIG. 6. Possible promoter/enhancer elements immediately upstream of the testis-specific exon. These potential transcription factor-binding sites were identified with the aid of a computer. The sequences of the elements found in the CaM kinase IV DNA are on the second line. The consensus sequence for C/EBP, CRE, and TRE elements are shown at the bottom.

potential CRE/TRE elements at −50 and −75. Whereas the usual CRE elements contain 8 nt and the middle two are CG, the ones at −50 in the CaM kinase IV DNA possess two C residues whereas the one at −75 contains TG. We have also discovered a testis-specific mRNA that would encode an amino-terminal truncated form of troponin C. The DNA sequence unique to this mRNA is also located in an intron of the rat cardiac troponin C gene. This testis-specific RNA appears at the same stage of spermatogenesis and in the same cells as does the calspermin mRNA (Means *et al.*, 1991; see Section V). The DNA that could represent a promoter/enhancer also contains a TGACCTCA sequence at about −50 relative to the CAP site. Preliminary experiments by Gayle Slaughter indicate that a fragment of DNA containing this putative CRE will band shift proteins present in nuclear extracts of spermiogenic cells. This affords the exciting possibility that the element could be one component required for stage-specific transcriptional initiation during spermatogenesis. This is a testable prediction that we are actively pursuing.

A precedent exists for the use of a testis-specific promoter that exists within an intron of a larger gene to regulate production of a testis-specific isoform of angiotensin-converting enzyme (ACE). In the mouse, the $NH_2$-terminal 66 amino acids of the testis isoform are unique whereas the remaining sequences are shared, with the much larger enzyme expressed in somatic tissues (Cushman and Cheung, 1971). Using a strategy similar to that we have used, Howard *et al.* (1990) found what they predicted to be a testis-specific promoter nested within the twelfth intron of the somatic ACE gene. These sequences were more than 7000 bp downstream of the somatic ACE promoter. Langford *et al.* (1991) ligated a 698-bp fragment of DNA suspected to contain this intragenic testis-specific promoter to a gene encoding $\beta$-galactosidase and generated transgenic mice. Histochemical analysis revealed that expression of the transgene was restricted to postmeiotic testicular germ cells. Thus the 598-bp DNA fragment is both necessary and sufficient to orchestrate male germ cell-specific expression. It is likely that a similar mechanism is employed to generate a testis-specific form of transferrin (Stallard *et al.*, 1991). This protein, called hemitransferrin, is identical to the COOH-terminal half of the protein produced in somatic tissues. However, the molecular details of this gene have yet to be published.

## V. Changes in Calspermin and CaM Kinase IV during Germ Cell Development

Northern hybridization had suggested that the 2.0-kb CaM kinase IV mRNA existed in brain, testis, and spleen. Antibodies to calspermin that cross-react with the kinase were used to examine whether the protein was also produced in these RNA-positive tissues (Means *et al.*, 1991). Cerebellum, cerebrum, spleen, and testis contained an immunoreactive 61-kDa protein. Cerebellum and testis also demonstrated a second immunoreactive protein that was about 64 kDa as judged from migration on SDS–PAGE. Epididymis contained only the 64-kDa species. Because this tissue does not exhibit cross-hybridizing mRNA to probes derived from CaM kinase IV, either the 64-kDa protein is encoded by a distinct mRNA from a separate gene or it is found in epididymal spermatozoa. Adult epididymis was used that contained sperm, thus either explanation is equally reasonable. On the basis of analysis of the *in vitro*-translated product of CaM kinase IV mRNA, we suspect that this gene product is the 61-kDa protein. Although cerebellum contains an additional 3.4-kb mRNA that cross-hybridizes with CaM kinase IV cDNA, testis does not. Thus it seems probable that the 64-kDa protein may well be translated from a mRNA that cannot be detected by the CaM kinase IV cDNA but does cross-react with our antibodies to calspermin. Relevant to this issue is the study of Frangakis *et al.* (1991b), who have isolated CaM kinase IV as a doublet from cerebellum but only as the single smaller protein from the cerebrum. A similar situation exists for the $\alpha$ and $\beta$ isoform of CaM kinase II in rat brain (Schulman, 1988). Whereas an $\alpha$ cDNA cannot recognize the $\beta$ mRNA, some antibodies identify both species and the proteins are reasonably similar at the amino acid level. In this case the

GENE EXPRESSION DURING SPERMATOGENESIS    93

cerebrum is the most abundant source of $\alpha$ and $\beta$ whereas cerebellum contains primarily the $\alpha$ isoform.

What is clear from our experiments is that testis contains both mRNA and protein for CaM kinase IV and calspermin. To determine which testicular cells express each mRNA, we used nucleic acid probes specific for either calspermin or the kinase to carry out *in situ* hybridization on sections of adult testis (Means *et al.*, 1991). Both mRNAs appeared to be restricted to germ cells but the pattern of expression was considerably different. Type A spermatogonia did not contain either of the mRNAs and a similar negative result was found for type B spermatogonia. The kinase mRNA first appeared at low concentrations in preleptotene primary spermatocytes and reached maximal concentrations by the midpachytene stage. Levels then declined in cells at later stages of differentiation and reached a low but stable level in spermatids. The concentration of kinase mRNA was similar in spermatids at all stages of germ cell development examined. Calspermin mRNA was first detected at very low levels in leptotene spermatocytes. Increased levels were observed in cells at each successive development step. The highest levels of calspermin mRNA were found in elongating stermatids of stage VIII seminiferous tubules; these cells have nearly completed differentiation and are soon to be released into the lumen of the tubule to begin the transit out of the testis and into the epididymis. The quantitative results show a distinct distribution of kinase and calspermin mRNA during germ cell development. The kinase mRNA appears first and is in greater abundance in meiotic cells, whereas calspermin mRNA, after getting a later start, reaches the highest concentration in the most advanced stages of postmeiotic germ cells. Preliminary electron microscopic studies reveal calspermin to be present in mature spermatozoa isolated from the caudal epididymis. This protein is also detected in ejaculated sperm by radioimmunoassay. Although we have been unable to detect CaM kinase IV activity in isolated sperm, we have yet to develop reagents that will specifically detect this protein. Therefore we must entertain the possibility that sperm could contain both proteins. Whereas promoter/reporter genes can be used to pinpoint transcriptional activity from the kinase or calspermin promoter in transgenic mice, it will be much more difficult to follow changes quantitatively in the kinase protein during germ cell development.

## VI.  Concerning the Functions of Calspermin and CaM Kinase IV

Equally difficult will be to determine what functional roles CaM kinase IV and calspermin may play during germ cell development. Considerable evidence currently exists to suggest that a $Ca^{2+}$/calmodulin-dependent

94 ANTHONY R. MEANS AND FRANCISCO CRUZALEGUI

protein kinase may be involved in the regulation of meiosis in sea urchin eggs (Batinger *et al.*, 1990) and *Xenopus* oocytes (Waldman *et al.*, 1990). It is very tempting to suggest a similar role for CaM kinase IV in regulation of germ cell meiosis. Germ cell-specific promoters now exist that become active in spermatocytes or spermatids (Stewart *et al.*, 1988; Robinson *et al.*, 1989; Langford *et al.*, 1991). Such promoters can be used to overexpress the kinase in transgenic mice to determine whether the timing or duration of meiosis is affected. It might also be possible to identify germ cell-specific substrates for CaM kinase IV. If so, then these molecules could be cloned. Analysis of the substrate sequence could provide clues to the function of the kinase. One interesting possibility is that the kinase could phosphorylate a CREB-like protein that would interact with the putative promoter/enhancer region of the gene that results in production of calspermin mRNA. If this were true, then the product of a parent gene could regulate transcriptional activation of a portion of the same contiguous piece of DNA. It has been reported by Jensen *et al.* (1991) that CaM kinase IV is found in the nucleus of cerebellar granule cells as determined by immunocytochemistry. A similar localization in spermatocytes would position the enzyme correctly to posttranslationally modify transcription factor(s). The granule cell enzyme also revealed axonal localization. A similar localization in the developing spermatid could indicate a role in axoneme formation or flagellar function. Finally, it has recently been reported by LeMagueresse-Battistoni *et al.* (1991) that spermatid differentiation can proceed from spermatocytes *in vitro*. If these events can be accomplished in culture, we will have a very promising system in which to study the function of proteins whose concentrations have been specifically altered in transgenic mice. At least, because CaM kinase IV is a protein Ser/Thr kinase, its functions will undoubtedly involve protein phosphorylation cascades.

Predicting potential functions for calspermin is quite another matter. From the sequence, one can only assign the calmodulin-binding region as a functional domain. One ponders the wisdom of producing large quantities of a calmodulin-binding protein in a differentiated and highly specialized cell such as the spermatozoon. The action of $Ca^{2+}$ in these cells is paradoxical. On the one hand, high concentrations of $Ca^{2+}$ inhibit sperm motility (Gibbons and Gibbons, 1980). On the other hand, low concentrations of $Ca^{2+}$ (Brokaw *et al.*, 1974), calmodulin (Brokaw and Nagayama, 1985), and a $Ca^{2+}$/calmodulin-dependent protein phosphatase (Tash *et al.*, 1988) appear to be vital to sperm motility (Garbers and Kopf, 1980). Maybe calspermin serves to position calmodulin in a way that would allow it to function without major changes in intracellular $Ca^{2+}$. This may be the case for calmodulin regulation of a myosin I molecule present in brush-border

## GENE EXPRESSION DURING SPERMATOGENESIS

epithelial cells (Swanljung-Collins and Collins, 1991). Equally plausible is that calspermin is present to protect the sperm from harmful effects of high concentrations of $Ca^{2+}$/calmodulin. Changes in intracellular $Ca^{2+}$ could result in binding to calmodulin, which, in turn, would interact with calspermin preferentially because of the high affinity of this interaction. At the very least we now have the molecular reagents to begin to investigate such possibilities. It certainly appears that germ cell-specific transcription and production of germ cell-specific proteins will be intimately involved in germ cell development and differentiation.

### ACKNOWLEDGMENTS

The authors are grateful to our laboratory colleagues, who contributed work described in this paper. Particular thanks are extended to Dr. Gayle Slaughter and Ms. Lauren McBride, who remain at Baylor College of Medicine. We also appreciate Mrs. Elizabeth Fletcher for competence and patience in preparing the typescript. The research reported in this paper was supported by research grants from the NIH to A.R.M. (HD-07503 and GM-33967).

### REFERENCES

Batinger, C., Alderton, J., Poenie, M., Schulman, H., and Steinhardt, R. A. (1990). *J. Cell Biol.* **111**, 1763–1773.

Beebe, S. J., Øyen, O., Sandberg, M., Frøysa, A., Hansson, V., and Jahnsen, T. (1990). *Mol. Endocrinol.* **4**, 465–475.

Bellvé, A. R. (1979). *Oxford Rev. Reprod. Biol.* **1**, 159–261.

Boer, P. H., Adra, C. N., Lau, Y.-F., and McBurney, M. W. (1987). *Mol. Cell. Biol.* **7**, 3107–3112.

Braun, R., Behringer, R., Peschon, J., Brinster, R., and Palmiter, R. (1989). *Nature (London)* **327**, 373–376.

Brokaw, C. J., Josslin, R., and Bobrow, L. (1974). *Biochem. Biophys. Res. Commun.* **58**, 795–800.

Brokaw, C. J., and Nagayama, S. M. (1985). *J. Cell Biol.* **100**, 1875–1883.

Cho, K.-O., Wall, J. B., Pugh, P. C., Ito, M., Mueller, S. A., and Kennedy, M. B. (1991). *Neuron* **7**, 439–450.

Clermont, Y. (1972). *Physiol. Rev.* **52**, 198–236.

da Cruz e Silva, E. F., and Cohen, P. T. W. (1987). *FEBS Lett.* **220**, 36–42.

Cunliffe, V., Koopman, P., McLaren, A., and Trowsdale, J. (1990). *EMBO J.* **9**, 197–205.

Cushman, D. W., and Cheung, H. S. (1971). *Biochim. Biophys. Acta* **250**, 261–265.

Distel, R. J., Kleene, K. C., and Hecht, N. B. (1984). *Science* **224**, 68–70.

Dixon, G. H. (1972). *Acta Endocrinol. (Suppl.)* **168**, 128–154.

Dym, M., and Fawcett, D. W. (1970). *Biol. Reprod.* **3**, 308–319.

Ernoult-Lange, M., Kress, M., and Hamer, D. (1990). *Mol. Cell. Biol.* **10**, 418–421.

Feig, L. A., Bellvé, A. R., Erickson, N. H., and Klagsbrun, M. (1980). *Proc. Natl. Acad. Sci. U.S.A.* **77**, 4774–4778.

Fischman, K., Edman, J. C., Shackleford, G. M., Turner, J. A., Rutter, W. J., and Nir, U. (1990). *Mol. Cell. Biol.* **10**, 146–153.

96 ANTHONY R. MEANS AND FRANCISCO CRUZALEGUI

Fong, Y-L., Taylor, W. L., Means, A. R., and Soderling, T. R. (1989). *J. Biol. Chem.* **264,** 16759–16763.

Frangakis, M. V., Chatila, T., Wood, E. R., and Sahyoun, N. (1991a). *J. Biol. Chem.* **266,** 17592–17596.

Frangakis, M. V., Ohmstede, C-A., and Sahyoun, N. (1991b). *J. Biol. Chem.* **266,** 11309–11316.

Garbers, D. L., and Kopf, G. S. (1980). *Adv. Cyclic Nucleotide Res.* **13,** 251–307.

Gibbons, B. H., and Gibbons, I. R. (1980). *J. Cell Biol.* **84,** 13–27.

Guerriero, V., Jr., Russo, M. A., Olson, N. J., Putkey, J. A., and Means, A. R. (1986). *Biochemistry* **25,** 8372–8381.

Hanks, S. K. (1989). *Mol. Endocrinol.* **3,** 110–116.

Hanks, S., Quinn, A., and Hunter, T. (1988). *Science* **241,** 42–52.

Hanley, R. M., Means, A. R., Ono, T., Kemp, B. E., Burgin, K., Waxham, N., and Kelly, P. T. (1987). *Science* **237,** 293–297.

Hanson, P. I., Kapiloff, M. S., Lau, L. L., Rosenfeld, M. G., and Schulman, H. (1989). *Neuron* **3,** 59–70.

Herring, B. P., Stull, J. T., and Gallagher, P. J. (1990). *J. Biol. Chem.* **265,** 1724–1730.

Housey, G. M., Johnson, M. D., Hsiao, W. L., O'Brian, C. A., Murphy, J. P., Kirschheimer, P., and Weinstein, I. B. (1988). *Cell (Cambridge, Mass.)* **52,** 343–354.

Howard, T. E., Shai, S-Y., Langford, K. G., Martin, B. M., and Bernstein, K. E. (1990). *Mol. Cell. Biol.* **10,** 4294–4302.

Huckins, C. (1972). *In* "Biology of Reproduction: Basic and Clinical Studies" (J. T. Velardo and B. A. Kasprow, eds.), pp. 395–421. PanAmerican Congress of Anatomy, New Orleans, LA.

Jensen, K. F., Ohmstede C-A., Fisher, R. S., and Sahyoun, N. (1991). *Proc. Natl. Acad. Sci. U.S.A.* **88,** 2850–2853.

Johnson, P. A., Bunick, D., and Hecht, N. B. (1991). *Biol. Reprod.* **44,** 127–134.

Jones, S. W., Erikson, E., Blenis, J., Maller, J. L., and Erikson, R. L. (1988). *Proc. Natl. Acad. Sci. U.S.A.* **85,** 3377–3381.

Kim, E., Waters, S. H., Hake, L. E., and Hecht, N. B. (1989). *Mol. Cell. Biol.* **9,** 1875–1881.

Kleene, K. C., Distel, R. J., and Hecht, N. B. (1984). *Dev. Biol.* **105,** 71–79.

Koide, Y., Ono, T., Ishinami, C., and Yamashita, K. (1986). *Am. J. Physiol.* **250,** C299–C305.

Koide, Y., Ono, T., and Yamashita, K. (1985). *Cell Calcium* **7,** 329–338.

Langford, K. G., Shai, S-Y., Howard, T. E., Kovac, M. J., Overbeek, P. A., and Bernstein, K. E. (1991). *J. Biol. Chem.* **266,** 15559–15562.

Leblond, C., and Clermont, Y. (1952). *Am. J. Anat.* **906,** 229–253.

LeMagueresse-Battistoni, B., Gérard, N., and Jégou, B. (1991). *Biochem. Biophys. Res. Commun.* **179,** 1115–1121.

Letwin, K., Yee, S-P., and Pawson, T. (1988). *Oncogene* **3,** 621–627.

Lin, C. R., Kapiloff, M. S., Durgerian, S., Tatemoto, K., Russo, A. F., Hanson, P., Schulman, H., and Rosenfeld, M. G. (1987). *Proc. Natl. Acad. Sci. U.S.A.* **84,** 5962–5966.

Matsushime, H., Jinno, A., Takagi, N., and Shibuya, M. (1990). *Mol. Cell. Biol.* **10,** 2261–2268.

Means, A. R. (1981). *Recent Prog. Horm. Res.* **37,** 333–368.

Means, A. R. (1988). *Recent Prog. Horm. Res.* **44,** 223–262.

Means, A. R., Cruzalegui, F., LeMagueresse, B., Needleman, D. S., Slaughter, G. R., and Ono, T. (1991). *Mol. Cell. Biol.* **11,** 3960–3971.

Means, A. R., Fakunding, J. L., Huckins, C., Tindall, D. J., and Vitale, R. (1976). *Recent Prog. Horm. Res.* **32,** 477–527.

## GENE EXPRESSION DURING SPERMATOGENESIS

Means, A. R., LeMagueresse, B., and Ono, T. (1990). In "Neuroendocrine Regulation of Reproduction" (S. S. C. Yen and W. W. Vale, eds.), pp. 143–154. Sorono Symposia, Norwell, MA.

Miller, S. G., Patton, B. L., and Kennedy, M. B. (1988). *Neuron* **1**, 593–604.

Monesi, V. (1964). *J. Cell Biol.* **22**, 521–532.

Ohmstede, C-A., Bland, M. M., Merrill, B. M., and Sahyoun, N. (1991). *Proc. Natl. Acad. Sci. U.S.A.* **88**, 5784–5788.

Ohmstede, C-A., Jensen, K. F., and Sahyoun, N. (1989). *J. Biol. Chem.* **264**, 5866–5875.

Ono, T., Koide, Y., Arai, Y., and Yamashita, K. (1984). *J. Biol. Chem.* **259**, 9011–9016.

Ono, T., Koide, Y., Arai, Y., and Yamashita, K. (1985). *J. Biochem. (Tokyo)* **98**, 1455–1461.

Ono, T., Koide, Y., Arai, Y., and Yamashita, K. (1987). *Arch. Biochem. Biophys.* **255**, 102–108.

Ono, T., Slaughter, G. R., Cook, R. G., and Means, A. R. (1989). *J. Biol. Chem.* **264**, 2081–2087.

Pausch, M. H., Kain, D., Kunisawa, R., Admon, A., and Thorner, J. (1991). *EMBO J.* **10**, 1511–1522.

Payne, M. E., Fong, Y., Ono, T., Colbran, R. J., Kemp, B. E., Soderling, T. R., and Means, A. R. (1988). *J. Biol. Chem.* **263**, 7190–7195.

Perey, B., Clermont, Y., and Leblond, C. P. (1961). *Am. J. Anat.* **108**, 47–77.

Ponzetto, C., and Wolgemuth, D. J. (1985). *Mol. Cell. Biol.* **5**, 1791–1794.

Propst, F., Rosenberg, M. P., and VandeWoude, G. F. (1988). *Trends Genet.* **4**, 183–187.

Robinson, M. O., McCarrey, J. R., and Simon, M. I. (1989). *Proc. Natl. Acad. Sci. U.S.A.* **86**, 8437–8441.

Sahyoun, N., McDonald, O. B., Farrell, F., and Lapetina, E. G. (1991). *Proc. Natl. Acad. Sci. U.S.A.* **88**, 2643–2647.

Schulman, H. (1988). *Adv. Sec. Mess. Phosphoprotein Res.* **22**, 39–116.

Schworer, C. M., Colbran, R. J., Keefer, J. R., and Soderling, T. R. (1988). *J. Biol. Chem.* **263**, 13486–13489.

Sikela, J. M., and Hahn, W. E. (1987). *Proc. Natl. Acad. Sci. U.S.A.* **84**, 3038–3042.

Sikela, J. M., Law, M. L., Kao, F. T., Hartz, J. A., Wei, Q., and Hahn, W. E. (1989). *Genomics* **4**, 21–27.

Stallard, B. J., Collard, M. W., and Giswold, M. D. (1991). *Mol. Cell. Biol.* **11**, 1448–1453.

Stewart, T. A., Hecht, N. B., Hollingshead, P. G., Johnson, P. A., Leong, J. C., and Pitts, S. L. (1988). *Mol. Cell. Biol.* **8**, 1748–1755.

Sullivan, K. F., Machlin, P. S., Ratrie, H., and Cleveland, D. W. (1986). *J. Biol. Chem.* **261**, 13317–13332.

Swanljung-Collins, and Collins, J. H. (1991). *J. Biol. Chem.* **266**, 1312–1319.

Tanaka, S., and Fujimoto, H. (1986). *Biochem. Biophys. Res. Commun.* **136**, 760–766.

Tash, J. S., Krinks, M., Patel, J., Means, R. L., Klee, C. B., and Means, A. R. (1988). *J. Cell Biol.* **106**, 1625–1633.

Thiel, G., Czernik, A. J., Gorelick, F., Nairn, A. C., and Greengard, P. (1988). *Proc. Natl. Acad. Sci. U.S.A.* **85**, 6337–6341.

Tindall, D. J., Vitale, R., and Means, A. R. (1975). *Endocrinology* **97**, 636–648.

Uhler, M. D., Chrivia, J. C., and McKnight, G. S. (1986). *J. Biol. Chem.* **261**, 15360–15363.

Waites, G. M. H. (1970). In "The Testis" (A. D. Johnson, W. R. Gomes, and N. L. Vandemark, eds.), Vol. I, pp. 241–279. Academic Press, New York.

Waldman, R., Hanson, P. I., and Schulman, H. (1990). *Biochemistry* **29**, 1679–1684.

Waxham, M. N., Aronowski, J., Westgate, S. A., and Kelly, P. T. (1990). *Proc. Natl. Acad. Sci. U.S.A.* **87**, 1273–1277.

Wolgemuth, D. J., and Watrin, F. (1991). *Mammalian Genome* **1**, 283–288.

# Retinoid Receptors

DAVID J. MANGELSDORF,* STEVEN A. KLIEWER,* AKIRA KAKIZUKA,*
KAZUHIKO UMESONO,*,† AND RONALD M. EVANS*,†

* The Salk Institute for Biological Studies, and †Howard Hughes Medical Institute, La
Jolla, California, 92037

## I. Introduction

One of the fundamental questions in hormonal signaling is how a group of closely related and simple compounds can mediate diverse and complex responses. Multicellular organisms have evolved two basic strategies for mediating the translation of an extracellular signal into a transcriptional response at the target cell (Fig. 1). In one mechanism the stimulus (e.g., a small peptide or growth factor) binds to a cell surface receptor. Transduction of this signal to the nucleus is then accomplished by one of a myriad of second messages. The many examples of signals that operate via a second messenger are demonstrated by the high degree of genetic and phenotypic variation of their receptors and pathways (Herschman, 1989). A second mechanism that cells have evolved to process an extracellular stimulus is through an intracellular receptor. In this pathway the blood-borne signals are small lipophilic molecules, such as steroids, which are either actively or passively transported through the cell membrane, where they bind cytosolic or nuclear receptors. The liganded receptor complex is a functionally active transcription factor that binds specifically to the regulatory region of target genes. In contrast to the cell surface receptors, the intracellular receptors belong to one highly conserved family of proteins (Evans, 1988; Green and Chambon, 1988). Thus, from the characterization of one of these receptor pathways a unified mechanism can be derived by which the whole class of intracellular receptors can be studied. To date, nuclear receptors have been discovered for steroids, thyroid hormone, and, recently, the retinoids (Evans, 1988; Green and Chambon, 1988).

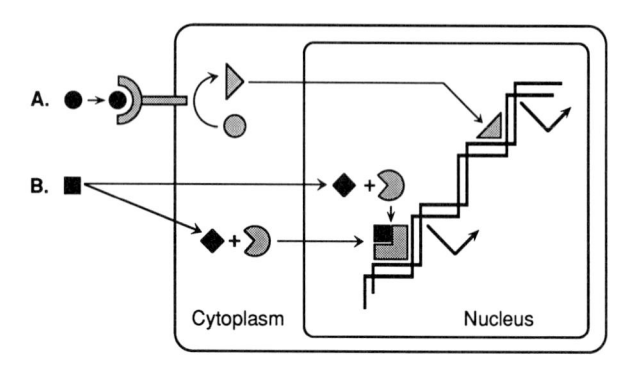

FIG. 1. Two pathways of hormonal signaling. (A) Water-soluble hormones (e.g., peptides) bind to their receptors on the target cell surface. Ligand binding triggers a series of enzymatic events, such as a change in intracellular calcium or the phosphorylation state of a protein, which in turn may act as second messengers to the nucleus. (B) Lipid-soluble hormones such as steroids, vitamin D, thyroid hormone, and retinoic acid penetrate the cell membrane and bind to an intracellular receptor. In this pathway the liganded receptor becomes the signal to the nucleus, where it directly interacts with the transcriptional machinery.

## II. Retinoid X Receptors

### A. A NOVEL RETINOID SIGNALING PATHWAY

Vitamin A derivatives, or retinoids [e.g., retinoic acid (RA)], are a group of signaling molecules that are essential for several life processes, including growth, differentiation, epithelial homeostasis, formation of the developing nervous system, and limb morphogenesis (Goodman, 1984; Sporn *et al.*, 1984; Dencker *et al.*, 1987; Durston *et al.*, 1989; Wagner *et al.*, 1990; Wedden *et al.*, 1988; Maden, 1982; Eichele, 1989). The heterogeneity of these responses suggests the existence of complex signaling pathways to account for the diverse regulatory roles of retinoids. In the past several years two distinct classes of nuclear hormone receptors were identified that mediate RA-dependent transcription. The first class is composed of the $\alpha$, $\beta$, and $\gamma$ RA receptors (RAR$\alpha$, RAR$\beta$, and RAR$\gamma$), each of which bind all-*trans*-RA with high affinity and share a high degree of structural conservation (Giguere *et al.*, 1987; Petkovich *et al.*, 1987; Brand *et al.*, 1988; Zelent *et al.*, 1989; Krust *et al.*, 1989; Ishikawa *et al.*, 1990). Recently, we described the existence of a second class of receptor that also responds to retinoids but, compared to the RARs, is substantially different in primary structure and ligand specificity (Mangelsdorf *et al.*, 1990). This new receptor was termed the retinoid X receptor (RXR).

RETINOID RECEPTORS 101

The discovery in RXR of a second retinoid transduction pathway has led us to investigate its functional properties and determine its relationship to the RARs. An understanding of the differences and similarities of the RAR and RXR systems requires knowledge of the diversity of the family members, their patterns of expression, and their pharmacology in response to cognate ligands. Recently, we reported the isolation of cDNAs encoding three mouse RXR proteins—mRXRα, mRXRβ, and mRXRγ (Fig. 2) (Mangelsdorf *et al.*, 1992). These proteins are closely related to each other both in their DNA-binding and ligand-binding domains. Their homologies indicate that these receptors are likely to regulate common target sequences and respond to common ligands. However, they differ markedly in their amino-terminal domains, which could confer distinct trans-activation functions. Interestingly, although there is no apparent insect homologue of the RARs, a *Drosophila* homologue of the RXR gene that has been identified maps to the *ultraspiracle* (*usp*) locus (Oro *et al.*, 1990). Although *usp* does not respond to retinoids, its homology to RXR indicates the ancient evolutionary origin of this gene family and raises the question as to whether RXRs might represent the original retinoid signaling system. In this respect, it will be of interest to discover the ligand for *usp*.

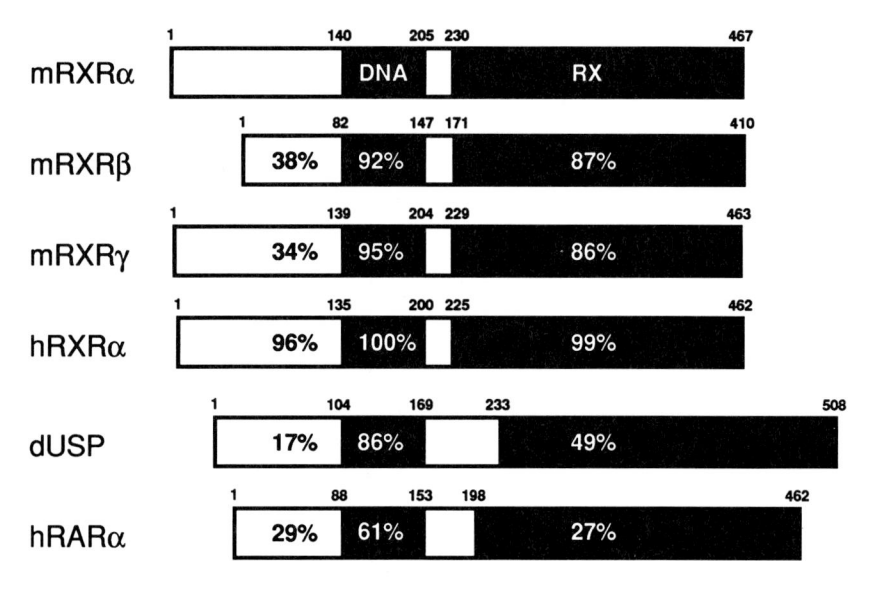

FIG. 2. Amino acid comparison between RXR and the other members of the retinoid receptor family. Shown are the retinoid receptors from the mouse (mRXRs), human (hRXRα and hRARα), and *Drosophila ultraspiracle* (dUsp).

## B.  A NEW HORMONE: 9-*cis*-RETINOIC ACID

A striking observation to come from the original characterization of the RXR family was the apparent dissimilarity in sequence to that of the RARs, providing evidence for an evolutionarily distinct retinoid response pathway. Human RXRα, like the RARs, was originally identified by its ability to respond to all-*trans*-RA; similarly, murine RXR subtypes are also activated by all-*trans*-RA. The initial observation that RA is the most potent inducer of both RAR and RXR suggested that it is also the naturla ligand for both of these receptors. However, there is some evidence to indicate that more potent ligands may exist which would discriminate between RAR and RXR. Cotransfection analyses have shown that although RAR and RXR respond with a rank order of potency similar to that of the naturally occurring metabolites of vitamin A (retinoic acid > retinal > retinal acetate > retinol), there are distinct pharmacological differences with respect to the synthetic retinoids (Mangelsdorf *et al.*, 1990). For example, the benzoic acid retinoid derivative, TTNPB, is known to be a potent agonist of RA activity (Keeble and Maden, 1989). In the cotransfection assay, however, TTNPB shows RA-like activity with human RARα (hRARα), but only 10% of this response with hRXRα. Furthermore, in a comparison of dose responses, hRARα has at least a 10-fold higher sensitivity to RA than does hRXRα, whose response has not yet saturated at levels as high as $10^{-5}$ $M$ RA. The idea that RA may not be the highest affinity RXR ligand is further supported by binding analyses. To date we have been unable to demonstrate high-affinity binding of [$^3$H]RA to RXR extracted from cells transfected with an RXR expression plasmid. In contrast, under similar conditions RA does bind with high affinity ($K_d$ = 4 n$M$) and specificity to the RAR proteins (Ishikawa *et al.*, 1990; Yang *et al.*, 1991). In the case of RXR, one explanation for the lack of high-affinity binding may be that RA is a lower affinity precursor that is metabolized to a more active form. Such a metabolic pathway would be reminiscent of vitamin D action, in which it originally was postulated that 25-hydroxyvitamin $D_3$ is the active principle. It was not until the discovery of a more polar metabolite specifically bound to chromatin (Haussler *et al.*, 1968) that it was realized 1,25-dihydroxyvitamin $D_3$ was the true hormone. Biologically active retinoic acid metabolites are known to exist. Recently, Thaller and Eichele (1990) have demonstrated the RA derivative 3,4-didehydroretinoic acid also has potent morphogenetic effects in the developing chick limb bud, where it is present at a concentration six times that of RA. Moreover, didehydroretinoic acid can differentially activate RXRs and RARs in the *Xenopus* system (Blumberg *et al.*, 1992). The observation that two distinct receptors respond with reciprocal specificity to similar ligands is not without precedent. An example is the estrogen and androgen receptors. The female sex steroid, estrogen, is produced via the aromatization of the male sex steroid, testosterone. Although these two hormones are closely related, the ligand-binding domains of their respective receptors share only marginal (25%) identity (Lubahn *et al.*, 1988). Based on this analogy, it is not unreasonable to presume that RARs and RXRs recognize chemically similar or identical molecules even though their ligand-binding domains are only distantly related. Based on these assumptions we devised a strategy based on the cotransfection assay to identify retinoid X. The implicit concept of this strategy was that all-*trans*-RA may be serving as a prohormone that would be metabolized to retinoid X. Accordingly, we have now demonstrated that 9-*cis*-RA, which is converted from all-*trans*-RA in cells, is a naturally occurring, high-affinity ligand for the human RXRα as determined by a combination of binding and activation studies (Heyman *et al.*, 1992). While not previously seen in living organisms, this work indicates that 9-*cis*-RA is an apparently new and potentially widely used vertebrate hormone. 9-*cis*-RA transactivates RXRα up to 40 times more efficiently than all-*trans*-RA and binds to RXR with high affinity ($K_d$ = 11.7 n$M$). We also confirmed that all-*trans*-RA shows no detectable binding affinity for RXRα. In addition, we have recently demonstrated that 9-*cis*-RA is also retinoid X for the mouse RXRα, RXRβ, and RXRγ subtypes (Mangelsdorf *et al.*, 1992). Each of these receptors is activated by 9-*cis*-RA with both increased potency and efficacy relative to all-*trans*-RA (Fig. 3). Though RXRs may interact with RARs (see below), they may also function independently. Expression in *Drosophila* Schneider cells is sufficient to reconstitute retinoid responsiveness. Thus, not only do the RXRs bind and respond to 9-*cis*-RA, but they are able to activate target genes in the absence of other retinoic acid receptor gene products. Furthermore, the RXRs are capable of activating through the recently described response element in the CRBPII gene, which is not induced by the RARs (Fig. 3) (Mangelsdorf *et al.*, 1991). These results indicate that the RXRs are bona fide members of the nuclear receptor family, and, together with the discovery of the new vertebrate hormone, 9-*cis*-RA, define a second retinoid signaling pathway.

A point of potential physiological significance is that 9-*cis*-RA also binds to and trans-activates both RXRs and RARs and may thus serve as a common or "bifunctional" ligand. Conversion of the all-trans to the 9-cis isomer could provide a novel means for differential cell-specific regulation of the activity of these retinoid pathways. The hypothesis that 9-*cis*-RA may be functionally distinct from its all-trans precursor raises the interesting possibility that regulation of its isomerization could be a key step in retinoid physiology. It is unknown whether this reaction is catalyzed by an enzyme. However, a precedence for this may be the dark-cycle reaction in the visual system, wherein a membrane-bound isomerase converts the all-trans isomer into the photolabile 11-cis isomer (Bernstein et al., 1987; Rando, 1990).

## C. RXR PATTERNS OF EXPRESSION

Though the RXRs can regulate target genes distinct from those regulated by the RARs, they may also regulate common target genes. To understand the physiologic role of 9-cis-RA, we have begun to analyze the target tissues for RXRs (Mangelsdorf et al., 1992). In aggregate, in the adult organism the RXRs are widely expressed. In general, this pattern reflects a perdurance of that observed in situ in the embryo. Although there is no absolute correlation between the individual RAR and RXR subtypes, there are several interesting similarities. For example, both RARα and RXRβ are found in almost all tissues, whereas RARβ and RARγ and RXRα and RXRγ are much more restricted. Although RXRβ shows no striking pattern, there is increased expression in the anterior extensions of the spinal cord and hind brain (Mangelsdorf et al., 1992). RXRα shows abundant expression in liver, kidney, spleen, and a variety of visceral tissues (Mangelsdorf et al., 1992) that are in marked distinction to the RAR subtypes, which are found in very low levels in the corresponding sites (Zelent et al., 1989; Brand et al., 1988; Kastner et al., 1990; Giguere et al., 1987). This pattern led us to propose that RXRα may be involved in retinoid metabolism, a hypothesis that was further bolstered by the demonstration that the CRBPII gene (Mangelsdorf et al., 1991) is RXR responsive. CRBPII helps to absorb dietary retinol and assist in its transfer to chylomicrons and ultimate transport to the liver. Because high levels of retinoids are teratogenic, especially to the early embryo, the abundant early expression of RXRα in the decidua and placenta may indicate RXRα has a role in feedback regulation of retinoid uptake and storage in the embryo. The expression of RXRα in vital organs suggests an additional role for retinoids in physiologic homeostasis. Interestingly, RXRα and RARγ are both abundantly expressed in the epidermis (Mangelsdorf et al., 1992; Kastner et al., 1990), indicating that these two receptors are likely to be responsible for the dermatologic effects of retinoids.

The most restricted of the three receptors is RXRγ, which also shows a lower level of expression in the embryo and the adult (Mangelsdorf et al., 1992). It is particularly common in the muscle and the brain. Thus, one might anticipate a role for RXRγ in aspects of muscle and CNS differentiation. In the chicken, a homologue of RXRγ has been shown to be expressed in the developing peripheral nervous system (Rowe et al., 1991). The CNS pattern of RXRγ expression in the mouse is particularly interesting, showing prominence in the embryonic enlage to the caudate

RETINOID RECEPTORS

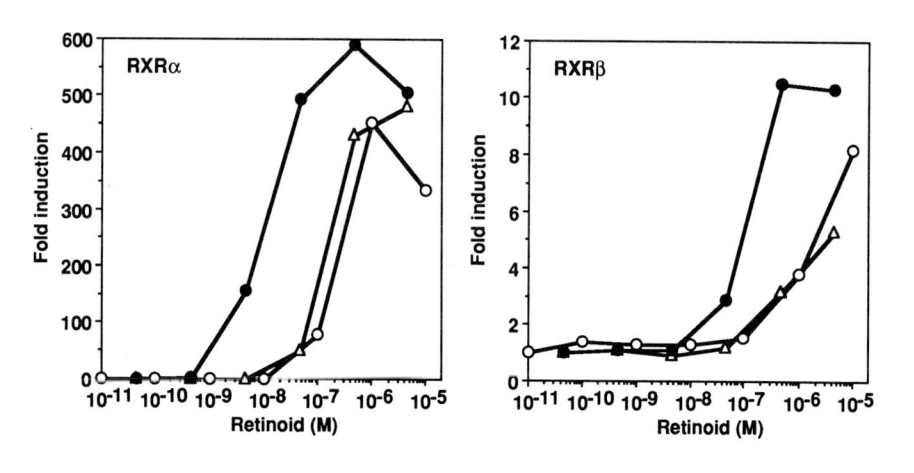

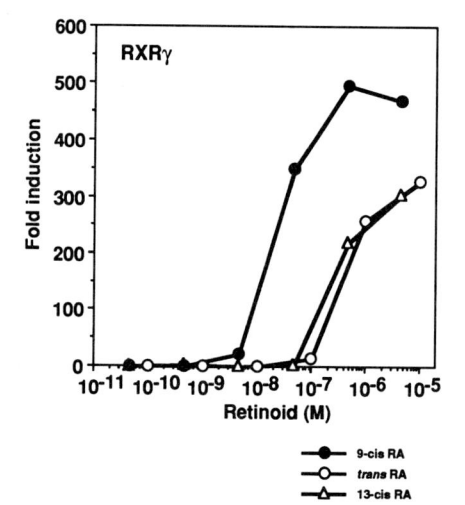

FIG. 3. 9-*cis*-RA is a high-affinity ligand for RXRs. Schneider cells cotransfected with receptor expression plasmids for mRXRα (a), mRXRβ (b), or mRXRγ (c) and the CRBPII–RXRE luciferase reporter plasmid were incubated with increasing amounts of 9-*cis*-RA (closed circles), all-*trans*-RA (open circles), or 13-*cis*-RA (open triangles). Transactivation is expressed as induction of luciferase activity (times control levels) in retinoid-treated cells compared with solvent-treated control cells and represents the average of triplicate assays.

putamen (basal ganglia). This is the major target of dopaminergic innervation by the substantia nigra. Interestingly, both RXRγ and RARβ are coexpressed in the motor neurons of the spinal cord, again suggesting a selective parity between these two receptor systems. We find prominent expression of RXRγ in the pituitary, suggesting a potential role for reti-

106                    DAVID J. MANGELSDORF ET AL.

noids in the regulatory cascade associated with hypophyseal differentiation.

These studies suggest that the RXR family plays critical roles in diverse
aspects of development, from embryo implantation to organogenesis and
central nervous system differentiation, as well as in adult physiology.

### D. THE 3–4–5 RULE PROVIDES A STRATEGY FOR CLASSIFYING KNOWN AND ORPHAN RECEPTORS

Members of the receptor superfamily modulate target gene expression
by binding as either homo- or heterodimers to hormone response elements
(HREs). We recently described the properties of direct repeats of the
consensus half-site sequence AGGTCA as HREs for nuclear receptors
(Umesono et al., 1991). Receptor specificity for binding and activation
was shown to be conferred through the number of nucleotides separating
the two half-sites. Spacers of 3, 4, or 5 nucleotides were originally shown to
serve as optimal REs for the vitamin D receptor (VDR), thyroid hormone
receptor (TR), and retinoic acid receptor (RAR), respectively, referred to
as the "3–4–5 rule" (Fig. 4) (Umesono et al., 1991). More recently,
as described above, we characterized an HRE present in the upstream
regulatory region of the CRBPII gene that confers selective responsiveness
to the RXR (Mangelsdorf et al., 1991). Termed the CRBPII–RXRE, this
RXR-specific HRE consists of four tandemly arranged AGGTCA repeats
separated by a single nucleotide, filling one of the spacing options left
unoccupied by the 3–4–5 rule.

In addition to the receptors that function in response to characterized
ligands, a large number of structurally related proteins, termed orphan
receptors, have been isolated that lack identified ligands (Evans, 1988).
Due to this lack of ligand information, the regulatory roles of the orphan
receptors have remained largely uncharacterized. The relationship between receptors and their cognate HREs expounded in the 3–4–5 rule
provides, in principle, a simple, systematic approach for classifying receptors, including the orphan receptors, on the basis of their half-site spacing
preferences. We have recently exploited this approach (Kliewer et al.,
1992a) to demonstrate functional interactions between RXR and COUP-
TF, an orphan receptor originally characterized via its interaction with a
response element present in the chicken ovalbumin gene promoter (Sagami
et al., 1986). The binding specificities of a variety of nuclear receptors
were examined through gel mobility shift analysis using a series of synthetic HREs consisting of two tandemly arranged AGGTCA repeats separated by a spacer ranging in length from zero to five nucleotides (DRs 0–5).
Bacterially expressed RXR and COUP-TF were both found to display a

RETINOID RECEPTORS 107

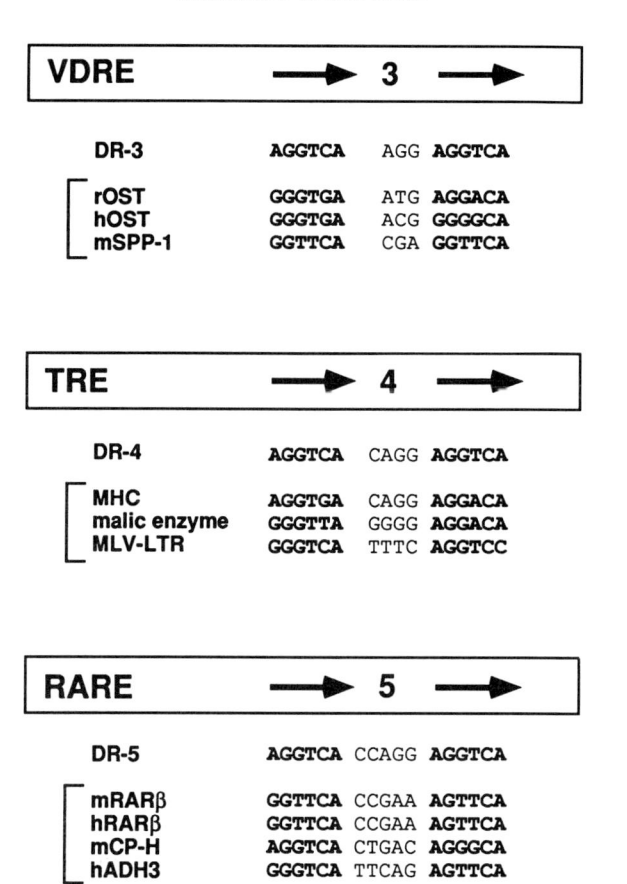

FIG. 4. The 3–4–5 rule for nuclear receptors. Shown are the sequence comparisons of the direct-repeat hormone response elements for vitamin D (VDRE), thyroid hormone (TRE), and retinoic acid (RARE). The direct-repeat (DR) sequence motif is based on the naturally occurring HREs described by Umesono *et al.* (1991).

marked preference for binding to DR-1 relative to the other spacing options (Fig. 5). These results suggested that the two receptors might modulate gene expression through an overlapping set of natural HREs. Interestingly, gel mobility shift experiments in which bacterially expressed COUP-TF and RXR were mixed revealed the preferential formation of COUP-TF–RXR heterodimers relative to the formation of either homodimer, further suggesting the possibility of regulatory "cross-talk" between the two receptors.

The potential for functional interactions between COUP-TF and RXR

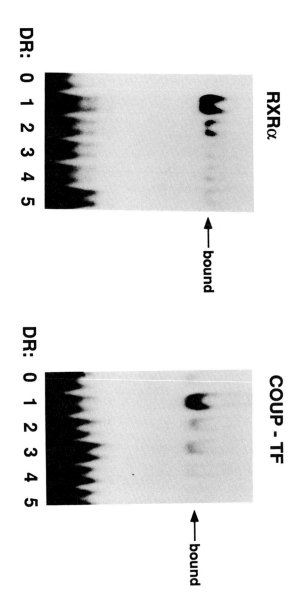

FIG. 5. Bacterially expressed RXR and COUP-TF preferentially interact with DR-1. Gel mobility shift assays were done with bacterially expressed RXR or COUP-TF and $^{32}$P-labeled, synthetic HREs containing two AGGTCA direct repeats separated by a spacer of 0–5 nucleotides (DR-0 through DR-5, respectively).

*in vivo* suggested by the *in vitro* studies was examined via cotransfection analysis. RXR was found to activate gene expression through the previously defined COUP-TF response element present in the promoter of the chicken ovalbumin gene. Although COUP-TF failed to activate gene expression through the CRBPII–RXRE, cotransfection of expression plasmids for COUP-TF and RXR revealed that COUP-TF was a potent repressor of RXR-mediated transaction via the CRBPII–RXRE (Fig. 6), suggesting a novel negative regulatory function for the orphan receptor. Thus, as predicted by their identical half-site spacing preferences, COUP-TF and RXR display a large degree of cross-regulation.

These studies demonstrate that knowledge of a receptor's half-site spacing preference can be exploited as part of a strategy to elucidate functional interactions between members of the superfamily, including the orphan receptors. In addition, by searching data bases, genes containing sequences related to HREs of distinct half-site spacing in their regulatory

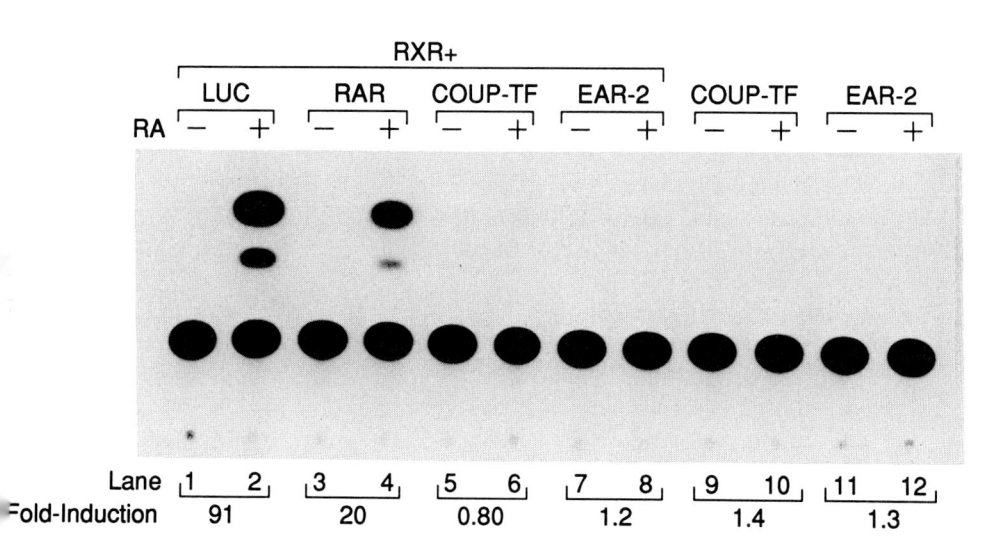

FIG. 6.   COUP-TF and related receptor EAR2 repress RXR-mediated transactivation through the CRBPII promoter. F9 cells were cotransfected in duplicate with reporter construct pCRBPII–CAT and RS–hRXRα plus either the control RS–LUC (lanes 1 and 2), RS–hRARα (lanes 3 and 4), RS–COUP-TF (lanes 5 and 6), or RS–EAR-2 (lanes 7 and 8). Cotransfections performed with the reporter pCRBPII–CAT and either RS–COUP-TF (lanes 9 and 10) or RS–EAR-2 (lanes 11 and 12) in the absence of RS–hRXRα are also shown. Cells were treated with either ethanol (−) or 10 μM RA (+) for 30 hours and the cell extracts were subsequently assayed for CAT activity.

110                    DAVID J. MANGELSDORF ET AL.

regions can be identified. Thus, this approach can provide insights into the complex, physiologic networks regulated by hormonal signals.

## E. RXR INTERACTS DIRECTLY WITH THE VDR, TR, AND RAR

Experiments from several laboratories have shown that accessory factors present in nuclear extracts are necessary for high-affinity binding of the RAR, TR, and VDR to their cognate HREs (Glass *et al.*, 1990; Liao *et al.*, 1990; Murray and Towle, 1989; Yang *et al.*, 1990). These accessory factors have been proposed to be receptor family members that function to stimulate receptor binding through heterodimer formation (Glass *et al.*, 1990). Remarkably, we have recently found that RXR functions as a common heterodimeric partner for the VDR, TR, and RAR (Kliewer *et al.*, 1992b), as well as for COUP-TF. These results emerged from our initial finding of a functional interaction between RAR and RXR. Cotransfection of RAR, as in the case of COUP-TF, was shown to efficiently repress RXR-mediated transactivation through the CRBPII–RXRE (Mangelsdorf *et al.*, 1991) (Fig. 6). Cotransfection experiments performed with a variety of RAR mutants revealed that this repression of RXR activity was mediated through the C-terminal region of RAR, which includes the dimerization domain, suggesting the possibility of a physical interaction between RAR and RXR (Kliewer *et al.*, 1992b). Indeed, *in vitro* studies showed that RAR and RXR form stable solution heterodimers and interact cooperatively in binding to target HREs, including the RARE of the RARβ promoter and the CRBPII–RXRE (Fig. 7). Thus, the signaling pathways mediated by the two families of retinoid receptors converge through direct RAR–RXR interactions.

We have also observed similar interactions between RXR and the VDR and TR (Kliewer *et al.*, 1992b). As in the case of RAR, RXR interacts cooperatively with the VDR and TR in binding to target HREs and forms stable solution heterodimers with the two receptors (Fig. 8). This ability of RXR to interact with receptors responsive to a variety of ligands establishes a central role for RXR in modulating multiple hormonal pathways. Recently, two additional subtypes of RXR, termed RXRβ and RXRγ, were isolated (Mangelsdorf *et al.*, 1992; see above). Thus, the interaction of multiple forms of RXR with nuclear receptors responsive to a diversity of signaling molecules likely provides the repertoire of transcription factor activities necessary to regulate the battery of hormone-responsive genes. It remains unclear why the VDR, TR, and RAR interact with a common partner, particularly because the actions of vitamin D and thyroid hormone do not appear to be retinoid dependent. It seems likely that RXR is able to exert some of its regulatory effects in the absence of ligand.

RETINOID RECEPTORS                                                   111

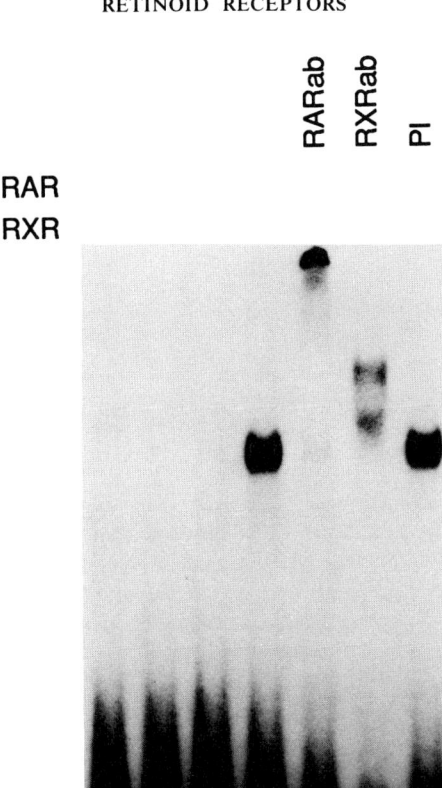

FIG. 7. Direct interactions between RAR and RXR. Gel mobility shift assays were performed using *in vitro*-synthesized RAR and/or RXR as indicated and $^{32}$P-labeled CRBPII–RXRE oligonucleotide. Polyclonal antisera prepared against either RAR (RARab) (lane 5) or RXR (RXRab) (lane 6) or preimmune serum (PI) (lane 7) were included in the reactions as indicated.

## F. FUSION RARα IN APL

As far back as 1977 it was first reported that the diseased blood cells of patients with acute promyelocytic leukemia (APL) demonstrated an abnormal karyotype, which manifests as a reciprocal translocation be-

112                    DAVID J. MANGELSDORF ET AL.

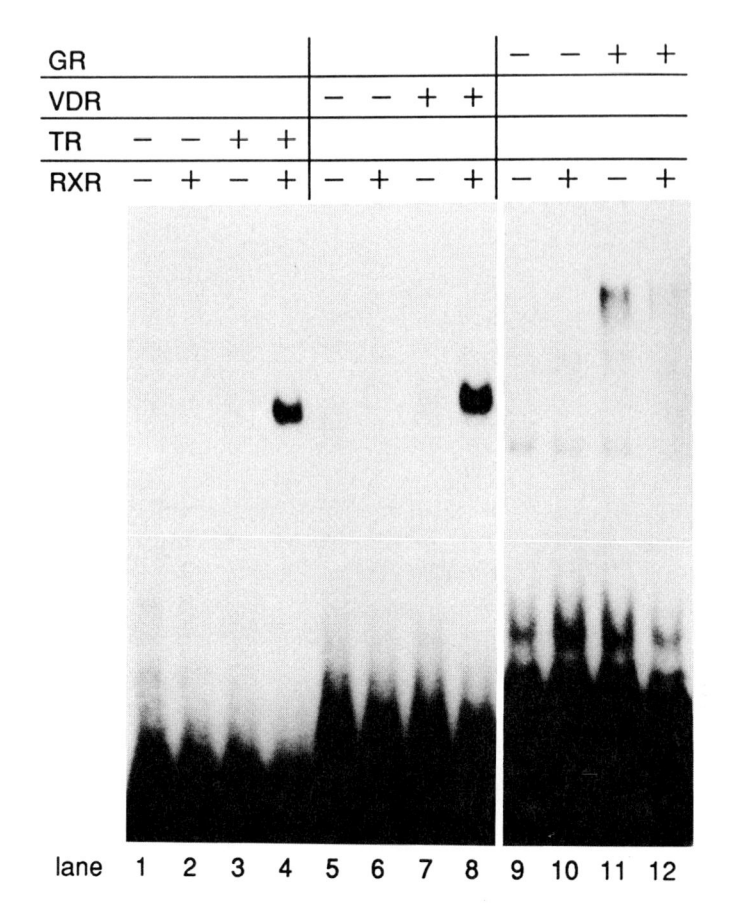

FIG. 8.   RXR interacts cooperatively with TR and VDR in DNA binding. Gel mobility
shift assays were performed using *in vitro*-synthesized RXR, TR, VDR, and GR as indicated
and $^{32}$P-labeled oligonucleotides encoding the murine leukemia virus long terminal repeat
TRE (lanes 1–4), the mouse osteopontin VDRE (lanes 5–8), or the palindromic GRE (lanes
9–12).

tween chromosome 15 and 17, t(15;17)(q22–q12–21), predicting a responsible gene(s) at either or both chromosomal break points (Rowley *et al.*,
1977). This particular translocation is detected in as many as 90% of APL
patients and has become a definitive marker for the disease. In 1988, it
was shown that high-dose RA therapy will induce these patients into
complete clinical remission (Huang *et al.*, 1988). The chromosome 17
break point in APL was recently mapped to the *retinoic acid receptor α*
*(RARα)* gene, almost exclusively in an intron between the first and second

RETINOID RECEPTORS 113

coding exons (Fig. 9) (de The *et al.*, 1990; Borrow *et al.*, 1990; Longo *et al.*, 1990). Thus, the translocation results in a fusion between the *RARα* gene and a region on chromosome 15 referred to as *PML*, which originally was named *myl* (de The *et al.*, 1990). Note also that the translocation is balanced and reciprocal and is expected to generate two novel abnormal gene-producing *RAR–PML* and *PML–RAR* fusion mRNAs. Northern analysis of APL RNA, probed with human *RARα* cDNA, gives rise to additional bands that are likely to represent the predicted transcripts of the translocated genes (Kakizuka *et al.*, 1991).

cDNAs have been identified coding for both reciprocal fusion mRNAs (*PML–RAR* and *RAR–PML*) as well as wild-type *RARα* and *PML* mRNAs from APL cells (Kakizuka *et al.*, 1991). Two isoform proteins of *PML–RAR* (type A is the shorter form, and type B is the longer form) are predicted from the cDNA sequences, which result from distinct fusion points in the PML protein (Kakizuka *et al.*, 1991; de The *et al.*, 1991; Pandolfi *et al.*, 1991). There are no differences clinically among patients who have either type A or type B fusion, although type B patients seems to be slightly more common than the ones with type A. Both isoforms

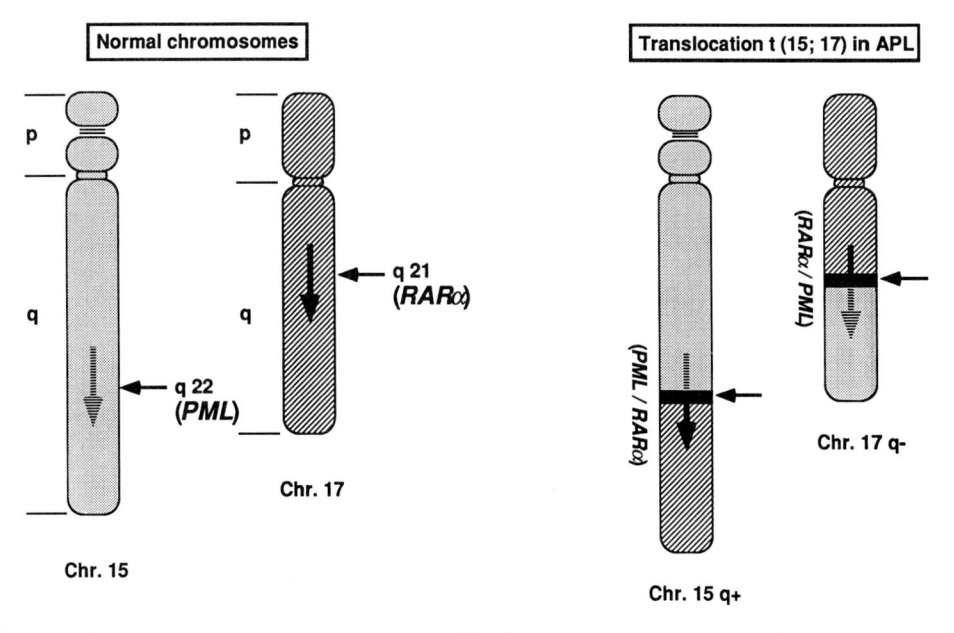

FIG. 9. Chromosomal translocation in APL. The arrows depict the normal chromosomal location for the wild-type *PML* and *RARα*, and the t(15;17) reciprocal translocation for the mutant fusion genes *PML–RARα* and *RARα–PML*.

114                    DAVID J. MANGELSDORF ET AL.

contain the functional DNA-binding and ligand-binding domains of the
RAR fused to most of the wild-type PML protein (Fig. 10). The reciprocal
fusion (RAR–PML) fuses only the first 59 amino acids of RARα to the
residual carboxy-terminal end of PML. Because of this observation, and
as mentioned below, it is more likely that the PML–RAR fusion contributes to the disease state.

The wild-type *PML* mRNAs show heterogeneity produced by alternative splicing, and several distinct cDNAs have been isolated (Kakizuka *et
al.*, 1991; de The *et al.*, 1991; Pandolfi *et al.*, 1991). One of the PML
proteins, PML-1 (Kakizuka *et al.*, 1991), consists of 560 amino acids and
contains four structurally interesting regions, including a proline-rich N
terminus, a cysteine-rich portion, a potential α-helical region, and a serine-
rich C terminus region (Fig. 11). There is no signal peptide sequence or a
transmembrane region, suggesting that PML-1 is an intracellular protein.
A segment of α-helical-rich domain (amino acids 280–327) shows homology to the leucine zipper region of the fos family. Because the fos leucine
zipper mediates dimerization, the presence of this related sequence in
PML raises the further possibility that PML or PML–RAR protein may
be part of a homo- or heterodimer complex. The translocation occurs at
amino acid 395 in type A and at 553 in type B. As mentioned above,
PML–RAR fusions would include most of these potentially important
regions, whereas the reciprocal fusion would only include a highly truncated segment of the PML protein.

Among the four structural regions, the most striking sequence is the
cysteine-rich motif, which we have referred to as the "cysteine-chapel

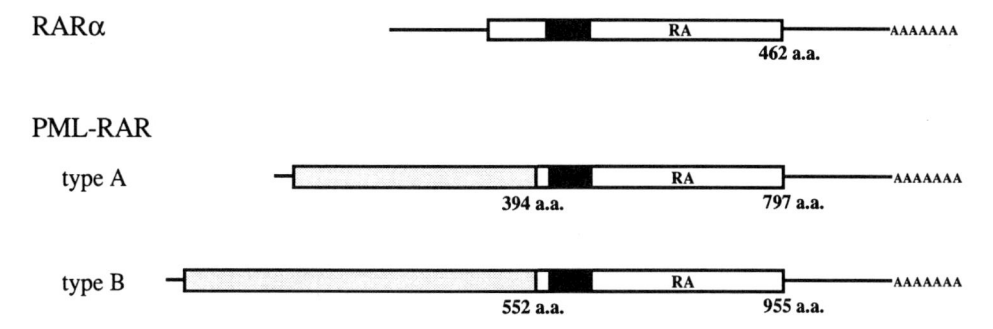

FIG. 10.  Schematic structures of *RARα* and *PML–RAR* mRNAs. The open reading
frames are shown as open boxes, and untranslated regions as lines. The fusion portions from
the *PML* gene are stippled. Regions encoding the DNA-binding domains derived from RARα
are in black. The RA-binding domains are marked as RA. Type A and type B represent
shorter and longer forms of fusion *PML–RAR* mRNA, respectively.

RETINOID RECEPTORS    115

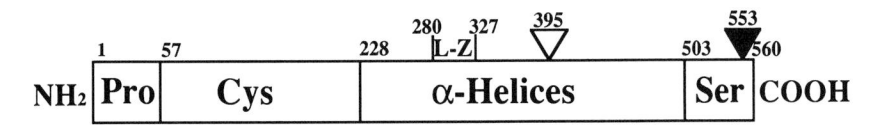

FIG. 11. Schematic structure of the PML-1 protein. The PML-1 protein contains four regions of interesting structure. They are denoted as follows: Pro, proline-rich region; Cys, cysteine-rich region; $\alpha$-helices, $\alpha$-helix-rich region; Ser, serine-rich C-terminus region. The open and closed triangles show the fusion points seen in type A and type B PML–RAR, respectively. L-Z shows the position of a putative leucine zipper. The numbers above the boxes represent the amino acids at the junction of the assigned regions.

motif.'' Data base searches that focused on this cysteine-rich region have identified several protein sequences with almost perfect conservation of all the cysteines found in the first of three clusters that occur in PML (Fig. 12). Furthermore, one histidine at amino acid 74 is also perfectly conserved. Cysteine and histidine conservation is reminiscent of the zinc finger motif seen in several different classes of transcription factors. Several of these proteins are in fact DNA-binding proteins and transcription factors. Although the function of this specific cysteine-rich domain is not yet clear, the presence of this motif is required for the ability of at least one of these proteins, Mel-18, to bind to DNA cellulose (Tagawa *et al.*, 1990). Likewise, another protein, LMCV-Z, binds $Zn^{2+}$ (Salvato and Shimomaye, 1989), supporting the possibility that the other members of this family may form cysteine–histidine fingers by chelating $Zn^{2+}$. Such analyses suggest that PML-1 may bind DNA through its cysteine-rich domain, thus making it a strong candidate for a transcription factor. This protein family also includes the recombination-activating gene product, RAG-1 (Schatz *et al.* 1989), and bmi-1 (van Lohuizen *et al.* 1991; Haupt *et al.* 1991), a protein that has shown to contribute to B cell lymphoma in the mouse and can collaborate with the *myc* oncogene, suggesting that other proteins that belong to the PML family can also contribute to oncogenesis in other types of tumors.

To address the question of how the t(15;17) translocation contributes to APL, the transcriptional profiles of the wild-type RAR$\alpha$ and the PML–RAR were compared. In one typical experiment, plasmids driving expression of either wild-type RAR$\alpha$ or PML–RAR were transfected into HL-60 cells and assayed for activation of two different reporters that contain retinoic acid response elements (RAREs) (Fig. 13). In the absence of RA, neither protein activated transcription. However, when RA was added, both RAR$\alpha$ and PML–RAR responded strongly to RA. Such experiments indicate PML–RAR is a hormone-responsive transcription factor

```
                        |─────────── cluster I ───────────|
PML      (  54) FLRCQQCQ------AEAKCPK-----LLPCLHTLCSGCLEAS--------GMQCPICQAPWPLGADTPA- (103)
RFP      (  13) ETTCPVCL------QYFAEPM-----MLDCGHNICCACLARCWGTAET----NVSCPQCRETFPQRHMRPNR (  69)
RPT-1    (  12) EVTCPICL------ELLKEPV-----SADCNHSFCRACITLNYESNRNTDG-KGNCPVCRVPYPFGNLRPNL (  71)
MEL-18   (  15) HLMCALCG------GYFIDAT-----TIVECLHSFCKTCIVRYLET-------NKYCPMCDVQ          (  59)
BMI-1    (  15) HLMCVLCG------GYFIDAT-----TIIECLHSFCKTCIVRYLET-------SKYCPICDVQ          (  59)
RING1           ELMCPICL------DMLKNTM-----TTKECLHRFCSDCIVTALRSG-----NKECPTCRKK
SS-A/Ro         EVTCPICL------DPFVEPV-----SIECGHSFCQECISQVGKGG------GSVCAVCRQR
PE38     (  83) KFECSVCL------ETYSQQS--NDTCPFLIPTTCDHGFCFKCVINLQSNAMNIPHSTVCCPLCNTQ (141)
RAG-1    ( 290) SISCQICE------HILADPV-----ETNCKHVFCRVCLIRCLKVM------GSYCPSCRYP         (334)
RAD18    (  25) LLRCHICK------DFLKVPV-----LTPCGHTFCSLCIRTHLNN------QPNCPICLFE         (  68)
L1-REP   (   7) YGMCAVCR------EPWAEGAV----ELLPCRHVFCTACVVQ---------RWRCPSCQRR         (  48)
IE110    ( 113) GDVCAVCT------DEIAPHL-----RCDTFPCMHRFCIPCMKTWMQL----RNTCPLCNAK        ( 159)
VZ61     (  16) DNTCTICM------STVSDLG-----KTMPCLHDFCEVCIRAWTST------SVQCPICRCP        (  60)
CG30     (   5) KLQCNICFSVAEIKNYFLQPIDRLTIIPVLELDTCKHQICSMCIRKIRKRK---KVPCPLCRVE       (  65)
LCMV-Z   (  29) PLSCKSCW------QKFDSLV------RC-------HDYICRHCLNLLLSV------SDRCPLCKYP    (  71)
                                                |─────── cluster II ───────|
PML      ( 104) -LDNVFFESLQRRLSVYRQIVDAQAVCTRCKESADFWCFECEQLLCAKCFEA--HQWFLKHEARPLAELRNQSVRE (176)
RFP      (  70) HLANVTQLVKQLRTERPSGPGGEMGVCEKHREPLKLYCEEDQMPICVVCDRSREHRG--HSVLPLEE-AVEGFKE (141)
RPT-1    (  72) HVANIVERLKGFKS-IP-EEEQKVNICAQHGEKLRLFCRKDMMVICWLCERSQEHRG---HQTALIEE-VDQEYKE (141)
                                    |─────── cluster III ───────|
PML      ( 177) FLDGTRKTNNIFCSNPNHRTPTLTSIYCRGCSKPLCCSCALLDSSHSELKC (227)
```

FIG. 12. The PML protein family. Amino acid alignment of proteins with a novel cysteine-rich motif are shown. Three cysteine clusters are indicated by arrows. Conserved amino acids, including cysteines/histidines, are denoted in bold type. Dashes represent gaps in amino acid alignment. The numbers represent the position of the first and the last amino acids in the respective protein sequences. References to protein sequences can be found in Kakizuka et al. (1991).

RETINOID RECEPTORS

117

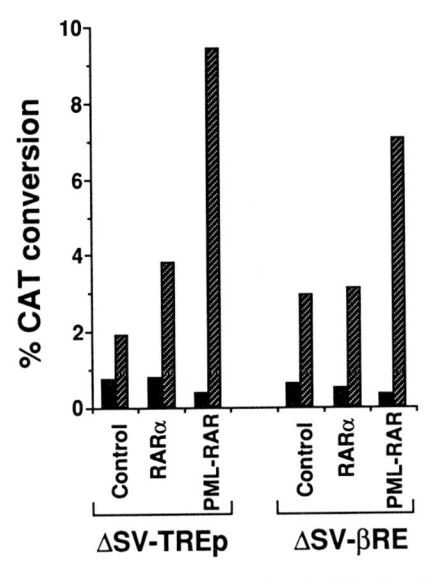

FIG. 13. *RARα* and *PML–RARα* trans-activation in HL-60 cells. Receptor expression plasmids containing control DNA, *RARα* cDNA, or *PML–RARα* cDNA were cotransfected with the CAT reporter plasmids ΔSV-TREp or ΔSV-βRE into HL-60 cells. Trans-activation was then monitored by measuring percent CAT conversion in cells treated with (shaded boxes) or without (solid boxes) $10^{-6}$ M retinoic acid.

and may be an equal or even better activator than wild-type RAR. These results suggest that PML–RAR may function to modulate the RA genetic network in promyelocytes. Furthermore, the comparison of the binding affinity to [³H]RA of RARα and PML–RAR expressed in COS cells does not show any significant difference between the two receptors (unpublished results). Given these results, the question therefore becomes how PML–RAR might contribute to APL. Its ability to function similarly to the RAR is evidence that it does not seem to contribute to APL through the RAR pathway. This conclusion is tempered, however, by the observation that PML–RAR displays marked cell type-specific variation in its activation profiles relative to wild-type RARα. The basis for this variation is not known. Though there are several possibilities for how the fusion might create an oncogene, each must take into account two clinical observations: (1) greater than 90% of APL patients have the (15;17) translocation, suggesting that the mutant RARα locus is a new oncogene, and (2) patients with APL can be induced into remission by high-dose RA treatments. Considering the structural features of PML–RAR, it is most likely that this fusion product interferes either with the RAR or the PML transcriptional pathway. For example, PML–RAR may block either one of these pathways in the absence of RA. In the presence of RA, the fusion protein could be activated and the block would be released. As discussed above, because the PML–RAR does not seem to interfere with the endogenous RAR pathway, we propose a model in which PML–RAR functions as a dominant-negative inhibitor of the endogenous PML pathway, and thereby blocks promyelocytic differentiation (Fig. 14). This idea is not without precedent. It has previously been shown that heterologous transcription factors fused to the ligand-binding domain of steroid receptors lose their normal activation properties and now become hormone dependent. For example, the transforming activity of the myc protein becomes estrogen-dependent when fused to the estrogen receptor (Eilers *et al.*, 1989). For PML–RAR, the t(15;17) translocation may place the PML product under control of the RAR ligand-binding domain. The abnormal fusion protein might block function of wild-type PML by forming nonproductive heterodimers or by blocking target DNA sites. Addition of RA would reverse this inhibition by releasing the block or transforming the fusion receptor to an activated state. This model provides a simple explanation as to why APL patients are responsive to RA treatment. A conclusive test will require the identification of PML–RAR target genes crucial for leukemogenesis.

In addition to clarifying the molecular origins of APL, the identification of the nucleotide sequences of two types of fusion mRNAs has provided an accurate and sensitive technique in which to diagnose APL by employing the reverse transcription/polymerase chain reaction (RT/PCR) technique. Recent clinical data from this method have demonstrated that

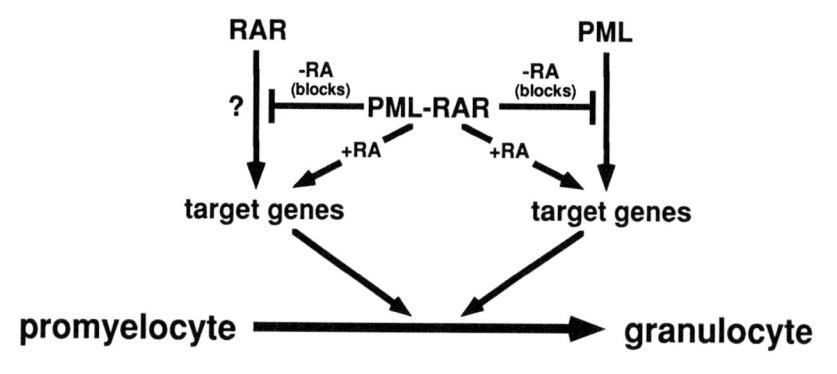

FIG. 14. Proposed model for APL. This model predicts how the aberrant PML–RAR fusion protein found in APL patients may contribute to the genesis of the disease. For discussion, see text.

## RETINOID RECEPTORS

there is a strong correlation between the existence of PML–RAR mRNA (either type A or type B) and clinical responsiveness to RA treatment; thus, patients who have the PML–RAR fusion can be predicted to benefit from RA treatment (Miller *et al.*, 1992). Interestingly, another clinical examination by Southern blot analyses for APL DNAs has demonstrated that in some cases only one of the PML and RARα genes is rearranged (Biondi *et al.*, 1991), suggesting that these cases of APL may not be responsive to RA. More detailed analysis will classify even morphologically undistinguishable APLs into at least two further subtypes.

### ACKNOWLEDGMENTS

Parts of this manuscript were compiled from edited versions of Umesono *et al.* (1991) Mangelsdorf *et al.* (1991, 1992), Kakizuka *et al.* (1991), and Kliewer *et al.* (1992a,b). We thank Elaine Stevens for manuscript preparation. S.A.K. is as Fellow of the Jane Coffin Childs Memorial Fund for Medical Research. D.J.M. and K.U. are Research Associates and R.M.E. is an Investigator of the Howard Hughes Medical Institute at the Salk Institute for Biological Studies. This work was supported in part by the Howard Hughes Medical Institute, National Institutes of Health, National Cancer Institute, and the Mathers Foundation.

### REFERENCES

Bernstein, P. S., Law, W. C., and Rando, R. R. (1987). *Proc. Natl. Acad. Sci. U.S.A.* **84**, 1849–1853.

Biondi, A., Rambaldi, A., Alcalay, M., Pandolfi, P. P., LoCoco, F., Diverio, D., Rossi, V., Mencarelli, A., Longo, L., Zamgrilli, D., Masera, G., Barbui, T., Mandelli, F., Grignani, F., and Pelicci, P. G. (1991). *Blood* **77**, 1418–1422.

Blumberg, B., Mangelsdorf, D. J., Dyck, J. A., Bittner, D. A., Evans, R. M., and De Robertis, E. (1992). *Proc. Natl. Acad. Sci. U.S.A.,* **89**, 2321–2325.

Borrow, J., Goddard, A. D., Sheer, D., and Solomon, E. (1990). *Science* **249**, 1577–1580.

Brand, N., Petkovich, M., Krust, A., Chambon, P., de The, H., Marchio, A., Tiollais, P., and Dejean, A. (1988). *Nature (London)* **332**, 850–853.

de The, H., Chomienne, C., Lanotte, M., Degos, L., and Dejean, A. (1990). *Nature (London)* **347**, 558–561.

de The, H., Lavau, C., Marchio, A., Chomienne, C., Degos, L., and Dejean, A. (1991). *Cell (Cambridge, Mass.)* **66**, 675–684.

Dencker, L., D'Arby, R., Danielson, B. R. G., Ghantous, H., and Sperber, G. O. (1987). *Dev. Pharmacol. Ther.* **10**, 212–223.

Durston, A. J., Timmermans, J. P. M., Hage, W. J., Kendriks, H. F. J., de Vries, N. J., Heideveld, M., and Nieuwkoop, P. D. (1989). *Nature (London)* **340**, 140–144.

Eichele, G. (1989). *TIGS* **5**, 246–251.

Eilers, M., Picard, D., Yamamoto, K. R., and Bishop, J. M. (1989). *Nature (London)* **340**, 66–68.

Evans, R. M. (1988). *Science* **240**, 889–895.

Giguere, V., Ong, E. S., Segui, P., and Evans, R. M. (1987). *Nature (London)* **330**, 624–629.

Glass, C. K., Devary, O. V., and Rosenfeld, M. G. (1990). *Cell* **63**, 729–738.

Goodman, D. S. (1984). *N. Engl. J. Med.* **310**, 1023–1031.

120 DAVID J. MANGELSDORF ET AL.

Green, S., and Chambon, P. (1988). *TIGS* **4**, 309–314.

Haupt, Y., Alexander, W. S., Barri, G., Klinken, S. P., and Adams, J. M. (1991). *Cell (Cambridge, Mass.)* **65**, 753–763.

Haussler, M. R., Myrtle, J. F., and Norman, A. W. (1968). *J. Biol. Chem.* **243**, 4055–4064.

Herschman, H. R. (1989). *TIBS* **14**, 455–458.

Heyman, R. A., Mangelsdorf, D. J., Dyck, J. A., Stein, R., Eichele, G., Evans, R. M., and Thaller, C. (1992). *Cell (Cambridge, Mass.)*, **68**, 397–406.

Huang, M. E., Ye, Y. C., Chen, S. R., Chai, J. R., Lu, J. X., Zhoa, L., Gu, L. J., and Wang, Z. Y. (1988). *Blood* **72**, 567–572.

Ishikawa, T., Umesono, K., Mangelsdorf, D. J., Aburtani, H., Stanger, B. Z., Shibasaki, Y., Imawari, M., Evans, R. M., and Takaku, F. (1990). *Mol. Endocrinol.* **4**, 837–844.

Kakizuka, A., Miller, W. H., Jr., Umesono, K., Warrell, R. P., Jr., Frankel, S. R., Murty, V. V. V. S., Dmitrovsky, E., and Evans, R. M. (1991). *Cell (Cambridge, Mass.)* **66**, 663–674.

Kastner, P., Krust, A., Mendelsohn, C., Garnier, J. M., Zelent, A., Leroy, P., Staub, A., and Chambon, P. (1990). *Proc. Natl. Acad. Sci. U.S.A.* **87**, 2700–2704.

Keeble, S., and Maden, M. (1989). *Dev. Biol.* **132**, 26–34.

Kliewer, S. A., Umesono, K., Heyman, R., Mangelsdorf, D. J., Dyck, J. A., and Evans, R. M. (1992a). *Proc. Natl. Acad. Sci. U.S.A.*, **89**, 1448–1452.

Kliewer, S. A., Umesono, K., Mangelsdorf, D. J., and Evans, R. M. (1992b). *Nature (London)*, **355**, 446–449.

Krust, A., Kastner, P., Petkovich, M., Zelent, A., and Chambon, P. (1989). *Proc. Natl. Acad. Sci. U.S.A.* **86**, 5310–5314.

Liao, J., Ozono, K., Sone, T., McDonnell, D. P., and Pike, J. W. (1990). *Proc. Natl. Acad. Sci. U.S.A.* **87**, 9751–9755.

Longo, L., Pandolfi, P. P., Biondi, A., Rambaldi, A., Mencarelli, A., LoCoco, F., Diverio, D., Pegoraro, L., Avanzi, G., Talilio, A., Zangrilli, D., Alcalay, M., Donti, E., Grignani, F., and Pelicci, P. G. (1990). *J. Exp. Med.* **172**, 1571–1575.

Lubahn, D. B., Joseph, D. R., Madhabananda, S., Tan, J., Higgs, H. N., Larson, R. E., *et al.* (1988). *Mol. Endocrinol.* **2**, 1265–1275.

Maden, M. (1982). *Nature (London)* **295**, 672–675.

Mangelsdorf, D. J., Borgmeyer, U., Heyman, R. A., Zhou, J., Ong, E. S., Oro, A. E., Kakizuka, A., and Evans, R. M. (1992). *Genes Dev.*, **6**, 329–344.

Mangelsdorf, D. J., Ong, E. S., Dyck, J. A., and Evans, R. M. (1990). *Nature (London)* **345**, 224–229.

Mangelsdorf, D. J., Umesono, K., Kliewer, S. A., Borgmeyer, U., Ong, E. S., and Evans, R. M. (1991). *Cell (Cambridge, Mass.)* **66**, 555–561.

Miller, W. H., Jr., Kakizuka, A., Frankel, S. R., Warrell, R. P., Jr., DeBlasio, A., Levine, K., Evans, R. M., and Dmitrovsky, E. (1992). *Proc. Natl. Acad. Sci. U.S.A.*, **89**, 2694–2698.

Murray, M. B., and Towle, H. C. (1989). *Mol. Endocrinol.* **3**, 1434–1442.

Oro, A. E., McKeown, M., and Evans, R. M. (1990). *Nature (London)* **347**, 298–301.

Pandolfi, P. P., Grignani, F., Alcalay, M., Mencanelli, A., Biondi, A., LoCoco, F., Grignani, F., and Pelicci, P. G. (1991). *Oncogene* **6**, 1285–1292.

Petkovich, M., Brand, N. J., A. Krust, A., and Chambon, P. (1987). *Nature (London)* **330**, 444–450.

Rando, R. R. (1990). *Angew. Chem. Int. Ed. Engl.* **29**, 461–480.

Rowe, A., Eager, N. S. C., and Brickell, P. M. (1991). *Development* **111**, 771–778.

Rowley, J. D., Golomb, H. M., and Dougherty, C. (1977). *Lancet* **1**, 549–550.

## RETINOID RECEPTORS

Sagami, I., Tsai, S. Y., Wang, H., Tsai, M.-J., and O'Malley, B. W. (1986). *Mol. Cell. Biol.* **6**, 4259–4267.

Salvato, M. S., and Shimomaye, E. M. (1989). *Virology* **173**, 1–10.

Schatz, D. G., Oettinger, M. A., and Baltimore, D. (1990). *Cell (Cambridge, Mass.)* **59**, 1035–1048.

Sporn, M. B., Roberts, A. B., and Goodman, D. S. (eds.) (1984). "The Retinoids," Vols. 1, 2. Academic Press, New York.

Tagawa, M., Sakamoto, T., Shigemoto, K., Matsubara, H., Tamura, Y., Ito, T., Nakamura, I., Okitsu, A., Imai, K., and Taniguchi, M. (1990). *J. Biol. Chem.* **265**, 20021–20026.

Thaller, C., and Eichele, G. (1990). *Nature (London)* **345**, 815–819.

Umesono, K., Murakami, K. K., Thompson, C. C., and Evans, R. M. (1991). *Cell (Cambridge, Mass.)* **65**, 1255–1266.

van Lohuizen, M., Verbeek, S., Scheijen, B., Wientjens, E., van der Gulden, H., and Berns, A. (1991). *Cell (Cambridge, Mass.)* **65**, 737–752.

Wagner, M., Thaller, C., Jessell, T., and Eichele, G. (1990). *Nature (London)* **345**, 819–822.

Wedden, S. E., Ralphs, J. R., and Tickle, C. (1988). *Development* **103**, 31–30.

Yang, N., Mangelsdorf, D. J., Schüle, R., and Evans, R. M. (1991). *Proc. Natl. Acad. Sci. U.S.A.* **88**, 3559–3563.

Zelent, A., Krust, A., Petkovich, M., Kastner, P., and Chambon, P. (1989). *Nature (London)* **339**, 714–717.

# The Growth Hormone/Prolactin Receptor Family

P. A. KELLY, S. ALI, M. ROZAKIS, L. GOUJON, M. NAGANO,
I. PELLEGRINI, D. GOULD, J. DJIANE, M. EDERY, J. FINIDORI,
AND M. C. POSTEL-VINAY

*INSERM Unité 344 Endocrinologie Moléculaire, Faculté de Médecine Necker-Enfants Malades, 75730 Paris Cedex 15, France; Laboratory of Molecular Endocrinology, Royal Victoria Hospital, McGill University, Montreal, Quebec, Canada H3A 1A1; and Unité d'Endocrinologie Moléculaire, Institut National de la Recherche Agronomique, 78350 Jouy en Josas, France*

## I. Introduction

The pituitary gland was first thought to be a vestigial structure, perhaps serving the function of lubrication of the nasal cavities. At the end of the nineteenth and beginning of the twentieth centuries, the importance of the pituitary in acromegaly, gigantism, and dwarfism was recognized. A direct role of the pituitary was confirmed by studies that showed that hypophysectomy resulted in the immediate cessation of growth in young dogs. Growth hormone (GH) was first isolated by Li *et al.* (1945) from bovine pituitaries. However, the bovine hormone was inactive in humans, thus the isolation of human GH (hGH) in the mid-1950s represented a major breakthrough in the treatment of growth-deficient children.

It was also early in the twentieth century that changes in the histology of the anterior pituitary gland of pregnant women was first noted. A lactogenic hormone was originally identified by Stricker and Grueter (1928) as a pituitary factor capable of inducing milk secretion in rabbits. Independently, American scientists made similar observations and also showed that the hormone that they named prolactin (PRL) was capable of stimulating the growth of pigeon crop sacs (Riddle *et al.*, 1932). Prolactin has now been shown to exist in all vertebrates.

There was some question in the past whether prolactin existed in humans, because hGH preparations were lactogenic in conventional bioassays, and all early attempts to separate GH and PRL activities failed. Clinical and histological observations, however, strongly suggested that the two hormones were, in fact, separate entities. Finally, human PRL was successfully isolated and purified, which led to numerous subsequent pathophysiological studies.

123

Copyright © 1993 by Academic Press, Inc.
All rights of reproduction in any form reserved.

Sequence analysis has confirmed that GH and PRL have selected regions of strong homology and, along with placental lactogen (PL) or chorionic somatotropin (CS), form a family of polypeptide hormones that appear to have arisin by duplication of an ancestral gene (Niall et al., 1971).

Hormones act by first binding to a specific receptor on or inside the cell. Two general classes of hormone receptors exist. The first class is known as DNA binding receptors, because the ligands for these receptors, which are usually small molecules [steroids, amino acids ($T_3$), vitamins, or retinoic acid], pass through the cell membrane and bind to cytoplasmic or nuclear receptors. The hormone–receptor complex in turn acts by regulating the level of transcription of specific target genes. The second class is known as membrane receptors, because they are located at the cell surface. Hormones bind to these cell surface receptors and transduce their message via a second-messenger system. Receptors in this class have been subdivided into four groups: multiple membrane-spanning receptors, coupled to G proteins (e.g., $\beta$-adrenergic, dopamine, substance K); multiple membrane-spanning receptors, coupled to ligand-gated channels [e.g., nicotinic acetylcholine, $\gamma$-aminobutyric acid-A (GABA), and glycine]; single membrane-spanning receptors, with tyrosine kinase activity [e.g., insulin, insulin-like growth factor, and platelet-derived growth factor (PDGF)]. Finally, there is a large fourth group of single membrane-spanning receptors with no known mechanism of action (e.g., GH, PRL, and cytokines). Although the recent cloning of the complementary DNAs encoding members of this fourth category of membrane-spanning receptors has permitted this correct classification based on structure, very little new information on their mechanism of action has been gained. The events responsible for the transfer of the hormonal message inside the cell occur just after the interaction of the ligand with the receptor; they remain, for the most part, unknown. The remainder of this article will deal with a description of the receptors for GH and PRL, actions of the two hormones, the expanded GH/PRL/cytokine receptor family, expression of receptor genes, what is known about the mechanism of action of the two hormones, and, finally, genetic diseases and models of receptor defects.

## II. Classical Biological and Biochemical Actions

### A. GROWTH HORMONE

Many actions of GH, such as on skeletal growth, are indirect and are mediated by insulin-like growth factor (IGF). GH has been shown to regulate directly IGF-I production in liver cells as well as in other cells in culture (Mathews *et al.,*1986). GH has other direct effects, such as the regulation of cytochrome P-450$_{15\beta}$ by liver cells (Tollet *et al.*, 1990) as well as glucose transport and glucose metabolism in adipocytes.

GH has a direct action on the differentiation of preadipocytes, where it is necessary for the initiation of the differentiation program for cells to become responsive to IGF-I and for its mitogenic effect (Green *et al.*, 1985). Rapid effects of GH in these cells include induction of c-*fos* and c-*jun* transcription (Gurland, *et al.*, 1990). A similar effect of GH on precursor cells could exist in several tissues, including adipose tissue, muscle, and cartilage.

Prenatal somatic growth appears to be largely independent of GH. Binding of GH to liver membranes from calf, lamb, and rat is minimal before birth and increases gradually during the first weeks of postnatal life (Maes *et al.*, 1983; Gluckman *et al.*, 1983). The ontogenesis of GH receptors probably does not follow the same pattern in all tissues.

## B. PROLACTIN

More than 85 distinct and diverse effects of prolactin have been described in the vertebrate subphylum (Nicoll and Bern, 1972). These actions can be subdivided into seven broad categories: (1) reproduction and lactation, (2) water and salt balance, (3) growth and morphogenesis, (4) metabolism, (5) behavior, (6) immunoregulation, and (7) effects on the ectoderm and the skin. Some of these actions are more characteristic of certain species and are less important in others.

In mammals, PRL is primarily responsible for the development of the mammary gland and lactogenesis. The best known effects of PRL are in mammary epithelial cells. PRL acts in association with insulin and glucocorticoids to stimulate milk protein gene expression at both the transcriptional and posttranscriptional levels, because the rate of gene transcription and the stability of milk protein mRNAs are increased under the influence of the hormone (Guyette *et al.*, 1979). PRL also regulates a number of other reproductive functions, but the kind and amplitude of these effects depend on the species examined. PRL exerts a marked influence on the gonads, by modulating the effects of gonadotropins (Armstrong *et al.*, 1970; Kelly *et al.*, 1980). Four principal actions have been documented: (1) stimulation of the number of luteinizing hormone (LH) receptors in the testis during puberty or in the stimulation of steroidogenesis, particularly in the corpus luteum; (2) stimulation of steroidogenesis, especially in the corpus luteum; (3) increase of high-density lipoprotein and low-density lipoprotein binding in membranes of the corpus luteum; and (4) stimulation of the growth of ovarian follicles.

The biological roles of the high levels of PRL receptors in the liver, particularly in the female rat, are not well understood. Administration of PRL to rats causes hepatic hypertrophy and increases ornithine decarboxylase activity, suggesting that PRL regulates hepatocyte renewal (Buckley et al., 1985; Richards, 1975). PRL has been shown to stimulate the expression of messenger RNA encoding the cytosolic form of phosphoenolpyruvate carboxykinase in liver of lactating rats and in primary cultures of hepatocytes (Zabala and Garcia-Ruiz, 1989). In addition, PRL induces a factor (synlactin) in the liver of pigeons or rats that acts synergistically with PRL to promote the growth of the crop sac or the mammary gland (Nicoll et al., 1985).

In humans, although it is reasonable to assume that milk protein genes are regulated by PRL, as in other species, the only gene that has actually been shown to be regulated by PRL is PRL-inducible protein (PIP). This protein was originally isolated from T-47D human breast cancer cells (Shiu and Iwasiow, 1985). PIP is a secreted glycoprotein inducible by androgens and PRL. Androgens regulate PIP transcription, whereas PRL has essentially posttranscriptional (mRNA stabilization) effects (Murphy et al., 1987b). PIP is present in several, but not all, human breast cancer cell lines and in benign and malignant breast tumor biopsies (Murphy et al., 1987a), where it is also known as gross cystic disease fluid protein (GCDFD-15) (Haagensen and Mazovjian, 1986). Interestingly, PIP is also expressed in some exocrine organs such as salivary, lacrimal, and sweat glands (Haagensen and Mazovjian, 1986), and in the secretions of these glands (Murphy et al., 1987a).

## III.   Growth Hormone, Prolactin, and the Immune System

Although the suppressive role of glucocorticoids on immune functions has been known for a long time, more recent evidence has shown that GH and PRL are immunostimulatory factors. These interactions have recently been reviewed (Gala, 1991).

Hypophysectomized rats have atrophic thymic glands and diminished immune function (Smith, 1930; Lunkin, 1960). Removal of the pituitary leads to reduced antibody production, diminished skin sensitivity to toxic substances and adjuvant arthritis, and prolonged skin allograft survival (Nagy and Berczi, 1978). Genetic dwarf mice models, such as the Snell or Ames strains, have also reinforced the concept of a role for GH and PRL in immune function. These animals, which are phenotypically dwarfs, lack not only GH but also PRL in their pituitary. Such dwarfs have atrophied thymus and lymphoid tissue and a depletion of bone marrow (Baroni, 1967). The administration of CB-154, a dopamine agonist, also leads to impaired immune responses (Palestine *et al.*, 1987). Hormone replacement studies have shown that either GH or PRL was able to restore immune function in animal models, such as hypophysectomy models or genetic dwarfs (Nagy *et al.*, 1983; Bernton *et al.*, 1988).

Several independent studies have reported *in vitro* effects of GH or PRL on lymphocytes. GH stimulated erythropoiesis in the presence of erythropoietin (Golde and Bersch, 1977). GH has also been shown to influence cytotoxic lymphocyte lysis (Snow *et al.*, 1981). Superoxide anion, which is responsible for death of intracellular pathogenic microbes, is increased by GH administration to mononuclear phagocytes (Edwards *et al.*, 1988). Many of the direct effects of GH appear to be mediated by a local production of IGF-I (see Gala, 1991).

Studies of direct effects of PRL on isolated lymphocytes have yielded variable results. Russel *et al.* (1984) reported PRL could increase ornithine decarboxylase activity in peripheral lymphocytes. Sometimes little or no direct effects of PRL can be seen, although antibodies to PRL inhibit lymphocyte proliferation (Hartman *et al.*, 1989). On the other hand, lymphocytes taken from spleens or lymph nodes of ovariectomized rats were shown to respond to PRL by an increase in cell proliferation (Mukherjee *et al.*, 1990). Also, the cylotoxic activities of natural killer (NK) cells are increased by PRL (Matera *et al.*, 1990).

The direct production of either GH (Weigent *et al.*, 1988) or prolactin (Hartmann *et al.*, 1989) by normal lymphocytes has been documented. In addition, DiMattia *et al.* (1990) have shown that a human $\beta$ lymphoblastoid cell line (IM-9-P) secreted hPRL, but no signal was observed in normal lymphocytes. On the other hand, we have found that a T cell line (JUR-KAT) and peripheral lymphocytes produce a mRNA specific for human PRL, which has, however, a different, nonpituitary exon 1, suggesting the regulation of PRL production by lymphocytes and lactotropic cells in the pituitary is regulated independently (Pellegrini *et al.*, 1992).

The fact that lymphocytes, and perhaps other GH or PRL target cells, are capable of producing these "hormones," suggests that they should also be considered as "growth factors," acting via classical paracrine or autocrine pathways.

## IV. Distribution and Regulation of Receptors

### A. GROWTH HORMONE RECEPTORS

Binding sites for GH were originally thought to be restricted to the liver. However, with the improvement in detection techniques, combined with analysis of mRNA transcripts, the number of tissues known to express

GH receptors has greatly increased (Table I). For some tissues, GH binding sites are not always detectable, although an mRNA transcript can be identified or an effect measured. This may be due to the low concentration of GH receptors in these tissues or to the fact that receptors may be found in a single cell type of a mixed cell population.

In rat and human liver, a higher proportion of GH receptors is localized in membranes associated with intracellular structures than in plasma membranes (Picard and Postel-Vinay, 1984; Hocquette et al., 1989). The presence of large numbers of receptors in the endosomal compartment probably reflects rapid receptor synthesis, because the half-life of the GH receptor is very short (Baxter, 1985).

The regulation of GH receptors has been studied under several different situations of growth failure in rats. Chronic renal failure (Finidori et al., 1980), fasting (Postel-Vinay et al., 1982), and streptozotocin-induced diabetes (Baxter et al., 1984) are associated with a decrease in GH receptors in liver. Four days of fasting results in a 50% decrease in the number of GH receptors with no change in affinity. Reduced receptor levels are found both at the cell surface (plasma membranes) and intracellularly (Golgi-endosomes). Normalization of receptor levels is rapid: in rats refed for 3 days, the number of GH binding sites returns to levels found in control animals (Postel-Vinay et al., 1982).

A clear relationship between the number of GH binding sites and a biological response to the hormone has been demonstrated in obese Zucker rat adipocytes. The fa/fa rat develops obesity with hyperinsulinemia. Adipocytes isolated from 30-day-old fa/fa rats have an increased number of GH receptors and are hyperresponsive to GH. In addition, a correlation between the number of GH binding sites and the GH-induced increase of glucose transport was demonstrated in obese rat adipocytes (Landron et al., 1989).

The role of GH in the regulation of its own receptor is variable. However, markedly reduced secretion of GH is associated with low receptor

TABLE I
*Distribution of GH Receptors in Mammals*

| | |
|---|---|
| Liver | Lung |
| Adipose tissue | Pancreas |
| Lymphatic and immune cells | Brain |
|     Lymphocytes and thymocytes | Cartilage |
| Intestine | Skeletal muscle |
| Heart | Corpus luteum |
| Kidney | Testis |

GROWTH HORMONE/PROLACTIN RECEPTORS

numbers, e.g., in streptozotocin-induced diabetes, fasting, and renal insufficiency (Finidori *et al.*, 1980; Baxter *et al.*, 1980). Adipocytes isolated from hypophysectomized rats have low GH binding, and treatment with bovine GH results in a twofold increase in receptor levels (Grichting and Goodman, 1986). A 7-day infusion (by minipump) of GH to either hypophysectomized or normal rats increases GH binding in liver membranes (Baxter and Zaltsman, 1984). For such studies, it is necessary to dissociate GH from its receptor, in order to measure the total number of GH binding sites. Finally, it has recently been shown that elevated levels of ovine GH expressed as a transgene in mice are capable of inducing hepatic GH receptors (Orian *et al.*, 1991).

However, a paradoxical effect occurs in hypophysectomized male rats. In these animals, which have an absence of circulating GH, an increased number of GH binding sites is observed in liver membranes (Picard and Postel-Vinay, 1984).

Homologous down-regulation of GH receptors has been shown in IM-9 lymphocytes. Exposure of cultured cells to GH results in a decrease in the number of binding sites (Lesniak and Roth, 1976). In addition, GH binding in liver is reduced 6 hours after a single injection of GH to male rats (Maiter *et al.*, 1988).

Estrogens also appear to play a role in the regulation of GH receptors. There is a twofold increase in GH binding to liver after puberty in female rats, whereas no change is observed in males (Maes *et al.*, 1983). Pregnancy results in a 10-fold increase in GH binding to liver membranes in rats and a twofold increase in rabbits (Hughes *et al.*, 1985).

Insulin also regulates the concentration of GH receptors. The reduced number of hepatic GH receptors in rats with streptozotocin-induced diabetes is partially restored by insulin treatment (Postel-Vinay *et al.*, 1982). In the fasted/refed rat model, the concentration of GH receptors in liver membranes parallels plasma insulin levels. In obese Zucker rats, the increase occurs in the number of liver GH receptors very shortly after the increase in plasma insulin concentrations, i.e., after day 18 (Postel-Vinay *et al.*, 1990).

## B. PROLACTIN RECEPTORS

Binding sites for PRL are widely distributed in mammalian tissues (see Table II and Hughes *et al.*, 1985). Human and other primate GHs, in addition to their conventional growth-related activities, are lactogenic (Forsyth *et al.*, 1965), and thus bind to both PRL and GH receptors. Recent studies demonstrate that zinc ions modulate the binding of hGH to the hPRL receptor (Cunningham *et al.*, 1990). PRL receptors have been

## TABLE II
### Distribution of PRL Receptors in Mammals

| | |
|---|---|
| Mammary gland | Pancreas |
| Normal | Islet of Langerhans |
| Milk[a] | Intestine |
| Tumor | Kidney |
| Ovary | Adrenal |
| Granulosa cell | Lymphatic and immune cells |
| Corpus luteum | Thymus |
| Uterus | Thymocyte |
| Placenta | Lymphocyte |
| Testis | Erythrocyte |
| Leydig cell | Neutrophil |
| Spermatid | Brain |
| Epididymis | Hypothalamus |
| Seminal vesicle | Substantia nigra |
| Prostate | Choroid plexus |
| Normal | Eye |
| Tumor | Choroid coat |
| Liver | Retinal photoreceptors |

[a] In this fluid, membrane-bound receptors and a soluble BP are found.

characterized in microsomal membrane preparations using the original method of Shiu and Friesen (1974) from an increasing number of tissues. The physiological function of PRL in some of these tissues is not known. The majority of studies performed have concentrated on receptors in mammary gland and liver. PRL receptors are found not only in plasma membranes, but also in endosomes, Golgi fractions, and lysosome-enriched preparations (Bergeron et al., 1978, 1986). In rat liver or lactating mammary gland, most of the receptors are localized in intracellular membranes. This probably reflects rapid receptor synthesis and degradation that occur even though the receptors are not occupied by PRL (Djiane et al., 1982). In addition, PRL receptors have been characterized in cytosol prepared from the liver of female rats or rabbit mammary gland (Amit et al., 1984; Ymer et al., 1987). More recently, a soluble form of PRL receptor has been found in milk (Postel-Vinay et al., 1991).

The hormonal regulation of PRL receptors is different depending on the target organ considered. In the mammary gland, PRL receptor numbers increase markedly early in lactation (Djiane et al., 1977). The high circulating levels of progesterone that occur during pregnancy are probably responsible for limiting PRL receptor numbers in the mammary gland (Djiane and Durand, 1977). In rat liver, PRL receptors are higher in females than

GROWTH HORMONE/PROLACTIN RECEPTORS 131

males and are increased by estrogens (Posner *et al.*, 1974). Testosterone increases PRL receptor levels in the prostate, whereas estrogens have the opposite effect (Kledzik *et al.*, 1976). In the liver, steroid hormones probably act indirectly, at least partially, via the increased secretion of prolactin. PRL, in most organs studied, is able to up- and down-regulate the level of its own receptor, depending on the concentration and duration of the exposure to the hormone. Down-regulation precedes up-regulation and is rapidly reversible, usually observed in the presence of high concentrations of PRL (Djiane *et al.*, 1979). This down-regulation is presumably a consequence of an acceleration of internalization of hormone–receptor complexes and their degradation in lysosomes (Djiane *et al.*, 1980). Up-regulation occurs after several days of exposure to moderate levels of PRL (Posner *et al.*, 1975) and is blocked by progesterone in the mammary gland (Djiane *et al.*, 1977). Administration of GH is also able to increase PRL receptor levels in rat liver (Baxter *et al.*, 1984). The expression of elevated levels of ovine GH as a transgene in mice is able to induce PRL receptors in liver (Orian *et al.*, 1991). Thus, both PRL and GH can up-regulate PRL receptors.

## V.  Receptor Structure—Identification of Multiple Forms

An almost identical approach was used to clone the cDNAs encoding the GH and PRL receptors. The purification to homogeneity of GH and PRL receptors by affinity chromatography (hormone affinity for GH, immunoaffinity for PRL) and preparative electrophoresis permitted the microsequence analysis of fragments of receptors. A purification factor of >100,000-fold was required to obtain homogeneous preparations. Receptor fragments were purified by reverse-phase HPLC and amino acid sequences were determined on a gas-phase sequencer. Oligonucleotide probes were prepared, based on the sequence of the receptor fragments, and were used to screen cDNA libraries.

## A.  GROWTH HORMONE RECEPTORS

Using a synthetic oligonucleotide probe, several clones were first identified in a rabbit cDNA library, and the sequence of a full-length cDNA was determined. The primary structure of the mature receptor, as deduced from this nucleotide sequence, resulted in a protein of 620 amino acids (aa). The receptor consists of an extracellular hormone-binding domain of 246 aa, a single transmembrane region, and a cytoplasmic domain of 350 residues (Fig. 1). The extracellular domain contains seven cysteine

132                    P. A. KELLY ET AL.

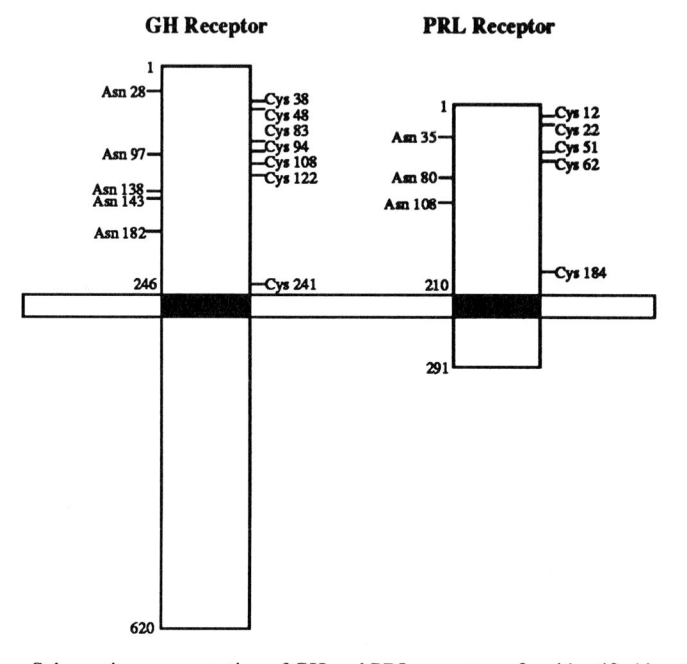

FIG. 1.  Schematic representation of GH and PRL receptors, first identified in rabbit and human liver (GH) and rat liver (PRL). The first and last amino acids of the mature protein, and the last amino acid of the extracellular domain, are indicated. Transmembrane domains are shown as black boxes. Extracellular cysteines (Cys) are shown at the right, and asparagines (Asn), which are potential sites of glycosylation, are at the left.


residues and five potential N-linked glycosylation sites (Leung *et al.,* 1987).

The purified GH receptor from rabbit liver was characterized by sodium dodecyl sulfate–polyacrylamide gel electrophoresis, and a single protein with an $M_r$ of 130,000 was identified. The $M_r$ of the deglycosylated receptor is 95,000, which is considerably greater than the $M_r$ of 70,000 calculated from the amino acid sequence. Thus far, no detailed explanation for the difference in apparent molecular weight of the glycosylated receptor (130,000) and the theoretical molecular weight (70,000) has been found, even when one considers that one molecule of ubiquitin appears to be covalently linked to the receptor molecule, because this only accounts for an additional molecular weight of approximately 10,000. A cDNA encoding the human GH receptor was cloned using the rabbit cDNA as a probe. There is a high degree of amino acid sequence identity (84%) between the rabbit and human receptors. The full-length rabbit receptor and a truncated form of the human receptor were expressed in COS-7 cells. The expressed

GROWTH HORMONE/PROLACTIN RECEPTORS 133

receptors showed the expected affinity and specificity for GH (Hughes *et al.*, 1985). Subsequently, mouse, rat, cow, and sheep GH receptors were cloned (Smith *et al.*, 1989; Brumbach *et al.*, 1989; Houser *et al.*, 1990; Adams *et al.*, 1990). The various GH receptors thus far cloned share approximately 70% overall amino acid similarity (Fig. 2).

Several studies have suggested the existence of multiple forms of the GH receptor, including GH binding data with rabbit liver membranes showing varying potencies of different forms of GH (rat GH, human 20K and 22K GH) (Hugues and Friesen, 1985), epitope mapping studies with several monoclonal antibodies to the GH receptor (Barnard *et al.*, 1985), and cross-linking studies, demonstrating multiple forms of GH receptor in mouse liver (Smith and Talamantes, 1987), in human liver (Hocquette *et al.*, 1990), and in RIN cells, a rat islet tumor cell line (Moldrup *et al.*, 1989). The mechanisms involved in the generation of the different forms of GH receptor and their physiological significance remain unknown.

The GH binding protein (BP), identified in mouse (Peeters and Friesen, 1977), rabbit (Ymer and Herington, 1985), and human (Baumann *et al.*, 1986) serum, is a soluble, short form of the liver GH receptor (Fig. 2). There is amino acid identity of the amino-terminal sequences of the receptor and the BP (Leung *et al.*, 1987). Two independent mechanisms have been proposed for the production of the GH BP: specific proteolysis of the membrane form of the receptor or translation from an alternatively spliced mRNA, produced from the same gene as the GH receptor. In man, cow, and rabbit, the serum GH BP probably results from proteolytic cleavage of the receptor, because only a single mRNA transcript of 4.5 kb has been clearly identified by Northern analysis. In mouse and rat liver, two mRNAs of approximately 4.5 and 1–1.5 kilobases (kb) are expressed, which encode the membrane receptor and the GH BP, respectively (Smith and Talamantes, 1987; Baumbach *et al.*, 1989). In rat and mouse, the GH BP is distinguished by a short hydrophylic C-terminal extension that is not found in the membrane receptor. Using an antibody raised against a synthetic peptide containing the 17-aa hydrophilic sequence, it was clearly shown that the GH BP in rat serum is derived almost entirely from the GH BP mRNA and not from proteolytic processing of the GH receptor (Sadeghi *et al.*, 1990), probably as a result of alternative splicing of a primary transcript.

## B. PROLACTIN RECEPTORS

The sequence of the mature PRL receptor, deduced from the cDNA, contained 291 aa (Fig. 1), with an extracellular region of 210 aa, a single transmembrane region of 24 aa, and a relatively short cytoplasmic domain of 57 aa (Boutin *et al.*, 1988). There are five extracellular cysteines, the

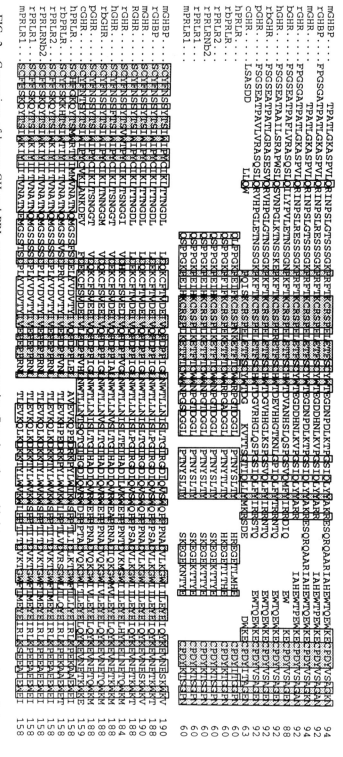

FIG. 2. Comparison of known GH and PRL receptor sequences across species. Gaps have been added to sequences, as necessary, to maximize alignment. Residues that are identical within the GH or PRL family are boxed; those identical between the two families are in shaded boxes. Abbreviations: GHBP, growth hormone binding proteins; GHR, growth hormone receptor; PRLR, prolactin receptor; m, mouse; r, rat; b, bovine; h, human; rb, rabbit; p, porcine; c, chicken.

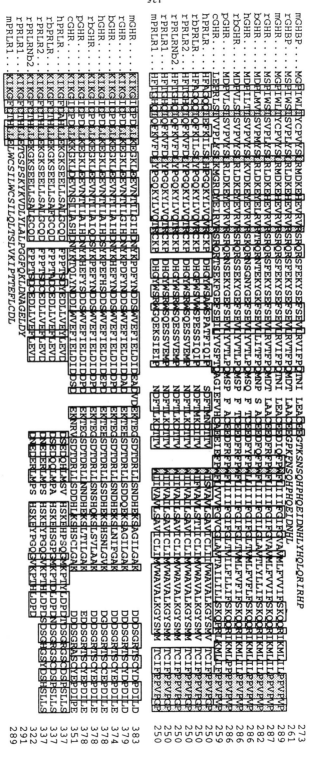

FIG. 2.—Continued

FIG. 2.—Continued

GROWTH HORMONE/PROLACTIN RECEPTORS 137

first four of which are located near the N-terminal region, with a fifth near the transmembrane domain. In addition, three potential N-linked glycosylation sites are present (Asn-35, Asn-80, and Asn-108), the first two of which have been confirmed by amino acid sequence analysis to be glycosylated.

Transfection of this cDNA and expression of the receptor in CHO or COS-7 cells confirmed that the expressed receptor had a specificity and an affinity for PRL identical to that already observed for PRL receptors in either rat liver or rabbit mammary gland.

1. A Second Form of PRL Receptor

Using the cDNA of the rat PRL receptor as a probe, a cDNA library prepared from rabbit mammary gland was screened and a cDNA was identified that encoded a mature receptor of 592 aa, with an extracellular and transmembrane region very similar to that of the rat PRL receptor (Fig. 2). The major difference is in the length of the cytoplasmic domain, 358 aa (Edery et al., 1989), which is much longer than that found for rat liver. Thus, a second class, representing a long form of the PRL receptor, has been identified.

The structure of the human PRL receptor was identified by screening cDNA libraries prepared from human hepatoma (HepG2) cells and a human breast cancer cell line (T-47D) (Boutin et al., 1989). The human PRL receptor is also a member of this second class of receptor (long form), containing 598 aa in the mature form.

Recently, two mouse liver cDNA sequences were reported (Davis and Linzer, 1989) that encode receptor proteins with a short cytoplasmic domain (39 and 50 aa vs. 57 aa in the rat). No such heterogeneity of the C-terminal portion of the short form of the PRL receptor was observed in the rat, as deduced from the oligonucleotide sequence of more than 12 cDNAs. However, the position at which the two mouse sequences diverge is identical to where the short and long forms of rat receptors differ (see below), supporting the hypothesis that the different mRNAs may result from alternative splicing of a single gene. A representation of the mouse PRL receptor containing the 50-aa cytoplasmic domain is illustrated in Figs. 2 and 3.

2. Multiple Forms of PRL Receptor in the Rat

The structure of the PRL receptor in the rat appears much more complex than either the human or rabbit receptor; only a single, long form has been identified, though initially only a short (291 aa) form was found. Because we had identified other forms in the rabbit and human, we decided to screen cDNA libraries from other rat tissues. With the short-form cDNA

138                        P. A. KELLY ET AL.

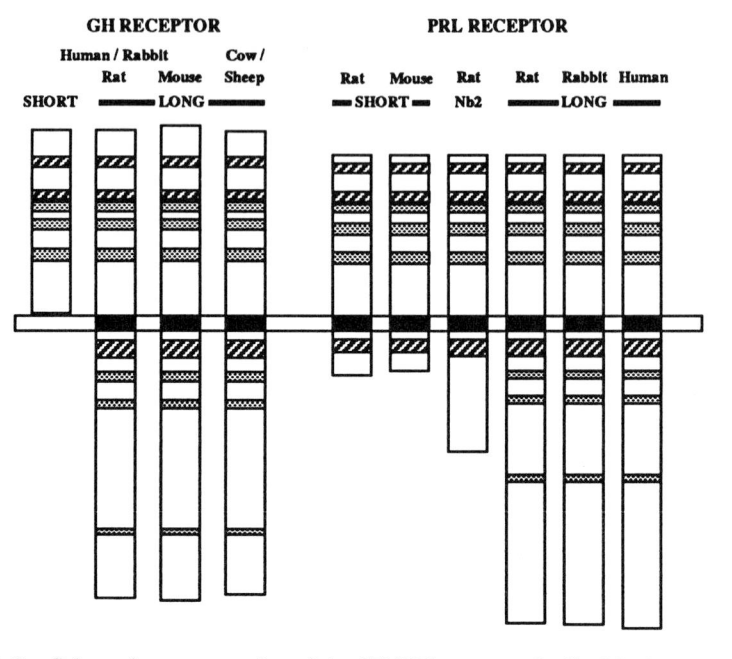

FIG. 3. Schematic representation of the GH/PRL receptor family. The long and short (binding protein) forms of the GH receptor in human, rabbit, cow, sheep, mouse, and rat are compared to the short form of PRL receptor in rat and mouse, the intermediate Nb2 form, and the long form in rat, rabbit, cow, and human. The transmembrane domains are shown in black. Regions of increased (>68%) amino acid identity are cross-hatched and those of moderate (40–60%) identity are stippled.

used as a probe, a second form of PRL receptor was identified in the ovary and liver. The cDNA encoded a PRL receptor in which the extracellular, transmembrane, and a portion of the cytoplasmic domain were identical to the short form of the PRL receptor already identified. The difference appeared at base pair (bp) 841 of the cDNA. The overall length of the mature form of the PRL receptor was thus 591 aa (Fig. 2), with the difference between the short and long forms appearing at aa 262 (Shirota et al., 1990). A long form of the PRL receptor in the rat ovary containing a consensus sequence for an ATP/GTP binding site has been reported (Zhang et al., 1990). Although, in general, the structure of the encoded proteins reported by Shirota et al. (1990) and Zhang et al. (1990) are very similar, we failed to confirm the nucleotide sequence encoding this putative ATP/GTP binding site in several cDNAs isolated from three independently prepared cDNA libraries. Because the presence of this ATP/GTP site

GROWTH HORMONE/PROLACTIN RECEPTORS 139

would confer kinase activity to the PRL receptor and represent an important element in the process of signal transduction, we wanted to clarify this discrepancy. To that end, three separate procedures capable of detecting a point mutation were performed: Southern analysis of reverse-transcribed PCR secretion products, allele-specific PCR, and S1 nuclease analysis. All results clearly indicate that the sequence for the ATP/GTP binding site does not exist in the PRL receptor (Nagano and Kelly, 1992).

Recently, we identified a mutant form of PRL receptor in the rat. The Nb2 lymphoma cell line is dependent on PRL for growth and contains high-affinity PRL receptors (Gout *et al.*, 1980; Shiu *et al.*, 1983). Biochemical studies have suggested that the receptor in these cells is different from the short and long forms already identified. Using the polymerase chain reaction of reverse-transcribed RNA and classical screening of a cDNA library, we have shown that the rat Nb2 PRL receptor is intermediate in size (393 aa) (Figs. 2 and 3) and appears to be due to a mutation in the PRL receptor gene, resulting in a loss of 594 bp in a region encoding a major portion of the cytoplasmic domain of the long form of the PRL receptor (Ali *et al.*, 1991).

The comparative structures of PRL and GH receptors are summarized in Fig. 3. Short and long forms of receptors exist for both hormones. For PRL, the short form is membrane bound, whereas for GH the short form is a soluble binding protein. Although, in general, the amino acid identity is approximately 30% between PRL and GH receptors, the identity increases to about 70% in certain domains of the extracellular regions, specifically between the first two pairs of cysteines, and in the cytoplasmic domain just inside the transmembrane segment. In addition, there are other extracellular and cytoplasmic regions of moderate (40–60%) identity. This increased structural identity led us to the conclusion that the receptors for PRL and GH form a new family of single membrane-spanning receptors (Boutin *et al.*, 1988).

### 3. Chromosomal Localization of PRL and GH Receptor Genes

Southern blot analysis of restriction fragments of genomic DNA suggested that the PRL and GH receptors are each encoded by a single gene. Chromosomal localization of the receptors was determined by *in situ* hybridization. The silver grain distribution for the receptor cDNAs in 100 metaphase spreads combined from two separate experiments confirmed a colocalization of PRL and GH receptor genes to chromosome 5p13 → p14 (Arden *et al.*, 1990). The GH receptor gene has been assigned to a similar location (chromosome 5p13.1 → p12) by another group (Barton *et al.*, 1989).

## C.  EXPANDED FAMILY OF GH/PRL/CYTOKINE RECEPTORS

The family that was originally identified to include the receptors for GH and PRL (Boutin *et al.*, 1988) has recently expanded. Interestingly, the receptors for a number of cytokines have also been shown to belong to this family (Bazan, 1989). As summarized in Fig. 4, in addition to GH and PRL, the family includes receptors for granulocyte colony-stimulating factor (Fukunaga *et al.*, 1990), erythropoietin (D'Andrea *et al.*, 1989), granulocyte–macrophage colony-stimulating factor (GM-CSF) (Gearing *et al.*, 1989), and the p75 or $\beta$ chain of interleukin-2 (IL-2) (Hatakeyama *et al.*, 1989), IL-3 (Itoh *et al.*, 1990), IL-4 (Mosley *et al.*, 1989), IL-5 (Takaki *et al.*, 1990), IL-6 (Yamasaki *et al.*, 1988), and IL-7 (Goodwin *et al.*, 1990). In addition, the IL-6 receptor is associated with a glycoprotein with an $M_r$ 130,000 (gp130) that appears to be associated with the signal transduction process (Taga *et al.*, 1989). Surprisingly, the gp130 is also a member of the cytokine/GH/PRL receptor family (Hibi *et al.*, 1990). Recently, it has been confirmed that, in humans, the receptors for GM-

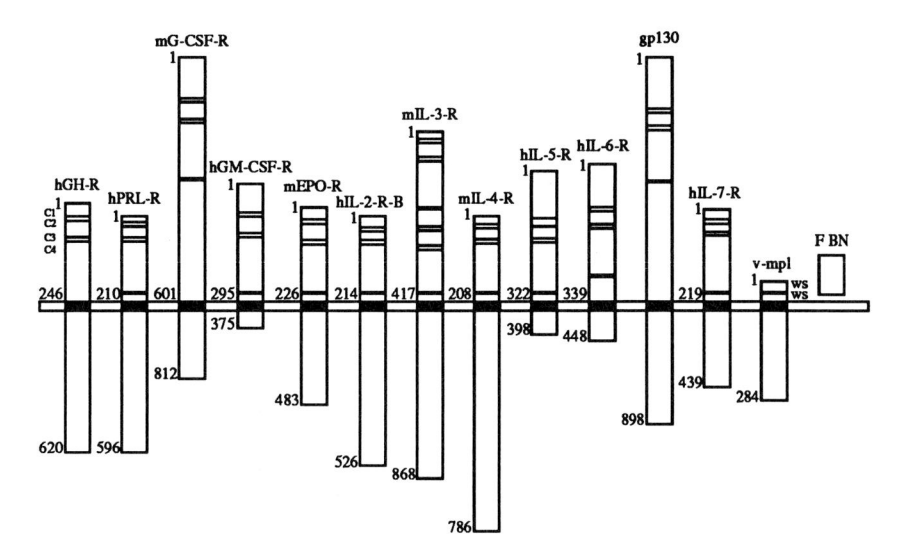

FIG. 4.  Schematic representation of members of the cytokine/GH/PRL receptor family. The abbreviations used can be found in the text. Homology with an approximately 90-aa type III domain of fibronectin (FBN) is also indicated. The first and last amino acid of the mature receptor, as well as the last amino acid of the extracellular region, are indicated. The transmembrane regions are shown as black boxes. Homology is restricted to approximately 200-aa regions in the extracellular domains that are limited N terminally by the two pairs of cysteines (C1, C2, C3, and C4, represented by thin black lines) and C terminally, by WSxWS motif (represented by thick black lines).

## GROWTH HORMONE/PROLACTIN RECEPTORS

CSF, IL-3, and IL-5 all share a commun $\beta$ subunit ($M_r$ 120,000). Both the $\alpha$ and $\beta$ subunits are members of this receptor family based on common structural features. High-affinity binding characteristics for these cytokines are observed only when the $\alpha$ and $\beta$ subunits are coexpressed (Gearing et al., 1989; Kitamura et al., 1991; Tavernier et al., 1991).

Over an approximately 200-aa residue region, although the level of amino acid identity is relatively low (14–25%), there are two characteristic features that are consistent throughout all members of this family. The first is the presence of two pairs of cysteines, usually in the N-terminal region of the molecule, which for the GH receptor have been shown to be linked sequentially by disulfide bonds, i.e., C1–C2 and C3–C4 (Fuh et al., 1990). These cysteines may be involved in forming ligand-binding pockets characteristic for each specific ligand. In addition, near the C-terminal extremity of this homologous region is a highly conserved WSxWS motif (tryptophan, serine, any amino acid, tryptophan, serine), which is found in all members of the family except the GH receptor. The approximately 200-aa residue region in the extracellular domain of members of this family can in fact be divided into two subdomains of about 100 aa (Bazan, 1990a; Thoreau et al., 1991). Fibronectin is a cell adhesion molecule that is a member of a larger family of adhesion or attachment molecules. It is a complex molecule composed of multiple units of type I, type II, and type III modules (Skorstengaard et al., 1986). The type III domain is an approximately 90-residue unit repeated 15 times. Homology of the second subdomain of members of the cytokine/GH/PRL receptor family with type III modules if fibronectin has been reported (Patthy, 1990). Portions of these domains may contribute to the interaction of receptors with their ligands or with other components of the receptor complex. Recently, a putative truncated form of a cytokine receptor gene that retains the WSxWS motif has been identified in a myeloproliferative leukemia virus (Souyri et al., 1990). It is proposed that myeloproliferative leukemia virus has transduced a truncated form of an as yet unidentified cytokine growth factor receptor.

Receptors for interferon-$\alpha$ (IFN-$\alpha$), IFN-$\beta$, and IFN-$\gamma$ have been proposed to form a subset of this family (Bazan, 1990a; Thoreau et al., 1991). In fact, this family of receptors is even more similar when one compares the secondary structures of the extracellular domains (Bazan, 1990b). Seven $\beta$ strands in conserved regions are found in each of the two approximately 100-aa domains. These two subdomains could each contribute to ligand binding as proposed by Bazan (1990b), or alternatively the first subdomain could be the primary site for interaction. The fact that a small proteolytic fragment of the PRL receptor of $M_r$ 15,000 retains binding activity (Dusanter-Fourt et al., 1990) is in favor of this second hypothesis.

142                    P. A. KELLY ET AL.

## D.  IDENTIFICATION OF LIGAND-BINDING DETERMINANTS
## OF THE GH AND PRL RECEPTORS

### 1.  Growth Hormone Receptor

Bass *et al.* (1991) reported the results if a systematic mutational analysis of the extracellular domain of the human GH receptor. They replaced charged and aromatic acids, as well as neighboring residues, with alanine. As shown in Fig. 5, residues within the cysteine-rich domain of the human GH BP form a patch when mapped on the structural motif proposed for members of this family (Bazan, 1990b; Thoreau *et al.*, 1991). Most of the alanine substitutions did not disrupt the overall three-dimensional

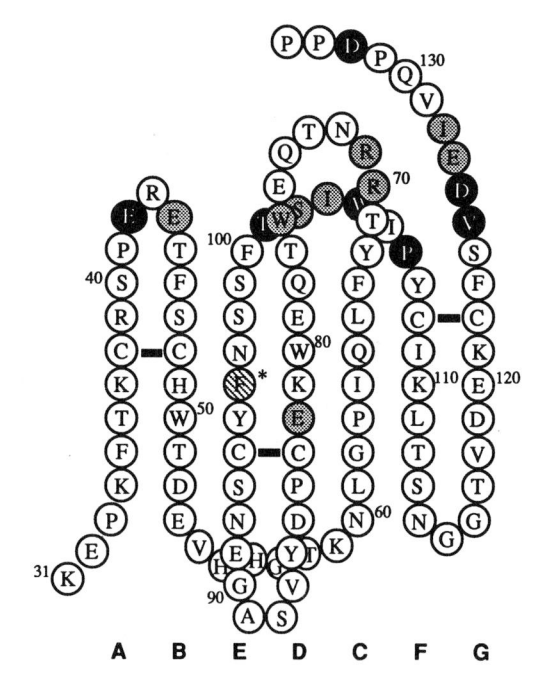

FIG. 5.   Residues of the first subdomain of the extracellular region of the GH receptor involved in binding of hGH. Seven antiparallel $\beta$ strands (A–G) make up a sandwich, with the G strand folded back to the A strand, where it hydrogen bonds. Residues causing >100-fold reduction in binding are indicated in black, stippled residues indicate a 2- to 99-fold reduction; disulfide bridges connecting cysteine residues are indicated by thick black lines (after Bass *et al.*, 1991). Note phenylalanine at position 96 (cross-hatched), which we find greatly affects binding, although probably indirectly (see text).

GROWTH HORMONE/PROLACTIN RECEPTORS 143

structure as determined by interaction with various monoclonal antibodies to hGH BP. Close inspection of Fig. 5 shows that the acidic residues E42 and E44 in the first cysteine loop, R70 and R71, W76, residues T101–W104, P106, residues V125–I128, and D132 are the amino acids that form the patch involved in receptor binding. Interestingly, E82, which is clearly outside this region, also appears to be part of the ligand-binding site.

A specific point mutation F96S, observed in a patient with Laron-type dwarfism (Amselem *et al.*, 1989) (see Section VIII), is indicated by an asterisk in Fig. 5. Bass *et al.* (1991) modified this phenylalanine to either an alanine or a serine. They failed to see any modification in the affinity of hGH for the soluble binding protein. On the other hand, we have made the F96S mutation in the cDNAs encoding the hGH BP and the full-length rabbit GH receptor and expressed the protein in COS-7 cells. We routinely failed to see any binding of the expressed mutants compared to the controls, although the protein is properly expressed, confirming that this phenylalanine is perhaps somehow important in providing structural and conformational constraints for agonist binding, perhaps rather than in providing a specific recognition site of hGH for its receptor (Edery *et al.*, 1992).

*2. Prolactin Receptor*

We recently demonstrated that modification of any of the first four conserved cysteines in the extracellular region of the rat PRL receptor, followed by transient transfection in COS-7 cells, completely abolished receptor binding, although the receptor protein is properly synthesized (Rozakis-Adock and Kelly, 1991). This confirmed that the cysteine-rich region was important for ligand binding. We have subsequently used homologue and alanine scanning within this region (residues 12–72) to determine those amino acids important in conferring specificity of binding. In addition, the effect of mutation of the WSxWS motif was examined (Rozakis-Adock and Kelly, 1992).

Based on the structural model of seven antiparallel $\beta$ strands (A–G) that form a $\beta$-sheet sandwich (Bazan, 1990b), and the secondary structure model proposed by Bass *et al.* (1991), we prepared a structural model for the first subdomain of the PRL receptor (Fig. 6A). However, closer examination of this model and prediction of the sequence of the $\beta$ turns not dictated by the disulfide bridges, based on the programs of Novotny and Auffray (1984) and Chou and Fasman (1978), led to a modification of the predicted sequence, shown in Fig. 6B. Note the altered position of the connecting loop between strands B and C that fold up for the PRL receptor, rather than lying down for the GH receptor.

144 P. A. KELLY ET AL.

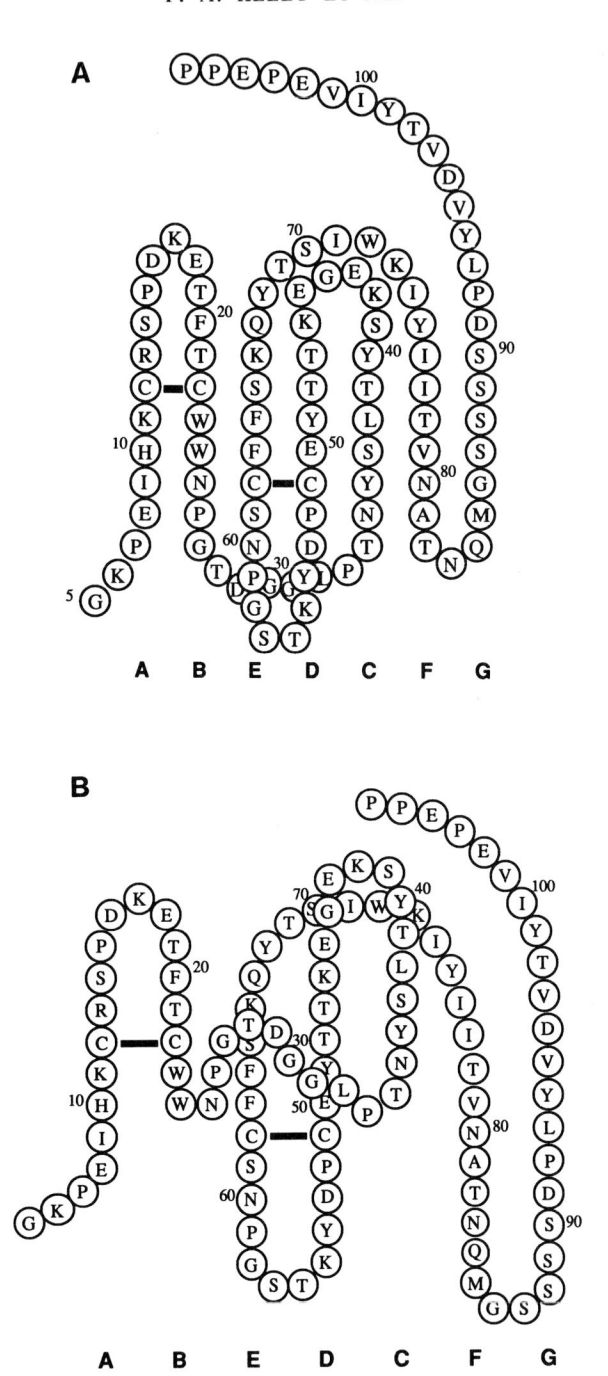

GROWTH HORMONE/PROLACTIN RECEPTORS   145

As shown in Fig. 7, a region within the first disulfide loop (residues R13, D16, and E18) and a set of lactogen-specific sequences between the two pairs of cysteines (residues N25–G30, G31–S37, and S41–Y49) converge to form a patch on the two-dimensional model (Rozakis-Adcock and Kelly, 1992).

Other amino acids could also play a role in determining ligand–receptor interactions. Notaby, residues W24, L38, Y40, and F64 are universally conserved in all members of the GH/PRL/cytokine receptor family (Bazan, 1990b), and G31, T69, S70, I71, and W72 are conserved in the subfamily of PRL/GH receptors (Fig. 2). We speculate that the side chains of these residues, which are buried in the hydrophobic interior $\beta$ sandwich, define structurally important determinants that contribute to the conformational integrity of these receptors. The phenylalanine at position 64, which corresponds to F96 in the GH receptor, is also important for the PRL receptor, as F64S failed to bind PRL (Edery et al., 1992; Rosakis-Adcock and Kelly, 1992).

Finally, the WSxWS motif appears to be an important structural element in regulating ligand binding. Selective mutations of these residues with alanine or substitution with the homologous residues of the GH receptor (YGEFs) resulted in a marked reduction in affinity for PRL. Because the structural integrity of the receptor (monoclonal antibody binding) was maintained, these results suggest that this motif might be involved in high-affinity binding (Rozakis-Adcock and Kelly, 1992). Recent studies with the IL-2 receptor demonstrated that normal binding is dependent on the presence of this motif in the $\beta$ subunit (Miyazaki et al., 1991). Taken together, these results suggest that the WSxWS motif may constitute the "floor" or hinge region of the ligand-binding crevice predicted by Bazan (1990a) for all cytokine receptors, and hence is intolerant to side chain substitutions. Alternatively, these residues may lie on the outer face of the ligand-binding pocket and provide a target site for the interaction of an accessory protein, important in the formation of a high-affinity binding complex, as has been illustrated for receptors for IL-2 (Szöllösi et al., 1987; Sharon et al., 1990l; Saragovi et al., 1990), IL-5 (Takaki et al., 1990), IL-6 (Hibi et al., 1990), and GM-CSF (Hayashida et al., 1990).

---

FIG. 6.   Proposed secondary structure of the prolactin receptor. (A) Structure based directly on model of immunoglobulins proposed by Bazan et al. (1990b). (B) Modified version, after taking into account the $\beta$ turns predicted by the programs of Novotny and Auffray (1984) and Chou and Fasman (1978).

146 P. A. KELLY ET AL.

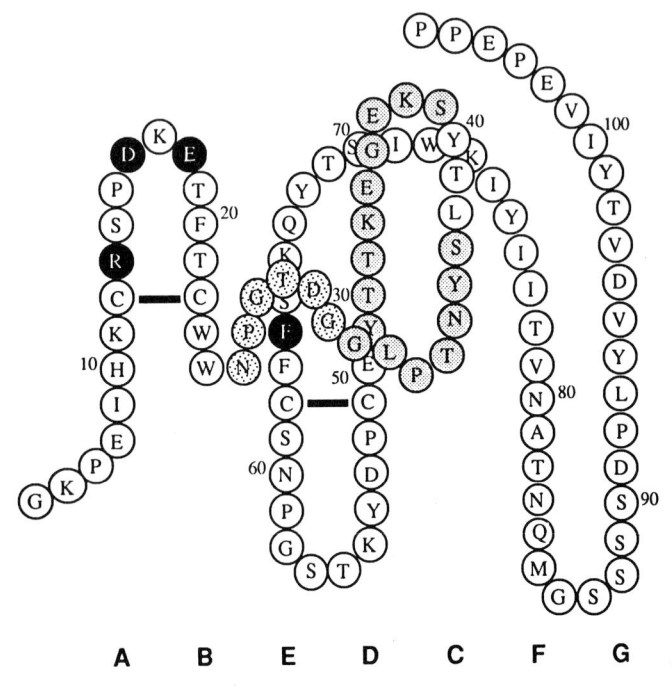

**A    B    E    D    C    F    G**

FIG. 7. Residues in the cysteine-rich domain of the rat PRL receptor involved in PRL binding. The seven antiparallel β strands (A–G) form a sandwich, as described in legend to Fig. 5. Individual amino acids demonstrated to be involved in reducing receptor affinity for PRL are shown in black; extended sequences, demonstrated by the preparation of PRL/GH receptor chimeras, are stippled (after Rosakis-Adcock and Kelly, 1992).


# VI.   Expression of Receptor Genes


## A.   GROWTH HORMONE RECEPTORS

GH receptor mRNA is present in some tissues in which no GH binding can be detected by radioligand binding. In rat intestine, heart, skeletal muscle, brain, and testis, GH receptor mRNA has been identified. As predicted from binding data, liver and adipose tissue contain the highest level of mRNA expression (Mathews *et al.*, 1989; Frick *et al.*, 1990). No GH receptor transcript is detected in rat spleen or thymus. Two mRNAs of 3.9–4.7 and 1–1.5 kb have been detected by Northern analysis in rat and mouse tissues (Baumbach *et al.*, 1989; Smith *et al.*, 1988). An additional low-abundance transcript of 2.6 kb has sometimes been found. Results from cell-free translation studies have suggested that the 1.2- and

GROWTH HORMONE/PROLACTIN RECEPTORS 147

3.9-kb transcripts encode the high-molecular-weight (receptor) and low-molecular-weight (BP) forms of the GH receptor, respectively (Smith *et al.*, 1988). In rabbit tissues and human liver, the approximately 4.5-kb transcript predominates, with a smaller transcript of about 1.2 kb being occasionally detected in low abundance (Leung *et al.*, 1987; Tiong *et al.*, 1989; Jammes *et al.*, 1991; and our unpublished results).

The ontogeny of GH receptor gene expression has been described in rat liver, kidney, and heart. In newborn rats, GH receptor expression is low, rising during postnatal development to a maximum at 5–8 weeks of age (Mathews *et al.*, 1989). In general, these results agree with actual GH receptor levels measured in liver membrane preparations. Similar levels of GH mRNA have been found in male and female rat liver; the increased concentration of GH mRNA in the liver of pregnant rats is consistent with the observed increase in GH binding (Mathews *et al.*, 1989). In a human hepatoma cell line, physiological levels of hGH increased GH receptor mRNA levels within 1 hour after addition of hormone, with a steady state being reached by 3–4 hours. Elevated hGH levels initially led to a down-regulation of GH receptor mRNA transcripts (at 3 hours), followed by a subsequent increase. These effects were due to changes in the rate of GH receptor gene transcription (Mullis *et al.*, 1991).

The concentration of GH receptor mRNA is not affected by hypophysectomy or GH treatment, suggesting that regulation of the number of GH receptors occurs at a posttranscriptional level in these situations. However, regulation could be tissue specific, because hypophysectomy increases GH receptor mRNA levels in muscle and decreases it in fat (Frick *et al.*, 1990).

The 1.2-kb transcript that encodes the GH/BP and that has been found in various rat tissues also appears to be developmentally regulated, similar to the 4.5-kb transcript (Carlsson *et al.*, 1990).

## B. PROLACTIN RECEPTORS

The expression of the PRL receptor gene has been studied in the rat, mouse, rabbit, and human. The sizes of the mRNAs are quite different depending on the species and the organs studied.

In the rat, the short form of the PRL receptor is encoded, almost entirely, by a transcript of 1.8 kb, whereas the long form is encoded by three transcripts of 2.5, 3, and 5.5 kb (Shirota *et al.*, 1990). The relative proportion of these transcripts differs from tissue to tissue. The liver expresses the majority of the 1.8-kb transcript, with the three transcripts in the ovary encoding, for the most part, the long form of PRL receptor. These different mRNAs are the result of alternative splicing of a primary transcript. This is supported by analysis of the structure of the PRL receptor gene in the rat (Banville *et al.*, 1992), which has revealed the existence of a single gene composed of at least 11 exons spanning greater than 70 kb. The eleventh exon appears to encode the cytoplasmic region specific to the short form of the receptor, whereas the tenth exon encodes the specific cytoplasmic region of the long form of the receptor.

PRL receptor mRNA levels have been studied in rat liver and mammary gland during development and during pregnancy and lactation. Receptor mRNA levels were undetectable in liver by Northern blot analysis in hypophysectomized animals, whereas mRNAs were easily identified in intact rats (Boutin *et al.*, 1988). Expression of PRL receptor mRNA was low or absent during neonatal periods, clearly increased at 40 days of age, just after puberty, and then stabilized until 70 days of age. PRL receptor levels measured by Scatchard analysis in the same tissue samples showed an almost identical pattern of development up to 40 days of age, after which receptor levels continued to increase while mRNA levels remained constant (Jolicoeur *et al.*, 1989).

After the first 3 days following a single injection of estradiol valerate (1 mg), a parallel increase in PRL receptor mRNA levels and receptor levels measured by Scatchard analyses was seen. After that time, receptor levels continued to increase while mRNA levels stabilized. These results, along with those described above during development, suggest that there is a dual regulation of PRL receptor expression, at the level of gene transcription as well as at a posttranscriptional level, either on the stabilization or the translation of mRNAs (Jolicoeur *et al.*, 1989).

More recently, expression of the short and long forms of PRL receptor mRNA has been measured in rat mammary gland and liver during pregnancy and lactation (Jahn *et al.*, 1991b). Because receptor levels are difficult to evaluate in the rat mammary gland using [125]I-labeled PRL, receptor mRNA levels were compared to receptor levels measured by the binding of an [125]I-labeled monoclonal antibody specific to the extracellular region of the PRL receptor (Okamura *et al.*, 1989). Although it was clear that the rat liver contained both the short and long forms of the PRL receptor, it was surprising to observe that the short form also predominated in the mammary gland. In fact, when quantitative measurements were made by Northern analyses using probes specific to the long and short forms of the receptor, we determined that the mammary gland contained approximately 70% short and 30% long forms of the receptor. The developmental pattern of the two forms of receptor was essentially similar within the same tissue but differed markedly in liver and mammary gland. That is, there was an increase during pregnancy of mRNAs for both forms of receptor in liver, with a significant reduction after parturition, whereas in

GROWTH HORMONE/PROLACTIN RECEPTORS 149

mammary gland, levels were relatively low during pregnancy but increased markedly during lactation. High progesterone levels during pregnancy are probably responsible for limiting PRL receptor gene expression in the mammary gland. Treatment with the progesterone antagonist (RU486) in late pregnancy induced an increase of PRL receptor mRNA in the mammary gland. PRL receptor levels, measured using [125]I-labeled U5 binding, paralleled the pattern of receptor mRNA (Jahn *et al.*, 1991b). Thus, there is a differential regulation of PRL receptors, both at the level of message and protein, between liver and mammary gland. Recently, we have shown that there is no modification in PRL receptor gene expression in the early stages of dimethylbenzanthracene-induced tumor transformation. However, an increased expression of the gene in established tumors is seen (Jahn *et al.*, 1991a). At this time, it is difficult to evaluate the importance of the two forms of the receptor. Obviously, the fact that there is so much short form in the mammary gland suggests that this form of receptor is transferred from mammary cells into milk, and this protein actually reaches the gut of newborns (Gonella *et al.*, 1989). Does this short form of receptor serve some transport function in the mammary gland?

In the rabbit, the situation is quite different, as analyzed by Dusanter-Fourt *et al.* (1991). Three major transcripts of 2.7, 3.4, and 10.5 kb and a minor transcript of 6.2 kb have been characterized. The four transcripts differ only in their 5' and 3' untranslated regions, and all encode a unique precursor corresponding to the long form of the PRl receptor. No specific transcript was found encoding the short form of PRL receptor in the rabbit. These four transcripts are identical in all organs tested. Among the various rabbit tissues, the male and female adrenals, the mammary gland, the ovaries, and the jejunum contain the highest level of PRl receptor mRNA. The PRL receptor gene was also expressed at moderate to weak levels in uterus, liver, kidney, pancreas, testes, and seminal vesicles.

In humans, PRL receptor transcripts are 2.5, 3, and 7.3 kb in T-47D breast cancer cells and chorion laeve, the only two tissues thus far examined (Boutin *et al.*, 1989). This is similar to the range of transcript sizes found in the rabbit. In the mouse, only a short form of PRL receptor with a cytoplasmic domain of 39 and 50 aa has thus far been identified (Davis and Linzer, 1989). It is likely that alternative splicing is also the explanation for these two transcripts observed in the mouse.

Taken together, these data suggest the existence of at least two species subgroups expressing the PRL receptor gene differently. In all species, the various mRNAs originate from a primary transcript by differential splicing, affecting exons in the translated and untranslated domains (rat an mouse) or only in the untranslated regions (rabbit and human). In the rat, the different forms of PRL receptor (short and long) are encoded by

150 P. A. KELLY ET AL.

specific mRNAs. This is not the case for the rabbit or perhaps for humans, in which the short form of the PRL receptor, if it exists, is probably generated by posttranslational mechanisms such as proteolytic processing.

## VII.  Mechanism of Action of Growth Hormone and Prolactin

### A.  IDENTIFICATION OF FUNCTIONAL DOMAINS OF RECEPTORS

In order to study structure–function relationships of cloned cDNAs for either the PRL or GH receptor, functional assays must be developed. Essentially, two basic approaches can be used. The first involves transfection of known target cells with receptor cDNA and the measurement of cellular responses in cells incubated in the presence of ligand. For GH, hepatic cells, preadipocytes, or islets of Langerhans are good potential systems. The responses to be measured are IGF-I expression, glucose transport, or insulin production.

The transfection of rat GH receptor cDNA into rat insulinoma (RIN) cells results in an increased expression of GH receptor that is accompanied by an increased responsiveness to GH in terms of insulin biosynthesis (Billestrup *et al.*, 1990). Similarly, when Chinese hamster ovary (CHO) cells, a receptor-negative cell line, were transfected with the rat GH receptor, GH was able to stimulate protein synthesis in the transfected cells. This short-term metabolic effect was dose and time dependent and was observed only under serum-free conditions (Emtner *et al.*, 1990).

The relative contribution of the intracellular domain of the GH receptor was examined using RIN cells transfected with GH receptor mutants. Moldrup *et al.* (1991) reported the expression of two truncation mutants that had either 6 or 169 intracellular amino acids, compared to the wild-type receptor with 350 amino acids inside the cell. In fact, only the wild-type receptor showed an increase in GH-stimulated insulin production, suggesting that the sequences in the C-terminal half of the cytoplasmic domain are required for GH signed transduction.

For PRL, normal mammary epithelial cells or tumor cells are good potential candidates. The synthesis of various milk proteins or PIP can be measured as a response to PRL. Thus far, in spite of numerous attempts using various human and murine cell lines, we have been unsuccessful in increasing the sensitivity to PRL of any of these transfected cell lines.

A second approach to the establishment of a functional assay involves cotransfection of a receptor cDNA along with a fusion gene containing a target gene for PRL ($\beta$-lactoglobulin, $\beta$-casein, $\alpha_{S1}$-casein, parathyroid-

GROWTH HORMONE/PROLACTIN RECEPTORS 151

like protein) or for GH (IGF-I, serine protease), coupled to a reporter gene such as that for chloramphenicol acetyltransferase (CAT). We have detected a clear response to PRL in CHO cells transiently transfected with the $\beta$-lactoglobulin gene promoter coupled to the gene for CAT and the long form of PRL receptor cDNA. A 4.6-fold increase in CAT activity was seen in these cells (Lesueur *et al.*, 1990). This effect was highly specific, with a maximal response at 14.6 n$M$ PRL in the incubation media. One drawback of these initial studies was the relatively high basal transcriptional activity occurring in cells transfected with the PRL receptor cDNA, even in the absence of PRL. Because CHO cells are usually grown in the presence of fetal calf serum and it is well known that this serum contains numerous growth factors, including lactogenic hormones, we tried to eliminate the potential lactogenic stimulation by removing serum from the incubation media. In fact, in more recent studies (Lesueur *et al.*, 1991), we have been able to almost completely eliminate the effect of serum and can routinely see a 12- to 20-fold increase in $\alpha$-lactoglobulin–CAT activity by PRL.

The establishment of this assay allowed a determination of the functional activity of different forms of PRL receptor occurring either naturally or constructed by mutagenesis. When CHO cells were cotransfected with an expression vector containing the cDNA of the long form of the PRL receptor and the $\beta$-lactoglobulin–CAT construct, a 17-fold induction of CAT activity was obtained in the presence of PRL in the culture media, whereas no induction was found when the cDNA of the short form was cotransfected. These results establish that only the long form of the PRL receptor is able to transduce the hormonal message to the milk protein genes (Lesueur *et al.*, 1991).

Using the same functional assay, we have determined the respective contribution of two domains of the PRL receptor that could be involved in the mechanism of signal transduction. The membrane-bound intracellular domain expressed alone was devoid of PRL binding activity, as expected, and failed to stimulate constitutive expression of the target gene. The extracellular domain, expressed alone as a soluble form, bound PRL with a 10-fold higher affinity than the membrane receptor and was also incapable of activating expression of the reporter gene in the presence of PRL. Interestingly, when the two constructs were transfected together, CAT activity was similar to that obtained with the full-length receptor cDNA (L. Lesueur, M. Edery, J. Paly, P. A. Kelly, and J. Djiane, submitted for publication). In contrast to what has been shown for several transmembrane receptors (e.g., epidermal growth factor, insulin), these results suggest that the membrane-anchored cytoplasmic domain of the PRL receptor has no independent activity and that the association of the extracellular and intracellular domains of the long form of the PRL receptor contributes to the transfer of the PRL signal to the milk protein gene.

We have recently developed a homologous functional system to test the activity of the two forms of PRL receptor as well as the mutant PRL receptor found in Nb2 cells, with respect to their ability to transmit a lactogenic signal. In this system, CHO cells were transiently transfected with a construct containing 2300 bp of the 5' flanking sequence of the rat $\beta$-casein gene (kindly provided by Dr. J. Rosen, Houston, Texas) fused to the chloramphenicol acetyltransferase gene and an expression vector containing the various forms of rat PRL receptor cDNA. Interestingly, the intermediate Nb2 form of the PRL receptor, which is missing 198 aa from the cytoplasmic domain, was fully able to transmit the lactogenic signal and stimulate the reporter gene (Ali *et al.*, 1992). Overall, these results suggest that the long and short forms of receptor are probably involved in mediating different biological functions of PRL. Also, it is clear that the amino acid sequences required for transduction of the lactogenic hormone signal are conserved in the Nb2 form of the PRL receptor. Thus the missing 198 aa must not be important, at least for this activity of PRL.

The development and application of functional tests as described above should allow the precise identification, by site-directed mutagenesis of the receptor cDNAs for GH and PRL, of the regions involved in the mechanism of action of these hormones. It should be remembered, however, that only one action is being studied with any specific test. Because multiple actions of GH and PRL have been reported, it is possible that there are a number of different signal pathways that must be identified.

## B. POSSIBLE SIGNAL TRANSDUCTION PATHWAYS

### 1. Growth Hormone

Very little is known about the early events that occur after the binding of GH to its receptor on the plasma membrane. As is true for PRL, no second messenger mediating the effects of GH has been identified. Attempts to link a GH effect with modifications of adenylate cyclase have resulted in conflicting results, indicating either increase, decrease, or no change in the levels of cAMP. One study has demonstrated that GH is able to increase guanylate cyclase activity in several tissues *in vitro* (Vesely, 1981).

The identification of the primary amino acid sequence of the GH receptor did not provide any clues into the mechanism of signal transduction. No consensus sequences in the GH receptor are found homologous to known tyrosine kinases. However, recent results suggest that a novel

GROWTH HORMONE/PROLACTIN RECEPTORS 153

tyrosine kinase or kinase activity may be associated with the GH receptor in several cell types. GH receptor complexes were partially purified from GH-treated 3T3F442A fibroblasts by immunoprecipitation using anti-GH antiserum. After incubation with $[^{32}P]ATP$, a protein with an $M_r$ of 121,000, phosphorylated on tyrosyl residues, was revealed. This protein was not observed when cells had not been treated with a control immunoglobulin G (Carter-Su et al., 1989). This kinase activity of or associated with the GH receptor has now been demonstrated in several other cell types, including human IM 9 lymphocytes, rat H-35 hepatoma cells, and freshly isolated rat adipocytes (Stred et al., 1990). The question of whether the tyrosine kinase is intrinsic to the GH receptor or is associated with a nonreceptor membrane protein remains to be resolved.

Protein kinase C activation could be involved in the mechanism of action of GH, as suggested by several recent reports. GH is able to stimulate phospholipase C activity in vitro with the production of inositol triphosphate and diacylglycerol in basolateral membranes of canine kidney (Rogers and Hammerman, 1989). In OB 1771 mouse preadipocyte cells, GH has also been shown to stimulate the production of diacylglycerol production by means of phosphatidylcholine breakdown, involving a phospholipase C coupled to the GH receptor in these cells (Catalioto et al., 1990). In isolated hepatocytes, GH and PRL are able to stimulate rapidly the production of diacylglycerol without changes in inositol phosphate concentration (Johnson et al., 1990). Diacylglycerol is known to activate protein kinase C; therefore, these data suggest that protein kinase could mediate at least some of the actions of GH in these models.

## 2. Prolactin

The events that occur after PRL binds to its receptor on the plasma membrane, events that are responsible for the intracellular responses to PRL, are poorly understood. The fact that several polyclonal and monoclonal antibodies to PRL receptors induce, in mammary gland (Djiane et al., 1985) and Nb2 cells (Shiu et al., 1983; Elberg et al., 1990), PRL-like effects suggests that PRL is probably not necessary other than for receptor binding, at least for Nb2 mitogenic activity and for milk protein gene stimulation, for which second-messenger mechanisms are most probably involved. Many studies have been designed in an attempt to identify such second messengers for PRL, but none has clearly demonstrated that the effects of PRL could be mimicked by any of these putative mediators. Because of the wide range of biological actions associated with PRL and the existence of various forms of PRL receptors, it is doubtful that one unifying mechanism of action will be found. A number of model systems have been studied, but most of the information on the mechanism of action

154                           P. A. KELLY ET AL.

of PRL has been gained using the Nb2 rat lymphoma cell line for mitogenic
effects and mammary explant cultures for the expression of milk protein
genes.

   *a.  PRL Mitogenic Activity in Nb2 Cells.*   PRL-mediated mitogenesis
does not involve modulation of intracellular cAMP concentrations (Kornberg and Liberti, 1989), but cAMP has, depending on the concentrations
used, stimulatory or inhibitory effects on PRL-induced mitogenesis
(Larsen and Dufau, 1988). The involvement of G proteins in PRL action
has been suggested by studies using pertussis toxin and cholera toxin,
known to effect ADP ribosylation of the $\alpha$ subunit of G proteins and which
have been shown to modulate PRL-stimulate mitogenesis (Barkey *et al.*,
1988). In Nb2 cells, cross-linking experiments have suggested that G proteins are associated with the PRL receptor (Too *et al.*, 1990). The possible
involvement of phospholipase $A_2$ and phospholipase C has been proposed
(Ofenstein and Rillema, 1987).

   Although phosphoinositide metabolism is increased during mitogenesis
in Nb2 cells, this is not an immediate consequence of the interaction of
PRL with its receptor (Gertler and Friesen, 1986). One report (Ko *et al.*,
1986) indicated that phosphatidylcholine levels are increased after PRL
action in Nb2 cells, but such modulations have not been related to second-
messenger release. Phorbol esters can induce a significant stimulation of
Nb2 cell proliferation, suggesting that activation of protein kinase C may
be involved in the effects of lactogenic hormones (Buckley *et al.*, 1986).
A rapid translocation of protein kinase C activity to the nucleus has
recently been reported (Ganguli *et al.*, 1990). In fact, the regulation of IL-
2-driven T lymphocyte proliferation by PRL has been shown to involve
translocation of PRL, taken up from the medium by cell surface receptors,
into the nucleus (Clevenger *et al.*, 1990). In rat liver, it has also been shown
that PRL induces a rapid activation of protein kinase C by interaction with
a putative nuclear receptor (Buckley *et al.*, 1988).

   Early events after PRL binding to its receptor may include activation
of a $Na^+/H^+$ exchange (Too *et al.*, 1987) and activation of $Ca^{2+}$ uptake. It
has recently been shown that PRL induces $Ca^{2+}$ influx within 20–100 sec,
which can be inhibited by specific voltage-gated $Ca^{2+}$ channel blockers
(Buckley *et al.*, 1986). Finally, the expression of several genes is rapidly
stimulated after the binding of PRL to Nb2 cells, including c-*myc* and
actin, interferon regulatory factor-1, ornithine decarboxylase (ODC), and
heat-shock protein-70-like genes (Fleming *et al.*, 1985; Yu-Lee, 1990; de
Toledo *et al.*, 1987).

   *b.  PRL Stimulation of Milk Protein Genes.*   As described above, the
long form of the PRL receptor expressed in mammalian host cells (CHO)

following transfection of the corresponding cDNA, after interaction with PRL, stimulates the promoter of milk protein genes cotransfected in the same cell (Lesueur *et al.*, 1990, 1991). This result indicates that the long form alone possesses both the functions of recognition of the hormone and of transduction of the hormonal message. However, the fact that no homology was found in the primary structure of the PRL receptor with any receptor for which the mechanism of signal transduction is known suggests that either the mechanism involved is completely distinct from what is known for other hormone receptors or that the transduction is carried out by some other associated membrane protein. This "transducer" protein could be widely distributed in different cell types, including CHO cells. This possibility has been documented for the IL-6 receptor, for which a membrane protein (gp130) appears to be directly responsible for the transduction of the hormonal signal after interaction of IL-6 with its receptor (Taga *et al.*, 1989).

Numerous studies have been performed using mammary cells to identify PRL second messengers. Chloroquine, which interferes with processing of hormone–receptor complexes within the cell and their degradation in lysosomes, does not block PRL effects in the mammary gland (Houdebine and Djiane, 1980), suggesting that degradation of hormone–receptor complexes in lysosomes does not generate an intracellular mediator for PRL. cAMP has been shown to have a slight inhibitory effect on casein synthesis, whereas cGMP has a slight stimulatory effect (Matusik and Rosen, 1980). However, the effects of these cyclic nucleotides are too small to be considered as a potential mediator of PRL action. Prostaglandins ($E_2$ and $F_2$) also have some PRL-like effects in the mammary gland (RNA synthesis), but do not mimic PRL action on casein synthesis (Vesely, 1984; Rillema, 1980). That they are able to amplify the effects of PRL in combination with polyamines is supported by the fact that PRL is able to stimulate ODC (Etindi and Rillema, 1987). Even if the participation of phospholipase C cannot be excluded, activation of the metabolism of phosphatidylinositol diphosphate in mammary membranes does not seem to be related to the immediate action of PRL (Etindi and Rillema, 1988). Recent studies indicate that diacylglycerol formation may be an early signaling event occurring after stimulation by PRL or GH (Johnson *et al.*, 1990). However, in the mammary gland, protein kinase C activation mimics PRL action only on ODC (Rillema and Whale, 1988) and not on milk protein genes. Stimulators, such as phorbol esters, or inhibitors, such as $H_7$ of protein kinase C, do not modify PRL action on casein synthesis (Devinoy *et al.*, 1989).

## VIII. Genetic Diseases and Models of Receptor Defects

The cloning of the genes encoding the receptors for PRL and GH and the subsequent chromosomal localization of other genes have opened the field of the study of PRL and GH resistance syndromes. Using molecular genetic approaches, it should be possible to determine whether resistance syndromes are due to receptor defects or are associated with postreceptor events.

### A. GROWTH HORMONE RECEPTORS

A rare autosomal recessive syndrome known as Laron-type dwarfism, characterized by resistance to GH, was identified a number of years ago (Laron et al., 1966, 1971). The patients had severe growth failure and low circulating levels of IGF-I, in spite of elevated levels of hGH, which appeared to be biologically active. Two groups have independently shown that this disease may be due to mutations in the GH receptor gene (Godowski et al., 1989; Amselem et al., 1989). Figure 8 summarizes these mutations. For one group, genetic analysis revealed the deletion of a large portion of the extracellular domain of the GH receptor; specifically exons 3, 5, and 6 were missing from two affected patients (Godowski et al., 1989). In another study, a single point mutation changes phenylalanine to a serine at position 96 of the extracellular region of the mature form of the GH receptor (Amselem et al., 1989). This region is highly conserved in receptors for GH and PRL. We have recently demonstrated that the substitution of the phenylalanine residue at position 96 of the GH receptor or the homologous phenylalanine in the extracellular domain of the PRL receptor abolishes ligand binding (M. Edery, M. Rosakis, L. Lesueur, J. Pay, L. Goujon, M. C. Postel-Vinay, J. Djiane, and P. A. Kelly, manuscript in preparation). Two additional nonsense mutations within the GH receptor gene have been identified recently, at positions corresponding to the amino-terminal extremity. These mutations would result in truncated proteins of 37 and 42 aa, deleting a large portion of the GH binding domain as well as the transmembrane and the cytoplasmic domains (Amselem et al., 1991).

For a large number of patients with Laron dwarfism (>80%), however, no defects in the GH receptor gene have been identified. Other modifications that could occur in such patients affect the successful translation of the mRNA into a protein and/or the insertion of the protein into the membrane. It is possible that other receptor mutations that interfere with the process of signal transduction could be involved in this disease. Obviously, more studies must be performed to resolve this question.

GROWTH HORMONE/PROLACTIN RECEPTORS 157

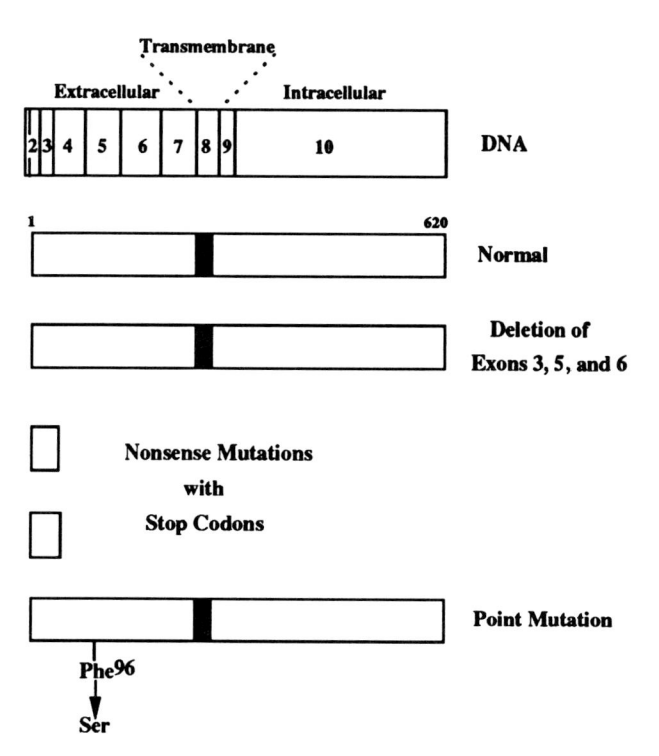

FIG. 8. Schematic representation of the human GH receptor gene and the various receptor mutations identified in Laron-type dwarfism. The numbers in the enclosed rectangles representing the receptor DNA indicate the corresponding exons, and the black vertical lines represent the introns. The "normal" receptor consisting of 620 aa is shown under the DNA. The transmembrane regions are shown in black.

## B. PROLACTIN RECEPTORS

No human or animal model of a genetic defect of the PRL receptor has thus far been reported. However, two animal models, the nude mouse and the hairless rat, appear to be interesting candidates for study. Homozygote females are unable to lactate after delivery, and in order for the young to survive, they must be placed with a nursing mother. Hairless rats secrete enough bioactive PRL to stimulate lactation (Guillaumot *et al.,* 1988a,b). In addition, normal PRL binding has been measured in the mammary gland of these animals. Thus the absence of lactation in the presence of normal PRL and PRL binding could be due either to a defect in the signal transduction mechanism in the cytoplasmic domain of the PRL receptor, or to a postreceptor defect.

158 P. A. KELLY ET AL.

As previously described, a genetic mutation appears to be responsible for the deletion resulting in the form of PRL receptor found in Nb2 cells (Ali *et al.*, 1991). Nb2 cells are uniquely dependent on PRL for growth. It will, therefore, be interesting to evaluate whether the modification in the cytoplasmic domain of the Nb2 PRL receptor is responsible for conferring the exaggerated mitogenic response of these cells.

## REFERENCES

Adams, T. E., Baker, L., Fiddes, R. J., and Brandon, M. R. (1990). *Mol. Cell. Endocrinol.* **73,** 135–145.

Ali, S., Edery, M., Pellegrini, I., Lesueur, L., Paly, J., Djiane, J., and Kelly, P. A. (1992). *Mol. Endocrinol.* **6,** 1242–1249.

Ali, S., Pellegrini, I., and Kelly, P. A. (1991). *J. Biol. Chem.* **266,** 20110–20117.

Amit, T., Barkey, R. J., Gavish, M., and Youdim, M. B. H. (1984). *Endocrinology (Baltimore)* **114,** 545–552.

Amselem, S., Duquesnoy, P., Bousnina, S., Postel-Vinay, M-C., and Goossens, M. (1989). *N. Engl. J. Med.* **231,** 989–995.

Amselem, S., Sobrier, M. L., Duquesnoy, P., Rappaport, R., Postel-Vinay, M. C., Gourmelen, M., Dallapiccola, B., and Goossens, M. (1991). *J. Clin. Invest.* **87,** 1098–1102.

Arden, K. C., Boutin, J. M., Kelly, P. A., Djiane, J., and Cavenee, W. K. (1990). *Cytogenet. Cell. Genet.* **53,** 161–165.

Armstrong, O. T., Knudsen, K. Z., and Miller, L. S. (1970). *Endocrinology (Baltimore)* **86,** 634–641.

Banville, D., Boie, Y., Shirota, M., Stocco, R., and Kelly, P. A. (1992). Submitted for publication.

Barkey, R. J., Calvo, J. C., and Dufau, M. L. (1988). *Biochem. Biophys. Res. Commun.* **156,** 776–782.

Barnard, R., Bundesen, P. G., Rylatt, D. B., and Waters, M. J. (1985). *Biochem. J.* **231,** 459–468.

Baroni, C. (1967). *Experientia* **23,** 282–283.

Barton, D. E., Foellmer, B. E., Wood, W. I., and Francke, U. (1989). *Cytogenet. Cell Genet.* **50,** 137–141.

Bass, S. H., Mulkerrin, M. G., and Wells, J. A. (1991). *Proc. Natl. Acad. Sci. U.S.A.* **88,** 4498–4502.

Baumann, G., Stolar, M. W., Amburn, K., Barsano, C. P., and Devries, B. C. (1986). *J. Clin. Endocrinol. Metab.* **62,** 134–141.

Baumbach, W. R., Horner, D. I., and Logan, J. S. (1989). *Genes Dev.* **3,** 1199–1205.

Baxter, R. C. (1985). *Endocrinology (Baltimore)* **117,** 650–655.

Baxter, R. C., Bryson, J-M., and Turtle, J. R. (1980). *Endocrinology (Baltimore)* **107,** 1176–1181.

Baxter, R. C., and Zaltsman, Z. (1984). *Endocrinology (Baltimore)* **115,** 2009–2014.

Baxter, R. C., Zaltsman, Z., and Turtle, J. R. (1984). *Endocrinology (Baltimore)* **114,** 1893–1901.

Bazan, F. J. (1989). *Biochem. Biophys. Res. Commun.* **164,** 788–795.

Bazan, F. J. (1990a). *Immunol. Today* **11,** 350–354.

Bazan, F. J. (1990b). *Proc. Natl. Acad. Sci. U.S.A.* **87,** 6934–6938.

## GROWTH HORMONE/PROLACTIN RECEPTORS 159

Bergeron, J. J. M., Posner, B. I., Josefsberg, Z., and Sikstrom, R. (1978). *J. Biol. Chem.* **253,** 4058–4066.

Bergeron, J. J. M., Searle, N., Khan, M. N., and Posner, B. I. (1986). *Biochemistry* **25,** 1756–1764.

Bernton, E. W., Meltzer, M. S., and Holaday, J. W. (1988). *Science* **239,** 401–404.

Billestrup, N., Moldrup, A., Serup, P., Mathews, L. S., Norstedt, G., and Nielsen, J. H. (1990). *Proc. Natl. Acad. Sci. U.S.A.* **87,** 7210–7214.

Boutin, J. M., Edery, M., Shirota, M., Jolicoeur, C., Leseur, L., Ali, S., Gould, D., Djiane, J., and Kelly, P. A. (1989). *Mol. Endocrinol.* **3,** 1455–1461.

Boutin, J. M., Jolicoeur, C., Okamura, H., Gagnon, J., Edery, M., Shirota, M., Banville, D., Dusanter-fourt, I., Djiane, J., and Kelly, P. A. (1988). *Cell (Cambridge, Mass.)* **53,** 69–77.

Buckley, A. R., Crowe, P. D., and Russell, D. H. (1988). *Proc. Natl. Acad. Sci. U.S.A.* **85,** 8649–8653.

Buckley, A. R., Montgomery, D. W., Kipler, R., Putnam, C. W., Zukoski, C. F., Gout, P. W., Been, C. T., and Russell, D. M. (1986). *Immunopharmacology* **12,** 37–51.

Buckley, A. R., Putnam, C., and Russel, D. H. (1985). *Life Sci.* **37,** 2569–2575.

Carlsson, B., Billig, H., Rymo, L., and Isaksson, O. G. P. (1990). *Mol. Cell. Endocrinol.* **73,** R1–R6.

Carter-Su, C., Stubbart, J. R., Wang, X., Stredt, S. E., Argelsinger, L. S., and Shafer, J. A. (1989). *J. Biol. Chem.* **264,** 18654–18661.

Catalioto, R. M., Ailhaud, G., and Negrel, R. (1990). *Biochem. Biophys. Res. Commun.* **173,** 840–848.

Chou, P. Y., and Fasman, G. D. (1978). *Adv. Enzymol.* **47,** 45–148.

Clevenger, C. V., Russell, D. H., Appasamy, P. M., and Prystowsky, M. B. (1990). *Proc. Natl. Acad. Sci. U.S.A.* **87,** 6460–6466.

Cunningham, B. C., Bass, S., Fuh, G., and Wells, J. A. (1990). *Science* **250,** 1709–1712.

D'Andrea, A. D., Lodish, H. F., and Wong, G. G. (1989). *Cell (Cambridge, Mass.)* **57,** 277–285.

Davis, J. H., and Linzer, D. I. H. (1989). *Mol. Endocrinol.* **3,** 674–680.

deToledo, S. M., Murphy, L. J., Hatton, T. H., and Friesen, H. G. (1987). *Mol. Endocrinol.* **1,** 430–434.

Devinoy, E., Jolivet, G. Thepot, D., and Houdebine, L. M. (1989). In "Development and Function of Reproductive Organs" (N. Josso, ed.), Rev. 21, p. 21. ARES Seron Symposia, Rome.

Djiane, J., Clauser, H., and Kelly, P. A. (1979). *Biochem. Biophys. Res. Commun.* **90,** 1371–1379.

Djiane, J., Delouis, C., and Kelly, P. A. (1982). *Mol. Cell. Endocrinol.* **25,** 163–170.

Djiane, J., and Durand, P. (1977). *Nature (London)* **266,** 641–643.

Djiane, J., Durand, P., and Kelly, P. A. (1977). *Endocrinology* **100,** 1348–1356.

Djiane, J., Dusanter-Fourt, I., Katoh, M., and Kelly, P. A. (1985). *J. Biol. Chem.* **260,** 11430–11435.

Djiane, J., Kelly, P. A., and Houdebine, L. M. (1980). *Mol. Cell. Endocrinol.* **18,** 87–91.

DiMattia, G. E., Gellersen, B., Duckworth, M. L., and Friesen, H. G. (1990). *J. Biol. Chem.* **265,** 16412–16421.

Doglio, A., Dani, C., Grimaldi, P., and Ailhaud, G. (1989). *Proc. Natl. Acad. Sci. U.S.A.* **86,** 1148–1152.

Dusanter-Fourt, I., Berthon, P., Belair, L., Kelly, P. A., and Djiane, J. (1990). *J. Receptor Res.* **9,** 479–493.

160   P. A. KELLY ET AL.

Dusanter-Fourt, I., Gaye, P., Belair, L., Petridou, B., Kelly, P. A., and Djiane, J. P. (1991). *Mol. Cell. Endocrinol.* **77**, 181–192.

Edery, M., Jolicoeur, C., Levi-Meyrueis, C., Dusanter-Fourt, I., Pétridou, B., Boutin, J. M., Lesueur, L., Kelly, P. A., and Djiane, J. (1989). *Proc. Natl. Acad. Sci. U.S.A.* **86**, 2112–2116.

Edery, M., Rozakis-Adcock, M., Goujon, L., Finidori, J., Levi-Meyrueis, C., Paly, J., Djiane, J., Postel-Vinay, M. C., and Kelly, P. A. (1993). *J. Clin. Invest.,* in press.

Edwards, C. K., Ghiasuddin, S. M., Schepper, J. M., Yunger, L. M., and Kelley, K. W. (1988). *Science* **239**, 764–771.

Elberg, G., Kelly, P. A., Djiane, J., Binder, L., and Gertler, A. (1990). *J. Biol. Chem.* **265**, 14770–14776.

Emtner, M., Mathews, L. S., and Norstedt, G. (1990). *Mol. Endocrinol.* **4**, 2014–2020.

Etindi, R. N., and Rillema, J. A. (1987). *Biochim. Biophys. Acta* **929**, 345–349.

Etindi, R. R., and Rillema, J. A. (1988). *Biochim. Biophys. Acta* **968**, 385–391.

Finidori, J., Postel-Vinay, M. C., and Kleinknecht, C. (1980). *Endocrinology (Baltimore)* **106**, 1960–1965.

Fleming, W. H., Murphy, P. R., Murphy, L. J., Hatton, T. W., Matusik, R. J., and Friesen, H. G. (1985). *Endocrinology (Baltimore)* **117**, 2547–2549.

Forsyth, I. A., Folley, L. S. J., and Chadwick, A. (1965). *J. Endocrinol.* **31**, 115–126.

Frick, G. P., Leonard, J. L., and Goodman, H. M. (1990). *Endocrinology (Baltimore)* **126**, 3076–3082.

Fuh, G., Mulkerrin, M. G., Bass, S., McFarland, N., Brochier, M., Bourell, J. H., Light, D. R., and Wells, J. A. (1990). *J. Biol. Chem.* **265**, 3111–3115.

Fukunaga, R., Ishizaka-Ikeda, E., Seto, Y., and Nagata, S. (1990). *Cell (Cambridge, Mass.)* **61**, 341–350.

Gala, R. R. (1991). *Proc. Soc. Exp. Biol. Med.* **198**, 513–527.

Ganguli, S., Hu, L. B., Menke, P. T., Freeman, J. J., McGrath, M. E., Collier, R. J., and Gertler, A. (1990). "Program of the 72nd Annual Meeting of the Endocrine Society, Atlanta, GA," p. 91 (Abstract 268).

Gearing, D. P., King, J. A., Gough, N. M., and Nicola, N. A. (1989). *EMBO J.* **8**, 3667–3676.

Gertler, A., and Friesen, H. G. (1986). *Mol. Cell. Endocrinol.* **48**, 221–228.

Gluckman, P. D., Butlet, J. H., and Elliott, T. B. (1983). *Endocrinology (Baltimore)* **112**, 1607–1612.

Godowski, P. J., Leung, D. W., Meacham, L. R., Galgani, J. P., Hellmiss, R., Keret, R., Rotwein, P. S., Parks, J. S., Laron, Z., and Wood, W. I. (1989). *Proc. Natl. Acad. Sci. U.S.A.* **86**, 8083–8087.

Golde, D. W., and Bersch, N. (1977). *Science* **196**, 1112–1113.

Gonella, P. A., Harmatz, P., and Walker, W. A. (1989). *J. Cell. Physiol.* **140**, 138–149.

Goodwin, R. G., Friend, D., Ziegler, S. F., Jerzy, R., Falk, B. A., Gimpel, S., Cosman, D., Dower, S. K., March, C. J., Namen, A. E., and Park, L. S. (1990). *Cell (Cambridge, Mass.)* **60**, 941–951.

Gout, P. W., Beer, C. T., and Noble, R. L. (1980). *Cancer Res.* **40**, 2433–2436.

Green, H., Morikawa, M., and Nixon, D. (1985). *Differentiation* **29**, 195–198.

Grichting, G., and Goodman, H. M. (1986). *Endocrinology (Baltimore)* **119**, 847–854.

Grinwich, D. L., Hichens, M., and Behrman, M. R. (1976). *Biol. Reprod.* **14**, 212–218.

Guillaumot, P., Sabbagh, I., Bertrand, J., and Cohen, H. (1988a). *Mol. Cell. Endocrinol.* **58**, 17–24.

Guillaumot, P., Sabbagh, I., Bertrand, J., and Cohen, H. (1988b). *Mol. Cell. Endocrinol.* **58**, 25–29.

## GROWTH HORMONE/PROLACTIN RECEPTORS 161

Gurland, G., Ashcom, G., Cochran, B. H., and Schwartz, J. (1990). *Endocrinology (Baltimore)* **127**, 3187–3195.

Guyette, W. A., Matusik, R. J., and Rosen, J. M. (1979). *Cell (Cambridge, Mass.)* **17**, 1013–1023.

Haagensen, D. E., and Mazovjian (1986). *In* "Diseases of the Breast" (C. D. Haagensen, ed.), pp. 474–500. Saunders, Philadelphia.

Hartmann, D. P., Holaway, J. W., and Bernton, E. W. (1989). *FASEB J.* **3**, 194–202.

Hatakeyama, M., Tsudo, M., Minamoto, S., Kono, T., Doi, T., Miyata, T., Miyasaka, M., and Taniguchi, T. (1989). *Science* **244**, 551–556.

Hauser, S. D., McGrath, M. F., Collier, R. J., and Krivi, G. G. (1990). *Mol. Cell. Endocrinol.* **72**, 187–200.

Hayashida, K., Kitamura, T., Gorman, D. M., Arai, K., Yokota, T., and Miyasima, A. (1990). *Proc. Natl. Acad. Sci. U.S.A.* **87**, 9655–9659.

Hibi, M., Murakami, M., Saito, M., Hirano, T., Taga, T., and Kishimoto, T. (1990). *Cell (Cambridge, Mass.)* **63**, 1149–1157.

Hocquette, J. F., Postel-Vinay, M. C., Djiane, J., Tar, A., and Kelly, P. A. (1990). *Endocrinology (Baltimore)* **127**, 1665–1672.

Hocquette, J. F., Postel-Vinay, M. C., Kayser, C., Hemptinne, B., and Amar-Costesec, A. (1989). *Endocrinology (Baltimore)* **125**, 2167–2173.

Houdebine, L. M., and Djiane, J. (1980). *Mol. Cell. Endocrinol.* **17**, 1–15.

Hughes, J. P., Elsholtz, H. P., and Friesen, H. G. (1985). *In* "Polypeptide Hormone Receptors" (B. I. Posner, ed.), pp. 157–190.

Hugues, J. P., and Friesen, H. G. (1985). *Annu. Rev. Physiol.* **47**, 469–482.

Itoh, N., Yonehara, S., Schreurs, J., Gorman, D. M., Maruyama, K., Ishii, A., Yahara, I. K., Arai, K-I., and Miyajima, A. (1990). *Science* **247**, 324–327.

Jahn, G. A., Diolez-Bojda, F., Belair, L., Kerdlhué, B., Kelly, P. A., Djiane, J., and Edery, M. (1991a). *Biomed. Pharmacol.* **45**, 15–23.

Jahn, G. A., Edery, M., Belair, L., Kelly, P. A., and Djiane, J. (1991b). *Endocrinology (Baltimore)* **128**, 2976–2984.

Jammes, H., Gaye, P., Belair, L., and Djiane, J. (1991). *Mol. Cell. Endocrinol.* **75**, 27–35.

Johnson, R. M., Napier, M. A., Cronin, M., and King, K. (1990). *Endocrinology (Baltimore)* **127**, 2099–2103.

Jolicoeur, C., Boutin, J. M., Okamura, H., Raguet, S., Djiane, J., and Kelly, P. A. (1989). *Mol. Endocrinol.* **3**, 895–900.

Kelly, P. A., Seguin, C., Cusan, L., and Labrie, F. (1980). *Biol. Reprod.* **23**, 924–928.

Kitamura, T., Sato, N., Arai, K. I., and Miyajima, A. (1991). *Cell (Cambridge, Mass.)* **66**, 1175–1184.

Kledzik, G. S., Marshall, S., Campbell, A. A., Gelato, M., and Meites, J. (1976). *Endocrinology (Baltimore)* **98**, 373–379.

Ko, K. W. S., Cook, H. W., and Vance, D. E. (1986). *J. Biol. Chem.* **261**, 7846–7852.

Kornberg, L. J., and Liberti, J. P. (1989). *Biochim. Biophys. Acta* **1011**, 205–211.

Landron, D., Guerre-Millo, M., Postel-Vinay, M. C., and Lavau, M. (1989). *Endocrinology (Baltimore)* **124**, 2305–2313.

Laron, Z., Pertzelan, A., Karp, M., Kowadlo-Silbergeld, A., and Daughaday, W. H. (1971). *J. Clin. Endocrinol. Metab.* **33**, 332–342.

Laron, Z., Pertzelan, A., and Mannhyeimer, S. (1966). *Isr. J. Med. Sci.* **2**, 152–155.

Larsen, J. L., and Dufau, M. L. (1988). *Endocrinology (Baltimore)* **123**, 438–444.

Lesniak, M. A., and Roth, J. (1976). *J. Biol. Chem.* **251**, 3720–3729.

Lesueur, L., Edery, M., Ali, S., Paly, J., Kelly, P. A., and Djiane, J. (1991). *Proc. Natl. Acad. Sci. U.S.A.* **88**, 824–828.

162    P. A. KELLY ET AL.

Lesueur, L., Edery, M., Paly, J., Clark, J., Kelly, P. A., and Djiane, J. (1990). *Mol. Cell. Endocrinol.* **71,** R7–R12.

Leung, D. W., Spencer, S. A., Cachianes, G., Hammonds, R. G., Collins, C., Henzel, W. J., Barnard, R., Waters, M. J., and Wood, W. I. (1987). *Nature (London)* **330,** 537–543.

Li, C. H., Evans, H. M., and Simpson, M. E. (1945). *J. Biol. Chem.* **159,** 353–366.

Lunkin, P. M. (1960). *Acta Pathol. Microbiol. Scand.* **48,** 351–355.

Maes, M., De Hertogh, R., Watrin-Granger, P., and Ketelsgers, J. M. (1983). *Endocrinology (Baltimore)* **113,** 1325–1332.

Maiter, D., Underwood, L. E., Maes, M., and Ketelsgers, J. M. (1988). *Endocrinology (Baltimore)* **122,** 1291–1296.

Maiera, L., Cesano, A., Muceloli, G., and Veglia, F. (1990). *Int. J. Neurosci.* **51,** 265–267.

Mathews, L. S., Enberg, B., and Norstedt, G. (1989). *J. Biol. Chem.* **264,** 9905–9910.

Mathews, L. S., Norstedt, G., and Palmiter, R. D. (1986). *Proc. Natl. Acad. Sci. U.S.A.* **83,** 9343–9347.

Matusik, R. J., and Rosen, J. M. (1980). *Endocrinology (Baltimore)* **106,** 252–259.

Miyazaki, T., Maruyama, M., Yamada, G., Hatakeyama, M., and Taniguchi, T. (1991). *EMBO J.* **10,** 3191–3197.

Moldrup, A., Allevato, G., Dryberg, T., Nielsen, J. H., and Billestrup, N. (1991). *J. Biol. Chem.* **266,** 17441–17445.

Moldrup, A., Billestrup, N., Thorn, N. A., Lernmark, A., and Nielsen, J. H. (1989). *Mol. Endocrinol.* **3,** 1173–1182.

Mosley, B., Beckmann, M. P., March, C. J., Idzerda, R. L., Gimpel, S. D., VandenBos, T., Friend, D., Alpert, A., Anderson, D., Jackson, J., Wignall, J. M., Smith, C., Gallis, B., Sims, J. E., Urdal, D., Widmer, M. B., Cosman, D., and Park, L. S. (1989). *Cell (Cambridge, Mass.)* **59,** 335–348.

Mukherjee, P., Mastro, A. M., and Hymer, W. C. (1990). *Endocrinology (Baltimore)* **126,** 88–94.

Mullis, P. E., Lund, T., Patel, M. S., Brook, C. G. D., and Brickell, P. M. (1991). *Mol. Cell. Endocrinol.* **76,** 125–133.

Murphy, L. C., Lee-Wing, M., Goldenberg, G. J., and Shiu, R. P. C. (1987a). *Cancer Res.* **47,** 4160–4164.

Murphy, L. C., Tsuyuki, D., Myal, Y., and Shiu, R. P. C. (1987b). *J. Biol. Chem.* **262,** 15236–15241.

Nagano, M., and Kelly, P. A. (1992). *Biochem. Biophys. Res. Commun.* **183,** 610–618.

Nagy, E., and Berczi, I. (1978). *Acta Endocrinol.* **89,** 530–537.

Nagy, E., Berczi, I., and Friesen, H. G. (1983). *Acta Endocrinol.* **102,** 351–357.

Niall, H. D., Hogan, M. L., Sayer, R., Rosenblum, I. Y., and Greenwood, F. C. (1971). *Proc. Natl. Acad. Sci. U.S.A.* **68,** 866–869.

Nicoll, C. S., and Bern, H. A. (1972). *In* "Lactogenic Hormones" (G. W. E. Wolstenholme and J. Knight, eds.), pp. 299–317.

Nicoll, C. S., Herbert, M. J., and Russel, S. M. (1985). *Endocrinology (Baltimore)* **116,** 1449–1453.

Novotny, J., and Auffray, C. (1984). *Nucleic Acids Res.* **12,** 243–255.

Ofenstein, J. P., and Rillema, J. A. (1987). *Proc. Soc. Exp. Biol. Med.* **185,** 147–152.

Okamura, H., Raguet, S., Bell, A., Gagnon, J., and Kelly, P. A. (1989). *J. Biol. Chem.* **264,** 5904–5911.

Orian, J. M., Snibson, K., Stevenson, J. L., Brandon, M. R., and Herington, A. C. (1991). *Endocrinology (Baltimore)* **128,** 1238–1246.

## GROWTH HORMONE/PROLACTIN RECEPTORS

Palestine, A. G., Muellenberg-Coulombre, C. G., Kim, M. K., Gelato, M. C., and Mussenblatt, R. B. (1987). *J. Clin. Invest.* **79,** 1078–1081.

Patthy, L. (1990). *Cell (Cambridge, Mass.)* **61,** 13–14.

Peeters, S., and Friesen, H. G. (1977). *Endocrinology (Baltimore)* **101,** 1164–1179.

Pellegrini, I., Ali, S., Lebrun, J. J., and Kelly, P. A. (1992). *Mol. Endocrinol.* **6,** 1023–1031.

Picard, F., and Postel-Vinay, M-C. (1984). *Endocrinology (Baltimore)* **114,** 1328–1333.

Posner, B. I., Kelly, P. A., and Friesen, H. G. (1974). *Proc. Natl. Acad. Sci. U.S.A.* **71,** 2407–2410.

Posner, B. I., Kelly, P. A., and Friesen, H. G. (1975). *Science* **188,** 57.

Postel-Vinay, M-C., Belair, L., Kayser, C., Kelly, P. A., and Djiane, J. (1991). *Proc. Natl. Acad. Sci. U.S.A.* **88,** 6687–6690.

Postel-Vinay, M-C., Cohen-Tanugi, E., and Charrier, J. (1982). *Mol. Cell. Endocrinol.* **28,** 667–669.

Postel-Vinay, M-C., Durand, D., Lopez, S., Kayser, C., and Lavau, M. (1990). *Horm. Metab. Res.* **22,** 7–11.

Richards, J. F. (1975). *Biochem. Biophys. Res. Commun.* **63,** 292–299.

Riddle, O., Bates, R. W., and Dykshorn, S. W. (1932). *Proc. Soc. Exp. Biol. Med.* **29,** 1211–1212.

Rillema, J. A. (1980). *Fed. Proc.* **39,** 2593–2598.

Rillema, J. A., and Whale (1988). *Proc. Soc. Exp. Biol. Med.* **187,** 432–434.

Rogers, S. A., and Hammerman, M. R. (1989). *Proc. Natl. Acad. Sci. U.S.A.* **86,** 6363–6366.

Rozakis-Adcock, M., and Kelly, P. A. (1991). *J. Biol. Chem.* **266,** 16472–16477.

Rozakis-Adcock, M., and Kelly, P. A. (1992). *J. Biol. Chem.*, in press.

Russel, D. H., Matrisian, L., Kibler, R., Larson, D. F., Paulos, B., and Magun, B. E. (1984). *Biochem. Biophys. Res. Commun.* **121,** 899–906.

Sadeghi, H., Wang, B. S., Lumanglas, A., Logan, J. S., and Baumbach, W. R. (1990). *Mol. Cell. Endocrinol.* **4,** 1799–1805.

Sharon, M., Gnarra, J. R., and Leonard, W. J. (1990). *Proc. Natl. Acad. Sci. U.S.A.* **87,** 4869–4873.

Shirota, M., Banville, D., Ali, S., Jolicoeur, C., Boutin, J. M., Edery, M., Djiane, J., and Kelly, P. A. (1990). *Mol. Endocrinol.* **4,** 1136–1142.

Shiu, R. P. C., Elsholtz, H. P., Tanaka, T., Friesen, H. G., Gout, P. W., Beer, C. T., and Noble, R. L. (1983). *Endocrinology (Baltimore)* **113,** 159–165.

Shiu, R. P. C., and Friesen, H. G. (1974). *Biochem. J.* **140,** 301–311.

Shiu, R. P. C., and Iwasiow, B. M. (1985). *J. Biol. Chem.* **360,** 11307–11313.

Skorstengaard, K., Jensen, M. S., Sahl, P., Petersen, T. E., and Magnusson, S. (1986). *Eur. J. Biochem.* **161,** 441–453.

Smith, P. E. (1930). *Anat. Rec.* **47,** 119–129.

Smith, W. C., Kunioyoshi, J., and Talamantes, F. (1989). *Mol. Endocrinol.* **3,** 984–990.

Smith, W. C., Linzer, D. I. H., and Talamantes, F. (1988). *Proc. Natl. Acad. Sci. U.S.A.* **85,** 9576–9579.

Smith, W. C., and Talamantes, L. F. (1987). *J. Biol. Chem.* **262,** 2213–2219.

Snow, E. C., Feldbush, T. L., and Ouks, J. A. (1981). *J. Immunol.* **126,** 161–164.

Souyri, M., Vigon, I., Penciolelli, J-F., Heard, J-M., Tambourin, P., and Wendling, F. (1990). *Cell (Cambridge, Mass.)* **63,** 1137–1147.

Stred, S. E., Stubbart, J. R., Argelsinger, L. S., Shafer, J. A., and Carter-Su, C. (1990). *Endocrinology (Baltimore)* **127,** 2506–2516.

Stricker, P., and Gruter, F. (1928). *C. R. Soc. Biol.* **99,** 1978–1980.

164          P. A. KELLY ET AL.

Szöllösi, J., Damjanovich, S., Goldman, C. K., Fulwyler, M., Aszalos, A. A., Goldstein, G., Rao, P., Talle, M. A., and Waldman, T. A. (1987). *Proc. Natl. Acad. Sci. U.S.A.* **84,** 7246–7251.

Taga, T., Hibi, M., Hirata, Y., Yamasaki, K., Yasukawa, K., Matsuda, T., Hirano, T., and Kishimoto, T. (1989). *Cell (Cambridge, Mass.)* **58,** 573–581.

Tavernier, J., Devos, R., Cornelis, S., Tuypens, T., Van der Heyden, J., Fiers, W., and Plaecinck, G. (1991). *Cell (Cambridge, Mass.)* **66,** 1175–1184.

Thoreau, E., Petridou, B., Kelly, P. A., Djiane, J., and Mornon, J. P. (1991). *FEBS Lett.* **282,** 26–31.

Tiong, T. S., Freed, K. A., and Herington, A. (1989). *Biochem. Biophys. Res. Commun.* **158,** 141–148.

Tollet, P., Enberg, B., and Mode, A. (1990). *Mol. Endocrinol.* **4,** 1934–1942.

Takaki, S., Tominaga, A., Hitoshi, Y., Mita, S., Sonada, E., Yamaguchi, N., and Takatsu, K. (1990). *EMBO J.* **9,** 4367–4374.

Too, C. K. L., Shiu, R. P. C., and Friesen, H. G. (1990). *Biochem. Biophys. Res. Commun.* **173,** 48–52.

Too, C. K. L., Walker, A., Murphy, P. R., Cragoe, E. J., Jacobs, H. K., and Friesen, H. G. (1987). *Endocrinology (Baltimore)* **121,** 1503–1511.

Vesely, D. L. (1981). *Am. J. Physiol.* **240,** E79–D82.

Vesely, D. L. (1984). Biochem. Biophys. Res. Commun. **123,** 1084–1090.

Wiegent, D. A., Baxter, J. B., Wear, W. E., Smith, C. R., Bost, K. L., and Blalock, J. E. (1988). *FASEB J.* **2,** 2812–2818.

Yamasaki, K., Taga, T., Hirata, Y., Yawata, H., Kawanishi, Y., Seed, B., Taniguchi, T., Hirano, T., and Kishimoto, T. (1988). *Science* **241,** 825–828.

Ymer, S. I., and Herington, A. C. (1985). *Mol. Cell. Endocrinol.* **41,** 153–161.

Ymer, S. I., Kelly, P. A., Herington, A. C., and Djiane, J. (1987). *Mol. Cell. Endocrinol.* **53,** 67–73.

Yu-Lee, L. Y. (1990). *Mol. Cell. Endocrinol.* **68,** 21–28.

Zabala, M. T., and Garcia-Ruiz, J. P. (1989). *Endocrinology (Baltimore)* **125,** 2587–2593.

Zhang, R., Buczko, E., Tsai-Morris, C-H., Hu, Z-Z., and Dufau, M. (1990). *Biochem. Biophys. Res. Commun.* **168,** 415–422.

RECENT PROGRESS IN HORMONE RESEARCH, VOL. 48

# Molecular Genetics of Laron-Type GH Insensitivity Syndrome

MICHEL GOOSSENS, SERGE AMSELEM, PHILIPPE DUQUESNOY, AND MARIE-LAURE SOBRIER

*Institut National de la Santé et de la Recherche Médicale (I.N.S.E.R.M), U 91, Hôpital Henri Mondor, 94010 Créteil, France*

## I. Introduction

Defective human growth hormone (hGH) production or an absence of peripheral action can cause metabolic alterations and growth failure. Though hGH deficiency is usually idiopathic, several genetic disorders or syndromes are associated with defective hGH secretion or action. These genetic diseases can be due to alterations of the hGH gene, of distant loci that act through epistatic effects, or of genes that affect the response to hGH (Phillips, 1989). The GH insensitivity syndrome initially described by Laron *et al.* (1966) falls into this latter category.

Growth hormone acts either directly at sites such as muscle, adipose tissue, and bone, or indirectly by inducing the liver to produce insulin-like growth factor-I (IGF-I), also called somatomedin C, thereby promoting growth. GH binding to specific liver receptors is considered to be the initial event in its biological action. In theory, GH may fail to act peripherally because growth-promoting factors such as somatomedins are not generated, or because target tissues do not respond to these promoting factors. In 1966, Laron and associates reported a new form of hereditary dwarfism in which patients show clinical signs of severe isolated GH deficiency (IGHD), with delayed growth, facial abnormalities, high-pitched voice, and small genitalia in males. However, in contrast to IGHD patients, these patients have high levels of circulating GH and very low serum levels of IGF-I that do not respond to injections of exogenous GH (Laron *et al.*, 1971). These features, combined with other experimental data recently reviewed (Amselem et al., 1991a), suggest that the primary defect in this disease, also called Laron-type GH insensitivity syndrome, could be abnormal GH receptors (GH-R).

At the end of 1987, the cloning of a putative growth hormone receptor cDNA by Leung *et al.* (1987) opened the way to the characterization of the gene abnormalities responsible for this autosomal recessive GH-resistance syndrome. Cloning was based on the isolation of a protein via

165

Copyright © 1993 by Academic Press, Inc.
All rights of reproduction in any form reserved.

its ability to bind growth hormone, and the subsequent screening of a rabbit liver cDNA library using a synthetic oligonucleotide probe derived from one of the tryptic peptides. A putative human GH-R cDNA clone was subsequently isolated from a human liver cDNA library by means of the rabbit cDNA probe obtained. These cloned sequences were shown to encode proteins containing a single membrane-spanning domain, sharing sequence homology with prolactin receptors isolated from a variety of species (Boutin et al., 1988, 1989; Davis and Linzer, 1989; Edery et al., 1989) and sequence homologies with members of a new cytokine receptor family (Cosman et al., 1990). Analysis of human genomic DNA revealed that the corresponding gene contains at least 10 exons spanning 87 kb on chromosome 5 (Godowski et al., 1989).

Given the multiple biological effects of GH (Isaksson et al., 1985; Edwards et al., 1988), there was a chance that the cloning strategy used would lead to the isolation of GH receptors specific for one or other action of the hormone; it was thus necessary to demonstrate that the protein characterized by Leung et al. was involved in bone growth promotion. One way to address this issue is to perform linkage studies between GH-R gene markers and the Laron phenotype; another way is to reconstitute a functional system in vitro, making use of the cloned GH receptor in cell transfection experiments. To this end, a system for expressing the full-length human GH-R cDNA in vitro was developed in our laboratory. Using these two approaches, we have now demonstrated the responsibility of the GH-R gene in the Laron phenotype and, as a first step toward understanding receptor structure–function relationships, we have started to characterize the spectrum of GH-R mutations.

## II.   The Growth Hormone Receptor Gene and Laron's Syndrome

When a gene is suspected of being involved in a hereditary disease, two approaches can be used. Deleterious gene abnormalities present only in patients or carriers but not in normal subjects provide clear-cut proof of cause, and, of these abnormalities, deletions are the easiest to detect. When this direct approach is unsuccessful, genetic linkage between the trait and DNA markers located within or close to the candidate gene can be sought.

### A.   LINKAGE STUDIES BETWEEN THE GH-R GENE AND THE LARON PHENOTYPE

Linkage analysis is based on the simple concept that a DNA marker located close to a disease gene is more likely to be inherited with the gene through multiple meiotic events than is one located at a more distant site

LARON-TYPE SYNDROME GENETICS 167

on the same or another chromosome. This approach requires appropriate DNA markers.

To characterize such markers, GH-R gene segments amplified by means of the polymerase chain reaction (PCR) (Saiki *et al.*, 1988) were analyzed for the presence of nucleotide sequence polymorphisms using denaturing gradient polyacrylamide gel electrophoresis (DGGE), a powerful method based on the principle that the mobility of double-stranded DNA is retarded by the denaturation (melting) of any portion of the fragment under study (Myers *et al.*, 1989). When DNA migrates through a gel with an increasing denaturant concentration, its mobility will thus depend on the base composition of the first domain that melts. As a single base pair change within that domain modifies the mobility of the fragment, variant or mutant DNA fragments are readily detected.

Using this technique, we analyzed an intervening sequence (IVS) located in part of the GH-R gene that encodes the intracellular domain of the receptor. PCR products containing this 333-bp IVS were tested for DNA melting polymorphisms and several patterns were observed (Figs. 1 and 2), suggesting the presence of multiallelic polymorphisms. Six intra-

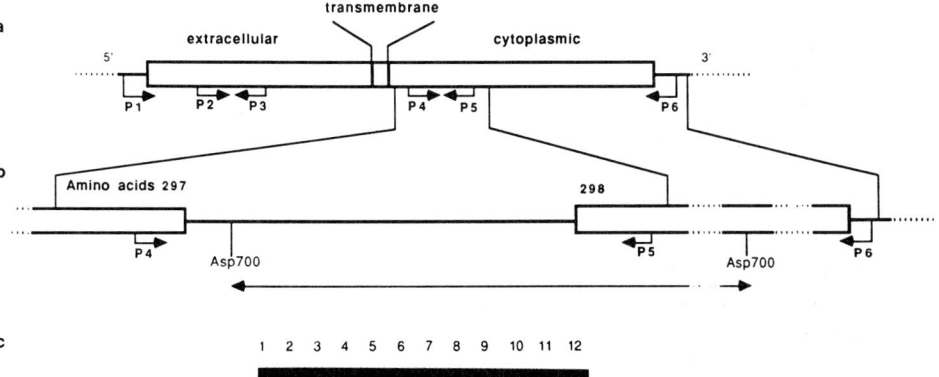

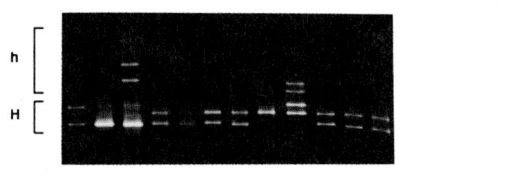

FIG. 1. Schematic representation of the putative growth hormone receptor cDNA sequence. (a) Arrows indicate the position and orientation of the oligonucleotide primers used in the polymerase chain reaction assays and sequencing reactions. (b) Intronic sequence (horizontal line) interrupting the cytoplasmic coding domain (open boxes). (c) Analysis of amplified DNA samples from 12 control subjects by electrophoresis on a denaturing gradient gel stained with ethidium bromide. The PCR products (P4–P6) were digested with the restriction enzyme *Asp*-700 before analysis. Each sample (lanes 1–12) displays homoduplex (H) and in some instances heteroduplex (h) bands.

168                    MICHEL GOOSSENS ET AL.

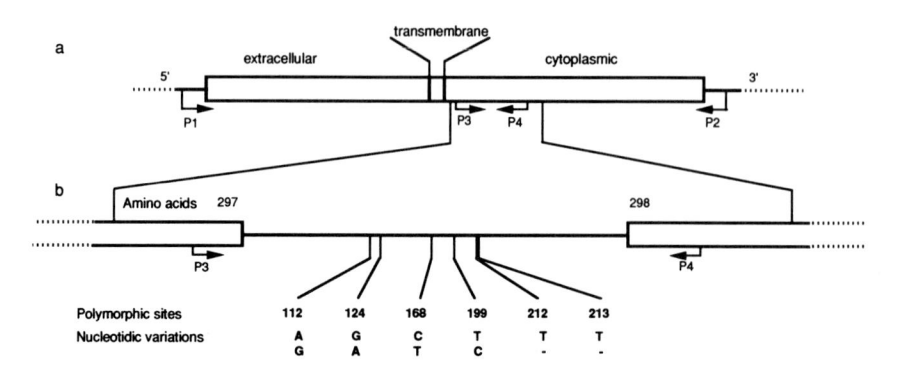

FIG. 2. Schematic representation of the GH-R cDNA sequence (a) and of an intron interrupting the GH-R gene in the cytoplasmic domain (b). The arrows indicate the position of the oligonucleotide primers used in the PCR assays and sequencing reactions. The position and nature of the nucleotide polymorphisms identified are indicated within the intervening sequence.

genic polymorphic sites were shown by direct DNA sequencing to be variously associated in several haplotypes (Table I) (Amselem *et al.*, 1989).

Linkage analysis is based on the confrontation, *a posteriori*, of the phenotype (normal or affected) with the genotype (pattern of inheritance of polymorphic DNA markers) in each family member. The parents of affected children must have different alleles of the marker so that it is possible to distinguish those inherited, respectively, from the father and the mother. We have been able to study two consanguineous families with two and three children; the markers identified were informative, i.e., present in the heterozygous state in family members, and analysis of their transmission showed that the affected children had inherited the same pair of GH-R alleles, whereas the phenotypically normal siblings had not (Fig. 3). Using "lod" score analysis (a mathematical test computing the logarithm of the odds that a particular polymorphic DNA marker is linked to the disease gene), we found a 10,000:1 probability of linkage between the GH-R DNA markers and the disease (Amselem *et al.*, 1989). These genetic findings established that the sequence isolated by Leung *et al.* mediates the growth-promoting effects of GH, and that Laron's syndrome most likely results from abnormalities in the GH-R gene. At that time, a missense mutation replacing phenylalanine by serine at position 96 (F96S) of the amino acid sequence of the extracellular domain of the growth hormone receptor was identified in one of the two families.

TABLE I
*GH Receptor Gene Haplotypes in Healthy Controls*

| Haplotypes | Location of polymorphic sites | | | | | | | Chromosomes studied | |
|---|---|---|---|---|---|---|---|---|---|
| | IVS nt 112 | IVS nt 124 | IVS nt 168 | IVS nt 198 | IVS nt 211 | IVS nt 212 | | Number | Percent |
| I | A | G | C | T | T | T | | 37 | 52.9 |
| II | A | G | C | T | — | — | | 14 | 20 |
| III | G | G | C | T | — | — | | 12 | 17.1 |
| IV | A | G | C | T | — | — | | 5 | 7.1 |
| V | A | A | C | T | T | — | | 1 | 1.4 |
| VI | A | G | T | T | T | T | | 1 | 1.4 |
| | | | | | | | Total | 70 | 100 |

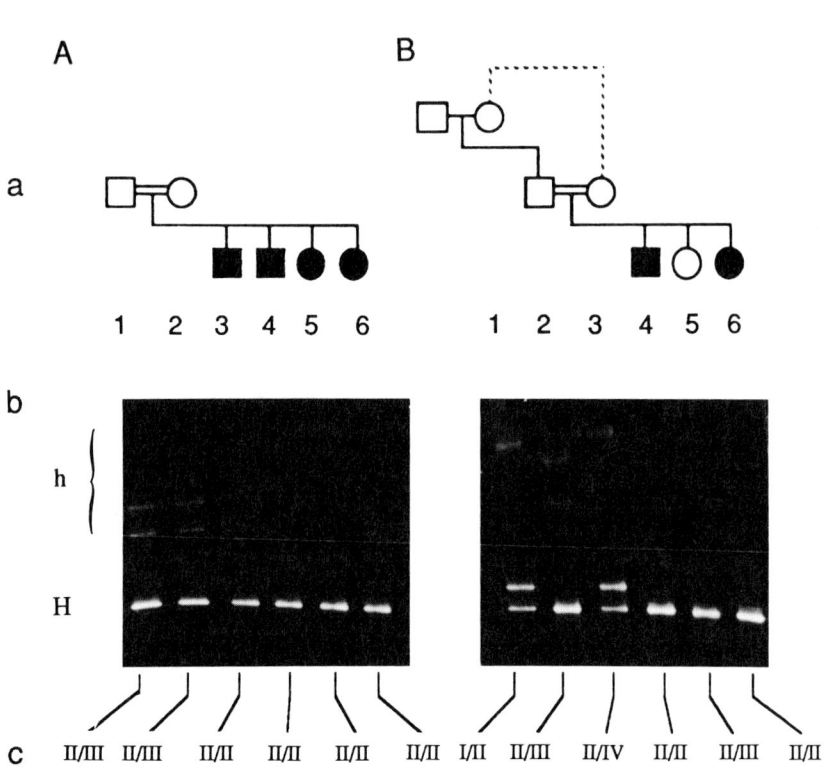

FIG. 3. (a) Pedigrees of two families (A and B) with Laron's syndrome; (b) denaturing gradient gel electrophoresis of amplified DNA samples as shown in Fig. 1; (c) GH-R gene haplotypes established by direct sequencing. The two families are informative for the DNA polymorphisms: the parents are heterozygous and have heteroduplexes (both families) or two different homoduplexes (family B). Only affected children (homozygous for a mutant allele) have patterns with no heteroduplexes (h) and with a single major band (i.e., homoduplex H). In both families, sequence analyses of the gene fragment studied indicated that the same haplotype (II) (Table I) consegregated with the Laron phenotype.

## B. IDENTIFICATION OF GH-R STOP MUTATIONS IN PATIENTS WITH LARON'S SYNDROME

Another direct way of determining the involvement of GH-R in Laron's syndrome is based on detecting gross rearrangements or deletions at the GH-R locus with the use of GH-R cDNA probes in Southern blotting experiments. Using this approach, Godowski et al. (1989) showed that, following hybridization with a GH-R-specific probe, genomic DNA from two patients with Laron's phenotype had partial or complete deletion of exons 3, 5, and 6 corresponding to parts of the extracellular domain of the receptor probably involved in hormone binding. However, because alternative splicings can occur during expression of the GH-R gene (Leung *et al.*, 1987; Godowski *et al.*, 1989), it remained possible that a shorter mRNA could give rise to a functional receptor. To obtain direct evidence that GH-R is involved in this disease, we used a strategy based on the identification of unambiguous molecular defects, i.e., stop mutations altering the coding sequence of the gene. Because recurrent mutations occur frequently at CpG sites (Cooper and Krawczak, 1989), we screened the potentially hypermutable CpG dinucleotides by focusing on CGA (arginine) codons, which are prone to generate TGA stop codons (cytosine can be methylated and subsequently deaminate spontaneously to thymine). Only two exons (4 and 7) in the GH-R coding sequence contain such a codon (Fig. 4). We found the same point mutation, an arginine 43 → stop defect (R43X), in two unrelated patients, whereas another stop mutation (C38X) that did not involve a CpG dinucleotide was observed in a third affected family (Amselem *et al.*, 1991b). The nature and position of these mutations preclude the expression of functional putative GH-R, which contains 620 residues, because codons 38 and 43 lie early in the amino terminus of the extracellular domain of GH-R and a nonsense mutation at

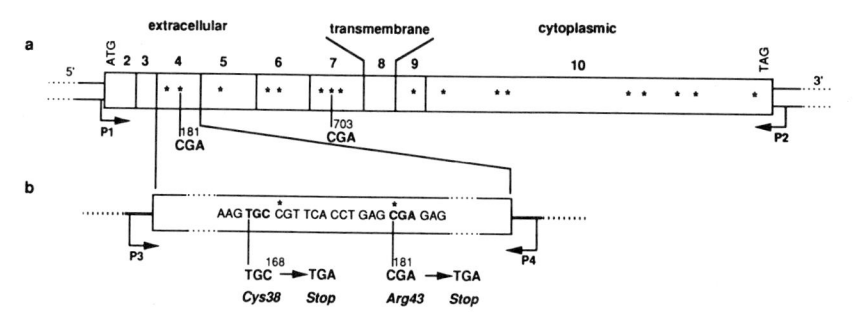

FIG. 4. (a) Diagrammatic representation of the human growth hormone receptor cDNA. (b) Nucleotide and amino acid changes found within exon 4 in patients with Laron-type insensitivity syndrome. The open reading frame is indicated by an open box, the untranslated mRNA region by double lines, and introns by thick black lines. The numbers above the cDNA indicate the corresponding exons according to Godowski *et al.* (1989). The arrows indicate locations and directions of PCR primers used in DNA amplifications. The position of each CpG dinucleotide is marked by an asterisk within the coding sequence. Two CpG doublets (nucleotides 181–182 and 703–704) lie in a CGA codon below the cDNA in exons 4 and 7, respectively. A partial sequence of exon 4 is represented within the open box at the bottom.

these positions would result in a truncated protein and delete a large portion of both the GH binding domain and the full transmembrane and intracellular domains. Identification of such genetic defects in patients with Laron's syndrome confirmed the role of this GH-R in the transduction of the growth signal.

## C.  MOLECULAR HETEROGENEITY ASSOCIATED WITH GH INSENSITIVITY

Analysis of the intragenic DNA markers associated with the different mutant alleles provided evidence that the molecular defects originated independently. Indeed, different mutations were found to be associated with the same chromosomal background. Two mutations were recurrent: in two instances, the same mutation was observed in unrelated patients with different GH receptor DNA markers. One of them, R43X, involves a CpG dinucleotide; this observation supports the hypothesis that these sequences represent hot spots for mutations in the GH-R gene. Overall, five molecular defects of the GH-R gene located within the coding sequence have been identified in Laron-type GH insensitivity (Table II). In addition to the four abnormalities cited above, a fifth mutation that alters RNA splicing has recently been described in an inbred population in southern Ecuador (Berg and Francke, 1991).

Finally, it should also be borne in mind that Laron-type GH insensitivity might also be caused by alterations of genes other than that encoding GH-R, e.g., those controlling GH-R gene expression. This possibility can only be tested by analysis of the genetic linkage between the disease and GH-R gene markers in affected families.

TABLE II

*Molecular Defects in the GH-R Gene from Patients with Laron's Syndrome*

| Molecular defect | Number of patients | Origin | Consanguinity | GH-R haplotype |
|---|---|---|---|---|
| Exons 3, 5, and 6 deleted | 2 | Middle East | + | ? |
| Arg 43 → stop | 2 | Mediterranean | + | II |
| | | Mediterranean | + | III |
| Cys 38 → stop | 1 | Northern European | + | I |
| Phe 96 → Ser | 1 | Mediterranean | + | II |
| Glu 180 → Glu (GAA → GAG) | 1 | South American | + | ? |

## III. Decoding Structure–Function Relationships of GH-R

Though nonsense mutations provide no information as to the structure–function relationships of the different domains of GH-R, missense mutations should be more informative. In one of the families with Laron's syndrome that we studied, a mutation replacing phenylalanine by serine was found at position 96, i.e., within the extracellular domain of GH-R (Amselem et al., 1989). The fact that this phenylalanine is conserved through evolution among the members of the transmembrane receptor family, in particular the growth hormone and prolactin receptors of the different species studied (Leung et al., 1987; Boutin et al., 1988, 1989; Baumbach et al., 1989; Davis and Linzer, 1989; Edery et al., 1989; Mathews et al., 1989; Smith et al., 1989; Adams et al., 1990; Cioffi et al., 1990; Hauser et al., 1990), strongly suggested that this F96S mutation was deleterious (Fig. 5). In order to determine its relevance to the GH-resistance phenotype, we studied COS-7 cells transiently transfected with wild-type or mutant receptor cDNAs encoding the full amino acid sequence of the molecule (Duquesnoy et al., 1991). To determine whether the GH-R proteins expressed could bind GH, plasma membranes of transfected COS-7 cells were incubated with $^{125}$I-labeled hGH and assayed for specific GH-binding activity. A high level of $^{125}$I-labeled hGH binding was observed in cells transfected with phGHRwt, an expression vector harboring the full-length wild-type hGH-R cDNA (Fig. 6). Scatchard analysis showed that membrane extracts from transfected COS-7 cells bound $^{125}$I-labeled hGH with an affinity ($K_a = 1.5 \times 10^9 \ M^{-1}$) similar to that reported for human GH-R. Because growth hormone binding protein (GH-BP) may derive from the extracellular hormone-binding domain of the receptor (Leung et al., 1987) and given that GH-binding activity is absent from the serum of patients who bear the F96S mutation—a finding in agreement with observations made in several cases investigated previously (Daughaday and Trivedi, 1987)—we expected this alteration of GH-R structure to impair GH binding. Indeed, plasma membranes of the COS-7 cells expressing the mutant full-length GH-R cDNA showed no GH-binding activity, but this was difficult to reconcile with the report by Bass and co-workers (Bass et al., 1991) that, when expressed in Escherichia coli, the extracellular domain of the mutant hGH-R has the same affinity as the wild-type binding protein. Nonetheless, the expression of a truncated molecule could have modified the influence of the variant amino acid in Bass's study, and proteins expressed in bacteria may not undergo the posttranslational modifications that may be involved in normal receptor binding. This latter point was tested indirectly by Fuh et al. (1990), who found that E. coli was able to express a correctly folded extracellular GH-

174 MICHEL GOOSSENS ET AL.

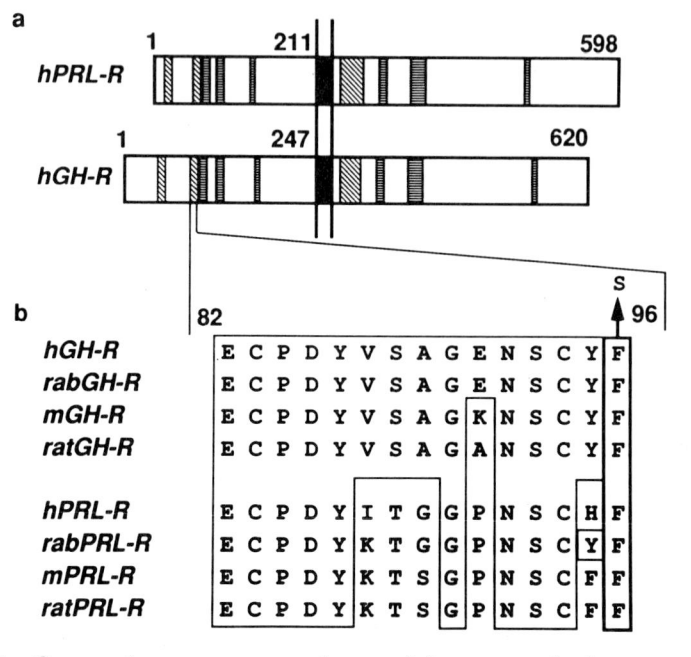

FIG. 5. Conserved sequences among the growth hormone–prolactin receptor family. (a) Diagrammatic representation of the human growth hormone receptor (hGH-R) and human prolactin receptor (hPRL-R). Numbers indicate the positions of the amino-terminal amino acid, the first amino acid of the transmembrane region (black), and the last amino acid of the mature receptor. Diagonal stripes indicate regions of high similarity between the two molecules (>68%), and horizontal stripes indicate those of moderate similarity (38–60%). (b) Sequence comparisons of a conserved domain (amino acids 82–96) between the human GH-R (*hGH-R*), rabbit GH-R (*rabGH-R*), mouse GH-R (*mGH-R*), rat GH-R (*ratGH-R*), human PRL-R (*hPRL-R*), rabbit PRL-R (*rabPRL-R*), mouse PRL-R (*mPRL-R*), and rat PRL-R (*ratPRL-R*). Sequences that exactly match the hGH-R are boxed and residues are numbered according to the hGH-R sequence. The Phe residue that is substituted (Phe → Ser) in a patient with Laron's syndrome is invariant at position 96.

R domain, as judged by its ability to bind hGH. Indeed, the unglycosylated extracellular domain had binding properties virtually identical to those of its natural glycosylated counterpart isolated from human serum, suggesting that glycosylation was not essential for GH binding.

The discrepancy between Bass's results and our own could be explained by a postranslational modification of the substituted Ser-96, which can only occur in eukaryotic cells and which, in turn, would be responsible for the lack of GH-binding activity observed. To test this hypothesis, we investigated the effect of substituting alanine at this position. Alanine is

# LARON-TYPE SYNDROME GENETICS

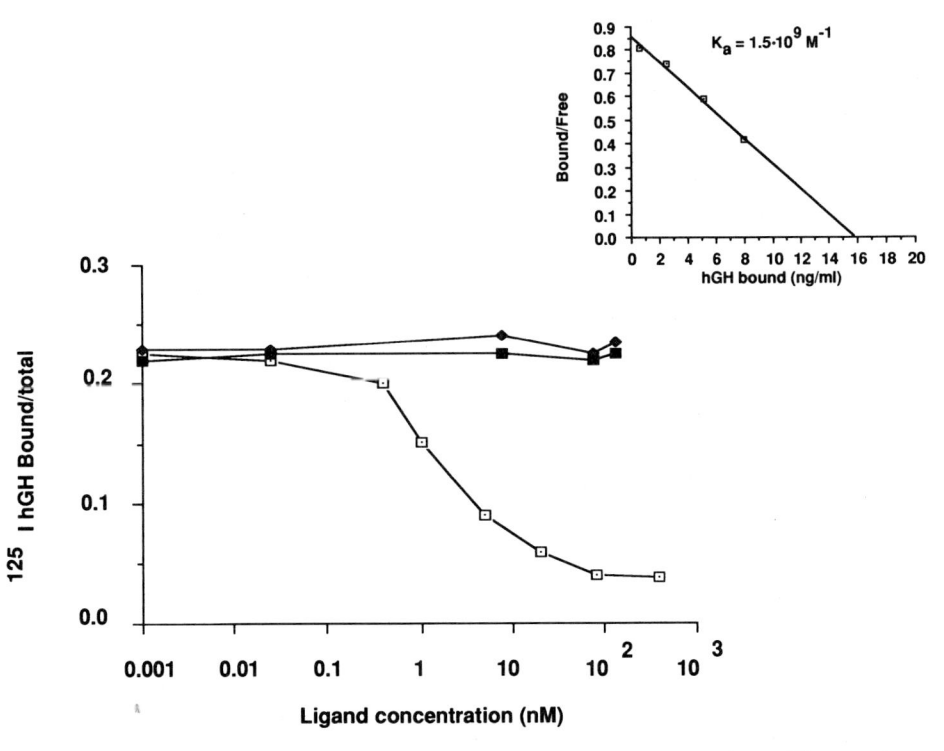

FIG. 6. Expression of the hGH-R cDNA clone in COS-7 cells. Binding of [125]I-labeled hGH to membranes from COS-7 cells transfected with phGHRwt was measured in the presence of increasing amounts of unlabeled hGH (open boxes), bGH (filled boxes), and hPRL (diamonds). No specific binding was observed in nontransfected COS-7 cells. The results are expressed as the percentage of total dpm bound per 100 μg of protein. *Inset:* Scatchard plot of the competition assay with unlabeled hGH.

structurally very similar to serine but lacks the hydroxyl radical, which is a potential target for posttranslational modification. Interestingly, this substitution did not restore GH-binding activity to COS-7 cell membranes. Therefore, given the existence of specific eukaryotic protein transport systems, we investigated whether the mutant GH-R was present or absent at the plasma membrane. Using specific monoclonal antibodies directed against the GH-R transiently expressed in COS-7 cells, we showed that specific immunostaining was detected in the cytosol only. In contrast, in cells transfected with the wild-type GH-R, both the cytosol and the plasma membrane were found to exhibit specific immunostaining (Fig. 7).

Taken together, these preliminary results strongly suggest that Phe-96, which is conserved among the GH, prolactin, and erythropoietin recep-

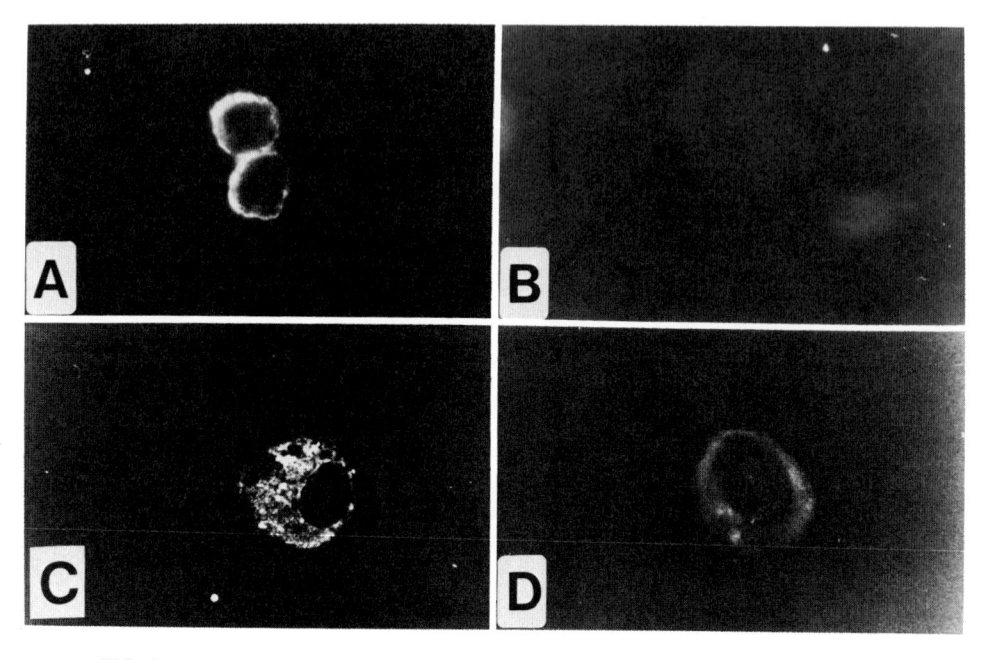

FIG. 7. Immunolocalization of wild-type and mutant hGH-R. Indirect immunofluorescence staining with a monoclonal antibody specific to hGH-R. COS-7 cells were transfected with phGHRwt (A and C) or phGHR (F96S) (B and D) and labeling was visualized by immunofluorescence microscopy before (A and B) and after cell permeabilization (C and D).

tors, is essential for normal intracellular GH-R trafficking. Further studies will be necessary to dissect the complex process of intracellular trafficking and identify the step at which the F96S mutation interferes.

In conclusion, progress in elucidating the molecular mechanisms involved in Laron's syndrome has been rapid. To date, five molecular defects have been characterized. Although the nature of the mutation is different (i.e., deletions, nonsense, missense, or splicing mutations), all alter GH binding. Other mutations that would alter the domains of GH-R involved in signal transduction can be anticipated and may be found in rare patients who have detectable circulating GH-BP (Buchanan *et al.*, 1991). The results so far obtained should be helpful in the genetic counseling of affected families and provide insight into receptor structure–function relationships. Ultimately, they will help to decode growth signal regulation and transduction.

## LARON-TYPE SYNDROME GENETICS

### REFERENCES

Adams, T. E., Baker, L. B., Fiddes, R. J., and Brandon, M. R. (1990). *Mol. Cell. Endocrinol.* **73**, 135–145.

Amselem, S., Duquesnoy, P., Attree, O., Novelli, G., Bousnina, S., Postel-Vinay, M.-C., and Goossens, M. (1989). *N. Engl. J. Med.* **321**, 989–995.

Amselem, S., Duquesnoy, P., and Goossens, M. (1991a). *Trends Endocrinol. Metab.* **2**, 35–40.

Amselem, S., Sobrier, M.-L., Duquesnoy, P., Rappaport, R., Postel-Vinay, M.-C., Gourmelen, M., Dallapiccola, B., and Goossens, M. (1991b). *J. Clin. Invest.* **87**, 1098–1102.

Bass, S. H., Mulkerrin, M. G., and Wells, J. A. (1991). *Proc. Natl. Acad. Sci. U.S.A.* **88**, 4498–4502.

Baumbach, W. R., Horner, D. L., and Logan, J. S. (1989). *Genes Dev.* **3**, 1199–1205.

Berg, M. A., and Francke, U. (1991). *In* Proceedings, 8th Int. Soc. Hum. Genet., Washington, p. 182.

Boutin, J. M., Edery, M., Shirota, M., Jolicoeur, C., Lesueur, L., Ali, S., Gould, S., Djiane, J., and Kelly, P. A. (1989). *Mol. Endocrinol.* **3**, 1455–1461.

Boutin, J. M., Jolicoeur, C., Okamura, H., Gagnon, J., Edery, M., Shirota, M., Banville, D., Dusanter-Fourt, I., Djiane, J., and Kelly, P. A. (1988). *Cell (Cambridge, Mass.)* **53**, 69–77.

Buchanan, C. R., Maheshwari, H. G., Norman, M. R., Morrell, D. J., and Preece, M. A. (1991). *In* "Proceedings 2nd International IGF Symposium, San Francisco," Abstract D18.

Cioffi, J. A., Wang, X., and Kopchick, J. J. (1990). *Nucleic Acids Res.* **18**, 6451.

Cooper, D. N., and Krawczak, M. (1989). *Hum. Genet.* **83**, 181–188.

Cosman, D., Lyman, S. D., Idzerda, R. L., Beckmann, M. P., Park, L. S., Goodwin, R. G., and March, C. J. (1990). *Trends Biochem. Sci.* **15**, 265–270.

Daughaday, W. H., and Trivedi, B. (1987). *Proc. Natl. Acad. Sci. U.S.A.* **84**, 4636–4640.

Davis, J. A., and Linzer, D. I. H. (1989). *Mol. Endocrinol.* **3**, 674–680.

Duquesnoy, P., Sobrier, M. L., Amselem, S., and Goossens, M. (1991). *Proc. Natl. Acad. Sci. U.S.A.* **88**, in press.

Edery, M., Jolicoeur, C., Levi-Meyrueis, C., Dusanter-Fourt, I., Pétridou, B., Boutin, J.-M., Lesueur, L., Kelly, P. A., and Djiane, J. (1989). *Proc. Natl. Acad. Sci. U.S.A.* **86**, 2112–2116.

Edwards, C. K., Ghiasuddin, S. M., Schepper, J. M., Yunger, L. M., and Kelley, K. W. (1988). *Science (Washington, D.C.)* **239**, 769–771.

Fuh, G., Mulkerrin, M. G., Bass, S., McFarland, N., Brochier, M., Bourell, J. H., Light, D. R., and Wells, J. A. (1990). *J. Biol. Chem.* **265**, 1–5.

Godowski, P. J., Leung, D. W., Meacham, L. R., Galgani, J. P., Hellmiss, R., Keret, R., Rotwein, P. S., Parks, J. S., Laron, Z., and Wood, W. I. (1989). *Proc. Natl. Acad. Sci. U.S.A.,* **86**, 8083–8087.

Hauser, S. D., McGrath, M. F., Collier, R. J., and Krivi, G. G. (1990). *Mol. Cell. Endocrinol.* **72**, 187–200.

Isaksson, O. G. P., Eden, S., and Jansson, J.-O. (1985). *Annu. Rev. Physiol.* **47**, 483–499.

Laron, Z., Pertzelan, A., and Mannheimer, S. (1966). *Isr J. Med. Sci.* **2**, 152–155.

Leung, D. W., Spencer, S. A., Cachianes, G., Hammonds, R. G., Collins, C., Henzel, W. J., Barnard, R., Waters, M. J., and Wood, W. I. (1987). *Nature (London)* **330**, 537–543.

Mathews, L. S., Enberg, B., and Norstedt, G. (1989). *J. Biol. Chem.* **264**, 9905–9910.

Myers, R. M., Sheffield, V. C., and Cox, D. R. (1989). *In* "PCR Technology: Principles and Applications for DNA Amplification" (H. A. Erlich, ed.), pp. 71–88. Stockton Press, New York.

Phillips, J. A. (1989). *In* "The Metabolic Basis of Inherited Disease" (C. R. Scriver, A. L. Beaudet, W. S. Sly, and D. Valle, eds.), 6, II, pp. 1965–1983. McGraw–Hill, New York.

Saiki, R., Gyllensten, U., and Erlich, H. (1988). *In* "Genome Analysis, a Practical Approach" (K. Davies, ed.), pp. 11–152. IRL Press, Oxford.

Smith, W. C., Kuniyoshi, J., and Talamantes, F. (1989). *Mol. Endocrinol.* **3,** 984–990.

RECENT PROGRESS IN HORMONE RESEARCH, VOL. 48

# Conventional and Nonconventional Uses of Growth Hormone

BARBARA M. LIPPE AND JON M. NAKAMOTO

*Department of Pediatrics, Division of Endocrinology and Metabolism, UCLA School of Medicine, Los Angeles, California 90024*

## I. Introduction

Human growth hormone (GH) is a 191-amino acid protein that is synthesized in, and secreted from, the anterior pituitary somatotrope under the dual regulation of GH-releasing hormone and somatostatin. It is released in a pulsatile fashion and transported in serum by GH-binding proteins, the primary one of which is related to the GH receptor. Its physiologic regulation and multiple actions, both direct and indirect through the complex insulin-like growth factors (IGFs) and their respective binding proteins, constitute a vast field of study that is continuously being enriched by new research. This discussion will focus on the clinical uses of GH in humans in the context of its conventional use, to promote growth in children, and its developing nonconventional use, to improve metabolic status in children and adults. These can be summarized under the broad headings outlined in Table I.

## II. Development of GH as a Pharmacologic Agent

### A. PITUITARY GROWTH HORMONE

Gemzell and Heijkenskjöld (1956), Li and Papkoff (1956), and Raben (1957) all successfully isolated and purified human pituitary GH, with the latter two groups recognizing the species-specific nature of the molecule. Reports of its short-term use in the treatment of human pituitary dwarfism followed quickly thereafter (Raben, 1958; Escamilla *et al.,* 1961; Soyka *et al.,* 1964; Seip and Trygstad, 1966; Moe, 1968).

Early studies were designed not only to document efficacy of growth promotion but also to assess the metabolic effects of the therapy (Shepard *et al.,* 1960; Raben, 1959; Prader *et al.,* 1968; Zachmann, 1969; Fernandez *et al.,* 1970; Novak *et al.,* 1972; Collipp *et al.,* 1973). Decreases in total body fat and extracellular water, increase in nitrogen balance, alterations

179

Copyright © 1993 by Academic Press, Inc.
All rights of reproduction in any form reserved.

## TABLE I
### Uses of Growth Hormone

Conventional uses to increase stature
  Classical growth hormone deficiency
  Non-growth hormone-deficient short stature
  Turner syndrome
  Chronic renal failure
  Syndromes with associated short stature
  Medical conditions with associated short stature

Nonconventional uses to improve metabolic status
  Growth hormone-deficient adult
  Obesity
  Aging
  Ovulation induction
  Lactation failure
  Critical illness or injury
  Glucocorticoid therapy
  Modulation of immunologic status

in amino acid concentrations, decreased skin fold thickness, and increased lean body mass were all demonstrated. Later reports (to be discussed below) focused on the longer term efficacy of growth hormone to promote linear growth.

In many of these early studies the diagnosis of pituitary dwarfism was made clinically, because assay techniques for the measurement of circulating growth hormone followed its therapeutic use by several years. It should be pointed out that the first clinically applicable assay, a hemagglutination inhibition method, was reported by Read and Bryan (1960). This was followed by application of the radioimmunoassay technique, which remained the standard for more than 35 years (Glick *et al.,* 1963).

It must be emphasized that the supply of human pituitary GH for clinical use was dependent on the availability of cadaveric pituitaries, the collection of which required the cooperation of a large number of pathologists. In the United States the task of coordinating the collection and distribution was undertaken by the NIH-supported National Pituitary Agency (NPA), with other countries having similar pituitary programs under their respective health councils. Supply was always limited and criteria for distribution of material were established to ensure treatment of the most severely affected individuals with the most efficient dosing schedule. Thus, save for short-term studies, the major use of pituitary GH between 1960 and 1985 was in the treatment of GH-deficient children.

USES OF GROWTH HORMONE          181

The diagnostic laboratory criteria for what constituted GH deficiency included serum GH responses to provocative stimuli (reviewed by Frasier, 1974), which were developed without long-term prospective data that correlated the peak response with therapeutic outcome. In addition, the assay methods for GH concentrations in serum began to vary in the early 1980s, as the methods changed from using the traditional polyclonal GH antibodies and standards prepared by the National Hormone and Pituitary Program (formerly the NPA) or the WHO, to commercial methods employing new polyclonal or monoclonal antibodies and new detection methods, such as the immunoradiometric technique (Reiter et al., 1988). Treatment practices were also somewhat arbitrary because there were few pharmacokinetic data to support the accepted treatment practice of giving the hormone by intramuscular injection three times a week (Frasier et al., 1969). Finally, the accepted dose–response curve for efficacy in the first year was a function of the minimum effective dose employed as well as the maximum doses that were distributed by the National Pituitary Agency at the time (Frasier et al., 1981).

## B.  BIOSYNTHETIC GROWTH HORMONE

With the successful application of molecular biologic techniques to the methodology necessary to express and then produce human GH by recombinant DNA technology (Goeddel et al., 1979), the current era of plentiful GH availability began. The biosynthetic material first contained an additional methionyl residue (met-hGH), which was a function of the methodology to effect the cloning and synthesis strategy in Escherichia coli (Flodh, 1986). No significant metabolic differences or growth-promoting differences between this synthetic hormone, somatrem, and human pituitary GH were demonstrated (Rosenfeld et al., 1982; Skottner et al., 1984). Subsequently, methodology progressed in E. coli to produce a biosynthetic human GH (rhGH) that lacked the methionyl residue, somatropin (Fryklund, 1987). Other strategies were used to produce this same hormone in mammalian cells (Facklam et al., 1988). However, the initial preparations from E. coli were highly antigenic. Though this led to extensive investigation into the question of safety and efficacy, antibodies to any of the recombinant preparations proved to be of little pharmacologic or clinical significance, save for the known development and growth attenuation in familial cases of the GH gene deletion (Jørgensen, 1991; Nishi et al., 1984) and the occasional example of what appeared to be growth-attenuating antibodies with the early preparations in an idiopathic GH-deficient child (Okada et al., 1987). Thus, the question of which therapeutic agent is currently employed in which study is less relevant than the experi-

182    BARBARA M. LIPPE AND JON M. NAKAMOTO

mental design, and we will therefore use GH as a designation for growth hormone regardless of its origin or method of synthesis.

It is beyond the scope of this progress report to assess all of the human metabolic and endocrine studies that have been undertaken with either human pituitary or biosynthetic GH. Similarly, it is impossible to review all the studies that assess efficacy of these growth hormones on short-term and long-term growth promotion [with much of this covered by Frasier (1983) and Jørgensen (1991)]. Instead, we will primarily discuss those data that reflect our cumulative progress to date and those data that may direct us to future areas of investigation.

Finally, it must be pointed out that the rapid transition from human GH to biosynthetic GH was not determined primarily by availability or efficacy, but rather by the events that followed the report that human pituitary GH was the likely source of transmission of the agent that caused Creutzfeldt–Jacob disease (CJD) in a GH recipient (Koch *et al.*, 1985; Committee on GH use report, 1985). The human material was withdrawn from distribution, first in the United States in early 1985, and then, over time, from most of the countries where it was available. Commercially available biosynthetic hormone, under clinical trials in the United States and in many other countries, was approved for sale within the year. At present, seven neuropathologically confirmed cases of CJD have been detected in the epidemiologic follow-up of 6284 recipients of GH distributed through the NHPP in the United States (Fradkin *et al.*, 1991), six CJD deaths reported among 1908 registered GH recipients in the United Kingdom (Buchanan *et al.*, 1991), and between 4 and 6 other cases documented or suspected. The data from the United States epidemiologic study indicate that all seven cases started therapy before 1970 and were treated with a median duration of 100 months (which was significantly longer than the 41-month median of all patients treated before 1970). Because the incubation period appears to range from 10 to over 20 years, and low-level infectivity among multiple hormone preparations may have occurred randomly over many years, there is a real possibility that a large number of patients, worldwide, remain at risk.

### III.    Conventional Uses of Growth Hormone: To Increase Stature

### A.    CLASSICAL GROWTH HORMONE DEFICIENCY

As stated earlier, the treatment of patients with so-called pituitary dwarfism was begun before there were means to assess circulating GH concentrations. Pituitary dwarfism was recognized as a clinical entity and

## USES OF GROWTH HORMONE

metabolic studies were performed prior to and after treatment (Shepard *et al.*, 1960). Following the introduction of assay systems for circulating GH, upper limits for the GH response to provocative stimuli were added to the laboratory diagnostic criteria (Frasier, 1974). The present authors contend (Lippe and Frasier, 1989) that classical growth hormone deficiency (GHD), whether idiopathic (hypothalamic, pituitary, or genetic) or organic (hypothalamic or pituitary), isolated, or a part of a multiple hormone deficiency, does exist as classically and clinically defined (Kaplan, 1989). We acknowledge that overlapping with these classically deficient individuals there may be children with a continuum or spectrum of growth hormone secretion patterns that may influence their growth response (Hindmarsh *et al.*, 1987) and be correlated with their stature, in that very tall children tend to secrete more GH than very short children (Albertsson-Wikland *et al.*, 1983). We also acknowledge that the spectrum of etiology for classic growth hormone deficiency is expanding, as an increasing number of children are surviving therapy for childhood malignancy and developing degrees of hypopituitarism as a consequence of their treatments. Nevertheless, that such groups may exist need not, and should not, alter the clinical classification of classical GHD.

The data on the efficacy of GH to promote growth in classical GHD divide along the lines of short-term effects (usually 1 year of treatment), longer duration effects (2 or more years of treatment), and very long-term effects (usually 4 or more years of treatment and/or reports to final height). Data assessment often suffers from the differences in treatment protocols, including dose, schedule, and concomitant medications, as well as clinical differences, including age, growth velocity, bone age, and pubertal status on entry. Thus, the first large-scale attempt to assess the effects of GH on growth included patients from 11 different centers, treated with seven different preparations of GH, and in patients varying in age from 2 to 17 years (Henneman, 1968). Few conclusions could be drawn from that early experience save that growth responsiveness appeared to wane over time. This was confirmed in subsequent early reports (Soyka *et al.*, 1970; Shizume *et al.*, 1970). The first report of a United States prospective collaborative study of growth hormone efficacy appeared in 1972 (Aceto *et al.*, 1972). Seventy-one patients were studied and several conclusions emerged. There did not appear to be a significant difference between the response of the patients with idiopathic or organic causes for the deficiency, an observation that is not found in all reports (and may depend, among studies, on variables such as age, bone age, dose of corticosteroids, and disease treatment such as radiation or chemotherapy). There was a very apparent dose effect (between the groups receiving two different dose schedules); what was described as a negative correlation with age, height, and weight may well have all been dose related, because the majority of the patients received a fixed dose, not adjusted for weight. Therefore, the fixed-dose feature made it very difficult to evaluate the relative predictiveness of auxologic variables. Following this report, few large-scale prospective studies of well-characterized and similarly treated patients were reported and usually when this occurred it was in the context of some therapeutic manipulation, such as intermittent treatment (Preece and Tanner, 1977), or with the same confounding feature of a fixed dose (Ranke et al., 1979; Wit et al., 1986). Thus, in the era of human pituitary GH treatment there was a significant gap between initial studies and final height assessments, and the questions of optimum treatment schedules and variables predictive of long-term response were not answered.

The long-term final outcome reports (Burns et al., 1981; Joss et al., 1983; Dean et al., 1985; Bourguignon et al., 1986; Bundak et al., 1988; Hibi and Tanaka, 1989; Libber et al., 1990) do shed some light, retrospectively, on the efficacy of treatment and the predictors of outcome. No study has a treated patient population that reached a mean height of either the population mean or the mean adjusted for midparental height, with the average mean height achieved being slightly more than $-2$ SD ($-12$ to $-14$ cm) below the midparental mean. Almost all investigators concluded that the degree of height loss at the onset of therapy [as determined according to Ht standard deviation score (SDS) for bone age] correlated negatively with final outcome (stressing the importance of early diagnosis and treatment) and that the initial height gained in the early years of treatment represented the major catch-up to be attained with treatment. The issue of the timing and duration of the pubertal growth phase and its effect on final height was addressed in several reports (Burns et al., 1981; Joss et al., 1983; Bourguignon et al., 1986; Hibi et al., 1989a). The consensus appears to be that the later puberty begins, the taller the final height. This is supported, in part, by data from untreated patients with idiopathic hypopituitarism whose final height is not much shorter than the heights in the reports cited above, but achieved at the expense of a protracted pubertal growth period (Van Der Werff Ten Bosch and Bot, 1990). In addition, there is a suggestion that when puberty develops spontaneously in GH-deficient individuals being treated with GH that the duration is shorter and the height gain less than in normal individuals (Hibi et al., 1989b, Darendeliler et al., 1990). Almost nothing additional can be said about dose and frequency from the studies with human GH due to the limitations and protocol constraints described above.

The results of the outcome data from patients treated with pituitary GH provided very clear implications for future studies with biosynthetic growth hormone. Efficacy in growth promotion was the first question

## USES OF GROWTH HORMONE

asked, and within several years of biosynthetic GH availability, the literature was replete with reports from countries throughout the world (only a few of which will be cited) of large multicenter studies of the efficacy of somatrem (Kaplan et al., 1986; Takano et al., 1986; Bierich, 1986; Milner, 1986) and somatropin (Albertsson-Wikland, 1987; Ranke and Bierich, 1987; Frasier et al., 1992). In all reports, first-year growth responses were either equal to or greater than those achieved with pituitary growth hormone. In the earliest report of longer term efficacy (Kaplan et al., 1986) there was some waning effect between year one and two, but good maintenance of the response in the third and fourth years. Dose, route of administration, and frequency of injection were the next issues addressed, and there, too, results were rapid in their appearance and have been recently reviewed (Jorgensen, 1991). Therefore, we will emphasize the conceptual results rather than the specifics of each report.

The first concept is that the dose requires some uniformity in expression. When extracted pituitary GH was assessed, a bioassay of activity was used and the results were given in standardized International Units (IU). Batches thus varied in potency due to extraction methods and IU dosing was used to account for these differences. Nevertheless, bioconversions are estimations, at best, and make early dose comparisons difficult. Recombinant GH can be accurately quantified, in milligrams, so that once a transition is made in determining biopotency, milligram dosing could be standard. The first preparations of somatrem were assessed to have a biopotency of 2 IU/mg; the biopotency of the product currently used clinically in the United States is given as approximately 2.6 IU/mg (on the product label for Protropin). For the somatropin preparation used in the United States (Humatrope) the conversion is also said to calculate to 2.6 IU/mg, whereas some of the somatropin preparations available in Europe (Genotropin, Norditropin, and Saizen) are listed as between 2.7 and 3 IU/mg. Second, dosing comparisons need also to be made based on whether body weight is used or whether the dose is given on a weight/height basis (surface area in square meters). Regression equations and reasonable approximation equations have been derived by Preece (1990). Finally, dosing should have some relationship to physiology; that is, whereas the pharmacokinetics of injected GH will be very different from pulsatile endogenous secretion, some relationship between production rate and replacement dose might be helpful. The approximation for production rate used by K. Albertsson-Wikland (personal communication) from data synthesized from her clinical studies and the literature is 2–3 IU/24 hours in the prepubertal child and 2–12 IU/24 hours in the pubertal child. When corrected for surface area, this production rate ranges from 0.1 IU/kg/day in the prepubertal child to 0.4 IU/kg/day during maximum secre-

186    BARBARA M. LIPPE AND JON M. NAKAMOTO

tion in puberty. When corrected for the dosage equivalent used at the time
of her studies (2.0 IU/mg), the rate can vary from 0.05 to 0.20 mg/kg/
day. Injection dose does not necessarily translate into efficacy because
absorption kinetics, degradation, and even injection volume will influence
bioavailability (Jørgensen, 1991). Therefore, even the newest data on the
roles of dose, frequency, and method of administration in therapeutic
efficacy must be evaluated with these considerations in mind.

Nevertheless, recent reports do suggest that increased frequency of
injection of the same weekly GH dose does result in increased growth
response. The group at the Middlesex Hospital in London (Smith *et al.*,
1988) evaluated the contribution of dose and frequency in the first-year
growth response of GH-insufficient children. They concluded from their
data that the combination of the highest dose tested ($>15$ IU/m$^2$/week)
and a frequency of at least six or seven injections per week (daily), rather
than three times a week, gave the best growth response. We have
found that both frequency and dose contribute significantly to the prediction of first-year growth response (Blethen *et al.*, 1991). Though the frequency component of this recommendation has been confirmed by others
(Albertsson-Wikland *et al.*, 1986; Kikuchi *et al.*, 1988), the dose component has yet to be assessed in long-term multicenter trials. Thus, the
reports of the efficacy of biosynthetic GH in the GHD individual confirm
a salutory effect on growth rate acceleration but do not yet define the
optimum treatment design to yield final heights in concert with genetic
target heights. These studies are currently under way.

## B.  NON-GH-DEFICIENT SHORT STATURE

The laboratory diagnosis of GH deficiency has been based primarily on
the acute GH secretory response to stimuli such as exercise, hypoglycemia
induced by insulin, or pharmacologic agents such as arginine, levodopa
(with or without propranolol), or clonidine. The definition of a normal GH
response is most commonly defined as a peak GH level $>10$ ng/ml, with
levels between 7 and 10 ng/ml often considered borderline. Such categori-
zation, while helpful, cannot separate with absolute certainty those who
will respond to GH treatment with accelerated growth from those who will
not; therefore, a number of studies have sought to identify the subset of
apparently non-GH-deficient patients who will respond to exogenous GH
therapy.

### 1.  Radiation-Induced GH Secretory Dysfunction

In 1982, Chrousos *et al.* studied monkeys that had received 4000 rad
(cGy) cranial irradiation and described blunted spontaneous GH secretion

over a 24-hour period despite normal GH responses to pharmacologic stimuli such as levodopa or arginine. Shortly thereafter, two studies reported similar findings in children (Blatt *et al.*, 1984; Romshe *et al.*, 1984) who were irradiated during therapy for acute lymphoblastic leukemia (ALL) or cranial tumors. Despite marked reduction in the number of spontaneous peaks of GH as well as the total amount secreted, at least two-thirds of the subjects in these studies achieved a peak GH of >7 ng/ml in response to either insulin-induced hypoglycemia, arginine, or levodopa. Thus, these studies led to the concept that the irradiation had damaged the pulsatile hypothalamic control of GH secretion, resulting in insufficient production for growth, but had left intact sufficient stimuli to synthesis that pharmacologic stimuli could elicit a response. In the nine subjects studied by Romshe *et al.* (1984), 1 year of GH treatment increased mean growth velocity from a baseline 2.1 cm/year to 6.5 cm/year. Notably, the GH response to arginine or levodopa, whether normal or low, did not predict the degree of response to GH. Similar findings have been reported by Kirk *et al.* (1987). More recently, Rappaport and Brauner (1989) have elegantly reviewed the spectrum of growth hormone response patterns that can be detected, over time, following various radiation protocols. Overall, it has been estimated that GH treatment of children after cranial irradiation can produce >6 cm height gain during the first year of treatment. Given this benefit, the known risk to the hypothalamus of >2400 rad of exposure, and the potential inability of pharmacologic testing to determine who will respond to GH treatment, we feel that a trial of GH treatment should be considered for any tumor-free subject with a poor growth rate after cranial irradiation, regardless of response to provocative testing. Most investigators feel that a 2-year tumor-free period should elapse before rhGH therapy is considered (Shalet *et al.*, 1988; Rappaport and Brauner, 1989), to avoid the period of greatest risk for relapse. Data on final height of irradiated patients treated with GH have shown minimal catch-up growth, in part due to less bone age delay than in classical GHD and an earlier onset of puberty in irradiated children; thus, combination therapy with GH and luteinizing hormone releasing hormone analogs (LHRHa) to delay puberty has been discussed (Sulmont *et al.*, 1990).

### 2. Idiopathic Short Stature

*a. Rationale for Use.* Most children brought to the attention of an endocrinologist for evaluation of height of less than the third percentile have no discernible organic etiology for their short stature. In a large subset of these patients with idiopathic short stature, the skeletal age is delayed relative to chronological age and growth rates are frequently marginal, i.e., less than the twenty-fifth percentile for age. Such findings are reminiscent of features seen in GH deficiency; however, growth hormone response to pharmacologic stimuli is typically >10 ng/ml. Nevertheless, subtle defects of GH secretion have been suspected. A parallel has been drawn (Spiliotis *et al.*, 1984; Bercu *et al.*, 1986) between some cases of idiopathic short stature and the GH secretory dysfunction seen after cranial irradiation (see Section III,B,1). Indeed, some studies have suggested that short children secrete less GH than those of normal height over a 24-hour period (Zadik *et al.*, 1985), albeit more recent data show significant night-to-night variability in GH secretion and overlap of values from hypopituitary and normal-statured children (Rose *et al.*, 1988; Costin *et al.*, 1989; Donaldson *et al.*, 1989; Lin *et al.*, 1989). The clinical relevance of neurosecretory dysfunction in children with idiopathic short stature remains controversial.

Though the existence of subtle defects in GH secretion in the short, slowly growing child remains uncertain, there is a growing body of evidence that GH can increase growth rate in a dose-dependent fashion. Experience with pituitary gigantism (Whitehead *et al.*, 1982) or overexpression of a human growth hormone gene in transgenic mice (Palmiter *et al.*, 1982) has demonstrated the ability of GH to promote growth beyond genetically determined potential. The likelihood of increased growth in children with idiopathic short nature, given an adequate rGH dose, has therefore appeared high, and growth rate increases have been demonstrated in early pilot studies (see Section III,B,2,b). The uncertainties concerning safety and final height outcome, however, have demanded large-scale clinical trials for more definitive results.

*b. Clinical Trials.* Most clinical trials have demonstrated that GH (pituitary derived or recombinant) can accelerate over the short-term the growth rate of some children with idiopathic short stature. Although an early trial by Tanner *et al.* (1971) showed little growth rate increase from administration of twice-weekly pituitary GH to four subjects with idiopathic short stature, most studies since have demonstrated significant growth rate increases in at least some subjects during the first year of treatment with GH. Growth rate increased dramatically in 9 of 20 short children treated for 6 months with pituitary GH by Rudman *et al.* (1981a), but those who responded had strikingly low somatomedin C (IGF-I) levels and a mean baseline growth rate of only 1.8 cm/year (prorated from 6-month data), well below the fifth percentile growth rate of 3.7 cm/year for subjects this age. A large number of studies since then have documented the ability of GH to improve growth rate during the first year of treatment in slowly growing (although not as slowly as those studied by Rudman *et al.*) short children (Frazer *et al.*, 1982; Lenko *et al.*, 1982; Plotnick *et al.*, 1983; Van Vliet *et al.*, 1983; Gertner *et al.*, 1984; Grunt *et al.*, 1984; Raiti *et al.*, 1987; Albertsson-Wikland, 1988; Bozzola *et al.*, 1988; Hermanussen *et al.*, 1989; Lin *et al.*, 1989; Genentech Collaborative Study Group, 1989; Rochiccioli *et al.*, 1989; Wit *et al.*, 1989; Walker *et al.*, 1990). A review of non-GH-deficient short stature by Cara and Johanson (1990) has summarized the results from many of these studies, and a modification and update of those data are given in Table II.

Complementing these studies of slowly growing short children is an investigation by Hindmarsh and Brook (1987) in which children growing at a rate no worse than $-0.8$ SDS (twenty-fifth percentile) below the mean were treated with GH; 16 children with normal baseline growth rates responded to GH therapy, with a mean growth rate $z$-score (standard deviation score) for age improving from $-0.44$ to $+2.2$ SDS score. Untreated controls showed no change during the study period.

No prestudy variable has been found that can reliably predict growth rate response, which varies in the above-mentioned studies from approximately 40% up to 100% of short subjects tested. The wide variability in response rates may result not only from differing response criteria, but also from the potential bias pointed out by Polychronakos *et al.* (1988): studies using subjects as their own controls may overestimate the true response rate, because children are more likely to enter a study during a period of slow growth, which may be transient, thus marking them as "responders" if their growth rates improve spontaneously during the study period. Early claims that IGF-I response to GH correlated with growth rate response (Rudman *et al.*, 1981a) were not supported by later researchers (Gertner *et al.*, 1984; Grunt *et al.*, 1984; Bozzola *et al.*, 1988). The degree of bone age delay does not correlate with growth response (Genentech Collaborative Study Group, 1989). Some studies have suggested a weak inverse correlation between pretreatment growth rate and response to GH treatment (Hindmarsh and Brook, 1987; Wit *et al.*, 1989), but this has not been supported by data from the larger multicenter Genentech Collaborative Study Group (1989). Some studies have suggested a weak inverse correlation between pretreatment growth rate and response to GH treatment (Hindmarsh and Brook, 1987; Wit *et al.*, 1989), but this has not been supported by data from the larger multicenter Genentech Collaborative Study Group (1989). Recent data presented by members of this same multicenter study (Hintz *et al.*, 1991) suggest that, while no prestudy characteristic can predict the initial growth response, the initial growth response during year 1 of therapy is a fair predictor ($r = 0.50$) of response in the next year and beyond.

Longer term data have shown that the acceleration of growth rate, while sustainable past 1 year of GH treatment, decreases with each additional year of treatment (Wit *et al.*, 1989; Albertsson-Wikland, 1988), unless the

TABLE II
Clinical Trials of GH Treatment for Non-GH-Deficient Short Stature[a,b]

| Investigators | Treatment criteria | Dose | Growth rate (cm/year) or SDS score (N) | | | | | |
|---|---|---|---|---|---|---|---|---|
| | | | Pre-Rx | 6 months | 1 year | 2 years | 3 years | 4 years |
| Tanner et al. (1971) | Hereditary short stature or growth delay | 20 U/week | 5.3 (4) | — | 6.3 (4) | — | — | — |
| Rudman et al. (1981a) | Ht < 3rd percentile, GR < 2.4 cm/year | 0.08 U/kg q day | 1.6–1.8 (20) | 2.6–10.1 (20) | — | — | — | — |
| Frazer et al. (1982) | Ht < 6 SD | 0.1 U/kg TIW × 6 months, then 0.2 U/kg TIW × 6 months | 4.2 (5) | — | 6.5 (5) | — | — | — |
| Lenko et al. (1982) | Ht < 3 SD | 4.6–18.8 mg/m²/week TIW | −1.8 SDS (14) | — | +0.9 SDS (14) | — | — | — |
| Plotnick et al. (1983) | Ht < 2 SD, GR < 3rd percentile | 0.075–0.15 U/kg TIW | 3.6 (16) | — | 7.1 (11) | — | — | — |
| Van Vliet et al. (1983) | Ht < 3 SD, GR < 5 cm/year | 0.1 U/kg TIW | 4.2 (15) | 6.1 (15) | — | — | — | — |
| Gertner et al. (1984) | Ht < 2.5 SD | 0.1 U/kg TIW | 4.3 (10) | 7.4 (10) | — | — | — | — |
| Grunt et al. (1984) | Ht < 4 SD | 2 U qod | 4.1 (7) | 8.6 (7) | — | — | — | — |

| Study | Criteria | Dose | | | | | |
|---|---|---|---|---|---|---|---|
| Hindmarsh and Brook (1987) | Ht < 2 SD, GR between 0 and −0.8 SD | 2 U SIW (12–21 U/m²/week) | −0.44 SDS (16) | — | +2.2 SDS (16) | — | — | — |
| Raiti et al. (1987) | Ht < 1st percentile, GR < 4 cm/year | 0.1 U/kg TIW | 3.4 (48) | 6.9 (48) | — | — | — | — |
| Bozzola et al. (1988) | Ht < 2 SD | 0.1 U/kg TIW | 3.7 (8) | 5.2 (8) | — | — | — | — |
| Lin et al. (1989) | Ht < 2 SD | 0.3–0.375 mg/kg/wk TIW | 4.0 (28) | 8.8 (28) | — | — | — | — |
| Albertsson-Wikland (1989) | Ht < 2 SD | 0.1 U/kg/day | 4.2 (24) | — | 8.1 (24) | 6.7 (24) | 6.0 (24) | 4.9 (24) |
| Wit et al. (1989) | Ht < 2.5 SD, GR < 25th percentile | 2–4 U/m²/day | 3.6–5.7 (17) | — | 6.9–7.6 (17) | 6.1–7.3 (17) | — | — |
| Rochiccioli et al. (1989) | 24-hour GH secretion < 3 ng/ml/minute | 0.4 U/kg/wk TIW or SIW | 4.9 (24) | 6.8 (24) | 6.2 (15) | — | — | — |
| Walker et al. (1990) | Ht < 3rd percentile | 30 U/m²/wk q day | −0.6 SDS (21) | +4.59 SDS (18) | +4.24 SDS (13) | — | — | — |
| Hintz et al. (Genentech Study Group) (1991) | Ht < 2 SD, GR < 50th percentile | 0.3 mg/kg/wk | 4.6 (121) | — | 8.0 (121) | 7.5 (103) | 6.9 (51) | — |
| Lesage et al. (1991) | Ht < 1st percentile | 0.3 U/kg q day | 4.0 (10) | — | 10.7 (10) | 8.8 (10) | — | — |

[a] Modified from Cara and Johanson (1990).
[b] Abbreviations: Ht, height; GR, growth rate; TIW, three times per week; SIW, six times per week; SDS, standard deviations scores.

192        BARBARA M. LIPPE AND JON M. NAKAMOTO

dose is increased (Gertner *et al.*, 1986) or the frequency of administration changed from 3 days a week to daily, at the same total weekly dose (Genentech Collaborative Study Group, 1990). The 3-year data from the Genentech Collaborative Study Group (Hintz *et al.*, 1991) have shown most of the improvement in predicted final height to occur in the first year, with a small degree in the second year, and no apparent further increase in predicted height $z$-score between years 2 and 3 of GH treatment. It should be noted that these data may be confounded by the admixture of young patients with older patients who are well into puberty. Nevertheless, even if these data are sustained, one needs to assess the issue of whether GH can be discontinued once height prediction no longer appears to be improving, given data such as reported by Raiti *et al.* (1987), which suggest that discontinuation of GH therapy was followed, in a third of cases, by deceleration of growth rate to less than the pretreatment rate (so called catch-down growth).

Overall, published abstracts from the longest running studies show some improvement in final height prediction after GH treatment of idiopathic short stature. The 3-year data from the Genentech Collaborative Study group (Hintz *et al.*, 1991) show mean predicted final height improving from $-2.9$ to $-1.7$, an increment of $+1.2$ SDS. In contrast, the Dutch Growth Hormone Working Group (Wit *et al.*, 1991) have presented 4-year data from a smaller study population showing a mean increment in predicted adult height $z$-score of only $+0.5$ SDS. These data may be more encouraging, however, because the untreated short controls achieve a final height $z$-score 0.5 SDS less than initial predictions. In what is one of the first of several long-awaited reports on final adult height achieved in these studies, Albertsson-Wikland and Karlberg (1991) have presented an abstract based on final heights in 33 normal short children treated with GH for extended periods of time; the mean final height $z$-score varied from $+0.1$ to $+0.7$ SDS greater than pretreatment predictions.

As mentioned in the discussion of classical GHD, the issue of the effect of GH on the tempo of puberty may be equally relevant to the non-GH-deficient child. Darendeliler *et al.* (1990) suggest that GH therapy may shorten the duration of puberty, which could potentially reduce any previously gained improvement in final height due to shortened duration of the pubertal growth spurt. These researchers studied retrospectively 134 children with isolated GH deficiency who were treated with pituitary or rhGH; time from first physical signs of puberty (breast or genitalia Tanner stage II) to breast or genitalia Tanner stage IV was recorded and compared to data from historical or contemporaneous surveys of pubertal development in normal children. Although the authors claimed that the median

## USES OF GROWTH HORMONE

193

duration of puberty of 1.5 years for both sexes were significantly different from that of comparison studies, large sources of variability must be considered, including significant interobserver variability in assigning pubertal stages, especially genital staging, which lacks a clear-cut dividing line between stages III and IV; visit intervals 3 months apart, which creates discontinuities in data points and reduces resolution (the ability to define exactly when a given change takes place); and differences between GH-deficient and nondeficient children. Nevertheless, given the known effects of GH and/or IGF-I on ovarian granulosa cells (see Section IV,D,1.), there is a need to assess further the hypothesis of GH acceleration of pubertal duration, perhaps in the current multicenter trials ongoing in the United States and Europe.

An interesting approach to GH treatment of idiopathic short stature, employing a high-dose, short-duration treatment protocol, has been taken by Lesage *et al.* (1991). Ten short prepubertal children with normal GH responses to pharmacologic tests were treated for 2 years with doses of rhGH that, at 2.1 U/kg/week (1.05 mg/kg/week), were much higher than the more typical maximum dose of 0.6 U/kg/week used in other studies. The age of treatment was chosen to be before, but close to the time of, puberty. Growth acceleration was remarkable, with mean growth velocity $z$-scores of $+6.9$ and $+3.1$ at 12 and 24 months, respectively; because treatment had been timed to precede puberty by approximately 2 years, growth rates 12 months after ending GH treatment remained significantly higher than baseline. Predicted adult height increased significantly, although final height data are not yet available. Fasting glucose and glucose profiles on oral glucose tolerance testing did not become abnormal. As expected, there was a significant rise in insulin concentrations and evidence of relative insulin resistance after 2 years of treatment, but this returned to normal when measured 12 months after GH therapy was stopped. No obvious acral changes were reported, although there are no specific measures, such as heel pad thickness, to support this contention. These positive data are consistent with the subgroup of patients included in the report of Albertsson-Wikland and Karlberg (1991) who were treated beginning 2 years before puberty and who appeared to have the best final height outcome.

In summary, there is now longer term evidence from multicenter studies that GH treatment can increase growth velocity in non-GH-deficient children, and may improve predicted adult height in some children with each study. Because there are few studies of true final height and no data on what pretreatment variables might correlate best with a positive outcome, further studies are indicated before recommendations for therapy can be developed.

## C. TURNER SYNDROME

In 1938 Henry Turner described seven women with the phenotype of sexual infantilism, webbed neck, and cubitus valgus; this condition now bears Turner's name (Turner, 1938). Though short stature was not mentioned in the title of this paper, Turner clearly recognized that short stature was a primary characteristic of the syndrome. Over the past 53 years, heterogeneity not only in the chromosomal karyotype (ranging from complete monosomy of the X to small deletions of either the short or long arm of the X), clinical phenotype (ranging from multiple stigmata to a normal physical appearance), and physiologic manifestations (including a spectrum of gonadal function from complete ovarian failure to documented fertility) have been described. The short stature, however, has remained an almost invariant finding (reviewed by Lippe in Kaplan, 1989).

### 1. Growth in Turner Syndrome

The short stature of Turner syndrome appears to be a result of the combination of impaired intrauterine growth and abnormally slow postnatal growth velocity between ages 3 and 13 years. This pattern, described by both cross-sectional height and longitudinal velocity data, was most comprehensively described by Ranke et al. (1983) from the analysis of a cohort of 150 girls who had not received any hormone or growth-promoting therapy. Final height data for four cohorts of European patients ($n = 366$) was reported to be 143.5 cm (Lyon et al., 1985). These investigators also developed cross-sectional growth charts, with height standard deviations for girls with Turner syndrome, and they introduced the concept that final height could be projected from earlier height data. They documented that there was an excellent correlation between the height standard deviation score on the Turner growth chart when first seen (between the ages of 3 and 12 years) and when last seen (ages 19–24 years). We (Lippe et al., 1991) have validated both the European growth data and the Lyon method of height projection for American girls with Turner syndrome.

The etiology of the short stature in Turner syndrome is clearly multifactorial, involving a possible intrinsic cellular mechanism secondary to chromosomal monosomy (Verp et al., 1988), documented intrauterine growth retardation, skeletal abnormalities, and lack of sex steroids at the time of puberty. Abnormalities of growth hormone secretion or action—as another factor in the growth failure—have been postulated by many investigators, but the data are subject to a variety of interpretations and a consensus on a pathophysiologic role has not been developed (Ross et al., 1985; Rappaport and Sauvion, 1989; Albertsson-Wikland and Rosberg, 1991; van Es, 1991). There are data showing that in normal children, more

USES OF GROWTH HORMONE                                    195

growth hormone is produced by the very tall ones than the short ones (Albertsson-Wikland and Rosberg, 1988), but in Turner syndrome this does not appear to be operative (Massarano et al., 1989). There also appears to be no correlation between the auxologic data before and during treatment with GH and the GH responses to pharmacologic stimuli (Massa et al., 1991). In addition, the data on whether low-dose sex steroids increase growth hormone secretion in Turner syndrome girls as part of their growth-promoting action are conflicting (Massarano et al., 1989; Mauras et al., 1990).

### 2. Treatment with Sex Steroids and Growth Hormone

Attempts to increase stature with both anabolic agents and growth hormone date back to the early 1960s (Whitelaw et al., 1962; Tzagournis, 1969). The efficacy of anabolic steroids alone is still being debated, with outcome data ranging from no significant effect on final height (Sybert, 1984) to a maximum of 5.2 cm increase over control patients (Joss and Zuppinger, 1984). The efficacy data for growth hormone treatment have been, until recently, equally unrewarding. In his comprehensive review, Wilton (1987) points out that some studies were very short term; others did not use comparable doses of GH and little is known about final height. In the one United States multicenter prospective collaborative study that used human pituitary growth hormone, there was an indication that it was effective in increasing height velocity in the first year of treatment (Raiti et al., 1986).

Because of the introduction of biosynthetic human GH, large numbers of patients are being studied in multiple countries in protocols that hope to clarify the role of GH therapy, either alone or in conjunction with androgens or estrogens (see Ranke and Rosenfeld, 1991). The combination of GH and androgen was first reported by Rudman et al. (1980). Their data suggested that while both hormones separately did increase growth velocity, the combination was most effective. This observation formed the basis for the first and currently the longest ongoing multicenter (including the authors' center) study of the effects of biosynthetic human growth hormone alone or in combination with oxandrolone in Turner syndrome (Rosenfeld et al., 1986, 1988, 1991). The mean cumulative height gained in the first 5 years of therapy was 15.3 cm more than expected based on the subjects' baseline Turner percentile in the growth hormone plus oxandrolone group ($n = 16$) and 9.1 cm more in the growth hormone-alone group ($n = 15$). Twenty-five girls have met the criteria for discontinuation and are at or near final height. Their height is $151.7 \pm 5.1$ cm, which is $8.6 \pm 4.4$ cm over their predicted adult height based on their original Turner height percentiles. We believe that the data are sufficiently compelling

196 BARBARA M. LIPPE AND JON M. NAKAMOTO

that we now recommend GH therapy for selected patients with Turner syndrome.

## D. CHRONIC RENAL FAILURE

Chronic renal failure is a significant cause of morbidity in childhood. The etiology includes developmental abnormalities of the kidney (dysplasia); destructive processes that begin *in utero* and/or develop postnatally (posterior urethral valves, obstructive uropathy); genetic disorders such as congenital nephrotic syndrome and cystinosis; acquired diseases as hemolytic–uremic syndrome, nephrosis, and nephritis; and acquired toxicity from drugs, chemotherapeutic agents, and radiation therapy. As the efficiency of medical treatment, including dialysis and transplantation, improves, an increasing number of affected children survive childhood and have the potential for normal adult lives.

### 1. Growth in Chronic Renal Failure

Growth failure is a major clinical complication of chronic renal failure (CRF) in infancy and childhood (Fine, 1984). Multiple factors have been implicated, including acidosis (McSherry and Morris, 1978), poor nutrition (Simmons *et al.*, 1971), lack of adequate vitamin D, hyperphosphatemia, and parathyroid hormone-induced metabolic bone disease (Arnaud, 1973; Korkor, 1987; Hudson *et al.*, 1983). Despite strict clinical attention to the correction of all these abnormalities, a significant number of children remain growth retarded and have markedly reduced adult stature.

Abnormalities in the growth hormone/IGF axis have been implicated in the growth retardation of CRF, based on studies demonstrating paradoxically elevated GH concentrations in uremia (Samaan and Freeman, 1970; Ramirez *et al.*, 1978), normal to high concentrations of IGF-I and IGF-II (Powell *et al.*, 1986; Hokken-Koelega *et al.*, 1990), and the apparent presence of an inhibitor of somatomedin/IGF-I (Phillips *et al.*, 1984). However, no clear hypothesis for what appeared to be a state of growth hormone resistance had emerged at the time significant quantities of growth hormone became available. Only in the last 3–4 years have the IGF-binding proteins been well characterized and their patterns determined in uremia (Lee *et al.*, 1989; Blum *et al.*, 1989).

### 2. Treatment with Growth Hormone

Initial studies of the effects of growth hormone administration to uremic rats were undertaken with the idea of using supraphysiologic doses to overcome the resistance (Mehls and Ritz, 1983; Mehls *et al.*, 1988), and before the current pathophysiologic model involving IGF and its binding proteins was developed (Mehls *et al.*, 1990). The results of these studies

USES OF GROWTH HORMONE 197

indicated that a salutory effect could be obtained and led to the first preliminary report demonstrating the initial efficacy of biosynthetic human growth hormone in increasing growth velocity in children with CRF not yet requiring dialysis (Lippe *et al.*, 1988). This report was soon followed by further data from the same center (Koch *et al.*, 1989) as well as from two other centers (Wilson *et al.*, 1989; Tönshoff *et al.*, 1989) confirming the efficacy of GH in increasing growth velocity in predialysis, patients, including two with cystinosis, as well as in patients already on dialysis. These studies, as well as the development of a radioimmunoassay for IGFBP-3 (Blum *et al.*, 1990), and the accumulating data supporting the hypothesis that the uremic resistance to endogenous growth hormone was a consequence of the increased IGFBP-3, leading to diminished secretion of IGF-I (Mehls *et al.*, 1990) and a decrease in biologically available IGF, have led to an increasing number of clinical studies of the effects of GH in uremia. These have ranged from short-term single-center studies including patients with a wide range of renal impairment, including dialysis and transplantation patients (Rees *et al.*, 1990), a placebo-controlled, double-blind crossover study (Hokken-Koelega *et al.*, 1991), and a multi-center European study, including uremic and posttransplant patients (Johansson *et al.*, 1990).

The growth response data have been uniformly favorable, albeit there are differences, depending on the age of the child, whether or not they have CRF without requiring dialysis, are on dialysis, or have received a renal transplant (and are, therefore, receiving corticosteroid therapy). Our own data on these three groups of patients—CRF (Fine *et al.*, 1991a), dialysis (Fine *et al.*, 1990), and following renal transplant (Fine *et al.*, 1991b)—confirm the efficacy of growth hormone treatment under all three conditions. However, it is clear that in the group with CRF, those who have not yet progressed to dialysis or transplantation exhibit the best first-year response, and continue to show catch-up growth into the second and third year of treatment. Thus, there are pathophysiologic and biologic data to strongly recommend that children with growth retardation associated with chronic renal disease be treated with recombinant human growth hormone at a young age, with the goal of restoring them to a normal genetic growth centile before the need for dialysis or transplantation.

## E.  OTHER NON-GH-DEFICIENT SYNDROMES WITH GROWTH FAILURE

A very large number of syndromes and conditions exist that are characterized by growth failure and for which a trial of GH treatment has been undertaken. Our goal is to discuss several of these categories, with the understanding that future research and clinical trials will be needed to determine the role that GH or any growth-promoting agent might have in the management of children with these conditions.

## 1. Intrauterine Growth Retardation/Silver–Russell Syndrome

Children with intrauterine growth retardation (IUGR) clearly comprise a heterogeneous group, given the wide variety of conditions that can produce a reduced weight or weight and length for a given gestational age. Regardless of cause, children with IUGR are at high risk for short stature in later childhood, much more so than are children with low birth weights due solely to prematurity (Binkin et al., 1988). Studies of the natural history of IUGR infants over the first year have shown that IGF-I levels are lower during the first week of life than in controls; those infants who have the least catch-up growth by 1 year of age also have the lowest IGF-I levels (Thieriot-Prevost et al., 1988). Though no consistent abnormalities of GH secretion in response to provocative tests have been demonstrated, three groups have used 24-hour or overnight multiple blood sampling to suggest that a low spontaneous secretion of GH may exist in subjects with IUGR (Albertsson-Wikland 1989; Stanhope et al., 1989; Rochiccioli et al., 1989). The earliest clinical trials (Tanner et al., 1971; Grunt and Enriquez, 1972) showed only minimal acceleration of growth with 6 to 12 months of pituitary GH treatment, though later studies (Foley et al., 1974; Lee et al., 1974; Lanes et al., 1979) documented some efficacy, with the acceleration of growth rate comparable to that seen, at the time, with GH treatment in idiopathic short stature. Foley et al. (1974) showed that the degree of growth rate acceleration was to some extent dose dependent, with a significant further increment of growth achieved by increasing the fixed daily GH dose from 2 units up to 5 units. Lanes et al. (1979) showed that growth rate acceleration could be sustained for at least 30 months of treatment, although the yearly growth increment decreased with time.

Three more recent studies confirm that GH can stimulate the growth rate of children with IUGR as well as those with Silver–Russell syndrome, often included as a specific form of IUGR when associated with pre- and postnatal growth failure, triangular facies, and skeletal asymmetry. Albertsson-Wikland (1989) studied 16 subjects and showed a treatment-induced acceleration of growth rate that was negatively correlated with the baseline spontaneous GH secretion. Stanhope et al. (1989) showed that the degree of growth rate acceleration was somewhat dose dependent. Rochiccioli et al. (1989) documented a doubling of mean growth velocity (3.5 to 7.0 cm/year) during the first 12 months of treatment. As is the case for children with idiopathic short stature, no firm conclusions about the efficacy of rhGH treatment in IUGR can be made until more data concerning final achieved height become available.

## USES OF GROWTH HORMONE 199

### 2. Noonan Syndrome

Noonan syndrome is a sporadic or dominantly inherited condition, occurring in both males and females, which shares some phenotypic similarity to Turner syndrome (short or webbed neck, ptosis) and is characterized by short stature. Mean childhood and final adult height are below the third percentile for both sexes (Ranke et al., 1988). Though peak GH response to pharmacologic provocative testing appears to be normal, the IGF-I levels have been reported as low (Cianfarani et al., 1987) or low to normal (Ahmed et al., 1991). Small, uncontrolled studies of the effects of GH have produced conflicting results: Cianfarani et al. (1987) found no increase in growth velocity with GH therapy of three patients with Noonan syndrome, whereas Ahmed et al. (1991), using more frequent administration and a higher total weekly dose, noted an increase in mean height velocity from 4.8 to 7.4 cm/year. In our own experience, two patients have shown a salutory response in the first year of GH treatment. Appropriate clinical trials appear warranted.

### 3. Down Syndrome

Childhood and adult short stature are characteristic of Down syndrome (trisomy 21). In five patients Anneren et al. (1986) documented normal GH secretion during sleep and/or in response to insulin–arginine stimulation tests. IGF-I levels were on the lower side of normal but not in the GH-deficient range. Nevertheless, six months of GH treatment appeared to double the baseline growth rate (Anneren et al., 1986; Anneren et al., 1990). Recently, Torrado et al. (1991) reported their experience with treating 13 Down syndrome patients with GH for 1 year. The mean growth rate in this study increased markedly, from 5.4 to 12.2 cm/year. Follow-up data are not yet available, and the same caveats we had about final height and the tempo of puberty apply to this and other syndromes. In addition, we note that Down syndrome individuals have a higher than normal incidence of leukemia (Robison et al., 1984). Until the issue of the relationship of GH therapy to the development of leukemia is completely resolved (see Section V,B,2) we are cautious about recommending clinical trials in this condition.

### 4. Prader–Willi Syndrome

Prader–Willi syndrome is characterized by short stature, appetite disturbances with resultant obesity, cryptorchidism, hypotonia in utero and in infancy, and frequent mental retardation. More than 50% of patients have a demonstrable deletion of the long arm of chromosome 15, and recent studies have demonstrated that the involved chromosome is of paternal origin (Butler *et al.*, 1986). The symptoms suggest possible hypothalamic dysfunction, but there are no consistent findings of abnormalities of GH secretion. Although GH response to levodopa is minimal, there is a response to arginine and insulin that is comparable to that of obese normal subjects (Bray *et al.*, 1983). Some studies have found decreased spontaneous GH secretion (Lee *et al.*, 1987a) regardless of body weight (Angulo *et al.*, 1991). Studies by Lee *et al.* (1987a,b) have shown acceleration of growth by either pituitary or recombinant GH in three of four subjects and Wu *et al.* (1988) found a significant growth response in three patients during the first 6 months of treatment. Preliminary results from a trial of GH in 19 patients with Prader–Willi syndrome indicate an increase in growth rate from 2.0 to 5.3 cm/year, with a net gain in final predicted height (Angulo *et al.*, 1991). Also noted during this study was a slowing of the excessive weight gain associated with this syndrome.

## 5. Skeletal Abnormalities and Other Congenital Anomalies

Patients with a number of osteochondrodystrophies (disorders of bone and cartilage) have undergone trials with GH. Older studies showed minimal effects on height in achondroplasia (Escamilla *et al.*, 1966) or hypochondroplasia (Soyka *et al.*, 1964). A more recent study of 84 patients with varying degrees of short-leggedness (termed hypochondroplasia by the authors) documented a significantly increased mean growth rate ($z$-score improved from $-1.66$ to $+1.62$) after 1 year of GH therapy (Appan *et al.*, 1990). Whether all these patients had a true genetic chondroldystrophy or, in fact, represented degrees of familial short stature, is not clear to us. In a prospective clinical trial of documented skeletal dysplasias including achondroplasia and hypochondroplasia, GH therapy was shown to have a small but consistent effect on increasing growth velocity, with almost all of the increased growth in the lower limbs (Tick *et al.*, 1991).

The observation that GH administration increases renal phosphate resorption (Gertner *et al.*, 1979) has served as the rationale for use of rhGH in hypophosphatemic rickets, in which defective renal and intestinal phosphate transport appear to be the primary defects. Short stature below the expected final adult height is a common problem in this syndrome as well. Though early small-scale studies did not show growth rate acceleration by pituitary GH (Gershberg *et al.*, 1964), some improvement in renal phosphorus handling was reported. The most recent study by Wilson *et al.* (1991a) in 11 children with hypophosphatemic rickets showed that 4 months of therapy with rhGH alone produced a significant rise in serum phosphorus that was not significantly different from levels achieved with standard calcitriol and oral phosphorus therapy, although the best results were achieved by combining both rhGH and standard therapy. In contrast

USES OF GROWTH HORMONE 201

to earier studies, this group demonstrated an increased growth velocity in all patients, with the mean growth velocity $z$-score increasing significantly from $-1.27 \pm 1.38$ to $+2.43 \pm 1.43$. The long-term effects on bone and mineral metabolism remain to be studied.

A study by Rotenstein *et al.* (1989) of seven slowly growing patients with neural tube defects (spina bifida or myelomeningocele) demonstrated significant short stature that could not be wholly explained by vertebral abnormalities such as scoliosis. The mean baseline growth rate was extremely low at 1.7 cm/year, due to the preselection for slow growth; after 6 months of rhGH treatment, the growth rate had increased to 7.9 cm/year. Bone age did not accelerate, suggesting a net gain in ultimate predicted height. Long-term studies are in progress.

## F. OTHER MEDICAL CONDITIONS ASSOCIATED WITH SHORT STATURE

Chronic illnesses such as cystic fibrosis, severe chronic anemias, bowel disorders resulting in long-standing malnutrition, and conditions requiring long-term high-dose corticosteroid therapy may all cause growth retardation. If protracted enough, many of these conditions can result in adult short stature. We recognize that, over time, studies of the efficacy of GH to ameliorate the growth retardation may be undertaken in any number of these and other conditions. We will discuss, briefly, only corticosteroid therapy.

### 1. Corticosteroid-Induced Growth Failure

There is little evidence, to date, that exogenous GH therapy can improve linear growth in children receiving chronic high-dose glucocorticoid therapy, albeit GH might have a role in reducing protein catabolic effects (see Section IV,G,2). Older studies reported no growth-stimulating effect of pituitary GH treatment of children receiving corticosteroid therapy for rheumatoid arthritis (Ward *et al.*, 1966), asthma, nephrotic syndrome (Morris *et al.*, 1968), or inflammatory bowel disease (McCaffery *et al.*, 1974). The study by Butenandt (1979) of corticosteroid-treated children with rheumatoid arthritis is difficult to interpret in light of the fluctuating course of the disease and steroid dosages. The general lack of GH effect may be due in part to costeroid interference with growth at several levels, including not only GH release (Miell *et al.*, 1991), also at the level of secretion and action of insulin-like growth factor I (Unterman and Phillips, 1985). We also speculate that the pharmacologic effect of the high-dose steroids may have a direct cytotoxic effect at the level of the prechondrocyte or subsequent differentiating cells. Though it is still possible that

202          BARBARA M. LIPPE AND JON M. NAKAMOTO

higher and more frequently administered doses of GH may partially over-come the growth-suppressive effects of corticosteroids, concerns about synergistic effects on glucose metabolism (insulin resistance and possible glucose intolerance) have limited the number of studies of children.

## IV. Non-Conventional Uses of GH: To Improve Metabolic Status

### A.  METABOLIC EFFECTS OF GH

In addition to stimulation of linear growth, GH has significant and complex effects on protein, fat, and carbohydrate metabolism (Isaksson *et al.*, 1985; Davidson, 1987). In general, the clinically significant effects of GH appears to be insulin-like (anabolic) for protein homeostasis and antiinsulin-like for fat (lipolytic) and carbohydrate (impaired glucose utilization) metabolism. Such a simplistic scheme fails to take into account the hormonal background on which GH exerts its effects; for example, GH can have a transient insulin-like effect on carbohydrate metabolism in tissues lacking recent exposure to GH or in hypophysectomized patients. Starvation, cortisol, epinephrine, or factors modulating the activity of adenylyl cyclase are only a few of the factors that may alter the metabolic response to GH. Still uncertain is the relative combination to these effects of GH versus mediators such as insulin-like growth factor I or insulin, induced *in vivo*, by GH treatment.

### 1.  Effects of GH on Protein Metabolism

The effect of GH on protein homeostasis is anabolic. In tissues from animals lacking endogenous GH, GH exerts a transient stimulatory effect on amino acid uptake and protein synthesis both *in vivo* and *in vitro* (Kostyo, 1968). This effect can be separated from the growth-promoting effect of GH; a variant form (20 kDa) of GH has significantly less effect on protein synthesis while retaining full growth-promoting ability (Kostyo, 1988). The early influence of GH on protein synthesis appears to be at the level of translation of mRNA into protein, as this effect is not inhibited by actinomycin D (Martin and Young, 1965). Clinically, administration of GH produces nitrogen retention (Ikkos *et al.*, 1958), which may take up to 2 days to become apparent and may not reach maximal values for several days. More recent studies using radiolabeled amino acids verify that GH stimulates protein synthesis in normal volunteers (Manson *et al.*, 1986; Horber and Haymond, 1990). Direct brachial artery infusion of GH ($0.14 \mu$/kg/minute) increases local skeletal muscle protein synthesis in the

USES OF GROWTH HORMONE 203

absence of any change in plasma amino acid or insulin levels (Fryburg *et al.*, 1991).

Evidence exists to support roles for both GH and IGF-I in anabolic effects on protein metabolism. Stimulation of protein synthesis by GH can occur without mediation by IGF-I: cultured Chinese hamster ovary (CHO) cells expressing GH receptors after transfection respond to GH with a generalized increase in protein synthesis but no detectable messenger RNA for IGF-I (Emtner *et al.*, 1991). On the other hand, IGF-I administered to patients lacking GH receptors produces positive nitrogen balance, suggesting that IGF-I alone is sufficient to exert anabolic effects (Walker *et al.*, 1991).

### 2. Effects of GH on Lipid Metabolism

GH can stimulate lipolysis, but the effect is of relatively slow onset and appears to be greatly influenced by the hormonal background that exists at the time of GH administration. For example, lipolytic effects of GH *in vitro* are often difficult to demonstrate unless other factors such as glucocorticoids or theophylline are present (Davidson, 1987). *In vivo*, release of free fatty acids induced by growth hormone is more clearly seen in the fasting state. Goodman *et al.* (1988) have suggested that the role of GH in lipolysis is to act as a permissive factor rather than an agonist, i.e., increasing sensitivity of adipocytes to lipolytic factors rather than directly inducing lipolysis. They suggest that GH may affect the guanine nucleotide-binding protein (G protein), $G_i$, that inhibits adenylyl cyclase.

### 3. Effects of GH on Carbohydrate Metabolism

The predominant clinical effects of GH on carbohydrate metabolism are antiinsulin-like (Davidson, 1987) and synergistic with the effects of glucocorticoids. Transient insulin-like effects can be seen with GH administration in GH-deficient subjects, but such effects are masked by prior recent exposure to GH. GH treatment increases mean fasting glucose concentrations despite a concurrent increase in insulin levels; further indications of insulin resistance include increased hepatic glucose production and decreased peripheral glucose uptake. Horber *et al.* (1991), using measurements of glucose, insulin, and C peptide in normal subjects, demonstrated a synergistic antiinsulin-like effect of rhGH and prednisone (a glucocorticoid) cotreatment; on the basis of this finding and differing effects on fat and protein metabolism, these workers suggest that GH and glucocorticoids induce insulin resistance via independent mechanisms.

## B. CLINICAL STUDIES IN GH-DEFICIENT ADULTS

### 1. Rationale for use in Adult GH Deficiency

More attention is now being paid to adult GH-deficient patients as possible candidates for hormonal replacement (Christiansen and Jorgensen, 1991). Contributing to this trend are increased appreciation of GH anabolic effects, availability of sufficient quantities of GH, and studies suggesting decreased physical and psychological well-being in GH-deficient adults. Retrospective analysis (Dean *et al.*, 1985) suggests that the social adjustment of GH-deficient adults is poorer than that of normal subjects, as measured by employment and marriage rates, and psychological questionnaires suggest that GH-deficient patients see themselves as less energetic, more anxious, more emotionally labile, and having a lower sense of well-being than matched normal controls (McGauley *et al.*, 1990). Anecdotally, there is a higher rate of somatic complaints, especially muscle weakness (Ranke, 1987). Body composition analysis showed that GH-deficient patients had lower muscle mass and more adipose tissue than normal subjects (Salomon *et al.*, 1989), and reduced muscle force per muscle area (Cuneo *et al.*, 1990). There has also been a suggestion of twofold increased mortality from vascular disease (myocardial infarction, congestive heart failure, or stroke) in hypopituitary adults, based on retrospective analysis of patients with hypopituitarism secondary to pituitary tumor surgery, but no GH replacement therapy over a 30-year period (Rosén and Bengtsson, 1990). This last study should be interpreted with caution, because many of the patients were not evaluated for GH deficiency postoperatively and there was relative undertreatment of hypogonadal women with estrogen.

### 2. Recent Clinical Trials

In 1989, results from two double-blind, placebo-controlled studies of GH treatment in adult GH deficiency were reported. In one study (Salomon *et al.*, 1989), 24 subjects with GH deficiency acquired in adulthood showed significantly increased lean body mass (averaging 6 kg by 6 months), decreased body fat (as much as 20%), and decreased waist-to-hip size ratio after 6 months of rhGH treatment. Side effects included transient fluid retention requiring reduction of dose in two patients. Blood glucose increased slightly along with C peptide levels, suggesting increased resistance to insulin. In the other study, Jørgensen *et al.* (1989), using a crossover design and functional outcome measurements, found after 4 months of GH therapy a significant increase in thigh muscle and decreased thigh fat volume by CT scan, which did not, however, reach the normal muscle to fat ratio (85 : 15%) of healthy adults. There was also an increase in

USES OF GROWTH HORMONE 205

exercise capacity by bicycle ergometer, and a trend toward increased isometric strength by dynamometer was noted, but did not reach significance ($p = 0.08$). Psychological questionnaires given to subjects in this study showed a clear preference for GH treatment over placebo treatment, based on feelings of increased well-being. Side effects included reversible peripheral edema in 1 of the 22 patients; glucose and insulin levels were not reported.

These initial findings have been further corroborated by additional studies. Degerblad *et al.* (1990) conducted a double-blind, placebo-controlled crossover study in which GH was administered for 3 months to six adult GH-deficient patients. The effects included decreased waist size, increased exercise endurance by bicycle ergometer, and an improved sense of well-being, but no change (during this brief treatment period) in isometric muscle force, bone mineral density, or scores on tests of cognitive function. In this study, one out of the six patients had clinically evident fluid retention; no differences in fasting blood glucose were noted during GH treatment (4 IU subcutaneously each day, 6–7 days/week = 0.5–0.6 IU/kg/week). Based on these encouraging early results, Christiansen *et al.* (1990) are conducting an open long-term trial of daily GH (2 IU/m² subcutaneously each evening). Out of 16 patients in this trial, 11 had participated in an earlier trial (Jørgensen *et al.*, 1989) and at 12 months showed significant further increases in thigh muscle volume by CT scan.

## C. MODIFICATION OF BODY COMPOSITION

The lipolytic and protein-sparing effects of GH have raised hopes of potential use in various conditions in which modification of body composition is desirable. Ever since the striking effects of GH treatment on body composition of patients with hypopituitarism were clinically demonstrated (Novak *et al.*, 1972), there has been much interest about the effect of GH in normal subjects. Two areas of interest in this regard have been (1) treatment of obesity, in an effort to preserve lean body mass while losing adipose tissue, and (2) reversing the tendency during the aging process to accumulate fat and lose lean body mass.

### 1. Use in Obesity

In 1987, Clemmons *et al.* reported results of a small placebo-controlled crossover study of the effects of 3 weeks of GH treatment in dieting obese subjects. During the treatment phase, positive nitrogen balance was achieved despite fasting, but body fat, determined by hydrostatic weighing, did not decrease any faster than with placebo. A larger study from the same group (Snyder *et al.*, 1988) examined effects of 12 weeks of GH treatment and found that the nitrogen-sparing effects of GH disappeared after 5 weeks, despite maintenance of increased plasma IGF-I levels. In this study as well, there was no acceleration of fat loss by GH treatment. In part, such findings may mirror the situation *in vitro*, wherein, as previously discussed, lipolytic effects of GH are dependent on the hormonal background present. In a different clinical context, wherein subjects are not under caloric restriction, GH treatment does seem to exert a lipolytic effect. A small double-blind, placebo-controlled study of obese women on a weight-maintaining diet showed a significant decrease in body fat (mean 38% versus baseline 40.3%) measured by hydrodensitometry after 4 weeks of rhGH treatment (Skaggs and Crist, 1991). Whether this effect can be extended beyond 4 weeks remains an open question.

## 2. Use in Aging Subjects

The changes in body composition associated with aging are well-recognized, with increased adipose tissue and decreased lean body mass not only from muscle, but also from bone, skin, and viscera, as summarized by Rudman (1985). Though not all studies concur (Dudl *et al.*, 1973), the demonstration by some investigators of decreased spontaneous GH secretion (Finkelstein *et al.*, 1972) and a tendency toward lower IGF-I levels (Rudman *et al.*, 1981b) with age has fueled speculation about possible beneficial effects of GH treatment in aging adults. It should be noted that the decrease in IGF-I levels is not uniform in all aging persons; rather, there is an increase with age in the percentage of subjects with IGF-I levels below that usually found in healthy 20-year olds, and those with lower IGF-I levels tend to have a greater percentage of adipose tissue (Rudman *et al.*, 1981b). The fall-off in spontaneous GH secretion and IGF-I levels appears to be particularly more prevalent after age 40 years (Florini *et al.*, 1985). The change is particularly apparent in women, and may be explained, in part, by the decrease in estrogen levels with menopause (Ho and Weissberger, 1990). Based on this finding, studies of the effects of estrogen administration on GH secretion in postmenopausal women have been conducted; interestingly, however, while GH secretion increased, IGF-I levels actually decreased (Ho and Weissberger, 1990) after estrogen therapy, perhaps due to suppression of hepatic IGF-I synthesis.

Other proposed explanations for lower GH secretion in aging include both a decrease in GH response to GH-releasing hormone (GHRH) and a decreased reserve of GHRH secreted by the hypothalamus. This latter hypothesis is supported by the data of Bando *et al.* (1991), which documented a decrease of both GH and GHRH in response to levodopa in elderly versus young men.

## USES OF GROWTH HORMONE

The first pilot clinical trial of GH in non-GH-deficient aging adults was that of Rudman et al. (1990). The 21 subjects were all males between 61 and 81 years of age, representing a group specifically selected for plasma IGF-I levels below the normal lower limit found in young men. These subjects were followed for a 6-month baseline period, then randomly assigned to a GH treatment or nontreatment group; no placebo was used. After 6 months of GH treatment, no adverse effects were noted, although mean fasting glucose rose slightly; insulin levels were not measured. Lean body mass calculated from total body potassium measurement increased significantly from baseline, although the difference between treated and untreated subjects was less striking. Subtraction of the lean body mass from the total body weight, which did not increase significantly, led to the inference that adipose tissue mass had decreased significantly. Unfortunately, the lack of data on the variability of total body potassium measurements does not allow full assessment of the results; the standard deviation of a single potassium determination measured by whole-body counting of the naturally occurring 40K isotope has been estimated at ±80 mmol (Bengtsson et al., 1990), which is equivalent to 1.2 kg lean body mass. Known effects of GH on sodium and fluid retention, lack of a placebo, and the nonblind nature of this study are additional sources of potential error. Therefore, the results of this study, while fully consistent with findings from placebo-controlled, double-blind studies in GH-deficient adults treated with rhGH (Salomon et al., 1989; Jørgensen et al., 1989), must be considered preliminary at best.

Marcus et al. (1990) undertook a short-term metabolic study of the effects of GH in both healthy elderly men and women. They documented the acute positive effects on nitrogen balance and demonstrated increases in serum organic phosphate, increased excretion of urinary calcium and hydroxyproline, and a rise in circulating parathyroid hormone (PTH), calcitriol, and osteocalcin. Because these data suggested a positive effect not just on nitrogen balance but also on bone metabolism, they undertook a longer placebo-controlled study of elderly women to determine if the GH affected overall calcium balance and bone mass or simply increased bone turnover. The 1-year data continued to confirm the increase in osteocalcin, urinary hydroxyproline, and calcium excretion and increased bone alkaline phosphatase activity. However, they were unable to document a change in body adiposity (by hydrostatic weighing) and could find no difference in mineralization of the spine or hip, as assessed by dual-energy X-ray absorptiometry (R. Marcus, personal communication). Thus, continued study of the effects of GH in both men and women needs to be done.

## D. GH AND INDUCTION OF OVULATION

### 1. Rationale for Use of GH

The effects of GH on ovarian physiology is believed to be largely mediated by IGF-I. A number of *in vitro* studies demonstrate an interaction of follicle-stimulating hormone (FSH) and IGF-I action on ovarian granulosa cells. FSH promotes the release of IGF-I in granulosa cells (Hsu and Hammond, 1987) and modulates release of IGF binding proteins (IGF-BPs) produced by granulosa cells (Adashi *et al.*, 1991). IGF-I enhances the replication of cultured mammalian granulosa cells (Savion *et al.*, 1981) and potentiates FSH action on the ovary at a postreceptor level by enhancing FSH stimulation of adenylyl cyclase and enhancing the effect of cyclic AMP on progesterone accumulation.

Although the majority of the evidence points toward IGF-I as the major mediator of GH action on the ovary, GH may have some direct effects or work indirectly via factors rather than IGF-I. Mason *et al.* (1990) have shown that GH can stimulate estradiol production from human granulosa cells in the absence of FSH; no immunoreactive IGF-I was detectable in the media. Insulin, which is known to be increased during GH treatment, can induce secretion of estradiol by granulosa cells (Peluso *et al.*, 1991).

That endogenous GH secretion may be having an *in vivo* effect on ovulation is suggested by the study of Menashe *et al.* (1990). While evaluating 25 anovulatory infertility patients, they found a correlation between the endogenous GH response to clonidine and the responsiveness to human menopausal gonadotropin (hMG) treatment for ovulation induction. Of 20 patients, 17 showed a "negative" GH response to clonidine (mean GH peak $2 \pm 1.2$ ng/ml) and required a mean total hMG dose three times that required in the 8 patients who responded well (average GH peak of $9.2 \pm 4.5$ ng/ml) to clonidine. The "negative" GH responders also had a much higher incidence of poor response to hMG therapy. However, there is good evidence that GH may not have an essential role in ovulation induction or fertility. Menashe *et al.* (1991) detected no evidence of infertility in two women with Laron dwarfism, a syndrome in which a genetic defect in GH receptor function leads to resistance to effects of GH and typically exhibits high serum GH and low IGF-I levels (Amsalem *et al.*, 1989). Stone and Marrs (1991) found no significant correlation of GH with estradiol or FSH either during normal or controlled ovarian hyperstimulation; however, this retrospective study was limited by reliance on single daily measurements of GH, which do not adequately assess the total daily GH secretory output.

## 2. Clinical Trials of GH Treatment for Ovulation Induction

In 1988, Homburg *et al.* reported that cotreatment with a short course of high-dose GH (approximately six times that used for childhood GH deficiency) decreased the total dose of human menopausal gonadotropin needed for ovulation induction in seven patients with hypogonadotropic hypogonadism. A lower GH dose also proved successful in combination with hMG treatment of a hypopituitary patient (Blumenfeld and Lunenfeld, 1989).

A prospective, placebo-controlled study in 16 patients with amenorrhea and anovulatory infertility (Homburg *et al.*, 1990) confirmed the effect of rhGH–hMG cotreatment. The mean total dose of hMG required was decreased 36% by cotreatment with high-dose rhGH. This decrease reflected both a shortened duration of hMG therapy as well as a reduction in the number of hMG ampules required per day to promote an ultrasound-detectable ovarian response. Ibrahim *et al.* (1991) observed a similar shortened duration of hMG therapy when GH was added to the *in vitro* fertilization treatment regimen of infertile women without evidence of pituitary insufficiency. These women, who were on a regimen of gonadotropin-releasing hormone agonist (GnRHa) therapy followed by hMG, showed a 27% mean increase in follicular growth rate when 12 days of rhGH was part of the treatment. The effect of GH was on rate of follicular growth only, as no difference in the number of large follicles ($>$14 mm) or estradiol levels was noted. Of the 10 women tested, all of whom had responded poorly to previous cycles of ovulation induction, 6 conceived during the cycle in which rhGH was added.

Burger *et al.* (1991) looked retrospectively at three patients, resistant to ovulation induction by hMG alone, who had received cotreatment with GH. Two interesting observations emerged: first, even a GH dose one-sixth of that used by Homburg *et al.* (1990) or Ibrahim *et al.* (1991) was effective in reducing the total dose of hMG required. Second, the effects of one cycle of cotreatment with rhGH appeared to carry over into subsequent cycles, with less total hMG required per cycle for as many as four cycles in one subject.

In contrast to these encouraging results, Owen *et al.* (1991) did not find an overall improvement in ovarian response to clomiphene citrate and hMG when GH was added to the regimen. Only a subset of patients with ultrasound-diagnosed polycystic ovaries showed any positive effect of GH cotreatment, primarily an increase in the number of larger follicles and an increased yield of oocytes collected. The treatment regimen differed from that of the previously mentioned studies in that the daily dose of hMG was fixed rather than being adjusted stepwise to a dose determined by ultrasound to be effective. This illustrates the difficulty in comparing studies in which patient selection, treatment regimens, and doses used may differ. Though it is possible that the effects of GH may be evident only in regimens that normally use higher doses of gonadotropins, further studies are needed to confirm the initial reports.

## E. LACTATION FAILURE

In ruminant animals, treatment with GH is lactogenic. Bovine GH (bGH) administered to dairy cows can increase mild yield 15–20% (Etherton, 1991). This occurs without affecting nutritional quality or exposing consumers to additional health risk (NIH Technology Assessment Conference, 1991; Juskevich and Guyer, 1990). Recently, studies of GH effects on lactation have been extended to primates, with data suggesting that GH treatment of rhesus monkeys during pregnancy and lactation can improve the weight gain of their nursing infants (Wilson *et al.*, 1991b). This effect may be due to both a slight increase in breast milk fat content and stimulation of mammary gland growth during pregnancy, leading to increased milk yield. Preliminary data from a double-blind, placebo-controlled trial of GH for 1 week in lactating women 2–4 months postpartum showed an increase in milk production that correlated with a rise in plasma IGF-I; no change in milk nutritional composition (including fat) was noted, and all but one sample had undetectable IGF-I concentrations (Milsom *et al.*, 1991). The apparent efficacy of GH in humans even when started late in the lactation phase suggests that it may have a role in treating lactation failure.

## F. DEFENSE AGAINST CATABOLISM OF BODY PROTEIN: CRITICAL ILLNESS OR INJURY

Breakdown of body protein stores is part of the physiologic response to trauma or critical illness, and is often exacerbated by a caloric intake inadequate to meet the increased demands of the postinjury hypermetabolic state. Such protein catabolism, if prolonged, can diminish the function of vital systems such as the heart, gastrointestinal system, respiratory muscles, and skin, and thus further compromise the critically ill individual (Baue, 1991). GH, with its known anabolic, nitrogen-sparing effects, may shift fuel utilization toward fat instead of protein, and thus has been under study for years as an adjunct to nutritional therapy in critical illness.

## USES OF GROWTH HORMONE

### 1. Early Studies of GH Treatment in Trauma

In 1941, Cuthbertson *et al.* observed that bovine pituitary extract could improve negative nitrogen balance in rats with femur fractures. Demonstration of the lipolytic activity of GH (Roe and Kinney, 1962) suggested an alternate source of substrates that might allow protein sparing. Early studies, however, provided mixed results: though adjunctive GH therapy was viewed as beneficial in burn patients (Prudden *et al.*, 1956; Liljidahl *et al.*, 1961; Wilmore *et al.*, 1974), Johnston and Hadden (1963) found no change in nitrogen balance in patients treated with hGH after hernia repair. The variability of results may have had to do with variability in patient selection and baseline nutritional status, the duration and dosage of treatment, and timing of treatment. For example, Soroff *et al.* (1967) demonstrated that the nitrogen-sparing effect of GH was evident during the later, convalescent phase of recovery from trauma but was ineffective in the acute catabolic phase, during which immunologic considerations appeared to be more important.

Major studies of GH effects prior to 1985 are few, primarily due to the relative scarcity of pituitary-derived GH in that period. Following availability of recombinant human growth hormone a number of pilot studies were undertaken that demonstrated the potential ability of GH to help protect body protein stores.

### 2. GH and Hypocaloric Nutrition in Normal Subjects

Previously it was believed (Wilmore *et al.*, 1974) that GH exerted its nitrogen-sparing effect only in nutritionally replete individuals. (Further work by Wilmore's group (Manson and Wilmore, 1986) showed that, for normal individuals in negative nitrogen balance from hypocaloric intake (30–50% of energy requirement), 1 week of GH treatment could restore positive nitrogen balance. In these individuals, analysis of $^{15}N$ turnover and urinary excretion of 3-methylhistidine, a muscle breakdown product, suggested that protein sparing was a result of increased protein synthesis rather than decreased protein breakdown (Manson *et al.*, 1988). Supporting this is a study in nutritionally depleted volunteers that demonstrates the ability of 6 hours of GH infusion to increase in skeletal muscle the messenger RNA levels for myosin heavy chain, a muscle protein subunit (Fong *et al.*, 1989). Effects of GH on adipose tissue are not reported in the study by Manson and Wilmore (1986); it would be interesting to see if lipolysis is more evident under conditions of nutritional repletion than under nutritional depletion, as was seen in studies of obese subjects (Clemmons *et al.*, 1987; Snyder *et al.*, 1988; Skaggs and Crist, 1991).

212                BARBARA M. LIPPE AND JON M. NAKAMOTO

## 3. GH and Critical Illness: Recent Clinical Trials

The demonstration that GH could be effective in states of hypocaloric nutrition resulted in several studies of GH in conditions in which nutrition might be considered adequate for usual demands but inadequate to meet hypermetabolic demands. In certain situations, such as recovery from major abdominal surgery, trauma, or sepsis, only the parenteral intravenous (piv) route may be available for intake of nutrients. Nutrition by piv in general cannot satisfy the high caloric demands imposed by a catabolic response to major injury. Central venous catheters can supply solutions of higher caloric density, but may involve technical demands for placement as well as higher risks for infection and embolization. In addition, even aggressive nutritional support does not always prevent protein loss in critically ill patients (Streat et al., 1987).

Most of the studies of the effects of GH fall into the following categories: (1) recovery from gastrointestinal surgery or other conditions preventing use of the gastrointestinal tract for nutritional intake, (2) adjunctive treatment of burns, (3) treatment of the catabolic state of sepsis, and (4) optimizing nutrition in long-term critical illness.

a. GH and Gastrointestinal Surgery. Several studies have documented the efficacy of GH as an adjunct to piv nutritional support in recovery from gastrointestinal surgery. During the first two postoperative weeks of piv alone, a weight loss in the range of 5–6% from preoperative weight is typical (Kinney et al., 1968; Jiang et al., 1989), with associated negative nitrogen balance (Ward et al., 1987; Manson et al., 1988; Ziegler et al., 1988). One week of daily GH treatment significantly improved nitrogen balance relative to placebo treatment, even on minimal (~20% of daily requirement) caloric intake without supplemental protein (Ward et al., 1987). Other studies with supplemental amino acids added to the (still hypocaloric) treatment regimen achieved positive nitrogen balance with addition of GH versus negative balance for control subjects treated with placebo (Ponting et al., 1988; Ziegler et al., 1988; Jiang et al., 1989). In contrast to studies in normal subjects (Manson et al., 1988), the protein-sparing effect in postabdominal surgery patients appeared to be due both to increased protein synthesis and decreased protein breakdown, although the former predominated somewhat (Jiang et al., 1989; Ponting et al., 1988). This effect could be maintained for at least 3 weeks with GH therapy and persisted even several days after stopping GH treatment (Ziegler et al., 1988). In the largest (n = 18) randomized, placebo-controlled, double-blind study of GH effects on recovery from abdominal surgery, the control group lost, on average, 5.3 + 2.9% of lean body mass and 11% body fat, while the GH-treated group lost essentially no lean body mass and 8.5 + 0.4% body fat (Jiang et al., 1989).

USES OF GROWTH HORMONE 213

Though GH appears to preserve body nitrogen stores and lean body mass, does it result in improved surgical outcome? Measurement of functional outcome is largely lacking, although Jiang *et al.* (1989) demonstrated a protective effect of short-term GH treatment on hand muscle grip force, one of the few functional outcome measurements available. Demonstration of lower complication rates and faster recovery awaits larger scale studies.

*b. GH and Burn Management.* GH appears to have potential as adjunctive therapy in patients with thermal injury, more likely in the anabolic recovery phase (Soroff *et al.,* 1967) than in the acute injury phase. In burn patients receiving adequate caloric intake, Wilmore *et al.* (1974) observed in 9 out of 10 subjects a significant decrease in nitrogen losses on a GH dose of 10 mg/day. In a recent study (Herndon *et al.,* 1990) of the effect of GH in children with burns, the authors demonstrated an accelerated rate of donor-site healing and shortened length of stay for equivalently sized severe burns in a placebo controlled, double-blind trial. The effect was substantial, as the calculated reduction of stay for a patient with a 60% body surface area burn was nearly 2 weeks. Gore *et al.* (1991) recently documented both decreased protein breakdown and increased protein synthesis in extremities of burn patients treated with GH. Elevation of circulating insulin levels by administration of exogenous insulin and glucose had a similar effect on protein turnover. Given concerns about potential glucose intolerance and possible hyperglycemia induced by high-dose GH therapy, it is tempting to speculate about simultaneous use of GH and insulin to maximize anabolic effect while balancing the risk of hyper versus hypoglycemia. Also of interest will be studies comparing the effects of GH versus recombinant IGF-I, administered either systemically or locally.

*c. GH and Its Role in the Treatment of Sepsis.* Studies of GH effects in septic states have been undertaken in the anticipation that the intensely catabolic stress of infection might be decreased. Results, however, have been largely disappointing, and it appears that septic patients may respond differently as a group to GH therapy than do patients with other catabolic conditions. Dahn *et al.* (1988) studied a small group of septic patients who received GH treatment for 2 days while receiving hypocaloric dextrose infusion; in contrast to normal subjects, none of these patients showed an increased IGF-I response, increase in splanchnic amino acid uptake, or decrease in urea excretion. This study suffers somewhat from the lack of a placebo-treated group of septic patients, who might well have shown more negative nitrogen balance in response to the hypocaloric stress under which the study was performed. An additional issue is the brief 2-day duration of GH therapy, given previous studies in postoperative patients showing that improvement in nitrogen balance may take anywhere from

214    BARBARA M. LIPPE AND JON M. NAKAMOTO

1 to 3 days to become evident, and up to a week to achieve a positive value (Jiang *et al.*, 1989). This may explain why Douglas *et al.* (1990) did see an improvement in nitrogen balance after 3 days of rhGH treatment in septic patients on more substantial parenteral nutrition, although they were not able to achieve a positive nitrogen balance. Notably, these changes were primarily due to increased protein synthesis rather than decreased protein catabolism. Thus, sepsis may decrease responsiveness to GH, which is consistent with the studies that suggest GH is more useful in the later stages of recovery than in the acute injury/illness phase.

*d. GH and Serious Long-Term Catabolic Conditions.* A number of serious but nonsurgical catabolic conditions also present problems in nutritional management. These include conditions such as chronic obstructive pulmonary disease (COPD), which will be used as a model for this discussion. Although clinical studies with GH are few, at present, we anticipate that they will be undertaken in the future. In patients with COPD, Suchner *et al.* (1990) found that 1 week of GH treatment increased fat oxidation and decreased protein and glucose oxidation, with a resultant decrease in respiratory quotient that allowed an increased number of calories to be delivered without further increasing the load of carbon dioxide. Positive nitrogen balance was achieved with a eucaloric rather than a hypercaloric diet (which increases the load of carbon dioxide and the work load of respiration). In this brief study there was little functional change in pulmonary function or muscle strength by dynamometer. A more recent article from Pape *et al.* (1991) describes an uncontrolled pilot study of seven patients with COPD who received 3 weeks of daily GH. Results included substantial weight gain, improved nitrogen balance, and improved pulmonary function based on measurements of maximal inspiratory pressure. Mean fasting serum glucose levels rose slightly and reversibly; in contrast to studies in other clinical situations, there was no evidence of fluid retention in these patients. More conclusive results in this and other catabolic medical conditions await placebo-controlled, double-blind studies.

*e. GH Effect on Wound Healing.* Several clinical reports have suggested that wound healing may be accelerated by GH treatment when administered systemically (Prudden *et al.*, 1958; Barbul *et al.*, 1978) or locally (Waago, 1987). Animal studies have suggested that GH can increase wound strength and healing (Pessa *et al.*, 1985; Jørgensen and Andreassen, 1988; Jørgensen *et al.*, 1989). In experimentally wounded rats, GH treatment led to significantly stronger wound strength (as measured by a tensiometer) than did placebo. However, this effect was abolished by the coexistence of a second, severe acute injury from burning (Belcher and Ellis, 1990), suggesting that GH may only be efficacious in mild injury or in the recovery phase of more severe injury. These effects

USES OF GROWTH HORMONE                                             215

on wound healing are likely to be mediated by IGF-I, given the known synergism between it and other growth factors in wound healing (Lynch *et al.*, 1989). In an animal model of wound healing, tissue DNA and protein content decreased after hypophysectomy but increased to control levels after local instillation of IGF-I (Steenfos *et al.*, 1989). However, the study by Herndon *et al.* (1990) in burn patients, which demonstrates an increase in the rate of donor-site healing, confirms that GH administration alone can affect the wound healing effect even in a moderately catabolic patient.

### 4. The Role of IGF-I Versus GH in Promotion of Nitrogen Retention

Most evidence points to IGF-I as the mediator of the protein-sparing effect of GH. Higher IGF-I levels correlate closely with positive nitrogen balance (Underwood *et al.*, 1986; Dahn *et al.*, 1988), although negative protein balance can occur under the stress of hypocaloric nutrition even prior to a decrease in IGF-I concentrations (Manson *et al.*, 1988). Clinical observations in acutely ill trauma patients or postsurgical patients in the intensive-care unit reveal increased GH levels and decreased serum IGF-I levels (Frayn *et al.*, 1984; Hawker *et al.*, 1987; Ross *et al.*, 1991), a pattern also seen in states of malnutrition or in fasting normal subjects (Phillips *et al.*, 1984; Ho *et al.*, 1988). As noted previously, critically ill septic patients show neither increased IGF-I levels nor decreased urea excretion in response to a brief course of GH administration (Dahn *et al.*, 1988). Infusion of recombinant IGF-I in normal men suppresses indices of protein breakdown such as radiolabeled leucine release (Elahi *et al.*, 1991).

Could IGF-I therefore take the place of rhGH in efforts to preserve body nitrogen stores, perhaps avoiding GH antagonism of insulin effects? Or is GH required as a permissive factor, as might be predicted by the "dual effector" theory of growth hormone action (Green *et al.*, 1985)? The increasing availability of recombinant IGF-I will no doubt help resolve this issue.

### G.   DEFENSE AGAINST CATABOLISM OF BODY PROTEIN: GLUCOCORTICOID THERAPY

### 1.   Overview: Metabolic Effects of Glucocorticoids

GH and glucocorticoids exert opposing effects on a number of metabolic processes. Glucocorticoids, in addition to antiinflammatory and immunosuppressive effects, exert catabolic effects such as increased muscle protein breakdown; decreased DNA and protein synthesis in muscle, heart, kidney, and liver; disruption of bone mineralization and inhibition of linear growth in children (Loeb, 1976). GH, in contrast, decreases efflux of amino

216    BARBARA M. LIPPE AND JON M. NAKAMOTO

acids from muscle, increases protein synthesis in muscle and major organs (Fong *et al.*, 1989), and stimulates osteoblastic activity (Stracke *et al.*, 1984).

In contrast to their antagonistic effects on protein turnover and growth, GH and glucocorticoids can produce carbohydrate intolerance by inducing resistance to the effects of insulin on glucose disposal. Horber *et al.* (1991) suggest that these effects may be via independent mechanisms, based on synergistic effects of combination therapy on insulin antagonism.

### 2. Clinical Studies

Preliminary evidence suggests that GH can prevent the protein wasting induced by glucocorticoid therapy, at least over the short term. Horber and Haymond (1990) used nitrogen balance and isotope dilution techniques to study protein metabolism in normal, adequately fed adult volunteers given 1 week of placebo, GH alone, prednisone alone, or a combination of GH and prednisone. By either technique, prednisone alone induced negative protein balance, GH alone induced significantly higher positive protein balance compared to controls, and a combination of prednisone–GH treatment induced comparable or slightly more positive protein balance relative to controls. The final clinical relevance of this study is uncertain, as it focuses on short-term, relatively high-dose glucocorticoid therapy in normal subjects. Still unresolved is the efficacy of GH over a longer term in patients on lower dose, more chronic glucocorticoid therapy. Also to be determined is the lowest effective dose, in an effort to minimize known complications such as carbohydrate intolerance and unknown risks.

## H.  BONE PHYSIOLOGY/OSTEOPOROSIS

Known effects of GH on bone and mineral homeostasis has raised hopes of potential therapeutic use in pathologic states such as osteoporosis. Indirect actions of GH include enhancement of calcium and phosphorus absorption from the intestine, either by increased production of 1,25-dihydroxyvitamin D (Spanos *et al.*, 1978) or increased intestinal sensitivity (Chipman *et al.*, 1980). Actions of GH on bone are believed to be both direct (Isaksson *et al.*, 1982) and via local generation of IGF-I (Isaksson *et al.*, 1987). GH or IGF-I secreted by bone cells can stimulate proliferation of osteoblasts, on which GH receptors have been identified (Slootweg *et al.*, 1988; Hock *et al.*, 1988). Clinically, bone turnover has been noted to be increased in acromegaly, as measured by [47]Ca tracer studies (Aloia *et al.*, 1972), and levels of osteocalcin, a marker of osteoblast activity, increase after GH-releasing hormone treatment in postmenopausal women

USES OF GROWTH HORMONE 217

(Franchimont et al., 1989). These findings have suggested that GH promotes bone turnover, and might have a role in treatment of osteroporosis.

Clinical trials of GH in postmenopausal osteoporosis have been, to date, disappointing. Previous clinical studies of 6 months of administration of human pituitary GH alone at low (0.2 U) or high dose (2 U) daily (Aloia et al., 1976) or on alternate days with calcitonin (Aloia et al., 1985) showed no improvement in bone mineral mass by single-photon absorptiometry (SPA) or bone biopsy, although there was some evidence that GH increased the area of bone undergoing active turnover. Notably, the higher doses of GH (2 U daily) were associated with increased urinary calcium and hydroxyproline, indices of increased bone resorption. A third protocol was attempted, with GH treatment for 2 months, followed by calcitonin treatment for 3 months (Aloia et al., 1987); although there was increased body calcium content and a trend toward increased bone mass by SPA, no change was evident on four paired bone biopsies. The data of Marcus et al. (1990) (Section IV,C,2) in elderly women confirm the evidence of active bone turnover without an apparent change in bone mineral content. Overall, at this time GH treatment appears to have little place in the treatment of most osteoporosis because it increases bone turnover by stimulating both osteoblastic and osteoclastic activity. However, the possibility remains that it may be useful for the less common "low-turnover" form of osteoporosis.

The patient population for whom GH may prove useful in terms of effects on bone is the GH-deficient adult. A pilot study by van der Veen and Netelenbos (1990) found that 8 weeks of GH treatment led to an increase in serum osteocalcin, urinary hydroxyproline, and bone mineral mass by dual photon absorptiometry of the lumbar spine. Whether this effect can be sustained or have clinical effect on osteoporosis in the GH-deficient adult is unknown at this time.


## I. MODULATION OF IMMUNOLOGIC STATUS

### 1. Rationale

A small number of studies suggest that GH deficiency in animals is associated with some degree of immunological abnormality and that GH treatment may at least partially correct this defect. In mice with autosomal recessive dwarfism (Fabris et al., 1971), chickens with sex-linked dwarfism (Marsh et al., 1984), and immunodeficient dwarf dogs (Roth et al., 1984), GH treatment enhanced immunologic function, including antibody responses and (in the chicken) bursal growth. GH deficiency after hypothalamic lesions has been shown to decrease natural killer (NK) cell function

218 BARBARA M. LIPPE AND JON M. NAKAMOTO

in rats (Cross *et al.*, 1984); deficient antibody responses have also been reported and can be corrected by administration of either GH or prolactin (Nagy *et al.*, 1983). GH treatment has been shown to enhance the activity of rat natural killer cells (Davila *et al.*, 1987) and prime neutrophils to secrete superoxide anion (Fu *et al.*, 1991). In a recent article, Murphy *et al.* (1992) reported preliminary findings of GH stimulation of reconstitution of T cells after thymocyte or bone marrow transplant into immunodeficient mice. Examination of the GH-treated mice showed significantly more T-cells in the thymus and lymph nodes than in untreated mice; >85% of the T cells were CD4+ (helper T cell)/CD3+. The clinical implications for bone marrow transplants and AIDS are intriguing.

Also interesting is a study by Edwards *et al.* (1991) showing significantly increased survival after *Salmonella typhimurium* infection in both intact and hypophysectomizd rats given natural or recombinant porcine GH compared to placebo. This effect was eliminated by heat inactivation of GH or by addition of anti-GH antibodies. Macrophages from hypophysectomized rats were significantly less effective than those from intact rats in an *in vitro S. typhimurium* killing test; killing ability was significantly increased by *in vivo* GH treatment prior to isolation of macrophages.

## 2. Clinical Studies

A survey of studies in humans reveals no clear-cut agreement on the immunologic status of patients with GH deficiency. Some authors have reported in GH-deficient patients normal immunologic function test values (Abassi and Bellanti, 1985; Peterson *et al.*, 1990; Spadoni *et al.*, 1991), while others have reported immunologic test abnormalities such as increased suppressor T cell number (Gupta *et al.*, 1983), decreased natural killer cell activity (Kiess *et al.*, 1986; Bozzola *et al.*, 1990), or decreased production of interleukin-2 by mononuclear cells (Casanova *et al.*, 1990). There has been no documentation of increased susceptibility to infection in GH deficiency. The apparent small increase in leukemia incidence reported in GH-deficient patients treated with GH (Fisher *et al.*, 1988) or untreated (Redman *et al.*, 1988) has provoked some speculation about the contribution of subtle immunologic abnormalities, but to date no documentation of specific immunodeficiency has been reported in GH-deficient patients with leukemia.

Conflicting results also have arisen from immunologic studies of GH-deficient patients receiving replacement GH therapy. Two groups have reported a transiently reduced B cell percentage of total lymphocytes (Petersen *et al.*, 1990; Spadoni *et al.*, 1991) after GH treatment of GH-deficient patients, while others have seen little or no change (Etzioni *et al.*, 1988; Church *et al.*, 1989); immunoglobulin levels do not change

USES OF GROWTH HORMONE 219

appreciably. Decreases in the ratio of CD4+ (T helper) to CD8+ (T suppressor) cells during GH therapy have been reported; it is interesting to note that these studies (Rapaport *et al.*, 1986; Bozzola *et al.*, 1989) employed intramuscular administration of GH, while other studies employing a subcutaneous route observed no significant change in the CD4+ to CD8+ ratio (Spadoni *et al.*, 1991; Church *et al.*, 1989; Etzioni *et al.*, 1988; Petersen *et al.*, 1990).

Stimulatory activity of GH on the immune system has been reported. In GH-deficient children, GH treatment corrects low mononuclear cell interleukin-1a (IL-1a) and IL-2 production (Casanova *et al.*, 1990). GH treatment increases NK cell number (Caruso-Nicoletti *et al.*, 1991) or activity in GH-deficient children (Bozzola *et al.*, 1990), adults with poor GH responses to provocative testing (Crist *et al.*, 1987), and normal adults (Crist and Kraner, 1990). Two patients with Laron dwarfism were noted to have higher (94th and 97th percentile) NK cell numbers/percentages and somewhat lower B cell numbers compared to controls and GH-deficient patients; such a report, while too preliminary to be conclusive, does raise the possibility of a route of GH action on the immune system through other than GH receptors or IGF-I action.

One further potential application of GH is to preserve weight and lean body mass in patients with AIDS. Krentz *et al.* (1991) have recently presented preliminary evidence from a small double-blind study of a 3-month GH treatment in men with AIDS. Prestudy weight loss was reversed, with increases coming from lean body mass and mild retention of fluid. Adipose tissue mass decreased. Muscular function also improved slightly, as measured by dynamometry. Despite animal data suggesting a positive effect of GH on the percentage of CD4+ cells, no changes in CD4+ cells were seen in this study.

## V. Potential Adverse Effects

Having discussed the growth-promoting and metabolic effects of GH in children and the metabolic effects in adults, it is obvious that GH (and for the purposes of this discussion, the IGF-I that is generated as a consequence) has multiple potent effects on numerous target tissues. It is beyond the scope of this report to review, for example, all of the clinical and metabolic alterations that occur in acromegaly. Some of the real or potentially adverse effects have been mentioned within the various sections of this report. Suffice it to say that investigators recognize that any GH treatment protocol should be designed and monitored to prevent the known adverse alterations of metabolism that can result from GH excess. Alterations in carbohydrate metabolism, thyroid hormone metabolism,

220                BARBARA M. LIPPE AND JON M. NAKAMOTO

renal function, salt and water metabolism (with the development of clinical
edema and the potential for hypertension), soft tissue enlargement (including the development of carpal tunnel syndrome), and arthralgia have all
been reported in recent clinical studies, although no serious or nonre-
versable effects have been noted. We will discuss, instead, only a few of
those known or potential adverse effects that appear either to be unique
to children or represent major areas of potential toxicity for children or
adults.

## A. CONCERNS FOR THE GROWING CHILD

### 1. Antibody Formation

Initial preparations of human pituitary GH were highly antigenic, although the actual number of documented patients who exhibited an attenuated growth response following an initial response, as a consequence of
the development of anti-GH antibodies, was small, save for patients with
a familial/genetic form of GH deficiency (reviewed by Frasier, 1983). With
the introduction of recombinant human GH, the question of purity of
preparation and incidence of antibody production was again raised, but as
the methods for removal of the *E. coli* proteins, which accompanied the
GH, improved, antigenicity declined, and the risk of growth-attenuating
antibody formation is essentially over (reviewed by Jørgensen, 1991).

### 2. Slipped Capital Femoral Epiphysis

The condition of slipped capital femoral epiphysis (SCFE) occurs in
growing children, most commonly in association with obesity or trauma
in the peripubertal period. It is also seen in conditions in which there is
marked delay of normal epiphyseal maturation, such as hypothyroidism
or GH deficiency. Thus, the finding of an increased incidence of this
condition in children with hypopituitarism who were subsequently treated
with GH (Rappaport and Fife, 1985) was interesting, but was not considered by most pediatric endocrinologists to be a specific side effect of the
GH treatment per se. Nevertheless, others have reported their experiences
with individual patients (Prasad *et al.*, 1990) treated with GH who had few
risk factors. Thus, the development of SCFE is listed as a potential risk
during GH therapy.

### 3. Hypoglycemia and Hypermetabolism

Hypoglycemia occurs in some infants and children with GH deficiency,
presumably as a consequence of both increased insulin sensitivity (lack of
the normal ''resistance'' to insulin present in the GH-sufficient individual)
and decreased ability to perform gluconeogenesis. The acute administra-

USES OF GROWTH HORMONE          221

tion of GH results in the rapid documentation of a degree of insulin resistance (Lippe *et al.*, 1981) and GH is indicated as part of the treatment of the hypoglycemia of hypopituitarism in childhood. However, when GH was administered either thrice weekly or every other day, hypoglycemia often recurred. Press *et al.* (1987) reported the occurrence of hypoglycemia 36–60 hours after GH treatment in three GH-deficient children and stressed the need for daily therapy to maintain plasma glucose levels in these at-risk children. In a child with a history of hypoglycemia, this caution should be noted by those who use GH daily, save for a "free" day once a week.

The lipolytic and anabolic effects of GH that are being assessed in the adult population have been recognized as occurring in children, although this is deemphasized in favor of reports of the growth-promoting effects. In the report of Walker *et al.* (1990) describing the responses of a group of short, slowly growing non-GH-deficient children, the authors stress the loss of fat mass and increase in lean body mass, as well as the increase in resting energy expenditure that occurred during the first 6 months of GH therapy. Although they present no direct evidence that this metabolic "stress" is adverse, and their treatment trail continues, this report has been cited in the lay press to indicate an adverse effect of GH. The authors do not concur with this interpretation but present the citation for completeness.

### 4. Therapeutic Failure

Should it occur, the psychologic implication of a given child's failure to respond well to the growth-promoting action of GH should be considered as a potential adverse effect of therapy. Coupled with the inherent message that treatment was undertaken because the child's stature was not acceptable, the failure to meet the expectations of family, physicians, and self could lead to significant psychologic consequences. Even treatment success must be viewed with caution. Increasing final adult stature with GH does not alter the genetic potential of the individual. Thus, his or her children will have the original parental height potential and be at risk not only for being short, but also for appearing shorter than one might expect given the observed parental heights.

### B. CONCERNS ABOUT ONCOGENESIS IN CHILDREN AND ADULTS

### 1. GH Effects on Existing Malignancies

Intracranial neoplasms may be associated with GH deficiency either directly by causing destruction of the hypothalamus or pituitary or indirectly as a result of the radiotherapy that has been used in their treatment.

222                BARBARA M. LIPPE AND JON M. NAKAMOTO

Similarly, high-dose cranial irradiation for CNS leukemia has also resulted
in GH deficiency (Shalet et al., 1976). In both cases GH has been considered as appropriate treatment when deficiency is present. However, the
issue of whether GH will cause a recurrence of the neoplasm as a consequence of its own mitogen activity has always been a concern to clinicians
(Rogers et al., 1977). The clinical attempts to answer the question about
leukemia and solid tumor recurrence take the form of retrospective assessments of recurrences in patients on treatment as compared to those never
treated (Arslanian et al., 1985; Clayton et al., 1987). The two studies cited
fail to show an increased recurrence risk in any category, albeit the number
of patients treated in both cohorts is small. Currently, most countries in
which recombinant GH is available maintain a registry of adverse events,
either through a governmental agency or through the commercial compa-
nies that manufacture the GH.

## 2.  GH and the Occurrence of New Malignancies

The known mitogenic potential of GH, coupled with data suggesting
that there is an increase in premalignant adenomatous polyps and gastrointestinal malignancies in patients with acromegaly (Ituarte et al., 1984), has
raised clinical concern about the oncogenic potential of GH treatment.
However, prior to a report of the association of GH treatment and leukemia
in five patients from Japan (Watanabe et al., 1988) and one from the
Netherlands (Delemarre-Van de Waal et al., 1988), the oncogenic potential
of GH had not been seriously considered as a potential adverse effect of
treatment. Discussions of this issue now appear in the literature (Fisher
et al., 1988; Stanke and Zeisel, 1989), but a causal link has yet to be
determined. As of May 1, 1991, 26 patients worldwide with GH deficiency
treated with GH are known to have developed de novo leukemia (report
of the Drugs and Therapeutics Committee, Lawson Wilkins Pediatric
Endocrine Society Annual Meeting, unpublished). While the occurrence
of the association exceeds the expected occurrence of leukemia alone in
children of the same age, there are several confounding variables that may
preclude implicating GH per se. These include, in many of the GH-treated
patients, the risk factors of a preceding malignancy, with or without radiation and chemotherapy (immunosuppressive therapy), conditions that
might predispose to a malignancy (syndromes associated with chromosomal breakage), and conditions with major metabolic abnormalities
(Kearns–Sayre syndrome). Of special note are the high numbers of patients who developed leukemia following a preceding brain tumor (two-
thirds of the United States cases). The question of the relationship of brain
tumors to the development of leukemia has been examined in a recent
report (Blatt et al., 1991). In the authors' review of the literature, the data

USES OF GROWTH HORMONE    223

appear to demonstrate an increased risk of leukemia developing in children or adults following an intracranial neoplasm, independent of the issue of GH treatment. One patient had documented GH deficiency as a consequence of the radiation treatment of the brain tumor but had not yet been treated with GH when the leukemia developed. The converse, the development of a second neoplasm after acute leukemia, with those being especially of the central nervous system in CNS-irradiated children (Neglia *et al.*, 1991), further confounds the issue. Not only does there appear to be an interrelationship between brain tumors and leukemia that is independent of GH, but there is also the risk that these second malignancies will develop in those cured leukemia patients who have been treated with GH. Finally, there is at least one published account of a GH-deficient child.

The concerns about the potential oncogenic effects of GH apply to the adult as well. We have discussed the issue of gastrointestinal malignancies in acromegaly, and therefore questions about monitoring for such occurrences in adults who might be part of long-term GH treatment protocols must be addressed. Because any discussion of the consequences of GH treatment must also include any potential adverse effects of high concentrations of either circulating or autocrine/paracrine IGF-I, the entire literature on the mitogenic actions of the IGFs could be cited. Suffice it to say that several human cancers have been reported to express IGF mRNA, protein, or receptors, and that studies of organs/tissues at high risk for adult malignancy (breast, prostate, uterus) will be of intense concern to investigators designing any long-term adult treatment protocol.

## VI.  Summary

Although GH has been available as a therapeutic agent for the GH-deficient child for more than 30 years, the conditions of its use have yet to be optimized. The availability of biosynthetic material has provided researchers with the opportunity to develop the protocols necessary to begin to finally answer the most fundamental questions pertaining to dose, frequency, and duration of treatment. It has also permitted the initiation of prospective trials in a large number of conditions that result in childhood short stature, with the expectation that some or many of them will be treated effectively and safely. Finally, it has opened the door to an entire spectrum of potentially new uses of GH and other growth factors for so-called nonconventional indications. That these have implications that range from the short-term rapid healing of a burngraft site, to the more efficient induction of ovulation, to the long-term preservation of lean body mass has excited the interest of investigators in many fields of medicine

224 BARBARA M. LIPPE AND JON M. NAKAMOTO

and physiology. Thus, the recent progress reported in this paper is really the beginning of the new research that will take place with GH and growth factors.

## ACKNOWLEDGMENTS

B.M.L. acknowledges the Genentech National Collaborative Study Investigators in the collection and preparation of some of these data, and Genentech, Inc., for the grant support for the studies in GH deficiency, Turner syndrome, idiopathic short stature, and chronic renal failure in which the author was a participant. B.M.L. also acknowledges the help of S. Douglas Frasier in the preparation of this manuscript. J.M.N. is supported by a K-11 Physician Scientist Award (DK01997) from the NIH.

## REFERENCES

Abassi, V., and Bellanti, J. A. (1985). *Pediatr. Res.* **19,** 299–301.
Aceto, T., Jr., Frasier, S. D., Hayles, A. B., Meyer-Bahlburg, H. F. L., Parker, M. L., Munschauer, R., and Di Chiro, G. (1972). *J. Clin. Endocrinol. Metab.* **35,** 483–496.
Adashi, E. Y., Resnick, C. E., Hurwitz, A., Ricciarelli, E., Hernandez, E. R., and Rosenfeld, R. G. (1991). *Endocrinology (Baltimore)* **128,** 754–760.
Ahmed, L., Foot, A. B. M., Edge, J. A., Lamkin, V. A., Savage, M. O., and Dunger, D. B. (1991). *Acta Pediatr. Scand.* **80,** 446–450.
Albertsson-Wikland, K. (1987). *Acta Pediatr. Scand. (Suppl.)* **331,** 28–34.
Albertsson-Wikland, K. (1988). *Acta Pediatr. Scand. (Suppl.)* **343,** 77–84.
Albertsson-Wikland, K. (1989). *Acta Pediatr. Scand. (Suppl.)* **349,** 35–41.
Albertsson-Wikland, K., and Karlberg, J. (1991). *Horm. Res.* **35**(Suppl. 2), 32 (Abstract 121).
Albertsson-Wikland, K., and Rosberg, S. (1988). *J. Clin. Endocrinol. Metab.* **67,** 493–500.
Albertsson-Wikland, K., and Rosberg, S. (1991). *In* "Turner Syndrome: Growth Promoting Therapies" (M.B. Ranke and R.G. Rosenfeld, eds.), pp. 23–28. Elsevier Science Publishers B.V., The Netherlands.
Albertsson-Wikland, K., Rosberg, S., Isaksson, O., and Westphal, O. (1983). *Acta Endocrinol. (Copenh.)* **103**(Suppl. 256), 72.
Albertsson-Wikland, K., Westphal, O., and Westgren, U. (1986). *Acta Pediatr. Scand.* **75,** 89–97.
Aloia, J. F., Roginsky, M. S., Jowsey, J., Dombrowski, C. S., Shukla, K. K., and Cohn, S. H. (1972). *J. Clin. Endocrinol. Metab.* **33,** 543–551.
Aloia, J. F., Vaswani, A., Kapoor, A., Yeh, J. K., and Cohn, S. H. (1985). *Metabolism* **34,** 124–129.
Aloia, J. F., Vaswani, A., Meunier, P. J., Edouard, C. M., Arlot, M. E., Yeh, J. K., and Cohn, S. H. (1987). *Calcif. Tissue Int.* **40,** 253–259.
Aloia, J. F., Zanzi, I., Ellis, K., Jowsey, J., Roginsky, M., Wallach, S., and Cohn, S. H. (1976). *J. Clin. Endocrinol. Metab.* **43,** 992–999.
Amsalem, S., Duquensoy, F., Attree, O., Novelli, G., Bousnina, S., Postel-Vinay, M. -C., and Goossens, M. (1989). *N. Engl. J. Med.* **321,** 989–995.
Angulo, M., Castro-Magana, M., Uy, J., and Rosenfeld, W. (1991). *Pediatr. Res.* **29,** 126A.
Anneren, G., Gustavson, K. -H., Sara, V. R., and Tuvemo, T. (1990). *Am. J. Med. Genet. (Suppl.)* **7,** 59–62.
Anneren, G., Sara, V. R., Hall, K., and Tuvemo, T. (1986). *Arch. Dis. Child.* **61,** 48–52.

## USES OF GROWTH HORMONE 225

Appan, S., Laurent, S., Chapman, M., Hindmarsh, P. C., and Brook, C. G. D. (1990). *Acta Pediatr. Scand.* **79**, 796–803.

Arnaud, C. D. (1973). *Kidney Int.* **4**, 89–95.

Arslanian, S. A., Becker, D. J., Lee, P. A., Drash, A. L., and Foley, T. P., Jr. (1985). *Am. J. Dis. Child.* **139**, 347–350.

Bando, H., Zhang, C., Takada, Y., Yamasaki, R., and Saito, S. (1991). *Acta Endocrinol. (Copenh.)* **124**, 31–36.

Barbul, A., Rettura, G., Prior, E., Levenson, S. M., and Seifter, E. (1978). *Surg. Forum* **29**, 93–95.

Baue, A. E. (1991). *Surg. Clin. North Am.* **71**, 549–565.

Belcher, H. J. C. R., and Ellis, H. (1990). *J. Clin. Endocrinol. Metab.* **70**, 939–943.

Bengtsson, B. -A., Brummer, R. -J., and Bosaeus, I. (1990). *Horm. Res.* **33**(Suppl. 4), 19–24.

Bercu, B. B., Shulman, D., Root, A. W., and Spiliotis B. E. (1986). *J. Clin. Endocrinol. Metab.* **63**, 709–716.

Bierich, J. R. (1986). *Acta. Pediatr. Scand. (Suppl.)* **325**, 13–18.

Binkin, N. J., Yip, R., Fleshood, L., and Trowbridge, F. L. (1988). *Pediatrics* **82**, 828–834.

Blatt, J., Bercu, B. B., Gillin, J. C., Mendelson, W. B., and Poplack, M. D., (1984). *J. Pediatr.* **104**, 182–186.

Blatt, J., Penchansky, L., Phebus, C., and Horn, M. (1991). *Pediatr. Hematol. Oncol.* **8**, 77–82.

Blethen, S. L., Lippe, B. M., August, G. P., Rosenfeld, R. G., Compton, P. G., Attie, K. M., Johanson, A. J., and the National Cooperative Growth Study. (1991). *Pediatr. Res.* **29**, 74A.

Blum, W. F., Ranke, M. B., Keitzmann, K., Gauggel, E., Zeisel, H. J., and Bierich, J. (1990). *J. Clin. Endocrinol. Metab.* **70**, 1292–1298.

Blum, W. F., Ranke, M. B., Kietzmann, K., Tonshoff, B., and Mehls, O. (1989). *In* "Insulin-like Growth Factor Binding Proteins" (S.L.S. Drop and R.L. Hintz, eds.), pp. 93–99. Elsevier, The Netherlands.

Blumenfeld, Z., and Lunenfeld, B. (1989). *Fertil. Steril.* **52**, 328–331.

Bourguignon, J. P., Vandeweghe, M., Vanderschueren-Lodeweyckx, M., Malvaux, P., Wolter, R., Du Caju, M., and Ernould, C. (1986). *J. Clin. Endocrinol. Metab.* **63**, 376–382.

Bozzola, M., Cisternino, M., Biscaldi, I., Maghnie, M., Valtorta, A., Moretta, A., and Severi, F. (1988). *Eur. J. Pediatr.* **147**, 248–251.

Bozzola, M., Cisternino, M., Valtorta, A., Moretta, A., Biscaldi, I., Maghnie, M., De Amici, M., and Schimpff, R. M. (1989). *Horm. Res.* **31**, 153–156.

Bozzola, M., Valtorta, A., Moretta, A., Cisternino, M., Biscaldi, I., and Schimpff, R. -M. (1990). *J. Pediatr.* **117**, 596–599.

Bray, G. A., Dahms, W. T., Swerdloff, R. S., Fiser, R. H., Atkinson, R. L., and Carrel, R. E. (1983). *Medicine* **62**, 59–80.

Buchanan, C. R., Preece, M. A., and Milner, R. D. G. (1991). *Br. Med. J.* **302**, 824–828.

Bundak, R., Hindmarsh, P. C., Smith, P. J., and Brook, C. G. D. (1988). *J. Pediatr.* **112**, 875–879.

Burger, H. G., Kovacs, G. T., Polson, D. M., McDonald, J., McCloud, P. I., Harrop, M., Colman, P., and Healy, D. L. (1991). *Clin. Endocrinol.* **35**, 119–122.

Burns, E. C., Tanner, J. M., Preece, M. A., and Cameron, N. (1981). *Eur. J. Pediatr.* **137**, 155–164.

Butenandt, O. (1979). *Eur. J. Pediatr.* **130**, 15–28.

Butler, M. G., Meaney, F. J., and Palmer, C. G. (1986). *Am. J. Med. Genet.* **23**, 793–809.

Cara, J. F., and Johanson, A. J. (1990). *Pediatr. Clin. North Am.* **37**, 1129–1254.

Caruso-Nicoletti, M., Mancuso, G. R., Sciotto, A., Spadaro, G., Guarcello, V., Farinella, Z., Lupo, L., and Schiliro, G. (1991). *J. Pediatr. Endocrinol.* **4,** 33–39.
Casanova, S., Repellin, A. M., and Schimpff, R. M. (1990). *Horm. Res.* **34,** 209–214.
Chipman, J. J., Zerwekh, J., Nicar, M., Marks, J., and Pak, C. Y. C. (1980). *J. Clin. Endocrinol. Metab.* **51,** 321–324.
Christiansen, J. S., and Jørgensen, J. O. L. (1991). *Acta. Endocrinol. (Copenh.)* **125,** 7–13.
Christiansen, J. S., Jørgensen, J. O., Pedersen, S. A., Møller, J., Jørgensen, J., and Skakkebœk, N. E. (1990). *Horm. Res.* **33**(Suppl. 4), 61–64.
Chrousos, G. P., Poplack, D., Brown, T., O'Neill, D., Schwade, J., and Bercu, B. B. (1982). *J. Clin. Endocrinol. Metab.* **54,** 1135–1139.
Church, J. A., Costin, G., and Brooks, J. (1989). *J. Pediatr.* **115,** 420–423.
Cianfarani, S., Spadoni, G. L., Finocchi, G., Ravet, P., Costa, F., Papa, M., Scirè, G., Manca Bitti, M. L., and Boscherini, B. (1987). *Minerva Pediatr.* **39,** 281–284.
Clayton, P. E., Shalet, S. M., Gattamaneni, H. R., and Price, D. A. (1987). *Lancet* **1,** 711–713.
Clemmons, D. R., Snyder, D. K., Williams, R., and Underwood, L. E. (1987). *J. Clin. Endocrinol. Metab.* **64,** 878–883.
Collipp, P. J., Curti, V., Thomas, J., Sharma, R. K., Maddaiah, V. T., and Cohn, S. H. (1973). *Metabolism* **22,** 589–595.
Committee on Growth Hormone Use of the Lawson Wilkins Pediatric Endocrine Society. (1985). *J. Pediatr.* **107,** 10–12.
Costin, G., Kaufman, F. R., and Brasel, J. A. (1989). *J. Pediatr.* **115,** 537–544.
Crist, D. M., and Kraner, J. C. (1990). *Metabolism* **39,** 1320–1324.
Crist, D. M., Peake, G. T., Mackinnon, L. T., Sibitt, W. L., Jr., and Kraner, J. C. (1987). *Metabolism* **36,** 1115–1117.
Cross, R. J., Markesbery, W. R., Brooks, W. H., and Roszman, T. L. (1984). *Immunology* **51,** 399–405.
Cuneo, R. C., Salomon, F., Wiles, C. M., and Sönksen, P. H. (1990). *Horm. Res.* **33**(Suppl. 4), 55–60.
Cuthbertson, D. P., Shaw, G. B., and Young, F. G. (1941). *J. Endocrinol.* **2,** 468–474.
Dahn, M. S., Lange, P., and Jacobs, L. A. (1988). *Arch. Surg.* **123,** 1409–1414.
Darendeliler, F., Hindmarsh, P. C., Preece, M. A., Cox, L., and Brook, C. G. D. (1990). *Acta Endocrinol. (Copenh.)* **122,** 414–416.
Davidson, M. B. (1987). *Endocr. Rev.* **8,** 115–131.
Davila, D. R., Brief, S., Simon, J., Hammer, R. E., Brinster, R. L., and Kelley, K. W. (1987). *J. Neurosci. Res.* **18,** 108–116.
Dean, H. J., McTaggart, T. L., Fish, D. G., and Friesen, H. G. (1985). *Am. J. Dis. Child.* **139,** 1105–1110.
Degerblad, M., Almkvist, O., Grunditz, R., Hall, K., Kaijser, L., Knutsson, E., Ringertz, H., and Thorén, M. (1990). *Acta. Endocrinol. (Copenh.)* **123,** 185–193.
Delemarre-Van de Waal, H. A., Odink, R. J. H., de Grauw, T. J., and de Waal, F. C. (1988). *Lancet* **1,** 1159.
Donaldson, D. L., Hollowell, J. G., Pan, F., Gifford, R. A., and Moore, W. V. (1989). *J. Pediatr.* **115,** 51–56.
Douglas, R. G., Humberstone, D. A., Haystead, A., and Shaw, J. H. F. (1990). *Br. J. Surg.* **77,** 785–790.
Dudl, R. J., Ensinck, J. W., Palmer, H. E., and Williams, R. H. (1973). *J. Clin. Endocrinol. Metab.* **37,** 11–16.
Edwards, C. K. III, Yunger, L. M., Lorence, R. M., Dantzer, R., and Kelley, K. W. (1991). *Proc. Natl. Acad. Sci. U.S.A.* **88,** 2274–2277.

## USES OF GROWTH HORMONE

Elahi, D., McAloon-Dyke, M., Fukagawa, N. K., Wong, G., Minaker, K. L., Seaman, J. J., Good, W. R., Vandepol, C. G., Shannon, R. S., Miles, J. M., and Wolfe, R. R. (1991). *In* "Modern Concepts of Insulin-like Growth Factors" (E. M. Spencer, ed.), p. 219–223. Elsevier, New York.

Emtner, M., Mathews, L. S., and Norstedt, G. (1991). *Mol. Endocrinol.* **4,** 2014–2020.

Escamilla, R. F., Hutchings, J. J., Deamer, W. C., Li, C. H., and Forsham, P. H. (1961). *J. Clin. Endocrinol. Metab.* **21,** 721–726.

Escamilla, R. F., Hutchings, J. J., Li, C. H., and Forsham, P. (1966). *Calif. Med.* **105,** 104–110.

Etherton, T. D. (1991). *J. Clin. Endocrinol. Metab.* **72,** 957A–957C.

Etzioni, A., Pollack, S., and Hochberg, Z. (1988). *Acta Pediatr. Scand.* **77,** 169–170.

Fabris, N., Pierpaoli, W., and Sorkin, E. (1971). *Clin. Exp. Immunol.* **9,** 227–240.

Facklam, T., Maillard, F., and Nguyen, D. (1988). *In* "Biosynthetic GH and GHRH: Basic and Clinical Aspects" (G. Chiumello and B. Di Natale, eds.), pp. 5–20. Ares-Serono Symposia via Ravenna, Italy.

Fernandez, A., Zachmann, M., Prader, A., and Illig, R. (1970). *Helv. Paediatr. Acta.* **25,** 566–576.

Fine, R. N. (1984). *In* "Clinical Dialysis" (A. R. Nissenson, R. N. Fine, and D. E. Gentile, eds.), pp. 661–670. Appleton and Lange, Norwalk.

Fine, R. N., Koch, V. H., Boechat, M. I., Lippe, B. M., Nelson, P. A., Fine, S. E., and Sherman, B. H. (1990). *Peritoneal Dialysis Int.* **10,** 209–214.

Fine, R. N., Pyke-Grimm, K., Nelson, P. A., Boechat, M. I., Lippe, B. M., Yadin, O., and Kamil, E. (1991a). *Pediatr. Nephrol.* **5,** 477–481.

Fine, R. N., Yadin, Ol, Nelson, P. A., Pyke-Grimm, K., Boechat, M. I. Lippe, B. M., Sherman, B. M., Ettenger, R. B., and Kamil, E. (1991b). *Pediatr. Nephrol.* **5,** 147–151.

Finkelstein, J. W., Roffwarg, H. P., Boyar, R. M., Kream, J., and Hellman, L. (1972). *J. Clin. Endocrinol. Metab.* **35,** 665–670.

Fisher, D. A., Job, J. -C., Preece M., and Underwood, L. E. (1988). *Lancet* **1,** 1159–1160.

Flodh, H. (1986). *Acta. Pediatr. Scand. (Suppl.)* **325,** 1–9.

Florini, J. R., Printz, P. N., Vitiello, M. V., and Hintz, R. L. (1985). *J. Gerontol.* **40,** 2–7.

Foley, T. P., Jr., Thompson, R. G., Shaw, M., Baghdassariam, A., Nissley, S. P., and Blizzard, R. M. (1974). *J. Pediatr.* **84,** 635–641.

Fong, Y., Rosenbaum, M., Tracey, K. J., Raman, G., Hesse, D. G., Matthews, D. E., Leibel, R. L., Gertner, J. M., Fischman, D. A., and Lowry, S. F. (1989). *Proc. Natl. Acad. Sci. U.S.A.* **86,** 3371–3374.

Fradkin, J. E., Schonberger, L. B., Mills, J. L., Gunn, W. J., Piper, J. M., Wysowski, D. K., Thomson, R., Durako, S., and Brown, P. (1991). *JAMA* **265,** 880–884.

Franchimont, P., Urbain-Choffray, D., Lambelin, P., Fontaine, M. -A., Frangin, G., and Reginster, J. -Y. (1989). *Acta Endocrinol. (Copenh.)* **120,** 121–128.

Frasier, S. D. (1974). *Pediatrics* **53,** 929–937.

Frasier, S. D. (1983). *Endocr. Rev.* **4,** 155–170.

Frasier, S. D., Costin, G., Ling, S. M., and Kaplan, S. A. (1969). *Pediatr. Res.* **3,** 557–561.

Frasier, S. D., Costin, G., Lippe, B. M., Aceto, T., Jr., and Bunger, P. F. (1981). *J. Clin. Endocrinol. Metab.* **53,** 1213–1217.

Frasier, S. D., Rudlin, C. R., Zeisel, H. J., Liu, H. H., Long, P. C., Senior, B., Finegold, D. N., Bercu, B. B., Marks, J. F., and Redmond, G. P. (1992). *Am. J. Dis. Child,* **146,** 582–587.

Frayn, K. N., Price, D. A., Maycock, P. F., and Carroll, S. M. (1984). *Clin. Endocrinol.* **20,** 179–187.

228    BARBARA M. LIPPE AND JON M. NAKAMOTO

Frazer, T., Gavin, J. R., Daughaday, W. H., Hillman, R. E., and Weldon, V. V. (1982). *J. Pediatr.* **101**, 12–15.

Fryburg, D. A., Gelfand, R. A., and Barrett, E. J. (1991). *Am. J. Physiol.* **260**, E499–E504.

Fryklund, L. (1987). *Acta. Pediatr. Scand. (Suppl.).* **331**, 5–8.

Fu, Y. -K., Arkins, S., Wang, B. S., and Kelley, K. W. (1991). *J. Immunol.* **146**, 1602–1608.

Gemzell, C. A., and Heijkenskjöld, F. (1956). *Endocrinology (Baltimore)* **59**, 681–687.

Genentech Collaborative Study Group. (1989). *J. Pediatr.* **115**, 713–719.

Genentech Collaborative Study Group. (1990). *Acta Pediatr. Scand. (Suppl.)* **366**, 24–26.

Gershberg, H., Neumann, L. L., and Mari, S. (1964). *Metabolism* **13**, 636–649.

Gertner, J. M., Genel, M., Gianfredi, S. P., Hintz, R. L., Rosenfeld, R. G., Tamborlane, W. V., and Wilson, D. M. (1984). *J. Pediatr.* **104**, 172–176.

Gertner, J. M., Horst, R. L., Broadus, A. E., Rasmussen, H., and Genel, M. (1979). *J. Clin. Endocrinol. Metab.* **49**, 185–188.

Gertner, J. M., Tamborlane, W. V., Gianfredi, S. P., and Genel, M. (1986). *J. Pediatr.* **110**, 425–428.

Glick, S. M., Roth, J., Yalow, R. S., and Berson, S. A. (1963). *Nature (London)* **199**, 784–787.

Goeddel, D. V., Heyneker, H. L., Hozumi, T., Arentzen, R., Itakura, K., Yansura, D. G., Ross, M. J., Miozzari, G., Crea, R., and Seeburg, P. H. (1979). *Nature (London)* **281**, 544–548.

Goodman, H. M., Gorin, E., and Honeyman, T. W. (1988). *In* "Human Growth Hormone: Progress and Challenges" (L. E. Underwood, ed.), pp. 75–111. Dekker, New York.

Gore, D. C., Honeycutt, D., Jahoor, F., Wolfe, R. R., and Herndon, D. N. (1991). *Arch. Surg.* **126**, 38–43.

Green, H., Morikawa, M., and Nixon, T. (1985). *Differentiation* **29**, 195–198.

Grunt, J. A., and Enriquez, A. R. (1972). *Pediatr. Res.* **6**, 664–674.

Grunt, J. A., Howard, C. P., and Daughaday, W. H. (1984). *Acta Endocrinol.* **106**, 168–174.

Gupta, S., Fikrig, S. M., and Noval, M. S. (1983). *Clin. Exp. Immunol.* **54**, 87–90.

Hawker, F. H., Stewart, P. M., Baxter, R. C., Borkmann, M., Tan, K., Caterson, I. D., and McWilliam, D. B. (1987). *Crit. Care Med.* **15**, 732–736.

Henneman, P. H. (1968). *JAMA* **205**, 828–836.

Hermanussen, M., Geiger-Benoit, K., Partsch, C. -J., and Burmeister, J. (1989). *Acta Pediatr. Scand.* **78**, 555–562.

Herndon, D. N., Barrow, R. E., Kunkel, K. R., Broemeling, L., and Rutan, R. L. (1990). *Ann. Surg.* **212**, 424–431.

Hibi, I., and Tanaka, T. (Committee for Treatment of Growth Hormone Deficient Children). (1989a). *Acta. Endocrinol.* **120**, 409–415.

Hibi, I., Tanaka, T., Tanae, A., Kagawa, J., Hashimoto, N., Yoshizawa, A., and Shizume, K. (1989b). *J. Clin. Endocrinol. Metab.* **69**, 221–226.

Hindmarsh, P. C., and Brook, C. G. D. (1987). *Br. Med. J.* **295**, 573–577.

Hindmarsh, P., Smith, P. J., Brook, C. G. D., and Matthews, D. R. (1987). *Clin. Endocrinol.* **27**, 581–591.

Hintz, R., Hopwood, N., and the Genentech Study Group. (1991). *Horm. Res.* **35** (Suppl. 2), 17 (Abstract 66).

Hock, J. M., Centrella, M., and Canalis, E. (1988). *Endocrinology (Baltimore)* **122**, 254–260.

Hodson, E. M., Shaw, P. F., Evans, R. A., Dunstan, C. R., Hills, E. E., Wong, S. Y. P., Rosenberg, A. R., and Roy, L. P. (1983). *J. Pediatr.* **103**, 735–740.

Ho, K. Y., Veldhuis, J. D., Johnson, M. L., Furlanetto, R., Evans, W. S., Alberti, K. G. M. M., and Thorner, M. O. (1988). *J. Clin. Invest.* **81**, 986–975.

Ho, K. Y., and Weissberger, A. J. (1990). *Horm. Res.* **33** (Suppl. 4), 7–11.

## USES OF GROWTH HORMONE 229

Hokken-Koelega, A. C. S., Hackeng, W. H. L., Stijnen, T., Wit, J. M., de Muinck Keizer-Schrama, S. M. P. F., and Drop, S. L. S. (1990). *J. Clin. Endocrinol. Metab.* **71**, 688–695.

Hokken-Koelega, A. C. S., Stijnen, T., de Muinck Keizer-Schrama, S. M. P. F., Wit, J. M., Wolff, E. D., deJong, M. C. J. W., Donckerwolcke, R. A., Abbad, N. C. B., Bot, A., Blum, W. F., and Drop, S. L. S. (1991). *Lancet* **338**, 585–590.

Homburg, R., Eshel, A., Abdalla, H. I., and Jacobs, H. S. (1988). *Clin. Endocrinol.* **29**, 113–117.

Homburg, R., West, C. Torresani, T., and Jacobs, H. S. (1990). *Fertil. Steril.* **53**, 254–260.

Horber, F. F., and Haymond, M. W. (1990). *J. Clin. Invest.* **86**, 265–272.

Horber, F. F., Marsh, H. M., and Haymond, M. W. (1991). *Diabetes* **40**, 141–149.

Hsu, C., and Hammond, J. M. (1987). *Endocrinology (Baltimore)* **121**, 1343–1348.

Ibrahim, Z. H. Z., Matson, P. L., Buck, P., and Lieberman, B. A. (1991). *Fertil. Steril.* **55**, 202–204.

Ikkos, D., Luft R., and Gemzell, C. A. (1958). *Lancet* **1**, 720–721.

Isaksson, O. G. P., Eden, S., and Jansson, J.-O. (1985). *Annu. Rev. Physiol.* **47**, 483–499.

Isaksson, O. G. P., Jansson, J. -O., and Gause, I. A. M. (1982). *Science* **216**, 1237–1239.

Isaksson, O. G. P., Lindahl, A., Nilsson, A., and Isgaard, J. (1987). *Endocr. Rev.* **8**, 426–438.

Ituarte, E. A., Petrini, J., and Hershman, J. M. (1984). *Ann. Int. Med.* **101**, 627–628.

Jiang, Z. -M., He, G. -Z., Zhang, S. -Y., Wang, X. -R., Yang, N. -F., Zhu, Y., and Wilmore, D. W. (1989). *Ann. Surg.* **210**, 513–525.

Johansson, G., Sietnieks, A., Janssens, F., Proesmans, W., Vanderschueren-Loedeweyckx, M., Holmberg, C., Sipila, I., Broyer, M., Rappaport, R., Albertsson-Wikland, K., Berg, U., Jodal, U., Rees, L., Rigden, S. P. A., and Preece, M. A. (1990). *Acta Pediatr. Scand. (Suppl.)* **370**, 36–42.

Johnston, I. D. A., and Hadden, D. R. (1963). *Lancet* **1**, 584–586.

Jørgensen, J. O. L. (1991). *Endocr. Rev.* **12**, 189–207.

Jørgensen, J. O. L., Pedersen, S. A., Thuesen, T., Jørgensen, J., Ingemann-Hansen, T., Skakkebaek, N. E., and Christiansen, J. S. (1989). *Lancet* **1**, 1221–1225.

Jørgensen, P. H., and Andreassen, T. T. (1988). *Horm. Metab. Res.* **20**, 490–493.

Jørgensen, P. H., Andreassen, T. T., and Jørgensen, K. D. (1989). *Acta Endocrinol. (Copenh.)* **120**, 767–772.

Joss, E., and Zuppinger, K. (1984). *Acta Pediatr. Scand.* **73**, 674–779.

Joss, E., Zuppinger, K., Schwarz, H. P., and Roten, H. (1983). *Pediatr. Res.* **17**, 676–679.

Juskevich, J. C., and Guyer, C. G. (1990). *Science* **249**, 875–884.

Kaplan, S. A. (1989). *In* "Clinical Pediatric Endocrinology" (S. A. Kaplan, ed.), pp. 1–62. Saunders, Philadelphia.

Kaplan, S. A., Underwood, L. E., August, G. P., Bell, J. J., Blethen, S. L., Blizzard, R. M., Brown, D. R., Foley, T. P., Hintz, R. L., Hopwood, N. J., Johansen, A., Kirkland, R. T., Plotnick, L. P., Rosenfeld, R. G., and Van Wyk, J. J. (1986). *Lancet.* **1**, 697–700.

Kiess, W., Doerr, H., Butenandt, O., and Belohradsky, B. H. (1986). *N. Engl. J. Med.* **314**, 321.

Kikuchi, K., Masakatsu, S., Miyamoto, A., Ohie, T., Mori, C., and Mikawa, H. (1988). *Acta Paediatr. Jpn.* **30**, 557.

Kinney, J. M., Long, C. L., Gump, F. E., and Duke, J. H., Jr. (1968). *Ann. Surg.* **168**, 459–474.

Kirk, J. A., Raghupathy, P., Stevens, M. M., Cowell, C. T., Menser, M. A., Bergin, M., Tink, A., Vines, R. H., and Silink, M. (1987). *Lancet* **1**, 190–193.

230 BARBARA M. LIPPE AND JON M. NAKAMOTO

Koch, T. K., Berg, B. O., De Armond, S. J., and Gravina, R. F. (1985). *N. Engl. J. Med.* **313,** 731–733.

Koch, V. H., Lippe, B. M., Nelson, P. A., Boechat, M. I., Sherman, B. M., and Fine, R. N. (1989). *J. Pediatr.* **115,** 365–371.

Korkor, A. B. (1987). *N. Engl. J. Med.* **316,** 1573–1577.

Kostyo, J. L. (1968). *Ann. N. Y. Acad. Sci.* **148,** 389–407.

Kostyo, J. L. (1988). In "Human Growth Hormone Progress and Challenges" (L. E. Underwood, ed.), pp. 63–73. Dekker, New York.

Krentz, A. J., Koster, F. T., Crist, D., Finn, K., Boyle, P. J., and Schade, D. S. (1991). *Clin. Res.* **39**(2). [abstract]

Lanes, R., Plotnick, L. P., and Lee, P. A. (1979). *Pediatrics* **63,** 731–735.

Lee, P. A., Blizzard, R. M., Cheek, D. B., and Holt, A. B. (1974). *Metabolism* **23,** 913–919.

Lee, P. D. K., Hintz, R. L., Sperry, J. B., Baxter, R. C., and Powell, D. R. (1989). *Pediatr. Res.* **26,** 308–315.

Lee, P. D. K., Wilson, D. M., Hintz, R. L., and Rosenfeld, R. G. (1987a). *J. Pediatr. Endocrinol.* **2,** 31–34.

Lee, P. D. K., Wilson, D. M., Rountree, L., Hintz, R. L., and Rosenfeld, R. G. (1987b). *Am. J. Med. Genet.* **28,** 865–871.

Lenko, H. L., Leisti, S., and Perheentupa, J. (1982). *Eur. J. Pediatr.* **138,** 241–249.

Lesage, C., Walker, J., Landler, F., Chatelain, P., Chaussain, J. L., and Bougnères, P. F. (1991). *J. Pediatr.* **119,** 29–34.

Li, C. H., and Papkoff, H. (1956). *Science* **124,** 1293–1294.

Libber, S. M., Plotnick, L. P., Johanson, A. J., Blizzard, R. M., Kwiterovich, P. O., and Migeon, C. J. (1990). *Medicine* **69,** 46–55.

Liljedahl, S. -O., Gemzell, C. A., Platin, L. O., and Birke, G. (1961). *Acta Chir. Scand.* **122,** 1–14.

Lin, T. -H., Kirkland, R. T., Sherman, B. M., and Kirkland, J. L. (1989). *J. Pediatr.* **115,** 57–63.

Lippe, B., Fine, R. N., Koch, V. H., and Sherman, B. M. (1988). *Acta Pediatr. Scand. (Suppl.)* **343,** 127–131.

Lippe, B., Frane, J., and the Genentech National Cooperative Study Group. (1991). In "Turner Syndrome Growth Promoting Therapies" (M. B. Ranke and R. G. Rosenfeld, eds.), pp. 59–65. Elsevier Science Publishers B. V., The Netherlands.

Lippe, B., Kaplan, S. A., Golden, M. P., Hendricks, S. A., and Scott, M. L. (1981). *J. Clin. Endocrinol. Metab.* **53,** 507–513.

Lippe, B. M., and Frasier, S. D. (1989). *J. Pediatr.* **115,** 585–587.

Loeb, J. N. (1976). *N. Engl. J. Med.* **295,** 547–552.

Lynch, S. E., Colvin, R. B., and Antoniades, H. N. (1989). *J. Clin. Invest.* **84,** 640–646.

Lyon, A. J., Preece, M. A., and Grant, D. B. (1985). *Arch. Dis. Child.* **60,** 932–935.

Manson, J. M., Smith, R. J., and Wilmore, D. W. (1988). *Ann. Surg.* **208,** 136–142.

Manson, J. M., and Wilmore, D. W. (1986). *Surgery* **100,** 188–197.

Marcus, R., Butterfield, G., Holloway, L., Gilliland, L., Baylink, D. J., Hintz, R. L., and Sherman, B. M. (1990). *J. Clin. Endocrinol. Metab.* **70,** 519–527.

Marsh, J. A., Gause, W. C., Sandhu, S., and Scanes, C. G. (1984). *Proc. Soc. Exp. Biol. Med.* **175,** 351–360.

Martin, T. E., and Young, F. G. (1965). *Nature (London)* **208,** 684–685.

Mason, H. D., Martikainen, H., Beard, R. W., Anyaoku, V., and Franks, S. (1990). *J. Endocrinol.* **126,** R1–R4.

Massa, G., Vanderschueren-Lodeweyckx, M., Craen, M., Vandeweghe, M., and van Vliet, G. (1991). *Eur. J. Pediatr.* **150,** 460–463.

USES OF GROWTH HORMONE 231

Massarano, A. A., Brook, C. G. D., Hindmarsh, P. C., Pringle, P. J., Teale, J. D., Stanhope, R., and Preece, M. A. (1989). *Arch. Dis. Child.* **64**, 587–592.

Mauras, N., Rogol, A. D., and Veldhuis, J. D. (1990). *Pediatr. Res.* **28**, 626–630.

McCaffery, T. D., Nasr, K., Lawrence, A. M., and Kirsner, J. B. (1974). *Dig. Dis.* **19**, 411–416.

McGauley, G. A., Cuneo, R. C., Salomon, F., and Sönksen, P. H. (1990). *Horm. Res.* **33**(Suppl. 4), 52–54.

McSherry, E., and Morris, R. C. (1978). *J. Clin. Invest.* **61**, 509–526.

Mehls, O., and Ritz, E. (1983). *Kidney Int.* **15**(Suppl.), S53–S60.

Mehls, O., Ritz, E., Hunziker, E-B., Eggli, P., Heinrich, U., and Zapf, J. (1988). *Kidney Int.* **33**, 45–52.

Mehls, O., Tönshoff, B., Blum, W. F., Heinrich, U., and Seidel, C. (1990). *Acta Pediatr. Scand. (Suppl.)* **370**, 28–34.

Menashe, Y., Lunenfeld, B., Pariente, C., Frenkel, Y., and Mashiach, S. (1990). *Fertil. Steril.* **53**, 432–435.

Menashe, Y., Sack, J., and Mashiach, S. (1991). *Human Reproduction* **6**, 670–671.

Miell, J. P., Corder, R., Pralong, F. P., and Gaillard, R. C. (1991). *J. Clin. Endocrinol. Metab.* **72**, 675–681.

Milner, R. D. G. (1986). *Acta. Pediatr. Scand. (Suppl.)* **325**, 25–28.

Milsom, S. R., Breier, B. H., Cox, V., Gallaher, B. W., and Blum, W. F. (1991). *In* Proceedings, Endocrine Society 73rd Annual Meeting, June 19–22, Washington, D. C.," Abstract 1312.

Moe, P. J. (1968). *Acta. Pediatr. Scand.* **57**, 300–304.

Morris, H. G., Jorgensen, J. R., Elrick, H., Goldsmith, R. E., and Subrayan, V. L. (1968). *J. Clin. Invest.* **47**, 436–451.

Murphy, W. J., Durum, S. K., and Longo, D. L. (1992). *Proc. Natl. Acad. Sci. U.S.A.* **89**, 4481–4485.

Nagy, E., Berczi, I., and Freisen, H. G. (1983). *Acta Endocrinol. (Copenh.)* **102**, 351–357.

Neglia, J. P., Meadows, A. T., Robison, L. L., Kim, T. H., Newton, W. A., Ruymann, F. B., Sather, H. N., and Hammond, G. D. (1991). *N. Engl. J. Med.* **325**, 1330–1336.

NIH Technology Assessment Conference Statement on Bovine Somatotropin. (1991). *JAMA* **265**, 1423–1425.

Nishi, Y., Aihara, K., Usui, T., Phillips III, J. A., Mallonee, R.L., and Migeon, C. J. (1984). *J. Pediatr.* **104**, 885–889.

Novak, L. P., Hayles, A. B., and Cloutier, M. D. (1972). *Mayo Clin. Proc.* **47**, 241–246.

Okada, Y., Taira, K., Takano, K., and Hizuka, N. (1987). *Endocrinol. Jpn.* **34**, 621–626.

Owen, E. J., West, C., Mason, B. A., and Jacobs, H. S. (1991). *Human Reproduction* **6**, 524–528.

Palmiter, R. D., Brinster, R. L., Hammer, R. E., Trumbauer, M. E., Rosenfeld, M. G., Birnberg, N. C., and Evans, R. M. (1982). *Nature (London)* **300**, 611–615.

Pape, G. S., Friedman, M., Underwood, L. E., and Clemmons, D. R. (1991). *Chest* **99**, 1495–1500.

Peluso, J. J., Delidow, B. C., Lynch, J., and White, B. A. (1991). *Endocrinology (Baltimore)* **128**, 191–196.

Pessa, M. E., Bland, K. I., Sitren, H. S., Miller, G. J., and Copeland, E. M., III (1985). *Surg. Forum* **36**, 6–8.

Peterson, B. H., Rapaport, R., Henry, D. P., Huseman, C., and Moore, W. V. (1990). *J. Clin. Endocrinol. Metab.* **70**, 1756–1760.

Phillips, L. S., Fusco, A. C., Unterman, T. G., and del Greco, F. (1984). *J. Clin. Endocrinol. Metab.* **59**, 764–772.

232    BARBARA M. LIPPE AND JON M. NAKAMOTO

Plotnick, L. P., Van Meter, Q. L., and Kowarski, A. A. (1983). *Pediatrics* **71**, 324–327.
Polychronakos, C., Abu-Srair, H., and Guyda, H. J. (1988). *Eur. J. Pediatr.* **147**, 582–583.
Ponting, G. A., Halliday, D., Teale, J. D., and Sim, A. J. W. (1988). *Lancet* **1**, 438–440.
Powell, D. R., Rosenfeld, R. G., Baker, B. K., Lu, F., and Hintz, R. L. (1986). *J. Clin. Endocrinol. Metab.* **63**, 1186–1192.
Prader, A., Zachmann, M., Poley, J. R., and Illig, R. (1968). *Acta Endocrinol.* **57**, 115–128.
Prasad, V., Greig, F., Bastian, W., Castells, S., Juan, C., and AvRuskin, T. W. (1990). *J. Pediatr.* **116**, 397–399.
Preece, M. A. (1990). *Acta Pediatr. Scand. (Suppl.)* **370**, 103–104.
Preece, M. A., and Tanner, J. M. (1977). *J. Clin. Endocrinol. Metab.* **45**, 169–170.
Press, M., Notarfrancesco, A., and Genel, M. (1987). *Lancet* **1**, 1002–1004.
Prudden, J. F., Nishihara, G., and Ocampo, L. (1958). *Surg. Gynecol. Obstet.* **107**, 481–482.
Prudden, J. F., Pearson, E., and Soroff, H. S. (1956). *Surg. Gynecol. Obstet.* **102**, 695–701.
Raben, M. S. (1957). *Science* **125**, 883–884.
Raben, M. S. (1958). *J. Clin. Endocrinol. Metab.* **18**, 901–903.
Raben, M. S. (1959). *Recent Prog. Horm. Res.* **15**, 71–105.
Raiti, S., Kaplan, S. L., Van Vliet, G., Moore, W. V., and the National Hormone and Pituitary Program Growth Hormone Committee. (1987). *J. Pediatr.* **110**, 357–361.
Raiti, S., Moore, W. V., Van Vliet, G., and Kaplan, S. L. (1986). *J. Pediatr.* **109**, 944–949.
Ramirez, G., O'Neill, W. M., Bloomer, A., and Jubiz, W. (1978). *Arch. Intern. Med.* **138**, 267–271.
Ranke, M. B. (1987). *Acta Pediatr. Scand. (Suppl.)* **331**, 80–82.
Ranke, M. B., and Bierich, J. R. (1987). *Acta Pediatr. Scand. (Suppl.)* **331**, 9–17.
Ranke, M. B., Heidemann, P., Knupfer, C., Enders, H., Schmaltz, A. A., and Bierich, J. R. (1988). *Eur. J. Pediatr.* **148**, 220–227.
Ranke, M. B., Pfluger, H., Rosendahl, W., Stubbe, P., Enders, H., Bierich, J. R., and Majewski, F. (1983). *Eur. J. Pediatr.* **141**, 81–88.
Ranke, M. B., and Rosenfeld, R. G. [eds.] (1991). "Turner Syndrome: Growth Promoting Therapies," pp. 189–269. Excerpta Medica, New York.
Ranke, M. B., Weber, B., and Bierich, J. R. (1979). *Eur. J. Pediatr.* **132**, 221–238.
Rapaport, R., Oleske, J., Ahdieh, H., Solomon, S., Delfaus, C., and Denny, T. (1986). *J. Pediatr.* **109**, 434–439.
Rappaport, E. B., and Fife, D. (1985). *Am. J. Dis. Child.* **139**, 396–399.
Rappaport, R., and Brauner, R. (1989). *Pediatr. Res.* **25**, 561–567.
Rappaport, R., and Sauvion, S. (1989). *Acta Pediatr. Scand. (Suppl.)* **356**, 82–86.
Read, C. H., and Bryan, G. T. (1960). *Recent. Prog. Horm. Res.* **16**, 187–212.
Redman, G. P., Shu, S., and Morris, D. (1988). *Lancet* **1**, 1335.
Rees, L., Rigden, S. P. A., Ward, G., and Preece, M. A. (1990). **65**, 856–860.
Reiter, E. O., Morris, A. H., MacGillivray, M. H., and Weber, D. (1988). *J. Clin. Endocrinol. Metab.* **66**, 68–71.
Robison, L. L., Nesbit, M. E., Jr., Sather, H. N., Level, C., Shahidi, N., Kennedy, A., and Hammond, D. (1984). *J. Pediatr.* **105**, 235–242.
Rochiccioli, P., Tauber, M., Moisan, V., and Pienkowski, C. (1989). *Acta Pediatr. Scand. (Suppl.)* **349**, 42–46.
Roe, C. F., and Kinney, J. M. (1962). *Surg. Forum* **13**, 369–371.
Rogers, P. C., Komp, D., Rogol, A., and Sabio, H. (1977). *Lancet* **2**, 434–435.
Romshe, C. A., Zipf, W. B., Miser, A., Miser, J., Sotos, J. F., and Newton, W. A. (1984). *J. Pediatr.* **104**, 177–181.
Rose, S. R., Ross, J. L., Uriarte, M., Barnes, K. M., Cassorla, F. G., and Cutler, G. B., Jr. (1988). *N. Engl. J. Med.* **319**, 201–207.

# USES OF GROWTH HORMONE 233

Rosén, T., and Bengtsson, B.-Å. (1990). *Lancet* **336**, 285–288.

Rosenfeld, R. G., Attie, K. M., Johanson, A. J., and the Turner Collaborative Group. (1991). *Pediatr. Res.* **28**(4), 85A.

Rosenfeld, R. G., Hintz, R. L., Johanson, A. J., Brasel, J., Burstein, S., Chernausek, S. D., Clabots, T., Frane, J., Gotlin, R. W., KJuntze, J., Lippe, B. M., Mahoney, P. C., Moore, W. V., New, M. I., Saenger, P., Stoner, E., and Sybert, V. (1986). *J. Pediatr.* **109**, 936–943.

Rosenfeld, R. G., Hintz, R. L., Johanson, A. J., Sherman, B., Brasel, J., Burstein, S., Chernausek, S., Compton, P., Frane, J., Gotlin, R. W., Kuntze, J., Lippe, B. M., Mahoney, P. C., Moore, W. V., New, M. I., Saenger, P., and Sybert, V. (1988). *J. Pediatr.* **113**, 393–400.

Rosenfeld, R. G., Wilson, D. M., Dollar, L. A., Bennett, A., and Hintz, R. L. (1982). *J. Clin. Endocrinol. Metab.* **54**, 1033–1038.

Ross, J. L., Long, L. M., Loriaux, D. L., and Cutler, G. B. (1985). *J. Pediatr.* **106**, 202–206.

Ross, R., Miell, J., Freeman, E., Jones, J., Matthews, D., Preece, M., and Buchanan, C. (1991). *Clin. Endocrinol.* **35**, 47–54.

Rotenstein, D., Reigel, D. H., and Flom, L. L. (1989). *J. Pediatr.* **115**, 417–420.

Roth, J. A., Kaeberle, M. L., Grier, R. L., Hopper, J. G., Spiegel, H. E., and McAllister, H. A. (1984). *Ann. J. Vet. Res.* **45**, 1151–1155.

Rudman, D. (1985). *J. Am. Geriatr. Soc.* **33**, 800–807.

Rudman, D., Feller, A. G., Nagraj, H. S., Gergans, G. A., Lalitha, P. Y., Goldberg, A. F., Schlenker, R. A., Cohn, L., Rudman, I. W., and Mattson, D. (1990). *N. Engl. J. Med.* **323**, 1–6.

Rudman, D., Goldsmith, M., Kutner, M., and Blackston, D. (1980). *J. Pediatr.* **96**, 132–135.

Rudman, D., Kutner, M. H., Blackston, R. D., Cushman, R. A., Bain, R. P., and Patterson, J. H. (1981a). *N. Engl. J. Med.* **305**, 123–131.

Rudman, D., Kutner, M. H., Rogers, C. M., Lubin, M. F., Fleming, G. A., and Bain, R. P. (1981b). *J. Clin. Invest.* **67**, 1361–1369.

Salomon, F., Cuneo, R. C., Hesp, R., and Sönksen, P. H. (1989). *N. Engl. J. Med.* **321**, 1797–1803.

Samaan, N. A., and Freeman, R. M. (1970). *Metabolism* **19**, 102–113.

Savion, N., Lui, G. -M., Laherty, R., and Gospodarowicz, D. (1981). *Endocrinology (Baltimore)* **190**, 409–420.

Seip, M., and Trygstad, O. (1966). *Acta. Pediatr. Scand.* **55**, 287–293.

Shalet, S. M., Beardwell, C. G., Pearson, D., and Morris Jones, P. H. (1976). *Clin. Endocrinol.* **5**, 287–290.

Shalet, S. M., Clayton, P. E., and Price, D. A. (1988). *Horm. Res.* **30**, 53–61.

Shepard, T. H., II, Nielsen, R. L., Johnson, M. L., and Bernstein, N. (1960). *Am. J. Dis. Child.* **99**, 90–96.

Shizume, K., Matsuzaki, F., Irie, M., and Osawa, N. (1970). *Endocrinol. Jpn.* **17**, 297–309.

Simmons, J. M., Wilson, C. J., Potter, D. E., and Holliday, M. A. (1971). *N. Engl. J. Med.* **285**, 653–656.

Skaggs, S. R., and Crist, D. M. (1991). *Horm. Res.* **35**, 19–24.

Skottner, A., Forsman, A., Löfberg, E., and Thorngren, K. G. (1984). *Acta Endocrinol.* **107**, 192–198.

Slootweg, M. C., van Buul-Offers, S. C., Hermann-Erlee, M. P. M., and Duursma, S. A. (1988). *Acta Endocrinol. (Copenh.)* **118**, 294–300.

Smith, P. J., Hindmarsh, P. C., and Brook, C. G. D. (1988). *Arch. Dis. Child.* **63**, 491–494.

Snyder, D. K., Clemmons, D. R., and Underwood, L. E. (1988). *J. Clin. Endocrinol. Metab.* **67**, 54–61.

234 BARBARA M. LIPPE AND JON M. NAKAMOTO

Soroff, H. S., Rozin, R. R., Mooty, J., Lister, J., Raben, M. S., MacAulay, A. J., and Paddock, A. B. (1967). *Ann. Surg.* **166,** 739–752.

Soyka, L. F., Bode, H. H., Crawford, J. D., and Flynn, F. J., Jr. (1970). *J. Clin. Endocrinol.* **30,** 1–14.

Soyka, L. F., Ziskind, A., and Crawford, J. D. (1964). *N. Engl. J. Med.* **271,** 754–764.

Spadoni, G. L., Rossi, P., Ragno, W., Galli, E., Cianfarani, S., Galasso, C., and Boscherini, B. (1991). *Acta Pediatr. Scand.* **80,** 75–79.

Spanos, E., Barrett, D., MacIntyre, I., Pike, J. W., Safilian, E. F., and Haussler, M. R. (1978). *Nature (London)* **273,** 246–247.

Spiliotis, B. E., August, G. P., Hung, W., Sonis, W., Mendelson, W., and Bercu, B. B. (1984). *JAMA* **251,** 2223–2230.

Stanhope, R., Ackland, F., Hamill, G., Clayton, J., Jones, J., and Preece, M. A. (1989). *Acta Pediatr. Scand. (Suppl.)* **349,** 47–52.

Stanke, N., and Zeisel, H. J. (1989). *Eur. J. Pediatr.* **148,** 591–596.

Steenfos, H., Spencer, E. M., and Hunt, T. K. (1989). *Surg. Forum* **40,** 68–70.

Stone, B. A., and Marrs, R. P. (1991). *Fertil. Steril.* **56,** 52–58.

Stracke, H., Schultz, A., Moeller, D., Rossol, S., and Schatz, H. (1984). *Acta Endocrinol. (Copenh.)* **107,** 16–24.

Streat, S. J., Beddoe, A. H., and Hill, G. L. (1987). *J. Trauma* **27,** 261–266.

Suchner, U., Rothkopf, M. M., Stanislaus, G., Elwyn, D. H., Kvetan, V., and Askanazi, J. (1990). *Arch. Intern. Med.* **150,** 1225–1230.

Sulmont, V., Brauner, R., Fontoura, M., and Rappaport, R. (1990). *Acta Pediatr. Scand.* **79,** 542–549.

Sybert, V. P. (1984). *J. Pediatr.* **104,** 365–369.

Takano, K., Shizume, K., Hizuka, N., Okuno, A., Umino, T., Kobayashi, Y., Kusano, S., Nakajima, H., Irie, M., Hibi, I., Kato, K., Suwa, S., Koshimizu, T., Ogawa, M., Sudo, M., Imura, H., Okada, Y., Kondo, T., Hashimoto, K., Miyao, M., Kohno, H., Iwatani, N., and Ono, S. (1986). *Endocrinol. Jpn.* **33,** 589–596.

Tanner, J. M., Whitehouse, R. H., Hughes, P. C. R., and Vince, F. P. (1971). *Arch. Dis. Child.* **46,** 745–782.

Thieriot-Prevost, G., Boccara, J. F., Francoual, C., Badoual, J., and Job, J. C. (1988). *Pediatr. Res.* **24,** 380–383.

Tick, D., Shohat, M., Baraket, S., Melmed, S., and Rimoin, D. (1991). *Clin. Res.* **39,** 54A.

Tönshoff, B., Mehls, O., Schauer, A., Heinrich, U., Blum, W., and Ranke, M. (1989). *Kidney Int.* **36**(Suppl 27), S201–S204.

Torrado, C., Bastian, W., Wisniewski, K. E., and Castells, S. (1991). *J. Pediatr.* **119,** 478–483.

Turner, H. H. (1938). *Endocrinology (Baltimore)* **23,** 566–574.

Tzagournis, M. (1969). *JAMA* **210,** 2373–2376.

Underwood, L. E., Clemmons, D. R., Maes, M., D'Ercole, A. J., and Ketelslegers, J. -M. (1986). *Horm. Res.* **24,** 166–176.

Unterman, T. G., and Phillips, L. S. (1985). *J. Clin. Endocrinol. Metab.* **61,** 618–626.

van der Veen, E. A., and Netelenbos, J. C. (1990). *Horm. Res.* **33**(Suppl. 4), 65–68.

Van Der Werff Ten Bosch, J. J., and Bot, A. (1990). *Clin. Endocrinol.* **32,** 707–717.

van Es, A., Massarano, A. A., Wit, J. M., Hindmarsh, P. C., Kamp, G. A., Brook, C. G. D., Preece, M. A., and Matthews, D. R. (1991). *In* "Turner Syndrome Growth Promoting Therapies" (M. B. Ranke and R. G. Rosenfeld, eds.), pp. 29–33. Elsevier Science Publishers B. V., The Netherlands.

Van Vliet, G., Styne, D. M., Kaplan, S. L., and Grumbach, M. M. (1983). *N. Engl. J. Med.* **309,** 1016–1022.

## USES OF GROWTH HORMONE

Verp, M. S., Rosinsky, B., Le Beau, M. M., Martin, A. O., Kaplan, R., Wallemark, C-B., Otano, L., and Simpson, J. L. (1988). *Clin. Genet.* **33**, 277–285.

Waago, H. (1987). *Lancet* **1**, 1485.

Walker, J. L., Ginalska-Malinowska, M., Romer, T. E., Pucilowska, J. B., and Underwood, L. E. (1991). *N. Engl. J. Med.* **324**, 1483–1488.

Walker, J. M., Bond, S. A., Voss, L. D., Betts, P. R., Wootton, S. A., and Jackson, A. A. (1990). *Lancet* **336**, 1331–1334.

Ward, D. J., Hartog, M., and Ansell, B. M. (1966). *Ann. Rheum. Dis.* **26**, 416–421.

Ward, H. C., Halliday, D., and Sim, A. J. W. (1987). *Ann. Surg.* **206**, 56–61.

Watanabe, S., Tsunematsu, Y., Fujimoto, J., and Komiyama, A. (1988). *Lancet* **1**, 1159.

Whitehead, E. M., Shalet, S. M., Davies, D., Enoch, B. A., Price, D. A., and Beardwell, C. G. (1982). *Clin. Endocrinol.* **17**, 271–277.

Whitelaw, M. J., Thomas, S. F., Graham, W., Foster, T. N., and Brock, C. (1962). *Am. J. Obstet. Gynecol.* **84**, 501–504.

Wilmore, D. W., Moylon, J. A., Jr., Bristow, B. F., Mason, A. D., Jr., and Pruitt, B. A., Jr. (1974). *Surg. Gynecol. Obstet.* **138**, 875–884.

Wilson, D. M., Lee, P. D. K., Morris, A. H., Reiter, E. O., Gertner, J. M., Marcus, R., Quarmby, V. E., and Rosenfeld, R. G. (1991a). *Am. J. Dis. Child.* **145**, 1165–1170.

Wilson, M. E., Gordon, T. P., Chikazawa, K., Gust, D., Tanner, J. M., and Rudman, C. G. (1991b). *J. Clin. Endocrinol. Metab.* **72**, 1302–1307.

Wilson, D. P., Jelley, D., Stratton, R., and Coldwell, J. G. (1989). *J. Pediatr.* **115**, 758–761.

Wilton, P. (1987). *Acta Pediatr. Scand.* **76**, 193–200.

Wit, J. M., Faber, A. J., and Van Den Brande, J. L. (1986). *Acta Pediatr. Scand.* **75**, 767–773.

Wit, J. M., Fokker, M. H., de Muinck Keizer-Schrama, S. M. P. F., Oostdijk, W., Gons, M., Otten, B. J., Delemarre-Van de Wall, H. A., Reeser, M., and Waelkens, J. J. J. (Dutch Growth Hormone Working Group). (1989). *J. Pediatr.* **115**, 720–725.

Wit, J. M., Kuilboer, M. M., de Muinck Keizer-Schrama, S. M. P. F., Oostdijk, W., Gons, M. H., Otten, B. J., Delemarre-Van de Waal, H. A., Reeser, M., and Waelkens, J. J. J. (1991). *Horm. Res.* **35**, 31A.

Wu, R. H., St. Louis, Y., Rubin, K., Cassidy, S. B., Thorpy, M. J., and Saenger, P. (1988). *Pediatr. Res.* **23**, 207A.

Zachmann, M. (1969). *Acta Endocrinol.* **62**, 513–520.

Zadik, Z., Chalew, S. A., Raiti, S., and Kowarski, A. A. (1985). *Pediatrics* **76**, 355–360.

Ziegler, T. R., Young, L. S., Manson, J. M., and Wilmore, D. W. (1988). *Ann. Surg.* **208**, 6–16.

RECENT PROGRESS IN HORMONE RESEARCH, VOL. 48

# Site-Directed Mutagenesis in the Mouse

ALLAN BRADLEY

*Institute for Molecular Genetics, Baylor College of Medicine, Houston, Texas 77030*

The correlation of genotype and phenotype is a central theme in the analysis of development. Indeed, the observational pathway has often been from phenotype to genotype and molecular analysis in organisms such as *Caenorhabditis elegans* and *Drosophila*. In the mouse this type of reverse genetics is constrained not only by the size of the murine genome, which makes positional cloning difficult, but also by the limited availability of mutations that can readily be identified and maintained (Green, 1989). Embryonic lethal mutations are particularly difficult to observe and maintain unless they are closely linked to a visible mutation. A number of mutations in mice have now been correlated with cloned genes (Balling *et al.*, 1988; Chabot *et al.*, 1988; Li *et al.*, 1990; Yoshida *et al.*, 1990). Mapping and attempting to identify mutations in candidate genes is clearly a worthwhile exercise, but its applicability is clearly limited by the availability of mutations.

Another approach to link phenotype to genotype is by using transgenic insertions that have mutated endogenous genes as a result of the integration of the transgene. The transgene is a molecular flag that can serve as a marker to clone the disrupted gene. A number of retroviral and microinjection transgenic insertions have led to the cloning of novel genes important to the developing organism (Schnieke *et al.*, 1983; Maas *et al.*, 1990).

Though each approach is uniquely important, one of the values of the genetic approach is the ability to generate an allelic series of mutations that functionally dissect a genetic locus. These are available for a limited set of genes in mice (such as *Steel* and *W*) (Chabot *et al.*, 1988; Copeland *et al.*, 1990), but for most cloned murine genes a mutant mouse does not exist.

Very recently it has become possible to generate mutant mice for virtually any cloned gene by implementing embryonic stem (ES) cell/gene targeting technologies (Fig. 1). The central discoveries that have enabled this technology were the isolation of permanent *in vitro* embryonic stem cell lines from preimplantation embryos (Evans and Kaufman, 1981) and

237

Copyright © 1993 by Academic Press, Inc.
All rights of reproduction in any form reserved.

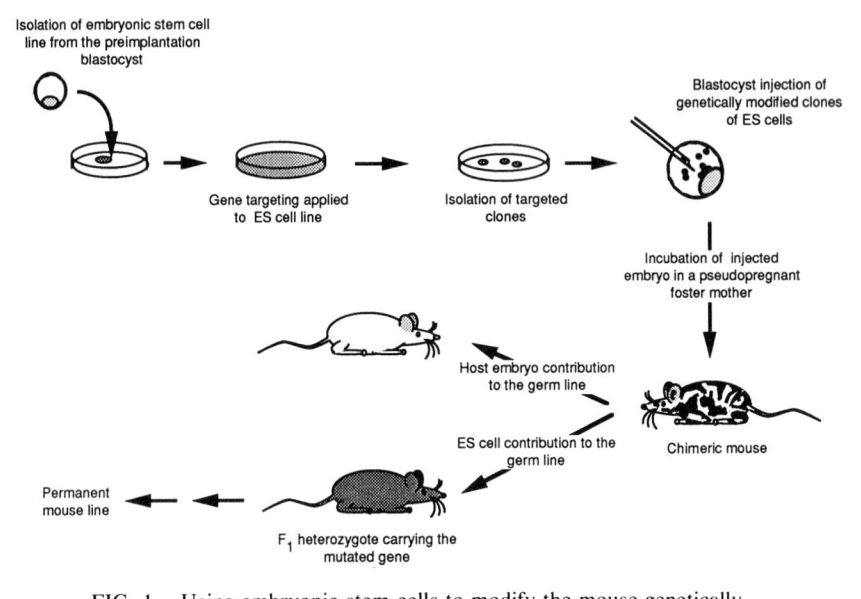

FIG. 1.    Using embryonic stem cells to modify the mouse genetically.

the demonstration that these cell lines are capable of recolonizing embryos, contributing to the germ line (Bradley *et al.*, 1984) despite considerable periods of *in vitro* growth. The most compelling feature offered by ES cells is the availability of large numbers of cells as an *in vitro* cell line. This presents an experimental opportunity to treat an entire population of cells. Thus it is possible to generate genetic modifications that may occur very infrequently, such as those that are the consequences of gene targeting procedures (Thomas and Capecchi, 1987). Furthermore, rare clones of cells that have been generated with a desired genetic change may be isolated from the population and used to generate a mouse, so that the modification may be established *in vivo* (Kuehn *et al.*, 1987; Schwartzberg *et al.*, 1989). Thus, unlike in *Drosophila* and *C. elegans*, the genetic dissection of the mouse, via gene targeting, need not be constrained by the generation of random mutations in functional areas of a genetic target.

The combination of ES cells and gene targeting techniques constitutes a very powerful genetic system. Our studies on vector–chromosome homologous recombination have enabled us to devise efficient strategies that facilitate the construction of site-directed mutations in nonselectable genes (Hasty *et al.*, 1991a; Davis *et al.*, 1992). The application of these methodologies to recently isolated ES cell lines (McMahon and Bradley, 1990)

MURINE SITE-DIRECTED MUTAGENESIS　239

has led to a much higher frequency of chimera formation and germ line transmission than was previously possible.

## I. ES Cell Lines: Clonal Heterogeneity and Maintenance of Totipotency

The ultimate assay for a normal ES cell is its ability to participate in embryogenesis after injection into a host blastocyst. The fate and germ line transmission of descendants of the introduced cells is determined by a variety of factors (Bradley, 1990), which include the differentiation state of the cell, the chromosomal state (aneuploid or euploid), the sex of both the host and ES cells, and the genotype of both the host and ES cell components.

The first *in vivo* differentiation experiments performed by injecting ES cells into the 3.5-day blastocyst used ES cells that had been expanded from isolated secondary clones from the primary explant of a single embryo (Bradley *et al.*, 1984). With the particular host–ES cell combination used at this time, germ line transmission was achieved. The significance of a host genotype was not initially realized (Bradley *et al.*, 1984; Robertson *et al.*, 1986), although studies of transmission from aggregation chimeras had previously detected specific biases with certain embryo combinations (Mullen and Whitten, 1977). The first systematic studies of the effect of the host embryo genotype in combination with microinjected 129-derived ES cells revealed a similar dependence on genotype (Schwartzberg *et al.*, 1989).

Although the reported successes of the injection of both parental and modified *populations* of ES cells into blastocysts has served as an inspiration for the field (Bradley *et al.*, 1984; Robertson *et al.*, 1986), the use of *populations* of ES cells in these experiments has greatly overscored the representation of true ES cells in some ES cell lines, due to the clonal selection of "normal" cells during differentiation *in vivo* (Bradley, 1990). Cloning of ES cell populations, a necessary procedure with most gene targeting experiments, has revealed the rapid "progression" of ES cell lines to a state in which only a relatively minor fraction of cells can contribute extensively to the somatic and germ lineages of chimeric mice (Schwartzberg *et al.*, 1989). In "progressed" cultures, the abnormal cells are morphologically indistinguishable from normal ES cells. In some cases the abnormality can be shown to be associated with the loss of euploidy (Kuehn *et al.*, 1987; Schwartzberg *et al.*, 1989), but often the differences between normal and abnormal clones are only discernible following blastocyst injection, because abnormal cells will not contribute extensively to chimeric mice nor will they transmit through the germ line.

Under optimal conditions, the routine culture of an ES cell line requires a complex mixture of growth factors experimentally provided by high concentrations of fetal serum and a mitotically inactive layer of fibroblast feeder cells (Robertson, 1987). Although many fibroblast subclones and serum batches may appear to exhibit growth-promoting activities with established ES cell lines, it will only become apparent on subcloning whether this proliferation was restricted to the normal or the abnormal cells, which will always be present in any *in vitro* cell line.

Some efforts have been made to grow ES cells as "feeder-free" cell lines using media conditioned by buffalo rat liver (BRL) cells (Smith and Hooper, 1987) or in the presence of recombinant leukemia inhibitory factor (LIF) (Williams *et al.*, 1988; Pease and Williams, 1990). Though cellular proliferation is readily accomplished under these culture conditions, there have been very few reports of the germ line transmission of *cloned* ES cells that were grown and isolated in the absence of feeder cells; in contrast, a large number of laboratories have successfully achieved germ line transmission of ES cell lines grown and *cloned* on feeder cells (see Bradley *et al.*, 1992, and references therein).

Feeder-free cultures can occasionally provide an appropriate balance for the proliferation of normal ES cells (Thompson *et al.*, 1989; Joyner *et al.*, 1991), but as a routine the use of feeder cells appears to be more reliable and can maintain ES cells with a high proportion of totipotent cells for long periods of time. We have recently described the AB1 (McMahon and Bradley, 1990) and AB2.1 (Soriano *et al.*, 1991) ES cell lines and the clonal SNL76/7 feeder cell line (McMahon and Bradley, 1990), which appear to represent a relatively stable culture system for ES cells because both the population and a substantial proportion of clones generated in the course of gene targeting experiments are transmitted through the germ line (Tables I and II). Furthermore, we have demonstrated that AB1 cells, which have been subcloned twice on SNL76/7 feeder cells during "hit and run" targeting experiments (Hasty *et al.*, 1991a), can still be transmitted

TABLE I

*Construction of Chimeras Using Embryonic Stem Cell Populations*

| Cell line | Percent chimeric | Extent of ES contribution to chimeras (%) | Frequency of germ line transmission (per animal) |
|---|---|---|---|
| CCE | 70 | 90 | 9/10 |
| AB1[a] and AB2.1 | 88 | 90 | 9/11 |

[a] Additional extensive germ line transmission data for AB1 populations have been reported previously by Friedrich and Soriano (1991).

# MURINE SITE-DIRECTED MUTAGENESIS

TABLE II

*Construction of Chimeras Using Embryonic Stem Cell Clones*

| Cell line | Number of clones tested | Number of clones showing germ line transmission |
|---|---|---|
| CCE | 14 | 3 |
| AB1 and AB2.1 | 38 | 29[a] |

[a] Includes clones with targeted mutations in eight different autosomal genes with a total of 11 different alleles. Eight of these clones have been reported previously (McMahon and Bradley, 1990; Soriano *et al.*, 1991; Donehower *et al.*, 1992).

through the germline (A. Bradley and Ramiro Ramirez-Solis, unpublished observations).

Tables I and II demonstrate that the assay of chimera formation divides clonal isolates from the CCE ES cell line (Robertson *et al.*, 1986) into two distinct subpopulations, one of which has properties analogous to the original population, while the others colonize embryos very poorly and do not show germ line transmission. The pattern of colonization of all of the clones derived from AB1 (McMahon and Bradley, 1990) and AB2.1 (Soriano *et al.*, 1991) cell lines resembles that of the parental population.

## II. Null Alleles Generated by Gene Targeting

One remarkable property of many mammalian cell lines, including ES cells (Thomas and Capecchi, 1987), is the ability of transfected DNA to locate and recombine with its homologous chromosomal counterpart in the genome (Smithies *et al.*, 1985). This process, commonly referred to as gene targeting, is being used in many laboratories to generate ES cells and, ultimately, mouse lines in which specific loci have been modified. A common rationale to disrupt gene function by gene targeting in ES cells is to construct a vector that is designed to undergo a homologous genetic exchange with an endogenous locus. These vectors are typically arranged so that they insert additional sequences, such as a positive selection marker, into the coding elements of the target, thereby ablating the function of the target gene (Thomas and Capecchi, 1987). However, when a targeting vector is introduced into an ES cell it may follow two different integrative pathways—either into a random chromosomal site or into its homologous site in the genome. Generally, the frequency of targeted integration events is lower than random integration events by several orders of magnitude (Thomas and Capecchi, 1987). The ratio of targeted to random integration events is important from an experimental point of view because it determines the ease with which it is possible to isolate the desired recombinant clone. There are many factors that may influence this ratio, including the length of homology (Thomas and Capecchi, 1987; Hasty *et al.*, 1991b), the target locus, the use of an insertion or replacement vector (Hasty *et al.*, 1991c), and the degree of polymorphic variation between the vector and the target.

Because gene targeting techniques are designed to generate specific recombinant alleles, features of the desired recombination locus have been used to select *in vitro* for targeted clones and against random integration events, thereby enhancing the representation of targeted clones in selected populations. Some of the strongest *in vitro* selection techniques for gene targeting events have utilized vectors wherein the positive selection marker lacks a promoter (Schwartzberg *et al.*, 1990) or polyadenylation (Donehower *et al.*, 1992) elements. The integration of such a vector into a random location in the genome will not, in most instances, lead to the efficient transcription of the selectable marker, resulting in the preferential survival of targeted clones under selective conditions specific for the positive selection marker cloned in the vector. It is also possible to select negatively against random integration events and thus partially enrich transfected populations in favor of targeted events (Mansour *et al.*, 1988). The integration structures of a targeting vector resulting from random insertions into the genome are typically not very precise; they may, for instance, often delete the negative selector gene. As a result the enrichments afforded by negative selection against random integration events are not as impressive as positive selection schemes. A summary of the types of selections that have been employed to enrich transfected populations for targeted events can be found in Bradley *et al.* (1992).

These techniques can be utilized separately or in combination to select for rare targeting events. However, their applicability is clearly limited to a subset of genetic changes: insertional inactivation or deletion of coding sequences to generate a null allele. They are not appropriate for directing very specific nucleotide alterations, which will ultimately be desirable for many genes. Targeting an expression cassette into a locus will introduce enhancers, promoters, and potential splicing artifacts, which may have a direct impact on both the pattern of expression and the translated product of a target gene. Analysis of mutations at a fine level, by definition, requires the exclusion of the selectable marker from the target locus. Thus, for directing minor modifications, selection cassettes should be excluded from the target locus. This introduces another complexity into attempts to generate subtle mutations, because the transfection events cannot be selected for directly. Given that the electroporation efficiency is of the order of $10^{-3}$, and considering a gene with a moderate targeting frequency of $10^{-3}$ targeted/illegitimate integration events, without positive selection

MURINE SITE-DIRECTED MUTAGENESIS 243

for transfection events the ratio would be $10^{-6}$ targeted events per cell transfected. The generation of a nonselectable mutation would thus become a very rare event. We describe three strategies to introduce nonselectable mutations: coelectroporation (Davis *et al.*, 1992), "hit-and-run" (Hasty *et al.*, 1991a), and "double-hit" gene targeting.

One of the basic rationales to which all gene targeting concepts are subject to is the belief that sequence modifications made in the homologous sequences in a vector will be faithfully transferred to the chromosomal target. Though the fidelity may not be a significant concern when generating a null allele, it is important in attempts to generate precisely defined changes in the genome. Although studies in a variety of gene targeting systems have observed a high frequency of error (Thomas and Capecchi, 1986; Brinster *et al.*, 1989), our recent survey of approximately 80 kb of junction fragments representing 44 targeted events revealed a very low error rate. We detected just two mutations in 80 kb: a deletion of a T and a T-to-G transversion (Zheng *et al.*, 1991). Thus, we are confident that minor sequence changes may be faithfully established in the genome by homologous recombination.

## A. COELECTROPORATION

Targeted recombination events may typically be generated at absolute frequencies of $10^{-5}$ to $10^{-6}$ per cell transfected (Thomas and Capecchi, 1987; Davis *et al.*, 1992; Soriano *et al.*, 1991; Hasty *et al.*, 1991a,b,c). Because only 1 in $10^3$ cells actually integrates the transfected DNA, this ratio can be enhanced by positive selection for the transfected cells. The relative frequency of targeted events in this subpopulation will be in the range of $10^{-2}$ to $10^{-3}$; thus, positive selection facilitates the direct isolation of targeted events. Positive selection of all transfection events, including legitimate as well as illegitimate recombinants, should be possible with an unlinked transfection marker (cotransfection), provided that the treated cells can take up and integrate more than one DNA molecule. For cotransfection to work with gene targeting, the additional proviso is that both a random and a targeted integration event must occur in the same cell.

To test if multiple-site integration is possible in ES cells, we transfected DNA at various concentrations into these cells. We routinely observed multiple insertions under a variety of conditions, and at least 15% of the G418$^r$ clones integrated the introduced DNA at two different chromosomal locations.

To test if we could obtain both a targeted and a nontargeted event in ES cells by coelectroporation, we chose to target the X-linked hypoxanthine phosphoribosyltransferase (*hprt*) gene (single copy in XY ES cells). Tar-

244                           ALLAN BRADLEY

geted recombinants generated in this locus can be designed to lose all
enzyme activity; these clones can thus be directly selected in 6-thioguanine
(6TG), regardless of the fidelity of the integration event.

We constructed a targeting vector that contains a 6.8-kb fragment of the
mouse *hprt* gene (Fig. 2). This fragment had been modified to include a 4-
bp frameshift insertion in exon 3 that destroyed a native *Xho*1 site and
created a diagnostic *Pvu*1 site. We cotransfected this vector at various
concentrations with a cassette carrying the *neomycin* gene into the CCE
ES cell line. We obtained 6TG$^r$ clones at a ratio of approximately 1 : 500
G418$^r$ clones, very similar to a control plasmid where the *neo* cassette was
cloned in the *Xho*1 site in exon 3.

Detailed molecular analysis of the integration structures in the 6TG$^r$
clones revealed that, although all the clones were targeted, only a subset
could be identified as having the desired integration pattern. Further analysis of these clones using the polymerase chain reaction verified the conversion of the endogenous *Xho*1 site to the desired *Pvu*1 site. The frequency
of bona fide clones was 8% of the 6TG$^r$ clones analyzed. Although this
figure is low, it is not unexpected, because the predicted double-crossover/

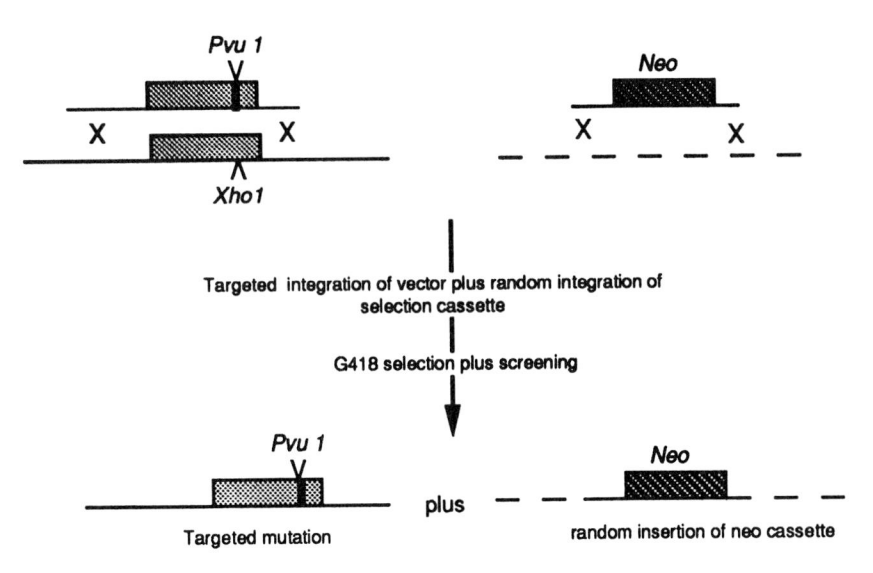

FIG. 2.   Generating nonselectable mutations by coelectroporation. A targeting vector
with the desired alteration (in this case represented by a *Pvu*1 site) and a separate unlinked
positive selection cassette are cointroduced into the same cell. In a subset of transfectants,
both a targeted and a random integration event occurs, rendering the cell selectable because
of the integration of a positive selectable marker. Targeted clones may be screened for with
conventional assays.

gene conversion integration event has occurred in only 10% of the 6TG$^r$ clones examined after transfection with the control vector (Hasty *et al.*, 1991c). In the cotransfection experiments, we are also requiring a second random integration event to generate a G418$^r$ clone.

Cotransfection enables one to generate specific mutations with the selectable marker located at a chromosomal site other than the target gene (Davis *et al.*, 1992). This makes it possible to target genes that suppress *neo* expression when integrated into the locus.

Any mutational event generated by the random insertion of the positive selectable marker in the genome may be separated from the targeted allele by meiosis following transmission through the germ line. The targeted genetic modifications generated by coelectroporation occur in a single step. This is important for many applications, because ES cells might lose their totipotent state during the prolonged culture that other targeting schemes require.

## B. HIT AND RUN

Arguably the best way to make minor changes to the genome is to modify the target locus by using two successive recombination events. This may be accomplished in two different ways, either using two rounds of single reciprocal recombination, such as that described in the hit-and-run strategy (Fig. 3) (Hasty *et al.*, 1991a; Valancius and Smithies, 1991), or using two rounds of double-reciprocal recombination, the double-hit strategy (Fig. 4). The rationale behind these double recombination events is to make the primary targeted clone relatively easy to isolate by conventional gene targeting selection/screens. Clones bearing this allele then serve as a substrate for the very inefficient ($10^{-5}$ to $10^{-7}$) secondary recombination event. Although these secondary events occur at a very low frequency, the recombinant clones may be readily isolated because they can be designed to exclude a negatively selectable marker included in the locus by the first recombination step (Hasty *et al.*, 1991a).

In the hit-and-run strategy, the first step is an insertional recombination of the vector into the target locus, which generates a duplication of the target homology separated by the vector, positive and negative selectable genes. Holliday junctions (Holliday, 1964) are formed during the initial integration event and these can migrate so that the region of the homologous sequence that differs between the vector and chromosomal target may form a region of heteroduplex. The differential repair of heteroduplexed DNA and the point of resolution of the Holliday junction will determine whether the introduced mutation ends up in the 5′, 3′, both, or neither of the duplicates.

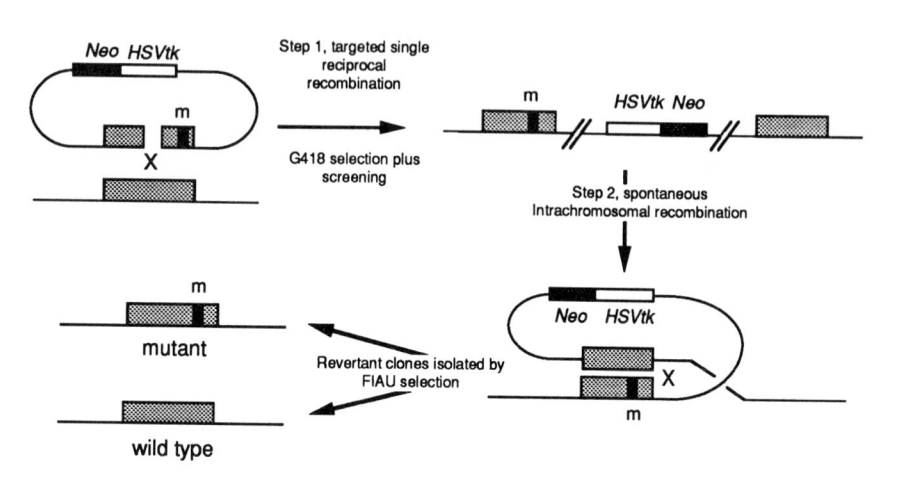

FIG. 3. Hit-and-run targeting to make subtle mutations. An insertion vector is used in the first step to target the desired locus. The insertional recombination event is stimulated by a double-strand break in the homologous sequences. The vector also carries the desired mutation, m, in the target homology. Recombinant clones are selected in G418 and screened by conventional methods for targeted events. For the second step, the targeted clones are expanded without selection. The duplication will spontaneously resolve by intrachromosomal recombination and revertant clones may be isolated by growing the cells in FIAU, because revertants will also lose the *HSVtk* gene and survive selection. A portion of the revertant clones will lose the mutation but it will be retained in others.

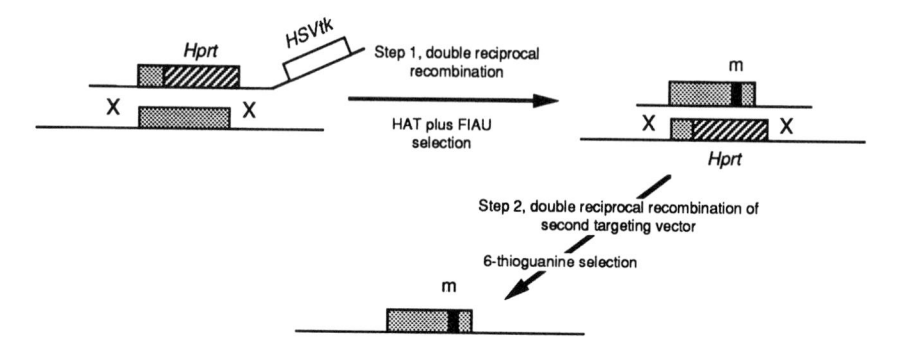

FIG. 4. Double-hit gene replacement to generate an allelic series rapidly. For the first step a standard replacement vector is used to target the desired genomic target in *hprt*-negative ES cells. Positive–negative selection for targeted clones is carried out in HAT media (for *hprt*-positive cells) plus FIAU (for cells that have lost the *HSVtk* gene. Double-resistant clones are screened for the desired recombination event. For the second step the clone targeted in the first step is transfected with a second vector, which carries target homology with the desired mutant allele. Targeted clones may be selected because they lose the *hprt* gene and become resistant to 6TG selection.

MURINE SITE-DIRECTED MUTAGENESIS 247

The direct repeat generated by the insertion vector will be resolved in the second step of the hit-and-run strategy. The reversion occurs by intrachromosomal recombination between duplicates on the same chromosome or by unequal sister chromatid recombination. The reversion also removes the negatively selectable gene, thus revertant clones may be directly selected in tissue culture. The position of the reversion crossover will determine if the revertant clones are wild type or have the desired modified allele.

At the *hprt* locus we have selectively isolated reversion events because the reverted cells regenerate a functional wild-type *hprt* gene and become resistant to hypoxanthine aminopterin thymidine (HAT). At this locus the frequency of reversion depends on the length of the duplicate. A 1.3-kb duplication showed a reversion frequency of $10^{-6}$ per cell generation, whereas a 6.8-kb duplication in the same region reverted at a frequency of $10^{-5}$ per cell generation. Introducing a small oligonucleotide mismatch into one unit of the duplication resulted in a 10-fold decrease in frequency. At the nonselectable *Hox-2.6* locus, we have selected reversion events by using the *HSVtk* gene. Revertant cells become resistant to the nucleoside analog 1-(2-deoxy-2-fluoro-$\beta$-D-arabinofuranosyl)-5-iodouracil (FIAU). Interestingly, the *Hox-2.6* locus displays a much higher reversion frequency compared with the *hprt* locus; with a 3.1-kb duplication and a single mutation, revertants could be selected at a frequency of $10^{-3}$ per cell division (Hasty *et al.*, 1991a).

At the *Hox-2.6* locus we examined the reversion events from six primary clones and were able to detect excision of the inserted vector in revertants from only two. From these clones, approximately 15% of the revertant cells retained the oligonucleotide mutation. Because these revertant clones were derived from a primary recombinant that had the mutation in the 5' duplicate, either the crossover occurred in just 0.5 kb or it occurred elsewhere in association with gene conversion (Bollag and Liskay, 1988).

The application of the hit-and-run technique at the *Hox-2.6* locus has enabled us to generate targeted clones that have a premature termination codon in the third helix of the homeodomain. Importantly, clones of ES clones that have been selectively isolated with this procedure have been shown to be transmitted through the germ line of mice (A.B. and Ramiro Ramirez-Solis, unpublished observations).

## C. DOUBLE-HIT GENE REPLACEMENT

The basis of this procedure is the introduction of a negatively selectable marker gene into the target site of interest. Clones of cells carrying this mutated allele become the target cells for secondary gene targeting experiments to perform gene replacement events at the allele that has already been targeted. The secondary gene replacement events are directly selectable because the negative marker should be removed by gene replacement, gene conversion (Fig. 4).

Negative selection is a very powerful means of selecting for the loss of gene activity at the target locus. However, for it to be effective, the background number of cells that survive selection must be $10^{-6}$ to $10^{-7}$ per selected cell, because secondary recombination events per transfected cell are likely to be generated at this frequency. We have targeted negatively selectable genes to both the *Hox-2.6* and c-*myc* loci (A.B., Hui Zheng, and Ann Davis, unpublished observations) and placed these targeted clones under negative selection. In both cases there is a very high background number of cells that survive negative selection. We have determined that the clones that survive selection have lost the negative selection cassette from the c-*myc* and *Hox-2.6* loci. The loss of this marker could be explained by gene conversion between the homologous chromosomes or by chromosome loss followed by reduplication of the remaining chromosome. The high background frequency is, however, a significant problem for isolating the secondary targeting events. Despite this, the double-hit gene replacement strategy has been demonstrated to work at the *Hox-2.6* locus (A.B. and Hui Zheng, unpublished observations).

## III. Discussion

We have described a second generation of gene targeting vectors and ES cells, which in combination enable us to modify the mouse genome in a variety of ways. Each type of vector will have a unique applicability, depending on the exact experimental situation, but all share the characteristic of being able to direct specific changes in a target locus at its normal chromosomal location. These vectors do not introduce other transcription units into a locus and thus constitute a major advance in our ability to interpret the consequences of the genetic changes we have engineered. Each experimental situation will dictate which of these strategies is most appropriate. The most powerful, double-hit gene replacement, relies on the exchange of a negative marker at the target site with vector sequences. Virtually any genetic change (deletions, insertions, exchange of control regions, and exchange of coding sequences) could potentially be made and isolated by selection. However, this particular strategy has yet to be demonstrated to work at high efficiency.

Coelectroporation can mediate the same types of genetic changes, namely gene replacement events. In this experimental situation, the frequency of the desired exchange is relatively modest, 1/6000 at the *hprt*

# MURINE SITE-DIRECTED MUTAGENESIS

locus, although it does not differ dramatically from that of gene replacement events that do not require coelectroporation in the same region of *hprt* (Hasty *et al.*, 1991c). For genes with a relatively high targeting efficiency, coelectroporation is a viable procedure. It will also be a very valuable technique where selective markers become silenced by the target locus. In situations where ES culture conditions are suboptimal, coelectroporation has the advantage of generating recombinants in a single step. This can be very important because the probability of isolating a totipotent clone through the two rounds of cloning required for the double-hit gene replacement approach may be low under some culture conditions.

The hit-and-run procedure utilizes the most efficient pathway for gene targeting in ES cells (Hasty *et al.*, 1991a). In experimental scenarios wherein considerable effort must be dedicated to screen/select rare recombinants from large populations of transfected cells, increasing the relative abundance of the required recombinant is desirable. Our results have shown that insertion vectors will form the predicted recombinant structures about 100-fold more efficiently than replacement vectors can mediate the desired replacement events (Hasty *et al.*, 1991c). The high frequency of targeting with insertion vectors was a necessary prerequisite for the success of the hit and run strategy. If the primary recombinants were difficult to detect, then the system would be more difficult to work with. Although the reversion events occurred at modest frequency, the ability to kill nonrevertant clones selectively negates the relative importance of the frequency of the reversion event. We have used the hit-and-run approach to introduce an oligonucleotide insertion (Hasty *et al.*, 1991a), but it may also be used for larger genome modifications.

It is now possible to modify virtually any gene and establish the mutation in the germ line. As developmental biologists attempting to unravel the genetic control of mouse development, our limitations are no longer at the level of designing the genotype, but understanding the phenotype!

### ACKNOWLEDGMENTS

This work was supported by grants from the NIH (A.B.), the Searle Scholars Program (Chicago Community Trust) (A.B.), and the Cystic Fibrosis Foundation.

### REFERENCES

Balling, R., Deutsch, U., and Gruss, P. (1988). *Cell (Cambridge, Mass.)* **55,** 531–536.
Bollag, R. J., and Liskay, R. M. (1988). *Genetics* **119,** 161–169.
Bradley, A. (1990). *Curr. Opinions Cell Biol.* **2,** 1013–1017.
Bradley, A., Evans, M., Kaufman, M. H., and Robertson, E. J. (1984). *Nature (London)* **309,** 255–256.

Bradley, A., Hasty, P., Davis, A., and Ramirez-Solis, R. (1992). *Biotechnology* **10**, 534–539.
Brinster, R. L., Braun, R. E., Lo, D., Avarbock, M. R., Oram, F., and Palmiter, R. D. (1989). *Proc. Natl. Acad. Sci. U.S.A.* **86**, 7087–7091.
Chabot, B., Stephenson, D. A., Chapman, V. M., Besmer, P., and Bernstein, A. (1988). *Nature (London)* **335**, 88–89.
Copeland, N. G., Gilbert, D. J., Cho, B. C., *et al.* (1990). *Cell (Cambridge, Mass.)* **63**, 175–183.
Davis, A. C., Wims, M., Bradley, A. (1992). *Mol. Cell. Biol.* **12**, 2769–2776.
Donehower, L. A., Harvey, M., Slagle, B., McArthur, M. J., Montgomery, C. A., Butel, J. S., and Bradley, A. (1992). *Nature (London)* **356**, 215–221.
Evans, M. J., and Kaufman, M. H. (1981). *Nature (London)* **292**, 154–156.
Friedrich, G., and Soriano, P. (1991). *Genes Dev.* **5**, 1513–1523.
Green, M. C. (1989). *In* "Genetic Variants and Strains of the Laboratory Mouse (M. F. Lyon and A. G. Searle, eds.), 2nd ed. Oxford Univ. Press, London.
Hasty, P., Ramirez-Solis, R., Krumlauf, R., and Bradley, A. (1991a). *Nature (London)* **350**, 243–246.
Hasty, P., Rivera-Perez, J., and Bradley, A. (1991b). *Mol. Cell. Biol.* **11**, 5586–5591.
Hasty, P., Rivera-Perez, J., Chang, C., and Bradley, A. (1991c). *Mol. Cell. Biol.* **11**, 4509–4517.
Holliday, R. (1964). *Genet. Res.* **5**, 282–304.
Kuehn, M. R., Bradley, A., Robertson, E. J., and Evans, M. J. (1987). *Nature (London)* **326**, 295–298.
Joyner, A. L., Auerbach, B. A., Davis, C. A., Herrup, K., and Rossant, J. (1991). *Science* **251**, 1239–1243.
Li, S., Crenshaw, E. B., Rawson, E. J., Simmons, D. M., Swanson, L. W., and Rosenfeld, M. C. (1990). *Nature (London)* **347**, 528–533.
Maas, R. L., Zeller, R., Woychik, R. P., Vogt, T. F., and Leder, P. (1990). *Nature (London)* **346**, 853–855.
Mansour, S. L., Thomas, K. R., and Capecchi, M. R. (1988). *Nature (London)* **336**, 348–352.
McMahon, A., and Bradley, A. (1990). *Cell (Cambridge, Mass.)* **62**, 1073–1085.
Mullen, R. J., and Whitten, W. K. (1977). *J. Exp. Zool.* **9**, 111–129.
Pease, S., and Williams, R. L. (1990). *Exp. Cell Res.* **190**, 209–211.
Robertson, E. J. (1987). *In* "Teratocarcinomas and Embryonic Stem Cells: A Practical Approach" (E. J. Robertson, ed.), pp. 71–112. IRL Press, Oxford.
Robertson, E. J., Bradley, A., Kuehn, M., and Evans, M. (1986). *Nature (London)* **323**, 445–448.
Schnieke, A., Harbers, K., and Jaenisch, R. (1983). *Nature (London)* **304**, 315–320.
Schwartzberg, P. L., Goff, S. P., and Robertson, E. J. (1989). *Science* **246**, 799–803.
Schwartzberg, P. L., Robertson, E. J., and Goff, S. P. (1990). *Proc. Natl. Acad. Sci. U.S.A.* **87**, 3210–3214.
Smith, A. G., and Hooper, M. L. (1987). *Dev. Biol.* **121**, 1–9.
Smithies, O., Gregg, R. G., Boggs, S. S., Koralewski, M. A., and Kucherlapati, R. S. (1985). *Nature (London)* **317**, 230–234.
Soriano, P., Montgomery, C., Geske, R., and Bradley, A. (1991). *Cell (Cambridge, Mass.)* **64**, 693–702.
Thomas, K. R., and Capecchi, M. R. (1986). *Mol. Cell. Biol.* **324**, 34–38.
Thomas, K. R., and Capecchi, M. R. (1987). *Cell (Cambridge, Mass.)* **51**, 503–512.
Thompson, S., Clarke, A. R., Pow, A., Hooper, M. L., and Melton, D. W. (1989). *Cell (Cambridge, Mass.)* **56**, 313–321.

Valancius, V., and Smithies, O. (1991). *Mol. Cell. Biol.* **11,** 1402–1408.
Williams, R. L., Hilton, D. J., Pease, S., Willison, T. A., Stewart, C. L., Gearing, D. P., Wagner, E. F., Metcalf, D., Nicola, N. A., and Gough, N. M. (1988). *Nature (London)* **336,** 684–686.
Yoshida, H., Hayashi, S.-I., Kunisada, T., Ogawa, M., Nishikawa, S., Okamura, H., Sudo, T., Shultz, L. D., and Nishikawa, S. I. (1990). *Nature (London)* **345,** 442–444.
Zheng, H., Hasty, P., Brenneman, M., Grompe, M., Gibbs, R., Wilson, J. H., and Bradley, A. (1991). *Proc. Natl. Acad. Sci. U.S.A.* **88,** 8067–8071.

# The Molecular Basis for
# Growth Hormone–Receptor Interactions

JAMES A. WELLS, BRIAN C. CUNNINGHAM, GERMAINE FUH, HENRY B.
LOWMAN, STEVEN H. BASS,* MICHAEL G. MULKERRIN,[†] MARK ULTSCH,
AND ABRAHAM M. DEVOS

*Departments of Protein Engineering, *Cell Genetics,
and [†]Medicinal and Analytical Chemistry, Genentech, Inc., South San Francisco,
California 94080*

## I. Introduction

Human growth hormone (hGH) is important for normal human growth
and development. This 22-kDa polypeptide hormone induces a variety of
biological effects, including linear growth, lactation, nitrogen retention,
lipolysis, a diabetogenic-like effect, macrophage activation, and others
(for review see Hughes and Friesen, 1985; Isaksson *et al.*, 1985). These
activities begin with the pulsatile release of hGH from the pituitary and
specific interactions with cellular receptors.

A molecular understanding of these phenomena is crucial to developing new therapies for disease states associated with hGH deficiency
(dwarfism) and excess (acromegaly), among others. Furthermore, such
work is fundamentally important for understanding the molecular basis
for hormone action. Here we review studies from our group designed to
characterize the molecular basis for growth hormone–receptor interactions.

## II. Expression and Biochemical Characterization of the
## hGH Binding Protein

The hGH receptor has been cloned and sequenced (Leung *et al.*,
1987). The receptor contains an extracellular hormone-binding domain
(~28 kDa), a single transmembrane domain, and an intracellular domain
(~35 kDa) of unknown function. The extracellular domain is found in
serum (Baumann *et al.*, 1986; Herington *et al.*, 1986) and is termed the
hGH binding protein (hGHbp). The hGHbp binds [125]I-labeled hGH about
three-fold weaker than the full-length receptor and is thought to be shed

254                    JAMES A. WELLS ET AL.

by proteolysis of the liver hGH receptor (Spencer *et al.*, 1988; Leung *et al.*, 1987).

hGH can be produced in large quantities as secretions of cultures of *Escherichia coli* (Chang *et al.*, 1987). The hormone is normally nonglycosylated and *E. coli*-derived hGH has the same receptor-binding properties as pituitary-derived hGH. We were able to express the hGHbp (residues 1–238) in a similar fashion in *E. coli* and to purify it in large quantities (Fuh *et al.*, 1990). Natural hGHbp is heavily glycosylated. Nonetheless, the *E. coli* expressed hGHbp has the same binding affinity and specificity for hGH as the natural hGHbp, suggesting glycosylation of the hGHbp is not critical for hGH binding. Having access to large quantities of the purified hGHbp allowed us to determine that the disulfide pairing consisted of three sequentially linked cystines (Cys-38–Cys-48, Cys-83–Cys-94, and Cys-108–Cys-122) (Fuh *et al.*, 1990). Furthermore, the highly purified binding protein simplified the binding analysis of hGH mutants and provided sufficient materials for biphysical studies and crystallography (see Section VII).

## III. Scanning Mutational Analysis of hGH

A two-tiered strategy was applied to identify important receptor-binding determinants in hGH (for review see Wells, 1991). The first level of analysis used an approach we call homologue-scanning mutagenesis (Cunningham *et al.*, 1989) (Fig. 1). hGH shares substantial sequence homology with several other hormones, such as prolactin (PRL) and placental lactogen (PL), which do not bind tightly to the hGH receptor. Because homologous proteins retain similar folds (for example, see Chothia and Lesk, 1986), we reasoned that segments substituted from functionally divergent but structurally similar molecules would be accommodated into the structure of hGH without disrupting the overall fold of the variant. By analyzing those chimeras that were most disruptive to binding we could infer areas of the molecule important in modulating binding affinity.

From a folding diagram of porcine growth hormone (pGH) reported from X-ray crystallographic studies (Abdel-Meguid *et al.*, 1987), it was possible to derive a simple structural model for hGH (Cunningham *et al.*, 1989). This model was very useful in designing segment substitutions that preferentially altered surface residues, and for mapping the mutational effects on the structure. From the analysis of 17 segment-substituted molecules that collectively altered 89 out of 191 residues in hGH, we inferred that residues in the N-terminal portion of helix 1, an extended loop region between residues 54–74 and the C-terminal portion of helix 4 were important for binding to the hGHbp (Fig. 2A).

The disruptive effects on binding were probably caused by local and not global changes in the structure of the variant hGH molecules. First, the

## Homologue scan

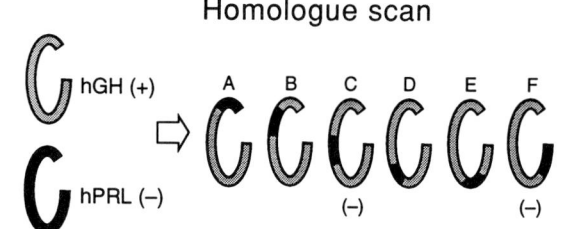

## Ala scan

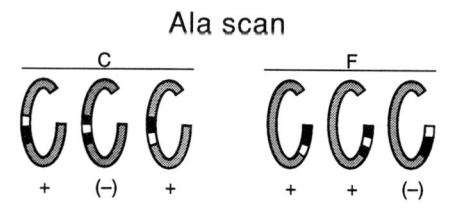

FIG. 1. Scanning mutational strategies for analysis of receptor binding determinants in hGH. Upper panel outlines homologue scanning in which segments (■) of nonbinding homologues of hGH (such as hPRL) are substituted throughout hGH. Mutants severely reduced in binding affinity to the hGHbp (−) identify putative areas of receptor binding. Lower panel shows alanine scanning in which side chains larger than alanine (□) in segments (■) identified by homologue scanning are mutated and tested for binding.

collection of disruptive variants formed a patch on the model of hGH despite being distantly separated in primary sequence (Fig. 2A). Second, homologue scanning effectively identified epitopes for eight different monoclonal antibodies (Mabs) (Cunningham *et al.*, 1989). Each of these Mabs sometimes overlapped each other on the hGHbp, but were always different in detail and usually discontinuous in character. Every segment-substituted mutant that disrupted the receptor binding by >10-fold retained the ability to bind to most of the other Mabs. Thus, the Mabs and hGHbp served as conformational probes showing that the segment-substituted mutants were not globally misfolded.

A second and higher resolution level of analysis used an approach called alanine-scanning mutagenesis (see Wells, 1991) (Fig. 1). Each side chain larger than alanine between residues 2–19, 54–74, and 164–191 was converted to alanine (Cunningham and Wells, 1989). There are a number of reasons for choosing alanine over other replacement residues. Alanine is the most common amino acid in proteins; it is found frequently in buried and exposed positions and does not break secondary structures. Finally, alanine does not impose new steric, electrostatic, or hydrogen-bonding schemes. In effect, this analysis "tests" the energetic or functional importance of side chains beyond the $\beta$-carbon for binding.

256 JAMES A. WELLS ET AL.

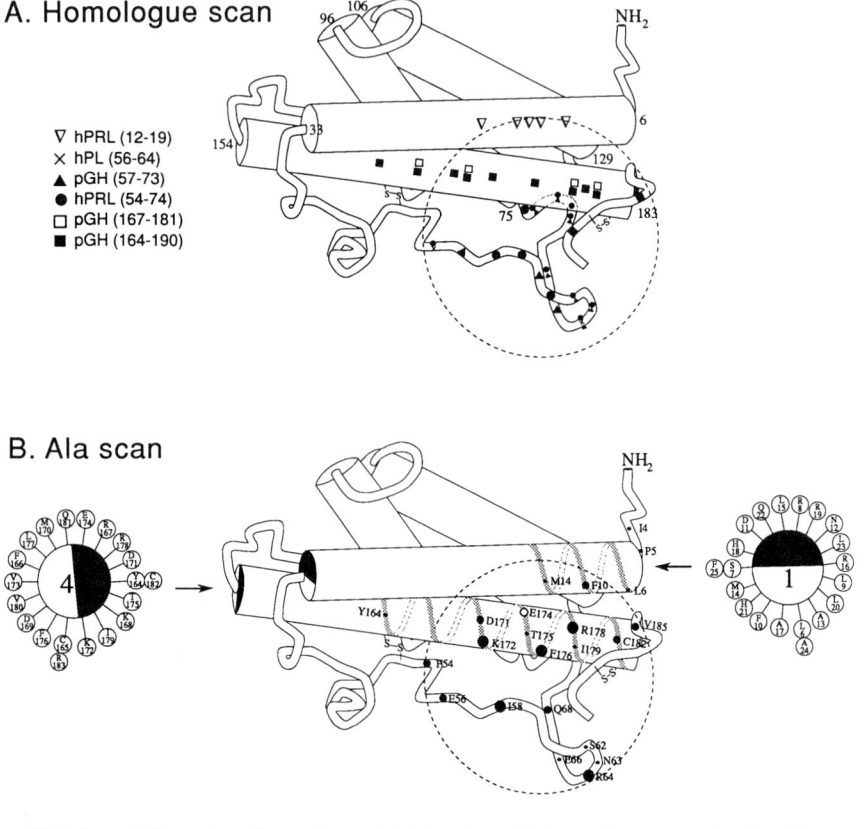

FIG. 2. (A) Location of mutations in hGH produced by homologue scanning that disrupt binding to the hGHbp by greater than 10-fold. The locations of specific mutations within each chimera are shown by a common symbol. The circle illustrates the approximate extent and location of the disruptive mutations (from Wells, 1991). (B) Receptor binding determinants are identified by alanine-scanning mutagenesis of hGH (Cunningham and Wells, 1989). Residues cause a 2- to 4-fold reduction (●), a 4- to 10-fold reduction (●), or enhancement (○), and a 10- to 100-fold reduction (●) in binding to the hGHbp. Helical wheel projections are shown for the amphipathic helices 1 and 4, with the hydrophilic surface blackened. From Cunningham and Wells (1991).

Binding analysis of the set of purified single alanine replacements showed that although most had little effect on binding, 10 variants caused 2- to 4-fold reductions in affinity, 7 analogs caused 4- to 10-fold reductions in affinity, and 5 variants caused 10- to 50-fold reductions in affinity (Fig. 2B). One substitution (E174A) actually had a 4-fold improvement in affinity, showing that binding affinity is not maximized between hGH and its receptor. These residues mapped to the same region in hGH as those

GROWTH HORMONE–RECEPTOR INTERACTIONS    257

revealed in the homologue scan. Moreover, the disruptive mutants on helix 4 were on the hydrophilic face of the amphipathic helix, suggesting that the important side chains point in the same direction. Similarly, the positions of disruptive alanine substitutions for helix 1 appear on the same side and point toward those in helix 4. This functional epitope is generally cationic and contains a number of important hydrophobic side chains. Some of these disruptive effects may result from indirect structural perturbations, but extensive Mab analysis indicates that the folded conformation of each mutated hormone is like wild-type hGH (Cunningham *et al.*, 1989; Cunningham and Wells, 1989). The alanine-scan data provide a high-resolution analysis of side chains in hGH capable of modulating binding affinity to the hGHbp. Recently, homologue and alanine scanning have been used to provide high resolution functional epitopes for binding 21 different Mabs to hGH (Jin *et al.*, 1992). Each functional epitope on hGH was unique, surface accessible, and discontinuous in character.

## IV.    Use of High-Resolution Functional Data on hGH

### A.    "RECRUITMENT" OF hPRL AND hPL TO BIND THE hGHbp

We reasoned if the alanine-scan data reflected functionally important interactions between hGH and the hGHbp, it may be possible to use this information to engineer nonbinding homologues (such as hPRL and hPL) to bind to the hGHbp even in the absence of an X-ray structure of the complex. hPRL shares only 23% sequence identity with hGH and contains different side chains at 17 positions in hGH where alanine substitutions caused a twofold or greater change in binding affinity (Fig. 3A). Using an iterative process of mutagenesis and binding analysis, we were able to construct a mutant of hPRL containing only eight substitutions that bound about sixfold weaker than hGH to the hGHbp (Cunningham *et al.*, 1990a). An abbreviated summary of this set of variants is shown in Table I.

We engineered hPL (Fig. 3B) by a similar process to bind tightly to the hGHbp (Lowman *et al.*, 1991a). hPL shares 85% sequence identity with hGH and binds over 2000-fold more weakly to the hGHbp. Following incorporation of five substitutions derived from alanine-scanning data on hGH we were able to construct an hPL variant that bound only 1.6-fold weaker than hGH (Table I). These studies provide confirmation that the structural scaffolds for these three homologous hormones were very similar and that residues identified by alanine scanning are indeed functionally important. Finally, these redesigned homologues were useful tools for distinguishing hormone binding and receptor activation (see Section VII,B).

## A. Important residues not conserved in hPRL

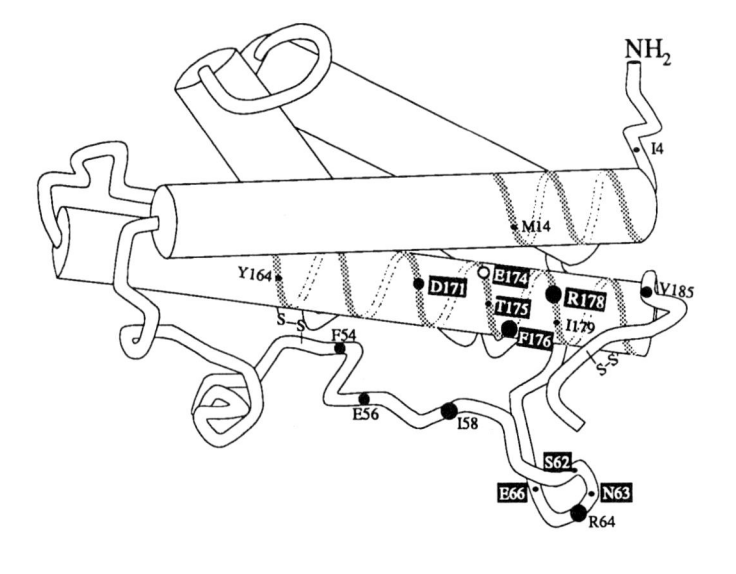

## B. Important residues altered in hPL

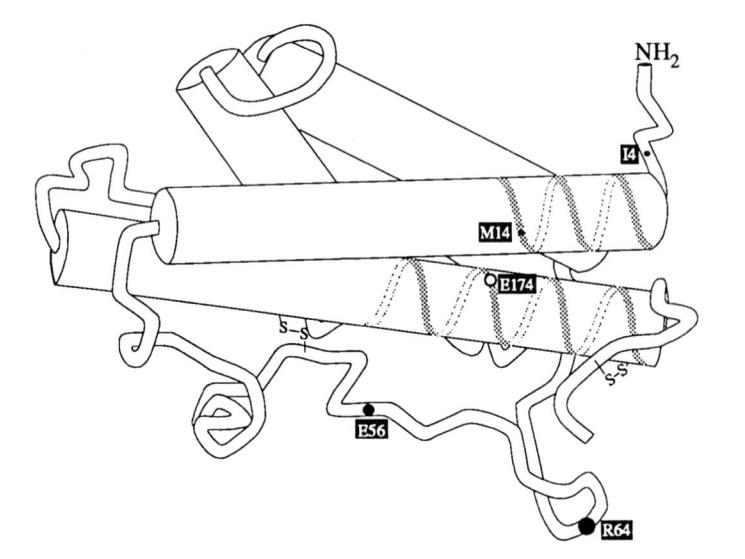

FIG. 3. Location of residues in hPRL (A) and hPL (B) that are different from hGH and important for binding of hGH to the hGHbp as determined by alanine scanning (Fig. 2B). Mutated residues in these nonbinding or weak-binding homologues that promoted binding to the hGHbp are shown enclosed in black squares.

GROWTH HORMONE–RECEPTOR INTERACTIONS 259

TABLE I

*Binding of hPRL and hPL Variants to the hGHbp*[a]

| Hormone ($K_d$) (mutations) | $K_d$(mut)/$K_d$(hGH) |
|---|---|
| hGH (0.3 n$M$) | 1 |
| hPRL (>30m$M$) | >100,000 |
| hPRL-A (H171D/N175T/Y176F) | 14,000 |
| hPRL-B (A plus K168R) | 660 |
| hPRL-C (B plus E174A) | 200 |
| hPRL-D (C plus E62S/D63N/Q66E) | 6.2 |
| hPL (770 n$M$) | 2300 |
| hPL-A (D56E) | 340 |
| hPL-B (A plus M64K/M179I) | 4.7 |
| hPL-C (B plus V4I/E174A) | 1.6 |

[a] Binding affinities were measured by competitive displacement of $^{125}$I-labeled hGH from hGHbp followed by immunoprecipitation of the hGHbp. Relative binding affinities were calculated from the ratio $K_d$ (mut)/$K_d$ (hGH). Data are compiled from Cunningham *et al.* (1990a) and Lowman *et al.* (1991a). Multiply mutated proteins are named by the parent hormone derivative followed by a series of single mutants. For example, hPRL-A is a triple mutant of hPRL in which His-171 is converted to Asp, Asn-175 is converted to Thr, and Tyr-176 is converted to Phe. hPRL-B contains these same mutations plus K168R.

## B. OPTIMIZATION OF BINDING AFFINITY OF hGH FOR THE hGHbp BY PHAGE DISPLAY

The fact that E174A hGH had a binding affinity about fourfold higher than hGH for the hGHbp suggested that this interface was not optimized for binding. To test the hypothesis that the binding affinity of hGH for the hGHbp could be enhanced further, we developed a random mutagenesis and *in vitro* selection method called monovalent phage display (Bass *et al.*, 1990). This technique allowed us to screen nearly 10$^6$ different hGH molecules for tight binding to the hGHbp (Lowman *et al.*, 1991b). In this method (Fig. 4), hGH was fused to the pili attachment protein (gene III) from the filamentous phage M13. On infection of *E. coli* cells bearing this plasmid with a helper phage, M13K07 (Vierra and Messing, 1987), phagemid particles were produced that packaged the hGH–gene III fusion plasmid and displayed hGH as a single-copy fusion protein. These phages bind to the hGHbp immobilized on polyacrylamide beads and can be enriched (10$^3$- to 10$^4$-fold) over non-hGH phages by a single binding and elution cycle (a binding selection). By displaying the hGH fusion protein as a single copy, we avoided "chelate" effects that can complicate binding

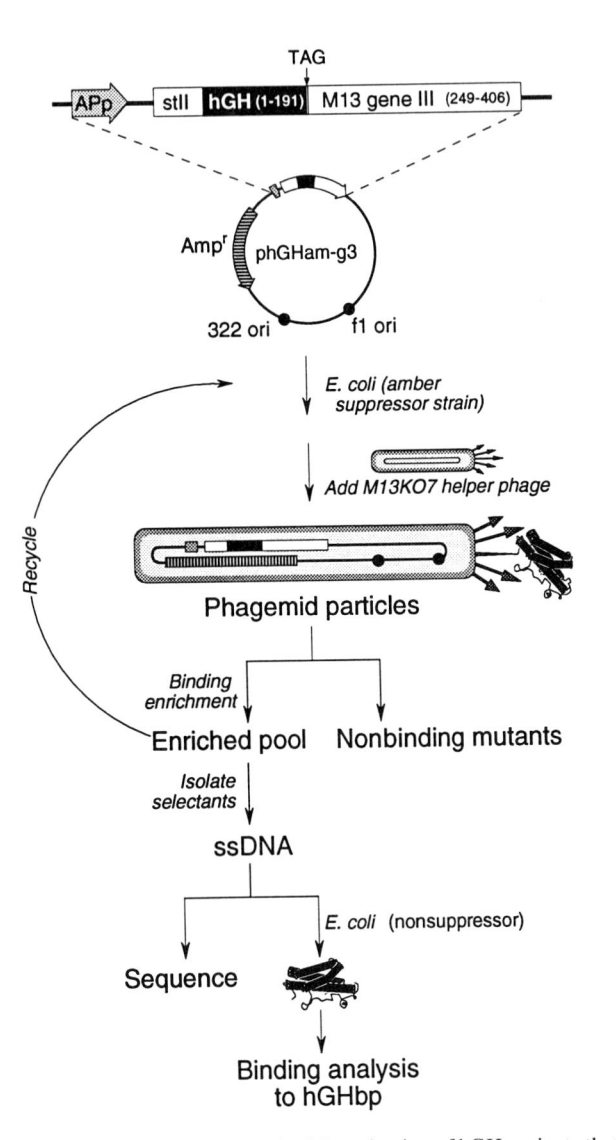

FIG. 4.  Monovalent phage display method for selection of hGH variants that bind tightly to the hGHbp. The plasmid, phGHam-g3, contains the alkaline phosphatase promoter (APp), stII signal sequence, and the hGH gene followed by codons 249–406 of the M13 gene III. After several cycles of binding selection to the hGHbp immobilized on polyacrylamide beads, individual phages are cloned and ssDNA is sequenced to determine what amino acid substitutions are present in hGH. Finally, the ssDNA is transformed into a nonsuppressor *E. coli* strain, in which the amber codon (TAG) terminates translation and free hormone is produced. See Lowman *et al.* (1991b) for details.

GROWTH HORMONE–RECEPTOR INTERACTIONS 261

selections in polyvalent display (for an example, see Cwirla *et al.*, 1990) by virtue of the phage binding through more than one ligand. Moreover, progeny phagemids from monovalent phagemid particles are much more easily propagated because each phage retains three to four copies of the wild-type gene III product for infection of *E. coli*. This method allowed us to sort tighter and weaker binding variants of hGH (Bass *et al.*, 1990).

We randomly mutated residues in hGH (four at a time) that were close to (or at) those shown to modulate binding affinity to the hGHbp by alanine scanning (Lowman *et al.*, 1991b). After three to six cycles of binding selection, "consensus" sequences developed for tight binding to the hGHbp. Such sequences often contained wild-type (or chemically similar) residues, if that residue was functionally important by alanine scanning. Residues not important by alanine scanning were usually sorted to something different from the wild-type hGH. The tightest binding hGH sequences from each of three different libraries had affinities three to five times tighter than hGH. These could be combined to produce even higher affinity variants (H. B. Lowman and J. A. Wells, unpublished data). We believe this is an extremely powerful method for producing high-affinity binding variants of hGH and other proteins. Moreover, such analyses should provide much information about how binding affinity and specificity can be modulated by amino acid changes at or near protein–ligand interfaces.

## C. DESIGN OF RECEPTOR-SELECTIVE VARIANTS OF hGH

hGH binds to both the hGH and hPRL receptors. To better define the molecular basis for the dual binding properties of hGH we wished to obtain a high-resolution functional map for the binding of hGH to the hPRL receptor. The cDNA for the prolactin receptor has been cloned and sequenced (Boutin *et al.*, 1989). We constructed a truncated version of the hRPL receptor (hPRLbp) containing the extracellular portion the receptor (residues 1–211) and secreted it in a binding-competent form from *E. coli* (Cunningham *et al.*, 1990b). Interestingly, on purification the hPRLbp lost its ability to bind hGH. Through a series of biochemical experiments it was determined that one equivalent of $Zn^{2+}$ was required for tight binding of hGH to hPRLbp. That is, hGH has a $K_d$ for the hPRLbp of 0.03 n$M$ in the presence of $Zn^{2+}$ compared to 270 n$M$ in the absence of zinc. In contrast, hGH binds to the hGHbp with $K_d$ values of 1.6 or 0.3 n$M$ in the presence or absence of $Zn^{2+}$, respectively. Thus, $Zn^{2+}$ is important for binding of hGH onto the hPRL receptor but not for binding to the hGH receptor. Moreover, other divalent metal ions present in serum do not substitute well for the role of $Zn^{2+}$ in binding hGH to the hPRLbp.

Homologue- and alanine-scanning mutagenesis elaborated a functional epitope (Fig. 5) for binding of the hPRLbp to hGH (Cunningham and Wells, 1991). This discontinuous epitope overlapped that for the hGHbp and contained three residues that were potential ligands for zinc coordination (His-18, His-21, and Glu-174). Equilibrium dialysis experiments on variants of hGH confirmed that these residues are important for direct binding of zinc in the hGH $\cdot$ Zn$^{2+}$ $\cdot$ hPRLbp ternary complex (Cunningham et al., 1990b).

Zinc typically coordinates four ligands in proteins. Because both hGH and the hPRLbp are required for tight binding of Zn$^{2+}$ and three ligands come from hGH, we reasoned that the fourth ligand may be donated from the hPRLbp. The hGHbp does not require zinc and therefore we targeted two His and one Cys residues in the hPRLbp for mutational studies because they were universally conserved in prolactin receptors and absent in growth hormone receptors. Indeed, such mutational analysis indicated that His-188 in the hPRLbp is the fourth zinc ligand (Cunningham et al., 1990b).

Although the epitopes for the hGHbp and hRPLbp overlapped, they were not identical (compare Figs. 2B and 5). Mutations at noninteracting sites in proteins exhibit additive free-energy effects on binding (Wells, 1990). Therefore, we reasoned by making multiple mutations at unshared binding determinants that we could produce variants of hGH exhibiting extreme selectivity for the hGHbp or hPRLbp. Also, by multiply mutating common residues it should be possible to produce variants of hGH that bind to neither receptor. Indeed, the double mutant K168A/E174A bound tightly to the hGHbp but not the hPRLbp, whereas the opposite was true for the R64A/D171A mutant (Table II). Furthermore, the K172A/E176A double mutant binds only very weakly to both receptors.

Such receptor-selective variants of hGH are useful probes to dissect the importance of binding to hGH or hPRL-like receptors for various biological activities. For example, the K168A/E174A variant is equipotent with hGH in promoting rat weight gain whereas the R64A/D171A and K172A/E176A variants are virtually inactive (R. Clark and D. Mortensen, unpublished results). Such data suggest that the hGH receptor and not the hPRL receptor is important in promoting rat weight gain. On the other hand, analysis of the receptor specificity on human neutrophils using these analogs indicates that the neutrophil receptor is activated by a prolactin-like epitope (Fu et al., 1992). We believe these receptor-selective analogs will be very useful in assigning specific receptor binding properties to the many biological activities of hGH.

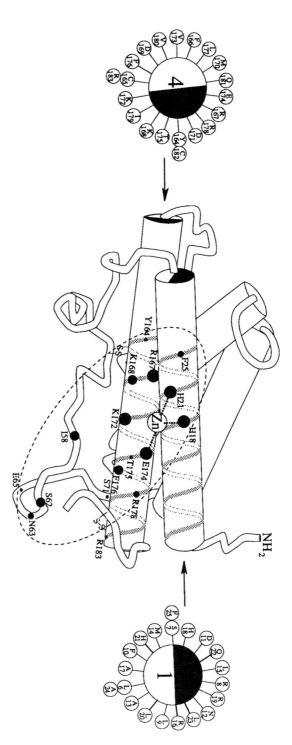

FIG. 5. Receptor binding determinants identified for the hPRLbp by alanine-scanning mutagenesis of hGH. Alanine substitutions at these residues cause a 2- to 4-fold reduction (●), a 4- to 10-fold reduction (●), a 10- to 100-fold reduction (●), or greater than 100-fold reduction (●) in binding to the hPRLbp. The locations of the putative zinc ligands and binding site are shown. From Cunningham and Wells (1991).

264                    JAMES A. WELLS ET AL.

TABLE II

*Construction of Receptor-Specific Variants*
*of hGH[a]*

| Hormone | $K_d$ (mut) $K_d$ (hGH) | |
|---|---|---|
| | hPRLbp | hGHbp |
| WT hGH | (1) | (1) |
| K168A/E174A | 9100 | 0.27 |
| R64A/D171A | 1.9 | 280 |
| K172A/F176A | 8400 | 560 |

## V.  Storage of hGH as a $Zn^{2+} \cdot$ hGH Dimer

Following the discovery that $Zn^{2+}$ is required for tight binding of hGH to the hPRLbp, we studied the binding of hGH to each component separately (Cunningham *et al.*, 1991a). Equilibrium dialysis experiments showed that while we could not detect binding of $Zn^{2+}$ to the hPRLbp, $Zn^{2+}$ bound to hGH with a moderate affinity ($\sim 1 \mu M$). Scatchard plots showed an upward deflection indicative of positive cooperativity. This was surprising because the stoichiometry of binding was one $Zn^{2+}$ per hGH. Subsequently, gel filtration and sedimentation equilibrium studies showed that $Zn^{2+}$ cooperativity induces formation of a dimeric complex $(Zn^{2+} \cdot hGH)_2$.

Mutational studies showed that His-18, His-21, and Glu-174 are important for binding of $Zn^{2+}$ and for dimer formation (Cunningham *et al.*, 1991a). These residues are the same ligands involved in forming the hGH $\cdot Zn^{2+} \cdot$ hPRLbp complex (Fig. 5). We believe that the ligand requirements for both zinc atoms are satisfied by monodentate coordination by two of the three residues plus bidentate bridging coordination from the third residue (probably Glu-174), which is donated to both $Zn^{2+}$ ions from each hGH in the dimer. Although other models are possible, such a bridged ligand model would help explain the cooperativity observed in dimer formation.

From sedimentation equilibrium experiments we estimate that in saturating $Zn^{2+}$ the hGH dimeric complex forms with a $K_d$ of $\sim 3 \mu M$. This is much above the peak concentrations of hGH found in serum ($\sim 2$ n$M$) (Thompson *et al.*, 1972) but much below the concentration of hGH as it is stored in somatotropic vesicles in the pituitary gland ($>2.5$ to 5 m$M$) (Daughaday, 1985; Cunningham *et al.*, 1991a). Moreover, the concentration of total $Zn^{2+}$ in such vesicles is estimated to be $\sim 4$ m$M$ (Thorlacius-Ussing, 1987). Thus, hGH and $Zn^{2+}$ are present in about equimolar amounts in somatotropic vesicles and thus probably exist in a dimeric

GROWTH HORMONE–RECEPTOR INTERACTIONS 265

complex. The $(Zn^{2+} \cdot hGH)_2$ complex is more stable than free hGH to denaturants (Cunningham *et al.*, 1991a). Thus, we believe zinc may serve to stabilize hGH prior to release from its storage granules in the pituitary. In addition, the dimerization of hGH by $Zn^{2+}$ covers important hGH receptor and hPRL receptor interfaces and may protect receptors proximal to the pituitary from the sudden efflux of hormone.

## VI.  Alanine Scanning of the hGHbp

We could not use homologue scanning to probe determinants in the hGHbp for hGH because hGH is capable of binding to all cloned GH and PRL receptors. Thus, we adopted a different strategy (Bass *et al.*, 1991). First, charged residues in the hGHbp were changed to alanines in clusters ranging from two to five residues. Charged residues are usually exposed and we had seen that several charged residues on hGH were important for binding the hGHbp. We dissected to single alanine mutants the clustered charge-to-alanine mutants that disrupted binding affinity to hGH. Although a number of clustered charge-to-alanine mutants did not express well in *E. coli*, we could usually obtain single alanine mutants at these positions. Overall we identified nine charged residues that, when changed to alanine, caused two- to eightfold reductions in binding affinity for hGH. Of these, the most disruptive ones were negatively charged residues, which complemented the fact that the most disruptive residues from hGh were positively charged. Furthermore, all of these important charged residues except one (K215A) were in the first half of the molecule—the disulfide-rich domain.

Fluorescence quenching experiments suggested that Trp residues in the hGHbp may become buried on binding of hGH (Bass *et al.*, 1991). Therefore, we mutated four Trp residues individually to Ala in the first half of the molecule. Indeed, one of these (W104A) caused a >2500-fold drop in binding affinity. A less drastic substitution (W104F) reduced affinity about 100-fold. Another variant, W76A caused a 2.5-fold drop in affinity. Finally, we mutated all residues within four of either side of those mutants causing a 2-fold or greater reduction in affinity. Ten of the most disruptive alanine-scan mutants were tested for binding with a panel of six conformationally sensitive Mabs. Each hGH variant bound to most or all Mabs with an affinity equal to that of wild-type hGH.

About the time that this analysis was complete, a structural prediction was made for the extracellular domain of the hematopoietic receptors (Bazan, 1990). Based on consensus sequences among cytokine receptors and the disulfide bonding pattern for the hGHbp, Bazan predicted the hGHbp would consist of two immunoglobulin-like domains. Indeed, when the set of disruptive alanine mutants in the first half of the molecule were mapped on this model, they formed a patch in the loop regions predicted for the hGHbp (Fig. 6). This is an example of how functional data from alanine scanning can provide support for structural models. Recently, Rozakis-Adcock and Kelly (1992) have identified a similar epitope by homologue and alanine scanning for binding of ovine PRL to a rat PRL receptor, suggesting that these homologous hormones interact with their receptors via topologically similar sites.

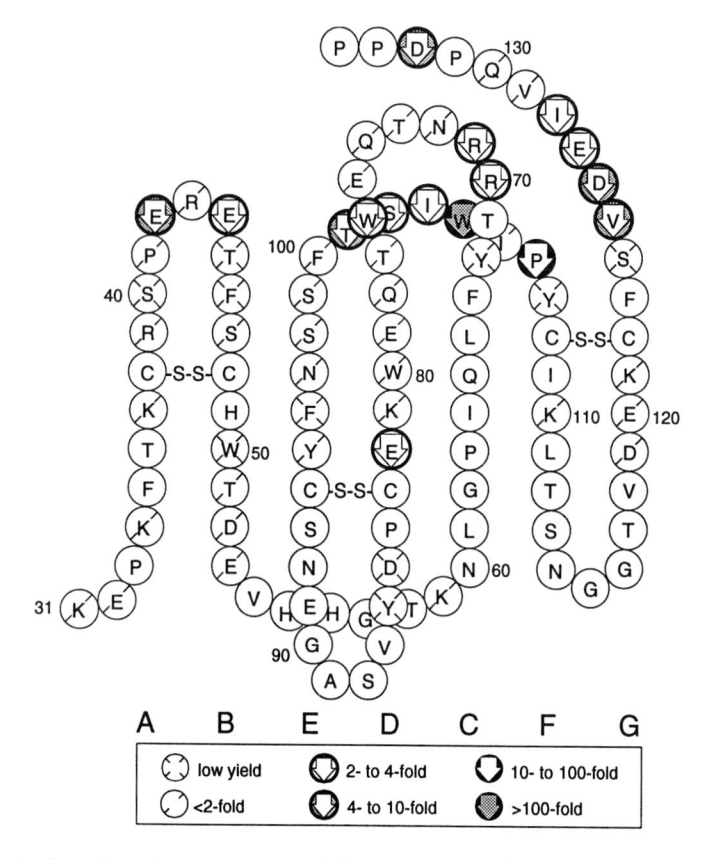

FIG. 6.   Location of residues in the hGHbp causing a 2-fold or greater effect on binding to hGH mapped on an immunoglobulin-like folding diagram predicted for the cysteine-rich domain of the hGHbp (Bazan, 1990). Seven antiparallel β strands (A–F) compose a β barrel by folding the G strand around the back until it hydrogen bonds to the A strand. Residues causing less than a 2-fold reduction, a 2- to 4-fold reduction, a 4- to 10-fold reduction, a 10- to 100-fold reduction, and >100-fold reduction in binding are indicated. Those residues that gave low expression yields are also indicated. Disulfides are shown by —S—S— and residues are numbered by tens. From Bass *et al.* (1991).

## VII.  One hGH Binds Two hGHbps

### A.  DETERMINATION OF THE STOICHIOMETRY OF THE COMPLEX

Cross-linking studies with dimethyl suberimidate suggested that hGH induced dimerization of the hGHbp (G. Fuh and J. Wells, unpublished results). However, such experiments are difficult to interpret because they are susceptible to artifacts based on the availability and differential chemical reactivity of groups near the binding interface. Crystals of hGH in complex with the hGHbp had been obtained, and when these were dissociated and analyzed by IIPLC, the ratio of hGII to hGIIbp were found to be 0.5 : 1 (Ultsch et al., 1991; Cunningham et al., 1991b). This striking result was not an artifact of crystal packing because titration calorimetry experiments showed that there was no further change in the heat of reduction ($\Delta H$) between hGH and the hGHbp once the ratio of hGH to hGHbp reached 0.5 (Cunningham et al., 1991b).

Gel filtration analysis of mixtures of hGH and the hGHbp showed that the complex had a molecular weight of ~75 kDa, consistent with formation of hGH·(hGHbp)$_2$ (Cunningham et al., 1991b). Moreover, at a 1 : 2 ratio of hGH to hGHbp we observed only a single 75-kDa complex peak whereas at different ratios we observed smaller peaks corresponding to free components. Thus, virtually all the hGH and hGHbp in our preparations were capable of forming an hGH·(hGHbp)$_2$ complex in solution. Furthermore, the hGHbp does not self-associate at the highest concentrations we have studied (~0.1 m $M$).

Competitive displacement of [125]-labeled hGH by hGH or hGH analogs from the hGHbp, followed by precipitation with the antireceptor antibody, Mab5 (Barnard et al., 1984), has been used extensively to study the binding of hGH to growth hormone receptors (Spencer et al., 1988; Cunningham et al., 1989). However, the stoichiometry of binding calculated from Scatchard analysis from these competitive displacement assays shows that only one hGH binds per hGHbp. We have shown that this is an artifact of the Mab5 used in these studies, because use of other Mabs (3B9 and 3D9) that were made to the E. coli-derived hGHbp gave a stoichiometry of hGh to hGHbp that was close to 1 : 2 (Cunningham et al., 1991b). Our data suggest that Mab5 blocks binding of the second hGHbp to the hGH·hGHbp complex. This could result if, for example, the contacts between the two hGHbp molecules in the hGH·(hGHbp)$_2$ complex are blocked by Mab5 (see section VII,D). Therefore, one should be cautious about deriving stoichiometries of binding using immunoprecipitation as a

268                    JAMES A. WELLS ET AL.

means of separating receptor-bound hormone from free hormone (especially if one employs polyclonal antibodies to the receptor).


## B.  FUNCTIONAL CHARACTERIZATION OF SITE 2 ON hGH

The immunoprecipitation assay using Mab5 "trapped" a monomeric complex (hGH·hGHbp) by blocking formation of the hGH·(hGHbp)$_2$ complex. In some respects this was a blessing because the results of the alanine scan (Fig. 2B) were readily interpretable as a single binding interface. However, the fact that hGH forms an hGH·(hGHbp)$_2$ complex suggested that there were two sites on hGH for binding the hGHbp.

We wanted to identify the determinants on hGH for binding a second hGHbp. For measuring the dimerization of the hGHbp by hGH, a fluorescence-based assay was developed that avoided the need for immunoprecipitation (Cunningham et al., 1991b). A single free thiol was introduced into the hGHbp at its penultimate residue by mutagenesis of Ser-237 to Cys. The S237C mutant of the hGHbp was stoichiometrically labeled with fluorescein-iodoacetamide. Fluorescein undergoes homoquenching when two molecules approach one another (Griep and McHenry, 1988). Indeed, the addition of hGH to the fluorescein-labeled hGHbp (F-hGHbp) caused fluorescence to be quenched with an EC$_{50}$ of 0.5 n$M$. Maximal quenching occurred when 0.5 equivalents of hGH had been added. Not surprisingly, alanine mutants tested within the epitope defined by the Mab5 immunoprecipitation assay (Fig. 2B) were reduced in their EC$_{50}$ value for dimerization in this fluorescence assay (B. Cunningham and J. Wells, unpublished results). Finally, by homologue- and alanine-scanning mutagenesis we identified mutants that disrupted the fluorescence dimerization assay but not the Mab5 immunoprecipitation assay (Cunningham et al., 1991b). In particular, specific residues at and near the N terminus as well as residues in helix 3 were very important. These mutants collectively formed a patch (which we call site 2) that is next to site 1 on hGH (Fig. 7).


## C.  THE hGHbp BINDS SEQUENTIALLY TO SITE 1 THEN SITE 2

Gel filtration experiments showed that when excess hGH was added to the hGH·(hGHbp)$_2$ complex it caused formation of the monomeric complex, hGH·hGHbp (Cunningham et al., 1991b). Indeed, when excess hGH was added to the preformed dimeric complex [hGH·(F-hGHbp)$_2$] it completely reversed the fluorescence quenching, as expected for formation of a monomeric complex (hGH·F-hGHbp). These data suggested that although the sites on hGH for binding the hGHbp are distinct, those on

GROWTH HORMONE–RECEPTOR INTERACTIONS 269

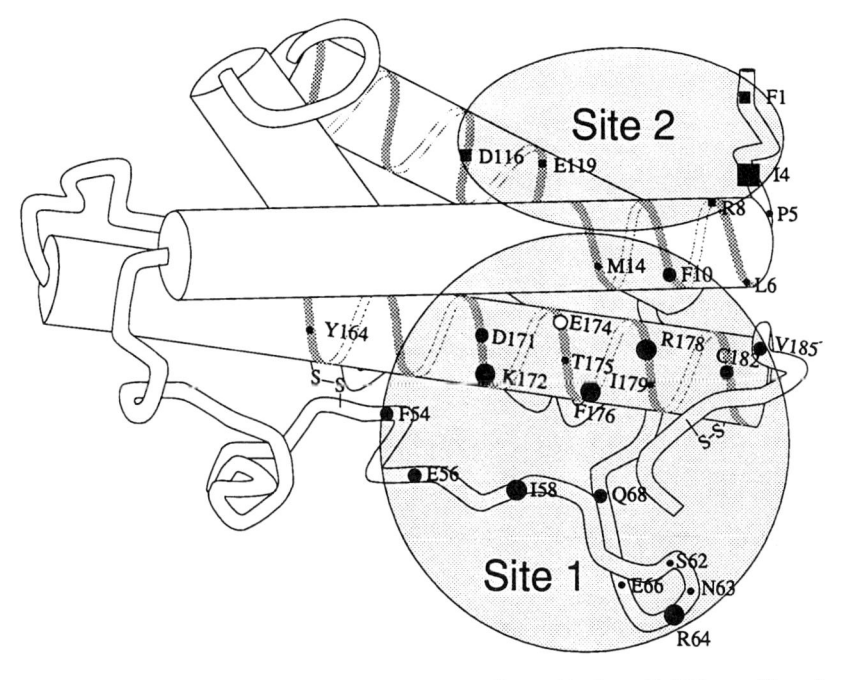

FIG. 7. Map of alanine substitutions in hGH that disrupt binding of hGHbp at either site 1 or site 2 (from Cunningham *et al.*, 1991b). The two sites are generally delineated by the large shaded circles. Residues for which alanine mutants reduce site 2 binding by 2- to 4-fold (∎), 4- to 10-fold (∎), 10- to 50-fold (∎), and greater than 50-fold (∎) are shown. Residues in site 1 are shown that represent sites where Ala mutations cause reductions of 2- to 4-fold (●) 4- to 10-fold (●), greater than 10-fold (●), or a 4-fold increase in binding affinity (○) for the hGHbp, using the Mab5 immunoprecipitation assay (Cunningham and Wells, 1989).

the hGHbp overlap. This would also explain why no higher molecular weight complexes have been found beyond hGH·(hGHbp)$_2$.

Next we evaluated what determinants were required on hGH for dissociating the hGH·(hGHbp)$_2$ complex to form the hGH·hGHbp complex. We can summarize these data by saying that analogs of hGH containing only a functional site 1 were effective at dissociating the hGH·(F-hGHbp)$_2$ complex, but those containing only a functional site 2 were incapable. This suggests that there is a sequential order to binding and dissociation. The hGHbp in the hGH·(hGHbp)$_2$ complex comes off site 2 on hGH first, and binds excess hGH through site 1. The fact that the Mab5 immunoprecipitation assay is only capable of detecting mutants in site 1 suggests that site 1 interactions occur first, followed by those in site 2. Moreover, mutations in site 2 do not affect formation of a 1:1 complex, whereas those in site 1 do.

## D. THREE-DIMENSIONAL STRUCTURE
## OF THE hGH·(hGHbp)₂ COMPLEX

Shortly after these functional data were obtained, a 2.8-Å-resolution X-ray structure of the hGH·(hGHbp)₂ complex was solved (de Vos *et al.*, 1992). This complex (Fig. 8) reveals for the first time the high-resolution structural details of a polypeptide hormone in complex with the extracellular domain of its receptor. The structure shows that the bound hGH is basically a four-helix bundle protein, as expected from the folding diagram of free pGH (Abdel-Meguid *et al.*, 1987). However, the region shown as an extended loop structure in pGH (corresponding to residues 5440–74 in

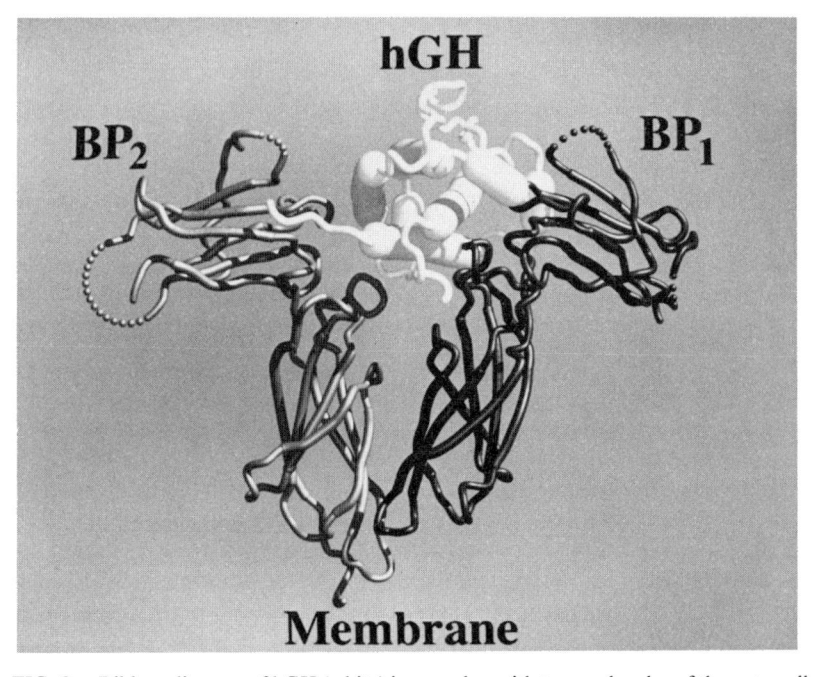

FIG. 8.   Ribbon diagram of hGH (white) in complex with two molecules of the extracellular domain of its receptor (hGHbp), shown as BP₁ (black) and BP₂ (gray). BP₁ and BP₂ use virtually the same binding determinants to recognize completely different determinants on hGH. There are also contacts between BP₁ and BP₂ that help to explain how hGH binds first to BP₁ and then BP₂. hGH is a four-helix bundle with N and C termini facing left and right, respectively. The hGHbp contains two β-sandwich domains with the N termini in the top domains and the C termini next to each other in the bottom domain, where the receptors would enter the membrane. The dashed portions of the hGHbp indicate loop regions that are not structurally well-defined. This illustration was produced by Germaine Fuh, Tom Hynes, and Jim Wells using crystallographic coordinates provided by Bart DeVos.

GROWTH HORMONE–RECEPTOR INTERACTIONS          271

hGH) actually contains two short pieces of $\alpha$ helix in hGH that contact one hGHbp at site 1. The structures of the two bound hGHbps are virtually identical and contain two immunoglobulin-like domains that are analogous to but not identical to the structural topology predicted by Bazan (1990). In particular, the receptor topology contains two $\beta$-sandwich domains that are more similar to the D2 domain of CD4 or PapD than an immunoglobulin constant domain (de Vos *et al.*, 1992).

The structural data corroborate much of the functional and mutgenesis data on the hormone (de Vos *et al.*, 1992). In general, there is good agreement between the sites elaborated by mutational analysis and those in contact. There are, however, some important differences that are currently under study. In particular, hGH has a larger contact surface than is revealed by the functional analysis. Residues 24, 39–42, and 167–168 make good contacts but appear to be of little or no functional consequence when mutated (Cunningham *et al.*, 1989; Cunningham and Wells, 1989). Moreover, Glu-174 is buried on binding, but when changed to alanine binding is improved, probably by relieving steric crowding. In site 2, Asn-12 appears as a contact but is functionally silent. Such discrepancies highlight the importance of combining high-resolution structural and functional studies in order to provide a full understanding of hormone–receptor interactions.

The functional analysis suggested that the binding sites overlapped on the hGHbp for sites 1 and 2 on hGH. In fact, the structure shows that the sets of residues used by the hGHbp for binding hGH are virtually identical! The only exception is Asn-218, which is buried in the hGHbp on binding site 1 but not site 2. Interestingly, Asn-218 is analogous to His-188 in the hPRLbp that is believed to function as a $Zn^{2+}$ ligand (Cunningham *et al.*, 1991b). Again, the structure corroborates much of the mutagenesis work on the hGHbp (Bass *et al.*, 1991) and further shows that Trp-169 (not tested in the mutational analyses) also contacts hGH. Moreover, Arg-43 from the hGHbp makes direct contact yet is functionally silent, whereas Glu-42 and Glu-44 are functionally important yet not in direct contact. The structure shows that in addition to hormone–receptor contacts, there are receptor–receptor contacts in the hGH·(hGHbp)$_2$ complex. Thus, it is not difficult to imagine how particular antibodies to the receptor (such as Mab5) may prevent formation of the dimeric complex.

The structure provides support for the sequential binding model. The area buried at site 1 on hGH is ~1200 $Å^2$ whereas that at site 2 is ~900 $Å^2$. However, the receptor–receptor interface buries about 500 $Å^2$. Thus, if hGH were to first bind through site 1 it would bury 1200 $Å^2$, followed by burial of 1400 $Å^2$ by binding the second hGHbp through site 2 (900 $Å^2$) plus receptor–receptor contacts (500 $Å^2$). Another important aspect of

272                    JAMES A. WELLS ET AL.

the hGH·(hGHbp)$_2$ structure (Fig. 8) is that the C termini of the hGHbp are brought close together in a way that one could envision would bring the transmembrane and intracellular domains into juxtaposition for signaling.

### E.   BIOLOGICAL RELEVANCE OF RECEPTOR DIMERIZATION

Hormone-induced receptor oligomerization has been proposed as a mechanism for signal transduction for the single transmembrane-containing class of tyrosine kinase receptors, including the EGF and PDGF receptors (for review see Ullrich and Schlessinger, 1990). We propose that receptor dimerization is critical for signaling in the hGH receptor. First, the hGHbp serves as a reasonable model for the full-length receptor because its affinity for $^{125}$I-labeled hGH is only threefold weaker (Spencer *et al.*, 1988). Thus, the fact that hGH induces dimerization of the hGHbp in solution strongly suggests that it will do the same for the full-length receptor *in vivo*. Second, the homologues of hGH (hPL and hPRL) that have been engineered to bind to the hGHbp via site 1 (but not site 2) (Table I) have been found to be inactive in the cell-based (G. Fuh and J. Wells, unpublished results) and rat weight-gain assays (R. Clark and D. Mortensen, unpublished results). Thus, analogs that do not dimerize the hGHbp *in vitro* are inactive *in vivo*. Finally, Mabs but not Fabs to a homologue of the hGH receptor, namely the prolactin receptor, are weak agonists in cell-based assays (Elberg *et al.*, 1990). We propose that dimerization of the extracellular portion of the hGH receptor by one hGH could bring the intracellular domains together so that they may interact. The dimerized receptor would present a "new" structure to cytosolic or membrane-bound components that would send a biochemical signal, which is currently under study.

### VIII.   Summary

High-resolution mutational and structural analyses of purified components have revealed a great deal about the molecular basis for growth hormone action. The structural and functional aspects of the interactions between hGH and its receptors have been largely elaborated. From these studies it has been possible to engineer homologues of hGH to bind to the hGH receptor and act as potential antagonists. Receptor-selective and high-affinity analogs have also been constructed based on a combination of alanine scanning and monovalent phage display.

From this molecular work much has been revealed about the biology of hGH (Fig. 9). Our data suggest that hGH is stored in the pituitary as a

# GROWTH HORMONE–RECEPTOR INTERACTIONS

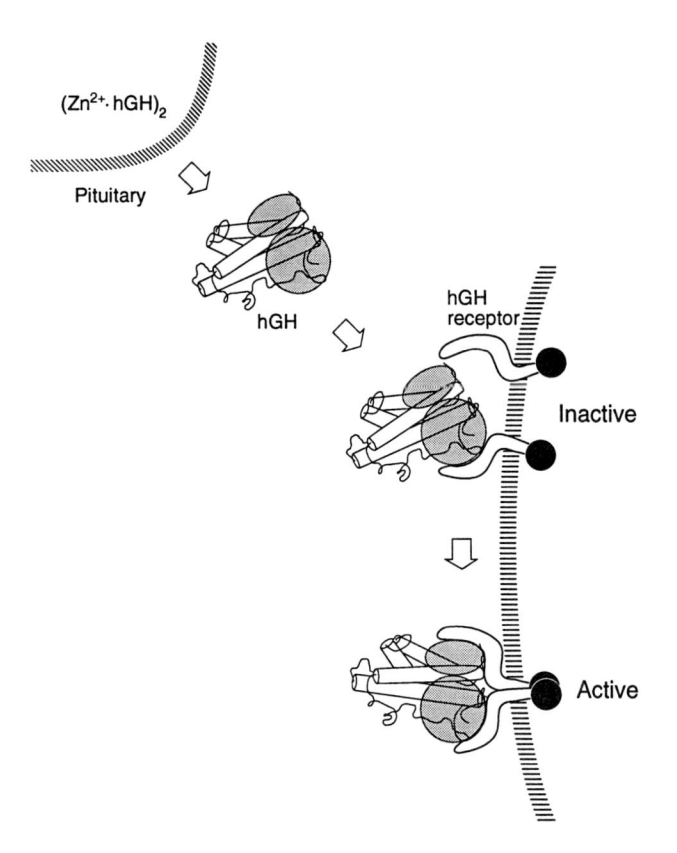

FIG. 9. Summary of molecular endocrinology of hGH. hGH is released as a $(Zn^{2+} \cdot hGH)_2$ complex from the pituitary. Upon dilution into the bloodstream, the dimer dissociates on dilution into a monomeric form, in which it is available to bind via site 1 to the hGH receptor (or with $Zn^{2+}$ to the hPRL receptor; not shown). The membrane-bound hGH then complexes a second receptor molecule using site 2 on hGH and binding determinants on the first receptor. Receptor dimerization by hGH initiates signal transduction.

$(Zn^{2+} \cdot hGH)_2$ complex. On release from somatotropic vesicles it dissociates into a monomeric form and reveals its primary receptor binding site (site 1). Free hGH can bind to the hGHbp in serum to form monomeric or dimeric complexes that slow the clearance of hGH (Moore *et al.*, 1989). However, because the affinity for the full-length receptor is greater, hGH can bind to it preferentially. Furthermore, the constitutive levels of the hGHbp (~0.5 to 1 n$M$) (Baumann *et al.*, 1986; Herrington *et al.*, 1986) are considerably below the levels of hGH after pulsatile release (~2 to 5 n$M$) (Thompson *et al.*, 1972). Our data indicate that hGH binds to the

274　　　　　　　JAMES A. WELLS ET AL.

hGH receptor on cell membranes through site 1 and subsequently forms dimers through site 2. We believe a similar process may occur for hGH to activate the hPRL receptor, except that $Zn^{2+}$ is required for site 1 association. Such receptor dimers are then activated and capable of interacting with other cellular components that may mediate the hGH "signal." Recently, based upon this proposed mechanism, we produced potent antagonists to the hGH receptor (Fuh *et al.*, 1992) and hPRL receptor (G. Fuh, P. Colosi, W. Wood, and J. Wells, unpublished results). These antagonists bind tightly to site 1 but are blocked in their ability to bind site 2 and dimerize the receptor. We believe these methods and discoveries will be relevant to the study of signaling by other hematopoietic hormones and receptors as well as other hormones and receptors.

## ACKNOWLEDGMENTS

We thank Ross Clark, Deborah Mortensen, Mike Cronin, and Jerry Moore for making animal data available prior to publication. We are also grateful to them, Tony Kossiakoff, and our colleagues in the Department of Protein Engineering for their enthusiastic support, and to Wayne Anstine for graphics.

## REFERENCES

Abdel-Meguid, S. S., Shieh, H. S., Smith, W. W., Dayringer, M. E., Violand, B. N., and Bentle, L. A. (1987). *Proc. Natl. Acad. Sci. U.S.A.* **84,** 6434–6437.

Barnard, R., Bundesen, P. G., Rylatt, D. B., and Waters, M. J. (1984). *Endocrinology* **115,** 1805–1813.

Bass, S., Greene, R., and Wells, J. A. (1990). *Proteins Struct. Funct. Genet.* **8,** 309–314.

Bass, S. H., Mulkerrin, M. G., and Wells, J. A. (1991). *Proc. Natl. Acad. Sci. U.S.A.* **88,** 4498–4502.

Baumann, G., Stolar, M. W., Ambarn, K., Barsano, C. P., and DeVries, B. C. (1986). *J. Clin. Endocrinol. Metab.* **62,** 134–141.

Bazan, F. (1990). *Proc. Natl. Acad. Sci. U.S.A.* **87,** 6934–6938.

Boutin, J. M., Edrey, M., Shirota, M., Jolicoeur, C., Lesueur, L., Ali, S., Guold, D., Djiane, J., and Kelly, P. A. (1989). *Mol. Endocrinol.* **3,** 1455–1461.

Chang, C. N., Rey, M., Bochner, B., Heyneker, H., and Gray, G. (1987). *Gene* **55,** 189–196.

Chothia, C., and Lesk, A. (1986). *Embo J.* **5,** 823–826.

Cunningham, B. C., Jhurani, P., Ng, P., and Wells, J. A. (1989). *Science* **243,** 1330–1336.

Cunningham, B. C., and Wells, J. A. (1989). *Science* **244,** 1081–1085.

Cunningham, B. C., Henner, D. J., and Wells, J. A. (1990a). *Science* **247,** 1461–1465.

Cunningham, B. C., Bass, S., Fuh, G., and Wells, J. A. (1990b). *Science* **250,** 1709–1712.

Cunningham, B. C., Mulkerrin, M. G., and Wells, J. A. (1991a). *Science* **253,** 545–548.

Cunningham, B. C., Ultsch, M., de Vos, A. M., Mulkerrin, M. G., Clauser, K. R., and Wells, J. A. (1991b). *Science* **254,** 821–825.

Cunningham, B. C., and Wells, J. A. (1991). *Proc. Natl. Acad. Sci. U.S.A.* **88,** 3407–3411.

Cwirla, S. E., Peters, E. A., Barrett, R. W., and Dower, W. J. (1990). *Proc. Natl. Acad. Sci. U.S.A.* **87,** 6378–6382.

GROWTH HORMONE–RECEPTOR INTERACTIONS 275

Daughaday, W. F. (1985). *In* "Textbook of Endocrinology" (J. D. Wilson and D. W. Foster, eds.), 7th ed., pp. 577–613. Saunders, Philadelphia.

de Vos, A. M., Ultsch, M., and Kossiakoff, A. A. (1992). *Science* **255**, 306–312.

Elberg, G., Kelly, P. A., Djiane, J., Binder, L., and Gertler, B. (1990). *J. Biol. Chem.* **265**, 14770–14776.

Fuh, G., Cunningham, B. C., Fukunaga, R., Nagata, S., Goeddel, D. V., and Wells, J. A. (1992). *Science* **256**, 1677–1680.

Fuh, G., Mulkerrin, M. G., Bass, S., McFarland, N., Brochier, M., Bourell, J. H., Light, D. R., and Wells, J. A. (1990). *J. Biol. Chem.* **265**, 3111–3115.

Fu, Y. -K., Arkins, S., Fuh, G., Cunningham, B. C., Wells, J. A., Fong, S., Cronin, M., Dantzer, R., and Kelley, K. W. (1992). *J. Clin. Invest.*, **89**, 451–457.

Griep, M. A., and McHenry, C. S. (1988). *Biochemistry* **27**, 5210–5215.

Herington, A. C., Ywer, S. I., and Stevenson, J. L. (1986). *J. Clin. Invest.* **77**, 1817–1823.

Hughes, J. P., and Friesen, H. G. (1985). *Annu. Rev. Physiol.* **47**, 469–482.

Isaksson, O. G. P., Eclen, S., and Jansson, J. O. (1985). *Annu. Rev. Physiol.* **47**, 483–499.

Jin, L., Fendley, B. M., and Wells, J. A. (1992). *J. Mol. Biol.* **226**, 851–865.

Leung, D. W., Spencer, S. A., Cachianes, G., Hammonds, G., Colius, C., Henzel, W. J., Barnard, R., Waters, M. J., and Wood, W. I. (1987). *Nature (London)* **330**, 537–543.

Lowman, H. B., Cunningham, B. C., and Wells, J. A. (1991a). *J. Biol. Chem.* **266**, 10982–10988.

Lowman, H. B., Bass, S. H., Simpson, N., and Wells, J. A. (1991b). *Biochemistry* **30**, 10832–10838.

Moore, J. A., Celniker, A., Fuh, G., Light, D., McKay, P., and Spenser, S. (1989). 71rst Meeting of the Endocrine Society, Abstract 1652. Seattle, WA.

Spencer, S. A., Hammonds, R. G., Henzel, W. J., Rodriguez, H., Waters, M. J., and Wood, W. I. (1988). *J. Biol. Chem.* **263**, 7862–7867.

Thompson, R. G., Rodriguez, A., Kowarski, A., and Blizzard, R. M. (1972). *J. Clin. Invest.* **51**, 3193.

Thorlacius-Ussing, O. (1987). *Neuroendocrinology* **45**, 233–241.

Ullrich, A., and Schlessinger, J. (1990). *Cell (Cambridge, Mass.)* **61**, 203–212.

Ultsch, M., de Vos, A. M., and Kossiakoff, A. A. (1991). *J. Mol. Biol.*, in press.

Vierra, J., and Messing, J. (1987). *In* "Methods in Enzymology" (R. Wu and L. Grossman, eds.), Vol. 153, pp. 3–11. Academic Press, San Diego.

Wells, J. A. (1990). *Biochemistry* **29**, 8509–8517.

Wells, J. A. (1991). *In* "Methods in Enzymology" (J. Langone, ed.), Vol. 202, pp. 390–411. Academic Press, San Diego.

RECENT PROGRESS IN HORMONE RESEARCH, VOL. 48

# Catecholamine Receptors: Structure, Function, and Regulation

## MARC G. CARON AND ROBERT J. LEFKOWITZ

*Howard Hughes Medical Institute, Departments of Cell Biology, Medicine and Biochemistry, Duke University Medical Center, Durham, North Carolina 27710*

## I. Introduction

The metabolic state of cells and tissues is highly controlled by the interplay of several extracellular signals in the form of hormones, neurotransmitters, and even sensory stimuli. These control phenomena are mediated by a series of complex transmembrane signaling events or intracellular processes that often lead to gene activation. Among the plasma membrane signaling systems, numerous stimuli utilize specific receptors coupled to guanine nucleotide regulatory proteins (G proteins). These systems are composed of three functional components, a receptor with seven hydrophobic transmembrane segments, a G protein that mediates the interactions between the receptor, and an effector component that can be either an enzyme such as adenylyl cyclase or an ion channel such as a $K^+$ or $Ca^{2+}$ channel (Lefkowitz and Caron, 1988; Casey and Gilman, 1988). Some of the best characterized G protein-coupled receptor systems include the adrenergic and dopaminergic receptors that mediate the actions of catecholamines and rhodopsin, which mediates light perception in the retina. In this article we will review briefly progress that has been made in the elucidation of the primary structure of this unexpectedly large family of receptors, the structure–function relationships for these receptors, and the mechanisms involved in the dynamic regulation of responsiveness of these systems.

## II. Structure and Heterogeneity of G Protein-Coupled Receptors

Pharmacological and biochemical characterizations of the receptors that mediate the actions of catecholamines suggest the existence of six distinct receptors: two $\beta$-adrenergic receptors, $\beta_1$ and $\beta_2$; two $\alpha$-adrenergic receptors, $\alpha_1$ and $\alpha_2$; and two dopaminergic receptors, $D_1$ and $D_2$. These receptors can bind ligands with different pharmacological profiles and couple

277

Copyright © 1993 by Academic Press, Inc.
All rights of reproduction in any form reserved.

to different signal transduction mechanisms, such as stimulation or inhibition of adenylyl cyclase or stimulation of phosphatidylinositol hydrolysis.

The approach we took to elucidate the structure of these receptors was to characterize these proteins biochemically in order to obtain primary sequence information. This task was accomplished by developing for each of these receptors specific affinity and photoaffinity probes to identify these proteins and to develop specific affinity chromatography matrices to purify them to apparent homogeneity (Dohlman *et al.*, 1987a; Caron *et al.*, 1988). The first clue to the primary structure of these receptors came with the sequencing of the $\beta_2$-adrenergic receptor, which eventually led to the cloning of its gene (Dixon *et al.*, 1986). The amino acid sequence deduced from the $\beta_2$-adrenergic receptor cDNA indicated that the protein contained about 415 amino acids (aa), with seven highly hydrophobic stretches of 20–28 aa that could represent transmembrane domains. Significant homology within these putative transmembrane domains of the $\beta_2$ receptor and those of rhodopsin was found. This observation and more direct studies on the topography of the $\beta_2$ receptor (Dohlman *et al.*, 1987b; Wang *et al.*, 1989) implied that the amino terminus of the protein resides on the outside of the cell membrane, with the C terminus on the cytoplasmic surface. The transmembrane domains are connected by various extra- and intracellular loops that vary in length.

Recent studies suggest that this seven-transmembrane motif is widely utilized in nature. In addition to being involved in the transduction of signals for many hormones and neurotransmitters, as well as light (vision), it is probable that other senses, such as olfaction and taste, may be mediated via such proteins (Buck and Axel, 1991). Receptor proteins with this motif are also found in lower organisms such as yeast and *Dictyostelium*, wherein they mediate responses to specific mating factors and chemotactic signals (Dohlman *et al.*, 1991).

Another interesting finding that was revealed by the molecular characterization of the seven-transmembrane domain receptors is that within individual families a much larger number of members exist than had been characterized initially by pharmacological and biochemical means. Thus, as shown in Table I for the adrenergic receptor family, three distinct $\beta$-adrenergic receptors exist, and three $\alpha_1$- as well as three distinct $\alpha_2$-adrenergic receptors. These receptors are all encoded by different genes. Whereas $\beta$ and $\alpha_2$ receptor genes are intronless, $\alpha_1$-adrenergic receptors are encoded by intron-containing genes. The encoded proteins contain 400–500 aa residues and range in molecular weight from 60,000–80,000. On expression of the various cDNAs in host cells, these receptors are all pharmacologically distinguishable (i.e., they bind ligands with unique patterns of specificity). Comparison of the amino acid sequence of these receptors reveals that overall they share 30–40% homology; however,

CATECHOLAMINE RECEPTORS

TABLE I

*Properties of the Cloned Adrenergic Receptor Subtypes*[a]

| Mammalian subtype | $M_r$ (kDa) | Peptide length (amino acids) | Topology identity (MSD) (%) | Introns | Chromosomal Location |
|---|---|---|---|---|---|
| $\beta_2$ | 64 | 413 | | 0 | 5 q31–q33 |
| $\beta_1$ | 64 | 477 | $71(\beta_2)$ | — | 10 q24–q26 |
| $\beta_3$ | — | 402 | $63(\beta_2)$ | 0 | — |
| $\alpha_2(C-10)$ | 64 | 450 | $42(\beta_2)$ | 0 | 10 q24–q26 |
| $\alpha_2(C-4)$ | 75 | 461 | $39(\beta_2), 75[\alpha_2(C-10)]$ | — | 4 |
| $\alpha_2(C-2)$ | — | 450 | $36(\beta_2), 74[\alpha_2(C-10)]$ | 0 | 2 |
| $\alpha_{1B}$ | 80 | 515 | $42(\beta_2)$ | 1 | 5 q32–q34 |
| $\alpha_{1C}$ | — | 466 | $42(\beta_2), 72(\alpha_{1B})$ | 1 | 8 |
| $\alpha_{1A}$ | — | 560 | $43(\beta_2), 73(\alpha_{1B})$ | — | 5 q23–q31 |

[a] The various adrenergic receptor subtypes are grouped in three families: the $\beta$, $\alpha_2$ and $\alpha_1$-adrenergic receptors. Three distinct subtypes exist within each subfamily. The table indicates the apparent molecular weight on SDS–PAGE, the peptide length in number of amino acids, the percent identity that exists within the membrane-spanning domains (MSD) with members of a subfamily in comparison to the $\beta_2$-adrenergic receptor or to another member of the subfamily; the minimum number of introns within the gene and the chromosomal localization of each gene.

within their transmembrane domains, the level of relatedness is much higher. Specifically, within a subfamily (i.e., subtypes of $\beta$ receptors) the identity between the various receptors in their transmembrane domains is 70–80%. This figure is significantly lower (40–45%) when the comparison is made between, for example, a $\beta$ and an $\alpha_1$ or an $\alpha_2$ receptor subtype. This property has been useful in developing a classification strategy for the various receptor subtypes (O'Dowd *et al.*, 1989a).

Table II presents a summary of the various subtypes of dopamine receptors that have been cloned and characterized. A level of heterogeneity similar to the adrenergic receptors has been found. Originally, two types of dopamine receptors had been defined on a pharmacological and biochemical basis. $D_1$ receptors were those coupled to stimulation of adenylyl cyclase whereas $D_2$ receptors inhibited the enzyme (Kebabian and Calne, 1979; Andersen *et al.*, 1990). As shown in Table II, two distinct subtypes of $D_1$ dopamine receptors have been characterized $D_{1A}$ and $D_{1B}$ (or $D_5$). These receptors share the same high level (75–80%) of identity within their transmembrane domains, characteristic of adrenergic receptor subtypes of the same subfamily. The genes for these two receptors, which positively

## TABLE II
### Properties of the Cloned Dopamine Receptor Subtypes[a]

| Subtype | Species | Coupling | Size (amino acids) | Exons | Introns | Chromosome |
|---------|---------|----------|--------------------|-------|---------|------------|
| $D_1$-like | | | | | | |
| $D_{1A}$ | Human Rat | ↑ AC | 446 | 1 | 0 | 5q35.1 |
| $D_5/D_{1B}$ | Rat | | 475 | | | |
| | Human | ↑ AC | 477 | 1 | 0 | 4p15.1–16.1 |
| $D_2$-like | | | | | | |
| $D_{2S}$ | Human Rat | | 415 | 6 | 5 | 11q22–23 |
| $D_{2L}$ | Bovine *Xenopus* | ↓ AC, ↑ K$^+$ | 444 | 7 | 6 | 11q22–23 |
| $D_3$ | Rat | | 446 | 6 | 5 | 3p13.3 |
| | Human | ? | 400 | ? | ? | 3p13.3 |
| $D_4$ | Human | ? | 387 | 5 | 4 | 11p |

[a] Two families of dopamine receptors are denoted $D_1$-like and $D_2$-like. Indicated in the table is the designation of each subtype, the spaces from which a cDNA or a gene has been characterized, the demonstrated coupling mechanism, the number of amino acids, the number of exons and introns contained within each gene, and the human chromosomal assignment. A question mark indicates that no information is available; AC, adenylyl cyclase; K$^+$, K$^+$ channel. Information is derived from work cited in the text.

couple to stimulation of adenylyl cyclase, are also intronless. On the other hand, three distinct members of the $D_2$-like dopamine receptor family have been characterized, the $D_2$, $D_3$, and $D_4$ dopamine receptors. When expressed in host cells, these three receptors bind ligands with generally similar yet distinct properties. Whereas it is obvious that $D_2$ receptors are biochemically coupled to inhibition of adenylyl cyclase and activation of K$^+$ channels through pertussis toxin-sensitive G proteins, the specific signal transduction mechanisms utilized by the $D_3$ and $D_4$ receptors have not been well characterized. Receptors for this subfamily of $D_2$-like receptors are all coded by intron-containing genes (Table II) (reviewed in Gingrich and Caron, 1993).

Another form of heterogeneity that is found with the $D_2$ dopamine receptor is the presence of two alternatively spliced forms of the receptor (reviewed in Sibley and Monsma, 1992). This situation is unprecedented within the G protein-coupled receptor family. The two isoforms of the $D_2$ receptor arise by the insertion of a 29-aa segment from an alternatively spliced exon within the third longer intracellular loop. Whereas the mRNA expression for the two receptor isoforms in various tissues and brain areas appear to vary (O'Malley *et al.*, 1990; Neve *et al.*, 1991a; Snyder *et al.*, 1991), no obvious functional differences in the expressed isoforms have yet been elucidated.

Thus, like the adrenergic receptor family, the family of dopamine receptors has grown beyond what could have been predicted on the basis

CATECHOLAMINE RECEPTORS                                    281

of functional properties of the originally defined receptors. Five distinct dopamine receptors have now been shown to exist. At this point in time, it is not completely apparent what specific physiological paradigms these various receptors will eventually subserve. This issue will obviously be important to resolve in future work.

The extensive heterogeneity of G protein-coupled receptors raises the question of the exact functional role these related yet distinct proteins might play in the organism. Several possible scenarios can be envisaged. In response to a given hormone or neurotransmitter it might be desirable to evoke distinct signal transduction responses, such as is the case for the binding of epinephrine to $\beta$-adrenergic receptors that elevate cAMP and $\alpha_1$-adrenergic receptors that activate the protein kinase C pathway. Receptor subtypes might be differentially regulated, as appears to be the case for the $\beta_2$- versus $\beta_3$-adrenergic receptors (Liggett $et$ $al.$, 1992). Receptor subtypes may need to be expressed preferentially or specifically in distinct tissues. One such example is provided by the two $D_1$-like dopamine receptors $(D_{1A}, D_{1B})$ that are expressed, respectively, most abundantly in the basal ganglia or virtually exclusively in the limbic system (Tiberi $et$ $al.$, 1990). Finally, receptor subtypes might serve to recognize the endogenous ligand with different affinity. Such a situation appears to prevail with the $D_3$ dopamine receptor that can interact with the endogenous ligand dopamine with at least a 10-fold higher affinity than the $D_2$ receptor (Sokoloff $et$ $al.$, 1991). Many other physiological circumstances can be envisaged in which receptor subtypes might dictate important properties of these signaling systems. Thus, understanding the various properties of receptor subtypes may help in elucidating their exact role in the control of physiological functions.

## III. Structure/Function Relationships of G Protein-Coupled Receptors

G protein-coupled receptors mediate a number of specific functions. First, they bind ligands with high affinity and selectivity, presumably leading to a change in conformation and activation of the receptor. Second, ligand-mediated activation of the receptor leads to the subsequent interaction of the receptor with its appropriate G protein. This latter interaction, which promotes nucleotide exchange and activation of the G protein, ultimately results in the modulation of the activity of an effector system such as adenylyl cyclase or phospholipase C. Mapping of the various domains on the receptor molecule responsible for these various functions should be helpful in understanding these various mechanisms.

Evidence obtained from mutational and chimeric receptor analysis of various cloned receptors has strongly suggested that the ligand binding site, at least for receptors that interact with small molecules such as the biogenic amines, resides within an area delineated by the transmembrane domains of these receptors (Kobilka *et al.*, 1988; Dohlman *et al.*, 1991). In the $\beta$-adrenergic receptor, for example, two serine residues within the fifth transmembrane segment and an aspartic acid residue in the third transmembrane domain have been implicated as interacting with the catechol hydroxyl and the amino nitrogen of catecholamines (Strader *et al.*, 1988, 1989). This motif is absolutely conserved within the 15 or so different catecholamine receptors that have been characterized to date. Similar evidence has been obtained for other receptors, for example, the dopamine receptors (Neve *et al.*, 1991b).

With respect to the identification of the regions of G protein-coupled receptors that interact with G proteins, much of the evidence has also been derived from the chimeric receptor approach, as well as mutational and deletional analysis. As one might have suspected, regions on the cytoplasmic surface of receptors have been implicated. In particular, sequence segments within the third intracellular loop in close proximity to the transmembrane segments 5 and 6 have been the regions most consistently identified in various receptors (Kobilka *et al.*, 1988; Strader *et al.*, 1987; Kubo *et al.*, 1988; Cotecchia *et al.*, 1990). In addition, a cytoplasmic region immediately adjacent to the seventh transmembrane domain has been inferred to play at least a modulatory role in the coupling process (O'Dowd *et al.*, 1989b). Moreover, work of Wong *et al.* (1990) with chimeric muscarinic and $\beta_1$-adrenergic receptors has suggested that determinants within the second intracellular loop may play an important role in the specificity of receptor–G protein interactions.

Recently, using the mutagenesis approach for one of these cytoplasmic regions (the carboxyl terminal of the third intracellular loop), we have obtained results that suggest a potential role for this region in the mechanism by which agonist binding induces an active conformation of the receptor capable of interacting with and activating G proteins (Cotecchia *et al.*, 1990; Kjelsberg *et al.*, 1992). A conservative substitution of the carboxyl end of the third intracellular loop of the $\alpha_{1B}$-adrenergic receptor with sequences derived from the $\beta_2$-adrenergic receptor was found to lead to a constitutive ability (i.e., in the absence of agonist) of the chimeric receptor to stimulate phosphatidylinositol hydrolysis (Cotecchia *et al.*, 1990) (Fig. 1). Within a stretch of seven amino acid residues in this region, the $\alpha_1$-adrenergic receptor differs from the $\beta_2$-adrenergic receptor in only three positions (Arg-288 Lys, Lys-290 His, and Ala-293 Leu). Individual mutations revealed that the Ala-293 residue contributed the most to the constitutively active phenotype. Consequently, this residue was systematically mutated with all 19 possible amino acid substitutions (Fig. 2). The resulting mutated receptors all possessed various degrees of constitutive

CATECHOLAMINE RECEPTORS 283

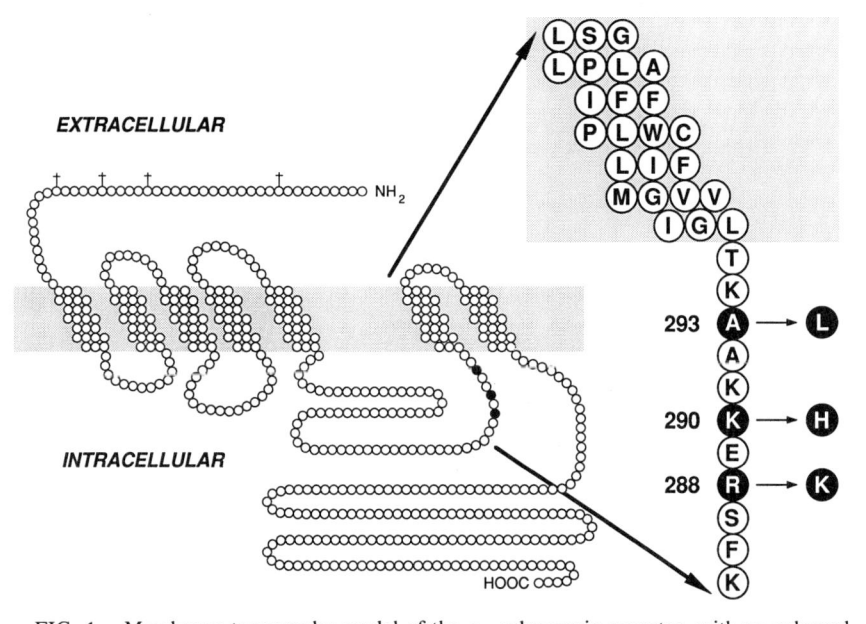

FIG. 1. Membrane topography model of the $\alpha_{1B}$-adrenergic receptor, with an enlarged region illustrating the carboxyl terminus of the third intracellular loop and the sixth transmembrane domain. Circles indicate individual amino acids. Solid circles indicate those amino acid residues that are divergent between the $\alpha_{1B}$- and $\beta_2$-adrenergic receptor within residues 285–295. A mutated chimeric receptor was produced by substituting residues 288, 290, and 293 with those shown on the right ($\beta_2$ receptor residues).

activity ranging from 21 to 211% above the level of inositol phosphates produced by the wild-type $\alpha_1$-adrenergic receptor in the absence of agonist. These levels of activity were as much as 50% as high as the maximal agonist-stimulated inositol phosphates obtained with the wild-type receptor (Kjelsberg *et al.*, 1992). Although agonist stimulation of the mutated receptors led to increased activity, the mutated receptors were often impaired in their biological response. This impairment was even more apparent when inositol phosphate levels were considered relative to the elevated basal.

Interestingly, these mutated receptors displayed increased affinity (3- to 175-fold higher) for agonists, with no significant changes in the affinities for antagonist binding when compared to the wild-type receptors. This increase in agonist affinity was translated into an increased potency of epinephrine to activate receptor-mediated phosphatidylinositol hydrolysis as compared with the wild-type receptors. The constitutive activity of

284          MARC G. CARON AND ROBERT J. LEFKOWITZ

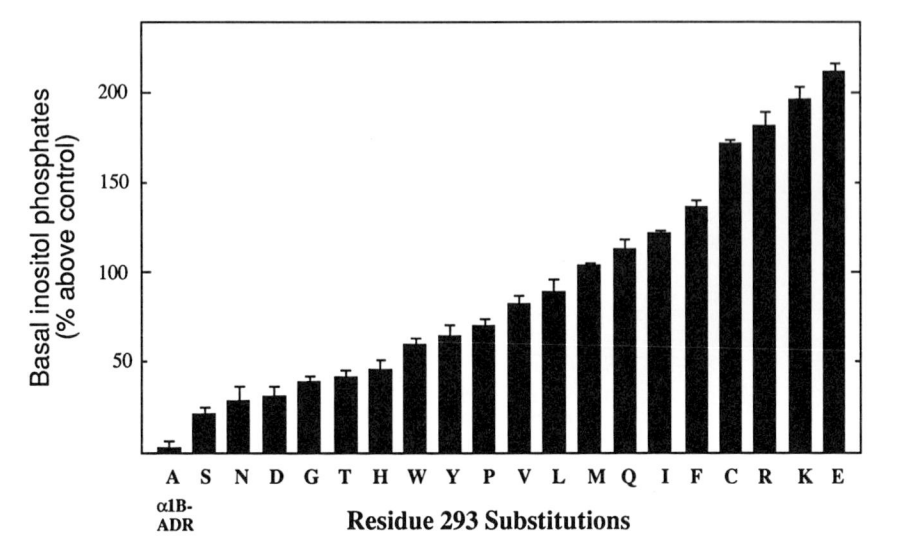

FIG. 2.    Constitutive activation of the $\alpha_{1B}$-adrenergic receptor from amino acid substitutions at position 293. The alanine residue (293) shown in Fig. 1 was substituted in the $\alpha_{1B}$-adrenergic receptor by site-directed mutagenesis with all possible 19 amino acid residues. Each mutated receptor was subcloned into the expression vector pBc12BI and transfected into COS-7 cells. The data represent the basal inositol phosphates produced in cells expressing the wild-type or the mutated receptors relative to the levels produced in cells transfected with the expression vector alone (control). From Kjelsberg *et al.* (1992).

the mutated receptors did not depend on the presence of low levels of endogenous agonists (Kjelsberg *et al.*, 1992).

The G protein activation or biological response and the increased affinity for agonists are two hallmarks of the activated conformation of G protein-coupled receptors. Normally, these properties are brought about by the binding of an agonist and the subsequent formation of the so-called ternary complex (hormone/receptor/G protein), wherein receptor displays an enhanced affinity for the agonist, which is stabilized by its high-affinity interaction with the G protein (DeLean *et al.*, 1980). Thus, the constitutive activity observed with the 19 amino acid mutants suggests that any mutations in this region of the receptor mimics the conformational change induced by agonist binding, which leads to agonist-mediated activation. In the wild-type receptor this region can be viewed as maintaining the receptor in an inactive (silent) state in the absence of agonist, which is then relieved (activated) on binding of an agonist. These results suggest that this region of the carboxyl terminal of the third intracellular loop may play a key role in mediating the change in conformation presumably responsible for the equilibrium between active and inactive receptors. This region of the third loop of G protein-coupled receptors had previously been implicated in receptor–G protein coupling (reviewed in Dohlman *et al.*, 1991). Synthetic peptides derived from this region are potent activators of G proteins *in vitro* (Palm *et al.*, 1989; Cheung *et al.*, 1991; Okamoto *et al.*, 1991).

A further demonstration of the constitutive activation of the mutated receptors was obtained when the original chimeric receptor containing the sequence of the $\beta_2$-adrenergic receptor (Arg-288 Lys, Lys-290 His, Ala-293 Leu) was transfected into fibroblasts and tested for the transforming phenotype of focus formation (Allen *et al.*, 1991). As opposed to the wild-type $\alpha_{1B}$-adrenergic receptor, which, when overexpressed in these cells, is capable of inducing formation of numerous foci in response to agonist stimulation, the mutated receptor was capable of inducing foci in the absence of agonist. These results strongly suggest that the mutated receptor possesses constitutive activity.

These findings suggest that the $\alpha_1$-adrenergic receptors can function as protooncogenes. Because the $\alpha_1$-adrenergic receptor activates a signal transduction pathway that regulates cellular growth mechanisms, spontaneously occurring mutations of the type described here could therefore subvert the normal function of these receptors and result in conditions associated with uncontrolled cell growth and proliferation. Such conditions might include neoplasia, tissue hyperplasia, and atherosclerosis.

## IV. Regulation of G Protein-Coupled Receptor Function by Phosphorylation Mechanisms

Another aspect of the structure–function relationships of G protein-coupled receptors is that the function of many of these receptors appears to be controlled by covalent modifications of the receptor proteins. The major posttranslational modification these receptors undergo is that of phosphorylation. In fact, several consensus sites for phosphorylation by various kinases are present on the cytoplasmic surface of G protein-coupled receptors. Most of the evidence of this mechanism of posttranslational modification has been obtained by examining the agonist-mediated desensitization of the $\beta_2$-adrenergic receptor.

The mechanism of agonist-mediated desensitization is complex. Several processes are distinguishable on a temporal basis. Following agonist occupancy of receptor, a rapid uncoupling (within seconds to minutes) of the response occurs. Within a few minutes, this uncoupling is followed by a process referred to as "sequestration," and at longer times "down-regulation" of receptors occurs (Collins *et al.*, 1992). The relationship between sequestration and the later process of down-regulation is not fully appreciated. However, it is quite obvious that the rapid uncoupling process of the response is primarily due to phosphorylation of the receptors (Hausdorff *et al.*, 1990).

Classically, the process of desensitization has been refered to as heterologous and homologous. Heterologous desensitization is when the response to various stimuli is diminished. On the other hand, homologous desensitization is defined as the situation when only the response to the specific stimulus is blunted. Whereas it is now obvious that these two modes of regulation are not totally separable on a temporal basis, they appear to have distinct biochemical mechanisms and subserve different physiological functions (Hausdorff *et al.*, 1990).

Heterologous desensitization is believed to involve protein kinase A phosphorylation of receptors. When agonist activation of the $\beta$ adrenergic receptor occurs and cAMP increases, protein kinase A is activated, and under these conditions the $\beta$ adrenergic receptor can be phosphorylated. In this process of phosphorylation, there is little or no dependence on agonist occupancy of the receptor. Thus, any receptor, $\beta$-adrenergic or other, occupied or not, can be effectively phosphorylated by the increased protein kinase A activity if the appropriate consensus phosphorylation site is present. Protein kinase A phosphorylation consensus sites are present in the $\beta$-adrenergic receptor within the regions involved in G protein coupling, thus suggesting that phosphorylation of these regions of the $\beta$ adrenergic receptor is involved in the alteration (diminution) of coupling to the G protein Gs. Mutagenesis and kinase inhibitor studies have suggested that this mechanism is an efficient means of desensitization, especially at low agonist concentrations (Hausdorff *et al.*, 1989; Lohse *et al.*, 1990a). These findings suggest that this process may be important at receptors that mediate the action of hormonal (low) levels of catecholamines (e.g., epinephrine in plasma) (Hausdorff *et al.*, 1990).

Homologous desensitization, on the other hand, is thought to involve phosphorylation of the $\beta$-adrenergic receptor by a specific receptor kinase ($\beta$ARK) and subsequent uncoupling of the ability of the receptor to activate Gs (Benovic *et al.*, 1988). In contrast to protein kinase A-mediated phosphorylation of the $\beta$ adrenergic receptor, $\beta$ARK-mediated $\beta$-adrenergic receptor phosphorylation is virtually totally dependent on $\beta$-adrenergic receptor activation or agonist occupancy of the receptor. At low (nanomolar) levels of hormonal activation of receptor (low occupancy), $\beta$ARK-mediated uncoupling is negligible because only a small portion of the receptors can be phosphorylated. Again, mutagenesis and kinase inhibitor studies have suggested that this mechanism is much more important at high agonist concentrations or high receptor occupancy (Hausdorff *et al.*, 1989; Lohse *et al.*, 1990a). This has led to the hypothesis

CATECHOLAMINE RECEPTORS 287

that this mechanism may be more important at synaptic locations where high levels of neurotransmitters are found in response to neuronal depolarization.

Recent evidence suggests that the tissue distribution of βARK mRNA and protein is more prominent in the brain and in highly innervated peripheral tissues (Benovic *et al.*, 1991) (Fig. 3). Moreover, immunocytochemistry has revealed that the enzyme βARK can be found both pre- and postsynaptically, thus positioning the activity in proximity to neurotrans-

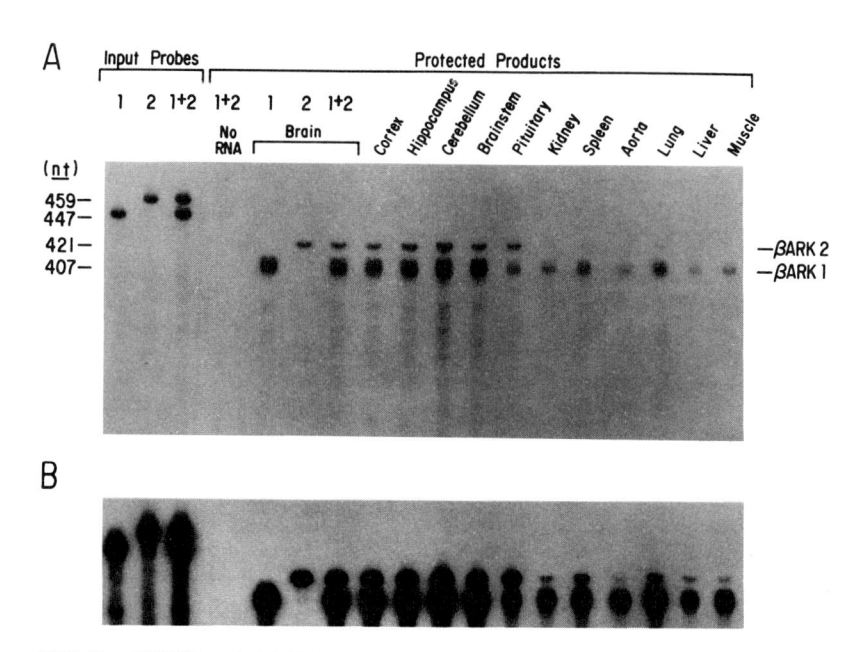

FIG. 3. βARK1 and βARK2 mRNAs in the nervous system and peripheral tissues. The relative levels of expression of βARK1 and βARK2 mRNAs were determined by a ribonuclease protection assay using the rat cDNA homologues of the bovine sequences. (A) The input cRNA probes synthesized from the rat homologues of βARK1 and βARK2 are shown to the left. These probes are 447 (1; βARK1) and 459 (2; βARK2) nucleotides (nt) in length, respectively, and include some sequences from the polylinker region of pGEM3. In control reactions, individual (1 or 2) or both (1 + 2) probes were hybridized with 20 μg of rat whole brain total RNA (Brain) or 20 μg of yeast tRNA (No RNA). Protected fragments corresponding to regions of complementary sequence (407 nt for βARK1 and 421 nt for βARK2) can be seen individually or in combination. In test reactions the combined probes for βARK1 and βARK2 were hybridized with 20 μg of rat total RNA from dissected regions of the brain or from peripheral tissues. Relative levels of expression can be seen in this 1-hour exposure. (B) βARK2 mRNA is detectable at low levels in peripheral tissues. This autoradiogram is an 18-hour exposure of the experiment in panel A. From Benovic *et al.* (1991).

mitter receptors (Arriza *et al.*, 1992). In addition, the distribution of $\beta$ARK within various neuronal systems in the brain has suggested a specificity that likely transcends the noradrenergic systems of the brain and indicates a much broader specificity of action for $\beta$ARK.

The molecular cloning of a cDNA for $\beta$ARK (Benovic *et al.*, 1989) has led to the identification of a second member of this enzyme family, referred to as $\beta$ARK2 (Benovic *et al.*, 1991). Both $\beta$ARK1 and $\beta$ARK2 share similar biochemical and anatomical properties, although, as shown in Fig. 3, $\beta$ARK1 and $\beta$ARK2 expression shows distinct patterns. The $\beta$ARK isozymes are members of a larger family that also includes rhodopsin kinase. Rhodopsin kinase is a retinal-specific enzyme involved in the phosphorylation of light-activated rhodopsin and the subsequent turn-off mechanism of light reception (Nathans, 1987). In their catalytic domains, $\beta$ARK1 and $\beta$ARK2 and rhodopsin kinase share a significant level of amino acid identity (45%) (Lorenz *et al.*, 1991). As in the light transduction system wherein another component called arrestin appears to prolong and augment the turn-off state of the system, two homologous arresting proteins called $\beta$arrestin1 and $\beta$arrestin2 have been characterized and implicated in the mechanism of $\beta$ARK-mediated desensitization of the $\beta_2$-adrenergic receptors (Lohse *et al.*, 1990b; Attramadal *et al.*, 1992). The tissue and subcellular distribution of $\beta$arrestin1 and $\beta$arrestin2 are such as to suggest that these proteins function in concert with the $\beta$ARK-mediated mechanisms of desensitization (Attramadal *et al.*, 1992).

The elucidation over the last 5 years of the structures of the G protein-coupled receptors has revealed a level of complexity and heterogeneity that could not have been predicted previously on the basis of biochemical and pharmacological properties of these systems. The availability of molecular tools has greatly facilitated approaches to understanding the mechanisms of action and regulation of this mode of signal transduction.

## REFERENCES

Allen, L. F., Lefkowitz, R. J., Caron, M. G., and Cotecchia, S. (1991). *Proc. Natl. Acad. Sci. U.S.A.* **88**, 11354–11358.

Andersen, P. H., Gingrich, J. A., Bates, M. D., Dearry, A., Falardeau, P., Senogles, S. E., and Caron, M. G. (1990). *Trends Pharmacol. Sci.* **11**, 231–236.

Arriza, J. L., Dawson, T. M., Simerly, R. B., Martin, L. J., Caron, M. G., Snyder, S. H., and Lefkowitz, R. J. (1992). *J. Neurosci.* **12**, 4045–4055.

Attramadal, H., Arriza, J. L., Aoki, C., Dawson, T. M., Codina, J., Kwatra, M. M., Snyder, S. H., Caron, M. G., and Lefkowitz, R. J. (1992). *J. Biol. Chem.* **267**, 17882–17890.

Benovic, J. L., Bouvier, M., Caron, M. G., and Lefkowitz, R. J. (1988). *Annu. Rev. Cell Biol.* **4**, 405–428.

Benovic, J. L., DeBlasi, A., Stone, W. C., Caron, M. G., and Lefkowitz, R. J. (1989). *Science* **246**, 235–240.

## CATECHOLAMINE RECEPTORS 289

Benovic, J. L., Onorato, J. J., Arriza, J. L., Stone, W. C., Lohse, M., Jenkins, N., Gilbert, D. J., Copeland, N. G., Caron, M. G., and Lefkowitz, R. J. (1991). *J. Biol. Chem.* **266,** 14939–14946.

Buck, L., and Axel, R. (1991). *Cell (Cambridge, Mass.)* **65,** 175–187.

Caron, M. G., Senogles, S. E., Amlaiky, N. and Berger, J. G. (1988). *In* "Dopamine Systems and Their Regulation" (P. M. Beart and D. M. Jackson, eds.), pp 151–158.

Casey, P. J., and Gilman, A. G. (1988). *J. Biol. Chem.* **263,** 2577–2580.

Cheung, A. H., Huang, R. R. C., Graziano, M. P., and Strader, C. D. (1991). *FEBS Lett.* **279,** 277–280.

Collins, S., Caron, M. G., and Lefkowitz, R. J. (1992). *Trends Biochem. Sci.* **17,** 37–39.

Cotecchia, S., Exum, S., Caron, M. G., and Lefkowitz, R. J. (1990). *Proc. Natl. Acad. Sci. U.S.A.* **87,** 2896–2900.

DeLean, A., Stadel, J. M., and Lefkowitz, R. J. (1980). *J. Biol. Chem.* **255,** 7108–7117.

Dixon, R. A. F., Kobilka, B. K., Strader, D. J., Benovic, J. L., Dohlman, H. G., Frielle, T., Bolanowski, M. A., Bennett, C. D., Rands, E., Diehl, R. E., Mumford, R. A., Slater, E. E., Sigal, I. S., Caron, M. G., Lefkowitz, R. J., and Strader, C. (1986). *Nature (London)* **321,** 75–79.

Dohlman, H. G., Caron, M. G., and Lefkowitz, R. J. (1987a). *Biochemistry* **26,** 2657–2664.

Dohlman, H. G., Bouvier, M., Benovic, J. L., Caron, M. G., and Lefkowitz, R. J. (1987b). *J. Biol. Chem.* **262,** 14282–14288.

Dohlman, H. G., Thorner, J., Caron, M. G., and Lefkowitz, R. J. (1991). *Annu. Rev. Biochem.* **60,** 653–688.

Gingrich, J. A., and Caron, M. G. (1993). *Annu. Rev. Neurosci.* **16,** 299–321.

Hausdorff, W. P., Bouvier, M., O'Dowd, B. F., Irons, G. P., Caron, M. G., and Lefkowitz, R. J. (1989). *J. Biol. Chem.* **264,** 12657–12665.

Hausdorff, W. P., Caron, M. G., and Lefkowitz, R. J. (1990). *FASEB J.* **4,** 2881–2889.

Kebabian, J. W., and Calne, D. B. (1979). *Nature (London)* **277,** 93–96.

Kjelsberg, M. A., Cotecchia, S., Ostrowski, J., Caron, M. G., and Lefkowitz, R. J. (1992). *J. Biol. Chem.* **267,** 1430–1433.

Kobilka, B. K., Kobilka, T. S., Daniel, K., Regan, J. W., Caron, M. G., and Lefkowitz, R. J. (1988). *Science* **240,** 1310–1316.

Kubo, T., Bujo, H., Akiba, I., Nakai, J., Mishina, M., and Numa, S. (1988). *FEBS Lett.* **241,** 119–125.

Lefkowitz, R. J., and Caron, M. G. (1988). *J. Biol. Chem.* **263,** 4993–4996.

Liggett, S. B., Schwinn, D. A., and Lefkowitz, R. J. (1992). *Clin. Res.* **40,** 252A.

Lohse, M. J., Benovic, J. L., Caron, M. G., and Lefkowitz, R. J. (1990a). *J. Biol. Chem.* **265,** 3202–3209.

Lohse, M. J., Benovic, J. L., Codina, J., Caron, M. G., and Lefkowitz, R. J. (1990b). *Science* **248,** 1547–1550.

Lorenz, W., Inglese, J., Palczewski, K., Onorato, J. J., Caron, M. G., and Lefkowitz, R. J. (1991). *Proc. Natl. Acad. Sci. U.S.A.* **88,** 8715–8719.

Nathans, J. (1987). *Annu. Rev. Neurosci.* **10,** 163–194.

Neve, K. A., Neve, R. L., Fidel, S., Janowsky, A., and Higgins, G. A. (1991a). *Proc. Natl. Acad. Sci. U.S.A.* **88,** 2802–2806.

Neve, K. A., Cox, B. A., Henningsen, R. A., Spanoyannis, A., and Neve, R. L. (1991b). *Mol. Pharmacol.* **39,** 733–739.

O'Dowd, B. F., Lefkowitz, R. J., and Caron, M. G. (1989a). *Annu. Rev. Neurosci.* **12,** 67–83.

O'Dowd, B. F., Hnatowich, M., Caron, M. G., Lefkowitz, R. J., and Bouvier, M. (1989b). *J. Biol. Chem.* **264,** 7564–7569.

Okamoto, T., Murayama, Y., Hayashi, Y., Inagaki, M., Ogata, E., and Nishimoto, I. (1991). *Cell (Cambridge, Mass.)* **67,** 723–730.

O'Malley, K. L., Mack, K. J., Gandelman, K. Y., and Todd, R. D. (1990). *Biochemistry* **29,** 1367–1371.

Palm, D., Münch, G., Dees, C., and Hekman, M. (1989). *FEBS Lett.* **254,** 89–93.

Sibley, D. R., and Monsma, F. J., Jr. (1992). *Trends Pharmacol. Sci.* **13,** 61–69.

Snyder, L. A., Roberts, J. L., and Sealfon, S. C. (1991). *Neurosci. Lett.* **122,** 37–40.

Sokoloff, P., Giros, B., Martres, M. P., Bouthenet, M. L., and Schwartz, J. C. (1990). *Nature (London)* **347,** 146–151.

Strader, C. D., Candelore, M. R., Hill, W. S., Sigal, I. S., and Dixon, R. A. F. (1989). *J. Biol. Chem.* **264,** 13572–13578.

Strader, C. D., Dixon, R. A. F., Cheung, H., Candelore, M. R., Blake, A. D., and Sigal, I. S. (1987). *J. Biol. Chem.* **262,** 16439–16443.

Strader, C. D., Sigal, I. S., Candelore, M. R., Rands, E., Hill, W. S., and Dixon, R. A. F. (1988). *J. Biol. Chem.* **263,** 10267–10271.

Tiberi, M., Jarvie, K. R., Silvia, C., Falardeau, P., Gingrich, J. A., Godinot, N., Bertrand, L., Yang-Feng, T. L., Fremeau, R. T., Jr., and Caron, M. G. (1991). *Proc. Natl. Acad. Sci. U.S.A.* **88,** 7491–7495.

Wang, H., Lipfert, L., Malbon, C. C., and Bahouth, S. (1989). *J. Biol. Chem.* **264,** 14424–14431.

Wong, S. K.-F., Parker, E. M., and Ross, E. M. (1990). *J. Biol. Chem.* **265,** 6219–6224.

# The Insulin Receptor and Its Substrate: Molecular Determinants of Early Events in Insulin Action

C. RONALD KAHN, MORRIS F. WHITE, STEVEN E. SHOELSON, JONATHAN M. BACKER, EIICHI ARAKI, BENTLEY CHEATHAM, PETER CSERMELY, FRANCO FOLLI, BARRY J. GOLDSTEIN, PEDRO HUERTAS, PAUL L. ROTHENBERG, MARIO J. A. SAAD, KENNETH SIDDLE, XIAO-JIAN SUN, PETER A. WILDEN, KAZUNORI YAMADA, AND STACY A. KAHN

*Research Division, Joslin Diabetes Center, Department of Medicine, Brigham and Women's Hospital, and Harvard Medical School, Boston, Massachusetts 02215*

Insulin is a potent metabolic and growth-promoting hormone that has pleiotropic effects at the level of the cell and within the intact organism. Insulin acts on cells to stimulate glucose, protein, and lipid metabolism, as well as RNA and DNA synthesis, by modifying the activity of a variety of enzymes and transport processes. The glucoregulatory effects of insulin at a whole body level are predominantly exerted by insulin action on liver, fat, and muscle. In liver, insulin stimulates glucose incorporation into glycogen and inhibits the production of glucose by glycogenolysis and gluconeogenesis. In muscle and fat, insulin stimulates glucose uptake, storage, and metabolism. In addition to these more classical effects, insulin also stimulates glucose metabolism in many other tissues that play little or no role in overall glucose homeostasis. In these nonclassical target tissues, insulin also often acts as a growth factor and in some manner modifies or augments the function of other regulators of metabolism of these cells.

Since the discovery of insulin over 70 years ago, considerable research has been devoted to attempting to understand the molecular mechanism of insulin action. The importance of understanding insulin action has been pointed out by its complex physiologic effects, as well as by the fact that altered insulin action, i.e., insulin resistance, plays important roles in the pathogenesis of many disorders, including obesity, diabetes mellitus, hypertension, and the glucose intolerance associated with many endocrine diseases (Caro *et al.*, 1986, 1987; DeFronzo, 1988; Reaven, 1988; Reddy and Kahn, 1988; Moller and Flier, 1991). It is only with the recent characterization of the insulin receptor as a tyrosine kinase (Kasuga *et al.*, 1982b; Shia and Pilch, 1983; Petruzzelli *et al.*, 1984) and identification of some of

292                    C. RONALD KAHN ET AL.

the related, downstream early components of the insulin response system
(White *et al.,* 1985; Sun *et al.,* 1991) that we have come to have some
understanding of the mechanism of insulin action at a molecular level. In
this review, we will focus our attention on the insulin receptor and its
primary cellular substrate, IRS-1, as the initial components of this insulin
action cascade. We will consider the structure and function of these proteins, the potential mechanisms by which they may be joined to the enzymes that regulate cellular metabolism, as well as how they may commu-
nicate with the actions of insulin at the level of the cell nucleus on RNA
and DNA.

## I. The Insulin Receptor

The actions of insulin at the cellular level are initiated by insulin binding
to its plasma membrane receptor. This receptor is present on almost all
mammalian cell types, ranging in number from <100 receptors per cell on
circulating erthrocytes to >300,000 receptors per cell on hepatocytes and
adipocytes (Ginsberg, 1977). In general, the concentration of the receptors
is higher in classical target tissues, although there are exceptions to this
rule in that muscle tends to have a relatively low concentration of insulin
receptors (Caro *et al.,* 1987). In cell types with high receptor content, such
as liver and adipose cells, classical metabolic effects occur with only a
small fraction of receptors occupied, i.e., there are "spare receptors" for
insulin action (Kahn *et al.,* 1981; Gammeltoft, 1984). There is no evidence,
however, that this is due to two structural classes of insulin receptors (one
of which is active and the other inactive); rather, the high concentration
of receptors on cells serves as a driving force for more rapid kinetics of
binding for insulin, because the circulating concentration of insulin ($10^{-10}$
to $10^{-9}$) is lower than the average affinity for binding (Gammeltoft, 1984).

## A.  INSULIN RECEPTOR STRUCTURE

The insulin receptor is a heterotetrameric protein consisting of two $\alpha$
subunits and two $\beta$ subunits linked together by disulfide bonds to form a
$\beta\alpha\alpha\beta$ structure (Fig. 1) (Czech, 1985, 1989; Avruch, 1989). The $\alpha$ subunits
have an apparent molecular weight ($M_r$) of 135,000 and the $\beta$ subunits
have an apparent $M_r$ of 95,000. Thus, the calculated molecular weight of
the holoreceptor is 460,000, although on SDS–gel electrophoresis under
nonreducing conditions the usual estimated size is ~350,000 (Massague *et
al.,* 1981; Kasuga *et al.,* 1982a). Both are glycosylated with complex N-
linked carbohydrates on the $\alpha$ subunit and with both N- and O-linked
carbohydrates on the $\beta$ subunit (Edge *et al.,* 1990). The role for the

INSULIN RECEPTOR—SUBSTRATE 293

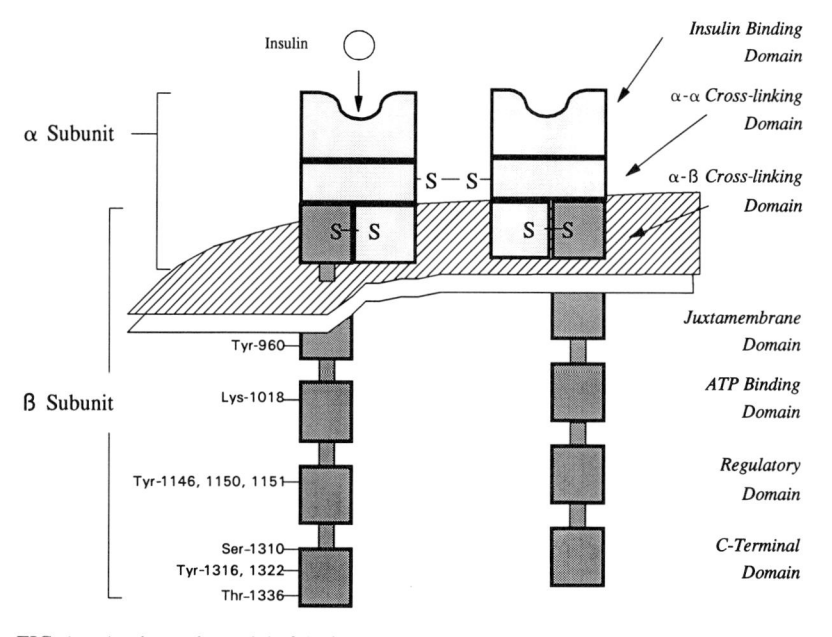

FIG. 1. A schematic model of the insulin receptor in its plasma membrane. The numbering system used in the figure represents that of the − exon 11 sequence. For the + exon 11 sequence, add 12 to each.

glycosylation of the insulin receptor is unknown. The α subunits of insulin receptors in brain and some neural tissues have much lower apparent molecular weights due to differences in glycosylation (Heidenreich *et al.*, 1988).

The two subunits of the insulin receptor are specialized to perform the two functions of this protein. The α subunit is entirely extracellular and contains the insulin binding site (Yip, 1992). The β subunit is a transmembrane protein and contains the tyrosine kinase signaling domain on its intracellular segment (White and Kahn, 1986). The insulin receptor is now recognized to be a member of the family of tyrosine protein kinase receptors for peptide hormones and growth factors. Based on overall structure, three classes of tyrosine kinase receptors have been identified (Yarden and Ullrich, 1988). One class represents monomeric proteins consisting of a single transmembrane polypeptide chain with a single continuous intracellular tyrosine kinase domain, as exemplified by the EGF receptor. A second class, similar to class I but with a discontinuous tyrosine kinase domain, is exemplified by the PDGF and CSF-1 receptors. The insulin receptor and the highly homologous IGF-1 receptor constitute class III.

## B. INSULIN RECEPTOR GENE AND RECEPTOR BIOSYNTHESIS

Although the insulin receptor is a tetrameric protein, the receptor is the product of a single gene located on the short arm of chromosome 19 (19p7) near the gene for the LDL receptor (Yang-Feng *et al.*, 1985). The structure of this gene has been elucidated by both restriction mapping (Muller-Wieland *et al.*, 1989) and genomic cloning (Seino *et al.*, 1989, 1990). The insulin receptor gene is about 150 kb in length and contains 22 exons separated by 21 introns (Fig. 2). Exons 1 through 11 are spread over 90 kb and code for the $\alpha$ subunit; exons 12 through 22 occur over a span of 30 kb and code for the $\beta$ subunit. Although the coding information is contained in about 4 kb of DNA, most cells possess four species of receptor mRNA ranging in length from 5.7 to 9.5 kb (Goldstein *et al.*, 1987; Goldstein and Kahn, 1989). Using RNase mapping, we have shown that the variability in length is primarily due to differences in the 3'-untranslated region with a relatively constant 5'-untranslated domain. Although differences in the 3'-untranslated length have been suggested to contribute to alterations in mRNA half-life, the $t_{1/2}$ for all species of insulin receptor mRNA is about 70 minutes. The concentration of receptor mRNA is greatest in brain and kidney, but is widely expressed in virtually all tissues.

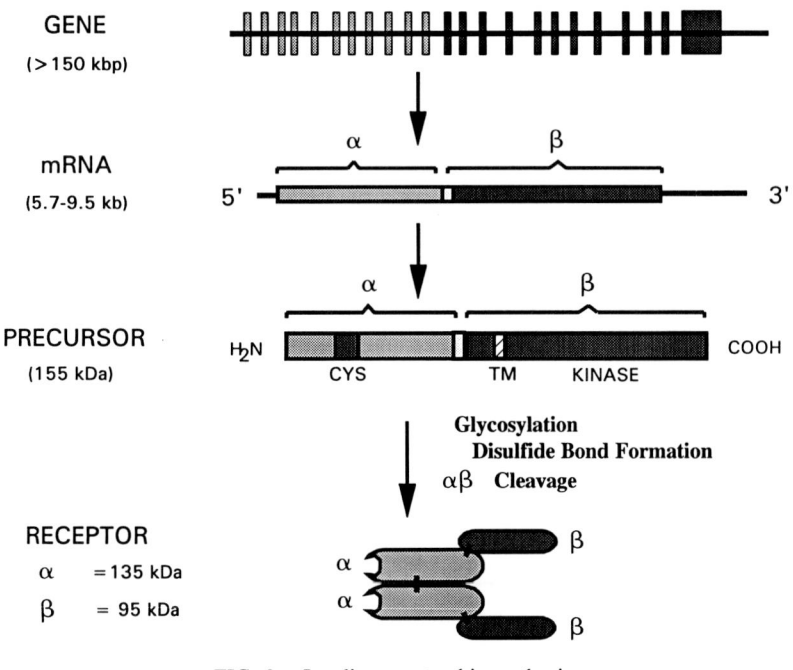

FIG. 2.   Insulin receptor biosynthesis.

INSULIN RECEPTOR–SUBSTRATE

The insulin receptor mRNAs code for a single-chain precursor of the insulin receptor that contains the entire coding sequence for both the $\alpha$ and $\beta$ subunits joined by a 4-amino acid connecting peptide (Fig. 2). Two forms of insulin proreceptor that differ in length by 12 amino acids near the C terminus of the $\alpha$ subunit are synthesized from these RNA species (Seino and Bell, 1989; Moller et al., 1989). This is based on alternative splicing of the 36-base pair (bp) exon 11 in the insulin receptor gene producing a larger form of insulin proreceptor containing 1382 amino acids and a smaller form of 1370 amino acids. These were recognized initially by differences in the cDNA cloning (Ullrich et al., 1985; Ebina et al., 1985). Both forms of the insulin receptor (+ and − exon 11) are expressed in most tissues, but the amount varies from tissue to tissue (Table I) (Seino and Bell, 1989; Goldstein and Dudley, 1990). The possible functional significance of these alternatively spliced forms is uncertain, but it appears that the − exon 11 form has a relatively higher affinity for both insulin and especially IGF-1 (Mosthaf et al., 1990; McClain, 1991; Moller et al., 1989; Benecke et al., 1992). In addition, this form is more rapidly internalized and down-regulated (Vogt et al., 1991; Yamaguchi et al., 1991). Mosthaf et al. (1991) have also suggested that more of the + exon 11 form is expressed in muscle of patients with non-insulin-dependent diabetes mellitus (NIDDM) as compared to normal individuals and thus may be a marker or may contribute to insulin resistance, but this finding has not been reproduced in other laboratories (Benecke et al., 1992) and thus remains controversial.

After synthesis in the rough endoplasmic reticulum the proreceptor is transported through the golgi apparatus, where it is rapidly glycosylated to give species of 180–210 kDa (Hedo et al., 1981; Hedo and Gorden 1985; Olson et al., 1988). The proreceptors are then disulfide linked, cleaved into $\alpha$ and $\beta$ subunits, and further glycosylated to give the final active

TABLE I
Tissue Distribution of Insulin Receptor mRNAs[a]

| Tissue | Relative concentration | + Exon 11 (%) |
|---|---|---|
| Brain | 78 | 0 |
| Kidney | 50 | 60 |
| Liver | 30 | 90 |
| Spleen | 22 | 0 |
| Placenta | n.d. | 50 |

[a] Tissue distribution of insulin receptor mRNA and relative concentration of + exon 11 variant; n.d., not determined. Adapted from Araki et al. (1992), Seino and Bell (1989), and Moller et al. (1989).

tetrameric receptor. The proreceptor contains a 27-amino acid signal peptide that helps guide its movement and processing in the cell. A small fraction of the fully glycosylated proreceptor may escape processing and appear on the cell surface; however, this unprocessed receptor form has low affinity for insulin and markedly reduced kinase activity (Williams *et al.*, 1990).

## C. DOMAIN STRUCTURE OF THE INSULIN RECEPTOR $\alpha$ SUBUNIT

As noted above, the two subunits of the insulin receptor are specialized to the two functions that this receptor must achieve. The $\alpha$ subunit is specialized for high-affinity insulin binding, and the $\beta$ subunit is specialized for transmembrane signal transduction. Each of the subunits can also be subdivided into a number of specific functional domains, which in many cases appear to follow the exonic structure of the receptor gene (Seino *et al.*, 1990).

The extracellular $\alpha$ subunit consists of an N-terminal domain (coded for by exons 1–2), a cysteine-rich domain (coded for by exons 3–5), a domain involved in $\alpha$–$\alpha$ disulfide bonding (of uncertain location), and a domain involved in $\alpha$–$\beta$ disulfide bonding (probably coded for by exon 10) (Fig. 1). Based on chemical modification and *in vitro* mutagenesis studies, the most critical domain for high-affinity insulin binding is that coded for by exons 1, 2, and 3 and consists of the first 298 amino acids of the receptor (Yip, 1992; DeMeyts *et al.*, 1990; Rafaeloff *et al.*, 1989). By replacing exons 1–3 of the insulin receptor with the corresponding sequences from the cDNA of the IGF-1 receptor, insulin binding affinity is significantly impaired, and the newly created chimeric receptor binds IGF-1 with high affinity (Andersen *et al.*, 1990; Gustafson and Rutter, 1990; Kjeldsen *et al.*, 1991). Within this region several critical residues have been identified, particularly Phe-89 (DeMeyts *et al.*, 1990). Mutation of this single residue to almost any other amino acid significantly reduces insulin binding to its receptor. If a chimeric receptor is prepared, replacing only exon 3 (corresponding approximately to amino acids 191 to 298) of the insulin receptor with that of the IGF-1 receptor, the resulting chimeric receptor retains high-affinity insulin binding and acquires high-affinity IGF-1 binding (Andersen *et al.*, 1990). These results suggest that these three N-terminal exons contain both positive and negative recognition elements, and that the regions coded for by exons 1 and 2 are most important for insulin recognition, whereas that for exon 3 may confer IGF-1 specificity to the insulin receptor.

Similar conditions have been reached by studies in which the binding domain has been mapped using insulin covalently cross-linked to the

INSULIN RECEPTOR–SUBSTRATE 297

receptor (Yip *et al.*, 1988; Waugh *et al.*, 1989; Wedekind *et al.*, 1989). It is likely, however, that the completely normal high-affinity binding involves a number of conformational determinants contributed by the entire $\alpha$ subunit, and perhaps even the $\beta$ subunit. Thus, when a solubilized form of receptor ectodomain, including the entire $\alpha$ and extracellular $\beta$ linked in tetrameric form, is expressed, it binds insulin with significantly lower affinity than the normal intact receptor (Schaefer *et al.*, 1990; Paul *et al.*, 1990). Free $\alpha$ subunits, as well as many receptors containing single point mutations, spread throughout the $\alpha$ subunit, bind insulin very poorly, and most truncated $\alpha$ subunits do not bind insulin at all (Taylor *et al.*, 1990a; Schaefer *et al.*, 1990). As noted above, near the C terminus of the $\alpha$ subunit is a site of alternative mRNA splicing that results in the addition or deletion of 12 amino acids to the sequence. This addition of 12 amino acids has effects on the affinity of the insulin receptor for both insulin and IGF-1, suggesting that in the three-dimensional structure of the receptor, the C-terminal domain of the $\alpha$ subunit may come into juxtaposition to the N-terminal regions, also of the $\alpha$ subunit involved in high-affinity insulin binding. Evidence for such a complex interaction of a peptide hormone with its receptor has recently been obtained for the human growth hormone receptor (see article by Wells *et al.*, 1993). This type of three-dimensional analysis for the insulin receptor will probably require an X-ray crystallographic analysis of the receptor and its subunits.

The intact insulin receptor, both in its solubilized, purified form and in its normal plasma membrane environment, exhibits negative cooperativity between insulin binding sites and curvilinear Scatchard plots (DeMeyts *et al.*, 1976). This means that the insulin receptor binds the first insulin molecule with high affinity, and binds the second insulin molecule with much lower affinity. The molecular mechanism underlying this phenomenon remains uncertain. Although the intact insulin receptor possesses two $\alpha$ subunits, some studies have suggested that only a single molecule of insulin binds with high affinity (Pang and Shafer, 1984; Sweet *et al.*, 1987). However, this property still persists in single tetrameric receptors and in isolated receptors reconstituted in liposomes such that there is no more than one receptor per liposome (Huertas and Kahn, 1992). In our hands, the stoichiometry of binding also suggests that a single tetrameric receptor can bind more than one insulin molecule. Isolated $\alpha\beta$ heterodimers obtained by mild reduction display linear Scatchard plots with a binding affinity intermediate between the high- and low-affinity states of the tetrameric receptor (Sweet *et al.*, 1987; Boni-Schnetzler *et al.*, 1987). Thus, negative cooperativity appears to be the property of a single tetrameric receptor, and is likely due to interactions and three-dimensional conformational changes within the receptor itself (Gu *et al.*, 1988). Changes in the negative cooperative properties have been observed with both natural

298                    C. RONALD KAHN ET AL.

mutations and *in vitro* mutations involving residues 460, 647, and the region 682–685 (Kadowaki *et al.*, 1990; Cheatham and Kahn, 1992).

All of the 36 cysteine residues in the $\alpha$ subunit are involved in either disulfide bonding or are otherwise covalently modified and thus contribute to important conformational features of the receptor molecule (Finn *et al.*, 1990). Treatment of the insulin receptor with increasing concentrations of reducing agent results initially in an increase in high-affinity binding, followed by a decrease in binding (Crettaz *et al.*, 1984). The exact residues involved in disulfide bonding between the two $\alpha$ subunits have not yet been identified, but these residues are more sensitive to reductant than those involved in $\alpha$–$\beta$ disulfide bonding (Czech, 1985). Although $\alpha$–$\alpha$ disulfide bonding occurs early in the course of proreceptor synthesis, when the receptor is in its intracellular (and insulin-free) environment (Olson *et al.*, 1988), when the receptor is split into its two half-receptors (or protomers) by mild reduction *in vitro*, insulin addition to the medium provides a driving force for reassociation of the protomers (Boni-Schnetzler *et al.*, 1988). The normal insulin receptor protomer may also form hybrids with the IGF-1 receptor or with mutant insulin receptors, a phenomenon that also occurs both *in vitro* and normally in the intact cell (Treadway *et al.*, 1989, 1991). Although the exact physiological significance of this is uncertain, it has been speculated that this phenomenon may account for some of the growth-promoting actions of insulin and may account for the "dominant-negative" inhibition of normal receptor function when cells also express mutant insulin receptors (Whittaker *et al.*, 1990).

Based on structural analysis of the insulin receptor subjected to mild tryptic proteolysis, the domain of the receptor involved in $\alpha$–$\beta$ disulfide bonding included the C-terminal 25 kDa of the $\alpha$ subunit (Shoelson *et al.*, 1988; Xu *et al.*, 1990). This fragment contains only four cysteine residues (the extracellular $\beta$ subunit also contains four) at positions 647, 682, 683, and 685. To determine which of these Cys residues might be involved in $\alpha$–$\beta$ bonding, we prepared and expressed two *in vitro* mutant receptors: one in which Cys-647 was converted to serine, and another in which Cys residues 682, 683, and 685 were converted to serine (Cheatham and Kahn, 1992). The triple mutant exhibited normal insulin binding and receptor structure. By contrast, the Cys-647 mutant exhibited an altered tetrameric structure, with most surface receptors consisting of an $\alpha_2$ structure noncovalently linked to the two $\beta$ subunits (Fig. 3). Interestingly, these receptor tetramers remained together on the surface of the cell, but could be dissociated with treatment by urea or acid pH. This noncovalent receptor structure was associated with an increase in receptor affinity and negative cooperativity. This noncovalently associated receptor was defective for insulin-stimulated receptor kinase activation, suggesting that normal disul-

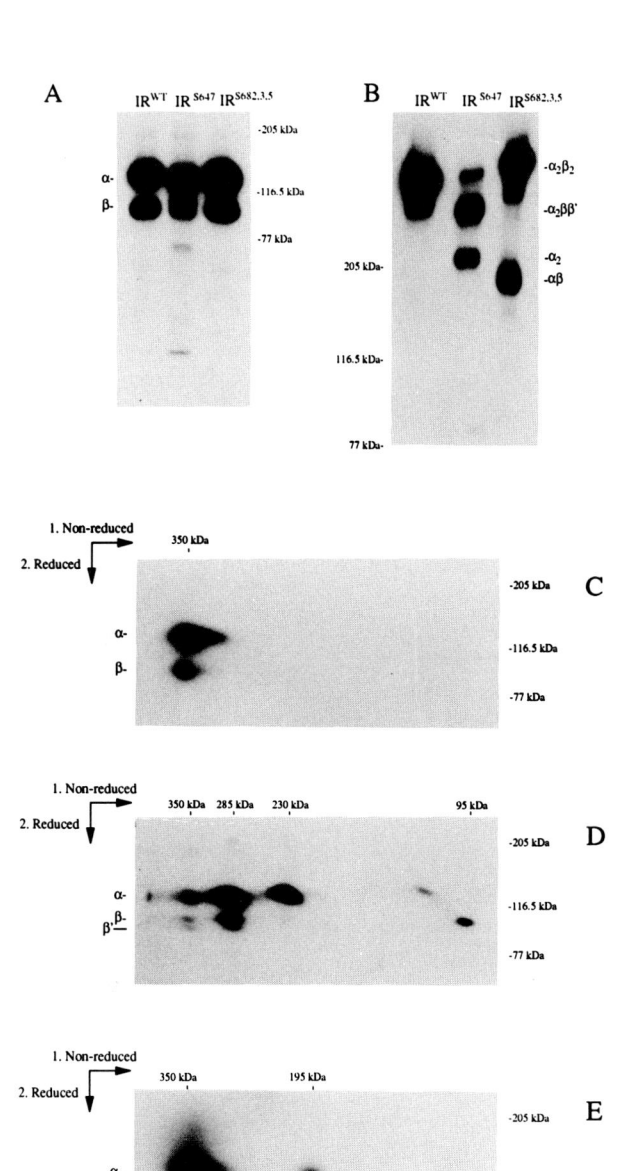

FIG. 3. Structure and function of the insulin receptor with a mutation of Cys-647 (IR$^{S647}$) and Cys-682, Cys-683, and Cys-685 (IR$^{S682,3,5}$). Left panel: CHO cells overexpressing wild-type (IR$^{WT}$) or mutant insulin receptors were surface labeled with Na[$^{125}$I]. Insulin receptors were immunoprecipitated and subjected to SDS–PAGE under reducing (A) or nonreducing (B) conditions. In C–E, receptors from B were subjected to a second electrophoresis under reducing conditions to separate individual components of the oligomeric forms. (C) IR$^{WT}$; (D) IR$^{S647}$; (E) IR$^{S682,3,5}$. Right panel: insulin-stimulated receptor autophosphorylation in intact cells following $^{32}$P labeling. Receptors were immunoprecipitated with antireceptor antibody. Adapted from Cheatham and Kahn (1992).

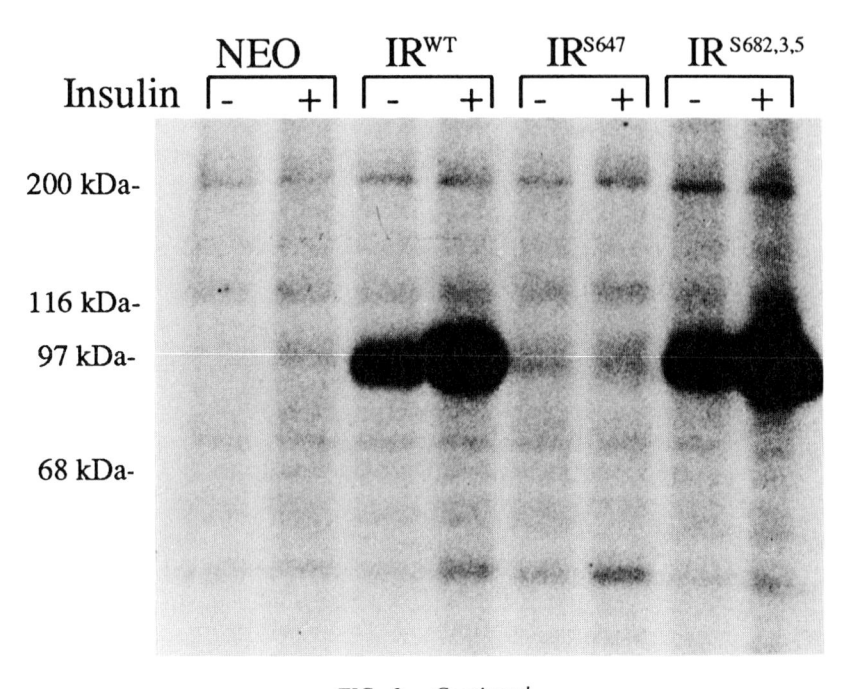

FIG. 3.—*Continued*

fide bond formation between $\alpha$ and $\beta$ subunits is required for normal signal transduction (Fig. 3).

## D.   DOMAIN STRUCTURE OF THE INSULIN RECEPTOR $\beta$ SUBUNIT

The $\beta$ subunit of the insulin receptor contains an extracellular domain of about 193 amino acids, a single transmembrane domain of 23 amino acids, and a complex intracellular domain of 402 amino acids. Little is known about the function of the extracellular domain of the $\beta$ subunit except that it contains the cysteine residues involved in interchain bonding between $\alpha$ and $\beta$ subunits (Cheatham and Kahn, 1992) and sites of both complex N-linked and O-linked glycosylation (Edge *et al.*, 1990). Of the four cysteine residues in the extracellular domain, it is not clear which are involved in the interchain disulfide with Cys-647 of the $\alpha$ subunit.

The transmembrane domain of the $\beta$ subunit is a single $\alpha$-helical segment of 23 amino acids that links the extracellular and intracellular domains of the receptor. Because this single $\alpha$-helical domain forms the only physical link between the extracellular and intracellular $\beta$ subunits, it should obviously play a critical role in signal transduction. In general, however, the transmembrane domains of tyrosine kinase receptors show little in the way of conservation of specific sequence, and a number of *in vitro* mutations may be made in this domain, including chimeric receptors, without affecting receptor function (Yarden and Ullrich, 1988; Riedel *et al.*, 1986; Frattali *et al.*, 1991). However, in the case of the insulin receptor, two types of studies suggest that the structure of this domain and its internalization with surrounding lipids may play an important role in transduction of the signal across the lipid bilayer, as well as the cellular trafficking of the receptor. These studies include reconstitution of receptors into liposomes of varying lipid composition and *in vitro* mutagenesis of the transmembrane domain itself.

## E. BEHAVIOR OF THE INSULIN RECEPTOR IN ARTIFICIAL LIPID MEMBRANES

Utilizing a system of affinity-purified human insulin receptors reconstituted in intact, unilamellar phospholipid vesicles we have evaluated the role of membrane thickness, lipid saturation, and cholesterol content in the binding and signal transduction of the insulin receptor. In unsaturated phosphatidylcholines, increasing fatty acyl chain length lowers the affinity of the receptor for insulin, and this is associated with a parallel decrease in insulin-stimulated autophosphorylation. These properties of the receptor, however, are dissociated by reconstitution of receptors into vesicles of saturated phosphatidylcholines. Thus, when receptors are inserted into liposomes composed of phosphatidylcholines with saturated fatty acids of chain length from $C_8$ to $C_{20}$, insulin binding remains intact up to a chain length of $C_{18}$, whereas above $C_{14}$ chain length there is a marked decrease in insulin-stimulated receptor autophosphorylation (Fig. 4). The coupling between binding and kinase activation can be restored in part by use of fatty acids with increasing degrees of unsaturation. These results are consistent with a model in which interactions between the receptor transmembrane domain and the hydrophobic membrane environment regulate binding and signal transduction by the insulin receptor. Modification of membrane phospholipids by enzymatic treatment of the cell also modifies receptor signaling (Zoppini and Kahn, 1992).

## F. *In Vitro* MUTANTS OF THE INSULIN RECEPTOR TRANSMEMBRANE DOMAIN

As an alternative approach to exploring the role of the transmembrane (TM) domain in signal transduction, we have recently prepared a series of *in vitro* mutants of the receptor in which the sequence of the TM domain

302                    C. RONALD KAHN ET AL.

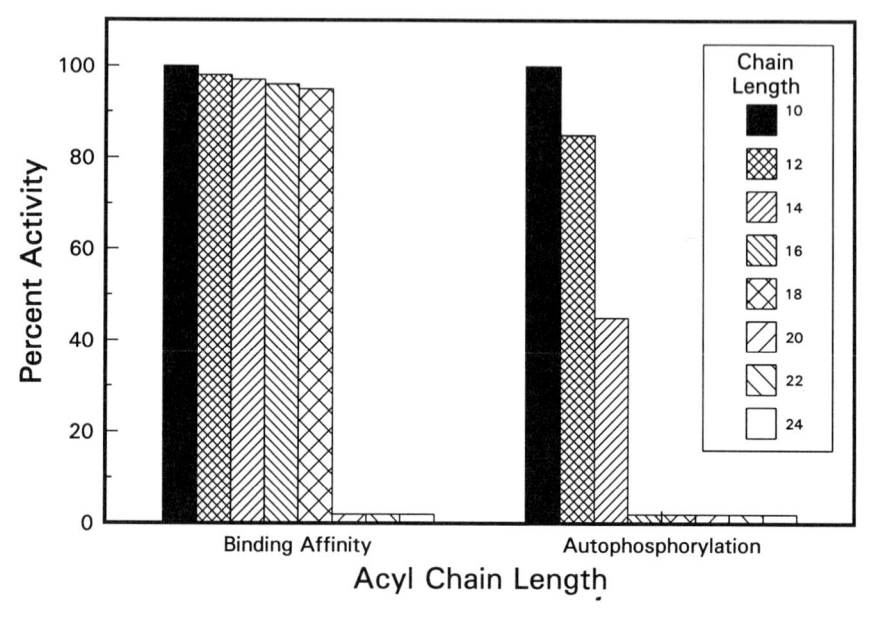

FIG. 4.   Effect of membrane lipids on insulin receptor function—activity in unsaturated phospatidylcholine liposomes. Purified placental insulin receptors were reconstituted in unilamellar liposomes consisting of unsaturated phosphatidylcholines of different fatty acyl chain lengths, varying from $C_{10}$ to $C_{24}$. Insulin binding and receptor autophosphorylation were then assessed using standard techniques.

has been modified in a variety of ways. These include substitution of other receptor TM domains into the insulin receptor, increasing TM length, point modifications of specific TM residues, and flipping the orientation of the TM domain with or without the associated charged residues that flank this domain on its exofacial and cytoplasmic surface. As has been observed with other receptor systems, the TM domain of the insulin receptor can be modified significantly with no apparent change in receptor properties. Substitution of the PDGF receptor TM domain, adding three amino acids to or deleting five amino acids from the TM domain, has no effect on binding or kinase properties (Frattali *et al.,* 1991; Yamada *et al.,* 1992). However, some changes in the TM domain do affect receptor function, indicating several important functional aspects of the TM domain.

One interesting series of mutants that we have produced is one in which the TM domain has a substitution of Val-938 with Asp, substitution of the entire TM domain with that of the c-*neu* protooncogene, or substitution of the TM with that from the *neu* oncogenic homologue (the c-*neu* TM domain containing a point mutation). This series is of interest because in

INSULIN RECEPTOR–SUBSTRATE                    303

the c-*neu* protooncogene (a relative of the EGF receptor), the tyrosine
kinase activity of the protein is normally very low. Modification of a single
Val in the c-*neu* TM domain to the acidic residue Glu results in the
constitutive activation of the tyrosine kinase and cellular transformation
(Bargmann *et al.*, 1986). In the insulin receptor, a Val-to-Glu or Val-to-
Asp point mutation alone has no effect on the receptor kinase activity
(Frattali *et al.*, 1991; Yamada *et al.*, 1992); however, substitution of the
complete c-*neu* TM domain in the insulin receptor will produce partial
activation of the kinase *in vitro* (Fig. 5) (Yamada *et al.*, 1992). Very
recently we found that substitution of the *neu* Val → Glu TM domain
produces an even greater, and perhaps fully constitutive, activation of the
insulin receptor kinase *in vivo* (B. Cheatham *et al.*, submitted for publication). The mechanism of this activation is uncertain, but in the case of the
*neu* oncoprotein, activation of the kinase has been ascribed to a tendency
of the protein to self-aggregation (Weiner *et al.*, 1989).

Other mutations in the insulin receptor TM may also alter the function
of the receptor. For example, almost all transmembrane proteins have one
basic amino acid on the extracellular face and three on the intracellular
face, which presumably serve as sequences that define the position of the

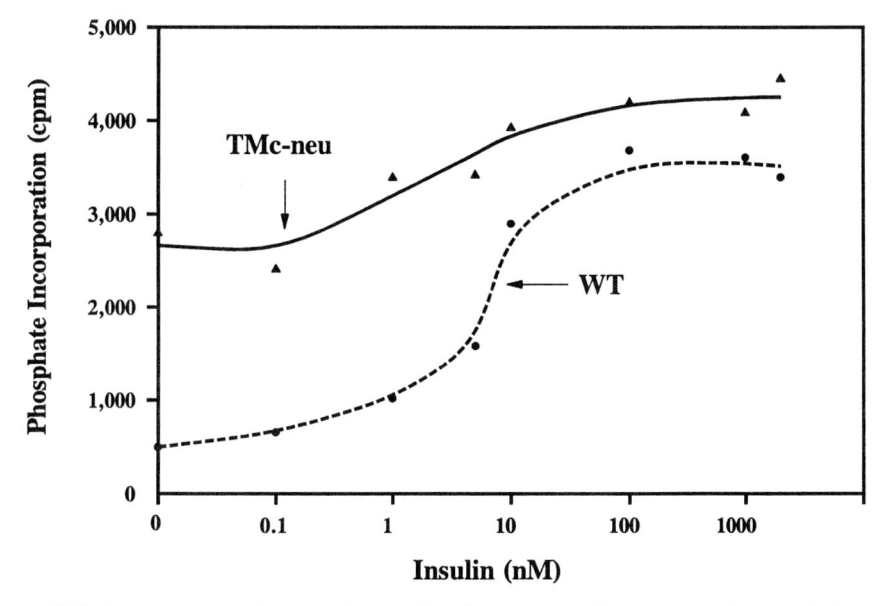

FIG. 5.    *In vitro* autophosphorylation of insulin receptors with a transmembrane substitution of the c-*neu* TM for the normal (WT) insulin receptor TM sequence. Note the basal,
i.e., constitutive, activation of the kinase. Adapted from Yamada *et al.* (1992).

304                    C. RONALD KAHN ET AL.

protein into the membrane. If an insulin receptor is constructed with a
TM containing the same amino acids in reversed direction, and if the
intracellular flanking charged amino acids are also moved to the extracellular face, the resultant receptor has markedly altered intracellular trafficking, with over 80% remaining in the golgi or endoplasmic reticulum of the
cell (K. Yamada et al., in preparation). The 10–20% of receptor that
reaches the plasma membrane is normally processed and binds insulin
normally, but fails to transmit a normal signal for kinase activation in
response to insulin binding.

The TM domain of most tyrosine kinase receptors is viewed as a single
$\alpha$-helical stretch of 21 to 24 amino acids (Yarden and Ullrich, 1988).
The insulin receptor TM is also predominantly $\alpha$-helical, but interestingly
contains a Gly and Pro sequence, which should act as "helix breakers,"
and this feature may also contribute to the function of the receptor. Thus,
when the insulin receptor TM is mutated to convert the Gly-Pro in the
normal sequence to Ala-Ala, amino acids that should promote increased
helicity, there is a threefold increase in mobility in the lipid matrix of
the membrane as assessed by fluorescent photobleaching and a similarly
increased rate of insulin-stimulated internalization (Goncalves et al.,
submitted). This is associated with a two- to threefold increase in cell-
associated insulin degradation and an increased rate of receptor down-
regulation. Thus, nature seems to have provided a "kink" in the insulin
receptor TM domain to slow its mobility in the membrane and slow the
rate of internalization and degradation of both ligand and receptor.


## G.   THE INTRACELLULAR REGION OF THE $\beta$ SUBUNIT

The intracellular or cytoplasmic region of the $\beta$ subunit consists of
402 amino acids, which can be functionally divided into four domains:
a juxtamembrane domain important in substrate binding and receptor
internalization, a kinase catalytic domain, a region of tyrosine autophosphorylation involved in regulation of kinase activity, and a C-terminal
domain containing two additional tyrosine phosphorylation sites, as well
as some sites of serine and threonine phosphorylation (Fig. 6). This structure was in large partially predicted prior to actual cloning of the receptor
molecule based on extensive characterization of its tyrosine kinase properties (Kasuga et al., 1982b; White et al., 1984; Avruch, 1989).

The most easily identified domain of the $\beta$ subunit was the catalytic
domain that contains the ATP binding site. In contrast to the extracellular
and transmembrane domains, this region contains the highest degree of
homology between the insulin receptor and all other members of the
protein tyrosine kinase family (Yarden and Ullrich, 1988; White and Kahn,

INSULIN RECEPTOR–SUBSTRATE                    305

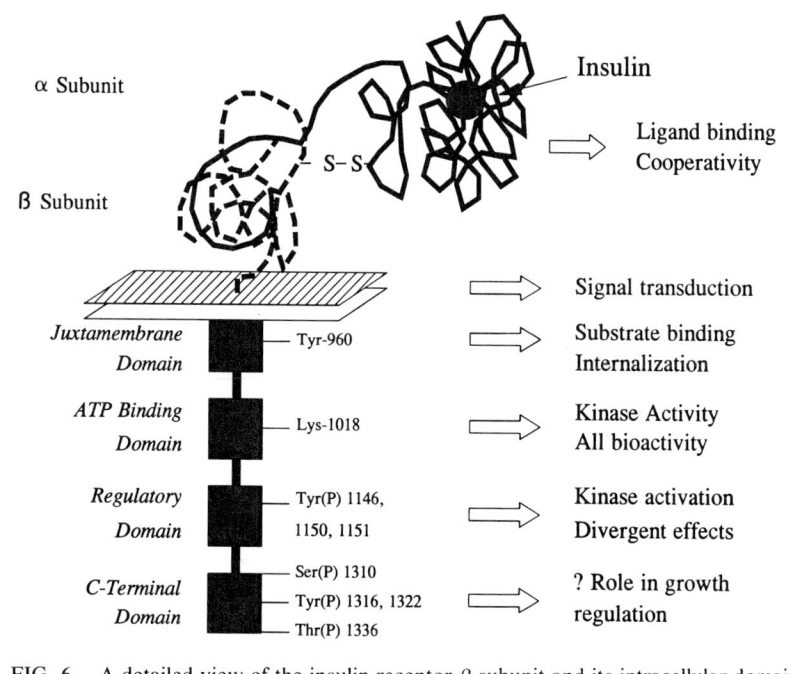

FIG. 6.   A detailed view of the insulin receptor $\beta$ subunit and its intracellular domains. The numbering system used for the amino acid residues refers to the − exon 11 sequence. For the + exon 11 sequence, add 12 to each.

1986). The distinctive motif Gly-X-Gly-X-X-Gly (where X is any amino acid) followed by a lysine residue 10 to 20 amino acids toward the C terminus forms the consensus amino acid sequence encoding the ATP binding domain. The lysine residue (number 1018 in the − exon 11 receptor and 1030 in the + exon 11 receptor) is absolutely required for kinase activity; substitution of this residue with any other amino acid invariably blocks autophosphorylation and kinase activity and most, if not all, biological response mediated by tyrosine kinase receptors (Chou *et al.*, 1987; Ebina *et al.*, 1987). A patient with the type A syndrome of insulin resistance and acanthosis nigricans has been described in which the Gly-X-Gly-X-X-Gly motif is modified by mutation of the final Gly to Val (Yamamoto-Honda *et al.*, 1990). This results in a kinase inactive receptor and a syndrome of severe insulin resistance.

About 110 amino acids toward the C terminus of the $\beta$ subunit is a cluster of three tyrosine residues (1146, 1150, and 1151 in the − exon 11 receptor and 1158, 1162, and 1163 in the + exon 11 variant) that form a major site of autophosphorylation and constitute the regulatory region of the kinase. As in other kinases, these tyrosine residues are near acidic amino acids (glutamic or aspartic acid) that increase their affinity as substrates of the receptor kinase. In the insulin receptor, autophosphorylation occurs on all three of these tyrosine residues within seconds after insulin stimulation (White et al., 1988b; Tornqvist and Avruch, 1988; Tornqvist et al., 1988; Tavare et al., 1988; Rosen, 1987), and this increases the activity of the kinase toward exogenous substrates dramatically (Fig. 7). As long as these three residues remain tyrosine phosphorylated, the kinase remains activated, even if insulin is removed from the insulin binding site. Not surprisingly, these three tyrosines also form the preferential site of dephosphorylation of the receptor by cellular phosphotyrosine phosphatases (King and Sale, 1990).

A second major region of phosphorylation of the insulin receptor is the C-terminal domain. This domain contains two tyrosine residues (1316, 1322, or 1328 and 1334 in the two receptor variants), both of which are autophosphorylated by the insulin receptor (White et al., 1988b). In addition, this C-terminal domain contains at least one threonine (1336/1348) and one serine (either 1293 or 1294 in the − exon 11 variant or 1305/1306 in the + exon 11 variant) site of receptor phosphorylation by exogenous Ser/Thr protein kinases (Lewis et al., 1990a,b). The exact role of these phosphorylations in insulin action remains debated. The tyrosine kinase activity of C-terminally truncated receptors is normal (Goren et al., 1987; Maegawa et al., 1988; Myers et al., 1991). Serine/threonine phosphorylation, on the other hand, may play a kinase regulatory role (see below). Several groups have studied receptors bearing C-terminal deletion mutations for other actions with varying results. In our hands, an in vitro mutant of the receptor with a deletion of the terminal 34 amino acids, including all of the proposed tyrosine, serine, and threonine sites of phosphorylation, exhibits a normal biological activity profile (Myers et al., 1991). By contrast, Thies et al. (1989) have reported that a similar mutant is defective in mediating metabolic effects of insulin while being superactive in stimulating growth effects. The reason for these very discordant results is not clear.

The fourth domain of the $\beta$ subunit of distinct function is the intracellular juxtamembrane region. Although we initially had proposed that this region was free of tyrosine sites of autophosphorylation (White et al., 1988a,b), more recent data from our laboratory and others (Tornqvist et al., 1988; Tavare et al., 1988) indicate that there are one or more sites of tyrosine phosphorylation in this domain as well. These sites of autophosphorylation, however, are not required for kinase activation, and account for less than 20% of the tyrosine phosphate in the $\beta$ subunit following insulin stimulation.

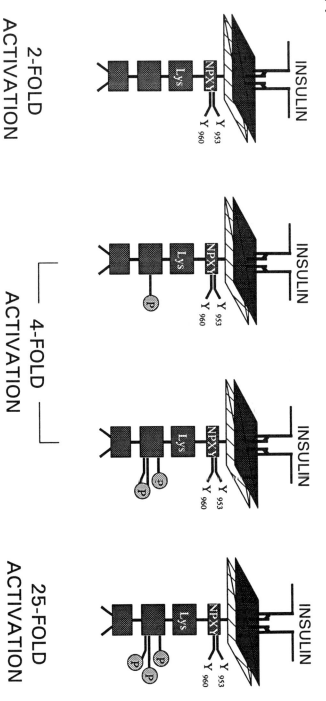

FIG. 7. (A) Schematic model of activation of the insulin receptor tyrosine kinase by the autophosphorylation cascade. (B) Behavior of the kinase toward exogenous (excess) substrates after activation by autophosphorylation. The dotted line represents the activity of the unactivated (basal) receptor; the dashed line represents the activity of the partially activated receptor; the solid line represents the activity of the fully activated (trisphosphorylated) receptor.

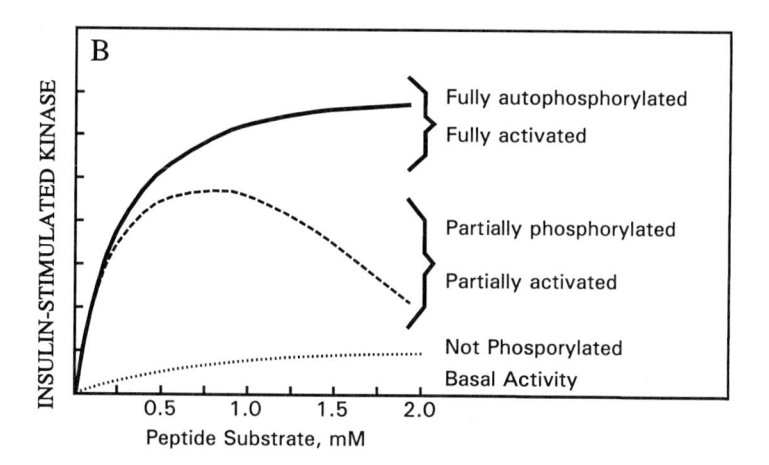

FIG. 7.—*Continued*

The juxtamembrane domain, however, appears to play an important role in several other functions of the receptor. Thus, mutation of the tyrosine at position 960 to phenylalanine (White *et al.*, 1988a), or deletion of a 12-amino acid peptide from this region (Rothenberg *et al.*, 1991), results in a receptor that is kinase active, but fails to phosphorylate the endogenous substrate of the insulin receptor pp185/IRS-1 (see below). Deletion mutations of this region also interfere with insulin-stimulated receptor internalization (Backer *et al.*, 1990; McClain *et al.*, 1987; Thies *et al.*, 1990). This latter effect has been ascribed to the fact that this region of the receptor contains a consensus sequence for internalization Asn-Pro-X-Tyr (NPXY), which is also present in a number of other receptors that exhibit high internalization rates (Chen *et al.*, 1990). This domain also contributes to an extended region of the kinase domain. As a result, the deletion mutant in the juxtamembrane region exhibits an altered $K_m$ for ATP (Backer *et al.*, 1991).

## H.  THE INSULIN RECEPTOR AS AN ALLOSTERIC ENZYME

Although the exact molecular changes that result in activation of the insulin receptor following insulin binding are uncertain, considerable data suggest that the insulin receptor behaves as a classic allosteric enzyme undergoing both conformational changes and modification by phosphorylation. Following insulin binding, the receptor undergoes a propagated conformation change. This conformational change results in alterations in the receiver's sensitivity to proteolysis (Donner and Yonkers, 1983; Lipson et al., 1986), susceptibility to chemical modification (Schenker and Kohanski, 1988; Wilden and Pessin, 1987; Waugh and Pilch, 1989), behavior on gel chromatography (Ginsberg et al., 1976), and sedimentation properties (Florke et al., 1990). This conformational change is propagated through both subunits and is associated with changes in antibody recognition to domains within the α and β subunits (Perlman et al., 1989; Baron et al., 1990; Prigent et al., 1990; Herrera and Rosen, 1986; Herrera et al., 1985). In the intact cell, this conformational change is also associated with increased receptor aggregation (Kahn et al., 1981).

Under normal circumstances, the major function of the α subunit is to suppress the kinase activity in the β subunit. Removal of all or part of the α subunit by either limited proteolysis (Shoelson et al., 1988) or in vitro mutagenesis (Ellis et al., 1988; Villalba et al., 1989) results in activation of the receptor in a manner similar to that induced by insulin binding. In the intact receptor, the insulin induced conformational and/or aggregational changes somehow reduce the normal effect of the α subunit to suppress the kinase activity in the β subunit. As a result, there is an increase in the kinase catalytic activity. This in turn results in autophosphorylation of the receptor on multiple tyrosine residues, initially involving the cluster of three regulatory tyrosines, as well as the juxtamembrane tyrosines, and the C-terminal tyrosines. And this in turn activates the kinase toward exogenous substrates.

It is still uncertain whether the autophosphorylation reaction is a cis event, i.e., occurs within a single β subunit, or a trans event between β subunits. Based on experiments with tryptically produced isolated receptor halves, we have shown that a cis-phosphorylation reaction can definitely occur, but trans phosphorylation is kinetically favored (Shoelson et al., 1988). On the other hand, autophosphorylation of the isolated intracellular domain of the β subunit suggests a concentration-dependent intermolecular (i.e., trans) reaction (Cobb et al., 1989). Pessin and colleagues have recently presented evidence for an even more complex form of trans reaction in which the two kinase domains in the receptor tetramer phosphorylate each other in a "ping-pong" fashion (Treadway et al., 1991). In IGF-1/insulin receptor hybrids, IGF-1 will induce autophosphorylation of both the insulin and IGF-1 receptor β subunits (Moxham et al., 1989). In either case, phosphorylation of the three regulatory tyrosines produces a further conformational change and activates the receptor kinase toward other cellular substrates, such as the endogenous substrate pp185/IRS-1 described below. Thus begins the chain of events that ultimately culminates in insulin action.

# I. REGULATION OF THE RECEPTOR MULTISITE PHOSPHORYLATION

Like many allosteric enzymes, the insulin receptor undergoes multisite phosphorylation, which contributes to its overall regulation. The most important component of this multisite phosphorylation is the tris-phosphorylation of tyrosines 1158, 1162, and 1163 (1146, 1150, and 1151 in the − exon 11 variant), which activates the kinase toward exogenous substrates. The importance of these phosphorylations in insulin action is illustrated by a series of seven *in vitro* mutants that we have prepared; in each of these, tyrosines have been mutated singly, doubly, or all together to phenylalanine, and then expressed stably in Chinese hamster ovary cells. Modification of even a single tyrosine to phenylalanine in this regulatory domain results in about a 50% decrease in receptor autophosphorylation and kinase activity (Fig. 8). Mutation of two tyrosines to phenylalanine produces a 70% reduction in kinase activity and mutation of all three tyrosines to phenylalanine causes an almost complete loss of kinase activity to a level similar to that seen in the ATP binding site mutant A1030 (Wilden *et al.*, 1992b). A similar change is also observed if the three tyrosines are changed to three serines, even though the serines can serve as potential phosphate acceptors. These changes in autokinase activity are paralleled by a similar decrease in phosphorylation of the endogenous substrate pp185/IRS-1 (see below).

These mutations produce alterations in downstream biological responses to insulin, but the exact extent to which these responses appear to be coupled varies to a considerable degree (Table II). For example, the ability of insulin to stimulate glucose incorporation into glycogen (a metabolic response) and thymidine incorporation into DNA (a growth response) tends to decrease in proportion to the decrease in receptor autophosphorylation and kinase activity (Wilden *et al.*, 1992a). By contrast, even a single tyrosine-to-phenylalanine mutation in the regulatory region causes a complete loss of insulin-stimulated receptor internalization; thus this insulin effect seems to be most tightly coupled to a normal kinase and autophosphorylation cascade. At the other extreme is insulin-stimulated receptor serine and threonine phosphorylation. This effect is mediated via cellular kinases distinct from the receptor, although the specific mechanism is unknown (Lewis *et al.*, 1990b). Although this insulin-stimulated effect is lost in the ATP binding site mutant (A1030), it is observed normally even in cells containing mutations of all three regulatory tyrosines. Thus, insulin-induced receptor serine phosphorylation is very insensitive to any change in the tyrosine kinase activity of the receptor or receptor autophosphorylation.

INSULIN RECEPTOR–SUBSTRATE   311

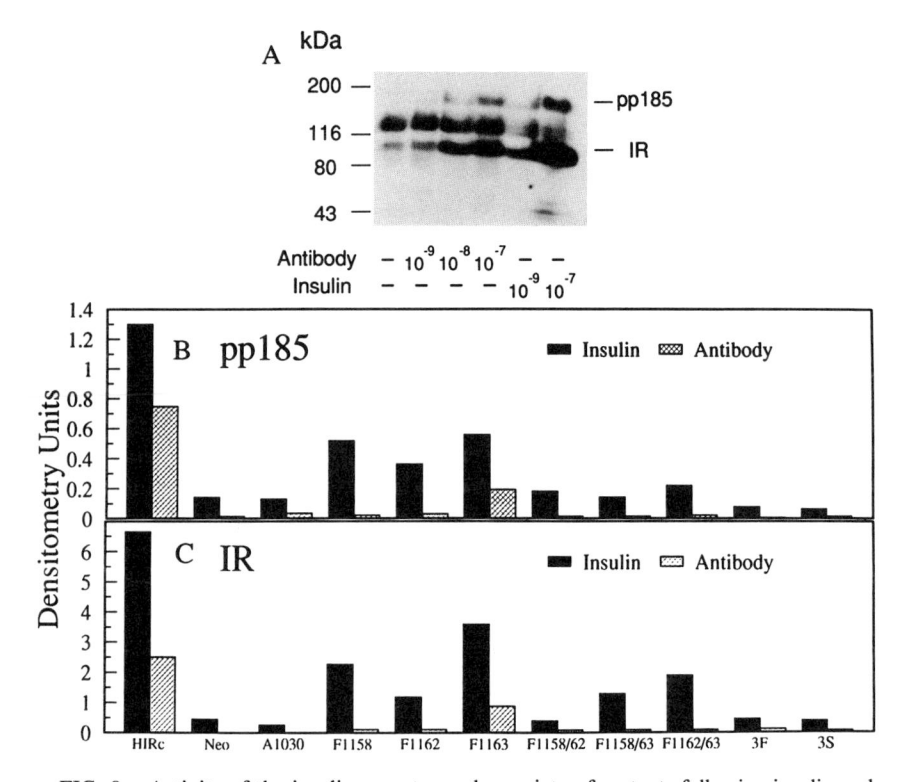

FIG. 8. Activity of the insulin receptor and a variety of mutants following insulin and antibody stimulation. CHO cells transfected with the insulin receptor or mutants were stimulated with insulin or antireceptor antibody 83-14 at a concentration of 100 n$M$; cells were extracted and subjected to SDS–gel electrophoresis and blotting with antiphosphotyrosine antibody. (A) A typical autoradiogram of insulin and antibody stimulation in cells expressing the wild-type human receptor. (B and C) These data are quantitated and compared with results of cells expressing mutant receptors or control vectors. Abbreviations: HIRc, normal human insulin receptor; Neo, neomycin control vector; A1030, ATP binding site mutant; and F1158, F1162, and F1163, receptors with changes in tyrosine to phenylalanine at residues 1158, 1162, and 1163 (the three regulatory tyrosines). The numbering system used in this figure represents that of the + exon 11 variant receptor.

Work in our laboratory and others has suggested that there may be a differential loss of some bioactivities in different single and double tyrosine mutants. For example, mutation of Tyr-1158 to phenylalanine resulted in a receptor that tends to lose growth-promoting activity more than metabolic activity (Wilden *et al.*, 1990). Debant *et al.* (1989) have reported that mutation of the other two regulatory tyrosines (1162 and 1163) tends to result in a preferential loss of metabolic activity over growth-promoting action. Neither of these findings has been reproducible in all studies (Zhang

# C. RONALD KAHN ET AL.

## TABLE II
### Summary of Insulin Receptor Mutant Function[a]

| Receptor | Serine phosphorylation | Kinase active | pp185 phosphorylation | DNA synthesis | Glycogen synthase | Internalization |
|---|---|---|---|---|---|---|
| Wild type | + | + | + | + | + | + |
| C terminal deletion | + | + | + | + | + | + |
| Juxtamembrane | + | + | − | − | − | + |
| Regulatory 1-Phe | + | < | < | < | < | − |
| Regulatory 2-Phe | + | << | << | << | << | − |
| Regulatory 3-Phe | + | − | − | − | − | − |
| Kinase deficient | − | − | − | − | − | − |

[a] The activities studied include insulin stimulation of receptor serine phosphorylation, kinase activity, pp185 phosphorylation, DNA synthesis, glycogen synthesis, and receptor internalization. The receptor mutants include the C-terminal deletion; the juxtamembrane (F960) mutant; mutants of 1, 2, or 3 tyrosines to phenylalanine in the regulatory region, and the kinase-deficient (ATP binding site) mutant.

et al., 1991; Wilden et al., 1992a). Whether the small differences in biological responses observed in in vitro mutants are significant is not clear, but similar differential biological response losses have been observed in some natural mutants of the insulin receptor that have been reported in syndromes of insulin resistance (Moller et al., 1990a,b).

In contrast to the activation of the receptor kinase produced by tyrosine autophosphorylation, insulin also stimulates the serine and threonine phosphorylation of the insulin receptor (Kasuga et al., 1982b). Indeed, in intact cells this accounts for more than half of the total phosphate incorporated into the receptor on prolonged insulin stimulation. In general, the stimulation of tyrosine phosphorylation is very rapid (maximal within 15 seconds), whereas the stimulation of serine/threonine phosphorylation is more gradual and occurs over several minutes to 1 hour (White et al., 1984). In cell-free systems, serine/threonine phosphorylation of the receptor is observed with protein kinase C and to a variable degree with cAMP protein kinase (Bollage et al., 1986; Tanti et al., 1987). In intact cells, serine phosphorylation is stimulated by phorbol esters and cAMP analogs, suggesting that these may be the kinases physiologically involved in this process (Takayama et al., 1988; Stadtmauer and Rosen, 1986; Häring et al., 1986). An insulin receptor-associated serine kinase has also been observed in partially purified receptor preparations, which appears to be distinct from these kinases (Czech et al., 1988; Lewis et al., 1990b). Thus far, however, this insulin-stimulated receptor serine kinase (IRSK) has not been isolated. The physiological importance of serine/threonine phosphorylation is uncertain. Receptors isolated from phorbol ester-treated cells or receptors phosphorylated in vitro by protein kinase C exhibit decreased tyrosine

INSULIN RECEPTOR–SUBSTRATE 313

kinase activity and some decreased biological responses (Takayama *et al.,* 1988). Thus, it appears that the insulin receptor can be positively regulated by tyrosine phosphorylation and negatively regulated by serine phosphorylation, producing a highly controlled kinase activity.

## II.  Substrates of the Insulin Receptor

Transmission of an insulin signal following activation of the receptor kinase could potentially involve several different pathways. In the most direct model of signal transmission, the activated kinase would catalyze the phosphorylation of an endogenous substrate or substrates (Fig. 9). Initial attempts to identify such substrates relied primarily on reconstitution experiments in which tyrosine phosphorylation of potential substrates was assessed by incubation of the partially purified receptor with candidate proteins *in vitro*. Under these conditions, the receptor will phosphorylate a large number of proteins, including histones, calmodulin, and lipocortins; however, the physiological significance of these events is impossible to ascertain (Rosen, 1987). Alternatively, attempts have been made to identify substrates by SDS–gel electrophoresis using insulin-stimulated [32]P-labeled cells and phosphoamino acid analysis. Using this approach, some relatively abundant proteins that have a low level of insulin-induced tyrosine phosphorylation may be identified; however, this approach is gener-

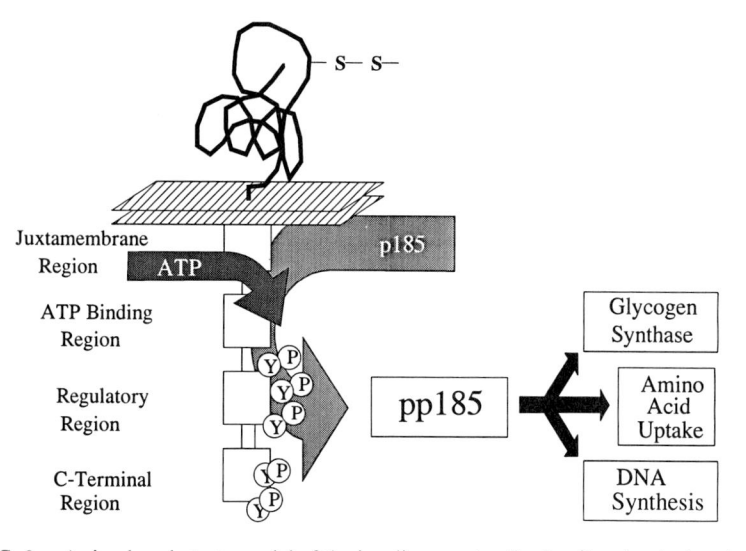

FIG. 9.   A simple substrate model of the insulin receptor β subunit and a single substrate leading to signaling in all pathways.

314                        C. RONALD KAHN ET AL.

ally both cumbersome and insensitive. A major step toward identifying the important physiological substrates of the insulin receptor was the development of antibodies to phosphotyrosine (Pang *et al.*, 1985). These antibodies can then be used to immunoprecipitate the P-Tyr-containing proteins of $^{32}$P-labeled cell extracts or immunoblot proteins from cell extracts before and after insulin stimulation. Because cells also contain many phosphotyrosine phosphatases, it is important to include a variety of inhibitors of these enzymes.

An example of a study using antiphosphotyrosine antibodies to study insulin-induced tyrosine phosphorylation is shown in Fig. 10. In this study, extracts of normal rat liver were made before and after insulin injection into the portal vein; the extracts were then subjected to SDS–gel electrophoresis, transferred to nitrocellulose paper, and immunoblotted with an antiphosphotyrosine antibody. In the absence of insulin stimulation, normal rat liver contains only two significant phosphotyrosine-containing proteins: one has an apparent molecular mass of 120 kDa and the other has a molecular mass of about 175 kDa. The 120-kDa protein (pp120) is observed as the major phosphotyrosine-containing protein of most cell

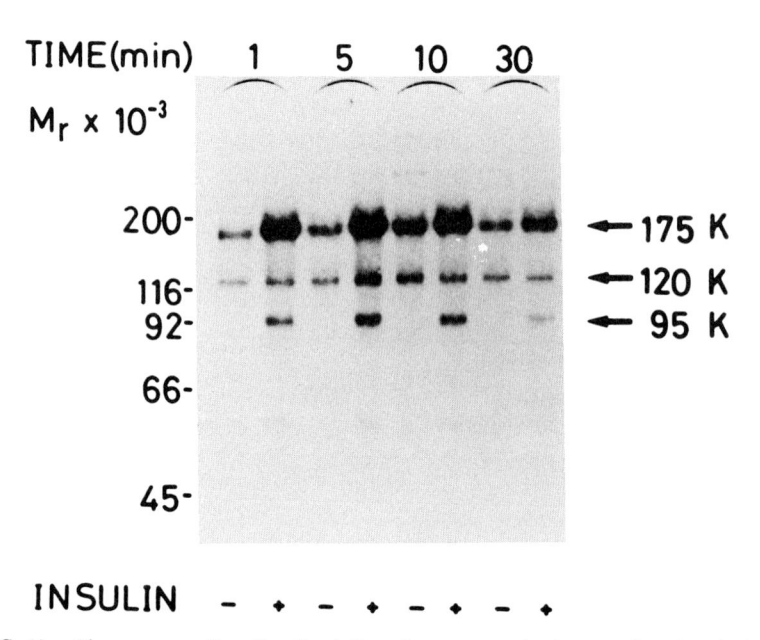

FIG. 10.   Time course of insulin stimulation of receptor and substrate phosphorylation in normal rat liver. Rats were injected in the portal vein and at various times their livers were extracted, SDS gels were performed, and the resulting blots were reacted with an antiphosphotyrosine antibody.

INSULIN RECEPTOR–SUBSTRATE                                    315

types. Recent work in our laboratory and that of Schaller *et al.* (1992) suggested that this protein corresponds to an endogenous tyrosine kinase associated with focal adhesion placques. The 175-kDa protein corresponds to the EGF receptor and can be precipitated and blotted with anti-EGF receptor antibody.

Following insulin stimulation, there is a rapid appearance of a two new phosphotyrosine-containing bands. One corresponds to the 95-kDa $\beta$ subunit of the insulin receptor; the other is a broad band between 165 and 185 kDa that overlaps the EGF receptor and is strongly insulin stimulated. However, based on a number of criteria, including antibody recognition, this higher molecular mass band is not the EGF receptor, but instead represents the major endogenous substrate of the insulin receptor. We initially termed this protein pp185 based on its electrophoretic mobility (White *et al.*, 1985); more recently we have renamed it insulin receptor substrate-1, or IRS-1, because its actual molecular mass is much less than the observed 185 kDa (Sun *et al.*, 1991).

In addition to pp185/IRS-1, adipocytes contain at least two other proteins, pp60 and pp46, which show some degree of insulin-stimulated tyrosine phosphorylation (Rothenberg *et al.*, 1991). Also, a protein of 15 kDa can be observed to undergo tyrosine phosphorylation in 3T3-L1 cells following insulin stimulation if the cells are first treated with phenylarsene oxide (Hresko *et al.*, 1988). This protein has been subsequently identified as the abundant fatty acid binding protein (AP2 or 422) (Bernier *et al.*, 1987). In liver, insulin can also be shown to stimulate phosphorylation of endogenous lipocortin 2, but this is best observed after induction of the protein by glucocorticoid administration (Karasik *et al.*, 1988). Interestingly, very few other proteins that undergo insulin-stimulated tyrosine phosphorylation are observed in cells or tissues using this technique (Rosen, 1987). Because pp185/IRS-1 is the major endogenous substrate in most cell types, it has received the most extensive characterization.

## A. INSULIN RECEPTOR SUBSTRATE-1/pp185

Some of the characteristics of pp185/IRS-1 are listed in Table III. As noted above, this protein is observed in virtually all tissues and cell types studied. This includes rat adipocytes (Monomura *et al.*, 1988; Mooney *et al.*, 1989; DelVecchio and Pilch, 1989), rat liver (Rothenberg *et al.*, 1991; Tobe *et al.*, 1990), cultured rat hepatoma cell lines (White *et al.*, 1985, 1987; Witters *et al.*, 1988; Tornqvist *et al.*, 1988; Tashiro-Hashimoto *et al.*, 1989), rat myoblasts (Beguinot *et al.*, 1989; Burdett *et al.*, 1990), mouse 3T3-L1 adipocytes (Gibbs *et al.*, 1986; Madoff *et al.*, 1988; Witters *et al.*, 1988), mouse fibroblasts (Pasquale *et al.*, 1988; Brindle *et al.*, 1990), mouse

TABLE III

*Characteristics of Insulin Receptor Substrate-1 (IRS-1/pp185)*

The major substrate of insulin in most cells
Is rapidly phosphorylated and dephosphorylated on tyrosine residues in response to insulin stimulation
Phosphorylation correlates with insulin action in cells expressing mutant insulin receptors
Coprecipitates with insulin-stimulated PI 3-kinase activity
Phosphorylated on serine and threonine in basal state and on tyrosine after insulin stimulation
Phosphorylation in regulated in physiologic and pathologic states

neuroblastoma cells (Shemer *et al.*, 1987), Chinese hamster ovary cells (White *et al.*, 1987; Chou *et al.*, 1987), human epidermoid carcinoma cells (Kadowaki *et al.*, 1987) and human adipocytes (Thies *et al.*, 1990). The kinetics of phosphorylation are rapid and reversible. In cultured cells pp185 phosphorylation is near maximal within 15–30 seconds after insulin stimulation (White *et al.*, 1985), and the time course is similar in intact tissues following intraportal injection of insulin (Rothenberg *et al.*, 1991) (Fig. 10). When insulin is removed from the cell, there is also a rapid dephosphorylation of pp185. Thus, the kinetics of response are consistent with the rapid kinetics required for insulin signaling of cells. On phospho-peptide and phosphoamino acid analysis, pp185 is multiply phosphorylated (White *et al.*, 1985; X.-J. Sun *et al.*, submitted for publication). In the basal state, most, if not all, of the phosphorylation is on serine and threonine; however, following insulin stimulation there is significate phosphorylation on tyrosine residues as well. Stimulation of tyrosine phosphorylation of IRS-1 (identified as pp160 in this tissue) is also observed in a differentiation-dependent manner in 3T3-L1 preadipocytes, consistent with their increasing insulin sensitivity (Keller *et al.*, 1991). IRS-1 phosphorylation is also observed in a number of cell types following incubation with IGF-1 (Beguinot *et al.*, 1989; Madoff *et al.*, 1988; Izumi *et al.*, 1987; Kadowaki *et al.*, 1987; Shemer *et al.*, 1987; Condorelli *et al.*, 1989). In many of these cases IGF-1 is probably acting through its own receptor, which is highly homologous to the insulin receptor. IRS-1 phosphorylation, however, is not observed following stimulation by other growth factors that work through other receptor tyrosine kinases, such as EGF or PDGF.

In addition to these characteristics, studies with cells expressing *in vitro* mutated insulin receptors suggested that pp185 was an important physiological substrate of the receptor. Thus, in cells expressing insulin receptors in which the ATP binding sites of the receptors or the regulatory tyrosines have been mutated, there is a decrease in pp185 phosphorylation that closely parallels the decrease in receptor autophosphorylation and induction of insulin action (Fig. 8) (Wilden *et al.*, 1992b).

INSULIN RECEPTOR–SUBSTRATE     317

The most important evidence that pp185 might be important in insulin action, however, came from studies of the insulin receptor that was mutated in the juxtamembrane region. As noted above, there are multiple tyrosine residues in the insulin receptor $\beta$ subunit that may undergo autophosphorylation, including at least two (Tyr-953 and Tyr-960) in the region just inside the membrane of the cell. In an effort to determine the role of these tyrosines in signal transduction, two mutant receptors were made, one in which Tyr-960 was converted to phenylalanine (IR-F960) and the other in which a 12-amino acid region, including Tyr-953 and Tyr-960 was deleted (IR-$\Delta$960). When these were expressed in cells, they were found to have normal kinase activity and to undergo insulin-stimulated tyrosine autophosphorylation in a normal or very near-normal manner (Fig. 11, left). Despite this, these two mutant receptors were unable to phosphorylate pp185 on tyrosine residues (Fig. 11, right) (White *et al.*, 1988b). This was not because of an absence of pp185/IRS-1 in these cells, because the protein could be observed on immunoblots using anti-IRS-1 antibodies. Most importantly, when insulin action was studied, these mutant insulin receptors were found to be inactive in supporting insulin-stimulated biological effects, including activation of glycogen synthase, stimulation of

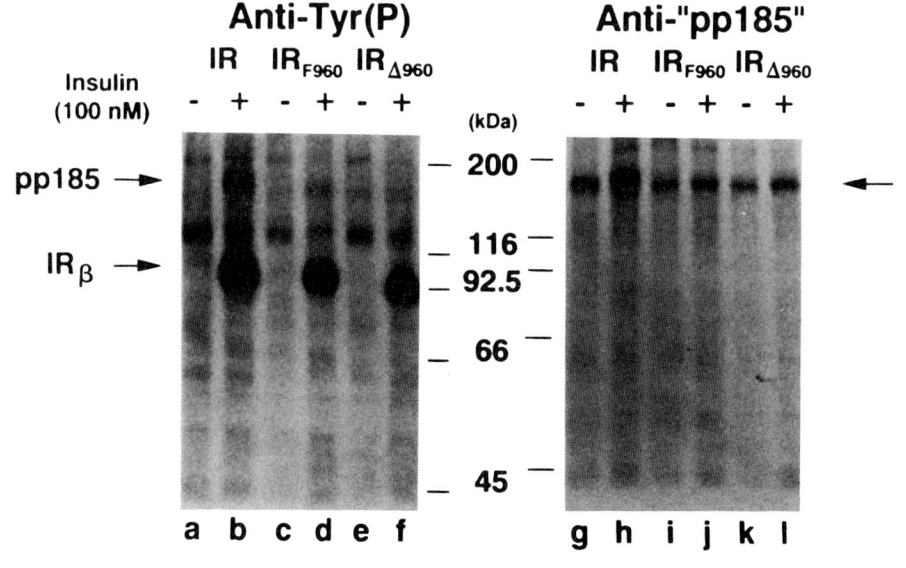

FIG. 11. Phosphorylation of the insulin receptor (pp185) and its substrate in CHO cells expressing normal and mutant insulin receptors in the juxtamembrane domain after $^{32}$P labeling. Proteins were immunoprecipitated with antiphosphotyrosine or anti-pp185 antibodies.

amino acid uptake, and thymidine incorporation into DNA (Fig. 12). Thus, receptors that are kinase active but cannot phosphorylate pp185 fail to transmit an insulin signal to more distal events. Hence, IRS-1/pp185 appears to play a critical role in the action of insulin.

Based on these findings, our laboratory undertook the task of purification of this protein. This was very difficult and challenging for several

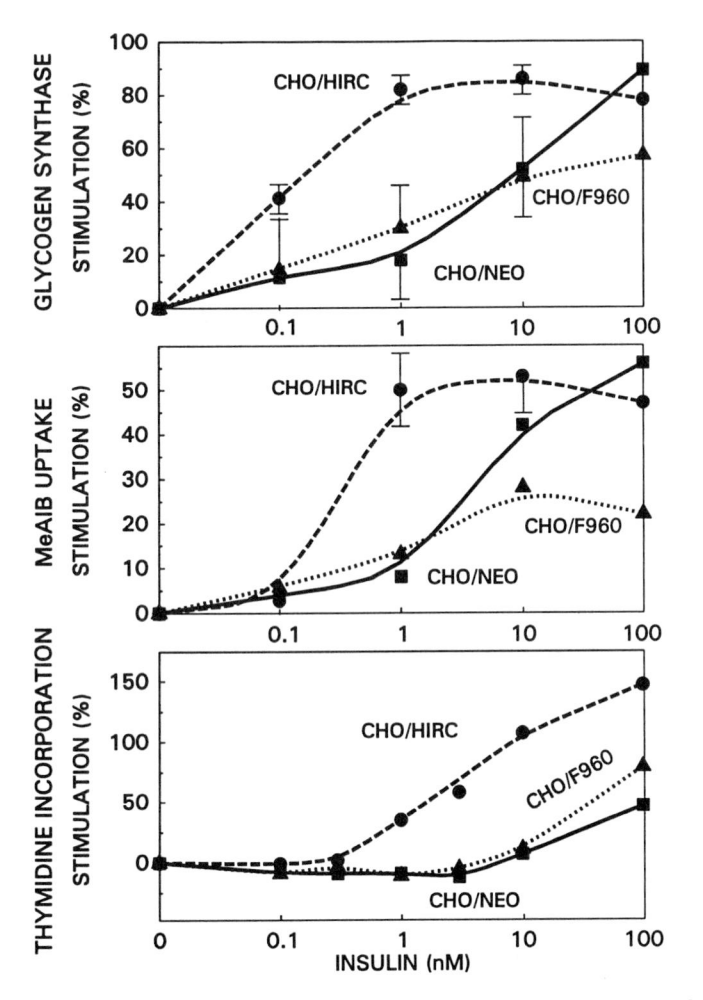

FIG. 12. Biological effects in cells expressing the normal insulin receptor (HIRC) or the Phe-960 (F960) mutant and control cells transfected with the neomycin (NEO) vector only. Top: Insulin stimulation of glycogen synthase. Middle: Insulin stimulation of (methyl-AIB) amino acid uptake. Bottom: Thymidine incorporation into DNA. Note that in every case the normal insulin receptor enhances insulin sensitivity whereas the F960 mutant does not.

INSULIN RECEPTOR–SUBSTRATE 319

reasons. First, pp185 is a very nonabundant protein in cells (about the same order of abundance as the insulin receptor), and no cell types were observed with very high levels of expression. Second, the only tools available initially for purification were antiphosphotyrosine antibodies, and pp185 was known to be very easily and rapidly tyrosine dephosphorylated even in the presence of high concentrations of inhibitors of PTPases. Finally, rat liver (the tissue chosen for purification), contains another protein of similar molecular weight and isoelectric point, carbamyl phosphate synthase (CPS), which is 100 to 1000 times more abundant than pp185. Thus, even using antiphosphotyrosine affinity columns to purify cell extracts, there was always significant contamination of the preparation with CPS.

To deal with these problems, a strategy was developed in which 50–100 rats were injected in the portal vein with insulin (a similar number of controls were injected with saline), the livers were excised at 1 minute and frozen in liquid nitrogen, the proteins were extracted in buffers containing high levels of PTPase inhibitors, the extracts were purified on anti-P-Tyr antibody columns and analyzed by SDS–gel electrophoresis, and the protein obtained in both the basal and insulin-stimulated states were subjected to tryptic peptide mapping by reverse-phase HPLC. Over 140 peptides were obtained from the tryptic digest of the pp185 band from the basal and stimulated samples, and 10 of these were new or increased following insulin injection (Rothenberg *et al.*, 1991). Each of these 10 was then subjected to microsequencing. In most cases, two amino acids were identified at each cycle. The more abundant one represented an amino acid from a peptide of CPS that contaminated the preparation; the less abundant amino acid represented a new peptide that was assumed to be from IRS-1. These amino acid sequences could then be used to make antipeptide antibodies and to predict the sequence of the mRNA, which should code IRS-1.

Using oligonucleotides based on this sequence information, two rat liver cDNA libraries were screened and ultimately sufficient clones were obtained to derive a full-length sequence for IRS-1 (Sun *et al.*, 1991). The complete IRS-1 cDNA was 5.4 kb in length and contained a single open reading frame of 3.4 kb. Northern blot analysis revealed that IRS-1 has a single dominant mRNA of ~9.5 kb. This coded for a hydrophilic protein of 1235 amino acids with a predicted molecular mass of 131 kDa. The difference between the predicted size of the protein and the apparent size on SDS gels is due to two factors. First, the protein exhibits anomalous migration. Following *in vitro* translation and SDS–gel electrophoresis, the apparent $M_r = 160,000$. In addition, the migration is further retarded due to phosphorylation on serine and threonine residues, as well as on tyrosine residues.This migration property explains the varied molecular weight estimates in studies from different laboratories. In view of the discrepancy between the true molecular mass and the apparent size on SDS gels, we now favor the name insulin receptor substrate-1, or IRS-1, for this protein. Recently we cloned the human muscle IRS-1 molecule (Araki *et al.*, submitted for publication) and find this protein to be very highly conserved (88% at the amino acid level) across species and tissues.

IRS-1 is shown schematically in Fig. 13 and appears to represent a new class of signal transduction molecules (Sun *et al.*, 1991). Consistent with its cytoplasmic location, IRS-1 is a hydrophilic molecule and contains no stretches of hydrophobic amino acids sufficiently long to transverse the membrane of the cell. Chou–Fasman analysis suggests that it has less than 5% helical content and is composed primarily of $\beta$ sheets, $\beta$ turns, and random-coil domains. Compared to a typical globular protein with 30–70% $\alpha$ helix, IRS-1 has much higher $\beta$-sheet content. Inspection of the sequence reveals a number of potential functionally important motifs. Near the N terminus of the molecule is a potential ATP (or GTP) binding site, beginning with a glycine-rich motif (Gly-X-Gly-X-X-Gly) and followed by an essential lysine residue 14 amino acids away in the sequence Ala-X-Lys. However, IRS-1 lacks the typical Asp-Phe-Gly and Ala-Pro-Glu motifs diagnostic of a protein kinase. There is also another potential nucleotide binding site near the C terminus in rat liver IRS-1, but this site is not fully conserved in the human muscle IRS-1 molecule, suggesting that it is less likely to be functionally important (Araki *et al.*, submitted).

IRS-1 contains many potential phosphorylation sites. Based on the typical motifs for cyclic AMP-dependent protein kinase, protein kinase C, casein kinase II, and cdc2 kinase, rat liver IRS-1 contains a total of 35 potential sites of serine or threonine phosphorylation distributed throughout the molecule (Fig. 13). Due to the slight differences in amino acid sequence, in human IRS-1 there are even more potential Ser/Thr phosphorylation sites (>50). Although it is not yet known which of these sites are phosphorylated in the intact cell, IRS-1 is highly serine and threonine phosphorylated, and this phosphorylation is increased following treatment of cells with phorbol esters, suggesting that IRS-1 is a substrate of protein kinase C (White *et al.*, 1985).

The most striking features of the molecule are the potential tyrosine phosphorylation sites, at least 10 of which can be identified based on the presence of tyrosine residues close to the acidic amino acids, glutamic or aspartic acid. Six of these are in a repetitive motif Tyr-Met-X-Met (YMXM), three have the sequence Tyr-X-X-Met (YXXM), and one has the sequence Glu-Tyr-Tyr-Glu (EYYE). For several of the YMXM sites there appears to be an even larger type of consensus motif in which this

INSULIN RECEPTOR–SUBSTRATE                                    321

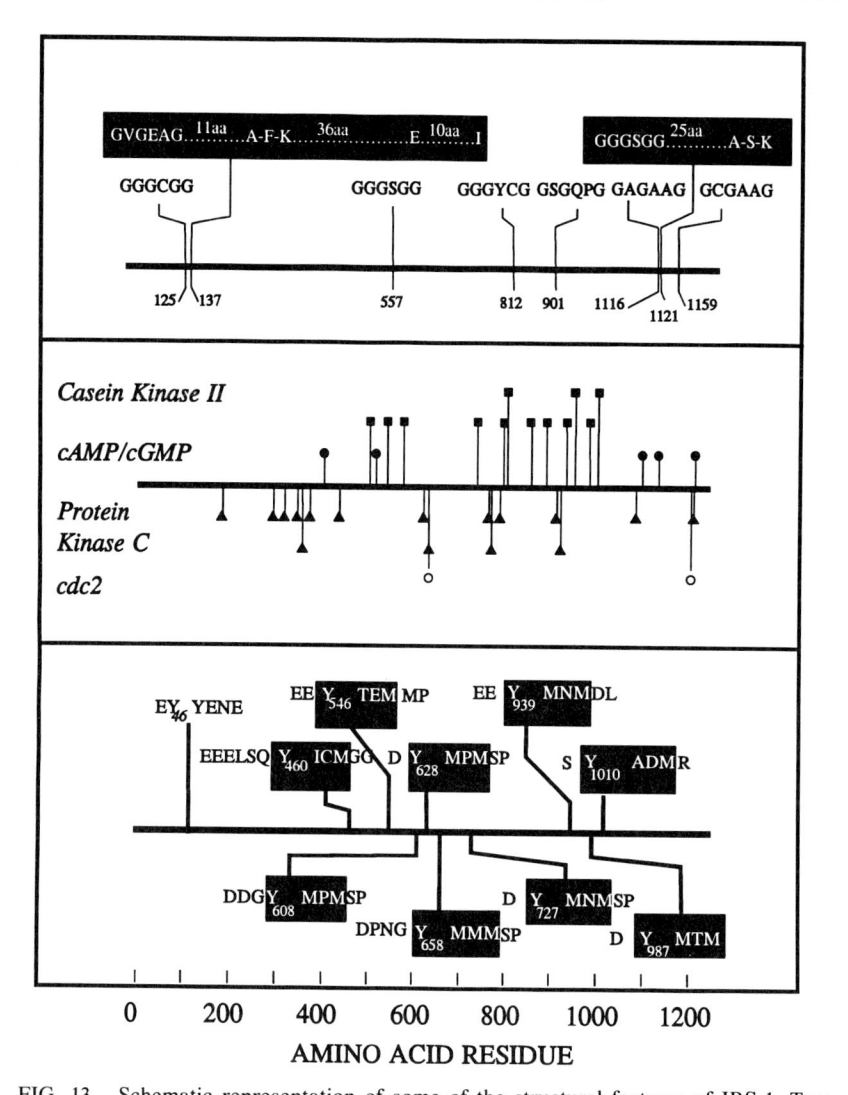

FIG. 13. Schematic representation of some of the structural features of IRS-1. Top: Potential nucleotide binding sites. Middle: Potential sites of serine or threonine phosphorylation. Bottom: Potential sites of tyrosine phosphorylation.

sequence is preceded by a Gly-Asp and followed by a Ser-Pro (SP). Thus, a longer overall consensus GDYMXMSP motif exists. According to the predicted secondary structure, all of the YMXM and YXXM motifs exist in a β sheet. Synthetic peptides containing these YMXM motifs are excellent high-affinity substrates of the insulin receptor kinase *in vitro*, with $K_m$

values in the low micromolar range (Shoelson *et al.*, 1992). Substitution of the methionine residues with other amino acids reduces the substrate efficiency of these peptides, suggesting that the YMXM motif is a key determinant for substrate recognition. The occurrence of this type of repetitive motif suggests some type of gene duplication event; however, a specific pattern consistent with this type of event (if this occurred) is not apparent at the nucleotide level. All 10 of these potential tyrosine phosphorylation sites are conserved exactly in the human IRS-1 molecule. Interestingly, several potential Ser/Thr phosphorylation sites exist very close to our overlapping YMXM or YXXM motifs, but the significance of this relationship is unknown at present.

Comparison of the IRS-1 sequence with sequences in the gene data banks reveals no similar proteins. The YMXM tyrosine phosphorylation motifs, however, are observed in several other tyrosine kinases and tyrosine kinase-related proteins, including two in the kinase intervening sequence of the PDGF receptor and one each in the EGF receptor, polyoma middle T protein, c-KIT, c-FMS, and several other oncogene tyrosine kinases (Table IV). In addition, several tyrosine kinases, including the insulin receptor itself, have a YXXM type sequence.

Recent studies have suggested that YMXM (and perhaps YXXM) phosphorylation motifs serve a special function in signal transduction by acting as intracellular recognition sites for other signaling proteins. In this model, when the YMXM motif is phosphorylated on the tyrosine (Y) residue, the

TABLE IV
*Proteins with YMXM Motifs*

| Protein | Tyr | Motif |
|---|---|---|
| IRS-1 | 608 | S N L H T D D G **YMP M** S P G V A P V |
| Middle T antigen | 298 | T Q A E R E N E **YMP M** A P Q I H L Y |
| PDGF-A Receptor | 731,742 | G D **YMD M** K N A D T T Q **Y VP M** L E |
| PDGF-B Receptor | 740,751 | G G **YMD M** S L D E S V D **Y VP M** L D |
| EGF Receptor | 920 | P P I C T I D V **YMI M** V K C W M I D |
| C-KIT | 718 | P S C D S S N E **YMD M** K P G V S Y V |
| C-FMS | 721 | F S S Q G V D T **Y VE M** R P V S T S S |
| cAMP-Dependent K | 118 | S F K D N S N L **YMVM** E Y V A G G E |
| Insulin Receptor | 1322 | R S Y E E H I P **YTHM** N G G K K N G |
| IRS-1 | 546 | S S V V S I E E **YTEM** M P A A Y P P |

affinity of this domain for a specific recognition domain in other signaling molecules is increased (Fig. 14). This $Y^{(P)}MXM$ recognition domain has been termed an SH2 domain because it was first recognized as a domain homologous to a region of the *src* oncogene product (Sugimoto *et al.*, 1984; Whitman *et al.*, 1985, 1987). Many oncogene tyrosine kinases and several normal intracellular enzymes contain SH2 domains (Table V). The SH2 domain contains enzymes, including a phosphatidylinositol 3-kinase (Carpenter *et al.*, 1990; Shibasaki *et al.*, 1991; Escobedo *et al.*, 1991; Skolnik *et al.*, 1991; Otsu *et al.*, 1991), phospholipase $C_\gamma$ (Margolis *et al.*, 1990), the G protein activating protein GAP (Anderson *et al.*, 1990; Kazlauskas *et al.*, 1990), and at least one phosphotyrosine phosphatase (Shen *et al.*, 1991). When the phosphorylated YMXM domain binds the SH2 domain of these enzymes, the enzymes may be regulated in their activity or changed in their subcellular localization. Thus, IRS-1 may be viewed as a "docking protein" with multiple YMXM motifs, several of which are phosphorylated and couple this protein to downstream enzymes involved in the insulin action cascade (Fig. 14).

There are two implications of this model. The first is that phosphorylation of IRS-1 results in an association of this protein with one or more molecules containing SH2 domains and this somehow changes their relative activity in the cell. Second, one might predict that it would be possible to mimic or block this interaction using isolated YMXM phosphopeptides. Thus far the only enzymatic activity definitely shown to associate with IRS-1 is the phosphatidylinositol 3-kinase (PI 3-kinase). Indeed, when insulin binds to its receptor and activates the receptor kinase, resulting in IRS-1 phosphorylation, there is an association of the regulatory (85 kDa)

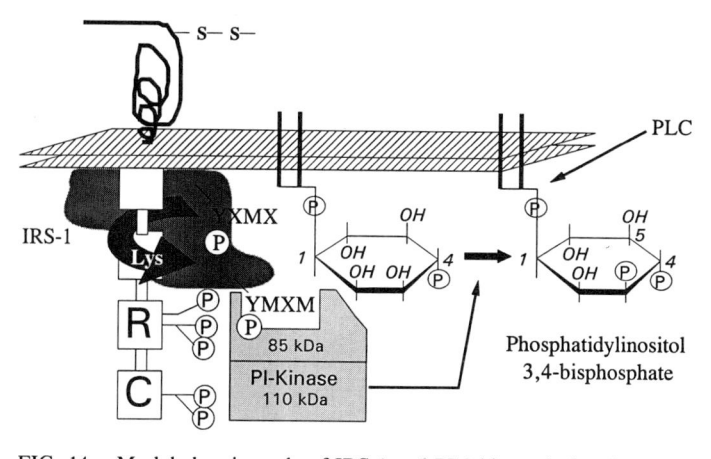

FIG. 14. Model showing role of IRS-1 and PI 3-kinase in insulin action.

TABLE V
Cellular Ligands and SH2 Domains

| SH2 domain ligands | SH2 domain proteins |
| --- | --- |
| | Tyrosine kinases |
| PDGF receptor | v-fps |
| EGF receptor | v-src |
| Polyoma middle-T antigen | v-crk |
| CSF-1 receptor | v-fgr |
| pp185/IRS-1 | v-abl |
| | v-yes |
| | $PLC_\gamma$ |
| | PI 3′-kinase |
| | G protein activating protein (GAP) |
| | Phosphotyrosine phosphatase(s) |

subunit of the PI 3-kinase with IRS-1, which can be detected by immunoblotting and a 5- to 10-fold stimulation of the IRS-1-associated PI 3-kinase activity (Fig. 15) (Ruderman et al., 1990; Backer et al., 1991, 1992; Kapeler et al., 1991). This results in phosphorylation of phosphatidylinositol (PI), PI 4-phosphate, and PI 4,5-bisphosphate on the 3-position to form PI-3-P, $PI-3,4P_2$, and $PI-3,4,5-P_3$ (Ruderman et al., 1990; Endemann et al., 1990).

A similar model has been derived for the interaction of other tyrosine kinases and their substrates with SH2-containing molecules (Anderson et al., 1990; Koch et al., 1991). The exact role of PI-3-P and related compounds in cellular metabolism is not yet known. However, PI 3-kinase is also activated by several other growth factor receptors (Cantley et al., 1991), and this activation can be observed in vivo at physiological concentrations of insulin (Folli et al., in press), suggesting that this reaction may play a role in the growth-promoting actions of these peptide hormones and growth factors. Cells expressing mutant insulin receptors that fail to phosphorylate IRS-1, such as the F960 and Δ960 mutants, do not activate the PI 3-kinase and do not transmit a downstream insulin signal (Kapeler et al., 1991; Backer et al., 1992).

It is still unclear exactly how tyrosine phosphorylation of IRS-1 is linked to other insulin actions. A theoretical model is shown in Fig. 16 (also see below). In addition to IRS-1 and PI 3-kinase, Maasen and colleagues have shown that insulin stimulates GTP loading of ras (Burgering et al., 1989, 1991), and GAP (which forms complexes with ras) is known to contain SH2 domains (Anderson et al., 1990; Kaplan et al., 1990; Kazlauskas et al., 1990). Thus far, however, we have been unable to find any association between the insulin receptor or IRS-1 and the ras–GAP complex (B.

INSULIN RECEPTOR–SUBSTRATE 325

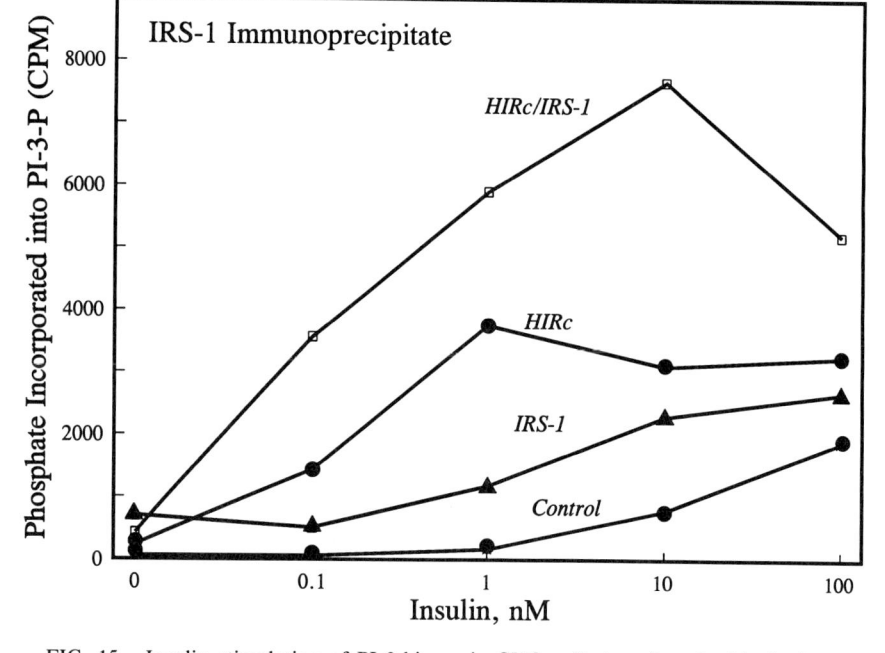

FIG. 15. Insulin stimulation of PI 3-kinase in CHO cells transfected with the human insulin receptor (HIRc), rat IRS-1 (IRS-1), both (HIRc/IRS-1), or neither (control). In each case the PI 3-kinase activity was measured in IRS-1 immunoprecipitates after insulin stimulation of the intact cells.

Cheatham *et al.*, unpublished data). Also there is no evidence of association of phospholipase $C_\gamma$ (Meisenhelder *et al.*, 1989) or a phosphotyrosine phosphatase (Shen *et al.*, 1991) with IRS-1, although thus far studies along this line have been relatively limited.

Because the insulin action cascade ultimately involves a number of serine/threonine phosphorylation events, it has been suggested that one of the molecules that might associate with IRS-1 via SH2 domains would be a serine/threonine kinase, which could act as a "switch kinase," converting the signal system from a tyrosine to serine phosphorylation cascade (Czech, 1989). Mitogen-activated protein (MAP) kinase is a candidate for such an intermediate because this enzyme appears to be central to the action of a number of growth factors and is activated by insulin (Ray and Sturgill, 1988; Rossomando *et al.*, 1989). MAP kinase requires both tyrosine and threonine phosphorylation for activity, a property expected for a switch kinase (Blenis, 1991). In the cytoplasm, MAP kinase has been shown to play a role in activation of the ribosomal S6 protein kinase (RSK)

326                    C. RONALD KAHN ET AL.

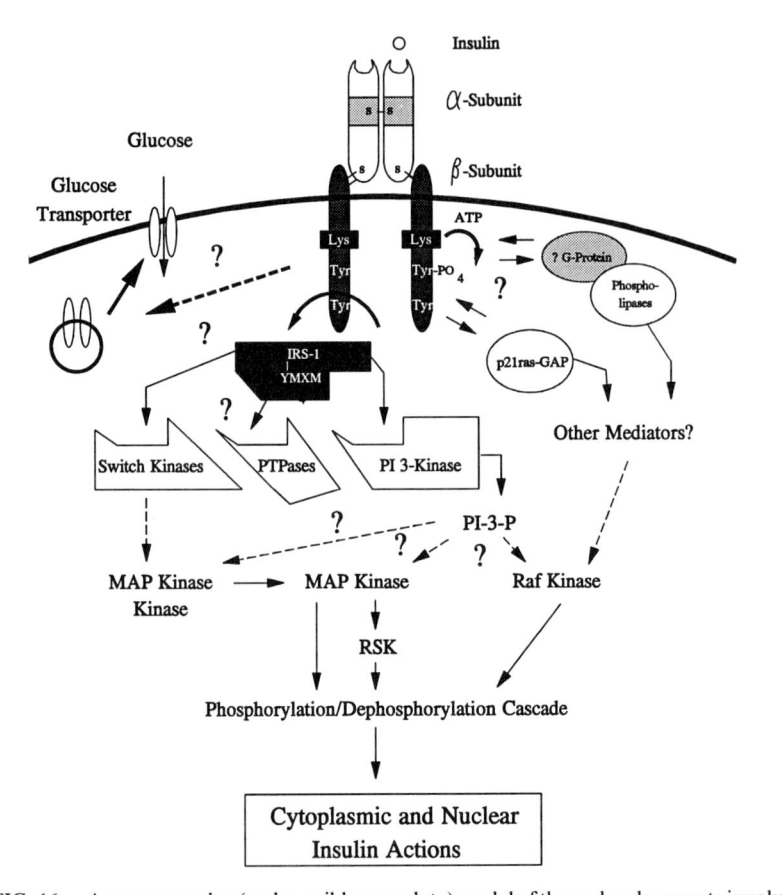

FIG. 16.   A more complex (and possibly complete) model of the molecular events involved in early steps of insulin action.

kinase, another component of the insulin action cascade (Rosen, 1987). Furthermore, MAP kinase can translocate between the cytoplasm and nucleus and could mediate some of the insulin actions at both levels. We have shown that insulin stimulation can affect protein serine phosphorylation at both the cytoplasmic and nuclear level (Csermely and Kahn, in press).

The *raf*-1 kinase, which has been shown to be capable of both tyrosine and serine/threonine phosphorylation and is a substrate for other tyrosine kinases, would also be candidate for this type of intermediate (Li *et al.*, 1991). Indeed, insulin has been shown to stimulate the phosphorylation of *raf* kinase; however, this is on serine residues and is not due to an IRS-1 or insulin receptor association (Kovacina *et al.*, 1990; Blackshear *et al.*, 1990). Thus, it seems likely that at least one additional step in the cascade

INSULIN RECEPTOR–SUBSTRATE 327

of early events in insulin action needs to be defined to complete the linkage between insulin-stimulated tyrosine and serine/threonine phosphorylations.

## B. IS TYROSINE PHOSPHORYLATION INVOLVED IN INSULIN ACTION?

Although the insulin receptor is clearly insulin-stimulated tyrosine kinase and considerable evidence suggests that the kinase activity is required for insulin action, the exact role of the kinase activity in insulin action continues to be debated. Evidence for the role of the receptor kinase activity in insulin action is substantial and includes the following points:

1. The insulin receptor is an insulin-stimulated tyrosine protein kinase, and thus far, no other activities have been directly associated with the receptor.

2. Overexpression of kinase-active insulin receptors in cells increases the sensitivity of the cell to all measurable insulin-stimulated actions, whereas overexpression of kinase-inactive insulin receptors has no effect (Ebina et al., 1987; Chou et al., 1987).

3. Both spontaneously occurring natural receptor mutants (Taylor et al., 1990b) and in vitro-produced receptor mutants with a defective ATP binding site or altered sites of autophosphorylation (Wilden et al., 1992a) are ineffective in transmitting a normal insulin signal when expressed in cells.

4. Injection of cells with antibodies to the insulin receptor β subunit (Morgan and Roth, 1987) or to phosphotyrosine inhibits the actions of insulin. Likewise, microinjection of Xenopus oocytes with a phosphotyrosine phosphatase will inhibit insulin action (Cicirelli et al., 1990).

There are several reported findings, however, that indicate that insulin action may involve more than simple kinase activation. First, although mechanistically unexplained, several investigators have reported insulin-stimulated responses in cells overexpressing kinase-inactive receptors (Gottschalk, 1991) or receptors in which all three regulatory tyrosines have been mutated (Sung et al., 1989; Rafaeloff et al., 1991). This includes insulin stimulation of pyruvate dehydrogenase, S6 kinase, and even amino acid uptake. What is confusing is that many of these same effects have been found to be defective when studied by other investigators using similar receptor constructs (Chou et al., 1987; Wilden et al., 1992b). Whether this indicates that similar insulin actions in different cells utilize different pathways or that some methodological differences account for the varied results has not yet been determined. Perhaps only a small amount of kinase activation is needed to imitate postreceptor events, and the differences lie in the sensitivity of the assays used to study the kinase activation.

A second finding has been used to suggest that kinase activity is not required for insulin action, i.e., that antiinsulin receptor antibodies are capable of mimicking insulin action without stimulating the receptor kinase (Forsayeth *et al.*, 1987) or can mimic insulin actions in cells expressing phosphorylation-defective receptors that fail to respond to insulin (Debant *et al.*, 1989). In our experience, as well as in the work of others, neither of these facts is uniformly reproducible. In many studies antiinsulin receptor antibodies do stimulate the insulin receptor kinase (Brindle *et al.*, 1990; Wilden *et al.*, 1992b; Steele-Perkins and Roth, 1990), although they are slightly less potent than insulin on a molar basis (Fig. 15). Also, we have failed to see any effect of antireceptor antibody to stimulate a classical insulin action, such as glycogen synthesis or DNA synthesis, in cells expressing mutant receptors that fail to respond to insulin (Wilden *et al.*, 1992a). The reasons for these discrepancies in the data have not been resolved.

## C.  ALTERNATIVE MECHANISMS OF INSULIN RECEPTOR SIGNALING

Although there is good evidence that insulin action involves stimulation of the receptor tyrosine kinase activity, several additional or alternative mechanisms of action in which the receptor may interact with other signaling molecules covalently or noncovalently may also participate in the full spectrum of insulin mediated events (Fig. 16). As noted above, there is good evidence that the phosphorylated receptor undergoes a conformational change and this change could result in noncovalent interactions of the receptor with other signaling molecules. Several studies have suggested a noncovalent interaction between the insulin receptor and some class of G proteins with insulin-modulating toxin-induced ADP ribosylation (Houslay *et al.*, 1989; Rothenberg and Kahn, 1988; Pyne *et al.*, 1989). Furthermore, toxins that catalyze ADP ribosylation and inactive G proteins block some of insulin's actions (Luttrell *et al.*, 1988; Ciaraldi and Maisel, 1989; Moises and Heidenreich, 1990).

In addition, several investigators have suggested that insulin action may stimulate a specific phospholipase C that is involved in the generation of one or more low-molecular-weight mediators of insulin action (Low and Saltiel, 1988; Saltiel, 1990). In this scheme, insulin stimulates a phosphatidylinositol glycan-specific phospholipase C (perhaps via a G protein),

INSULIN RECEPTOR–SUBSTRATE 329

which would hydrolyze a precursor in the membrane of the cell to produce a PI glycan and 1,2-diacylglycerol, both of which could have regulatory effects on intracellular enzymes. Such mediators have been proposed to play a role in insulin stimulation of pyruvate dehydrogenase, cAMP phosphodiesterase, adenylate cyclase, and phospholipid methyltransferase. Diacylglycerol, a potent activator of protein kinase C, has also been proposed to play a role as an insulin second messenger (Farese and Cooper, 1989). Although support for an important role for any of these alternative mechanisms in insulin action is limited, as noted above studies have suggested that cells expressing mutant insulin receptors with markedly reduced kinase activity may be active for signal transduction, at least along some insulin action pathways. Most likely, insulin signaling involves activation of several pathways. However, we are still rather far from a full understanding of this very complex scheme, and indeed still do not know even where in these action pathways the various components of insulin signaling diverge.

## III. Regulation of Early Steps of Insulin Action in Physiologic and Pathologic States

Very soon after methods became available for the study of the insulin receptor, it became clear that insulin receptors were regulated in a variety of physiologic and pathologic states. In fact, at least three previous presentations at The Laurentian Hormone Conference have dealt in some detail with these regulatory events (Roth et al., 1975; Kahn et al., 1981; Taylor et al., 1990a). These studies have indicated that the insulin receptor can be subjected to genetic mutations and regulated in its expression, turnover, affinity, and tyrosine kinase activity and that these changes may play a role in the pathophysiology of a number of disease states. These studies have also indicated that there can be differential regulation of the binding and kinase functions of the receptor. A detailed review of these studies is beyond the scope of this article, but is the subject of many excellent reviews, in addition to those listed above (Häring and Obermaier-Kusser, 1989; Kahn and Goldstein, 1989; Moller and Flier, 1991). Some of the findings of these studies have been very relevant to increasing our understanding of the molecular aspects of insulin by further elucidating some of the structure/function relationships within the receptor and the potential for regulation of function by both genetic and environmental variables. Recently we have begun to extend these studies to the IRS-1 molecule, which could also serve as a site for physiologic regulation of insulin action.

## A. REGULATION OF IRS-1 IN PATHOPHYSIOLOGIC STATES

The basic approach for studies of IRS-1 physiology and pathophysiology thus far is analysis of protein tyrosine phosphorylation in liver and muscle using antiphosphotyrosine and anti-IRS-1 antibodies after injection of insulin into the portal vein of mice or rats. This is similar to the technique used initially to identify IRS-1 in tissues of intact animals and allows an analysis of the function of both the receptor and its substrate in the *in situ* condition.

An example of such a study in fed and fasted rats is shown in Fig. 17. As in the previous studies, the major constitutive phosphotyrosine-containing protein in liver and muscle is the pp120 protein. Following insulin stimulation there is rapid stimulation of tyrosine phosphorylation

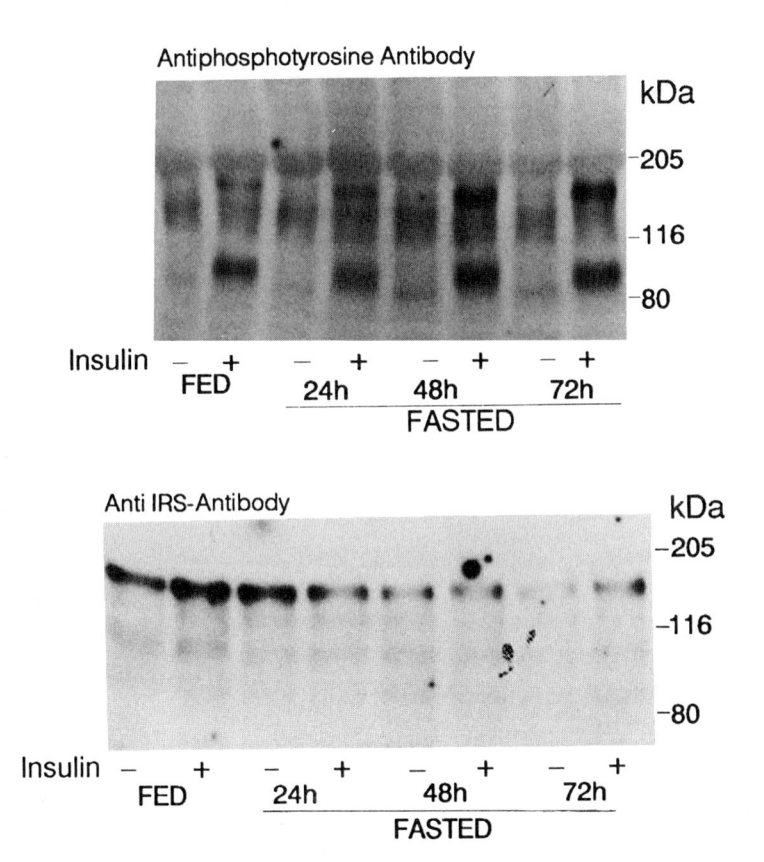

FIG. 17.   Insulin receptor and IRS-1 phosphorylation in muscle of fed and fasted rat. The basic technique was as given in Fig. 10; however, in this case skeletal muscle was extracted and immunoblotted with antiphosphotyrosine and anti-IRS-1 antibodies. Adapted from Saad *et al.* (submitted).

of the $\beta$ subunit of the insulin receptor (95 kDa) and the IRS-1 molecule (pp185). During fasting, there is a progressive increase in the insulin-stimulated phosphorylation of the insulin receptor. This is in part due to the increase in receptor number that occurs as insulin levels fall and receptors are "up-regulated." There is a parallel and even greater increase in the insulin-stimulated phosphorylation of IRS-1. In fact, we had made this observation even prior to beginning the purification of IRS-1, and as a result used fasted rats for the initial purification. Subsequently, when anti-IRS-1 antibodies became available, it became clear that although IRS-1 phosphorylation increased during fasting by almost twofold in liver, this was associated with a *decrease* in the actual amount of IRS-1 protein as determined by immunoblotting (Fig. 17). A similar increase in insulin receptor and IRS-1 phosphorylation is also observed in muscle following fasting, but in this tissue the levels of the IRS-1 protein do not change or may increase slightly.

We have also studied the regulation of IRS-1 in two forms of diabetes: the insulin-deficient diabetes of the streptozotocin (STZ)-treated rat and the insulin-resistant diabetes of the ob/ob mouse. In STZ-diabetic rats there is an increase in insulin receptor number and receptor autophosphorylation. The increase in receptor number again represents up-regulation, which occurs in this hypoinsulinemic state. Despite this increase in receptor number, receptor kinase activity decreases; however, overall there is an increase in receptor autophosphorylation, because the increase in receptor number outweighs the decrease in kinase activity. IRS-1 phosphorylation increases in parallel with the increased availability of the total active receptor kinase pool. As in fasting, IRS-1 protein levels decrease in liver and tend to increase in muscle.

The ob/ob mouse is an obese insulin-resistant rodent that is often used as an animal model of noninsulin-dependent diabetes in humans. In this animal there are very high circulating insulin levels, and down-regulation of the insulin receptor occurs in both liver and muscle. There is also a decrease in receptor kinase activity and overall receptor autophosphorylation is decreased about 50%. In contrast to the findings in fasting and STZ diabetes, in this case there is a significant decrease in IRS-1 phosphorylation, reflecting the overall decrease in receptor kinase activity. At the protein level, the changes are more complex with a decrease in IRS-1 in liver and an increase in IRS-1 in muscle. These changes are summarized in Table VI.

Several conclusions can be derived from these studies. First, IRS-1 phosphorylation is regulated and tends to parallel the changes in insulin receptor phosphorylation, indicating that the enzyme (i.e., the receptor) is rate limiting in the overall reaction in intact cells and tissues. Second,

## TABLE VI
### Regulation of Insulin Receptor and IRS-1 in Muscle and Liver in Physiologic and Pathologic States[a]

| | Insulin Receptor | | | | IRS-1/pp185 | | | |
| | Binding | | Phosphorylation | | Protein | | Phosphorylation | |
| | *Muscle* | *Liver* | *Muscle* | *Liver* | *Muscle* | *Liver* | *Muscle* | *Liver* |
|---|---|---|---|---|---|---|---|---|
| *Fasting* | ↑↑ | ↑↑ | ↑↑ | ↑ | ↓↓ | ↑ | ↑↑ | ↑↑ |
| *STZ Diabetes* | ↑↑ | ↑↑ | ↑ | ↑ | ↓ | ↑ | ↑↑ | ↑ |
| *ob/ob Mice* | ↓↓ | ↓↓ | ↓↓ | ↓ | → | ↓↓ | ↓↓ | ↓↓ |

[a] Regulation of the insulin receptor and IRS-1 expression and phosphorylation in liver and muscle in various physiologic and pathological states. Adapted from Saad *et al.* (submitted).

the level of IRS-1 protein is regulated, but there are different patterns of regulation in the liver and muscle. In the liver, IRS-1 regulation is similar to insulin receptor regulation, being increased in hypoinsulinemic states and decreased in hyperinsulinemic states. In muscle, the converse is observed with decreased IRS-1 in hypoinsulinemic states and increased IRS-1 in hyperinsulinemic states. Further studies will be needed to determine if and how these changes in IRS-1 protein and phosphorylation play a role in the overall physiology of these disorders.

## IV. Conclusions and Future Directions

Over the past 10 years, a great deal of progress has been made in defining many of the steps in insulin action at the cellular and molecular levels. In this review, we have focused our attention on the two earliest molecular events, binding and activation of the insulin receptor kinase and phosphorylation and subsequent signal transduction by the insulin receptor substrate IRS-1. Defining these events has helped add a whole new dimension to our understanding of insulin action. At the same time, studies defining the molecular events at the end of the insulin action cascade, such as activation of glucose transport and regulation of gene expression, have also begun to clarify the specific components required for these signaling events. Although a "black box" remains between the early and late events in insulin action, it is becoming smaller. With further study over the next several years we hope to fill in even more components of this important pathway and ultimately use this knowledge to elucidate the pathophysiology underlying diabetes and to develop new and better therapeutic approaches.

## INSULIN RECEPTOR–SUBSTRATE

### ACKNOWLEDGMENTS

The authors wish to thank Terri-Lyn Bellman for excellent secretarial assistance. This work was supported in part by National Institutes of Health Grants DK 31036 (to CRK) and DK 33201 (to CRK), the Marilyn M. Simpson Research Program in Diabetes, and the Mary K. Iacocca Professorship. EA is the Mary K. Iacocca Senior Research Fellow at the Joslin Diabetes Center.

### REFERENCES

Andersen, A. S., Kjeldsen, T., Wiberg, F. C., Christensen, P. M., Rasmussen, J. S., Norris, K., Moller, K. B., and Moller, N. P. H. (1990). *Biochemistry* **29**, 7363–7366.

Anderson, D., Koch, C. A., Grey, L., Elis, C., Moran, M. F., and Pawson, T. (1990). *Science* **982**, 979–982.

Araki, E., Sun, X. J., Hoag, III, B. L., Chuang, L. M., Zhang, Y., Yang-Feng, T. L., White, M. F., and Kahn, C. R. (1992). Submitted for publication.

Avruch, J. (1989). *In* "Insulin Action," pp. 65–77. A. R. Liss, New York.

Backer, J. M., Kahn, C. R., Cahill, D. A., Ullrich, A., and White, M. F. (1990). *J. Biol. Chem.* **265**, 16450–16454.

Backer, J. M., Schroeder, G. G., Cahill, D. A., Ullrich, A., Siddle, K., and White, M. F. (1991). *Biochemistry* **300**, 6366–6372.

Backer, J. M., Schroeder, G. G., Kahn, C. R., Myers, M. G., Wilden, P. A., Cahill, D. A., and White, M. F. (1992). *J. Biol. Chem.* **267**, 1367–1374.

Bargmann, C. L., Hung, M., and Weinberg, R. A. (1986). *Cell (Cambridge, Mass.)* **45**, 649–657.

Baron, V., Gautier, N., Komoriya, A., Hainaut, P., Scimeca, J.-C., Mervic, M., Lavielle, S., Dolais-Kitabgi, J., and Van Obberghen, E. (1990). *Biochemistry* **29**, 4634–4641.

Beguinot, F., Kahn, C. R., Moses, A. C., White, M. F., and Smith, R. J. (1989). *Endocrinology (Baltimore)* **125**, 1599–1605.

Benecke, H., Flier, J. S., and Moller, D. (1992). *J. Clin. Invest.* **89**, 2066–2070.

Bernier, M., Laird, D. M., and Lane, M. D. (1987). *Proc. Natl. Acad. Sci. U.S.A.* **84**, 1844–1848.

Blackshear, P. J., Haupt, D. M., App, A., and Rapp, U. R. (1990). *J. Biol. Chem.* **265**, 12131–12134.

Blenis, J. (1991). *Cancer Cells* **2**, 445–448.

Bollage, G. E., Roth, R. A., Beaudoin, J., Mochly-Rosen, D., and Koshland, D. E. (1986). *Proc. Natl. Acad. Sci. U.S.A.* **83**, 5833–5824.

Boni-Schnetzler, M., Kaligian, A., DelVecchio, R., and Pilch, P. F. (1988). *J. Biol. Chem.* **263**, 6822–6828.

Boni-Schnetzler, M., Scott, W., Waugh, S. E., DiBella, E., and Pilch, P. F. (1987). *J. Biol. Chem.* **262**, 8395–8401.

Brindle, N. P. J., Tavare, J. M., Dickens, M., Whittaker, J., and Siddle, K. (1990). *Biochem. J.* **268**, 615–620.

Burdett, E., Mills, G. B., and Klip, A. (1990). *Am. J. Physiol.* **258**, C99–C108.

Burgering, B. M. T., Medema, R. H., Maasen, J. A., Wetering, M. L., Eb, A. J., McCormick, I., and Bos, J. L. (1991). *EMBO J.* **10**, 1103–1109.

Burgering, B. M. T., Snijders, A. J., Maasen, A., Eb, A., and Bos, J. L. (1989). *Mol. Cell. Biol.* **9**, 4312–4322.

Cantley, L. C., Auger, K. R., Carpenter, C., Duckworth, B., Kapeller, R., and Soltoff, S. (1991). *Cell (Cambridge, Mass.)* **64**, 281–302.

334 C. RONALD KAHN ET AL.

Caro, J. F., Ittoop, O., Pories, W. J., Meelheim, D., Flickinger, E. G., Thomas, F., Jenquin, M., Silverman, J. F., Khazanie, P. G., and Sinha, M. D. (1986). *J. Clin. Invest.* **78**, 249–258.

Caro, J. F., Sinha, M. K., Raju, S. M., Ittoop, D., Pories, W. J., Flickinger, E. G., Meelheim, D., and Dohm, G. L. (1987). *J. Clin. Invest.* **79**, 1330–1337.

Carpenter, C. L., Duckworth, B. C., Auger, K. R., Cohen, B., Schaffhausen, B. S., and Cantley, L. C. (1990). *J. Biol. Chem.* **265**, 19704–19711.

Cheatham, B., and Kahn, C. R. (1992). *J. Biol. Chem.* **267**, 7108–7115.

Cheatham, B., Shoelson, S. E., Yamada, K., Goncalves, E., and Kahn, C. R. (1992). Submitted for publication.

Chen, W.-J., Goldstein, J. L., and Brown, M. S. (1990). *J. Biol. Chem.* **265**, 3116–3123.

Chou, C. K., Dull, T. J., Russell, D. S., Gherzi, R., Lebwohl, D., Ullrich, A., and Rosen, O. M. (1987). *J. Biol. Chem.* **262**, 1842–1847.

Ciaraldi, T. P., and Maisel, A. (1989). *Biochem. J.* **264**, 389–396.

Cicirelli, M. F., Tonks, N. K., Diltz, C. D., Weiel, J. E., Fischer, E. H., and Krebs, E. G. (1990). *Proc. Natl. Acad. Sci. U.S.A.* **87**, 5514–5518.

Cobb, M. H., Sang, B.-C., Gonzalez, R., Goldsmith, E., and Ellis, E. (1989). *J. Biol. Chem.* **264**, 18701–18706.

Condorelli, G., Formisand, P., Villone, G., Smith, R. J., and Beguinot, F. (1989). *J. Biol. Chem.* **264**, 12633–12638.

Crettaz, M., Jialal, I., Kasuga, M., and Kahn, C. R. (1984). *J. Biol. Chem.* **259**, 11543–11549.

Csermely, P., and Kahn, C. R. (1992). *Biochemistry,* in press.

Czech, M. P. (1985). *Annu. Rev. Physiol.* **47**, 357–381.

Czech, M. P. (1989). *Cell (Cambridge, Mass.)* **59**, 235–238.

Czech, M. P., Klarlund, J. K., Yagaloff, K. A., Bradford, A. P., and Lewis, R. E. (1988). *J. Biol. Chem.* **263**, 11017–11020.

Debant, A., Ponzio, G., Clauser, E., Contreras, J. O., and Rossi, B. (1989). *Biochemistry* **28**, 14–17.

DeFronzo, R. A. (1988). *Diabetes* **37**, 667–687.

DelVecchio, R. L., and Pilch, P. F. (1989). *Biochim. Biophys. Acta* **986**, 41–46.

DeMeyts, P., Bianco, A. R., and Roth, J. (1976). *J. Biol. Chem.* **251**, 1877–1888.

DeMeyts, P., Gu, J.-L., Shymko, R. M., Kaplan, B. E., Bell, G. I., and Whittaker, J. (1990). *Mol. Endocrinol.* **4**, 409–416.

Donner, D. B., and Yonkers, K. (1983). *J. Biol. Chem.* **258**, 9413–9418.

Ebina, Y., Araki, E., Taira, M., Shimada, F., Mori, M., Craik, C. S., Siddle, K., Pierce, S. B., Roth, R. A., and Rutter, W. J. (1987). *Proc. Natl. Acad. Sci. U.S.A.* **84**, 704–708.

Ebina, Y., Ellis, L., Jarnagin, K., Edery, M., Graf, L., Clauser, E., Ou, J.-H., Masiarz, F., Kan, Y. W., Goldfine, I. D., Roth, R. A., and Rutter, W. J. (1985). *Cell (Cambridge, Mass.)* **40**, 747–758.

Edge, A. S. B., Kahn, C. R., and Spiro, R. G. (1990). *Endocrinology (Baltimore)* **127**, 1887–1896.

Ellis, L., Levitan, A., Cobb, M. H., and Ramos, P. (1988). *J. Virol.* **62**, 1634–1639.

Endemann, G., Yonezawa, K., and Roth, R. A. (1990). *J. Biol. Chem.* **265**, 396–400.

Escobedo, J. A., Navankasattusas, S., Kavanaugh, W. M., Milfay, D., Fried, V. A., and Williams, L. T. (1991). *Cell (Cambridge, Mass.)* **65**.

Farese, R. V., and Cooper, D. R. (1989). *Diab. Metab. Rev.* **5**, 455–474.

Finn, F. M., Ridge, K. D., and Hofmann, K. (1990). *Proc. Natl. Acad. Sci. U.S.A.* **87**, 419–423.

Florke, R.-R., Klein, H. W., and Reinauer, H. (1990). *Eur. J. Biochem.* **191**, 473–482.

## INSULIN RECEPTOR–SUBSTRATE 335

Folli, F., Saad, M. J. A., Backer, J. M., and Kahn, C. R. (1992). *J. Biol. Chem.*, in press.
Forsayeth, J. R., Caro, J. F., Sinha, M. K., Maddux, B. A., and Goldfine, I. D. (1987). *Proc. Natl. Acad. Sci. U.S.A.* **84**, 3448–3451.
Frattali, A. L., Treadway, J. L., and Pessin, J. E. (1991). *J. Biol. Chem.* **266**, 9829–9834.
Gammeltoft, S. (1984). *Physiol. Rev.* **64**, 1322–1378.
Gibbs, E. M., Allard, W. J., and Lienhard, G. E. (1986). *J. Biol. Chem.* **261**, 16597–16603.
Ginsberg, B. H. (1977). *In* "Biochemical Actions of Hormones," pp. 313–349. Academic Press, New York.
Ginsberg, B. H., Kahn, C. R., Roth, J., and DeMeyts, P. (1976). *Biochem. Biophys. Res. Commun.* **73**, 1068–1074.
Goldstein, B. J., and Dudley, A. L. (1990). *Mol. Endocrinol.* **4**, 235–244.
Goldstein, B. J., and Kahn, C. R. (1989). *Biochem. Biophys. Res. Commun.* **159**, 664–669.
Goldstein, B. J., Muller-Wieland, D., and Kahn, C. R. (1987). *Mol. Endocrinol.* **1**, 759–766.
Goncalves, E., Yamada, K., Thatte, H. S., Backer, J. M., Golan, G. E., Kahn, C. R., and Shoelson, S. E. (1992). Submitted for publication.
Goren, H. J., White, M. F., and Kahn, C. R. (1987). *Biochemistry* **26**, 2374–2381.
Gottschalk, A. K. (1991). *J. Biol. Chem.* **266**, 8814–8819.
Gu, J. L., Goldfine, I. D., Forsayeth, J. R., and DeMeyts, P. (1988). *Biochem. Biophys. Res. Commun.* **150**, 694–701.
Gustafson, T. A., and Rutter, W. J. (1990). *J. Biol. Chem.* **265**, 18663–18667.
Häring, H., Kirsch, D., Obermaier, B., Ermel, B., and Machicao, F. (1986). *J. Biol. Chem.* **261**, 3869–3875.
Häring, H., and Obermaier-Kusser, B. (1989). *Diab. Metab. Rev.* **5**, 431–441.
Hedo, J. A., and Gorden, P. (1985). *Horm. Metab. Res.* **17**, 487–490.
Hedo, J., Kasuga, M., VanObberghan, E., Roth, J., and Kahn, C. R. (1981). *Proc. Natl. Acad. Sci. U.S.A.* **78**, 4791–4795.
Heidenreich, K. A., deVellis, G., and Gilmore, P. R. (1988). *J. Neurochem.* **51**, 878–887.
Herrera, R., Petruzzelli, L., Thomas, N., Bramson, H. N., Kaiser, E. T., and Rosen, O. M. (1985). *Proc. Natl. Acad. Sci. U.S.A.* **82**, 7899–7903.
Herrera, R., and Rosen, O. M. (1986). *J. Biol. Chem.* **261**, 11980–11985.
Houslay, M. D., Pyne, N. J., O'Brien, R. M., Siddle, K., Strassheim, D., Palmer, T., Spence, S., Woods, M., Wilson, A., Lavan, B., Murphy, G. J., Saville, M., McGregor, M., Kilgour, E., Anderson, N., Knowles, J. T., Griffiths, S., and Milligan, G. (1989). *Biochem. Soc. Trans.* **17**, 627–629.
Hresko, R. C., Bernier, M., Hoffman, R. D., Flores-Riveros, J. R., Liao, K., Laird, D. M., and Lane, M. D. (1988). *Proc. Natl. Acad. Sci. U.S.A.* **85**, 8835–8839.
Huertas, P., and Kahn, C. R. (1992). Unpublished observation.
Izumi, T., White, M. F., Kadowaki, T., Takaku, F., Akanuma, Y., and Kasuga, M. (1987). *J. Biol. Chem.* **262**, 1282–1287.
Kadowaki, T., Kadowaki, H., Cama, A., Marcus-Samuels, B., Rovira, A., Bevins, C. L., and Taylor, S. I. (1990). *J. Biol. Chem.* **265**, 21285–21296.
Kadowaki, T., Koyasu, S., Nishida, E., Tobe, K., Izumi, T., Takaku, F., Sakai, H., Yahara, I., and Kasuga, M. (1987). *J. Biol. Chem.* **262**, 7342–7350.
Kahn, C. R., Baird, K. L., Flier, J. S., Grunfeld, C., Harmon, J. T., Harrison, L. C., Karlsson, F. A., Kasuga, M., King, G. L., Lang, U. C., Podskalny, J. M., and Van Obberghen, E. (1981). *Recent Prog. Horm. Res.* **37**, 477–538.
Kahn, C. R., and Goldstein, B. J. (1989). *Science* **245**, 13.
Kapeler, R., Chen, K., Yoakim, M., Schaffhausen, B. S., Backer, J. M., White, M. F., Cantley, L. C., and Ruderman, N. B. (1991). *Mol. Endocrinol.* **5**, 769–777.

336     C. RONALD KAHN ET AL.

Kaplan, D. R., Morrison, D. K., Wong, G., McCormick, F., and Williams, L. T. (1990). *Cell* (*Cambridge, Mass.*) **61**, 125–133.

Karasik, A., Pepinsky, R. B., Shoelson, S. E., and Kahn, C. R. (1988). *J. Biol. Chem.* **263**, 11862–11867.

Kasuga, M., Hedo, J. A., Yamada, K. M., and Kahn, C. R. (1982a). *J. Biol. Chem.* **257**, 10392–10399.

Kasuga, M., Karlsson, F. A., and Kahn, C. R. (1982b). *Science* **215**, 185–186.

Kazlauskas, A., Ellis, C., Pawson, T., and Cooper, J. A. (1990). *Science* **247**, 1578–1581.

Keller, S. R., Kitagawa, K., Aebersold, R., Lienhard, G. E., and Garner, C. W. (1991). *J. Biol. Chem.* **266**, 12817–12820.

King, M. J., and Sale, G. J. (1990). *Biochem. J.* **266**, 251–259.

Kjeldsen, T., Andersen, A. S., Wiberg, F. C., Rasmussen, J. S., Schaffer, L., Balschmidt, P., Moller, K. B., and Moller, N. P. H. (1991). *Proc. Natl. Acad. Sci. U.S.A.* **88**, 4404–4408.

Koch, C. A., Anderson, D., Moran, M. F., Ellis, C., and Pawson, T. (1991). *Science* **252**, 668–674.

Kovacina, A., Yonezawa, K., Brautigan, D. L., Tonks, N. K., Rapp, U. R., and Roth, R. A. (1990). *J. Biol. Chem.* **265**, 12115–12118.

Lewis, R. E., Cao, L., Perregaux, D., and Czech, M. P. (1990a). *Biochemistry* **29**, 1807–1813.

Lewis, R. E., Wu, G. P., MacDonald, R. G., and Czech, M. P. (1990b). *J. Biol. Chem.* **265**, 947–954.

Li, P., Wood, K., Mamon, H., Haser, W., and Roberts, T. (1991). *Cell* (*Cambridge, Mass.*) **64**, 479–482.

Lipson, K. E., Yamada, K., Kolhatkar, A. A., and Donner, D. B. (1986). *J. Biol. Chem.* **261**, 10833–10838.

Low, M. G., and Saltiel, A. R. (1988). *Science* **239**, 268–275.

Luttrell, L. M., Hewlett, E. L., Romero, G., and Rogol, A. D. (1988). *J. Biol. Chem.* **263**, 6134–6141.

Madoff, D. H., Martensen, T. M., and Lane, M. D. (1988). *Biochem. J.* **252**, 7–15.

Maegawa, H., McClain, D. A., Friedenberg, G., Olefsky, J. M., Napier, M., Lipari, T., Dull, T. J., Lee, J., and Ullrich, A. (1988). *J. Biol. Chem.* **263**, 8912–8917.

Margolis, B., Li, N., Koch, A., Mohammadi, M., Hurwitz, D. R., Zilberstein, A., Ullrich, A., Pawson, T., and Schlessinger, J. (1990). *EMBO J.* **9**, 4375–4380.

Massague, J. P., Pilch, P. F., and Czech, M. P. (1981). *Proc. Natl. Acad. Sci. U.S.A.* **77**, 7137–7141.

McClain, D. A. (1991). *Mol. Endocrinol.* **5**, 734–739.

McClain, D. A., Maegawa, H., Lee, J., Dull, T. J., Ullrich, A., and Olefsky, J. M. (1987). *J. Biol. Chem.* **262**, 14663–14671.

Meisenhelder, J., Suh, P. G., Rhee, S. G., and Hunter, T. (1989). *Cell* (*Cambridge, Mass.*) **57**, 1109–1122.

Moises, R. A., and Heidenreich, K. (1990). *J. Cell. Physiol.* **144**, 538–545.

Moller, D. E., and Flier J. S. (1991). *N. Engl. J. Med.* **325**, 938–948.

Moller, D. E., Yokota, A., Caro, J. F., and Flier, J. S. (1989). *Mol. Endocrinol.* **3**, 1263–1269.

Moller, D. E., Yokota, A., Ginsberg-Fellner, F., and Flier, J. S. (1990a). *Mol. Endocrinol.* **4**, 1183–1191.

Moller, D. E., Yokota, A., White, M. F., Pazianos, A. G., and Flier, J. S. (1990b). *J. Biol. Chem.* **265**, 14979–14985.

Monomura, K., Tobe, K., Seyama, Y., Takaku, F., and Kasuga, M. (1988). *Biochem. Biophys. Res. Commun.* **155**, 1181–1186.

Mooney, R. A., Bordwell, K. L., Lukowskyj, S., and Casnelli, J. E. (1989). *Endocrinology* (*Baltimore*) **124**, 422–429.

INSULIN RECEPTOR–SUBSTRATE 337

Morgan, D. O., and Roth, R. A. (1987). *Proc. Natl. Acad. Sci. U.S.A.* **84**, 41–45.

Mosthaf, L., Grako, K., Dull, T. J., Coussens, L., Ullrich, A., and McClain, D. A. (1990). *EMBO J.* **9**, 2409–2413.

Mosthaf, L., Vogt, B., Häring, H. U., and Ullrich, A. (1991). *Proc. Natl. Acad. Sci. U.S.A.* **88**, 4728–4730.

Moxham, C. P., Duronio, V., and Jacobs, S. (1989). *J. Biol. Chem.* **264**, 13238–13244.

Muller-Wieland, D., Taub, R., Tewari, D. S., Kriauciunas, K. M., Reddy, S. S. K., and Kahn, C. R. (1989). *Diabetes* **38**, 31–38.

Myers, M. G., Backer, J. M., Siddle, K., and White, M. F. (1991). *J. Biol. Chem.* **266**, 10616–10623.

Olson, T. J., Bamberger, M. J., and Lane, M. D. (1988). *J. Biol. Chem.* **263**, 7342–7351.

Otsu, M., Hiles, I., Gout, I., Fry, M. J., Ruis-Larrea, F., Panayotou, G., Thompson, A., Dhand, R., Hsuan, J., Totty, N., Smith, A. D., Morgan, S. J., Courtneidge, S. A., Parker, P. J., and Waterfield, M. D. (1991). *Cell (Cambridge, Mass.)* **65**, 91–104.

Pang, D. T., and Shafer, J. A. (1984). *J. Biol. Chem.* **259**, 8589–8596.

Pang, D. T., Sharma, B. R., Shafer, J. A., White, M. F., and Kahn, C. R. (1985). *J. Biol. Chem.* **260**, 7131–7136.

Pasquale, E. B., Maher, P. A., and Singer, S. J. (1988). *J. Cell. Physiol.* **137**, 146–156.

Paul, J. I., Tavare, J. M., Denton, R. M., and Steiner, D. F. (1990). *J. Biol. Chem.* **265**, 13074–13083.

Perlman, R., Bottaro, D. P., White, M. F., and Kahn, C. R. (1989). *J. Biol. Chem.* **264**, 8946–8950.

Petruzzelli, L., Herrera, R., and Rosen, O. M. (1984). *Proc. Natl. Acad. Sci. U.S.A.* **81**, 3327–3331.

Prigent, S. A., Stanley, K. K., and Siddle, K. (1990). *J. Biol. Chem.* **265**, 9970–9977.

Pyne, N. J., Heyworth, C. M., Balfour, N., and Houslay, M. D. (1989). *Biochem. Biophys. Res. Commun.* **165**, 251–256.

Rafaeloff, R., Maddux, B. A., Brunetti, A., Sbaccia, P., Sung, C. K., Patel, R., Hawley, D. M., and Goldfine, I. D. (1991). *Biochem. Biophys. Res. Commun.* **179**, 912–918.

Rafaeloff, R., Patel, P., Yip, C., Goldfine, I. D., and Hawley, D. M. (1989). *J. Biol. Chem.* **264**, 15900–15904.

Ray, L. B., and Sturgill, T. W. (1988). *Proc. Natl. Acad. Sci. U.S.A.* **85**, 3753–3757.

Reaven, G. M. (1988). *Diabetes* **37**, 1595–1607.

Reddy, S. S. K., and Kahn, C. R. (1988). *Diab. Med.* **5**, 621–629.

Riedel, H., Dull, T. J., Schlessinger, J., and Ullrich, A. (1986). *Nature (London)* **324**, 68–70.

Rosen, O. M. (1987). *Science* **237**, 1452–1458.

Rossomando, A. J., Payne, D. M., Weber, M. J., and Sturgill, T. W. (1989). *Proc. Natl. Acad. Sci. U.S.A.* **86**, 6940–6943.

Roth, J., Kahn, C. R., Lesniak, M. A., Gorden, P., DeMeyts, P., Megyesi, K., Neville, D. M., Jr., Gavin, J. R., III, Soll, A. H., Freychet, P., Goldfine, I. D., Bar, R. S., and Archer, J. A. (1975). *Recent Prog. Horm. Res.* **31**, 95–139.

Rothenberg, P. L., and Kahn, C. R. (1988). *J. Biol. Chem.* **263**, 15546–15552.

Rothenberg, P. L., Lane, W. S., Karasik, A., Backer, J. M., White, M. F., and Kahn, C. R. (1991). *J. Biol. Chem.* **266**, 8302–8311.

Ruderman, N., Kapeller, R., White, M. F., and Cantley, L. C. (1990). *Proc. Natl. Acad. Sci. U.S.A.* **87**, 1411–1415.

Saltiel, A. R. (1990). *Diab. Care* **13**, 244–256.

Schaefer, E. M., Siddle, K., and Ellis, L. (1990). *J. Biol. Chem.* **265**, 13248–13253.

Schaller, M. D., Borgman, C. A., Cobb, B. S., Vines, R. R., Reynolds, A. B., and Parsons, J. T. (1992). *Proc. Natl. Acad. Sci. U.S.A.* **89**, 5192–5196.

Schenker, E., and Kohanski, R. A. (1988). *Biochem. Biophys. Res. Commun.* **157**, 140–145.

338    C. RONALD KAHN ET AL.

Seino, S., and Bell, G. I. (1989). *Biochem. Biophys. Res. Commun.* **159**, 312–316.

Seino, S., Seino, M., and Bell, G. I. (1990). *Diabetes* **39**, 129–133.

Seino, S., Seino, M., Nishi, S., and Bell, G. I. (1989). *Proc. Natl. Acad. Sci. U.S.A.* **86**, 114–118.

Shemer, J., Adamo, M., Wilson, G. L., Heffez, D., Zick, Y., and LeRoith, D. (1987). *J. Biol. Chem.* **262**, 15476–15482.

Shen, S.-H., Bastien, L., Posner, B. I., and Chretien, P. (1991). *Nature (London)* **352**, 736–739.

Shia, M. A., and Pilch, P. F. (1983). *Biochemistry* **22**, 717–721.

Shibasaki, F., Homma, Y., and Takenawa, T. (1991). *J. Biol. Chem.* **266**, 8108–8114.

Shoelson, S. E., Chatterjee, S., Chaudhuri, M., and White, M. F. (1992). *Proc. Natl. Acad. Sci. U.S.A.* **89**, 2027–2031.

Shoelson, S. E., White, M. F., and Kahn, C. R. (1988). *J. Biol. Chem.* **263**, 4852–4860.

Skolnik, E. Y., Margolis, B., Mohammadi, M., Lowenstein, E., Fischer, R., Drepps, A., Ullrich, A., and Schlessinger, J. (1991). *Cell (Cambridge, Mass.)* **65**, 83–90.

Stadtmauer, L., and Rosen, O. M. (1986). *J. Biol. Chem.* **261**, 3402–3407.

Steele-Perkins, G., and Roth, R. A. (1990). *J. Biol. Chem.* **265**, 9458–9463.

Sugimoto, Y., Whitman, M., Cantley, L. C., and Erikson, R. L. (1984). *Proc. Natl. Acad. Sci. U.S.A.* **81**, 2117–2121.

Sun, X.-J., Rothenberg, P., Kahn, C. R., Backer, J. M., Araki, E., Wilden, P. A., Cahill, D. A., Goldstein, B. J., and White, M. F. (1991). *Nature (London)* **352**, 73–77.

Sung, C. K., Maddux, B. A., Hawley, D. M., and Goldfine, I. D. (1989). *J. Biol. Chem.* **264**, 18951–18959.

Sweet, L. J., Morrison, B. D., and Pessin, J. E. (1987). *J. Biol. Chem.* **262**, 6939–6942.

Takayama, S., White, M. F., and Kahn, C. R. (1988). *J. Biol. Chem.* **263**, 3440–3447.

Tanti, J. F., Gremeaux, T., Rochet, N., Van Obberghen, E., and LeMarchand-Brustel, Y. (1987). *Biochem. J.* **245**, 19–26.

Tashiro-Hashimoto, Y., Tobe, K., Kashio, O., Izumi, T., Takaku, F., Akanuma, Y., and Kasuga, M. (1989). *J. Biol. Chem.* **264**, 6879–6885.

Tavare, J. M., O'Brien, R. M., Siddle, K., and Denton, R. M. (1988). *Biochem. J.* **253**, 783–788.

Taylor, S. I., Kadowaki, T., Accili, D., Cama, A., Kadowaki, H., McDeon, C., Moncada, V., Marcus-Samuels, B., Bevins, C., Ojamaa, K., *et al.* (1990a). *Recent Prog. Horm. Res.* **46**, 185–213.

Taylor, S. I., Kadowaki, T., Kadowaki, H., Accili, D., Cama, A., and McKeon, C. (1990b). *Diab. Care* **13**, 257–279.

Thies, R. S., Ullrich, A., and McClain, D. A. (1989). *J. Biol. Chem.* **264**, 12820–12825.

Thies, R. S., Webster, N. J., and McClain, D. A. (1990). *J. Biol. Chem.* **265**, 10132–10137.

Tobe, K., Koshio, O., Tashiro-Hashimoto, Y., Takaku, F., Adanuma, Y., and Kasuga, M. (1990). *Diabetes* **39**, 528–533.

Tornqvist, H. E., and Avruch, J. (1988). *J. Biol. Chem.* **263**, 4593–4601.

Tornqvist, H. E., Gunsalus, J. R., Nemenoff, R. A., Frackelton, A. R., Pierce, M. W., and Avruch, J. (1988). *J. Biol. Chem.* **263**, 350–359.

Treadway, J. L., Morrison, B. D., Goldfine, I. D., and Pessin, J. E. (1989). *J. Biol. Chem.* **264**, 21450–21453.

Treadway, J. L., Morrison, B. D., Soos, M. A., Siddle, K., Olefsky, J., Ullrich, A., McClain, D. A., Pessin, J. E. (1991). *Proc. Natl. Acad. Sci. U.S.A.* **88**, 214–218.

Ullrich, A., Bell, J. R., Chen, E. Y., Herrera, R., Petruzzelli, L. M., Dull, T. J., Gray, A., Coussens, L., Liao, Y. C., Tsubokawa, M., Mason, A., Sccburg, P. H., Grunfeld, C., Rosen, O. M., and Ramachandran, J. (1985). *Nature (London)* **313**, 756–761.

Villalba, M., Wente, S., Russell, D. S., Ahn, J., Reichelderfer, C. F., and Rosen, O. M. (1989). *Proc. Natl. Acad. Sci. U.S.A.* **86**, 7848–7852.

## INSULIN RECEPTOR–SUBSTRATE    339

Vogt, B., Carrascosa, J. M., Ermel, B., Ullrich, A., and Häring, H. U. (1991). *Biochem. Biophys. Res. Commun.* **177**, 1013–1018.

Waugh, S. M., DiBella, E., and Pilch, P. F. (1989). *Biochemistry* **28**, 3448–3455.

Waugh, S. M., and Pilch, P. F. (1989). *Biochemistry* **28**, 2722–2727.

Wedekind, F., Baer-Pontzen, K., Bala-Mohan, S., Choli, D., Zahn, H., and Brandenburg, D. (1989). *Biol. Chem. Hoppe-Seyler* **370**, 251–258.

Weiner, D. B., Liu, J., Cohen, J. A., Williams, W. V., and Green, M. I. (1989). *Nature (London)* **339**, 230–231.

Wells, J. A., Cunningham, B. C., Fuh, G., Lowman, H. B., Bass, S. H., Mulkerrin, M. G., Ultsch, M., and deVos, A. (1993). *Recent Progress Horm. Res.* **48**, 253–275.

White, M. F., Häring, H., Kasuga, M., and Kahn, C. R. (1984). *J. Biol. Chem.* **259**, 255–264.

White, M. F., Maron, R., and Kahn, C. R. (1985). *Nature (London)* **318**, 183–186.

White, M. F., and Kahn, C. R. (1986). *In* "The Enzymes" (P. Boyer and E. Krebs, eds.), pp. 247–310. Academic Press, New York.

White, M. F., Stegmann, E. W., Dull, T. J., Ullrich, A., and Kahn, C. R. (1987). *J. Biol. Chem.* **262**, 9769–9777.

White, M. F., Livingston, J. N., Backer, J. M., Lauris, V., Dull, T. J., Ullrich, A., and Kahn, C. R. (1988a). *Cell (Cambridge, Mass.)* **54**, 641–649.

White, M. F., Shoelson, S. E., Keutmann, H., and Kahn, C. R. (1988b). *J. Biol. Chem.* **263**, 2969–2980.

Whitman, M., Kaplan, D., Roberts, T., and Cantley, L. (1987). *Biochem. J.* **247**, 165–174.

Whitman, M., Kaplan, D. R., Schaffhausen, B., Cantley, L., and Roberts, T. M. (1985). *Nature (London)* **315**, 239–242.

Whittaker, J., Soos, M. A., and Siddle, K. (1990). *Diab. Care* **13**, 576–581.

Wilden, P. A., and Pessin, J. E. (1987). *Biochem. J.* **245**, 325–331.

Wilden, P. A., Backer, J. M., Kahn, C. R., Cahill, D. A., Schroeder, G., and White, M. F. (1990). *Proc. Natl. Acad. Sci. U.S.A.* **87**, 3358–3362.

Wilden, P. A., Siddle, K., Häring, E., Backer, J. M., White, M. F., and Kahn, C. R. (1992a). *J. Biol. Chem.* **267**, 13,719–13,727.

Wilden, P. A., Kahn, C. R., Siddle, K., and White, M. F. (1992b). *J. Biol. Chem.*, **267**, 16,660–16,668.

Williams, J. F., McClain, D. A., Dull, T. J., Ullrich, A., and Olefsky, J. M. (1990). *J. Biol. Chem.* **265**, 8463–8469.

Witters, L. A., Watts, T. D., Gould, G. W., Lienhard, G. E., and Gibbs, E. M. (1988). *Biochem. Biophys. Res. Commun.* **153**, 992–998.

Xu, Q.-Y., Paxton, R. J., and Fujita-Yamaguchi, Y. (1990). *J. Biol. Chem.* **265**, 18673–18681.

Yamada, K., Goncalves, E., Kahn, C. R., and Shoelson, S. E. (1992). *J. Biol. Chem.* **267**, 12,452–12,461.

Yamaguchi, Y., Flier, J. S., Yokota, A., Benecke, H., Backer, J. M., and Moller, D. E. (1991). *Endocrinology (Baltimore)* **4**, 2058–2066.

Yamamoto-Honda, R., Koshio, O., Tobe, K., Shibasaki, Y., Momomura, K., Odawara, M., Kadowaki, T., Takaku, F., Adanuma, Y., and Kasuga, M. (1990). *J. Biol. Chem.* **265**, 14777–14783.

Yang-Feng, T. L., Francke, U., and Ullrich, A. (1985). *Science* **228**, 728–731.

Yarden, Y., and Ullrich, A. (1988). *Annu. Rev. Biochem.* **57**, 443–478.

Yip, C. C., Hsu, H., Patel, R. G., Hawley, D. M., Maddux, B. A., and Goldfine, I. D. (1988). *Biochem. Biophys. Res. Commun.* **157**, 321–329.

Yip, C. (1992). *J. Cell. Biochem.* **48**, 19–25.

Zhang, B., Tavare, J. M., Ellis, L., and Roth, R. A. (1991). *J. Biol. Chem.* **266**, 990–996.

Zoppini, G., and Kahn, C. R. (1992). *Diabetologia* **35**, 109–115.

RECENT PROGRESS IN HORMONE RESEARCH, VOL. 48

# Thyrotropin-Releasing Hormone Receptor: Cloning and Regulation of Its Expression

## MARVIN C. GERSHENGORN

*Division of Molecular Medicine, Department of Medicine, Cornell University Medical College and The New York Hospital, New York, New York 10021*

## I. Introduction

Down-regulation of receptors is an important mechanism for diminishing cell responsiveness to many hormones, neurotransmitters, and growth factors. The number of receptors on the surface membrane of cells may be decreased by the action of the receptor ligand (homologous down-regulation) or by the action of other extracellular regulatory factors that interact with other cell receptors (heterologous down-regulation). Down-regulation has been shown to be affected at several steps in the synthesis and turnover of receptors (Collins *et al.*, 1991; Hadcock and Malbon, 1991), including posttranslational processes, such as internalization and subsequent intracellular degradation (Chen *et al.*, 1990; Valiquette *et al.*, 1990), and pretranslational processes, such as inhibition of receptor synthesis secondary to decreases in the level of receptor mRNA (Hadcock and Malbon, 1988; Freissmuth *et al.*, 1989; Burnstein *et al.*, 1990; Akamizu *et al.*, 1990; Izzo *et al.*, 1990; Birnbaumer, 1990). Studies of regulation of mRNA levels in eukaryotic cells have primarily focused on control of gene transcription, that is, mRNA synthesis. During the past few years, however, modulation of mRNA degradation or stability has been recognized as an important mechanism for regulation of expression of a number of proteins in different cells (Cleveland, 1988; Brawerman, 1989; Klausner and Harford, 1989; Nielsen and Shapiro, 1990), including several G protein-coupled receptors (Hadcock *et al.*, 1989; Bouvier *et al.*, 1989; Akamizu *et al.*, 1990; Saji *et al.*, 1991).

Down-regulation of the number of thyrotropin-releasing hormone receptors (TRH-Rs)[1] on anterior pituitary cells has been well-documented

---

[1] Abbreviations: TRH, thyrotropin-releasing hormone; TRH-R, TRH receptor; VIP, vasoactive intestinal peptide; PtdIns(4,5)P$_2$, phosphatidylinositol-4,5-bisphosphate; Ins(1,4,5)P$_3$, inositol-1,4,5-trisphosphate; H-7, 1-(5-isoquinolinesulfonyl)-2-methylpiperazine dihydrochloride; cAMP, adenosine 3′,5′-cyclic monophosphate; NEO, neomycin-resistance gene product; GAPDH, glyceraldehyde phosphate dehydrogenase.

341

Copyright © 1993 by Academic Press, Inc.
All rights of reproduction in any form reserved.

342                    MARVIN C. GERSHENGORN

(Hinkle, 1989). Homologous down-regulation by TRH (Hinkle and Tashjian, 1975; Gershengorn, 1978) as well as heterologous down-regulation
by other extracellular signaling molecules, such as the thyroid hormones
(Perrone and Hinkle, 1978; Gershengorn, 1978), epidermal growth factor (Hinkle *et al.*, 1991), and vasoactive intestinal peptide (VIP) (Imai
and Gershengorn, 1985; Fujimoto and Gershengorn, 1991), have been
demonstrated. Importantly, the decrease in the number of TRH-Rs
causes pituitary cells to exhibit a less marked response to TRH (Gershengorn, 1978; Mu *et al.*, 1982; Imai and Gershengorn, 1985; Ramsdell
and Tashjian, 1985, 1986; Winicov and Gershengorn, 1989). We and
others showed that down-regulation caused by TRH was not due to rapid
receptor internalization, but that TRH causes a slow decrease in
TRH-Rs with a half-time of approximately 12 hours. We therefore suggested that homologous down-regulation may be due, at least in part,
to decreased synthesis of TRH-Rs secondary to decreases in the level
of TRH-R mRNA. With our cloning of the mouse pituitary TRH-R
cDNA (Straub *et al.*, 1990) new reagents became available that have
permitted a more detailed and direct study of the regulation of TRH-R
mRNA.


## II.   Preliminary Studies in *Xenopus* Oocytes

Oocytes from the African frog *Xenopus laevis* injected with exogenous
mRNAs have become a widely used and important tool in the study of
mammalian receptors, ion channels, and transporters (see Oron and Lass,
1985; Snutch, 1988). Microinjection of mRNA from a variety of sources,
including tissue, cell lines, and *in vitro* transcription, leads to expression
of new proteins in *Xenopus* oocytes. Oocytes will translate, modify, and
transport receptor molecules so that they become functionally active at
the surface membrane. Because of their large size (1.2 mm in diameter),
which allows for easy access to the cell interior, oocytes are an excellent
system in which to manipulate the intracellular environment of intact,
living cells and thereby probe the molecular details of receptor activation.
In a series of studies, my colleagues and I injected *Xenopus* oocytes
with pituitary cell mRNA that encodes TRH-Rs and we characterized the
effects of activation of TRH-Rs expressed in these oocytes (Oron *et al.*,
1987a,b, 1988; Oron *et al.*, 1988b; Straub *et al.*, 1989a,b; Gershengorn *et al.*,
1990; Shapira *et al.*, 1990). Other investigators have performed related
studies (McIntosh and Catt, 1987; Meyerhof *et al.*, 1988). These studies
were necessary for our subsequent use of the oocyte system in an expression strategy to clone the pituitary TRH-R cDNA. They also provided

new insights into the mechanism of TRH action and the initial observation of regulation of TRH-R mRNA by TRH.

In mammalian pituitary cells, TRH-Rs are coupled to the PtdIns-$(4,5)P_2$–1,2-diacylglycerol–$Ca^{2+}$ signal transduction pathway (Gershengorn, 1985, 1986; Drummond, 1986). Previous studies in oocytes delineating the pathway used by endogenous muscarinic receptors (Dascal *et al.*, 1984, 1985; Oron *et al.*, 1985; Nadler *et al.*, 1986, Gillo *et al.*, 1987) and muscarinic and serotonin receptors acquired by the oocyte after injection of rat brain mRNA (Gundersen *et al.*, 1984; Dascal *et al.*, 1986) had shown that acquired receptors, which in their native tissue were coupled to PtdIns$(4,5)P_2$ hydrolysis, could utilize the same signaling pathway in the oocyte. We tested whether this could occur with the TRH-R by injecting mRNA from three sources, including a cloned rat pituitary cell line (GH$_3$ cells), a mouse thyrotropin (TSH)-producing pituitary tumor (TtT), and rat anterior pituitary tissue.

Uninjected oocytes exhibited no membrane electrical response to TRH but oocytes injected with RNA from these three sources showed a robust membrane depolarization on stimulation by TRH (Oron *et al.*, 1988b; Straub *et al.*, 1990). We showed that the acquired response to TRH in oocytes was mediated by the PtdIns$(4,5)P_2$–$Ca^{2+}$ signal transduction pathway and that the TRH-R was coupled to a phospholipase C via a guanine nucleotide-binding (G) protein (Oron *et al.*, 1978a,b, 1988; Oron *et al.*, 1988b; Straub *et al.*, 1989b). as in mammalian cells. The response elicited was a $Ca^{2+}$-mediated "opening" of chloride channels, which caused depolarization of the oocyte surface membrane. The data from one experiment that we performed that supports these conclusions are shown in Fig. 1. In this experiment, we showed that similar membrane depolarizations could be elicited by TRH or by intracellular injection of $Ca^{2+}$ or Ins$(1,4,5)P_3$ into oocytes injected 48 hours previously with GH$_3$ cell RNA. The sequence of molecular events that mediates TRH action in oocytes can, therefore, be summarized as follows:

$$\text{TRH} + \text{TRH-R} \rightarrow \text{TRH-R*} + G_p \rightarrow G_p^* + \text{PLC} \rightarrow \text{PLC*} + \text{PtdIns}(4,5)P_2 \rightarrow$$
$$\text{Ins}(1,4,5)P_3 \rightarrow Ca^{2+} \rightarrow I_{Cl} \rightarrow \text{depolarization}$$

where the asterisk denotes activated species, $G_p$ is a G protein that couples to PLC, PLC is a PtdIns$(4,5)P_2$-specific phospholipase C, and $I_{Cl}$ is a chloride conductance. Hence, injection of exogenous RNAs containing TRH-R mRNA leads to the acquisition of a functionally competent TRH-R on the surface of *Xenopus* oocytes that signals via the PtdIns$(4,5)$-$P_2$–InsP$_3$–$Ca^{2+}$ pathway and can be monitored as TRH-elicited electrical responses.

344                    MARVIN C. GERSHENGORN

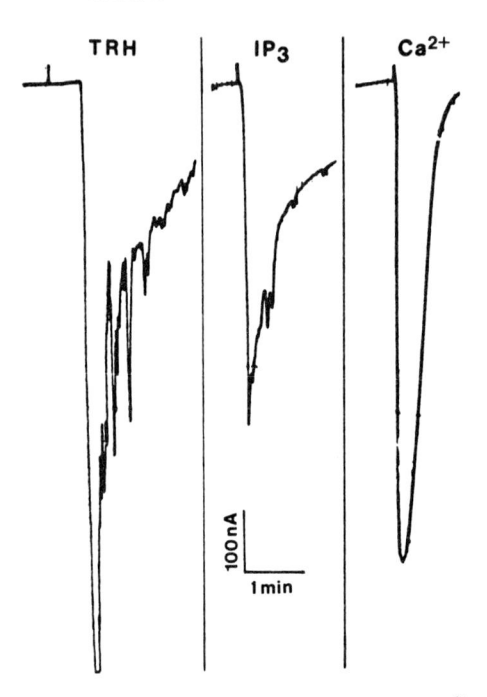

FIG. 1. Comparison of the effects of TRH, Ins(1,4,5)$P_3$, and $Ca^{2+}$ on chloride current responses in voltage-clamped *Xenopus laevis* oocytes injected with RNA from $GH_3$ cells. Oocytes were injected with 240 ng total cytosolic $GH_3$ cell RNA. The cells were stimulated with 1 $\mu M$ TRH or injected with 0.5 pmol Ins(1,4,5)$P_3$ or 10 pmol $Ca^{2+}$. TRH addition was started and Ins(1,4,5)$P_3$ and $Ca^{2+}$ injections were made at the marks on the tracings. These are typical responses in oocytes from a single frog. These data show that TRH, Ins(1,4,5)$P_3$, and $Ca^{2+}$ elicit similar responses in oocytes injected with RNA from $GH_3$ cells. From Straub *et al.* (1989b).

## III.  Cloning of a TRH-R Complementary DNA

Although there have been attempts in several laboratories (Johnson *et al.*, 1984; Sullivan *et al.*, 1987; Phillips and Hinkle, 1989), including our own (Winicov *et al.*, 1990), to solubilize, isolate, and purify the TRH-R over a number of years, no one has succeeded. At the time we decided to attempt to isolate a cDNA clone for the TRH-R there was no sequence information, which could be used to develop oligonucleotide probes, nor antibodies to use in screening a cDNA library. We elected, therefore, to use an expression strategy in *Xenopus* oocytes to clone a TRH-R cDNA (Straub *et al.*, 1990) in a manner similar to that employed in cloning a cDNA for the serotonin type 1C receptor (Julius *et al.*, 1988). After failing to detect a signal using four libraries constructed from cDNA synthesis of $GH_3$ cell RNA, we made a library using response-selected poly(A)$^+$ RNA from TtT tumors. Figure 2 illustrates the strategy that we employed. A

# TRH-R: CLONING AND EXPRESSION    345

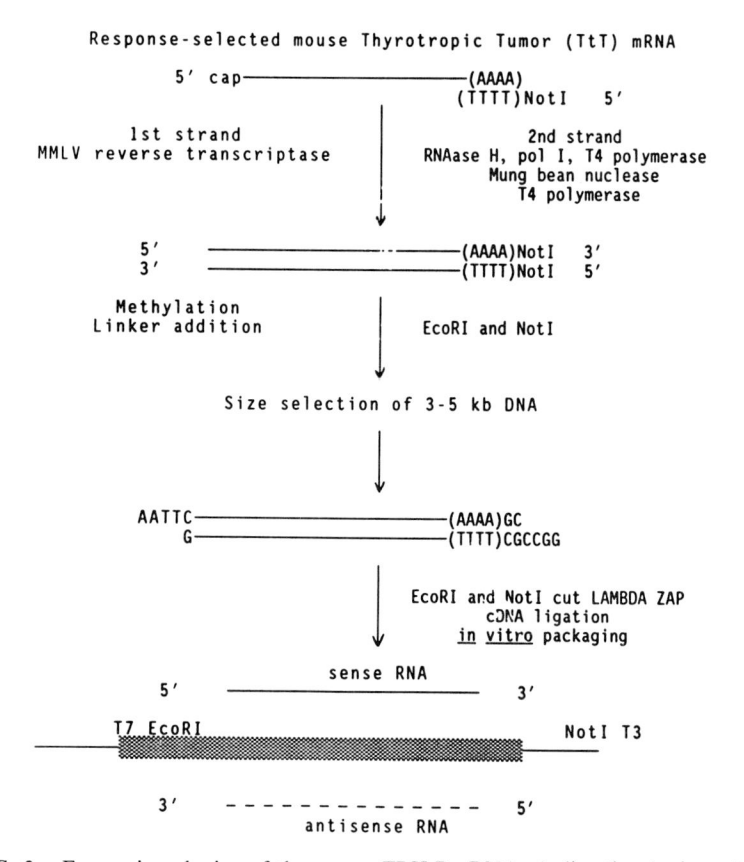

FIG. 2. Expression cloning of the mouse TRH-R cDNA. A directional, size-selected cDNA library was constructed. Poly(A)⁺ RNA was isolated from a mouse pituitary thyrotropic TtT tumor and fractionated on a sucrose gradient. The fraction that gave the highest specific TRH-R mRNA activity when injected into *Xenopus* oocytes was used for preparative cDNA synthesis. cDNAs of between 3 and 5 kb were purified by agarose gel electrophoresis and electroelution. The size-selected cDNAs were ligated directionally into phagemids such that transcription using T7 RNA polymerase would yield sense RNA. A library of $1.2 \times 10^6$ clones was constructed that was >95% recombinant. The primary library was immediately divided and amplified individually in 60 sublibraries of approximately 20,000 clones each. Phagemid DNA was prepared and digested with *Not*I and then proteinase K treated, extracted, and precipitated. *In vitro* transcription of sense RNA was carried out. Transcripts from 20 pools of 20,000 clones each were injected individually into *Xenopus* oocytes. Two or three days later the oocytes were placed under voltage clamp and 1 $\mu M$ TRH was administered in the bath. One pool gave a clear TRH-evoked current response in the oocyte but the other 19 pools were negative. The positive pool of 20,000 clones was reduced by division to pools of 2000, 200, 30, and 10, and finally a single "positive" clone was isolated that contained a 3.8-kb insert.

346 MARVIN C. GERSHENGORN

directional cDNA library of $1.2 \times 10^6$ clones was constructed in a phagemid using response-selected RNA. The library was screened by injecting RNA transcribed *in vitro* from pools of approximately 20,000 clones into *Xenopus* oocytes and measuring the membrane electrical response elicited by TRH. Serial division of the positive pool of clones yielded a single clone that conferred responsiveness to TRH. Figure 3 illustrates the response in oocytes injected with response-selected poly(A)$^+$ RNA from mouse thyrotropic tumors (which was the starting material for cDNA synthesis), RNA transcribed *in vitro* from the single positive clone, and RNA transcribed *in vitro* from another clone from the same library. The left tracing shows a typical response evoked by a maximally effective dose of TRH in voltage-clamped oocytes after injection of 20 ng of size-fractionated TtT poly(A)$^+$ RNA. The center tracing

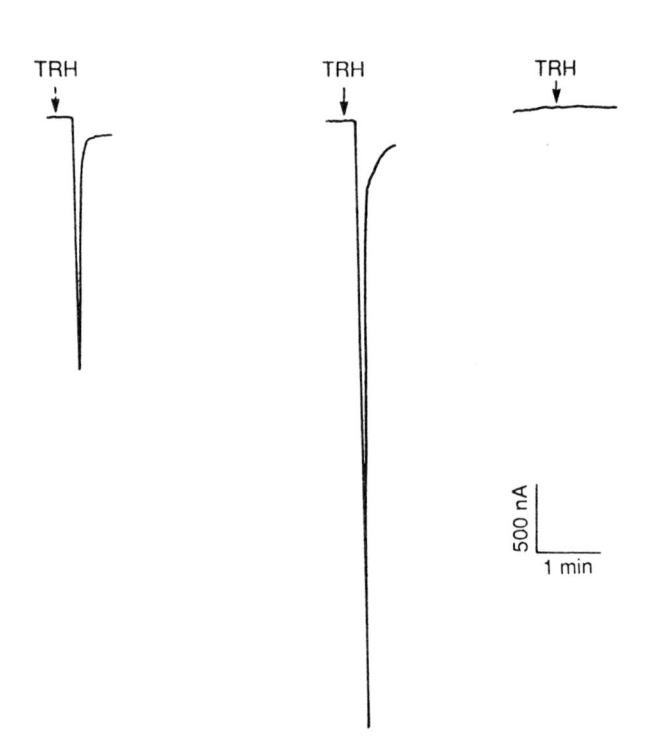

FIG. 3. Electrophysiological responses to TRH in *Xenopus* oocytes injected 48 hours previously with 20 ng of sucrose gradient-fractionated TtT poly(A)$^+$ RNA (left trace), 2 ng of RNA transcribed *in vitro* from the TRH-R cDNA (center trace), or 20 ng of RNA transcribed from a different clone from the same cDNA library (right trace). These data show that the positive clone encodes a protein that functions as a TRH-R when expressed in *Xenopus* oocytes. From Straub *et al.* (1990).

shows the response conferred by injection of 2 ng of RNA transcribed *in vitro* from a single clone isolated by serial division of this library. Responses could be obtained by injection of as little as 1 pg of RNA transcripts from this single clone. The right tracing is representative of RNA transcribed *in vitro* from any other individual clone isolated from the same library that did not confer TRH responsivity. Furthermore, we showed that oocytes injected with RNA transcribed *in vitro* from the positive clone bound methyl-TRH (a high-affinity analog of TRH), TRH, and chlordiazepoxide (an antagonist of TRH binding to its receptor) (Drummond, 1985; Gershengorn and Paul, 1986; Drummond *et al.*, 1989) with the appropriate affinities, and responded to TRH with the appropriate concentration dependency. Thus, the isolated cDNA encodes a functional TRH-R as assessed in *Xenopus* oocytes.

To show that the cloned mouse TRH-R cDNA sequence is similar to endogenous rat pituitary TRH-R mRNA, we studied whether antisense RNA transcribed *in vitro* from a portion of the TRH-R cDNA would inhibit TRH-R expression in oocytes injected with RNA isolated from normal rat anterior pituitary glands. The idea underlying this experiment is that homologous antisense RNA would specifically hybridize to the endogenous TRH-R mRNA and inhibit receptor synthesis, perhaps by inhibiting translation (Melton, 1985). Injection of RNA isolated from normal rat pituitaries led to acquisition of responses to TRH and carbachol, a muscarinic agonist (Fig. 4, upper panel, left trace); there were no intrinsic responses to TRH or carbachol in uninjected oocytes. In Fig. 4 (upper panel, right trace) are responses from an oocyte injected with RNA from normal rat pituitaries that had been incubated with antisense TRH-R RNA. The TRH response in this oocyte was abolished, but the carbachol response was unaffected. As shown in the compilation of data in Fig. 4, lower panel, when antisense RNA was allowed to hybridize to rat pituitary RNA prior to injection, the response to TRH was inhibited by 87% whereas the response to carbachol was not significantly inhibited. These data showed that the rat TRH-R nucleotide sequence is similar to that of the mouse and provided strong additional support for the conclusion that the isolated cDNA encodes a mouse TRH-R.

Final proof that the cDNA that we cloned encoded a TRH-R was obtained by expressing it in a mammalian system. We transiently transfected monkey kidney COS-1 cells with the TRH-R cDNA subcloned into a eukaryotic expression vector. No specific binding or TRH-stimulated inositol phosphate formation was detectable in untransfected COS-1 cells or cells transfected with vector alone. Cells transfected with pCDM8mTRH-R exhibited high-affinity, saturable binding of [$^3$H]methyl-TRH and displacement by TRH and chlordiazepoxide; the $K_i$ values for

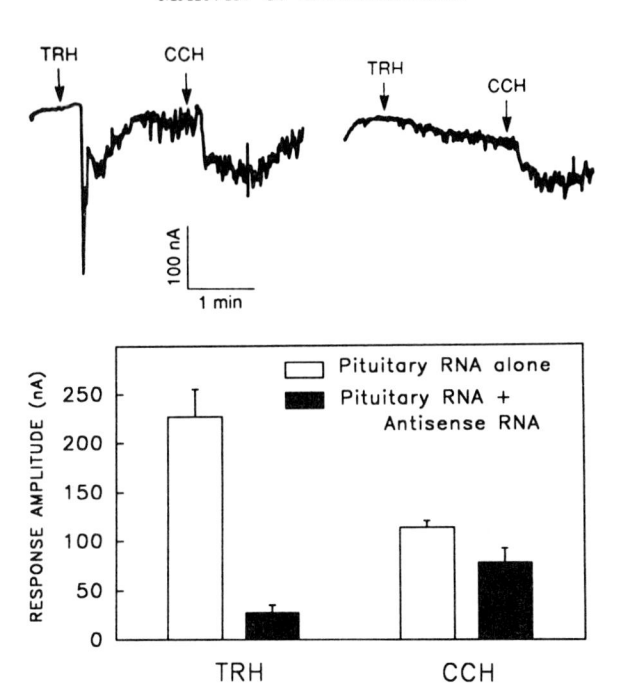

FIG. 4. Effect of antisense TRH-R RNA on the acquired TRH and carbachol (CCH) responses in *Xenopus* oocytes. Antisense RNA (5 ng) transcribed *in vitro* from a plasmid containing a portion of the TRH-R cDNA was incubated for 1 hour with 300 ng of RNA isolated from normal rat pituitary glands. The mixture was injected into oocytes and electrophysiological responses to 1 $\mu M$ TRH and 1 m$M$ CCH were assayed 3 days later. The traces at the top are representative responses in the absence (left trace) or presence (right trace) of antisense TRH-R RNA. The graph on the botton is a compilation of the antisense inhibition of the TRH but not the carbachol response. These data show that the cloned cDNA is highly homologous to rat TRH-R mRNA. From Straub *et al.* (1990).

TRH (10 n$M$) and for chlordiazepoxide (20 $\mu M$) were appropriate. TRH-stimulated inositol phosphate formation exhibited an $EC_{50}$ of approximately 10 n$M$, which is similar to the $EC_{50}$ found in $GH_3$ cells. These data showed conclusively that we had cloned a cDNA that encodes a mouse pituitary TRH-R.

## IV. Structure of the TRH-R

Figure 5 illustrates the primary structure and putative topology in the cell surface membrane of the mouse pituitary TRH-R. The cDNA contained an open reading frame of 1179 base pairs (bp) from nucleotides 259 to 1437 [out of 3498 nucleotides plus the poly(A)$^+$ tail] that encodes a protein of

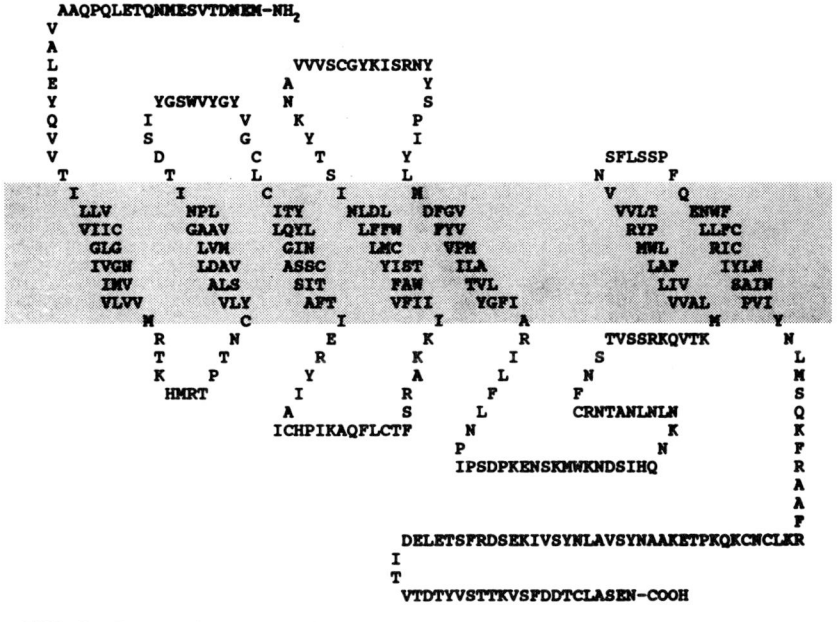

FIG. 5. Proposed structure of the mouse pituitary TRH-R. The seven transmembrane helices are numbered I to VII from left to right (numbers not shown). The extracellular space is at the upper part of the figure, the cell surface membrane is shaded, and the cytoplasm is at the lower part of the figure. See text for discussion.

393 amino acids. The receptor is predicted to have seven transmembrane-spanning domains, in agreement with the topology proposed for other G protein-coupled receptors (Raymond *et al.*, 1990). There are two potential N-linked glycosylation sites with the N-X-T/S consensus sequence in the N terminus at positions 3 and 10. Two cysteines, which are homologous to those in the $\beta_2$-adrenergic receptor that have been proposed to engage in a disulfide bond (Dohlman *et al.*, 1990), are present in the second (Cys-98) and third (Cys-179) extracellular loops. These residues may also form a disulfide bond in the TRH-R because it has been shown that TRH binding is reduced when disulfide bonds are disrupted (Sharif and Burt, 1984). Fifteen of the last 52 amino acids are either serine or threonine residues that are potential sites for regulatory phosphorylation, which may be mediated by a $\beta$-adrenergic receptor kinase-like enzyme (Hausdorff *et al.*, 1990), and there are four sites in the C terminus (Ser-324, Thr-342, Ser-360, and Thr-379) that are candidates for phosphorylation by protein kinase C (Kikkawa *et al.*, 1989). The third intracellular loop has been shown to be important in the coupling of other receptors to G proteins and there are

350                          MARVIN C. GERSHENGORN

candidates for phosphorylation by protein kinase C in this intracellular loop also (Thr-250, Ser-259, and Ser-260). Phosphorylation at these sites may be involved in receptor desensitization (Perlman and Gershengorn, 1991) or down-regulation.

Hence, the overall structure of the TRH-R is similar to other members of the family of G protein-coupled receptors, but the exact sites of binding of TRH, of coupling to the G protein, and of regulation have not been demonstrated.

## V.  Down-Regulation of TRH-R Number

The number of available receptors on the cell surface may be decreased or increased by ligand occupancy (homologous down-regulation or up-regulation) or by other factors that do not interact with the receptor (heterologous down-regulation or up-regulation). In pituitary cells, TRH-Rs have been shown to exhibit homologous down-regulation (Hinkle and Tashjian, 1975; Gershengorn, 1978), heterologous down-regulation (Gershengorn, 1978; Perrone and Hinkle, 1978; Peck and Gershengorn, 1980; Imai and Gershengorn, 1985; Hinkle et al., 1991), and heterologous up-regulation (DeLean et al., 1977; Peck and Gershengorn, 1980). The mechanisms by which these changes in TRH-R number are affected are not known (Hinkle, 1989).

In TtT cells in culture, for example, we observed down-regulation of TRH-R number caused by TRH and by thyroid hormones (Gershengorn, 1978). Figure 6 illustrates these effects in an experiment in which TtT cells were incubated with various concentrations of TRH or L-triiodothyronine ($T_3$) for 48 hours and then TRH-Rs were measured. Down-regulation of TRH-Rs was initially observed in $GH_3$ cells and related cell lines (for review, see Hinkle, 1989b). Of note is that in these two instances, and in all of the other examples of TRH-R down-regulation in pituitary cells that have been reported, the half-time of disappearance of the receptor was greater than 12 hours. In fact, a more rapid receptor degradation caused by TRH was looked for but could not be found (Halpern and Hinkle, 1981; Hinkle and Kinsella, 1982). Thus, inhibition of receptor synthesis, which is predicted to require a longer time for its effects to become measurable, was viewed as a likely mechanism for TRH-R down-regulation.

The important complementary question was whether decreases in TRH-R density affected the responses of pituitary cells to stimulation by TRH. Figure 7 shows the results of an experiment in which stimulation by maximally effective doses of TRH and the muscarinic agonist, carbamylcholine, provided to control TtT cells and TtT cells in which TRH-Rs were down-regulated, were compared (Winicov and Gershengorn, 1989). In control cells, the number of muscarinic receptors was approximately 10%

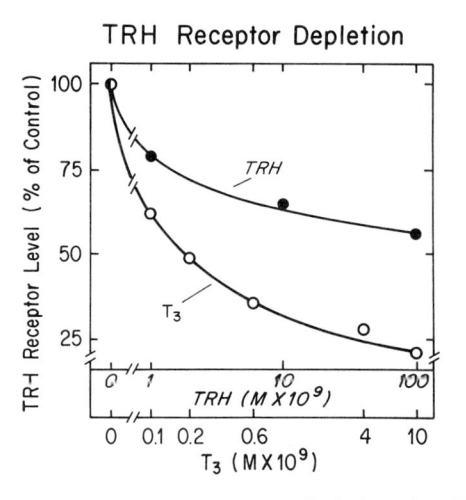

FIG. 6. Dose–response effects of TRH and L-triiodothyronine ($T_3$) on the number of TRH-Rs on TtT cells. TtT cells in culture were incubated with medium alone (control cells), with 1 to 100 nM TRH, or with 0.1 to 10 nM $T_3$. After 48 hours the TRH-R number was estimated using [$^3$H]TRH. Half-maximal receptor depletion occurred with approximately 1 nM TRH and 0.15 nM $T_3$. These data show homologous and heterologous down-regulation of TRH-Rs in TtT cells. From Gershengorn (1978).

of that of TRH-Rs, and the response to carbamylcholine, measured as increases in the intracellular mediators Ins(1,4,5)$P_3$ and cytoplasmic $Ca^{2+}$, was proportionally lower than that to TRH. In contrast, in cells in which the number of TRH-Rs were down-regulated to a level similar to that of muscarinic receptors, the responses to TRH and carbamylcholine were similar. We (Imai and Gershengorn, 1985) and others (Ramsdell and Tashjian, 1986; Mu et al., 1982; Ramsdell and Tashjian, 1985) found similar effects of TRH-R down-regulation on TRH responses in $GH_3$ and also in related cell lines. Thus, down-regulation of TRH-Rs is a process stimulated by several pituitary cell regulatory factors and the decrease in TRH-R number leads to a diminished response to TRH in pituitary cells.

## VI. Down-Regulation of Endogenous TRH-R mRNA

Based primarily on the slow kinetics of TRH-R down-regulation, we suggested that the decrease in TRH-R number was caused, at least in part, by decreased receptor synthesis that we thought might be secondary to decreases in the levels of receptor mRNA. In 1987, at a time prior to the cloning of the TRH-R cDNA, we developed evidence that homologous down-regulation was associated with effects on TRH-R mRNA (Oron et al., 1987b). In these experiments, TRH-evoked membrane electrical

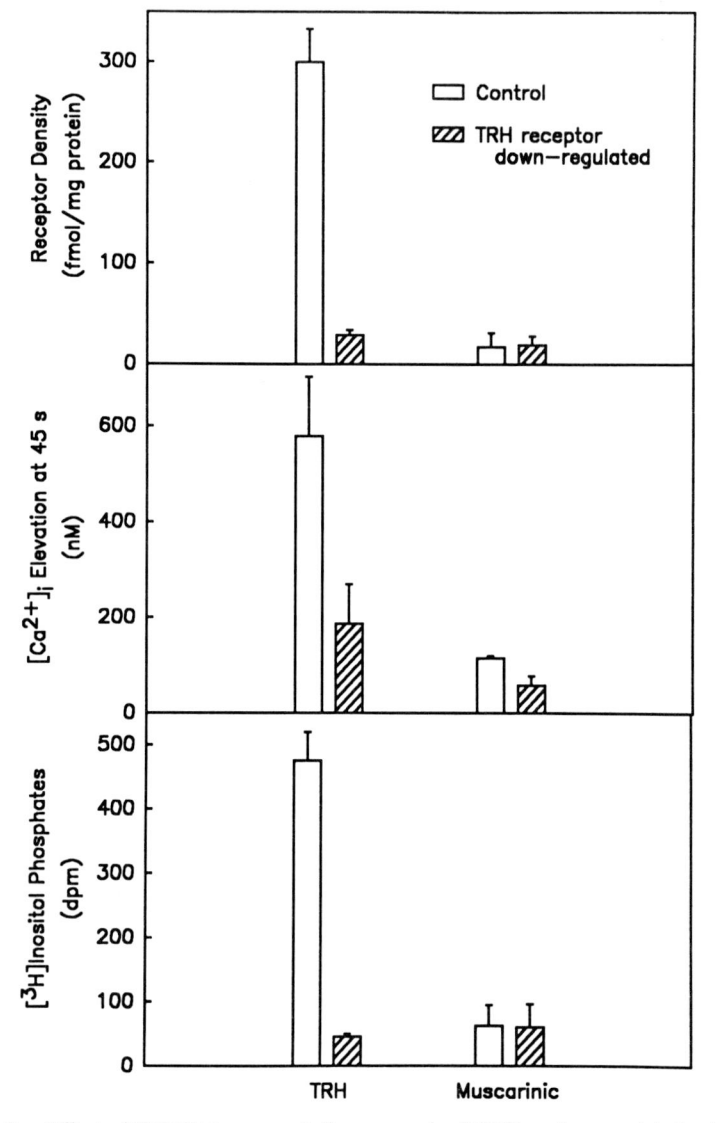

FIG. 7.   Effect of TRH-R down-regulation on maximal TRH- and muscarinic (carbamyl-choline)-stimulated elevation of cytoplasmic free $Ca^{2+}$ concentration ($[Ca^{2+}]_i$) and Ins(1,4,5)$P_3$ accumulation at 45 seconds in TtT cells. TtT cells were preincubated without (open bars) or with (striped bars) 100 n$M$ TRH for 16 hours. Top: $[^3]$MeTRH (1 n$M$) and $[^3H]$quinuclidinyl benzilate (3 n$M$) binding were measured in whole cells. $B_{max}$ was calculated using $K_d$ values measured in separate experiments. Middle: cells were loaded with the fluorescent $Ca^{2+}$ probe Quin 2, and $[Ca^{2+}]_i$ was measured after 45 seconds of acute stimulation with 100 n$M$ TRH or 1 m$M$ carbamylcholine. Basal levels of $[Ca^{2+}]_i$ (160 ± 24 n$M$, control; 140 ± 16 n$M$, TRH pretreated) have been subtracted. Bottom: cells were labeled with myo-$[^3H]$inositol for 24 hours prior to incubation without (control) or with 100 n$M$ TRH. Total inositol phosphate second-messenger molecule accumulation at 45 seconds (normalized to

TRH-R: CLONING AND EXPRESSION 353

responses in *Xenopus* oocytes were used as a bioassay. This assay could not distinguish between changes in mRNA level and changes in the translational efficiency of the mRNA, and we therefore referred to the measured value as "mRNA activity." TRH-R mRNA activity was measured as the amplitude of the membrane electrical response to a maximally effective dose of TRH in oocytes injected 48 to 72 hours previously with a known amount of pituitary cell RNA. In a typical experiment, RNA was isolated from control and TRH-treated pituitary cells and 240 ng of RNA was injected into *Xenopus* oocytes. We found that TRH-induced downregulation of TRH-Rs was preceded by a marked decrease in TRH-R mRNA activity. This was the initial evidence that TRH stimulation of pituitary cells is associated with effects on TRH-R mRNA.

After cloning the mouse TRH-R cDNA, we were able to develop reagents that permitted direct measurement of TRH-R mRNA. We have now used these probes to measure the effects of several types of regulatory factors on TRH-R mRNA levels. TRH-R mRNA levels in $GH_3$ cells were measured by Northern analysis and in ribonuclease protection assays. We measured the effects of TRH, elevation of cytoplasmic free $Ca^{2+}$ concentration, phorbol-12-myristate-13-acetate (which activates protein kinase C), and H-7 [1-(5-isoquinolinesulfonyl)-2-methylpiperazine dihydrochloride, an inhibitor of protein kinases]. These agents, in addition to TRH, were studied to gain insight into the mechanism of the TRH effect. As described above, signal transduction by TRH involves generation of $Ins(1,4,5)P_3$ and elevation of cytoplasmic free $Ca^{2+}$ concentration, which leads to activation of $Ca^{2+}$/calmodulin-dependent protein kinase, and generation of 1,2-diacylglycerol, which leads to activation of protein kinase C. Figure 8 shows that a maximally effective dose of TRH caused a marked, transient decrease in the level of TRH-R mRNA, which attained a nadir of 20 to 45% of control by 3 to 6 hours, increased after 9 hours, but was still below control levels after 24 hours. A maximally effective dose of phorbol-12-myristate-13-acetate (PMA) caused decreases in TRH-R mRNA that were similar in magnitude and in time course to the decreases induced by TRH. Figure 9 shows that H-7 blocked the effects of TRH and of PMA to lower TRH-R mRNA to similar extents. In contrast, we found that elevation of cytoplasmic free $Ca^{2+}$ concentration had no effect on TRH-R mRNA. Hence, our data showed that TRH and PMA decrease the levels of TRH-R mRNA in $GH_3$ cells, and we suggest that they are consistent

---

100,000 dpm [³H]inositol-labeled lipid) is shown. The basal levels in unstimulated cells, 655 ± 55 dpm (control) and 680 ± 63 dpm (down-regulated), have been subtracted from the data. These data show that the response of TtT cells to TRH, but not to muscarinic stimulation, is decreased in TRH-R down-regulated cells. From Winicov and Gershengorn (1989).

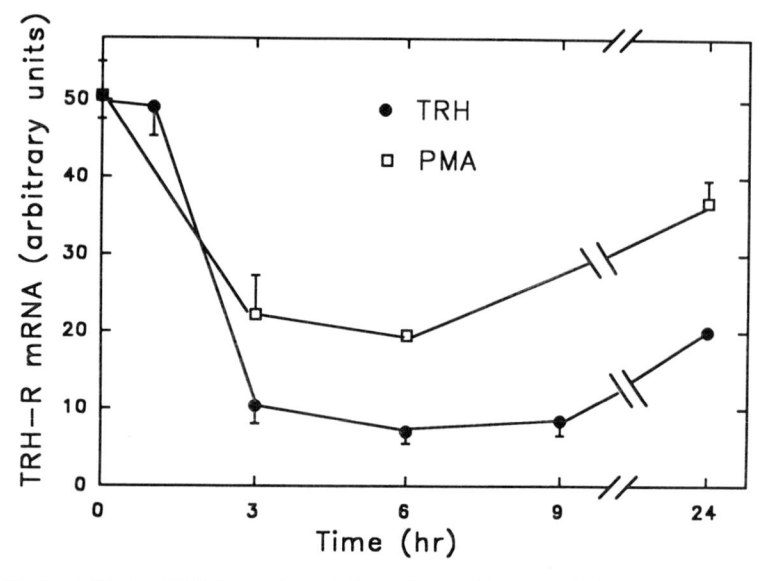

FIG. 8. Effects of TRH and phorbol-12-myristate-13-acetate (PMA) on TRH-R mRNA in GH₃ cells as measured in an RNase protection assay. GH₃ cells were incubated for the times indicated with 1 μM TRH (filled circles) or 1 μM PMA (open squares); total RNA was isolated and 50 or 100 μg were assayed. These data show that TRH and PMA down-regulate TRH-R mRNA in GH₃ cells. From Fujimoto *et al.* (1991).

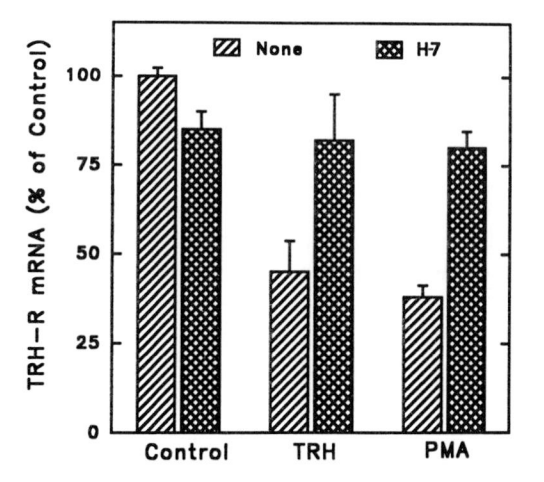

FIG. 9. Effect of the protein kinase inhibitor 1-(5-isoquinolinesulfonyl)-2-methylpiperazine dihydrochloride (H-7) on the decrease in TRH-R mRNA caused by TRH and PMA in GH₃ cells. GH₃ cells were preincubated with 20 μM H-7 for 1 hour (cross-hatched bars) or incubated in medium alone (hatched bars) and then were exposed to 1 μM TRH or 1 μM PMA for 3 hours. Total RNA was isolated and 50 or 70 μg were assayed. These data show that H-7 inhibits the effects of TRH and PMA to decrease TRH-R mRNA. From Fujimoto *et al.* (1991).

with the idea that the effect of TRH is via a protein kinase C-mediated mechanism.

Because TRH caused down-regulation of both TRH-R protein and mRNA, and because we had shown that agents that elevate cyclic AMP down-regulated TRH-R number, we studied whether VIP and other agents that elevate cellular cAMP also down-regulate TRH-R mRNA in $GH_3$ cells. We found that VIP caused a time- and concentration-dependent decrease in TRH-R mRNA. To determine whether this effect of VIP is mediated by cAMP, we measured the effects on TRH-R mRNA of several agents that elevate cellular cAMP levels. Figure 10 shows that the effect of VIP could be simulated by 8-(4-chlorophenylthio)-cAMP (a cAMP analog), forskolin (which directly activates adenylyl cyclase), cholera toxin (which stimulates adenylyl cyclase by activating the stimulatory G protein), and 1-methyl-3-isobutylxanthine (a cAMP phosphodiesterase inhibitor). Because the level of TRH-R mRNA could be lowered by elevation of cAMP and by activation of protein kinase C by TRH or PMA, we determined whether activating these two signaling pathways simultaneously would cause a greater decrease in TRH-R mRNA. Figure 11 illustrates that maximally effective doses of VIP or forskolin plus maximally effective doses of TRH or PMA caused greater decreases in TRH-R mRNA than any of these agents alone. When we preincubated cells with H-7 the effects of VIP, TRH, and PMA were inhibited. We suggested, therefore, that cAMP mediates down-regulation by VIP of TRH-R mRNA and, because this effect is inhibited by H-7, that it most likely involves protein kinase A.

TRH and VIP stimulate hormone secretion from pituitary cells, but the effects of TRH and VIP are mediated by different intracellular pathways that lead to activation of different protein kinases. We suggest that TRH

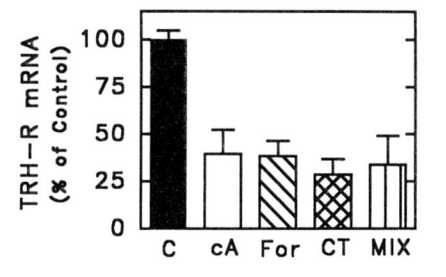

FIG. 10. Effects of pharmacological agents that elevate cAMP on TRH-R mRNA levels in $GH_3$ cells. $GH_3$ cells were incubated for 3 hours with 0.5 m$M$ 8-(4-chlorophenylthio)-cAMP (cA), 3 $\mu M$ forskolin (For), 50 n$M$ cholera toxin (CT), or 0.5 m$M$ methylisobutylxanthine (MIX); C, control. Total RNA was isolated and assayed. These data show that elevation of cAMP causes down-regulation of TRH-R mRNA in $GH_3$ cells. From Fujimoto and Gershengorn (1991).

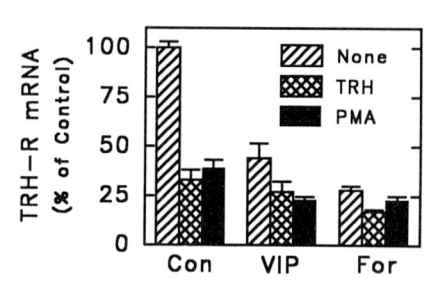

FIG. 11. Effects of simultaneous addition of VIP or forskolin (For) and TRH or phorbol-12-myristate-13-acetate (PMA) on TRH-R mRNA levels in $GH_3$ cells. $GH_3$ cells were incubated for 3 hours with no addition (Con), 1 $\mu M$ TRH, 1 $\mu M$ PMA, 100 n$M$ VIP, VIP and TRH, VIP and PMA, 3 $\mu M$ For, For and TRH, or For and PMA. Total RNA was isolated and assayed. These data are consistent with the idea that TRH and VIP cause down-regulation of TRH-R mRNA in $GH_3$ cells by activating different pathways that involve protein kinases C and A, respectively. From Fujimoto and Gershengorn (1991).

and VIP down-regulate TRH-R mRNA by activating protein kinases C and A, respectively.

## VII. Mechanism of Down-Regulation of TRH-R mRNA: Use of Transfected Cells

Although we were able to demonstrate that TRH, VIP, and various pharmacologic agents down-regulate endogenous TRH-R mRNA in $GH_3$ cells, we were not able to determine definitively whether these effects were caused by decreased gene transcription or increased mRNA turnover. We had suggested that TRH-induced decreases were caused by increased mRNA turnover based on preliminary evidence (Oron *et al.*, 1987b), but we could not measure turnover of the endogenous rat TRH-R mRNA in $GH_3$ cells. We therefore developed a $GH_3$ cell line that is stably transfected with mouse pituitary TRH-R cDNA (GH-mTRHR-1 cells). We thought that it would be possible to measure mRNA synthesis and turnover in these cells, reasoning that they might have higher levels of TRH-R mRNA than parental $GH_3$ cells because transcription of mouse TRH-R DNA would be driven by the strong cytomegalovirus promoter, which is the promoter in the expression plasmid that was used for transfection, and the DNA may be present in more copies than the endogenous gene. Because we cannot measure the absolute level of endogenous rat TRH-R mRNA in our heterologous ribonuclease protection assay, we cannot determine whether there actually is a higher level of mouse TRH-R mRNA in GH-mTRHR-1 cells compared to endogenous rat TRH-R mRNA in $GH_3$ cells. We can, however, measure mouse TRH-R mRNA in extracts of GH-mTRHR-1 cells without interference by endogenous rat mRNA by per-

# TRH-R: CLONING AND EXPRESSION 357

forming our ribonuclease protection assay under more stringent conditions. Using this more specific assay for mouse TRH-R mRNA, rat TRH-R mRNA is not detected in RNA from parental $GH_3$ cells nor from a cell line of $GH_3$ cells stably transfected with the vector alone.

We first characterized the TRH-Rs and responses to TRH in GH-mTRHR-1 cells. GH-mTRHR-1 cells were found to have 2.4 times as many TRH-Rs as the parent $GH_3$ cell line, 340 ± 50 fmol/mg protein as compared to 140 ± 30 fmol/mg, but there was no difference in the apparent equilibrium dissociation constants, which were 1.4 ± 0.1 and 1.4 ± 0.2 n$M$ for methyl-TRH for the parental cells and GH-mTRHR-1 cells, respectively. We are not certain whether GH-mTRHR-1 cells contain rat or mouse TRH-Rs, or both, because the binding characteristics of the mouse and rat TRH-Rs are too similar to permit this differentiation (Hinkle et al., 1974; Gershengorn, 1978) and no other means of distinguishing between them is available. In GH-mTRHR-1 cells, a maximally effective dose of TRH stimulated a 2.5-fold greater increase in formation of inositol phosphate intracellular mediator molecules than in parental $GH_3$ cells. Half-maximal stimulation of inositol phosphate accumulation occurred at the same concentration of TRH, approximately 7 n$M$, in both cell lines. Thus, as was found in nontransfected pituitary cells, there is a direct correlation between the number of TRH-Rs and the magnitude of the response to TRH in these transfected cells. Regulation by TRH of TRH-R number in GH-mTRHR-1 cells was similar to that in parental $GH_3$ cells. TRH caused a time- and concentration-dependent down-regulation of the number of TRH-Rs. We therefore concluded that GH-mTRHR-1 cells could be used to study regulation of TRH-R mRNA, because their response to TRH was similar to that of $GH_3$ cells.

We studied the effects of TRH on TRH-R mRNA in GH-mTRHR-1 cells. GH-mTRHR-1 cells contain 26 ± 1.6 molecules of mouse TRH-R mRNA per cell as compared to 230 ± 31 molecules of mRNA for the neomycin (NEO) resistance gene with which it was cotransfected. We used NEO mRNA and the endogenous mRNA for glyceraldehyde phosphate dehydrogenase (GAPDH) as control mRNAs. We showed that TRH caused a dose-dependent (Fig. 12), transient decrease in mouse TRH-R mRNA in GH-mTRHR-1 cells, with a nadir to 20% of control levels after 6 hours (Fig. 13). This effect is very similar to the decrease in endogenous rat TRH-R mRNA caused by TRH in parental $GH_3$ cells. In contrast, TRH did not affect NEO mRNA nor GAPDH mRNA. Thus, TRH caused a specific decrease in TRH-R mRNA that was similar in GH-mTRHR-1 cells to that in $GH_3$ cells.

We thought it unlikely that the decrease in TRH-R mRNA was caused by a decrease in the rate of transcription, given that transcription was driven by the powerful cytomegalovirus promoter. We tested this hypothe-

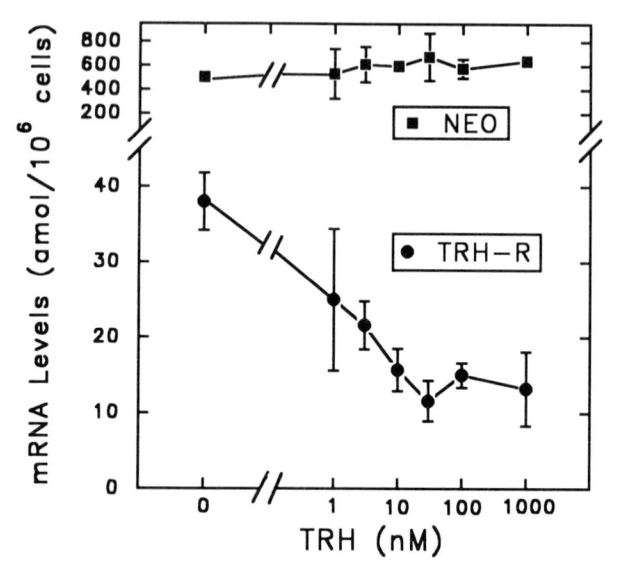

FIG. 12. Effects of TRH on mouse TRH-R and NEO mRNAs in GH-mTRHR-1 cells. GH-mTRHR-1 cells, which are GH₃ cells stably transfected with a plasmid containing the mouse TRH-R cDNA, were incubated without TRH or with the indicated concentrations of TRH for 3 hours. Total cell RNA was extracted and the levels of TRH-R and NEO mRNAs were measured. These data show that TRH specifically down-regulates mouse TRH-R mRNA in rat pituitary cells stably transfected with the mouse TRH-R cDNA. From Fujimoto *et al.* (1991).

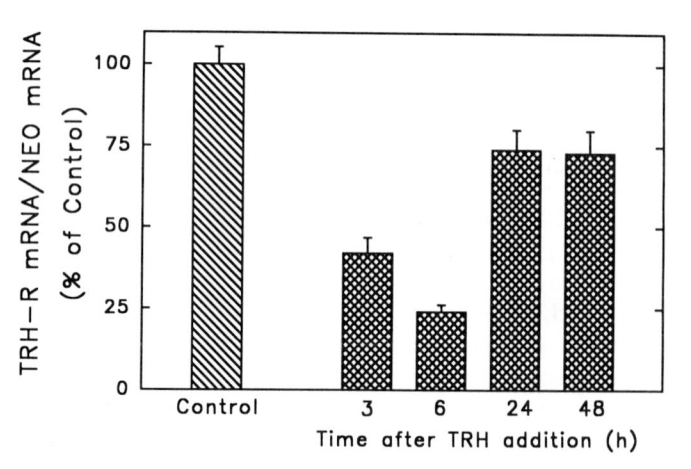

FIG. 13. Time course of the effect of TRH on TRH-R mRNA in GH-mTRHR-1 cells. GH-mTRHR-1 cells were incubated without (control) or with 1 $\mu M$ TRH for the times indicated; RNA was extracted and mouse TRH-R and NEO mRNAs were assayed. Data are presented as percent of control of the ratios of the amounts of mouse TRH-R mRNA divided by NEO mRNA in order to normalize the values for mouse TRH-R mRNA between experiments. These data show that the time course of down-regulation by TRH of TRH-R mRNA in GH-mTRHR-1 cells is similar to that of the endogenous rat TRH-R mRNA in parental GH₃ cells. From Fujimoto *et al.* (1991).

sis by measuring the effects of TRH on transcription of mouse TRH-R DNA. We found that TRH *stimulated* the rate of transcription of mouse TRH-R DNA by approximately twofold. Thus, the effect of TRH on TRH-R DNA transcription could certainly not account for its effect of *decreasing* TRH-R mRNA. We think the increase in transcription of TRH-R DNA caused by TRH in GH-mTRHR-1 cells is likely a manifestation of a "nonspecific" increase in transcription driven by the cytomegalovirus promoter (Yan *et al.*, 1991), rather than it being reflective of effects on the endogenous TRH-R gene, because the mouse TRH-R cDNA contains only 258 bases of 5′ untranslated sequence (Straub *et al.*, 1990) and, of course, does not contain the upstream gene regulatory elements.

Based on these data, it was likely that the effect of TRH was to increase the rate of turnover of TRH-R mRNA. We measured the rate of mRNA turnover after inhibiting transcription with actinomycin D. We compared the effects of TRH on the rates of turnover of mouse TRH-R mRNA and GAPDH mRNA (Fig. 14). TRH caused an increase in the turnover rate of TRH-R mRNA but had no effect on the rate of turnover of GAPDH

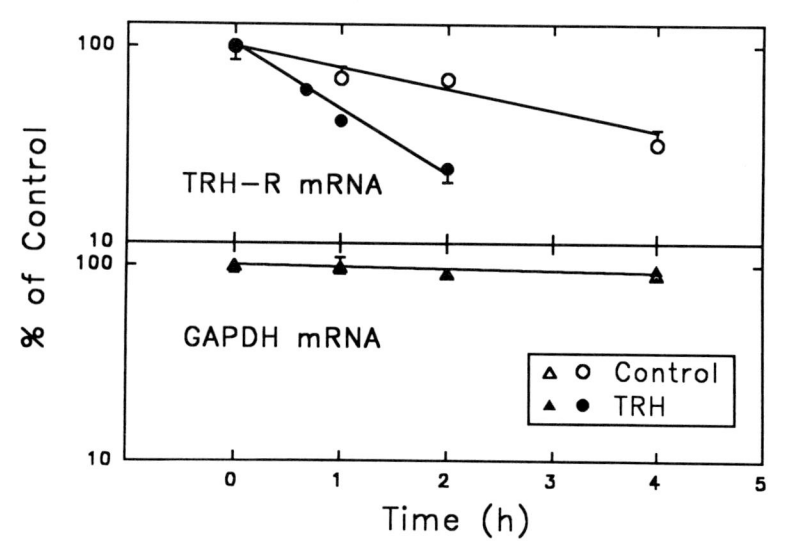

FIG. 14.   Effects of TRH on the rates of turnover of mouse TRH-R and GAPDH mRNAs in GH-mTRHR-1 cells. GH-mTRHR-1 cells were incubated without or with 1 μM TRH for 1.5 hours and then actinomycin D was added to a final concentration of 5 μg/ml, which inhibits transcription by more than 98%. At the times indicated, RNA was extracted and mouse TRH-R and GAPDH mRNAs were assayed. Data are presented as a percentage of the respective control. The level of mouse TRH-R mRNA was 88 ± 13% of the control level after exposure to TRH for 1.5 hours. These data show that TRH increases the rate of turnover of mouse TRH-R mRNA in GH-mTRHR-1 cells by fourfold, which represents a decrease in half-life from 3 hours to 0.75 hour. From Fujimoto *et al.* (1991).

mRNA. The half-life of mouse TRH-R mRNA was 3 hours in control cells and 0.75 hour in cells treated with a maximally effective dose of TRH for 1.5 hours; that is, TRH increased the rate of turnover of TRH-R mRNA fourfold.

We have begun to explore whether there may be regions within the TRH-R mRNA that are involved in regulation of its turnover. In our first approach we have created a series of truncation mutations of the 3' untranslated region (3'UTR) of the cDNA. We have generated stable transfectants of GH$_3$ cells with several of these mutant cDNAs. One such cell clone is transfected with a cDNA that is truncated at base 3356; that is, it is missing the last 143 nucleotides and those that encode the poly(A)$^+$ tail of the TRH-R mRNA. The mRNA in transfected cells would contain a poly(A)$^+$ tail because a tail is generated by plasmid sequences. We refer to this cell clone as GH-mTRHR-D9 cells. These cells respond to TRH and have levels of TRH-Rs and TRH-R mRNA similar to those in GH-mTRHR-1 cells. In contrast to our findings in GH-mTRHR-1 cells, which are transfected with the wild-type cDNA containing the full 3498 bases, the level of TRH-R mRNA in GH-mTRHR-D9 cells is not down-regulated by TRH and its half-life is not down-regulated by TRH. Based on these findings we suggest that there is a region within the 3' end of the TRH-R mRNA that is necessary for its turnover to be regulated by TRH. We postulate, therefore, that the increased turnover of TRH-R mRNA caused by TRH is secondary to an effect on the mRNA itself that leads to decreased mRNA stability. An increase in the activity of an RNAase that degrades the receptor mRNA occurs also (Narayanan et al., 1992). We have begun to study these last 143 nucleotides to determine which is (are) necessary for regulation by TRH.

We conclude that the predominant effect of TRH to lower mouse TRH-R mRNA levels in stably transfected, pituitary GH-mTRHR-1 cells is caused by an increase in the rate of mRNA turnover. We suggest that a similar mechanism is involved in TRH-induced decrease in endogenous TRH-R mRNA in parental GH$_3$ cells and in other pituitary cells.

## VIII. Speculations and Future Directions

Based on the data reviewed here I propose the following working hypothesis for the mechanism of homologous down-regulation of TRH-R mRNA in pituitary cells. TRH-R mRNA down-regulation is caused by TRH-induced increases in the rate of mRNA turnover. The central idea in this hypothesis is that increased mRNA turnover is secondary to decreased mRNA stability and to increases in the degrading activity of an RNAase. TRH-R mRNA turnover in unstimulated cells is determined by the intrinsic stability of the mRNA and the state of activity (activities) of the RNase(s) present. This rate of turnover is equal to the rate of TRH-R gene transcription at steady state. I propose that there is a domain within the 3' end of the TRH-R mRNA that contains an element for regulation of mRNA stability by TRH. This element may not be a specific base sequence but may be a secondary structural feature of the mRNA (Atwater et al., 1990). The regulatory element may act as a binding region for a protein that on binding leads to decreased mRNA stability. This idea is similar to the mechanism that has been delineated for regulation of transferrin receptor mRNA stability by an iron-binding regulatory protein (Klausner and Harford, 1989), except that binding in that system causes an increase in mRNA stability. I propose that the binding affinity of this protein to the response element in the TRH-R mRNA is increased on phosphorylation, which is caused directly by protein kinase C or by a kinase(s) that is distal in this signaling pathway. That is, the stability of TRH-R mRNA is regulated by phosphorylation of a trans-acting RNA binding protein. This is analogous to regulation of gene transcription by phosphorylation of a DNA binding transcription factor (Yamamoto et al., 1988). A similar mechanism could mediate the effects of VIP on TRH-R mRNA through a protein kinase A-mediated phosphorylation.

Our plans to test this hypothesis include using in vivo and in vitro systems to delineate the mechanism of regulation of TRH-R mRNA stability. Mutations within the 3'UTR of the mRNA will be made and these will be used to define the domain within the mRNA that confers regulatability in transfected $GH_3$ cells. In parallel, an in vitro system for RNA turnover has been developed, which we plan to use to isolate, purify, and clone the protein that regulates TRH-R mRNA stability in order to study the molecular details of its interaction with TRH-R mRNA. It is now clear that regulation of mRNA turnover is an important and widespread mechanism for the control of gene expression in eukaryotic cells. We think that regulation of TRH-R mRNA turnover will be an excellent model for studies of this type and that generally applicable ideas will be forthcoming from these studies.

### ACKNOWLEDGMENTS

I gratefully thank my colleagues, especially J. Fujimoto, C. S. Narayanan, Y. Oron, and R. E. Straub, for their contributions to the work described. I thank P. J. Deutsch for critical review of this manuscript.

### REFERENCES

Akamizu, T., Ikuyama, S., Saji, M., Kosugi, S., Kozak, C., McBride, O. W., and Kohn, L. D. (1990). Proc. Natl. Acad. Sci. U.S.A. **87,** 5677–5681.
Atwater, J. A., Wisdom, R., and Verma, I. M. (1990). Annu. Rev. Genet. **24,** 519–541.

362     MARVIN C. GERSHENGORN

Birnbaumer, L. (1990). *Annu. Rev. Pharmacol. Toxicol.* **30**, 675–705.
Bouvier, M., Collins, S., O'Dowd, B. F., Campbell, P. T., De Blasi, A., Kobilka, B. K., MacGregor, C., Irons, G. P., Caron, M. G., and Lefkowitz, R. J. (1989). *J. Biol. Chem.* **264**, 16786–16792.
Brawerman, G. (1989). *Cell (Cambridge, Mass.)* **57**, 9–10.
Burnstein, K. L., Jewell, C. M., and Cidlowski, J. A. (1990). *J. Biol. Chem.* **265**, 7284–7291.
Chen, W.-J., Goldstein, J. L., and Brown, M. S. (1990). *J. Biol. Chem.* **265**, 3116–3123.
Cleveland, D. W. (1988). *TIBS* **13**, 339–343.
Collins, S., Caron, M. G., and Lefkowitz, R. J. (1991). *Annu. Rev. Physiol.* **53**, 497–508.
Dascal, N., Gillo, B., and Lass, Y. (1985). *J. Physiol.* **366**, 299–313.
Dascal, N., Ifune, C., Hopkins, R., Snutch, T. P., Lubbert, H., Davidson, N., Simon, M. I., and Lester, H. A. (1986). *Brain. Res.* **387**, 201–209.
Dascal, N., Landau, E. M., and Lass, Y. (1984). *J. Physiol.* **352**, 551–574.
DeLean, A., Garon, M., Kelly, P. A., and Labrie, F. (1977). *Endocrinology (Baltimore)* **100**, 1505–1510.
Dohlman, H. G., Caron, M. G., Deblasi, A., Frielle, T., and Lefkowitz, R. J. (1990). *Biochemistry* **29**, 2335–2342.
Drummond, A. H. (1985). *Biochem. Biophys. Res. Commun.* **127**, 63–70.
Drummond, A. H. (1986). *J. Exp. Biol.* **124**, 337–358.
Drummond, A. H., Hughes, P. J., Ruiz-Larrea, F., and Joels, L. A. (1989). *Ann. N.Y. Acad. Sci.* **553**, 197–204.
Freissmuth, M., Casey, P. J., and Gilman, A. G. (1989). *FASEB J.* **3**, 2125–2131.
Fujimoto, J., and Gershengorn, M. C. (1991). *Endocrinology (Baltimore)* **129**, 3430–3432.
Fujimoto, J., Narayanan, C. S., Benjamin, J. E., Heinflink, M., and Gershengorn, M. C. (1992). *Endocrinology (Baltimore)* **130**, 1879–1884.
Fujimoto, J., Straub, R. E., and Gershengorn, M. C. (1991). *Mol. Endocrinol.* **5**, 1527–1532.
Gershengorn, M. C. (1978). *J. Clin. Invest.* **62**, 937–943.
Gershengorn, M. C. (1985). *Recent Prog. Horm. Res.* **41**, 607–653.
Gershengorn, M. C. (1986). *Annu. Rev. Physiol.* **48**, 515–526.
Gershengorn, M. C., and Paul, M. E. (1986). *Endocrinology (Baltimore)* **119**, 833–839.
Gershengorn, M. C., Oron, Y., and Straub, R. E. (1990). *J. Exp. Zool.* (Suppl. 4), 78–83.
Gillo, B., Lass, Y., Nadler, E., and Oron, Y. (1987). *J. Physiol. (London)* **392**, 349–361.
Gundersen, C. B., Miledi, R., and Parker, I. (1984). *Nature (London)* **308**, 421–424.
Hadcock, J. R., and Malbon, C. C. (1988). *Proc. Natl. Acad. Sci. U.S.A.* **85**, 5021–5025.
Hadcock, J. R., and Malbon, C. C. (1991). *TINS* **14**, 242–247.
Hadcock, J. R., Ros, M., and Malbon, C. C. (1989). *J. Biol. Chem.* **264**, 13956–13961.
Halpern, J., and Hinkle, P. M. (1981). *Proc. Natl. Acad. Sci. U.S.A.* **78**, 587–591.
Hausdorff, W. P., Caron, M. G., and Lefkowitz, R. J. (1990). *FASEB J.* **4**, 2881–2889.
Hinkle, P. M. (1989). *Ann. N.Y. Acad. Sci.* **553**, 176–187.
Hinkle, P. M., and Kinsella, P. A. (1982). *J. Biol. Chem.* **257**, 5462–5470.
Hinkle, P. M., Shanshala, E. D., II, and Yan, Z. (1991). *Endocrinology (Baltimore)* **129**, 1283–1288.
Hinkle, P. M., and Tashjian, A. H., Jr. (1975). *Biochemistry* **14**, 3845–3851.
Hinkle, P. M., Woroch, E. L., and Tashjian, A. H Jr. (1974). *J. Biol. Chem.* **249**, 3085–3090.
Imai, A., and Gershengorn, M. C. (1985). *J. Biol. Chem.* **260**, 10536–10540.
Izzo, N. J., Jr., Seidman, C. E., Collins, S., and Colucci, W. S. (1990). *Proc. Natl. Acad. Sci. U.S.A.* **87**, 6268–6271.
Johnson, W. A., Nathanson, N. M., and Horita, A. (1984). *Proc. Natl. Acad. Sci. U.S.A.* **81**, 4227–4231.
Julius, D., MacDermott, A. B., Axel, R., and Jessell, T. M. (1988). *Science* **241**, 558–564.
Kikkawa, U., Kishimoto, A., and Nishizuka, Y. (1989). *Annu. Rev. Biochem.* **58**, 31–44.

## TRH-R: CLONING AND EXPRESSION 363

Klausner, R. D., and Harford, J. B. (1989). *Science* **246**, 870–872.

McIntosh, R. P., and Catt, K. J. (1987). *Proc. Natl. Acad. Sci. U.S.A.* **84**, 9045–9048.

Melton, D. A. (1985). *Proc. Natl. Acad. Sci. U.S.A.* **82**, 144–148.

Meyerhof, W., Morley, S., Schwarz, J., and Richter, D. (1988). *Proc. Natl. Acad. Sci. U.S.A.* **85**, 714–717.

Mu, M.-C., P., Chakrabarti, S., Hanes, S. D., and Biswas, D. K. (1982). *Biochem. Biophys. Res. Commun.* **106**, 811–817.

Nadler, E., Gillo, B., Lass, Y., and Oron, Y. (1986). *FEBS. Lett.* **199**, 208–212.

Narayanan, C. S., Fujimoto, J., Geras-Raaka, E., and Gershengorn, M. C. (1992). *J. Biol. Chem.* **267**, 17,296–17,303.

Nielsen, D. A., and Shapiro, D. J. (1990). *Mol. Endocrinol.* **4**, 953–957.

Oron, Y., Dascal, N., Nadler, E., and Lupu, M. (1985). *Nature (London)* **313**, 141–143.

Oron, Y., Gillo, B., Straub, R. E., and Gershengorn, M. C. (1987a). *Mol. Endocrinol.* **1**, 918–925.

Oron, Y., Straub, R. E., Traktman, P., and Gershengorn, M. C. (1987b). *Science* **238**, 1406–1408.

Oron, Y., Gillo, B., and Gershengorn, M. C. (1988a). *Proc. Natl. Acad. Sci. U.S.A.* **85**, 3820–3824.

Oron, Y., and Lass, Y. (1985). *Rev. Clin. Basic Pharmacol.* **5**, 15S–24S.

Oron, Y., Gillo, B., Straub, R. E., and Gershengorn, M. C. (1988b). *Mol. Endocrinol.* **1**, 918–925.

Peck, V., and Gershengorn, M. C. (1980). *J. Clin. Endocrinol. Metab.* **50**, 1144–1146.

Perlman, J. H., and Gershengorn, M. C. (1991). *Endocrinology (Baltimore)* **129**, 2679–2686.

Perrone, M. H., and Hinkle, P. M. (1978). *J. Biol. Chem.* **253**, 5168–5173.

Phillips, W. J., and Hinkle, P. M. (1989). *Mol. Pharmacol.* **35**, 533–540.

Ramsdell, J. S., and Tashjian, A. H., Jr. (1985). *Mol. Cell. Endocrinol.* **43**, 173–180.

Ramsdell, J. S., and Tashjian, A. H., Jr. (1986). *J. Biol. Chem.* **261**, 5301–5306.

Raymond, J. R., Hnatowich, M., Caron, M. G., and Lefkowitz, R. J. (1990). *In* "ADP-Ribosylating Toxins and G Proteins" (J. Moss, and M. Vaughan, eds.), pp. 163–188. American Society for Microbiology Washington, DC.

Saji, M., Ikuyama, S., Akamizu, T., and Kohn, L. D. (1991). *Biochem. Biophys. Res. Commun.* **176**, 94–101.

Shapira, H., Lupu-Meiri, M., Gershengorn, M. C., and Oron, Y. (1990). *Biophys. J.* **57**, 1281–1285.

Sharif, N. A., and Burt, D. R. (1984). *J. Neurochem.* **42**, 209–214.

Snutch, T. P. (1988). *TINS* **11**, 250–256.

Straub, R. E., Frech, G. C., Joho, R. H., and Gershengorn, M. C. (1990). *Proc. Natl. Acad. Sci. U.S.A.* **87**, 9514–9518.

Straub, R. E., Oron, Y., and Gershengorn, M. C. (1989a). *In* "Methods in Neurosciences, Vol. 1, Gene Probes" (P. M. Conn, ed.), pp. 46–61. Academic Press, San Diego.

Straub, R. E., Oron, Y., Gillo, B., Thomson, R., and Gershengorn, M. C. (1989b). *Mol. Endocrinol.* **3**, 907–914.

Sullivan, N. J., Lautens, L. L., and Tashjian, A. H., Jr. (1987). *Mol. Endocrinol.* **1**, 889–898.

Valiquette, M., Bonin, H., Hnatowich, M., Caron, M. G., Lefkowitz, R. J., and Bouvier, M. (1990). *Proc. Natl. Acad. Sci. U.S.A.* **87**, 5089–5093.

Winicov, I., Cory, R. N., and Gershengorn, M. C. (1990). *Endocrinology (Baltimore)* **126**, 1668–1672.

Winicov, I., and Gershengorn, M. C. (1989). *J. Biol. Chem.* **264**, 9438–9443.

Yamamoto, K. K., Gonzalez, G. A., Biggs, W. H., III, and Montminy, M. R. (1988). *Nature (London)* **334**, 494–498.

Yan, G., Pan, W. T., and Bancroft, C. (1991). *Mol. Endocrinol.* **5**, 535–541.

RECENT PROGRESS IN HORMONE RESEARCH, VOL. 48

# Bombesin-Like Peptides: Of Ligands and Receptors

ELIOT R. SPINDEL,* ELIEZER GILADI,* THOMAS P. SEGERSON,[†]
AND SRINIVASA NAGALLA*

*Division of Neuroscience, Oregon Regional Primate Research Center, Beaverton,
Oregon 97006 and [†] Vollum Institute for Advanced Biomedical Research, Portland,
Oregon 97201*

## I. Introduction

The bombesin-like peptides comprise a large family of peptides common to frogs and man. The tetradecapeptide bombesin, isolated from the skin of *Bombina orientalis* by Anastasi and co-workers (1971), was the first family member isolated. Bombesin, when injected into rats, proved to have a wide variety of pharmacologic effects, including effects on body temperature, cardiac output, blood pressure, and the release of gastrointestinal hormones. These many effects prompted a search for mammalian homologues of the amphibian bombesin-like peptides. Subsequently (as is described in detail below), two mammalian bombesin-like peptides have been characterized, gastrin-releasing peptide (GRP), which is related to bombesin, and neuromedin B (NMB), which is related to ranatensin. A third class of bombesin-like peptides, the phyllolitorins, has been characterized in amphibians, and although mammalian phyllolitorins likely exist, they have not yet been isolated. The bombesin-like peptides have been divided into three subfamilies, as shown in Fig. 1, based on their penultimate and adjacent amino acids. The bombesin subfamily contains Leu as its penultimate amino acid, the ranatensin subfamily contains Phe as its penultimate amino acid, and the phyllolitorins occur in both a Leu (bombesin-like) and Phe (ranatensin-like) form, but substitute a serine adjacent to the penultimate residue.

When GRP and NMB were initially characterized in mammals, they were considered the mammalian homologues of bombesin and ranatensin, respectively. That is, it was believed that frogs would have bombesin or ranatensin and mammals would have GRP and NMB. Recent evidence, however, shows this is not true and that frogs have both GRP and bombesin. Thus the family of bombesin-like peptides is more complicated than previously conceived and it is likely that there may be four or more different mammalian bombesin-like peptides and an equivalent number of receptors for these peptides.

365

Copyright © 1993 by Academic Press, Inc.
All rights of reproduction in any form reserved.

366    ELIOT R. SPINDEL ET AL.

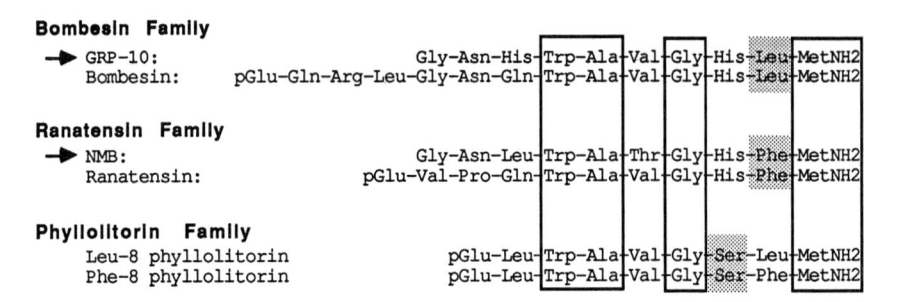

FIG. 1.    Bombesin-like peptide subfamilies. Representative members of each of the three subfamilies are shown. The shaded regions show the amino acid residue that defines each family. Amino acid residues that are conserved among all the subfamilies are boxed. The two horizontal arrows indicate the two mammalian bombesin-like peptides, NMB and GRP-10. GRP-10 is the C-terminal decapeptide of GRP and is highly conserved among mammalian species (see Fig. 2). pGlu denotes pyroglutamate; NH2 denotes the carboxyl-terminal amide.


## II.  Ligands


### A.  BOMBESIN/GRP

Bombesin was initially isolated from frog skin in 1971 (Anastasi *et al.*, 1971). In 1973, Bertaccini *et al.* (1973) demonstrated that bombesin induced gastrin release in canine stomach. Over the next few years, the ability of bombesin to stimulate secretion of pancreatic enzymes, cholecystokinin, and other gastrointestinal hormones was reported (Erspamer and Melchiorri, 1975; Erspamer *et al.*, 1974; Deschodt Lanckman *et al.*, 1976; Konturek *et al.*, 1976; Basso *et al.*, 1975). In 1977, Brown *et al.* reported that bombesin injected into the CNS had profound effects on thermoregulation (1977). In 1978, Moody *et al.* and Jensen *et al.* demonstrated specific binding of bombesin to brain and pancreatic membranes. These many potent effects of bombesin in mammals prompted a search for the mammalian homologue of bombesin. Such a peptide was isolated in 1979 by McDonald *et al.* using gastrin release as a bioassay. This peptide was named gastrin-releasing peptide and shares a C terminus with amphibian bombesin (Figs. 1 and 2). GRP has been isolated or cloned from a variety of mammalian and nonmammalian species (Orloff *et al.*, 1984; Reeve *et al.*, 1983; McDonald *et al.*, 1979, 1980; Shaw *et al.*, 1987; Lebacq-Verheyden *et al.*, 1988; Spindel *et al.*, 1984; Nagalla *et al.*, 1992; Conlon *et al.*, 1987), whereas bombesin has been characterized only in frogs of the *Bombina* genus. As shown in Fig. 2, all the GRPs share considerable homology in the C terminus. Full biological activity of GRP appears to reside in the C-

BOMBESIN-LIKE PEPTIDES

```
                     1                                                        13
1  Human GRP:     Val┤Pro-Leu-Pro├    -    ├Ala-Gly-Gly-Gly-Thr-Val-Leu-Thr-Lys─
2  Porcine GRP:   Ala┤Pro┤Val-Ser-    -     -Val├Gly-Gly-Gly-Thr-Val-Leu-Ala-Lys─
3  Rat GRP:       Ala┤Pro┤Val-Ser-Thr-Gly├Ala-Gly-Gly-Gly-Thr-Val-Leu-Ala-Lys─
4  Canine GRP:    Ala┤Pro┤Val┤Pro├    -     -Gly├Gly┤Gln├Gly-Thr-Val-Leu-Asp-Lys─
5  Avian GRP:     Ala┤Pro┤Leu┤Gln-    -     -Pro├Gly┤Gly┤Ser-Pro-Ala├Leu-Thr-Lys─
6  Shark GRP:     Ala┤Pro┤Val-Glu-    -    -   -    -Asn-Gln-Gly-Ser┤Phe├Pro├Lys─
7  Frog GRP:      Ser┤Pro┤Thr-Ser-Gln-Gln-His-Asn-Asp-Ala-Ala-Ser┤Leu┤Ser┤Lys─

                     14            17                                          27
1  Human GRP:     ┌Met-Tyr-Pro-Arg-Gly-Asn-His-Trp-Ala-Val-Gly-His-Leu-MetNH2
2  Porcine GRP:   │Met-Tyr-Pro-Arg-Gly-Asn-His-Trp-Ala-Val-Gly-His-Leu-MetNH2
3  Rat GRP:       │Met-Tyr-Pro-Arg-Gly-Ser-His-Trp-Ala-Val-Gly-His-Leu-MetNH2
4  Canine GRP:    └Met-Tyr-Pro-Arg-Gly-Asn-His-Trp-Ala-Val-Gly-His-Leu-MetNH2
5  Avian GRP:      Ile┤Tyr-Pro-Arg-Gly┤Ser├His-Trp-Ala-Val-Gly-His-Leu-MetNH2
6  Shark GRP:      Met┤Phe├Pro-Arg-Gly┤Ser├His-Trp-Ala-Val-Gly-His-Leu-MetNH2
7  Frog GRP:       Ile┤Tyr-Pro-Arg-Gly┤Ser├His-Trp-Ala-Val-Gly-His-Leu-MetNH2
8  Bombesin:      pGlu-Gln-Arg-Leu┤Gly├Asn-Gln┤Trp-Ala-Val-Gly-His-Leu-MetNH2
                  └─────────────────── GRP-10 ───────────────────┘
```

FIG. 2. Peptides in the bombesin/GRP subfamily. Numbering is relative to human GRP. Sequences are derived from protein sequencing or cDNA cloning as follows: human GRP (Orloff *et al.*, 1984; Spindel *et al.*, 1984), porcine GRP (McDonald *et al.*, 1979), canine GRP (Reeve *et al.*, 1983), avian GRP (McDonald *et al.*, 1980), rat GRP (Lebacq-Verheyden *et al.*, 1988), guinea pig GRP (not shown; it is the same as rat GRP) (Shaw *et al.*, 1987), frog GRP (Conlon *et al.*, 1991; Nagalla *et al.*, 1992), mud shark GRP (Conlon *et al.*, 1987), and amphibian bombesin (Anastasi *et al.*, 1971). Homologies with human GRP are boxed. The region of GRP-10 in the GRPs is as shown. pGlu denotes pyroglutamate; MetNH2 denotes methionine amide.

terminal decapeptide (GRP-10) of GRP, and this region is most homologous to bombesin. GRP-10 appears to be present in all species that have GRP. At present, all biological activities of GRP can be reproduced by GRP-10 and by bombesin. The significance of the N terminus of GRP remains to be determined.

GRP was considered mammalian bombesin until Nagalla *et al.* (1991, 1992) and Conlon *et al.* (1991) showed that frogs have both bombesin and GRP. As shown in Fig. 3, frogs express distinctly different GRP and bombesin mRNAs in stomach extracts. The homology between frog GRP and mammalian GRP prohormones is significantly higher (alignment scores = 190–200) than between the frog GRP and frog bombesin prohormones (alignment scores < 100). Thus GRP and bombesin separated quite early in the evolutionary scale. The relatively high levels of bombesin mRNA in frog brain (Fig. 3) make it likely that mammals, like frogs, will have a bombesin mRNA expressed in the CNS distinct from the GRP mRNA.

By RIA and immunohistochemistry, GRP has been shown to be widely distributed in mammals. In brain, GRP is widely distributed in neurons and has been implicated in the control of appetite, thermal regulation,

368                    ELIOT R. SPINDEL ET AL.

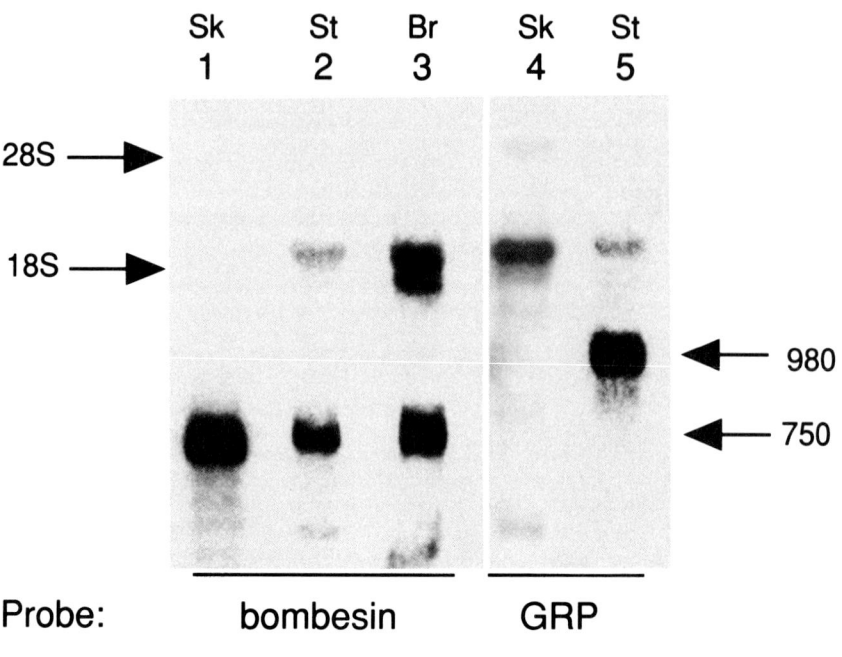

FIG. 3.  Northern blot analysis of bombesin and GRP mRNA expression in *Bombina orientalis;* 10 μg of total or poly(A)$^+$ RNA was resolved on 1.5% formaldehyde–agarose, transferred to a nylon membrane, and probed with the bombesin (lanes 1–3) or GRP cRNA (lanes 4 and 5) at 65°C in 50% formamide, with washing in 0.1× SSC, 65°C. Lane 1, 2.5 μg of total RNA from norepinephrine-stimulated dorsal skin (Sk); lanes 2 and 5, 10 μg of poly(A)$^+$ RNA from stomach (st); lane 3, 10 μg total RNA from brain (Br); lane 4, 10 μg of poly(A)$^+$ RNA from norepinephrine-stimulated dorsal skin. Because of the large differences in RNA levels, the autoradiogram was exposed for three different times. Lane 1, 15-minute exposure; lanes 2, 3, and 5, 24 hours; lane 4, 72 hours.


sympathetic output, blood sugar, and diverse other processes. Consistent with this, high levels of GRP are located in the hypothalamus and visceral nuclei of the brain stem (Wada *et al.*, 1990; Chronwall *et al.*, 1985; Roth *et al.*, 1982; Panula *et al.*, 1982).

GRP is also widely distributed in the gastrointestinal (GI) tract (Polak *et al.*, 1976; Dockray *et al.*, 1979; Price *et al.*, 1984), where it controls diverse gastrointestinal functions including motility (Bertaccini and Impicciatore, 1975), anion secretion (Kachur *et al.*, 1982) and stimulation of gastrointestinal hormone secretion. GRP is a potent secretagogue for gastrin, insulin, glucagon, vasoactive intestinal polypeptide, pancreatic polypeptide, cholecystokinin (CCK), and pancreatic exocrine enzymes, including trypsin and amylase (McDonald *et al.*, 1983; Basso *et al.*, 1975;

Knuhtsen *et al.*, 1985). Beyond these functions GRP is also a growth factor for GI tissues. Both parenteral and nonparenteral administration of GRP cause hyperplasia of GI epithelial cells and pancreatic hyperplasia (Lehy *et al.*, 1983, 1986; Puccio and Lehy, 1989). Interestingly, in light of the effects of GRP on GI growth, GRP has been detected in both human and bovine milk (Jahnke and Lazarus, 1984; Takeyama *et al.*, 1991), where it may act as a growth factor for the developing GI tract. In mammals, all GRP in the GI tract appears to be in the intrinsic neurons (Dockray *et al.*, 1979; Schultzberg and Dalsgaard, 1983), where it acts as a paracrine regulator of GI function. By contrast, in avian and amphibian species, GRP appears to be in endocrine cells (Vaillant *et al.*, 1979; Lechago *et al.*, 1981).

GRP is developmentally expressed in the lung. High levels of GRP are detected in fetal lung from approximately 10 to 30 weeks of gestation (Johnson *et al.*, 1982; Spindel *et al.*, 1987). In lung GRP is found in pulmonary neuroendocrine cells (Johnson *et al.*, 1982; Stahlman *et al.*, 1985). These cells are thought to be the progenitor cells of small cell lung carcinoma (SCLC) (Gould *et al.*, 1983; Sunday *et al.*, 1988) and, consistent with this, most small cell lung carcinomas express GRP mRNA (Erisman *et al.*, 1982; Moody *et al.*, 1981). Depending on the degree of neuroendocrine differentiation, the GRP mRNA will be translated and processed to mature GRP. Many large cell lung carcinomas also express GRP (Sunday *et al.*, 1991), and, again, whether mature GRP is expressed follows the degree of neuroendocrine differentiation (Carney *et al.*, 1985; Bepler *et al.*, 1989; Sunday *et al.*, 1991).

Both normal and neoplastic pulmonary cells show a mitogenic response to GRP (Cuttitta *et al.*, 1985; Weber *et al.*, 1985; Willey *et al.*, 1984). As some SCLCs both express and respond to GRP, Cuttitta *et al.* (1985) have proposed that GRP is an autocrine growth factor for SCLC. Consistent with this, antibodies to bombesin cause inhibition of SCLC growth both *in vitro* and *in vivo* (Cuttitta *et al.*, 1985). Similarly, GRP receptor antagonists can cause inhibition of the growth of some SCLC cell lines (Woll and Rozengurt, 1990; Mahmoud *et al.*, 1991). The exact function of GRP in normal lung development remains to be determined.

## B. RANATENSIN/NEUROMEDIN B

At about the same time that Anastasi *et al.* (1971) described the isolation of bombesin from the skin of *Bombina bombina*, Nakajima *et al.* (1970) independently described the isolation of the related peptide ranatensin from the skin of *Rana pipiens*. As shown in Fig. 1, bombesin and ranatensin are highly homologous, the most critical difference being the penultimate

370                    ELIOT R. SPINDEL ET AL.

residue, which is Leu in bombesin and Phe in ranatensin. Tests in mammals showed that ranatensin, like bombesin, had a variety of effects. Ranatensin had strong effects on smooth muscle and blood pressure (Clineschmidt *et al.*, 1971; Geller *et al.*, 1970) and also replicated many of bombesin's effects on GI and CNS function, but with less potency (Modlin *et al.*, 1981; Mukai *et al.*, 1987). Studies using antibodies to ranatensin with only minimal cross-reactivity to GRP or bombesin showed a distribution of ranatensin immunoreactivity distinct from that of GRP in rats (Chronwall *et al.*, 1985). Then, in 1983, Minamino *et al.*, isolated a peptide from porcine spinal cord that was highly homologous to ranatensin; this peptide, because it was the second peptide this research group had isolated from spinal cord, was named neuromedin B. The initial NMB isolated comprised 9 amino acids, but subsequently a 32-amino acid form of NMB (NMB-32) was isolated by Minamino *et al.* (1985) (Fig. 4). Thus NMB, like GRP, exists in both a short and long form.

cDNAs encoding NMB have been isolated from rat and human species (Krane *et al.*, 1988; Wada *et al.*, 1990). Like GRP, homology in the amino terminus of the large form of NMB (NMB-32) is striking (Fig. 4). As for GRP, the function of this highly conserved amino terminus is completely unknown.

NMB has been considered the mammalian homologue of ranatensin, but in light of the new knowledge as to the relation of GRP and bombesin, quite likely frogs will turn out to have both an NMB and a ranatensin.

```
            1                                                        16
1   Human NMB-32:   Ala-Pro-Leu-Ser-Trp-Asp-Leu-Pro-Glu-Pro-Arg-Ser-Arg-Ala-Ser-Lys-
2   Porcine NMB-32: Ala-Pro-Leu-Ser-Trp-Asp-Leu-Pro-Glu-Pro-Arg-Ser-Arg-Ala-Gly-Lys-
3   Rat NMB-32:     Thr-Pro-Phe-Ser-Trp-Asp-Leu-Pro-Glu-Pro-Arg-Ser-Arg-Ala-Ser-Lys-


            17          22                                          32
1   Human NMB-32:   Ile-Arg-Val-His-Ser-Arg-Gly-Asn-Leu-Trp-Ala-Thr-Gly-His-Phe-MetNH2
2   Porcine NMB-32: Ile-Arg-Val-His-Pro-Arg-Gly-Asn-Leu-Trp-Ala-Thr-Gly-His-Phe-MetNH2
3   Rat NMB-32:     Ile-Arg-Val-His-Pro-Arg-Gly-Asn-Leu-Trp-Ala-Thr-Gly-His-Phe-MetNH2
4   NMB:                            Gly-Asn-Leu-Trp-Ala-Thr-Gly-His-Phe-MetNH2
5   Rohdei-litorin:                 pGlu-Leu-Trp-Ala-Thr-Gly-His-Phe-MetNH2
6   Ranatensin-R:   Ser-Asn-Thr-Ala-Leu-Arg-Arg-Tyr-Asn-Gln-Trp-Ala-Thr-Gly-His-Phe-MetNH2
7   Ranatensin:                     pGlu-Val-Pro-Gln-Trp-Ala-Val-Gly-His-Phe-MetNH2
8   Litorin:                            pGlu-Gln-Trp-Ala-Val-Gly-His-Phe-MetNH2
```

FIG. 4. Peptides in the ranatensin/NMB subfamily. Sequences derived from protein sequencing or cDNA cloning as follows: human NMB (Krane *et al.*, 1988), porcine NMB (Minamino *et al.*, 1983, 1985), rat NMB (Wada *et al.*, 1990), and the amphibian peptides rohdei-litorin (Barra *et al.*, 1985), ranatensin-R (Yasuhara *et al.*, 1979), ranatensin (Nakajima *et al.*, 1970), and litorin (Anastasi *et al.*, 1975). Homologies with human NMB-32 are boxed. NMB (the C-terminal decapeptide of NMB-32) is identical in humans, pigs, and rats. pGlu denotes pyroglutamate; MetNH2 denotes methionine amide.

Whether mammals will also have two such peptides remains to be determined. The distribution of NMB is more localized than that of GRP and generally NMB levels are lower than GRP levels (Minamino *et al.,* 1984; Namba *et al.,* 1985; Sakamoto *et al.,* 1987). In brain, NMB is present, among other regions, in the olfactory bulb, hippocampus, brain stem, and substantia nigra (Wada *et al.,* 1990; Chronwall *et al.,* 1985). Principal differences in the CNS distribution of NMB compared to GRP consist of the presence of NMB in the olfactory bulb, substantia nigra, and the trigeminal sensory nucleus.

NMB is also found throughout the GI tract but, except in the esophagus, at lower levels than GRP. Because NMB cross-reacts with the GRP receptor (see Section II), it has been difficult to clearly differentiate the effects of GRP and NMB. Only recently was it discovered that the NMB receptor could be differentiated from the GRP receptor by two criteria: a higher affinity for NMB than GRP and an insensitivity to specific GRP receptor antagonists such as [D-Phe[6]]bombesin(6-13)propylamide and [Leu[13],$\psi$CH-$_2$NHLeu[14]]bombesin (von Schrenck *et al.,* 1990). By these criteria it has been shown that the esophagus expresses a fairly pure population of NMB receptors (von Schrenck *et al.,* 1988, 1989, 1990). Physiologic roles for NMB have not yet been clearly delineated, but with the availability of pharmacologic agents this should change.

Interestingly, it has recently been shown that many SCLCs also express NMB and NMB receptors (Corjay *et al.,* 1991; Cardona *et al.,* 1991a). The interaction of NMB with both NMB and GRP receptors can produce a mitogenic response, thus there may be a complicated interplay between these ligands and receptors.

## C. PHYLLOLITORINS

The phyllolitorins remain the mystery of the bombesin-like peptide family. These peptides (Fig. 1) are characterized by a Ser as the third-to-last amino acid, whereas the bombesin subfamily and the ranatensin subfamily have His. Interestingly, the phyllolitorins occur in a Leu-8 form similar to bombesin and a Phe-8 form similar to ranatensin. To date the phyllolitorins have only been isolated from the *Phyllomedusa* genus of tree frogs and it is not yet known if there are mammalian homologues. There are, however, several intriguing pieces of evidence to suggest that there may be mammalian homologues of the phyllolitorins. First, there have been several anecdotal reports of phyllolitorin-like immunoreactivity, although it is not clear to what extent these reports represented cross-reactivity with GRP and NMB-like peptides. Second, specific receptors for phyllolitorin, distinct from the receptors for GRP and NMB, have been reported in some SCLC cell lines and in WiDR cells (a human colon adenocarcinoma cell line) (Cardona *et al.*, 1991b). As to whether these receptors are truly different from the GRP and NMB receptors remains to be determined. Consistent with the idea that pharmacologically distinct phyllolitorins and their receptors exist, the pharmacologic activities of phyllolitorins are clearly distinct from the activities of peptides of the bombesin and ranatensin subfamilies (Falconieri Erspamer *et al.*, 1984; Broccardo and Cardamone, 1985; Negri *et al.*, 1988).

## III. Receptors for the Bombesin-Like Peptides

The number of bombesin-like peptides and the extent to which they bind to multiple receptors has made it clear that cloning of the receptors is necessary to understand the specific physiologic roles of each bombesin-like peptide. The expression of GRP and GRP receptors in neoplasia and the possibility that mutations in receptor structure could lead to oncogenic activation has made a molecular understanding of the bombesin receptors even more important.

Membrane bound receptors in mammals capable of binding amphibian bombesin were initially described by Jensen *et al.* (1978) on pancreatic acinar cells and by Moody *et al.* (1978) in rat brain. Westendorf and Schonbrunn demonstrated the presence of bombesin receptors in $GH_4$ cells (1983), which was consistent with reports by Rivier *et al.* (1978) on the effects of bombesin on prolactin and growth hormone secretion. Even in these early studies, the difference in affinities for bombesin-like and ranatensin-like peptides for membrane binding was noted. Subsequent to the discovery of bombesin-like peptide expression in SCLCs (Moody *et al.*, 1981; Erisman *et al.*, 1982), Moody *et al.* demonstrated the expression of bombesin-like peptide receptors in SCLCs and other neoplasms (1983, 1985).

The potential role of bombesin-like peptides as growth factors was reported initially by Rozengurt and Sinnett-Smith (1983), who showed bombesin was a mitogen for Swiss 3T3 cells, and by Lehy *et al.* (1983), who demonstrated that bombesin administration stimulated antral gastrin cell proliferation in the rat. The critical observation by Zachary and Rozengurt (1985) that Swiss 3T3 cells expressed greater than 100,000 bombesin receptors per cell provided a cell line to permit the eventual molecular characterization of the bombesin receptors.

In Swiss 3T3 cells, the mitogenic effect of bombesin and GRP was clearly demonstrated: both bombesin and GRP bound with a $K_d$ of $0.5–2.0 \times 10^{-9} M$ (Zachary and Rozengurt, 1985) and NMB bound with 5- to 10-fold less potency. In Swiss 3T3 cells it was demonstrated that bombesin-like peptides activated phospholipase C, resulting in increases

BOMBESIN-LIKE PEPTIDES 373

in inositol phosphates, diacylglycerol, and intracellular calcium (Heslop *et al.*, 1986; Takuwa *et al.*, 1987; Lopez-Rivas *et al.*, 1987). These effects were clearly mediated through a G protein-coupled receptor because non-hydrolyzable GTP analogs inhibited the binding of bombesin to membranes of cells expressing bombesin receptors (Fischer and Schonbrunn, 1988; Plevin *et al.*, 1990; Sharoni *et al.*, 1990). The G protein involved in the initial binding of bombesin agonists and in the activation of phospholipase C does not appear to be sensitive to cholera toxin or pertussis toxin (Fischer and Schonbrunn, 1988; Taylor *et al.*, 1988; Zachary *et al.*, 1987; Viallet *et al.*, 1990).

Though bombesin-stimulated secretory events and the initial binding of bombesin-like peptides are phospholipase C linked and are not inhibited by pertussis or cholera toxin, the mechanism of bombesin-induced mitogenesis is more complex. Although pertussis toxin and cholera toxin do not block initial G protein binding they do block the subsequent mitogenic response of bombesin (Letterio *et al.*, 1986; Zachary *et al.*, 1987; Viallet *et al.*, 1990). Additionally, down-regulation of phospholipase C with phorbol esters does not block the mitogenic response to bombesin (Zachary *et al.*, 1987; Brown *et al.*, 1990). Thus there appears to be dual pathway for the actions of bombesin-like peptides, partly through IP3 and partly through an as yet unknown mechanism. Though the known bombesin-like peptide receptors do not have any intrinsic tyrosine kinase activity (see below), there is apparent activation of tyrosine kinase activity with activation of the bombesin receptor (Cirillo *et al.*, 1986; Gaudino *et al.*, 1988; Zachary *et al.*, 1991). The significance of this phosphorylation remains unclear (Isacke *et al.*, 1986), though the phosphorylation, like the mitogenic effects of bombesin, is not blocked by down-regulation of protein kinase C (Zachary *et al.*, 1991).

## A. MOLECULAR CHARACTERIZATION OF THE GRP RECEPTOR

The high number of receptors in Swiss 3T3 cells made them the logical starting point for characterizing and cloning the receptors for the bombesin-like peptides. Early efforts to purify to homogeneity the bombesin receptor expressed in 3T3 cells proved unsuccessful, though SDS–PAGE of partially purified receptors showed a molecular weight of 70,000–80,000, which could be decreased to 43,000 by *N*-glycanase treatment (Kris *et al.*, 1987; Zachery and Rozengurt, 1987; Ghatei *et al.*, 1983). The success of Masu *et al.* (1987) and Julius *et al.* (1988) in cloning IP3-linked receptors by expression in *Xenopus* oocytes and the demonstration that bombesin receptors could be expressed in *Xenopus* oocytes (Moriarty *et al.*, 1988; Meyerhof *et al.*, 1988) prompted us to use oocyte expression cloning to clone the Swiss 3T3 bombesin receptor.

*Xenopus* oocytes injected with poly(A) RNA from Swiss 3T3 cells showed, in response to bombesin, a membrane depolarization that could be blocked by specific bombesin antagonists (Spindel *et al.*, 1989). A cDNA library from Swiss 3T3 cells was prepared in the vector λZAPII (Stratagene) and RNA transcribed from pools of 50,000 clones was expressed in oocytes. Screening of oocytes was initially performed by two-electrode voltage-clamp analysis and then by a simpler luminometric assay that measured the bombesin-induced increase in oocyte intracellular calcium (Sandberg *et al.*, 1988; Giladi and Spindel, 1991). For this method, *Xenopus* oocytes are coinjected with the RNA of interest and the calcium photoprotein aequorin (Blinks, 1989; Shimomura, 1991). The binding of bombesin to its receptor increases intracellular calcium in the oocyte, which causes the aequorin to emit light. This light can be detected in a luminometer, thus providing a simple, automated assay of receptor expression. A positive pool of 50,000 clones was identified and successive pools were subdivided; a single clone encoding the bombesin receptor was isolated (Fig. 5A). Oocytes expressing the bombesin receptor showed up to a 10,000-fold increase in light output (Spindel *et al.*, 1990).

Oocytes injected with 0.5–1.0 ng of RNA transcribed from this clone gave a log-linear response from $10^{-10}$ to $10^{-6}$ $M$ bombesin (Fig. 5B). The $ED_{50}$ of this response was approximately $3 \times 10^{-9} M$ bombesin, in good agreement with the published data for the binding of $^{125}$I-labeled bombesin to Swiss 3T3 cells (Rozengurt and Sinnett-Smith, 1983; Kris *et al.*, 1987; Zachary and Rozengurt, 1987). The effect of $3 \times 10^{-9}$ $M$ bombesin was blocked by $3 \times 10^{-8} M$ of the specific bombesin antagonist [D-Phe$^6$]bombesin(6-13)propylamide and by $3 \times 10^{-7}$ $M$ of the bombesin antagonist [Leu$^{13}$,ψCH$_2$NHLeu$^{14}$]bombesin, thus demonstrating the specificity of this response (Wang *et al.*, 1990a,b; Coy *et al.*, 1988; Dickinson *et al.*, 1988) (Fig. 5C). Note that these GRP antagonists, as do most GRP antagonists, demonstrate partial agonist activity. Oocytes expressing the 3T3 bombesin receptor showed similar responses to GRP and bombesin, and a slightly weaker response to NMB. Thus the 3T3 bombesin receptor meets the pharmacological criteria to be considered the GRP-preferring bombesin receptor subtype, or simply, the murine GRP receptor (mGRP-R) (Spindel *et al.*, 1990). As measured by two-electrode voltage-clamp analysis, oocytes injected with the mGRP-R cDNA showed a large inward depolarizing current, followed by a rapid return to baseline despite continued exposure to bombesin. Thereafter oocytes were refractory to continued exposure to bombesin, NMB, or GRP. This desensitization was also observed by luminescence assay. Treatment of oocytes with $10^{-10}$ $M$ bombesin blocked subsequent response to $10^{-8}$ $M$ bombesin for periods longer than 60 minutes.

The structure of the murine GRP-R was deduced from its DNA sequence

(Fig. 6). The GRP-R is 384 amino acids long with a predicted molecular weight of 43,200. Hydrophobicity analysis shows the presence of seven potential membrane-spanning domains. There are three potential N-linked glycosylation sites in the amino terminus and two in the carboxyl terminus. The carboxyl terminus of the receptor is rich in Ser, Thr, and Tyr residues, which may represent potential phosphorylation sites. Phosphorylation of these sites may play a role in the desensitization of the GRP-R, as has been proposed for the $\beta_2$-adrenergic receptor (Strader *et al.*, 1989; Hausdorf *et al.*, 1989). The cysteine at position 340 is a potential site for palmitoylation

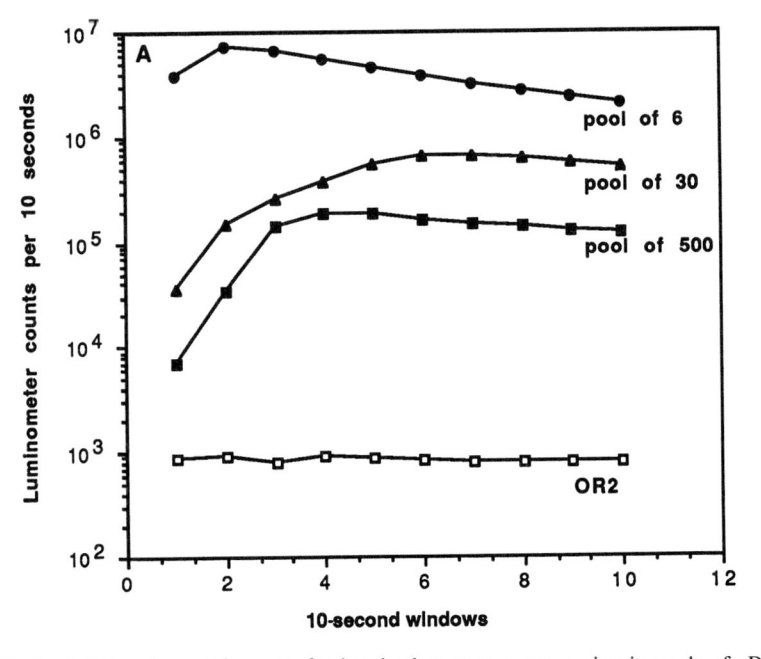

FIG. 5. (A) Luminometric assay for bombesin receptor expression in pools of cDNA clones. RNA was transcribed from pools of the size shown and coinjected with aequorin into oocytes. OR2 (open boxes) or 2 $\mu M$ bombesin (solid symbols) was injected into the luminometer and light output was counted in 10-second windows. Negative pools showed no response to bombesin. (B) Dose response of oocytes injected with 0.5 ng RNA transcribed from MBR1. For each concentration, four oocytes were individually tested first with OR2 (open boxes) then with bombesin (solid boxes). Total light output over 90 seconds is plotted logarithmically. The midpoint of the log-linear response to bombesin was $3 \times 10^{-9} M$ bombesin. (C) Effect of bombesin antagonists. Oocytes were injected with 0.5 ng RNA transcribed from MBR1 and treated with antagonist plus or minus $3 \times 10^{-9} M$ bombesin. Concentrations of antagonists are as shown. Phe6Bn, [D-Phe[6]]bombesin(6-13)propylamide; $\psi$Leu 13,14 Bn, [Leu[13],$\psi$CH$_2$NHLeu[14]]bombesin (Wang *et al.*, 1990a,b; Coy *et al.*, 1988). Values shown are means ± SEM of the responses of four individual oocytes. Four oocytes were tested per group. Fig. 5 continues.

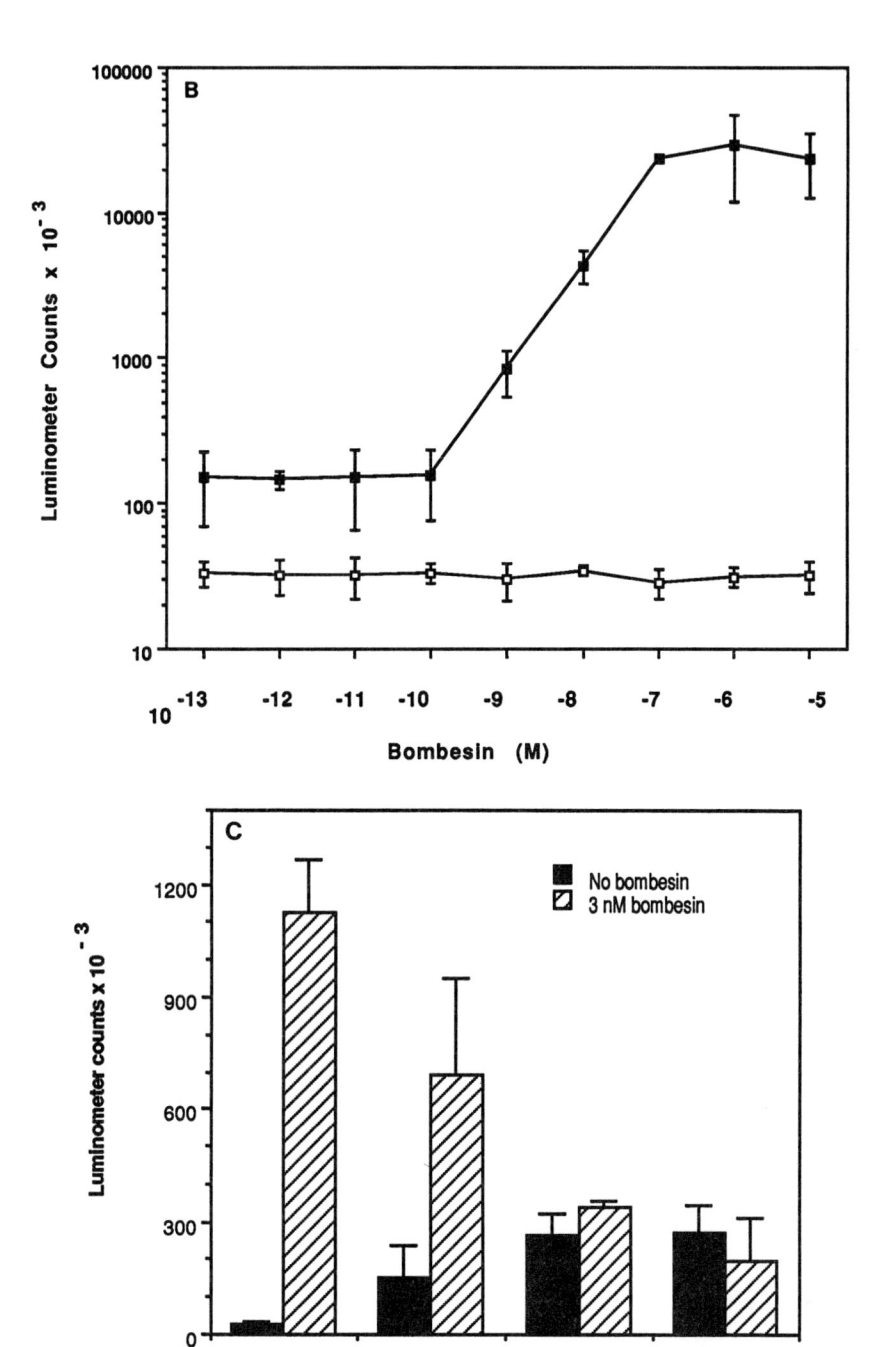

FIG. 5.—*Continued*

(ODowd *et al.*, 1989). Not including the NMB receptor, data base analysis shows greatest similarity between the mGRP-R and tachykinin receptors (Sasai and Nakanishi, 1989; Yokota *et al.*, 1989; Shigemoto *et al.*, 1990); the next closest is the rat $\beta_2$ receptor (Fig. 6). Homology is highest in the hydrophobic domains, and the aspartate in the second hydrophobic domain that has been linked to ligand binding is conserved in the GRP-R (Strader *et al.*, 1987). Homology between the GRP-R and the tachykinin receptors is particularly striking in the fourth and sixth membrane-spanning domains (Fig. 6). A GRP-R of identical sequence was also isolated by Battey *et al.* (1991) using a protein isolation scheme (Feldman *et al.*, 1990) to derive peptide sequence for cloning with mixed oligonucleotides.

No tyrosine kinase domains are present in the sequence of the GRP receptor. Thus bombesin-induced tyrosine phosphorylation events in Swiss 3T3 cells (Cirillo *et al.*, 1986; Zachary *et al.*, 1991) must occur through accessory proteins or downstream events. The role of these phosphorylation events remains to be determined.

Because of the oncogenic potential of the GRP receptor, it was crucial to determine the structure of the GRP receptor expressed in SCLCs. A cDNA encoding the human GRP receptor (hGRP-R) was isolated from a library prepared from the SCLC cell line H345. Sequence analysis of this clone revealed a 2600-base insert encoding a 384-amino acid protein with seven hydrophobic domains. Amino acid identity with the mouse and rat GRP receptors was greater than 90% (Fig. 7). Expressed in oocytes, the hGRP-R showed a response to bombesin similar to that of the murine GRP receptor. Consistent with its identity as the GRP-preferring subtype, when expressed in oocytes, the response of the hGRP-R to bombesin could be blocked by the specific GRP-R antagonist, [D-Phe$^6$]bombesin(6-13)propylamide (Wang *et al.*, 1990a,b; Wada *et al.*, 1991).

Northern blot analysis showed that the hGRP-R was encoded by a 9.0-kb mRNA and was similar in size to the murine GRP-R (mGRP-R). High levels of hGRP-R were found in stomach, colon, and duodenum (Fig. 8). Abundant hGRP-R mRNA was also present in PC3 prostate carcinoma U138 glioblastoma and T47d breast carcinoma cells. Curiously, very little GRP-R could be found in human fetal lung at a time when levels of GRP are quite elevated. Thus the issue of GRP-R expression in developing lung requires further study.

Whether there are any alterations in the structure of the SCLC GRP receptor was determined in two ways. First, the GRP-R RNA expressed in a panel of SCLC cell lines and in normal human brain and GI tissue was examined by RNase protection. Using three difference probes derived from the H345 GRP-R, no evidence for an altered GRP receptor was found in any of the SCLC cell lines, GI tract, or brain tissues examined.

```
                                                                       M1
                                                            50
Bn   ------MAPNNCSHL---NLDVDPFLSCNDT------FNQSLSPPKMDNWFHPGFIYVI-PAVYGLIIVIGLI-GNITLIKIFCTVKSMRNVPNLFISSL
SP   -----------MDNVLPMDSDLFPNISTN---------TSES------NQFVQPTWQIVLWAAAYTVIVVTSVV-GNVVVIWIILAHKRMRTVTNYFLVNL
A1   MNPDLDTGHNTSAPAQWGELKDANFTGPNQTS------SNSTLPQLDVTR------AISVGLVLGAFILFAIV-GNILVILSVACNRHLRTPTNYFIVNL
B2   --------------MEPHGNDSDFLLAPNGSRAPGHDITQ-----------ERDEAWVVGMAILMSVIVLAIVFGNVLVITAIAKFERLQTVTNYFITSL

       M2                            M3                                    M4
                 100                                    150
Bn   ALGDLLLVTCAPVDASKYLADRWLFGRIGCKLIPFIQLTSVGVSVFTLTALSADRYKAIVRPMDIQASHALMKICLKAALIWIVSMLLAI.PEAVFSDL
SP   AFAEACMAAFNTVVNFTYAVHNVWYYGLFYCKFHNFFPIAALFASIYSMTAVAFDRYMAIIHPLQPRLSATATKVVI..FVIWVLALLLAF.PQGYYSTT
A1   AIADLLLSFTVLPFSATLEVLGYWVLGRIFCDIWAAVDVLCCTASILSLCAISIDRYIGVRYSLQYPTLVTRRKAILALLSVWVLSTVISIGPLLGWKEP
B2   ACADLVMGLAVVPFGASHILMKMWNFGNFWCEFWTSIDVLCVTASIETLCVIAVDRYVAITSPFKYQSLLTKNKARVVILMVWIVSGL.....TSFLPIQ

                                 M5
                 200                                                                    250
Bn   HPFHVKDTNQTFISCAPYP.HSNELHPKIHSMASFLVFYVIPLAIISVYY...YFIA....RNLIQS.........AYNLP.VEGNIHVK.......KQI
SP   ETMPSR.....VVCMIEWPEHPNRTYEKAYHICVTVLIYFLPLLVIGYAY.......................TVVGITL..WASEIPGDSSDRYHEQVS
A1   APNDDKECGVT..........EEPFYALFSSLGSFY....IPLAVILVMYCRVYIVAKRTTKNLEAGVMKEMSNSKELTLRIHSKNFHEDTLSSTKAKGH
B2   MHWYRATHKQAIDCYAK.ETCCDFFTNQAYAIASSIVSFYVPLVVMVFVYSRVFQVAKRQLQKIDKS....EGRFHAQNLSQVEQDGR.......SGHGL

                       M6                                    M7
                                          300
Bn   ESR.............KRLAKTVLVFVGLFAFCWLPNHVIYLYRSYHYSEVDTSMLHFVTSIC..ARLLAFTNSCVNPFALYLLSKSFRKQFNTQLLCCQ
SP   AKR.............KVVKMMIVVVCTFAICWLPFHVFFLLP...YINPDLYLKKFIQQVYLASMWLAMSSTMYNPIIYCCLNDRFRLGFKHAFRCC.
A1   NPRSSIAVKLFKFSREKKAAKTLGIVVGMFILCWLPFFIALPLGSL.F.....STLKPPDAVFKVVFWLGYFNSCLNPIIYPCSSKEFKRAF.MRILGCQ
B2   RSSSKFC......LKEHKALKTLGIIMGTFTLCWLPFFIVNIVHVIRANLIPKEVYILLN.......WLGYVNSAFNPL.IYCRSPDFRIAFQ.ELLCLR

                              350                                  384
Bn   ..........................PGLMNRSHS.TG...RSTTCM....TSFKSTNPSATFSLINRNICHEGYV.....................
SP   .P..........................FISAGDYEGLEMKSTRYLQTQSSVYKVSRLETTISTVVGAHEEEPEEGPKATPSSLDLTSNGSSRSNSKTMT
A1   CRSGRRRRRRRLGACAYTYRPWTRGGSLERSQSRKDSLDDSGSCMSGSQRTLPSASPSPGYLGRGAQPPLELCAYPEWKSGALLSLPEPPGRRGLDSG
B2   RSSSKTYGNGYSSNSNGRTDYTGEQSAYQLGQEKENELLCEEAPGMEGFVNC.QGTVPSLSIDSQGRNCNTNDSPL.................

SP   ESSSFYSNMLA....................................
A1   PLFTFKLLGEPESPGTEGDASNGGCDATTDLANGQPGFKSNMPLAPGHF
```

BOMBESIN-LIKE PEPTIDES 379

Second, the DNA sequences of human GRP-R genomic clones were determined and the sequence of the hGRP-R encoded by the exonic segments of the genomic clones was found to be identical to that of the H345 cDNA. Thus, direct sequencing showed no evidence of altered GRP receptor structure in the H345 SCLC cell line.

In the hGRP gene, two introns were found to interrupt the coding sequence for the receptor: one intron between hydrophobic domains III and IV and one between domains V and VI (Fig. 7). Interestingly, the first intron interrupts the "DRY" sequence, one of the most conserved amino acid sequences among the seven membrane-spanning domain receptors. Though the adrenergic receptors lack introns, the presence of introns in the GRP receptor is similar to that observed for the tachykinin receptors (Graham et al., 1991; Gerard et al., 1990). Notably, the first intron of the neurokinin A/substance K receptor gene also identically interrupts the DRY sequence of that receptor (Graham et al., 1991; Gerard et al., 1990). This points to the evolutionary closeness of the bombesin and tachykinin families. The homology of these receptors and ligands is further demonstrated by the fact that some tachykinin receptor antagonists also function as GRP receptor antagonists (Bepler et al., 1988; Merali et al., 1988). Intron placement is identically conserved in the rat GRP-R (Spindel, unpublished) and in the NMB receptor genes (Corjay et al., 1991; Giladi et al., 1991a). Schantz et al. (1991) has localized the hGRP receptor gene to the X chromosome between p11 and q11.

Though we observed no evidence of altered GRP receptor structure in SCLCs, the oncogenic potential of hGRP receptor mutations is clear. Cotecchia et al. (1990) have described sequences in the third cytoplasmic loop of the hamster $\alpha_1$-adrenergic receptor believed to be important for G protein coupling. Mutations of these residues in the $\alpha_1$-receptor cause increased phosphoinositol production in response to a given dose of agonist (Cotecchia et al., 1990) and increased (Allen et al., 1991) tumorigenicity of the $\alpha_1$ receptor. Though these sequences are not perfectly conserved in the hGRP-R, two critical residues, Lys-290 and Ala-292 of the hamster $\alpha_1$-adrenergic receptor, are present in the hGRP-R. Just as in the $\alpha_1$ receptor, site-directed mutagenesis of these residues increases agonist-induced responses of the hGRP-R. The effect of mutation of Lys-290 to His in the hGRP-R is shown in Fig. 9. In oocytes, the His-260 GRP receptor shows

FIG. 6. Alignment of the murine 3T3 bombesin receptor (Bn) with the rat substance P receptor (SP) (Yokota et al., 1989; Hershey and Krause, 1990), rat $\beta_2$-adrenergic receptor (B2) (Gocayne et al., 1987), and the hamster $\alpha_1$-adrenergic receptor (A1) (Cotecchia et al., 1988). The putative membrane-spanning domains are indicated by solid lines and are numbered M1 to M7. Numbers above the sequence refer to amino acid positions within the bombesin receptor. Amino acids conserved (allowing conservative changes) between the bombesin receptor and the other receptors are outlined.

```
          1                                                                                    80
Hum GRP   Malndcf1Ln levdhfmh.. cnis.shsad lpvnddwshp gilyvIPavY gvIIiIGLiG NItLiKIFcT vksMRnVPNl
Rat GRP   Mdpnncsh1n levdpfls.. cnntfnqt1n ppkmdnwfhp gilyvIPavY gliivlGLiG NItLiKIFcT vksMRnVPNl
3T3 GRP   Mapnncshln ldvdpfls.. cndtfnqsls ppkmdnwfhp gflyvIPavY gliiviGLiG NItLiKIFcT vksMRnVPNl
Hum NMB   Mpskslsn1s vttganesgs vpegwerdfl pasdgtttel vircvIPs1Y llliltvGL1G NImLvKIFiT nsaMRsVPNi
Rat NMB   Mpprslpnls lpteasesel epewwendfl pdsdgttael vircVIPs1Y ll1isvGL1G NImLvKIF1T nstMRsVPNi
Consens   M-------L- ---------- ---------- ---------- ---VIP--Y  --II--GL-G NI-L-KIF-T ---MR-VPN-
                                                              ‾‾‾‾‾‾‾‾‾‾‾‾‾‾‾‾‾‾‾‾‾‾‾‾‾‾‾‾‾‾‾
                                                                           I

          81                                                              II                  160
Hum GRP   FISsLAiGDL L111iTCaPVD AsrY1aDrW1 FGrigCKLIP fIQLTSVGVS VFLTLTALSAD RYrAIVrPMD 1QaShalmki
Rat GRP   FISsLAiGDL L111vTCaPVD AsrY1aDrW1 FGrigCKLIP fIQLTSVGVS VFLTLTALSAD RYrAIVrPMD 1QaShalmki
3T3 GRP   FISsLAiGDL L111vTCaPVD AsKY1aDrWi FGrigCKLIP fIQLTSVGVS VFLTLTALSAD RYKAIVrPMD 1QaShalmki
Hum NMB   FISnLAaGDL L111TCvPVD  AsrYffDeWm FGkvGCKLIP vIQLTSVGVS VFLTLTALSAD RYrAIVnPMD mQtSgallrt
Rat NMB   FISnLAaGDL L111TCvPVD  AsrYffDeWv FGkiGCKLIP aIQLTSVGVS VFLTLTALSAD RYrAIVnPMD mQtSgvvlwt
Consens   FIS-LA-GDL LLL-TC-PVD  AS-Y---D-W- FG---GCKLIP -IQLTSVGVS VFLTLTALSAD RY-AIV-PMD -Q-S------
          ‾‾‾‾‾‾‾‾‾‾‾‾‾‾‾‾‾‾‾‾                                   ‾‾‾‾‾‾‾‾‾‾‾‾‾‾‾‾‾‾‾‾‾‾‾‾
                                                                         III         ▼
                                                                                     V

          161                      IV                                                         240
Hum GRP   clKAafIWil SmLLAiPEAV FSdlhpfhee stNqtFisCa PYPhsnELHP KIHSmasFLV fyvIPLsiIS vYYYfIAknL
Rat GRP   clKAaliWiv SmLLAiPEAV FSdlhpfhvk dtNqtFisCa PYPhsnELHP KIHSmasFLV fyiIPLsiIS vYYYfIArnL
3T3 GRP   clKAaliWiv SmLLAiPEAV FSdlhpfhvk dtNqtFisCa PYPhsnELHP KIHSmasFLV fyvIPLaiIS vYYYfIArnL
Hum NMB   cvKAmgIWvv SvLLAvPEAV FSevarissl d.NssFtaCl PYPqtdELHP KIHSvliFLV yfIIPLaiIS iyYYhiAktL
Rat NMB   s1KAvgIWvv SvLLAvPEAV FSevarigss d.NssFtaCi PYPqtdELHP KIHSvliFLV yfIIPLviIS iyYYhiAktL
Consens   --KA--IW-- S--LLA-PEAV FS-------- --N--F---C- PYP---ELHP KIHS---FLV ---IPL-iIS -YYY-IA--L
                     ‾‾‾‾‾‾‾‾‾‾‾‾‾‾‾‾‾‾‾                                    ‾‾‾‾‾‾‾‾‾‾‾‾‾‾‾‾‾
```

```
                    241                                VI                                                                              320
Hum GRP     IQSAYNLPVE gNlHvKKQlE sRKRLAktVL VFVGlFaFCW lPNHviYlYR SyhYsEvDtS mlHfvtslcA RLLaFtNSCV
Rat GRP     IQSAYNLPVE gNlHvKKQlE sRKRLAktVL VFVGlFaFCW lPNHviYlYR SyhYsEvDtS mlHfitslcA RLLaFtNSCV
3T3 GRP     IQSAYNLPVE gNlHvKKQlE sRKRLAktVL VFVGlFaFCW lPNHviYlYR SyhYsEvDtS mlHfvtslcA RLLaFtNSCV
Hum NMB     IkSAhNLPgE yNeHtKKQmE tRKRLAkiVL VFVGcFlFCW fPNHilYMYR SfnYnEldpS lgHmivtlvA RvLsFgNSCV
Rat NMB     IrSAhNLPgE yNeHtKKQmE tRKRLAkiVL VFVGcFvFCW fPNHilYlYR SfnYkEldpS lgHmivtlvA RvLsFsNSCV
Consens     I-SA-NLP-E -N-H-KKQ-E -RKRLAK-VL VFVG-F-FCW -PNH---Y-YR S---Y-E-D-S --H------A R-L-F-NSCV
                                                                                                                              
                    321                                                                                                392
Hum GRP     NPFAlYLlSk SFRkqFNtQL lCCqpgllIR shS..tgrst tcMTslKstn psvaTfS.Li Ngnlcheryv  *.
Rat GRP     NPFAlYLlSk SFRkqFNtQL lCCqpsllnR shS..tgrst tcMTsfKstn ps.aTfS.Li Ngnlchegyv  *.
3T3 GRP     NPFAlYLlSk SFRkqFNtQL lCCqpglmnR shS..tgrst tcMTsfKstn ps.aTfS.Li Nrnlchegyv  *.
Hum NMB     NPFAlYLlSe SFRrhFNsQL cCgrksyqeR gtSyllsssA vrMTslKSsna knmVTnSvtl Nghsmkqema  m*
Rat NMB     NPFAlYLlSe SFRkhFNsQL cCgqksyper stSyllsssA vrMTslKSsna knvVTnSvtl Nghstkqela  l*
Consens     NPFAlYLLS- SFR--FN-QL -C------R -S------S- -MTS-KS--- ----T-S-L- N---------- --
```

FIG. 7. Alignment of the human GRP receptor (Hum GRP) (Giladi et al., 1991a; Corjay et al., 1991) with the rat GRP receptor (Rat GRP) (Giladi et al., 1991b), the mouse GRP receptor (3T3 GRP) (Spindel et al., 1990), the human NMB receptor (Hum NMB) (Corjay et al., 1991), and the rat NMB receptor (Rat NMB) (Wada et al., 1991). Homology between all the GRP and NMB receptors is shown in caps and on the consensus line. Dashes represent gaps introduced into the sequence to optimize alignment. The hydrophobic domains are marked by heavy lines and are numbered. The positions of introns in the hGRP and hNMB receptor genes are indicated by the black triangles. Residues that are conserved between these receptors and the rat substance P receptor (Hershey and Krause, 1990) are underlined in the consensus line. Alignment was computed by the PileUp program (GCG version 7.0).

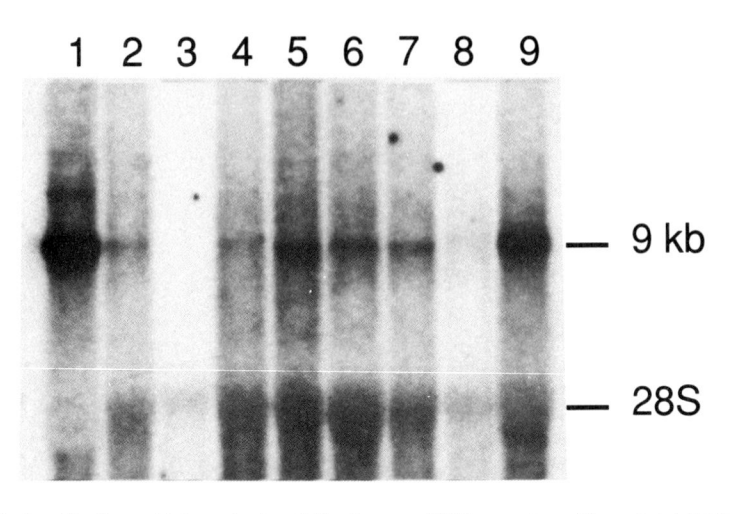

FIG. 8. Northern blot analysis of the human GRP receptor; 10 μg total RNA was loaded per lane, electrophoresed through 1.25% formaldehyde agarose, transferred to nylon membranes, and UV cross-linked. All RNAs are human except lane 1, which is murine. Lane 1, Swiss 3T3 cells; lane 2, forebrain; lane 3, esophagus; lane 4, stomach; lane 5, duodenum; lane 6, small bowel; lane 7, colon; lane 8, pancreas; lane 9, NCI H345 cells. Hybridization conditions were as described in Fig. 3. 28S indicates position of 28S RNA.

a larger increase in intracellular calcium in response to bombesin than does the unmutated receptor. Similarly, Rat-1 fibroblasts stably transfected with the His-260 GRP receptor showed increased uptake of tritiated thymidine in response to bombesin than did RAT-1 cells transfected with the normal receptor (Fig. 9). Thus, this demonstrates the potential of a single amino acid substitution to activate the hGRP-R.

## B. NEUROMEDIN B RECEPTOR

Early studies showed that in most tissues NMB and ranatensin bound with 5- to 10-fold less affinity than did GRP and bombesin. Von Schrenk *et al.* (1988, 1989) demonstrated that in esophagus NMB bound almost 100 times more potently than did GRP and that NMB binding could not be blocked by specific antagonists of the GRP receptor such as [D-Phe[6]]-bombesin(6-13)propylamide or [Leu[13],ψCH$_2$NHLeu[14]]bombesin (Qian *et al.*, 1989; Wang *et al.*, 1990a,b). This suggested that a distinct NMB-preferring bombesin receptor subtype, distinct from the GRP receptor subtype, was expressed in the esophagus. This was confirmed when Wada *et al.* (1991) and Giladi *et al.* (1991b) cloned cDNAs encoding the NMB-preferring receptor. Data base analysis showed that the NMB and GRP receptors were more closely related to each other than to any other receptors (Fig. 7). The next most closely related receptors were the tachykinin

BOMBESIN-LIKE PEPTIDES

**A**

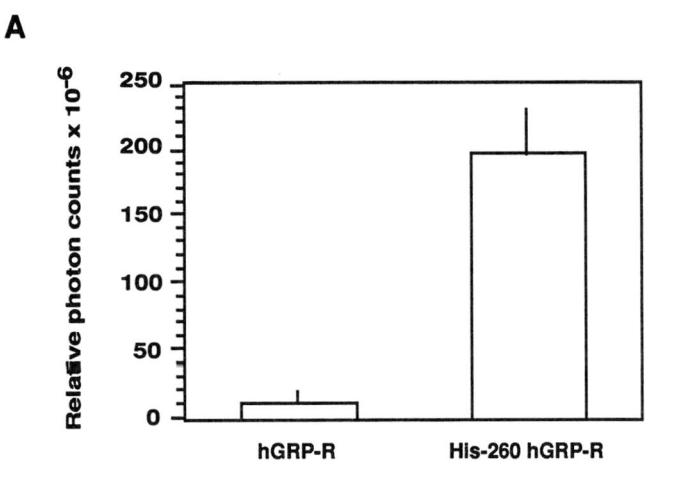

**B**

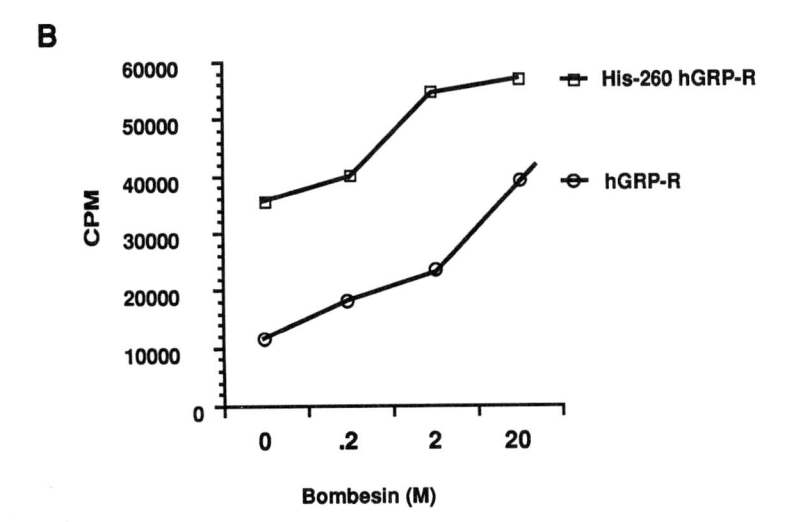

FIG. 9. Response of the His-260 GRP-R to bombesin. (A) Bombesin-evoked luminescence in *Xenopus* oocytes expressing the normal human GRP receptor or the His-260 hGRP
receptor. Oocytes were injected with 5 ng receptor RNA in 50 nl 0.1% aequorin; 18 hours
later the response to 10 n*M* bombesin was measured in a luminometer. Values shown are
the mean ± SEM of the total light output over 90 seconds of four individual oocytes.
(B) Mitogenic response of Rat-1 fibroblasts stably transfected with the normal receptor and
the His-260 hGRP receptor. Cells were growth arrested by 24 hours in DMEM + 0.3% serum
then exposed to bombesin at the doses shown for 24 hours. [³H]thymidine was added 4 hours
prior to harvest. Binding studies (not shown) indicated that both cell lines expressed an equal
number of receptors with equal affinities for bombesin and GRP.

receptors. Homology between the GRP and NMB receptors was strongest in the third hydrophobic domain, with 20 out of 21 amino acids identical (Wada *et al.*, 1991; Corjay *et al.*, 1991). Though the GRP and NMB receptors are closely related, they are encoded on different chromosomes—the hGRP receptor is on the X chromosome (Schantz *et al.*, 1991) and the NMB receptor has been localized by Naylor and co-workers to chromosome 6, q21-qter (Daly *et al.*, 1991).

Expression of the GRP and NMB receptor in *Xenopus* oocytes shows the pharmacologic differences between these two receptors. In terms of agonists, the GRP receptor responds relatively similarly to NMB and GRP, but the NMB receptor has greatly increased selectivity for NMB over GRP (Fig. 10). In terms of antagonists, consistent with the data of Von Schrenk *et al.* (1988, 1989), the GRP receptor is completely blocked by the series of GRP antagonists developed by Coy and co-workers (Qian *et al.*, 1989; Wang *et al.*, 1990a,b), but the NMB receptor is relatively less affected by these antagonists. Which domains of these receptors mediate agonist and antagonist binding remains to be determined by construction and expression of chimeric receptors.

RNA blot analysis showed that the NMB receptor was encoded by two mRNAs of 3.2 and 2.7 kb (Wada *et al.*, 1991), with highest levels present in the esophagus. In a careful mapping study by *in situ* hybridization, Wada *et al.* (1991) showed that the distribution of NMB receptor RNA was distinct from that of the GRP receptor. Levels of NMB receptor RNA were highest in the olfactory bulb and central thalamic regions. The localization of NMB receptor mRNA levels correlated well with studies on NMB ligand binding as reported by Ladenheim *et al.* (1990) and earlier studies on NMB distribution (Wada *et al.*, 1990; Chronwall *et al.*, 1985).

## IV. Perspectives

At present there are two mammalian bombesin-like peptides, GRP and NMB, and two known receptors, the GRP-preferring subtype and the NMB-preferring subtype. In that the GRP receptor responds almost equally to NMB and to GRP and in some regions both GRP and NMB are found, it is possible that GRP may be an endogenous ligand for either receptor. This may occur particularly in SCLCs. As shown in Table I, in SCLCs, all combinations of GRP, NMB, GRP receptor, and NMB receptor expression may occur. In the H209 cell line for example, there is abundant expression of GRP, but rather than the GRP receptor being expressed, the NMB receptor is expressed (Corjay *et al.*, 1991). Thus in some SCLCs, the mitogenic effects of bombesin-like peptides may be mediated by NMB rather than GRP. Likewise, in some areas of the brain, particularly in the hypothalamus and brain stem, it is not clear which

BOMBESIN-LIKE PEPTIDES                                    385

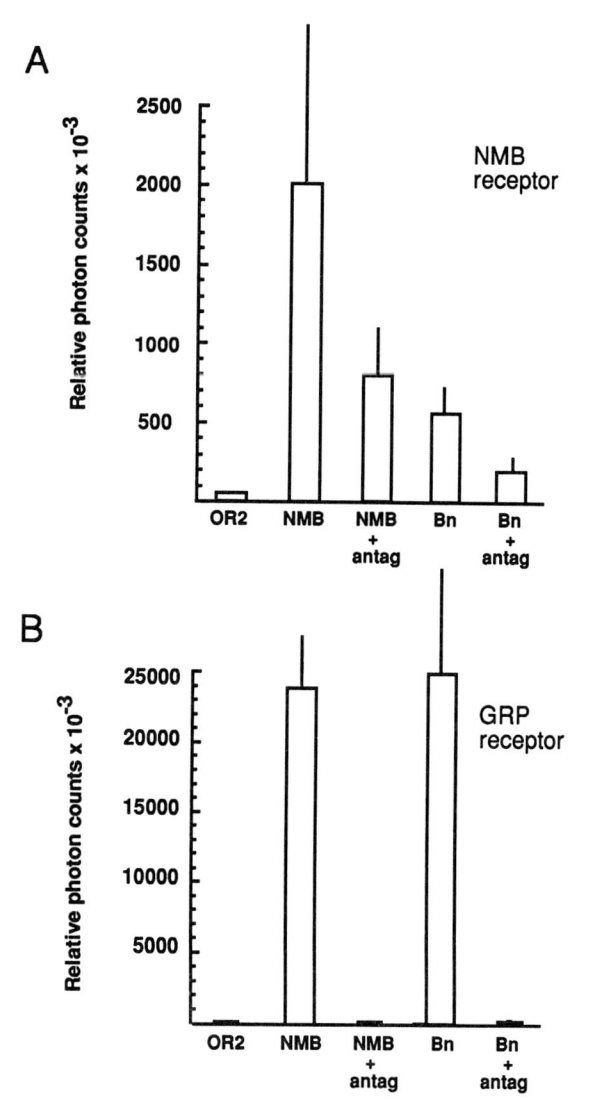

FIG. 10. Comparison of the luminescent response of oocytes injected with either the NMB receptor (A) or the GRP receptor (B). Oocytes were coinjected with aequorin and RNA was transcribed from either the NMB or the GRP receptor cDNAs; 18 hours later the luminescent response to 10 n$M$ bombesin (Bn) or NMB, with or without 100 n$M$ GRP antagonist {[D-Phe[6]]bombesin(6-13)propylamide} was recorded. Values shown are the mean ± SEM of the total light output over 90 seconds of four individual oocytes. OR2, Oocyte buffer. Note that the decreased photon counts from the oocytes injected with the NMB receptor compared to oocytes injected with GRP receptor reflect that a smaller amount of NMB receptor RNA was injected.

TABLE I

Expression of GRP, NMB, GRP-R, and NMB-R in SCLC Cells

| Cell line | GRP[a] (fmol/mg) | GRP-R[b] | NMB[a] (fmol/mg) | NMB-R[b] |
|-----------|------------------|----------|------------------|----------|
| H69 | 67 | − | 2.5 | − |
| H82 | <2 | − | ND[c] | Trace |
| H209 | 5533 | − | 11.3 | ++ |
| H345 | 5287 | ++ | 6.2 | ++ |

[a] From Cardona et al. (1991a).
[b] From Corjay et al. (1991).
[c] ND, Not determined.

effects of bombesin are mediated by the GRP receptor and which are mediated by the NMB receptor. Careful mapping studies can accurately localize bombesin-like ligands and their receptors to specific regions, but specific physiologic roles cannot be assigned to NMB, GRP, and their receptors without the use of specific antagonists. Though this is possible for the GRP receptor, at present no NMB receptor antagonists exist. Thus, when interpreting the literature on the many physiologic effects of bombesin, it is important to keep in mind that these effects may represent either endogenous ligand interacting with either receptor.

As discussed above, the situation is likely to become more complex as other bombesin-like peptides are described. Cardona et al. (1991b) have described evidence of a specific phyllolitorin receptor in SCLC cell lines and in WiDR colon carcinoma cells. We have also observed a response in Xenopus oocytes injected with WiDR RNA to phyllolitorin. Whether the responses to phyllolitorin observed by Cardona et al. (1991b) and by our laboratory represent a new bombesin receptor subtype or merely cross-reaction of phyllolitorin with the GRP or NMB receptors remains to be determined. In frog brain, both GRP and bombesin are expressed, which suggests that in frogs there are distinct receptors for bombesin and GRP. Whether mammals will also have two such receptors remains to be determined as does whether mammals will have a true bombesin distinct from GRP. Thus there are potentially at least two more mammalian bombesin-like peptides and two more bombesin receptor subtypes still to be characterized. Given the potential of crosstalk between these receptors and ligands, there is much complexity still to be unraveled to understand fully the physiologic and pathologic roles of the bombesin-like peptides.

NOTE ADDED IN PROOF. A third bombesin-receptor subtype has been characterized by Fathi et al. (1992) and Gorbulev et al. (1992) from SCLC cell lines, rat testis, and pregnant guinea pig uterus. Notably, this new receptor has 100-fold less affinity for GRP and NMB than the GRP and NMB receptors, and thus appears to be a receptor for an as yet uncharacterized mammalian bombesin-like peptide such as mammalian bombesin or mammalian phyllolitorin.

## BOMBESIN-LIKE PEPTIDES

387

## ACKNOWLEDGMENTS

This work was supported by NIH Grants CA39237 and CA53584 and by a grant from the Council on Tobacco Research.

## REFERENCES

Allen, L. F., Lefkowitz, R. J., Caron, M. G., and Cotecchia, S. (1991). *Proc. Natl. Acad. Sci. U.S.A.* **88**, 11354–11358.

Anastasi, A., Erspamer, V., and Bucci, M. (1971). *Experientia* **27**, 166–167.

Anastasi, A., Erspamer, V., and Endean, R. (1975). *Experientia* **31**, 510–511.

Barra, D., Falconieri Erspamer, G., Simmaco, M., Bossa, F., Melchiorri, P., and Erspamer, V. (1985). *FEBS Lett.* **182**, 53–56.

Basso, N., Giri, S., Improta, G., Lezoche, E., Melchiorri, P., Percoco, M., and Speranza, V. (1975). *Gut* **16**, 994–998.

Battey, J. F., Way, J. M., Corjay, M. H., Shapira, H., Kusano, K., Harkins, R., Wu, J. M., Slattery, T., Mann, E., and Feldman, R. I. (1991). *Proc. Natl. Acad. Sci. U.S.A.* **88**, 395–399.

Bepler, G., Bading, H., Heimann, B., Kiefer, P., Havemann, K., and Moelling, K. (1989). *Oncogene* **4**, 45–50.

Bepler, G., Zeymer, U., Mahmoud, S., Fiskum, G., Palaszynski, E., Rotsch, M., Willey, J., Koros, A., Cuttitta, F., and Moody, T. W. (1988). *Peptides* **9**, 1367–1372.

Bertaccini, G., Erspamer, V., and Impicciatore, M. (1973). *Br. J. Pharmacol.* **49**, 437–444.

Bertaccini, G., and Impicciatore, M. (1975). *Naunyn. Schmiedebergs Arch. Pharmacol.* **289**, 149–156.

Blinks, J. R. (1989). *In* "Methods in Enzymology" (S. Fleischer and B. Fleischer, eds.), Vol. 172, pp. 164–203. Academic Press, San Diego.

Broccardo, M., and Cardamone, A. (1985). *Peptides* **6**, 99–102.

Brown, K. D., Littlewood, C. J., and Blakeley, D. M. (1990). *Biochem. J.* **270**, 557–560.

Brown, M., Rivier, J., and Vale, W. (1977). *Science* **196**, 998–1000.

Cardona, C., Rabbitts, P. H., Spindel, E. R., Ghatei, M. A., Bleehen, N. M., Bloom, S. R., and Reeve, J. G. (1991a). *Cancer Res.* **51**, 5205–5211.

Cardona, C., Reeve, J. G., and Bleehen, M. (1991b). *Br. J. Cancer* 63(Suppl. 13), 14. [Abstract]

Carney, D. N., Gazdar, A. F., Bepler, G., Guccion, J. G., Marangos, P. J., Moody, T. W., Zweig, M. H., and Minna, J. D. (1985). *Cancer Res.* **45**, 2913–2923.

Chronwall, B. M., Pisano, J. J., Bishop, J. F., Moody, T. W., and O'Donohue, T. L. (1985). *Brain Res.* **338**, 97–113.

Cirillo, D. M., Gaudino, G., Naldini, L., and Comoglio, P. M. (1986). *Mol. Cell. Biol.* **6**, 4641–4649.

Clineschmidt, B. V., Geller, R. G., Govier, W. C., Pisano, J. J., and Tanimura, T. (1971). *Br. J. Pharmacol.* **41**, 622–628.

Conlon, J. M., Henderson, I. W., and Thim, L. (1987). *Gen. Comp. Endocrinol.* **68**, 415–420.

Conlon, J. M., O'Harte, F., and Vaudry, H. (1991). *Biochem. Biophys. Res. Commun.* **178**, 526–530.

Corjay, M. H., Dobrzanski, D. J., Way, J. M., Viallet, J., Shapira, H., Worland, P., Sausville, E. A., and Battey, J. F. (1991). *J. Biol. Chem.* **266**, 18771–18779.

Cotecchia, S., Exum, S., Caron, M. G., and Lefkowitz, R. J. (1990). *Proc. Natl. Acad. Sci. U.S.A.* **87**, 2896–2900.

Cotecchia, S., Schwinn, D. A., Randall, R. R., Lefkowitz, R. J., Caron, M. G., and Kobilka, B. K. (1988). *Proc. Natl. Acad. Sci. U.S.A.* **85**, 7159–7163.

388 ELIOT R. SPINDEL ET AL.

Coy, D. H., Heinz-Erian, P., Jiang, N.-.Y., Sasaki, Y., Taylor, J., Moreau, J.-.P., Wolfrey, W. T., and Gardner, J. D. (1988). *J. Biol. Chem.* **263**, 5056–5060.

Cuttitta, F., Carney, D. N., Mulshine, J., Moody, T. W., Fedorko, J., Fischler, A., and Minna, J. D. (1985). *Nature (London)* **316**, 823–826.

Daly, M. C., Spindel, E. R., Giladi, E., Grzeschik, K., and Naylor, S. L. (1991). *In* "Proceedings, 11th Annual Int. Human Gene Mapping Meeting." [Abstract]

Deschodt Lanckman, M., Robberecht, P., De Neef, P., Lammens, M., and Christophe, J. (1976). *J. Clin. Invest.* **58**, 891–898.

Dickinson, K. E., Uemura, N., Sekar, M. C., McDaniel, H. B., Anderson, W., Coy, D. H., and Hirschowitz, B. I. (1988). *Biochem. Biophys. Res. Commun.* **157**, 1154–1158.

Dockray, G. J., Vaillant, C., and Walsh, J. H. (1979). *Neuroendocrinology* **4**, 1561–1568.

Erisman, M. D., Linno, R. I., Hernandez, O., DiAngustine, R. P., and Lazarus, L. H. (1982). *Proc. Natl. Acad. Sci. U.S.A.* **79**, 2379–2383.

Erspamer, V., and Melchiorri, P. (1975). *In* "Gastrointestinal Hormones: A Symposium" (J. C. Thompson, ed.), pp. 575–589. Univ. of Texas Press, Austin.

Erspamer, V., Improta, G., Melchiorri, P., and Sopranzi, N. (1974). *Br. J. Pharmacol.* **52**, 227–232.

Falconieri Erspamer, G., Mazzanti, G., Farruggia, G., Nakajima, T., and Yanaihara, N. (1984). *Peptides* **5**, 765–768.

Fathi, Z., Corjay, M. H., Shapira, H., Wada, E., Benya, R., Jensen, R., Sausville, E. A., and Battey, J. F. (1992). *J. Biol. Chem.*, submitted for publication.

Feldman, R. I., Wu, J. M., Jenson, J. C., and Mann, E. (1990). *J. Biol. Chem.* **265**, 17364–17372.

Fischer, J. B., and Schonbrunn, A. (1988). *J. Biol. Chem.* **263**, 2808–2816.

Gaudino, G. G., Cirillo, D. M., Naldini, L., Rossino, P., and Comoglio, P. M. (1988). *Proc. Natl. Acad. Sci. U.S.A.* **85**, 2166–2170.

Geller, R. G., Govier, W. C., Pisano, J. J., Tanimura, T., and Clineschmidt, B. (1970). *Br. J. Pharmacol.* **40**, 605–610.

Gerard, N. P., Eddy, R. L., Jr., Shows, T. B., and Gerard, C. (1990). *J. Biol. Chem.* **265**, 20455–20462.

Ghatei, M. A., Sheppard, M. N., Henzen-Logman, S., Blank, M. A., Polak, J. M., and Bloom, S. R. (1983). *J. Clin. Endocrinol. Metab.* **57**, 1226–1232.

Giladi, E., Daly, M., Naylor, S. L., and Spindel, E. R. (1991a). *In* "Proceedings, 21st Annual Meeting, Society for Neuroscience," Abstract 79.1

Giladi, E., Palkki, H., Kelly, M., and Spindel, E. R. (1991b). *In* "Proceedings, 73rd Annual Meeting of the Endocrine Society," p. 110. [Abstract]

Giladi, E., and Spindel, E. R. (1991). *Biotechniques* **10**, 744–747.

Gocayne, J., Robinson, D. A., FitzGerald, M. G., Chung, F. Z., Kerlavage, A. R., Lentes, K. U., Lai, J., Wang, C. D., Fraser, C. M., and Venter, J. C. (1987). *Proc. Natl. Acad. Sci. U.S.A.* **84**, 8296–8300.

Gorbulev, V., Akhundova, A., Büchner, H., and Fahrenholz, F. (1992). *Eur. J. Biochem.* **208**, 405–410.

Gould, V. E., Linnoila, R. I., Memoli, V. A., and Warren, W. H. (1983). *Lab. Invest.* **49**, 519–539.

Graham, A., Hopkins, B., Powell, S. J., Danks, P., and Briggs, I. (1991). *Biochem. Biophys. Res. Commun.* **177**, 8–16.

Hausdorf, W. P., Bouvier, M., O'Dowd, B. F., Irons, G. P., Caron, M. G., and Lefkowitz, R. J. (1989). *J. Biol. Chem.* **264**, 12657–12665.

Hershey, A. D., and Krause, J. E. (1990). *Science* **247**, 958–962.

Heslop, J. P., Blakeley, D. M., Brown, K. D., Irvine, R. F., and Berridge, M. J. (1986). *Cell (Cambridge, Mass.)* **47**, 703–709.

BOMBESIN-LIKE PEPTIDES 389

Isacke, C. M., Meisenhelder, J., Brown, K. D., Gould, K. L., Gould, S. J., and Hunter, T. (1986). *EMBO J.* **5**, 2889–2898.

Jahnke, G. D., and Lazarus, L. H. (1984). *Proc. Natl. Acad. Sci. U.S.A.* **81**, 578–582.

Jensen, R. T., Moody, T., Pert, C., Rivier, J. E., and Gardner, J. D. (1978). *Proc. Natl. Acad. Sci. U.S.A.* **75**, 6139–6143.

Johnson, D. E., Lock, J. E., Elde, R. P., and Thompson, T. R. (1982). *Pediatr. Res.* **16**, 446–454.

Julius, D., MacDermott, A. B., Axel, R., and Jessel, T. M. (1988). *Science* **241**, 558–564.

Kachur, J. F., Miller, R. J., Field, M., and Rivier, J. (1982). *J. Pharmacol. Exp. Ther.* **220**, 449–455.

Knuhtsen, S., Holst, J. J., Jensen, S. L., Ulrich, K., and Nielsen, O. V. (1985). *Am. J. Physiol.* **248**, G281–G285.

Konturek, S. J., Krol, R., and Tasler, J. (1976). *J. Physiol.* **257**, 663–672.

Krane, I. M., Naylor, S. L., Chin, W. W., and Spindel, E. R. (1988). *J. Biol. Chem.* **263**, 13317–13323.

Kris, R. M., Hazan, R., Villines, J., Moody, T. W., and Schlessinger, J. (1987). *J. Biol. Chem.* **262**, 11215–11220.

Ladenheim, E. E., Jensen, R. T., Mantey, S. A., McHugh, P. R., and Moran, T. H. (1990). *Brain Res.* **537**, 233–240.

Lebacq-Verheyden, A. M., Krystal, G., Sartor, O., Way, J., and Battey, J. F. (1988). *Mol. Endocrinol.* **2**, 556–563.

Lechago, J., Crawford, B. G., and Walsh, J. H. (1981). *Gen. Comp. Endocrinol.* **45**, 1–6.

Lehy, T., Accary, J. P., Labeille, D., and Dubrasquet, M. (1983). *Gastroenterology* **84**, 914–919.

Lehy, T., Puccio, F., Chariot, J., and Labeille, D. (1986). *Gastroenterology* **90**, 1942–1949.

Letterio, J. J., Coughlin, S. H., and Williams, L. T. (1986). *Science* **234**, 1117–1119.

Lopez-Rivas, A., Mendoza, S. A., Nanberg, E., Sinnett-Smith, J., and Rozengurt, E. (1987). *Proc. Natl. Acad. Sci. U.S.A.* **84**, 5768–5772.

Mahmoud, S., Staley, J., Taylor, J., Bogden, A., Moreau, J.-P., Coy, D., Avis, I., Cuttitta, F., Mulshine, J. L., and Moody, T. W. (1991). *Cancer Res.* **51**, 1798–1802.

Masu, Y., Nakayama, K., Tamaki, H., Harada, Y., Kuno, M., and Nakanishi, S. (1987). *Nature (London)* **329**, 836–838.

McDonald, T. J., Ghatei, M. A., Bloom, S. R., Adrian, T. E., Mochizuki, T., Yanaihara, C., and Yanaihara, N. (1983). *Regul. Pept.* **5**, 125–137.

McDonald, T. J., Jornvall, H., Ghatei, M., Bloom, S. R., and Mutt, V. (1980). *FEBS Lett.* **122**, 45–48.

McDonald, T. J., Jornvall, H., Nilsson, G., Vagne, M., Ghatei, M., Bloom, S. R., and Mutt, V. (1979). *Biochem. Biophys. Res. Commun.* **90**, 227–233.

Merali, Z., Merchant, C. A., Crawley, J. N., Coy, D. H., Heinz Erian, P., Jensen, R. T., and Moody, T. W. (1988). *Synapse* **2**, 282–287.

Meyerhof, W., Morley, S. D., and Richter, D. (1988). *FEBS Lett.* **239**, 109–112.

Minamino, N., Kangawa, K., and Matsuo, H. (1983). *Biochem. Biophys. Res. Commun.* **114**, 541–548.

Minamino, N., Kangawa, K., and Matsuo, H. (1984). *Biochem. Biophys. Res. Commun.* **124**, 925–932.

Minamino, N., Sudoh, T., Kangawa, K., and Matsuo, H. (1985). *Biochem. Biophys. Res. Commun.* **130**, 685–691.

Modlin, I. M., Lamers, C. B., and Walsh, J. H. (1981). *Regul. Pept.* **1**, 279–288.

Moody, T. W., Bertness, V., and Carney, D. N. (1983). *Peptides* **4**, 683–686.

Moody, T. W., Carney, D. N., Cuttitta, F., Quattrocchi, K., and Minna, J. D. (1985). *Life Sci.* **37**, 105–113.

390 ELIOT R. SPINDEL ET AL.

Moody, T. W., Pert, C. B., Gazdar, A. F., Carney, D. N., and Minna, J. D. (1981). *Science* **214,** 1246–1248.

Moody, T. W., Pert, C. B., Rivier, J., and Brown, M. (1978). *Proc. Natl. Acad. Sci. U.S.A.* **75,** 5372–5376.

Moriarty, T. M., Gillo, B., Sealfon, S., Roberts, J., Blitzer, R. D., and Landau, E. M. (1988). *Mol. Brain Res.* **4,** 75–79.

Mukai, H., Kawai, K., Suzuki, Y., Yamashita, K., and Munekata, E. (1987). *Am. J. Physiol.* **252,** E765–E761.

Nagalla, S. R., Gibson, B. W., Reeve, J. R., Jr., and Spindel, E. R. (1991). *In* "Proceedings, 21st Annual Meeting, Society for Neuroscience," Abstract 219.2.

Nagalla, S. R., Gibson, B. W., Tang, D., Reeve, J. R., Jr., and Spindel, E. R. (1992). *J. Biol. Chem.* **267,** 6916–6922.

Nakajima, T., Tanimura, T., and Pisano, J. J. (1970). *Fed. Proc.* **29,** 282.

Namba, M., Ghatei, M. A., Bishop, A. E., Gibson, S. J., Mann, D. J., Polak, J. M., and Bloom, S. R. (1985). *Peptides* **6**(Suppl. 3), 257–263.

Negri, L., Improta, G., Broccardo, M., and Melchiorri, P. (1988). *Ann. N.Y. Acad. Sci.* **547,** 415–428.

ODowd, B. F., Hnatowich, M., Caron, M. G., Lefkowitz, R. J., and Bouvier, M. (1989). *J. Biol. Chem.* **264,** 7564–7569.

Orloff, M. S., Reeve, J. R., Jr., Ben-Avram, C. M., Shively, J. E., and Walsh, J. H. (1984). *Peptides* **5,** 865–870.

Panula, P., Yang, H-.Y. T., and Costa, E. (1982). *Regul. Pept.* **4,** 275–283.

Plevin, R., Palmer, S., Gardner, S. D., and Wakelam, M. J. O. (1990). *Biochem. J.* **268,** 605–610.

Polak, J. M., Bloom, S. R., Hobbs, S., Solcia, E., and Pearse, A. G. E. (1976). *Lancet* **1,** 1109–1110.

Price, J., Penman, E., Wass, J. A. H., and Rees, L. H. (1984). *Regul. Pept.* **9,** 1–10.

Puccio, F., and Lehy, T. (1989). *Am. J. Physiol.* **256,** G328–G334.

Qian, J. M., Coy, D. H., Jiang, N. Y., Gardner, J. D., and Jensen, R. T. (1989). *J. Biol. Chem.* **264,** 16667–16671.

Reeve, J. R., Jr., Walsh, J. H., Chew, P., Clark, B., Hawke, D., and Shively, J. E. (1983). *J. Biol. Chem.* **258,** 5582–5588.

Rivier, C., Rivier, J., and Vale, W. (1978). *Endocrinology (Baltimore)* **102,** 519–523.

Roth, K. A., Weber, E., and Barchas, J. D. (1982). *Brain Res.* **251,** 277–282.

Rozengurt, E., and Sinnett-Smith, J. (1983). *Proc. Natl. Acad. Sci. U.S.A.* **80,** 2936–2940.

Sakamoto, A., Kitamura, K., Haraguchi, Y., Yoshida, T., and Tanaka, K. (1987). *Am. J. Gastroenterol.* **82,** 1035–1041.

Sandberg, K., Markwick, A. J., Trinh, D. P., and Catt, K. J. (1988). *FEBS Lett.* **241,** 177–180.

Sasai, Y., and Nakanishi, S. (1989). *Biochem. Biophys. Res. Commun.* **165,** 695–702.

Schantz, L. J., Naylor, S. L., Giladi, E., and Spindel, E. R. (1991). *In* "Proceedings, 11th Annual Int. Human Gene Mapping Meeting." [Abstract]

Schultzberg, M., and Dalsgaard, C.-J. (1983). *Brain Res.* **269,** 190–195.

Sharoni, Y., Viallet, J., Trepel, J. B., and Sausville, E. A. (1990). *Cancer Res.* **5257,** 5262.

Shaw, C., Thim, L., and Conlon, J. M. (1987). *J. Neurochem.* **49,** 1348–1354.

Shigemoto, R., Yokota, Y., Tsuchida, K., and Nakanishi, S. (1990). *J. Biol. Chem.* **265,** 623–628.

Shimomura, O. (1991). *Cell Calcium* **12,** 635–643.

Spindel, E. R., Chin, W. W., Price, J., Rees, L. H., Besser, G. M., and Habener, J. F. (1984). *Proc. Natl. Acad. Sci. U.S.A.* **81,** 5699–5703.

Spindel, E. R., Giladi, E., Brehm, P., Goodman, R. H., and Segerson, T. P. (1990). *Mol.*

*Endocrinol.* **4,** 1956–1963.

Spindel, E. R., Segerson, T. P., Duckworth, B. C., Brehm, P., and Goodman, R. H. (1989). *In* "Proceedings, Annu. Meeting of the Endocrine Society," Abstract 1348

Spindel, E. R., Sunday, M. E., Hofler, H., Wolfe, H. J., Habener, J. F., and Chin, W. W. (1987). *J. Clin. Invest.* **80,** 1172–1179.

Stahlman, M. T., Kasselberg, A. G., Orth, D. N., and Gray, M. E. (1985). *Lab. Invest.* **52,** 52–60.

Strader, C. D., Sigal, I. S., and Dixon, R. A. F. (1989). *FASEB J.* **3,** 1825–1832.

Strader, C. D., Sigal, I. S., Register, R. B., Candelore, M. R., Rands, E., and Dixon, R. A. (1987). *Proc. Natl. Acad. Sci. U.S.A.* **84,** 4384–4388.

Sunday, M. E., Choi, N., Spindel, E. R., Chin, W. W., and Mark, E. J. (1991). *Hum. Pathol.* **22,** 1030–1039.

Sunday, M. E., Kaplan, L. M., Motoyama, E., Chin, W. W., and Spindel, E. R. (1988). *Lab. Invest.* **59,** 5–24.

Takeyama, M., Kondo, K., Takayama, F., Kondo, R., Murata, H., and Miyakawa, I. (1991). *Biochem. Biophys. Res. Commun.* **176,** 931–937.

Takuwa, N., Takuwa, Y., Bollag, W. E., and Rasmussen, H. (1987). *J. Biol. Chem.* **262,** 182–188.

Taylor, C. W., Blakeley, D. M., Corps, A. N., Berridge, M. J., and Brown, K. D. (1988). *Biochem. J.* **249,** 917–920.

Vaillant, C., Dockray, G. J., and Walsh, J. H. (1979). *Histochemistry* **64,** 307–314.

Viallet, J., Sharoni, Y., Frucht, H., Jensen, R. T., Minna, J. D., and Sausville, E. A. (1990). *J. Clin. Invest.* **86,** 1904–1912.

von Schrenck, T., Heinz-Erian, P., Moran, T., Coy, D., Gardner, J. D., and Jensen, R. T. (1988). *Gastroenterology* **94,** A482. [Abstract]

von Schrenck, T., Heinz-Erian, P., Moran, T., Mantey, S. A., Gardner, J. D., and Jensen, R. T. (1989). *Am. J. Physiol.* **256,** G747–G758.

von Schrenck, T., Wang, L.-H., Coy, D. H., Villanueva, M. L., Mantey, S., and Jensen, R. T. (1990). *Am. J. Physiol. Gastrointest. Liver Physiol.* **259,** G468–G473.

Wada, E., Way, J., Lebacq-Verheyden, A. M., and Battey, J. F. (1990). *J. Neurosci.* **10,** 2917–2930.

Wada, E., Way, J., Shapira, H., Kusano, K., Lebacq-Verheyden, A. M., Coy, D., Jensen, R., and Battey, J. (1991). *Neuron* **6,** 421–430.

Wang, L.-H., Coy, D. H., Taylor, J. E., Jiang, N.-Y., Kim, S. H., Moreau, J.-P., Huang, S. C., Mantey, S. A., Frucht, H., and Jensen, R. T. (1990a). *Biochemistry* **29,** 616–622.

Wang, L.-H., Coy, D. H., Taylor, J. E., Jiang, N.-Y., Moreau, J.-P., Huang, S. C., Frucht, H., Haffar, B. M., and Jensen, R. T. (1990b). *J. Biol. Chem.* **265,** 15695–15703.

Weber, S., Zuckerman, J. E., Bostwick, D. G., Bensch, K. G., Sikic, B. I., and Raffin, T. A. (1985). *J. Clin. Invest.* **75,** 306–309.

Westendorf, J. M., and Schonbrunn, A. (1983). *J. Biol. Chem.* **258,** 7527–7535.

Willey, J. C., Lechner, J. F., and Harris, C. C. (1984). *Exp. Cell Res.* **153,** 245–248.

Woll, P. J., and Rozengurt, E. (1990). *Cancer Res.* **50,** 3968–3973.

Yasuhara, T., Ishikawa, O., and Nakajima, T. (1979). *Chem. Pharm. Bull.* **27,** 492–498.

Yokota, Y., Sasai, Y., Tanaka, K., Fujiwara, T., Tsuchida, K., Shigemoto, R., Kakizuka, A., Ohkubo, H., and Nakanishi, S. (1989). *J. Biol. Chem.* **264,** 17649–17652.

Zachary, I., Millar, J., Nanberg, E., Higgins, T., and Rozengurt, E. (1987). *Biochem. Biophys. Res. Commun.* **146,** 456–463.

Zachary, I., and Rozengurt, E. (1985). *Proc. Natl. Acad. Sci. U.S.A.* **82,** 7616–7620.

Zachary, I., and Rozengurt, E. (1987). *J. Biol. Chem.* **262,** 3947–3950.

Zachary, I., Sinnett-Smith, J., and Rozengurt, E. (1991). *J. Biol. Chem.* **266,** 24126–24133.

RECENT PROGRESS IN HORMONE RESEARCH, VOL. 48

# Thyroid Hormone Regulation of Thyrotropin Gene Expression

WILLIAM W. CHIN, FRANCES E. CARR,[1] JOAN BURNSIDE,[2] AND DOUGLAS S. DARLING[3]

*Division of Genetics, Department of Medicine, Brigham and Women's Hospital, Howard Hughes Medical Institute and Harvard Medical School, Boston, Massachusetts 02115*

## I. Introduction

Thyrotropin (thyroid-stimulating hormone; TSH) is a pivotal hormone in the regulation of the hypothalamic–pituitary–thyroid axis (Chin, 1985) (Fig. 1). It is synthesized and secreted by the thyrotrope cell of the anterior pituitary gland of most vertebrates under the stimulation by thyrotropin-releasing hormone (TRH) and inhibition by somatostatin (SRIF), dopamine (DA), and thyroid hormones (Hershman and Pekary, 1985; Scanlon *et al.*, 1979). TSH stimulates the thyroid gland to produce and release thyroid hormones [tetraiodothyronine (T4) and triiodothyronine (T3)]. It is composed of two different, noncovalently associated glycoprotein subunits termed $\alpha$ and TSH$\beta$ (Pierce and Parsons, 1981). Each subunit is encoded by a separate gene located on different chromosomes in humans and mice (Chin, 1986). Coordinate regulation of expression of the TSH subunit genes results in nearly balanced production of the subunit proteins. Often, however, the $\alpha$ subunit is synthesized and released in excess. Only the heterodimer, TSH, can act in this endocrine system; unassociated TSH subunits are not known to possess biologic activity.

TSH is a member of a small polypeptide hormone family that shares similarities in the structures of both the hormone and their cognate receptors. The other members of this glycoprotein hormone family include pituitary lutropin (luteinizing hormone; LH), follitropin (follicle-stimulating hormone; FSH), and placental chorionic gonadotropin (CG in primates and horse) (Pierce and Parsons, 1981).

---

[1] Present address: Kyle Metabolic Unit Research Labs, Department of Clinical Investigation, Walter Reed Army Medical Center, Washington, DC 20307.

[2] Present address: Department of Animal Science, University of Delaware, Townsend Hall, Newark, Delaware 19717.

[3] Present address: Department of Biological and Biophysical Sciences, University of Louisville, Louisville, Kentucky 40292.

Copyright © 1993 by Academic Press, Inc.
All rights of reproduction in any form reserved.

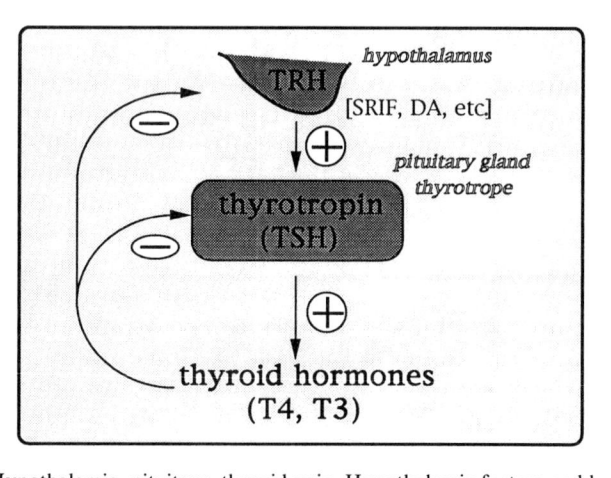

FIG. 1. Hypothalamic–pituitary–thyroid axis. Hypothalamic factors and hormones are delivered to the anterior pituitary gland via the hypophyseal portal system. One critical hormone is thyrotropin-releasing hormone (TRH), which stimulates thyrotropin synthesis and release; other factors, such as somatostatin (SRIF) and dopamine (DA), are inhibitors of thyrotrope function. Thyrotropin, in turn, stimulates the thyroid gland to produce and secrete thyroid hormones (T4 and T3). These thyroid hormones enter the general circulation to act at both the hypothalamic and the pituitary levels. In particular, thyroid hormones can negatively control thyrotropin synthesis and release directly in the thyrotrope.

## II.  Regulation of TSH Synthesis and Release by Thyroid Hormone

It is well known that TSH synthesis by the pituitary thyrotrope is directly regulated by thyroid hormone in a negative fashion (Hershman and Pekary, 1985). Early studies were not definitive with respect to the question of whether thyroid hormone regulation of TSH subunit genes was exerted at the transcriptional level. However, the advent of molecular approaches provided investigators with the ability to assess gene activity. In particular, the cDNAs encoding the subunits of TSH were isolated and characterized (Chin, 1986). These reagents allowed the measurement of steady-state TSH subunit mRNA levels in pituitary tissues using blot and liquid hybridization techniques (Carr *et al.*, 1985b; Carr and Chin, 1988) and indirect determinations of subunit gene transcriptional activities (Shupnik *et al.*, 1989).

## III.  α Subunit cDNA and Gene

The mouse α subunit cDNA was first identified and characterized from a mouse thyrotropic tumor cDNA library (Chin *et al.*, 1981). The cDNA revealed that the coding region of the mouse α subunit mRNA predicted an α subunit protein precursor of 120 amino acid residues. Northern blot analysis of total pituitary and thyrotropic tumor RNA showed that the $\alpha$ subunit mRNA is approximately 800 nucleotides in size. Further, Southern blot analysis of mouse genomic DNA revealed a simple arrangement suggesting a single $\alpha$ gene in the mouse. These data were consistent with results obtained in rats (Godine *et al.*, 1982), humans (Fiddes and Goodman, 1979), and cows (Erwin *et al.*, 1983; Nilson *et al.*, 1983), with the coding region possessing 74–97% similarity at the nucleotide and amino acid sequence levels.

In the mouse, the $\alpha$ subunit precursor consists of a 24-amino acid signal or leader peptide and a 96-amino acid apoprotein. The mouse $\alpha$ apoprotein is identical in size to the $\alpha$ subunits present in cows and rats, but different from the 92-amino acid homologue in humans. This difference is likely due to a loss of 12 nucleotides, representing four amino acid residues located near the $NH_2$ terminus of the apoprotein, during evolution in the $\alpha$ subunit mRNA in humans relative to that in cows and rats. Of note, this variation occurs at the site of the second intron, suggesting a splicing mechanism in the evolution of this difference (Chin *et al.*, 1983). Unlike their human counterpart, the $\alpha$ subunit mRNA in rodents and cows is not expressed in the placenta but rather is found largely in the anterior pituitary gland (Carr and Chin, 1985, 1988; Tepper and Roberts, 1984).

The $\alpha$ subunit gene in the rat has been isolated and characterized (Burnside *et al.*, 1989b) (Fig. 2). It is a single-copy gene, as suggested by the

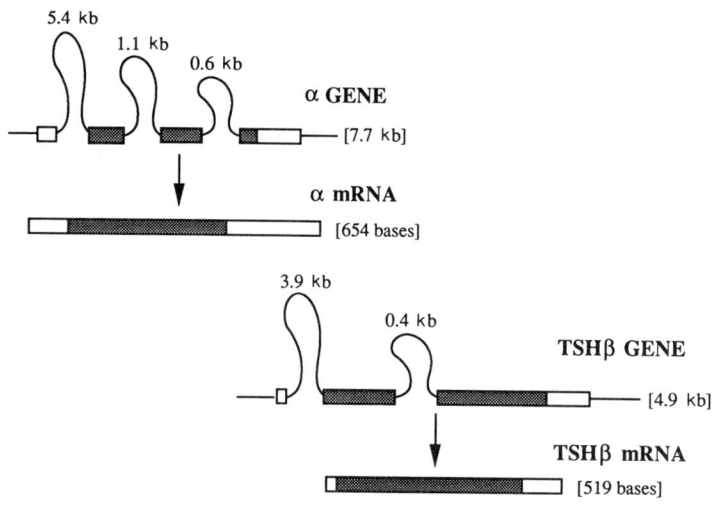

FIG. 2. Structures of the rat $\alpha$ and TSH$\beta$ mRNAs and genes. Schematic diagram of the rat TSH subunit mRNAs and genes. The boxes represent exons and the loops denote introns. The shaded regions encode protein.

396                    WILLIAM W. CHIN ET AL.

aforementioned Southern blots of genomic DNA, is approximately
7.7 kb in extent, and contains four exons and three introns. The exons are
interrupted by introns with sizes of 5.4, 1.1, and 0.6 kb. The first intron
is located in the 5' untranslated region and the second is found at codon
+6. The organization of the $\alpha$ subunit gene in humans (Boothby et al.,
1981; Fiddes and Goodman, 1981), cows (Goodwin et al., 1983) and mice
(Gordon et al., 1988) is nearly identical, with variations seen largely in
the sizes of the introns (especially the first intron). There is a single
transcriptional start site associated with a conventional TATA box-
containing promoter. The human homologue is located on the long arm of
chromosome 6 (Naylor et al., 1983).

## IV.  TSH$\beta$ Subunit cDNA and Gene

The TSH$\beta$ cDNAs have been cloned from rats (Chin et al., 1985a;
Croyle and Maurer, 1984), mice (Chin et al., 1985b; Gurr et al., 1983),
and cows (Maurer et al., 1984). The rat TSH$\beta$ cDNA encodes a protein
precursor that is 138 amino acid residues in size with a 20-amino acid
leader peptide and a 118-amino acid apoprotein (Chin et al., 1985a) (Fig.
2). The TSH$\beta$ cDNA detects a single mRNA in the rat anterior pituitary
gland of approximately 700 nucleotides in extent in RNA blot hybridization
analyses (Chin et al., 1985a; Croyle and Maurer, 1984). The TSH$\beta$ subunit
cDNAs from several species possess conserved DNA sequences, so that
the rat sequence compared to those of man, cow, and mouse has 80–91%
similarity at the nucleotide and amino acid sequence levels.

The rat TSH$\beta$ subunit gene has been cloned and analyzed (Carr et al.,
1987; Croyle et al., 1985). The gene is 4.9 kb in size, and contains three
exons and two introns (Fig. 2). The first intron is located near the junction
of the 5' untranslated and coding regions, and the second intron is found
between codons 34 and 35. Of note, there are two transcriptional start
sites in the rat TSH$\beta$ subunit gene that are located 43 nucleotides apart,
and each is associated with conventional promoter elements (Carr et al.,
1987). Thus, two TSH$\beta$ subunit mRNAs are generated in the pituitary
gland from a single transcriptional unit. The human TSH$\beta$ subunit gene
also contains three exons but has a single transcriptional start site (Hayashizaki et al., 1985; Tatsumi et al., 1988). The mouse TSH$\beta$ subunit gene,
however, differs slightly from this arrangement in that it contains five
exons. It, too, contains two transcriptional start sites but multiple TSH$\beta$
mRNAs are generated by alternative splicing of the first and two additional
exons encoding the 5' untranslated region (Gordon et al., 1988; Wolf et
al., 1987; Wood et al., 1987). The human TSH$\beta$ gene exists in the haploid
genome as a single copy and is located on chromosome 19 (Naylor et al.,

1986). Inasmuch as the regulation of the synthesis of TSH requires the balanced control of the synthesis of the subunits, the location of the two TSH subunit genes on separate chromosomal loci indicates that common modulatory elements must be present in the regulatory regions of these genes.

## V. Regulation of $\alpha$ and TSH$\beta$ Subunit Gene Expression by Thyroid Hormones

The effect of thyroid hormones on the synthesis of TSH at the pretranslational level was determined. In one model, the mouse thyrotropic tumor (TtT97), originally described by Furth (1955), was utilized. This biological system involves a transplantable mouse pituitary tumor, containing mostly thyrotropes, that grows only in a hypothyroid host. This tumor tissue produces TSH in large quantities in a thyroid hormone-responsive manner. Mice bearing TtT97 were treated with large doses of T4 or T3, and the steady-state levels of $\alpha$ and TSH$\beta$ subunit mRNAs in tumor and pituitary tissues were determined using a blot hybridization technique and cloned mouse $\alpha$ and TSH$\beta$ subunit cDNAs as probes (Chin et al., 1985b). First, consistent with previous results, thyroid hormones rapidly (within hours to a day) decreased serum levels of TSH and the separate free subunits, as determined by radioimmunoassays. Second, thyroid hormones decreased $\alpha$ and TSH$\beta$ subunit mRNAs in both tumor and pituitary tissues in a time- and dose-dependent fashion (Fig. 3). The mRNAs declined rapidly with statistically significant decreases observed within 4 hours for $\alpha$ subunit mRNA and within 1 hour for the TSH$\beta$ subunit mRNA. Furthermore, the extent of regulation of the TSH$\beta$ subunit was greater than that for the $\alpha$ subunit, at longer times. These results were consistent with those of Gurr and Kourides (1983).

In order to determine whether the regulation of TSH subunit mRNAs in thyrotropes was effected at the level of transcription, a series of experiments using an indirect measure of gene transcription rate was performed (Shupnik et al., 1983, 1985a). Employing the mouse thyrotropic tumor model in vivo and in vitro (tumor fragments in organ culture), the rates of transcription of the $\alpha$ and TSH$\beta$ subunit genes were assessed by the nuclear run-on approach (McKnight and Palmiter, 1979). In this method, nuclei from tissues in different experimental groups were isolated, partially purified, and allowed to complete the elongation of transcripts that had already been initiated at the time of sacrifice of the tissue in the presence of labeled nucleotides. The amount of specific, newly synthesized transcripts provides a measure of the RNA polymerase complexes on that gene at a given time.

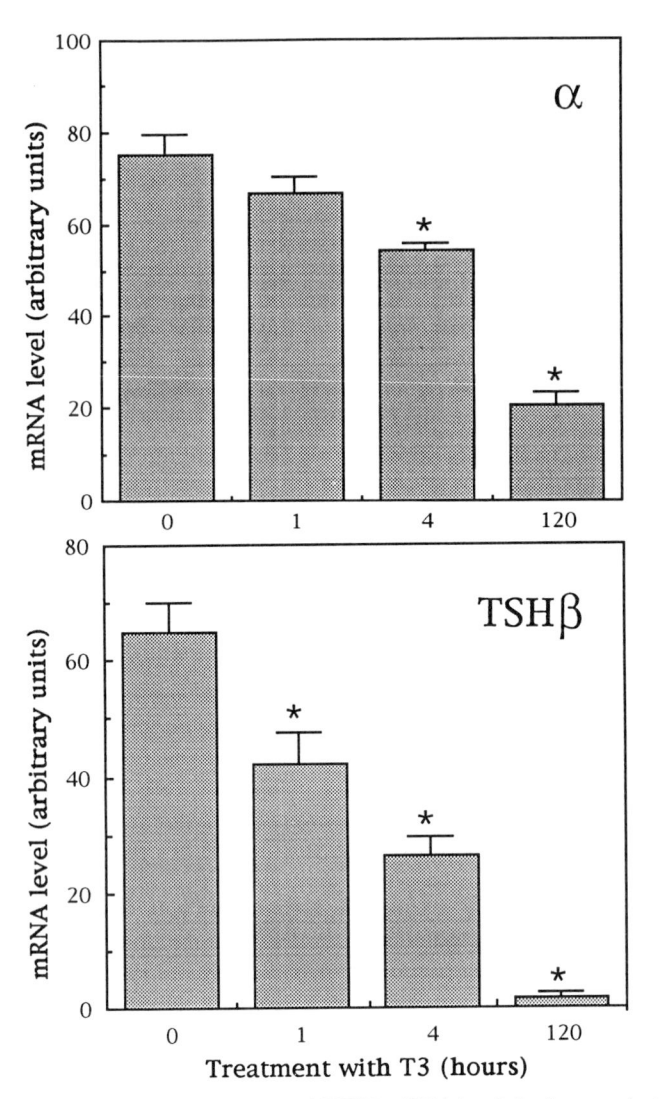

FIG. 3.   Regulation of steady-state $\alpha$ and TSH$\beta$ mRNA levels in thyrotropic tumors by thyroid hormones. Mice bearing thyrotropic tumors were treated with T3 (10 $\mu$g/100g/day) for varying times, and the steady-state TSH subunit mRNA levels in tumor tissues were determined by blot hybridization analyses. From Shupnik *et al.* (1985a), with permission.

Thyroid hormone, administered *in vivo* or *in vitro*, markedly and rapidly decreased the rates of transcription of the $\alpha$ and TSH$\beta$ subunit genes in the thyrotropic tissue (Fig. 4). The effects were evident in 1–2 hours and in 30 minutes for the $\alpha$ and TSH$\beta$ subunit genes, respectively (Shupnik *et al.*, 1985a). Thyroid hormones reduced the transcription rate for the TSH$\beta$

THYROTROPIN GENE EXPRESSION

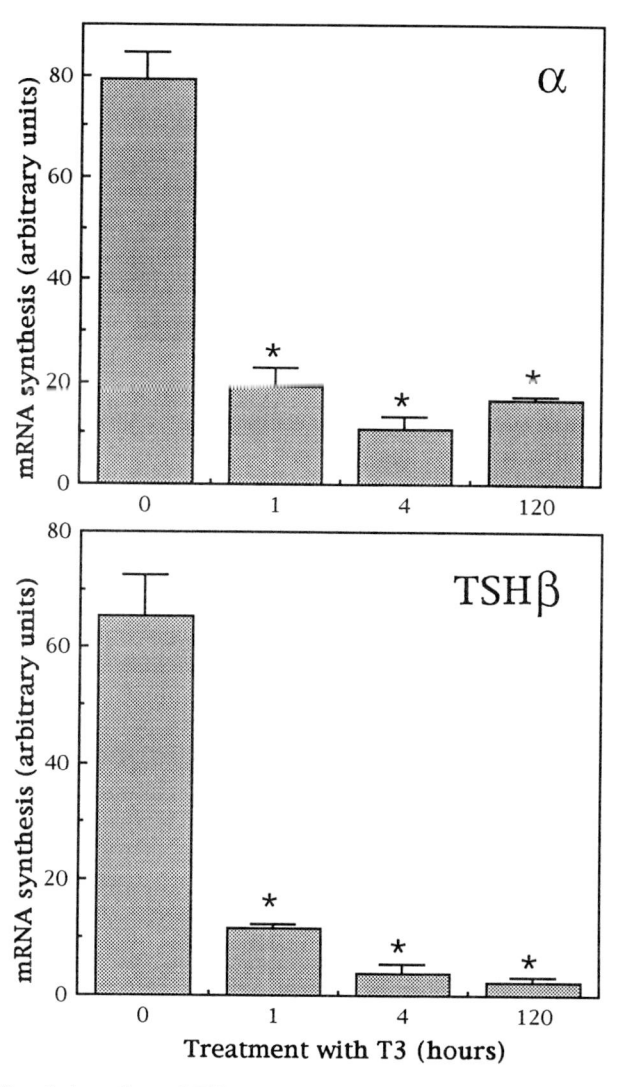

FIG. 4. Regulation of $\alpha$ and TSH$\beta$ gene transcription in thyrotropic tumors by thyroid hormones. Mice bearing thyrotropic tumors were treated with T3 (10 $\mu$g/100g/day) for varying times, and TSH subunit gene transcription rates were determined in tumor tissues by nuclear run-on approaches. From Shupnik *et al.* (1985a), with permission.

gene to <5% of the initial value at 4 hours. Thus, the effect of thyroid hormones on the synthesis of TSH is exerted, at least in part, at the transcriptional level. Further, thyroid hormones could inhibit $\alpha$ and TSH$\beta$ subunit gene expression even in the absence of new protein synthesis inasmuch as cycloheximide could not abrogate the negative effect of the thyroid hormones using thyrotropic tumor explant cultures (Shupnik *et al.*, 1986). The effects were shown to be directly correlated with the extent and time-course of the occupancy of nuclear thyroid hormone binding sites, nearly complete by 1 hour (Shupnik *et al.*, 1986). These results suggest strongly that thyroid hormones decrease TSH subunit gene expression by a *direct* effect, presumably via the binding of the ligand to the thyroid hormone receptor (TR), and the subsequent interaction of the complex with the regulatory region of the TSH subunit genes.

These results, which were also obtained by Gurr and Kourides (1985), provided confidence that putative cis-acting negative thyroid hormone-response elements (nTREs) could be located in the $\alpha$ and TSH$\beta$ genes. In this light, the previously isolated and characterized rat $\alpha$ and TSH$\beta$ subunit genes were used to define the nTREs by mutational and gene transfection approaches.

## VI. Definition of Putative nTREs in the Rat $\alpha$ Subunit Gene

The 5′ flanking region ($-246$ to $+36$) of the rat $\alpha$ subunit gene was fused to the chloramphenicol acetyltransferase (CAT) reporter gene, and transfected into $GH_3$ cells, a clonal rat pituitary cell line that permits the expression of the $\alpha$ subunit gene, using DEAE–dextran (Burnside *et al.*, 1989). After transfection, cells were treated with or without thyroid hormone for 48 hours, and the effect of hormone on the expression of the reporter gene was determined, as assayed by the activity of CAT in cellular extracts. The $-246$ to $+36$ fragment of the rat $\alpha$ subunit gene was sufficient to confer nearly 50% negative regulation of CAT expression by thyroid hormone. The effect was dose dependent with a half-maximal effective dose in the nanomolar range.

Further analysis of 5′ truncation mutants showed that a region between $-80$ and $+33$, containing sequences just upstream of the TATA box, was necessary for the thyroid hormone effect. In addition, DNA binding, as determined by the avidin–biotin complex DNA binding assay (ABCD) (Glass *et al.*, 1987), showed that the $-74$ to $-38$ region can bind to *in vitro*-synthesized TRs. Thus, the $-74$ to $-38$ region of the rat $\alpha$ subunit gene likely contains a putative nTRE (Fig. 5). Note that the sequence contains a set of direct repeats of the consensus TRE hexamer half-site [AGGT(c/a)A] (Brent *et al.*, 1989b). Other work in the human system has shown that a nTRE, in contrast, is located just downstream of the TATA box in the human $\alpha$ subunit gene (Chatterjee *et al.*, 1989), using similar criteria. Figure 6 shows a comparison of the proximal promoter regions of the rat and human $\alpha$ subunit genes. Remarkably, the region analogous to the putative nTRE in the human gene is apparently deleted in the rat.

THYROTROPIN GENE EXPRESSION 401

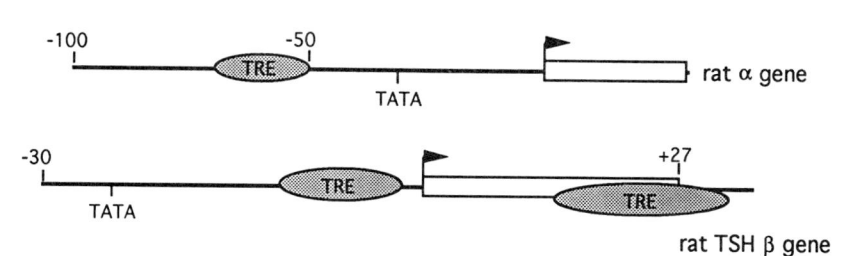

FIG. 5. Location of putative nTREs among the rat α and TSHβ gene promoters. Schematic diagram of the proximal 5′ flanking regions of the α and TSHβ genes in which the locations of the putative negative TREs are shown. TATA refers to the TATA box in the basal promoter of each gene. Only the downstream promoter of the rat TSHβ gene is shown. The bent arrow denotes the site of transcription initiation. The nucleotide numbers are indexed to the start site of transcription.

Whether rodent nTRE sequences have been translocated during evolution to this new site in the human gene is unclear.

## VII. Definition of Putative nTREs in the Rat TSHβ Subunit Gene

Using an approach similar to that used in the determination of the nTRE in the rat α subunit gene, we identified the nTRE in the rat TSHβ subunit gene (Carr *et al.*, 1989). In short, a large fragment (4 kb) of the 5′ flanking region of the rat TSHβ gene, including the first exon (+1 to +27) and a small portion of the first intron, was fused to the CAT reporter gene. This fusion gene was transfected into GH₃ cells and the activity of CAT determined as before. Surprisingly, the GH₃ cells supported the expression

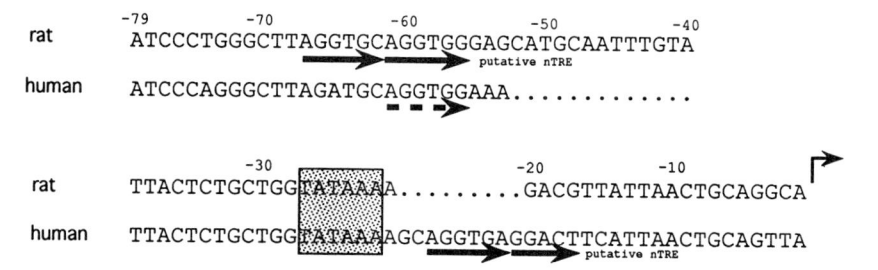

FIG. 6. DNA sequences of the putative nTREs in the rat and human α subunit gene promoters. A comparison of the proximal promoter regions of the rat and human α subunit genes is shown. Boldface arrows denote TRE consensus half-sites identified by functional and TR binding assays (Burnside *et al.*, 1989a; Chatterjee *et al.*, 1989). The broken arrow depicts a potential half-site in the human gene that mirrors the putative nTRE in the rat gene. The TATA boxes are shaded, and the bent arrow denotes the site of transcription initiation.

of the 5' flanking region of the TSHβ gene, even though the endogenous gene is not expressed (Mirell *et al.*, 1986). It has been shown recently that Pit-1, a pituitary-specific transcription factor, can activate growth hormone, prolactin, and TSHβ subunit genes (Ingraham *et al.*, 1990). Thus, it is possible that sufficient Pit-1 is expressed in GH$_3$ cells to allow some low level of exogenous TSHβ gene expression.

Transfection of the rat TSHβ 4-kb fragment fusion reporter resulted in CAT activity that could be modulated by thyroid hormone so that <50% of control levels could be observed after 48 hours (Carr *et al.*, 1989). Most surprising was the finding that further 5' deletion mutants could manifest thyroid hormone regulation even when the TATA box of the downstream promoter was deleted, and only the first seventeen 5' flanking region nucleotides, the first exon (27 base pairs), and a few intronic sequences (13 base pairs) (57 nucleotides total) remained! Thus, a cryptic promoter sequence in the vector provides enough basal gene expression to permit the detection of the putative nTRE(s) that was resident *downstream* of the TATA box. Remarkably, start-site analysis using RNase protection discovered authentic cap site use in this shortest construct. These results suggested that an nTRE(s) was present in this atypical region (Fig. 5).

To define further this 57-bp region with respect to TR binding, as expected in TREs, this DNA was tested for its ability to interact with *in vitro*-synthesized TRs (Darling *et al.*, 1989). Using the ABCD assay, two specific subregions were identified that each bound labeled TR. The DNA sequences so defined are shown in Fig. 7. Note again the presence of overlapping direct repeats in the downstream site. Similar findings in the mouse (Wood *et al.*, 1989) and human (Bodenner *et al.*, 1991; Wondisford

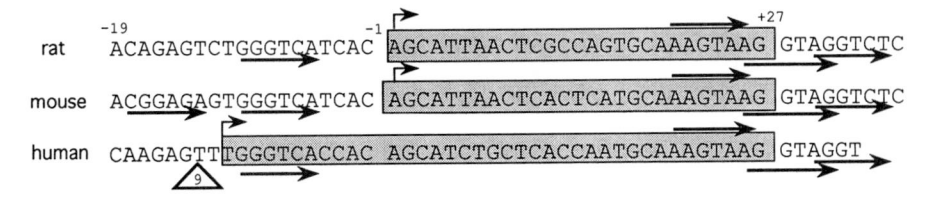

FIG. 7. DNA sequences of the putative nTREs in the rat, mouse, and human TSHβ subunit gene promoters. A comparison of the proximal promoter regions of the rat, mouse, and human TSHβ subunit genes is shown. The straight arrows denote TRE consensus half-sites identified by functional and TR binding assays (Bodenner *et al.*, 1991; Carr *et al.*, 1989; Darling *et al.*, 1989; Wondisford *et al.*, 1989; Wood *et al.*, 1989). The first exons (relative to the downstream promoter for the rat and mouse genes) are shaded, and the bent arrows denote the sites of transcription initiation. Note a nine-nucleotide deletion in the human gene relative to the rodent genes indicated by the triangle just 5' of the transcriptional start site.

*et al.*, 1989) TSH$\beta$ genes have been reported. A comparison of sequences in this region among the rat, mouse, and human genes shows that although the sequence of the downstream nTRE region is totally conserved, the upstream region is more variable. Of particular note, the mouse upstream nTRE consists of a pair of consensus TRE hexamer half-sites in a direct repeat arrangement separated by two nucleotides. Recently, Naar *et al.* (1991) provided evidence that such a TRE (direct repeat +2; DR+2) can subserve negative regulation by thyroid hormone in a heterologous promoter setting. Also, cotransfected TR cDNA with this DR+2 TRE results in increased expression of the reporter gene product in the *absence of ligand*. With thyroid hormone, the level of expression is reduced to levels below control values. While this observation remains to be confirmed, such sequences in the TSH$\beta$ gene also suggest the possibility that direct repeats in a specific context, or even single half-sites, may subserve the role of an nTRE.

The identification of putative nTREs downstream of the TATA box in the TSH$\beta$ gene promoters initially suggested the position-dependent nature of these elements. In particular, it indicated possible steric effects in the mechanism of negative regulation. However, several points and some experimental data argue against this explanation as the sole mechanism. First, as will be indicated, TRs can bind effectively to TREs in the absence of ligand. Because the negative regulation of TSH subunit gene expression is strictly thyroid hormone dependent, the role of the ligand is incorporated in this model with difficulty. Second, Brent *et al.* (1991b) have recently placed the downstream nTRE from the rat TSH$\beta$ gene and the nTRE from the rat $\alpha$ subunit gene in a location analogous to the native sites in the context of the rat growth hormone gene promoter, and observed negative regulation of the gene by thyroid hormone. Further, these putative nTREs, when placed at position −55 (upstream of the TATA box in the rat GH gene promoter), also mediated a negative thyroid hormone response (Fig. 8). Third, Carr *et al.* (1992) have shown that sequences +11 to +27 mediate negative regulation by thyroid hormone at positions −125, −50, −17, and +11, relative to a heterologous promoter. Mutation analysis revealed that sequences from +18 to +27 are essential for this negative regulation. This region functions only in the presence of TR, as evidenced by cotransfection studies with TR binding to the half-site motif. A second site, −17/+9, also binds TR and mediates negative regulation by thyroid hormone but is less effective, possibly reflecting the issue of context and requirement for additional DNA-binding proteins and coregulators/coactivators. In summary, these results indicate that the identified nTREs in the rat $\alpha$ and TSH$\beta$ subunit genes may not necessarily function in a position-specific manner.

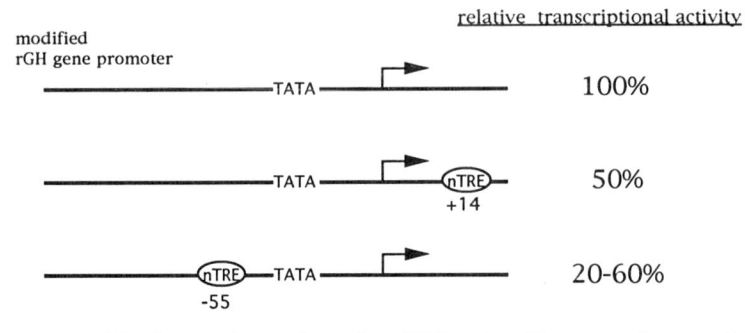

FIG. 8. Position independence of putative nTRE action. Placement of rat $\alpha$ and TSH$\beta$ nTREs in two different contexts within a modified rat growth hormone gene promoter results in continued negative regulation by thyroid hormones in a TR/reporter gene cotransfection paradigm. Note one context is the placement of the nTRE downstream of the transcription start site, and the other is the placement of the element just upstream of the TATA box. From Brent *et al.* (1989b), with permission.

## VIII. Posttranscriptional Regulation of TSH Gene Expression by Thyroid Hormones

It is possible that thyroid hormones may influence gene expression at loci other than transcription. Recently, evidence has been obtained suggesting that thyroid hormones can increase the turnover of TSH$\beta$ mRNA in cultured rat anterior pituitary cells (Krane *et al.*, 1991). In the presence of actinomycin D, the apparent half-life of TSH$\beta$ mRNA was shortened to 9 hours with thyroid hormone, compared to >24 hours in control cells. Further, it was noted that the poly(A) tail of the mRNA was also decreased in size (from 180–200 to 30 nucleotides, over an 8-hour period), as determined by an RNase H/oligo-dT procedure. Although it is not known whether the alteration in the poly(A) tail of the TSH$\beta$ mRNA is causal in effecting the change in mRNA half-life, these results emphasize the potential importance of posttranscriptional events in thyroid hormone action.

## IX. Mechanisms of Thyroid Hormone Action

In order to develop a model of thyroid hormone negative regulation of the TSH subunit gene expression, it is essential to appreciate the current knowledge of the mechanisms of thyroid hormone action (Brent *et al.*, 1991a; Chin, 1991; Lazar and Chin, 1990; Oppenheimer *et al.*, 1987; Samuels *et al.*, 1988). The recent identification of multiple thyroid hormone receptors and nuclear factors that interact with these transcription factors

THYROTROPIN GENE EXPRESSION            405

has greatly modified our view of the molecular events initiated by thyroid hormone.

Figure 9 depicts the current dogma regarding the details of the mode of thyroid action. Thyroid hormones (T4, T3) enter the cell via passive mechanisms. Also, T4 can be monodeiodinated to T3 in certain cell types. Thyroid hormones then move from the cytoplasm to the nucleus, where they interact with high-affinity nuclear receptors (TR; c-erbA) (Oppenheimer *et al.*, 1972). The TRs interact with target DNAs (thyroid hormone response element) as homodimers and monomers, as well as heterodimers, by association with retinoic acid receptors, retinoid X receptors, and TR auxiliary proteins (TRAPs). The combination of these TR/TR and other protein interactions result in activation or repression of genes that are regulated by thyroid hormones.

In late 1986, two key reports (Sap *et al.*, 1986; Weinberger *et al.*, 1986) detailed the identity of c-erbA, the protooncogene for the v-erbA, as the thyroid hormone receptor (TR). They showed that human placental and chicken embryo c-erbA can bind thyroid hormone and related analogs with relative affinities in the appropriate hierarchy of hormone potency. Like other members of the extensive steroid/thyroid hormone receptor superfamily (Evans, 1988), TRs possess DNA and ligand binding domains. The DNA binding domain contains two $Zn^{2+}$ fingers, with a P box (or proximal finger) that has homology with those of the retinoic acid recep-

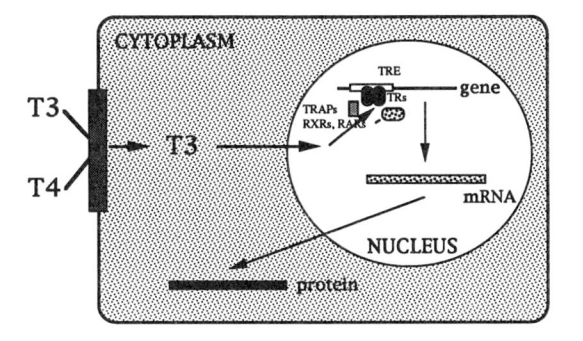

FIG. 9.   Thyroid hormone action. Schematic diagram of the action of thyroid hormones (T3 and T4) in their action in the regulation of gene expression. T3 and T4 enter the cell by a passive mode; T4 can be converted to T3 in certain cells by 5'-monodeiodination. T3, the more potent thyroid hormone, then enters the nucleus to interact with thyroid hormone receptors (TRs) and associated proteins such as TR auxiliary proteins (TRAPs), retinoid X receptors (RXRs), and retinoic acid receptors (RARs). The TR monomers, homodimers, and heterodimers then bind the thyroid hormone response element (TRE) in a thyroid hormone-regulated gene. Transcription of the gene to yield its cognate mRNA and protein then proceeds.

tor (RAR) and vitamin D receptor (VDR). Thus, TR along with RAR and VDR form a major subfamily of the nuclear receptor superfamily. Dimerization domains are found in the COOH-terminal portions of TR that superimpose with the ligand binding domain (Forman *et al.*, 1989). Finally, transactivation domains in TR have been poorly defined (O'Donnell and Koenig, 1990; Thompson and Evans, 1989).

Even with the original reports in 1986, it was immediately evident that at least two isoforms of c-erbA were present in vertebrates. Indeed, c-erbAα (TRα) and c-erbAβ (TRβ) are encoded by separate genes located on separate chromosomes in humans (chromosome 17 and 3, respectively) (Dayton *et al.*, 1984; Drabkin *et al.*, 1988; Weinberger *et al.*, 1986). Further studies revealed additional TR heterogeneity (Chin, 1991; Lazar and Chin, 1990). The rat TRα gene is transcribed to yield a transcript that is subjected to alternative mRNA splicing to produce at least two mRNAs that encode TRα-1 and c-erbAα-2 (Izumo and Mahdavi, 1988; Lazar *et al.*, 1988; Mitsuhashi *et al.*, 1988). TRα-1 is a bona fide TR based on its ability to bind TREs and thyroid hormones and to transactivate thyroid hormone-responsive genes. On the other hand, c-erbAα-2 cannot bind thyroid hormone due to an altered COOH-terminal end of the molecule along with an 82-amino acid extension. However, though it can bind TREs weakly, c-erbAα-2 is incapable of transactivation of thyroid hormone-responsive genes (Izumo and Mahdavi, 1988; Lazar *et al.*, 1988; Mitsuhashi *et al.*, 1988). Further, other experiments have shown that c-erbAα-2 can interfere with thyroid hormone action and hence exhibit dominant negative action, by yet unclarified mechanism(s) (Koenig *et al.*, 1989).

The single rat TRβ gene is transcribed from two promoters to yield two transcripts (including mandatory alternative RNA splicing at the 5' end). Each encodes a distinct TR isoform, TRβ-1 and TRβ-2 (Hodin *et al.*, 1989; Koenig *et al.*, 1988; Murray *et al.*, 1988; Yen *et al.*, 1991b). The two TR isoforms differ only by the NH₂-terminal regions, with identical DNA and ligand binding domains. Each is a bona fide TR by virtue of its ability to bind the appropriate DNAs and thyroid hormones as well as to transactivate TREs.

There is marked tissue-specific expression of the TR isoforms (Hodin *et al.*, 1989, 1990). TRα-1 and c-erbAα-2 are found in high abundance in the brain, skeletal and cardiac muscle, and brown fat, whereas TRβ-1 is expressed in most tissues, including the pituitary gland, liver, and kidney. TRβ-2 is highly unusual in that its expression is restricted to the anterior pituitary gland and to specific regions of the brain (Childs *et al.*, 1991; Cook *et al.*, 1992; Hodin *et al.*, 1989). Thus, TRα-1, c-crbAα-2, and TRβ-1 are widely expressed, although the relative amounts of these isoforms are different from tissue to tissue. TRβ-2 is the most spectacular example of this diversification, with highly limited tissue distribution. The functional ramification of this observation is still unclear because most studies have shown little difference in action among the bona fide TRs (Yen *et al.*, 1991a).

The nature of the TRE has been intensively studied. Analyses of the regulatory regions of two thyroid hormone-regulated genes, rat growth hormone (Brent *et al.*, 1991a; Crew and Spindler, 1986; Glass *et al.*, 1987; Samuels *et al.*, 1988) and rat α-myosin heavy chain (Izumo and Mahdavi, 1988), have revealed the presence of direct and inverted repeats of a putative TRE half-site. A consensus TRE half-site sequence, AGGT(c/A)A, has been described (Brent *et al.*, 1989b). Early in these efforts, a TRE palindrome (AGGTCATGACCT) was shown to be sufficient to confer thyroid hormone responses to a heterologous promoter (Glass *et al.*, 1988). More recently, several groups have demonstrated that direct repeats of the consensus TRE half-site with a gap of four nucleotides (AGGTCAN-NNNAGGTCA; direct repeat +4; DR+4) possess a similar ability (Naar *et al.*, 1991; Umesono *et al.*, 1991). In addition, TRE half-sites oriented in a head-to-head configuration (TGACCT $N_6$ AGGTCA) can also mediate thyroid hormone responses (Baniahmad *et al.*, 1990). Thus, TREs generally exist as pairs of half-sites, although the precise orientations of these component elements appear less important, as distinct from the steroid hormone response elements (Evans, 1988). The presence of paired TRE half-sites in functional TREs and the analogy with steroid hormone receptors suggest that TRs interact with TREs as dimeric molecules. Furthermore, three other features are apparently unique to TR/TRE interactions: (1) a single TRE half-site can bind TR as a monomer (Lazar and Berrodin, 1990); (2) unliganded TR can bind TREs (Darling *et al.*, 1991; Lazar and Berrodin, 1990), an observation consistent with the well-known association of TRs with the nucleus in general and chromatin in particular (Samuels *et al.*, 1988); (3) TRs cotransfected in the absence of thyroid hormones can possess functional activity, namely, repressive activity (Brent *et al.*, 1989a; Damm *et al.*, 1989; Graupner *et al.*, 1989; Sap *et al.*, 1989).

Another aspect of TR isoform expression is the regulation of TRs by thyroid hormones and other hormones and factors (Hodin *et al.*, 1990; Jones and Chin, 1991). Evidence that selective decreases in TRβ-2 expression result in altered thyroid hormone regulation of growth hormone gene expression (Jones and Chin, 1991) indicates that TRβ-2 may play a critical role in thyroid hormone action in the pituitary and selected regions of the brain. Such potential moment-to-moment modulation of the amounts of TR isoforms, along with their tissue-specific expression, highlights the likely importance of the relative amounts of the TR isoforms in determining ultimate thyroid hormone action. Of course, implicit to this statement is the assumption that the TR isoforms possess variable biological activities with diverse genes in different tissues. At this point, however, little evidence concerning such differential activities of the TR isoforms is available.

Finally, it is important to consider a set of rare heritable clinical disorders in which resistance of various tissues and organs to thyroid hormones is featured. Best studied are the syndromes of generalized resistance to thyroid hormones. The condition is inherited largely as an autosomal dominant disorder (Ono et al., 1991; Parilla et al., 1991; Refetoff, 1983). Recent genetic analyses have delineated clusters of point and other mutations in two "hot spots" limited to exons 9 and 10, located in the penultimate and COOH-terminal exons encoding the ligand-binding domain, of the TRβ gene (Takeda et al., 1991). The mutant alleles generally encode a TRβ that binds thyroid hormones with decreased affinity. The ability of a defect in a single TRβ allelic product to interfere with the function of "normal" TRs probably accounts for its dominant negative effect. Interestingly, a single family has been described that possesses an autosomal recessive inheritance pattern (Takeda et al., 1992). The genetic defect is a TRβ that is largely deleted; thus, a mutant TRβ that cannot interact with other receptors and related proteins results in a lack of dominant negative effect. These observations raise important issues with regard to the role of thyroid hormones, TRs, and other factors in thyroid hormone action.

## X. Role of TR Heterodimers in Thyroid Hormone Action

Glass et al. (1988, 1989) first described the potential of TRs to interact with other proteins, showing that TR enhanced the binding of RAR to TREs. Soon thereafter, Murray and Towle (1989) and Burnside et al. (1990), using different DNA binding assays, observed that nuclear extracts from nearly all mammalian tissues and cell lines could increase severalfold the binding of labeled, in vitro-translated TRα-1 and TRβ-1 to diverse TREs. This effect was shown to be a protein(s) with an apparent molecular size of 60–65 kDa (in pituitary tissue, with different sizes seen in other tissues) but is neither a TR nor a RAR(s). This factor is called TR auxiliary protein (TRAP).

Further characterization showed that the COOH-terminal region of TR was essential for the ability of TR to interact with TRAP (Darling et al., 1991) and that TRAP could bind weakly to specific DNA elements (Beebe et al., 1991). In addition, O'Donnell et al. (1991) and Lazar et al. (1991) showed potential close associations of TRs with TRAP by cross-linking studies. Several studies suggested that the formation of TR heterodimers possibly plays a role in thyroid hormone function. First, several point mutants in residues +285 to +300 of human TRβ shown by O'Donnell et

# THYROTROPIN GENE EXPRESSION

*al.* (1991) to fail to transactivate TREs, despite an ability to bind TREs and appropriate ligands, also fail to interact with TRAP (Darling *et al.,* 1991). Second, a cell line apparently lacking TRAP activity (HeLa tk⁻) also is incapable of permitting TR action (Zhang *et al.,* 1991). In summary, at that stage, it was likely that TRs interacted with a member(s) of the steroid/thyroid hormone receptor superfamily to form heterodimers that might be essential for thyroid hormone action.

Very recently, several groups simultaneously showed that a previously described orphan member of the nuclear receptor superfamily, retinoid X receptor (RXR), can form heterodimers with TRs and concomitantly augment the binding of TR to TREs (Hallenbeck *et al.,* 1991; Kliewer *et al.,* 1992; Leid *et al.,* 1992; Yu *et al.,* 1991; Zhang *et al.,* 1992). Furthermore, cotransfection of RXR and TR cDNAs was shown to synergize the transactivation ability of TR alone (Leid *et al.,* 1992; Zhang *et al.,* 1992). Thus, RXR is a TRAP, as previously defined. In addition, multiple RXRs (RXRα, RXRβ, and RXRγ) have been described, each with different tissue expression (Mangelsdorf *et al.,* 1992). Of great importance, it has been shown that 9-*cis*-retinoic acid is the natural ligand for the RXRs (Heyman *et al.,* 1992). Hence, an additional level of control of TR/RXR heterodimer formation and function may be possible. It is yet unclear whether other TRAPs exist in addition to the RXRs. Finally, it is important to note that RXRs can form heterodimers with RARs and VDRs with likely similar functions (Glass *et al.,* 1990; Liao *et al.,* 1990).

## XI. Ligand Decreases TR Homodimer Binding to TREs

Aside from the necessity of a ligand for full thyroid hormone action, the details of the role of a ligand in its interactions with TRs are limited. Recently, Yen *et al.* (1992) showed that thyroid hormones, in physiological concentrations (2–5 n*M*), can greatly decrease TR homodimer binding to various TREs, especially those with direct-repeat orientation. Most important, thyroid hormones had no effect on the formation of TR/TRAP heterodimers. For positively regulated TREs, these results indicate that thyroid hormones may initiate a ligand response by decreasing binding of TR homodimer to TREs, and hence diminish the unliganded TR repressive effect. Thereafter, thyroid hormones may engage TR/TRAP heterodimers in "true" activation by mechanisms still not elucidated.

## XII. Models of Negative Thyroid Hormone Regulation of TSH Subunit Gene Expression

Figure 10 depicts the possible models for the mechanisms by which the interactions of thyroid hormone, TRs, and nuclear factors with nTREs

## [1] Steric hindrance

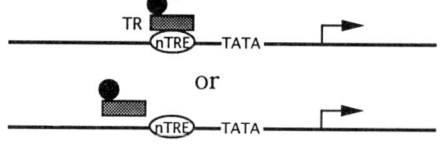

## [2] Squelching

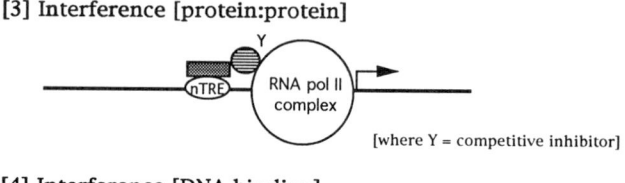

[where X = critical enhancer BP,
coactivator or coregulator]

## [3] Interference [protein:protein]

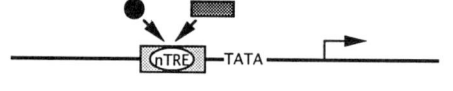

[where Y = competitive inhibitor]

## [4] Interference [DNA binding]

FIG. 10. Mechanisms for negative regulation of gene transcription. Four models for negative regulation of nTREs by thyroid hormones are illustrated. Shown in each example are the TATA box in the proximal promoter, the start of transcription (bent arrow), and the nTRE. The shaded rectangles represent TRs. Each model is described in the text. For Model 2, squelching of a critical enhancer binding protein (BP), coactivator, or coregulator (solid circle; X) is shown either on or off DNA. For Model 3, a shaded oval or Y is a competitive inhibitor that may interact directly with TR and the transcription complex. For Model 4, a composite nTRE/enhancer [nTRE within an enhancer (shaded rectangle)] is shown.

lead to negative regulation of gene expression by T3 and T4. To maintain perspective, the mechanisms for positive thyroid hormone regulation of TREs have not been fully delineated. Hence, the models presented here are highly speculative. First, an nTRE may be present in a position that sterically occludes efficient formation or transit of the RNA polymerase II complex (Model 1); however, the relative position independence of several nTREs would argue against this model as a sole mechanism. Second, an nTRE may bind TRs and associated proteins to form a complex that may squelch a critical factor such as a coregulator or coactivator, which may be important for efficient transcription (Model 2). Third, inter-

THYROTROPIN GENE EXPRESSION 411

ference may occur at the protein:protein level where the TR/nTRE interaction may bind an inhibitor (Model 3). Finally, interference may occur at the level of DNA binding such as may occur in a composite response element (Diamond *et al.*, 1990) (Model 4). Inherent in each model is the importance of the nTRE:TR:nuclear factor interaction in mediating a hormone effect different from that seen in TREs mediating a positive one. For instance, direct TRE half-site repeats with a two-nucleotide gap (DR+2) (Naar *et al.*, 1991) or a single TRE half-site, such as seen in overlapping sites found in the rat $\alpha$ and TSH$\beta$ subunit gene promoters, may dictate a different conformation of the bound TR, which may then allow differential interactions with other proteins, etc., in these models, as described above.

## XIII. Summary

Thyroid hormones suppress the synthesis and release of thyrotropin from thyrotropes in the anterior pituitary gland, a feature that is critical in the classic negative-feedback loop of the pituitary–thyroid endocrine axis. The major effect of thyroid hormones in this system is exerted at the transcriptional level. The molecular mechanisms by which there is negative regulation of TSH subunit gene expression by thyroid hormone have been elucidated. The TSH subunit genes have isolated and characterized. Structure–function analyses using fusion genes and DNA transfection approaches have defined the putative negative TREs among the promoters of the rat, mouse, and human $\alpha$ and TSH$\beta$ genes. These sequences are either largely overlapping direct TRE half-sites, TRE half-sites as direct repeats gapped by two nucleotides, or single TRE half-sites. These arrangements are distinct from those seen in positive TREs. Recent knowledge regarding the molecular mechanisms of thyroid action in general forces consideration of multiple TR isoforms, TR heterodimer partners (TRAPs), and thyroid hormones in the ultimate mechanisms of negative action. Several models have been proposed, but none has yet been proved. In addition, the role of thyroid hormone in the regulation of gene expression at the posttranscriptional level is beginning to be addressed. Future work should continue to illuminate these important facets of gene regulation.

### REFERENCES

Baniahmad, A., Steiner, C., Kohne, A. C., *et al.* (1990). *Cell (Cambridge, Mass.)* **61,** 505–514.
Beebe, J. S., Darling, D. S., and Chin, W. W. (1991). *Mol. Endocrinol.* **5,** 85–93.
Bodenner, D. L., Mroczynski, M. A., Weintraub, B. D., *et al.* (1991). *J. Biol. Chem.* **266,** 21666–21673.

Boothby, M., Ruddon, R. W., Anderson, C., et al. (1981). J. Biol. Chem. **256**, 5121–5127.
Brent, G. A., Dunn, M. K., Harney, J. W., et al. (1989a). New Biologist **1**, 329–336.
Brent, G. A., Harney, J. W., Chen, Y., et al. (1989b). Mol. Endocrinol. **3**, 1996–2004.
Brent, G. A., Moore, D. D., and Larsen, P. R. (1991a). Annu. Rev. Physiol. **53**, 17–35.
Brent, G. A., Williams, G. R., Harney, J. W., et al. (1991b). Mol. Endocrinol. **5**, 542–548.
Burnside, J., Darling, D. S., Carr, F. E., et al. (1989). J. Biol. Chem. **264**, 6886–6891.
Burnside, J., Buckland, P. R., and Chin, W. W. (1988). Gene **70**, 67–74.
Burnside, J., Darling, D. S., and Chin, W. W. (1990). J. Biol. Chem. **265**, 2500–2504.
Carr, F. E., Burnside, J., and Chin, W. W. (1989). Mol. Endocrinol. **3**, 709–716.
Carr, F. E., and Chin, W. W. (1985). Endocrinology (Baltimore) **116**, 1151–1157.
Carr, F. E., and Chin, W. W. (1988). Mol. Endocrinol. **2**, 667–673.
Carr, F. E., Kaseem, L. L., and Wong, N. C. W. (1992). J. Biol. Chem., **267**, 18689–18694.
Carr, F. E., Need, L. R., and Chin, W. W. (1987). J. Biol. Chem. **262**, 981–987.
Carr, F. E., Ridgway, E. C., and Chin, W. W. (1985). Endocrinology (Baltimore) **117**, 1272–1278.
Chatterjee, V. K., Lee, J. K., Rentoumis, A., et al. (1989). Proc. Natl. Acad. Sci. U.S.A. **86**, 9114–9118.
Childs, G. V., Taub, K., Jones, K. E., et al. (1991). Endocrinology (Baltimore) **129**, 2767–2773.
Chin, W. W. (1985). In "The Pituitary Gland" (H. Imura, ed.), pp. 103–125. Raven Press, New York.
Chin, W. W. (1986). In "Molecular Cloning of Hormone Genes" (J. F. Habener, ed.), pp. 137–172. Humana Press, Clifton, NJ.
Chin, W. W. (1991). In "Nuclear Hormone Receptors" (M. Parker, ed.), pp. 77–102. Academic Press, New York/London.
Chin, W. W., Kronenberg, H. M., Dee, P. C., et al. (1981). Proc. Natl. Acad. Sci. U.S.A. **78**, 5329–5333.
Chin, W. W., Maizel, J. V., Jr., and Habener, J. F. (1983). Endocrinology (Baltimore) **112**, 482–485.
Chin, W. W., Muccini, J. A., and Shin, L. (1985a). Biochem. Biophys. Res. Commun. **128**, 1152–1158.
Chin, W. W., Shupnik, M. A., Ross, D. S., et al. (1985b). Endocrinology (Baltimore) **116**, 873–878.
Cook, C. B., Kakucska, I., Lechan, R. M., et al. (1992). Endocrinology (Baltimore) **130**, 1077–1079.
Crew, M. D., and Spindler, S. R. (1986). J. Biol. Chem. **261**, 5018–5022.
Croyle, M. L., Battacharya, A., Gordon, D. F., et al. (1986). DNA **5**, 299–304.
Croyle, M. L., and Maurer, R. A. (1984). DNA **3**, 231–236.
Damm, K., Thompson, C. C., and Evans, R. M. (1989). Nature (London) **339**, 593–597.
Darling, D. S., Beebe, J. S., Burnside, J., et al. (1991). Mol. Endocrinol. **5**, 73–84.
Darling, D. S., Burnside, J., and Chin, W. W. (1989). Mol. Endocrinol. **3**, 1359–1368.
Dayton, A. I., Selden, J. R., Laws, G., et al. (1984). Proc. Natl. Acad. Sci. U.S.A. **81**, 4495–4499.
Diamond, M. I., Miner, J. W., Yoshinaga, S. K., et al. (1990). Science **249**, 1266–1272.
Drabkin, H., Kao, F. T., Hartz, J., et al. (1988). Proc. Natl. Acad. Sci. U.S.A. **85**, 9258–9262.
Erwin, C. R., Croyle, M. L., Donelson, J. E., et al. (1983). Biochemistry **22**, 4856–4860.
Evans, R. M. (1988). Science **240**, 889–895.
Fiddes, J. C., and Goodman, H. M. (1979). Nature (London) **281**, 351–355.
Fiddes, J. C., and Goodman, H. M. (1981). J. Mol. Appl. Genet. **1**, 3–18.
Forman, B. M., Yang, C. R., Au, M., et al. (1989). Mol. Endocrinol. **3**, 1610–1620.
Furth, J. (1955). Recent Prog. Horm. Res. **11**, 221–255.

## THYROTROPIN GENE EXPRESSION    413

Glass, C. K., Devary, O. V., and Rosenfeld, M. G. (1990). *Cell (Cambridge, Mass.)* **63**, 729–738.

Glass, C. K., Franco, R., Weinberger, C., *et al.* (1987). *Nature (London)* **329**, 738–741.

Glass, C. K., Holloway, J. M., Devary, O. V., *et al.* (1988). *Cell (Cambridge, Mass.)* **54**, 313–323.

Glass, C. K., Lipkin, S. M., Devary, O. V., *et al.* (1989). *Cell (Cambridge, Mass.)* **59**, 697–708.

Godine, J. E., Chin, W. W., and Habener, J. F. (1982). *J. Biol. Chem.* **257**, 8368–8371.

Goodwin, R. G., Moncman, C. L., Rottman, F. M., *et al.* (1983). *Nucleic Acids Res.* **11**, 6873–6882.

Gordon, D. F., Wood, W. M., and Ridgway, E. C. (1988). *DNA* **7**, 17–26.

Graupner, G., Wills, K. N., Tzukerman, M., *et al.* (1989). *Nature (London)* **340**, 653–656.

Gurr, J. A., Catterall, J. F., and Kourides, I. A. (1983). *Proc. Natl. Acad. Sci. U.S.A.* **80**, 2122–2126.

Gurr, J., and Kourides, I. (1983). *J. Biol. Chem.* **258**, 10208–10211.

Gurr, J. A., and Kourides, I. A. (1985). *DNA* **4**, 301–308.

Hallenbeck, P., Marks, M., Mitsuhashi, T., *et al.* (1991). *Thyroid* **1**, S60.

Hayashizaki, Y., Miyai, J., Kato, K., *et al.* (1985). *FEBS Lett.* **188**, 394–400.

Hershman, J. M., and Pekary, A. E. 1985. *In* "The Pituitary Gland," (H. Imura, ed.), pp. 149–188. Raven Press, New York.

Heyman, R. A., Mangelsdorf, D. J., Dyck, J. A., *et al.* (1992). *Cell (Cambridge, Mass.)* **68**, 397–406.

Hodin, R. A., Lazar, M. A., and Chin, W. W. (1990). *J. Clin. Invest.* **85**, 101–105.

Hodin, R. A., Lazar, M. A., and Wintman, B. I., *et al.* (1989). *Science* **244**, 76–79.

Ingraham, H. A., Albert, V. R., Chen, R. P., *et al.* (1990). *Annu. Rev. Physiol.* **52**, 773–791.

Izumo, S., and Mahdavi, V. (1988). *Nature (London)* **334**, 539–542.

Jones, K. E., and Chin, W. W. (1991). *Endocrinology (Baltimore)* **128**, 1763–1768.

Kliewer, S. A., Umesono, K., Mangelsdorf, D. J., *et al.* (1992). *Nature (London)* **355**, 446–449.

Koenig, R. J., Lazar, M. A., Hodin, R. A., *et al.* (1989). *Nature (London)* **337**, 659–661.

Koenig, R. J., Warne, R. L., Brent, G. A., *et al.* (1988). *Proc. Natl. Acad. Sci. U.S.A.* **85**, 5031–5035.

Krane, I. M., Spindel, E. R., and Chin, W. W. (1991). *Mol. Endocrinol.* **5**, 469–475.

Lazar, M. A., and Berrodin, T. J. (1990). *Mol. Endocrinol.* **4**, 1627–1635.

Lazar, M. A., Berrodin, T. J., and Harding, H. P. (1991). *Mol. Cell. Biol.* **11**, 5005–5015.

Lazar, M. A., and Chin, W. W. (1990). *J. Clin. Invest.* **86**, 1777–1782.

Lazar, M. A., Hodin, R. A., Darling, D. S., *et al.* (1988). *Mol. Endocrinol.* **2**, 893–901.

Leid, M. P., Kastner, R., Lyons, H., *et al.* (1992). *Cell (Cambridge, Mass.)* **68**, 377–395.

Liao, J., Ozono, K., Sone, T., *et al.* (1990). *Proc. Natl. Acad. Sci. U.S.A.* **87**, 9751–9755.

Mangelsdorf, D. J., Borgmeyer, U., Heyman, R. A., *et al.* (1992). *Genes Dev.* **6**, 329–344.

Maurer, R. A., Croyle, M. L., and Donelson, J. E. (1984). *J. Biol. Chem.* **255**, 5024–5025.

McKnight, G. S., and Palmiter, R. D. (1979). *J. Biol. Chem.* **254**, 9050–9058.

Mirell, C. J., Lau, R., Huaco, M., *et al.* (1986). *Mol. Cell. Endocrinol.* **47**, 145–151.

Mitsuhashi, T., Tennyson, G. E., and Nikodem, V. M. (1988). *Proc. Natl. Acad. Sci. U.S.A.* **85**, 5804–5808.

Murray, M. B., Liz, N. D., McCreary, N. L., *et al.* (1988). *J. Biol. Chem.* **263**, 12770–12777.

Murray, M. B., and Towle, H. C. (1989). *Mol. Endocrinol.* **3**, 1434–1442.

Naar, A. M., Boutin, J. M., Lipkin, S. M., *et al.* (1991). *Cell (Cambridge, Mass.)* **65**, 1267–1279.

Naylor, S. L., Chin, W. W., Goodman, H. M., *et al.* (1983). *Somat. Cell. Mol. Genet.* **9**, 757–770.

414 WILLIAM W. CHIN ET AL.

Naylor, S. L., Sakaguchi, A. Y., MacDonald, L., *et al.* (1986). *Somat. Cell. Mol. Genet.* **12**, 307–311.

Nilson, J. H., Thomason, A. R., Cserbak, M. T., *et al.* (1983). *J. Biol. Chem.* **258**, 4679–4682.

O'Donnell, A. L., and Koenig, R. J. (1990). *Mol. Endocrinol.* **4**, 715–720.

O'Donnell, A. I., Rosen, E. D., Darling, D. S., *et al.* (1991). *Mol. Endocrinol.* **5**, 94–99.

Ono, S., Schwartz, I. D., Mueller, O. T., *et al.* (1991). *J. Clin. Endocrinol. Metab.* **73**, 990–994.

Oppenheimer, J. H., Koerner, D., Schwartz, H. L., *et al.* (1972). *J. Clin. Endocrinol. Metab.* **35**, 330–333.

Oppenheimer, J. H., Schwartz, H. L., Mariash, C. N., *et al.* (1987). *Endocr. Rev.* **8**, 288–308.

Parilla, R., Mixson, A. J., McPherson, J. A., *et al.* (1991). *J. Clin. Invest.* **88**, 2123–2130.

Pierce, J. G., and Parsons, T. F. (1981). *Annu. Rev. Biochem.* **50**, 465–495.

Refetoff, S. (1983). *Am. J. Physiol.* **243**, E88–E98.

Samuels, H. H., Forman, B. M., and Horowitz, Z. D. (1988). *J. Clin. Invest.* **81**, 957–967.

Sap, J., Munoz, A., Damm, K., *et al.* (1986). *Nature (London)* **324**, 635–640.

Sap, J., Munoz, A., Schmitt, J., *et al.* (1989). *Nature (London)* **340**, 242–244.

Scanlon, M. F., Weightman, D. R., Shale, D. J., *et al.* (1979). *Clin. Endocrinol. (Oxford)* **10**, 7–15.

Shupnik, M. A., Ardisson, L. J., Meskell, M. J., *et al.* (1986). *Endocrinology (Baltimore)* **118**, 367–371.

Shupnik, M. A., Chin, W. W., Habener, J. F., *et al.* (1985a). *J. Biol. Chem.* **260**, 2900–2903.

Shupnik, M. A., Ridgway, E. C., and MacVeigh, M. S. (1985b). *Endocrinology (Baltimore)* **117**, 1940–1946.

Shupnik, M. A., Chin, W. W., Ross, D. S., *et al.* (1983). *J. Biol. Chem.* **253**, 15120–15124.

Shupnik, M. A., Ridgway, E. C., and Chin, W. W. (1989). *Endocr. Rev.* **10**, 459–475.

Takeda, K., Balzano, S., Sakurai, A., *et al.* (1991). *J. Clin. Invest.* **87**, 496–502.

Takeda, K., Sakurai, A., DeGroot, L. J., *et al.* (1992). *J. Clin. Endocrinol. Metab.* **74**, 49–55.

Tatsumi, K., Hayashizaki, Y., Hiraoka, Y., *et al.* (1988). *Gene* **73**, 489–497.

Tepper, M. A., and Roberts, J. L. (1984). *Endocrinology (Baltimore)* **115**, 385–391.

Thompson, C. C., and Evans, R. M. (1989). *Proc. Natl. Acad. Sci. U.S.A.* **86**, 3494–3498.

Umesono, K., Murakami, K. K., Thompson, C. C., *et al.* (1991). *Cell (Cambridge, Mass.)* **65**, 1255–1266.

Weinberger, C., Thompson, C. C., Ong, E. S., *et al.* (1986). *Nature (London)* **324**, 641–646.

Wolf, O., Kourides, I. A., and Gurr, J. A. (1987). *J. Biol. Chem.* **262**, 16596–16603.

Wondisford, F. E., Farr, E. A., Radovick, S., *et al.* (1989). *J. Biol. Chem.* **264**, 14601–14604.

Wood, W. M., Gordon, E. F., and Ridgway, E. C. (1987). *Mol. Endocrinol.* **1**, 875–883.

Wood, W. M., Kao, M. Y., Gordon, D. F., *et al.* (1989). *J. Biol. Chem.* **264**, 14840–14847.

Yen, P. M., Darling, D. S., Carter, R. L., *et al.* (1992). *J. Biol. Chem.* **267**, 3565–3568.

Yen, P. M., Darling, D. S., and Chin, W. W. (1991a). *Endocrinology (Baltimore)* **129**, 3331–3336.

Yen, P. M., Sunday, M. E., Darling, D. S., *et al.* (1991b). *Endocrinology (Baltimore)* **130**, 1539–1546.

Yu, V. C., Delsert, C., Andersen, B., *et al.* (1991). *Cell (Cambridge, Mass.)* **67**, 1251–1266.

Zhang, X. K., Hoffmann, B., Tran, P. B. V., *et al.* (1992). *Nature (London)* **355**, 441–446.

Zhang, X. K., Tran, P. B. V., and Pfahl, M. (1991). *Mol. Endocrinol.* **5**, 1909–1920.

RECENT PROGRESS IN HORMONE RESEARCH, VOL. 48

# Prohormone Structure Governs Proteolytic Processing and Sorting in the Golgi Complex

LINDA J. JUNG, THANE KREINER, AND RICHARD H. SCHELLER

*Howard Hughes Medical Institute, Department of Molecular and Cellular Physiology, Beckman Center, Stanford University, Stanford, California, 94305*

## I. Introduction

Peptides, along with biogenic amines and amino acids, are used as chemical messengers in the nervous system. Many of these neuropeptides are synthesized as parts of larger prohormones that may encode multiple physiologically active peptides (Douglass *et al.*, 1984). As the prohormones travel through the secretory pathway, the neuropeptides are groomed for action via a variety of processing events. In the trans-Golgi network (TGN), neuropeptides are segregated from other secretory proteins and are packaged into dense-core vesicles (DCVs) (Griffiths and Simons, 1986; Orci *et al.*, 1988). These vesicles are then transported to the nerve terminal and their contents are released by exocytosis when the appropriate signal is received by the neuron. The neuropeptides are released at synaptic sites, or into the circulation where they may travel to distant target tissues.

Processing begins with the removal of the signal sequence as the nascent peptide enters the endoplasmic reticulum (ER) (Walter and Blobel, 1981). The physiologically active peptide is frequently released from the prohormone molecule by proteolytic cleavage at basic residues flanking the neuropeptide unit (Fig. 1) (Docherty and Steiner, 1982). The carboxy-terminal basic residue(s) are then trimmed by a carboxypeptidase-like activity, and if the remaining carboxy-terminal amino acid is glycine, it is reduced to an amide group. Hence, elucidating the molecular and cellular mechanisms governing these processing events is key to fully understanding the regulation of neuropeptide biosynthesis.

In this review on neuropeptide processing, we will discuss (1) the structural features of the prohormone important for processing, (2) the processing enzymes, and (3) how the processing sequence of the egg-laying hormone (ELH) prohormone influences the sorting of peptides derived from this molecule.

415

Copyright © 1993 by Academic Press, Inc.
All rights of reproduction in any form reserved.

FIG. 1. Steps in neuropeptide maturation. The first step in prohormone processing is usually the cleavage at the carboxy-terminal site of a recognition site, frequently a pair of basic residues. In yeast, this endoproteolytic cleavage is performed by the *KEX2* gene product during the maturation of the α mating factor in yeast. Next, the basic residues are trimmed by the carboxypeptidase E enzyme, or the *KEX1*-encoded product in yeast. If the peptide ends in glycine, the carboxy terminus is amidated by the enzyme peptidyl-glycine-α-amidating monooxygenase (PAM).

## II. Prohormone Structure

Although there is great diversity among the primary amino acid sequences of prohormones, a basic organizational scheme exists. The structures of several prohormones are diagrammed in Fig. 2. All prohormones have signal sequences that direct them into the secretory pathway, where they can be further processed and packaged. The neuropeptides can be encoded as single discrete segments of a precursor, or in disulfide-linked units that are processed to mature hormones, as is in the case for proinsulin (Ullrich *et al.*, 1977). Here, the A and B peptides are separated by a C (connecting) peptide that is not present in mature insulin. In some cases, the mature peptides are in tandem, whereas in other precursors they are separated by spacer regions. Spatial organization of the physiologically active units may also be significant to processing. For example, in the ELH prohormone, the bag cell peptides (BCPs) are encoded on an amino-terminal intermediate and ELH is encoded on a carboxy-terminal intermediate. This organization allows the BCPs to be sorted into a compartment separate from that of ELH (see Section IV) (Fisher *et al.*, 1988).

As mentioned earlier, neuropeptide units are flanked by sets of basic residues, usually dibasic combinations of Lys-Arg or Arg-Arg. The neuropeptides are released by endoproteolytic cleavage at the carboxy terminus of the last basic residue of the set. However, not all basic residues are sites for cleavage. For example, the von Willebrand factor contains 28

PROHORMONE STRUCTURE EFFECTS 417

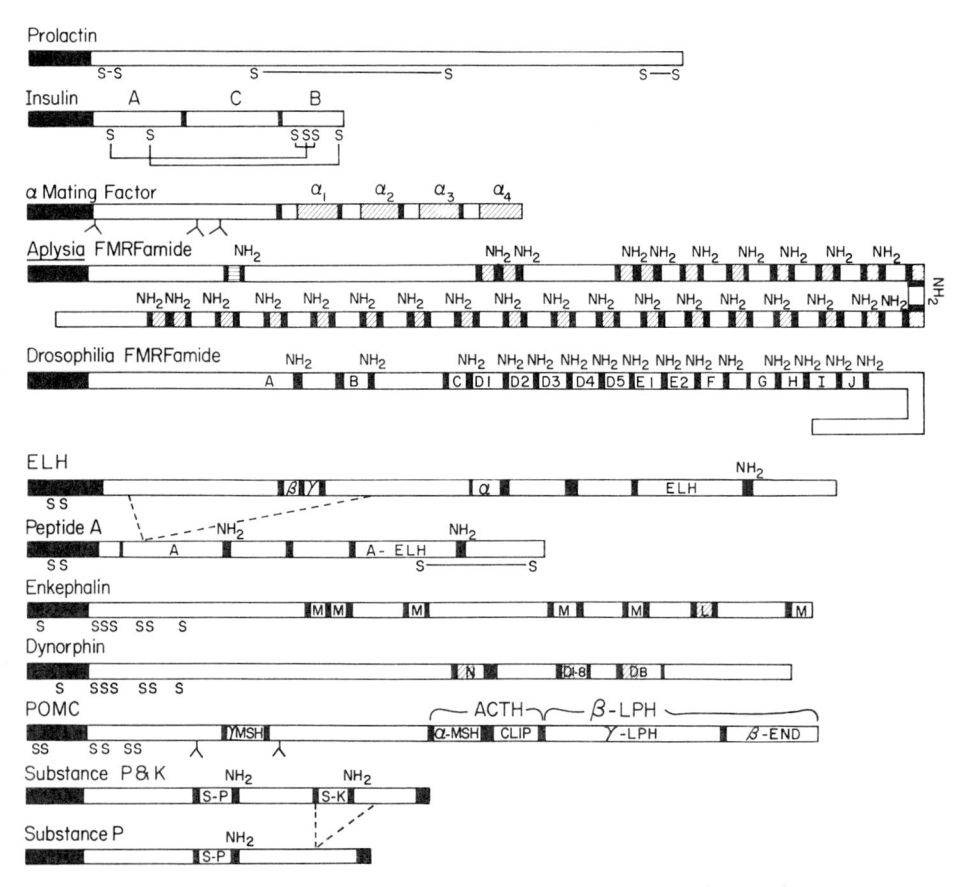

FIG. 2. Peptide hormone and neuropeptide precursor structures. Schematic representations of several hormone and neuropeptide precursors are illustrated. Each precursor contains a hydrophobic signal sequence (□). Internal cleavages at basic residues are indicated by the vertical bars, and the active peptides are labeled or shaded. Cysteine (S) and sugar ( λ ) residues are indicated below the schematic. Deletions between members of gene families or alternative RNA processing between two similar precursors are indicated by broken lines.

dibasic residue pairs, but only one is utilized. Thus, basic residues are necessary but are not a sufficient signal for endoproteolytic cleavage.

Clearly, structural features other than the basic residues of the processing site or prohormones help to constitute recognition sites for enzymatic cleavage. In some cases, a single point mutation outside of the processing site may render a prohormone resistant to cleavage. A proinsulin-encoding gene isolated from a patient with hyperinsulinoma was found to have a histidine-to-aspartate change at position B10; the proinsulin was not processed at the dibasic cleavage site (Carroll *et al.*, 1988). Other studies have shown that if a certain set of four amino acids proximal to the C peptide of proinsulin is deleted, proinsulin is not cleaved. (Gross *et al.*, 1989). Interestingly, these four amino acids were selected for mutation because they are conserved among proinsulins from various species. Similarly, a group of six amino acids proximal to the processing site in nerve growth factor is conserved and is likewise necessary for correct proteolytic processing (Suter *et al.*, 1991). Theoretical calculations indicate that dibasic residues in an $\alpha$ helix are not accessible for cleavage (Rholam *et al.*, 1986). Thus, the secondary and tertiary structure of the prohormone is likely to be an important factor in processing. Consequently, many studies have focused on using site-directed mutagenesis to localize structural determinants essential in processing.

## III. Processing Enzymes

If we are to fully understand the biosynthesis of neuropeptides, prohormone-processing enzymes must also be isolated and their structures and mechanisms characterized. For many years, various groups have tried to biochemically isolate the endoproteases involved in dibasic cleavages of prohormones (reviewed by Loh *et al.*, 1984). However, this has turned out to be a very difficult task due to the paucity of individual enzymes and the propensity of different enzymes that can perform dibasic cleavages. Furthermore, it is likely that more than one enzyme is responsible for internal cleavages of proinsulin (Davidson *et al.*, 1988) and prosomatostatin (Mackin and Noe, 1987). In addition, proteolytic cleavages occur in tissue-specific patterns, and therefore certain enzymes may be expressed specifically in different cell types. For example, in the anterior lobe of the pituitary, the proopiomelanocortin (POMC) precursor is cleaved at the basic residues flanking ACTH to yield ACTH and $\beta$-LPH (reviewed by Smith and Funder, 1988). However, in the intermediate lobe of the pituitary, ACTH and $\beta$-LPH are further processed by cleavage at additional internal basic sites to yield $\alpha$-MSH and $\beta$-endorphin, respectively (Fig. 2).

Until recently, the only characterized prohormone-specific endoprotease was the product of the *KEX2* gene isolated from yeast (Julius *et al.*, 1984; Fuller *et al.*, 1989a). Yeast deficient at the *KEX2* locus are unable to cleave the $\alpha$ mating prohormone and killer toxin precursor at the appropriate dibasic sites. This provides clear genetic confirmation that the *KEX2* locus encodes an endoproteolytic processing enzyme. This enzyme, kex2p, a glycoprotein of approximately 100 kDa, is a calcium-dependent serine protease with a neutral pH optimum (Fuller *et al.*, 1989a). The amino acid sequence deduced from the *KEX2* gene reveals homology to

PROHORMONE STRUCTURE EFFECTS 419

bacterial subtilisin enzymes. Cotransfection of *KEX2* and POMC cDNAs into a mammalian cell line incapable of processing POMC enables the cells to process the precursor (Thomas *et al.*, 1988). In fact, POMC was cleaved at proper dibasic sites to produce ACTH. These results prompted many groups to hypothesize that the mammalian dibasic processing enzymes would be homologous to kex2p.

Use of the *KEX2* gene as a hybridization probe in an attempt to find the mammalian homologue was unsuccessful. Furin, the first identified mammalian homologue of kex2p, was found by doing a computer search for proteins with similarity to kex2p (Fuller *et al.*, 1989b). Originally, furin was isolated as a gene upstream of the fes/fps oncogenes and was thought to be a receptor (Roebroek *et al.*, 1986). Furin has 50% amino acid identity to kex2p in a 300-amino acid region of the catalytic domain. When expressed in COS cells, proteins of 100 and 90 kDa are detected by Western analysis of the transfected cells. Furin has a configuration similar to that of kex2p, including a signal sequence, a glycosylated and hydrophobic amino-terminal domain, a transmembrane domain for anchoring in the membrane, and a highly charged carboxy-terminal region (Breshnahan *et al.*, 1990). In kex2p, it is proposed that certain residues in the carboxy-terminal tail are crucial to localize and retain the protein in the Golgi. Similarly, light-level immunocytochemistry showed that furin is also Golgi localized. Furin can cleave growth factors such as pro-$\beta$ nerve growth factor (Breshnahan *et al.*, 1990) and von Willebrand factor (van de Ven *et al.*, 1990; Wise *et al.*, 1990). Mutational studies to determine the specific cleavage site for furin indicated that the consensus sequence is Arg-X-Lys/Arg-Arg, with the first and fourth Arg being crucial for cleavage. It was also proposed that the third basic residue is not necessary for furin processing; this can be confirmed by cotransfection of furin with substrates containing different processing-site sequences. Interestingly, the Arg-X-Arg/Lys-Arg consensus sequence is more common to precursors of receptors and growth factors than to prohormones (Barr, 1991). Because furin is found to be expressed in a broad range of tissues (Barr *et al.*, 1991), it is not necessarily specific for prohormone processing.

Assuming that other potential endoproteolytic enzymes have similarity to the subtilisin-like active site of kex2p and furin, two additional potential mammalian endoproteolytic processing hormones were isolated. These are denoted as PC2 (Seidah *et al.*, 1990; Smeekens and Steiner, 1990) and PC1/PC3 (Seidah *et al.*, 1990; Smeekens *et al.*, 1991). As expected, each is similar to both kex2p and furin, particularly at the active-site region. Although PC2 and PC1/PC3 do not contain transmembrane domains, the carboxy-terminal regions are similar to amphipathic helices, which may be used for membrane anchoring. The ability of these enzymes to cleave prohormones was tested by coexpression of genes encoding PC1/PC3 or PC2 with the POMC gene into BSC-40 cells (Benjannet et al., 1991; Thomas et al., 1991). BSC-40 cells are epithelial in origin and contain a constitutive but not a regulated secretory pathway. Therefore, they should not contain endogenous prohormone-processing enzymes. POMC was cleaved by PC1/PC3 at the Lys-Arg flanking ACTH; the Lys-Arg in $\beta$-LPH was less efficiently cleaved, which is reminiscent of the processing pattern seen in the anterior lobe of the pituitary. PC2, but not PC1/PC3, selectively cleaved POMC to give $\beta$-LPH and also cleaved at the internal diabasic site in ACTH. Together, PC1/PC3 and PC2 cleaved POMC to yield the peptides generated in the intermediate lobe of the pituitary. The messages for PC1/PC3 and PC2 are preferentially expressed in neuroendocrine tissue such as the pituitary gland, the hypothalamus, and pancreatic insulin secretory cells. This result strengthens the hypothesis that PC1/PC3 and PC2 are actual prohormone cleavage enzymes. Subcellular localization studies will provide interesting information regarding the sites of these processing events in the secretory pathway.

Trimming of the carboxy-terminal basic residues is performed by a carboxypeptidase-like enzyme (Fig. 1). This activity is crucial to the maturation of hormones. Yeast deficient at the KEX1 locus are unable to remove the basic residues from the $\alpha$ mating factor, rendering it inactive; the yeast therefore display a sterile phenotype. As expected, KEX1 encodes a carboxypeptidase. (Fuller et al., 1988). A mammalian version, carboxypeptidase E (CE), was the first prohormone-processing enzyme to be isolated (Fricker and Snyder, 1983). CE was isolated biochemically from granules in the bovine adrenal medulla and the amino acid sequence was deduced from its cDNA (Fricker et al., 1986). CE is a 50-kDa glycoprotein with a pH optimum of 5 to 6 and it is stimulated by cobalt. It is inhibited by metal chelators unlike lysosomal carboxypeptidases. No striking homology is observed between the amino acid sequence of CE and carboxypeptidases A or B, which have broader substrate specificities and favor aromatic or bulky aliphatic side chains. CE, like neuropeptides, is synthesized as part of a larger precursor, and it is likely that processing of the CE precursor generates active CE. In the cell, CE exists in both soluble and membrane-associated forms. The soluble form is 2 kDa smaller than the membrane form and may result from cleavage at a dibasic site near the carboxy terminus. In situ hybridization experiments have shown that CE is localized preferentially in peptidergic cells (MacCumber et al., 1990).

Further modification of neuropeptides occurs if glycine is the carboxy-terminal amino acid (Fig. 1) (Eipper et al., 1983). These glycines are frequently hydrolyzed, leaving a carboxyamide; often the unamidated peptides are inactive. Carboxy-terminal amidation is also thought to confer stability to neuropeptides. The amidating enzyme, known as peptidyl-glycine-$\alpha$-amidating-monooxygenase (PAM), was isolated from bovine neurointermediate pituitary as two forms of 53 and 39 kDa, denoted as PAM-A and PAM-B, respectively. PAM requires molecular oxygen, ascorbate, and copper; it has an alkaline pH optimum and is inhibited by divalent ion chelators (Murthy et al., 1986). Its mechanism of action is similar to the oxidative enzyme dopamine hydroxylase, which also requires copper (Bradbury and Smyth, 1991). The cDNA for PAM was isolated using probes to amino acid sequences from the isolated enzyme (Eipper et al., 1987). Curiously, the cDNA encodes a 108-kDa protein that contains both the 53- and 39-kDa forms isolated biochemically. PAM is responsible only for the formation of the peptidyl-$\alpha$-hydroxyglycine; the conversion to the $\alpha$-amidated peptide can occur spontaneously at neutral pH. A recently identified enzyme activity catalyzes the conversion of the peptidyl-$\alpha$-hydroxyglycine to the $\alpha$-amidated peptide; this enzyme is denoted as a peptidylhydroxyglycine-$\alpha$-amidating lyase (PAL) (Katapodis et al., 1990; Takahashi et al., 1990). PAL was isolated as a 43-kDa protein and is encoded in the 108-kDa PAM precursor. Thus, the two enzymatic activities involved in peptide amidation are encoded in the PAM precursor (Perkins et al., 1990).

PAM is widely distributed throughout different tissues; unlike CE, it is found in nonneuronal tissues such as heart, thyroid, and bone. In the cell PAM is membrane associated in the ER but is soluble when packaged into DCVs. Its presence in DCVs is validated by its cosecretion with peptides (Mains and Eipper, 1984). However, it is not known at what stage of granule formation amidation occurs.

Much excitement is anticipated in future studies of prohormone-processing enzymes, especially for those active at basic site residues. Understanding the localization of these enzymes in the secretory pathway, the determination of structural sorting signals, and the maturation and regulation of the enzymes are important problems to solve. Evidence is mounting that internal basic residue cleavages occur both in the TGN, which has a neutral pH, and as the immature granule, which has a more acidic pH, is forming (Davidson et al., 1988; Lepage-Lezin et al., 1991). This contradicts earlier evidence that processing occurs in the mature DCVs (Orci et al., 1985; Tooze et al., 1987). Perhaps the site of cleavage is specific to the prohormone or cell or both. Isolation of more endoproteo-lytic enzymes in other peptidergic tissues will reveal how many different enzymes are involved in the processing of a particular prohormone, and if each prohormone has its own specific set of enzymes. The regulation of the enzymes themselves will also be of great interest and crucial to understanding neuropeptide biosynthesis.

## IV. ELH Prohormone Structure and Processing

The ELH prohormone is synthesized in the bag cell neurons of the marine snail *Aplysia*, and constitutes approximately 50% of the total cellular protein. These bag cells are grouped in two clusters of 400 neurons each and are situated just rostral to the abdominal ganglion (Frazier *et al.*, 1967; Kandel, 1976; Pinsker and Dudek, 1977; Kaczmarek *et al.*, 1979). The ELH prohormone encodes multiple biologically active peptides, including the bag cell peptides (BCPs) and egg-laying hormone (Scheller *et al.*, 1982). The BCPs are encoded in an amino-terminal intermediate, and ELH is encoded in a carboxy-terminal intermediate (Fig. 3). These peptides are involved in eliciting the stereotypic reproductive behavior of egg laying (Mayeri *et al.*, 1979a,b). The bag cell peptides act in an autocrine fashion as well as on the abdominal ganglion (Rothman *et al.*, 1983; Kauer *et al.*, 1987; Brown and Mayeri, 1989). ELH also acts on abdominal ganglion neurons, but additionally is released into the circulation, where it can travel to distant target sites. Hence, understanding the processing and sorting of the ELH prohormone is integral to elucidating the cellular regulation of the egg-laying behavior.

Biochemical and cellular studies on the processing and sorting of the peptides derived from the ELH prohormone have provided extensive

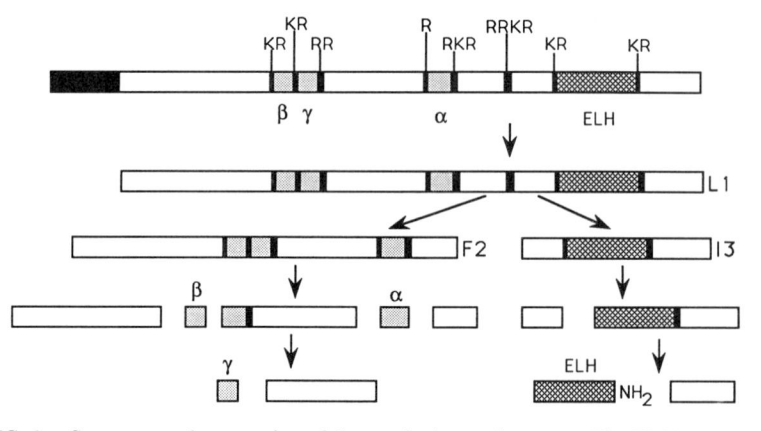

FIG. 3. Structure and processing of the egg-laying prohormone. The ELH prohormone is depicted. This prohormone contains a hydrophobic signal sequence (□) that is cleaved in the ER. The physiologically active $\alpha$, $\beta$, and $\gamma$ bag cell peptides and ELH are labeled. The internal cleavage sites are indicated by the vertical bars, with the corresponding basic residue sequence shown above. After the signal sequence cleavage, the first internal cleavage occurs at the unique tetrabasic site, resulting in amino- and carboxy-terminal intermediates, labeled F2 and I3 respectively. Further processing of F2 yields the $\alpha$, $\beta$, and $\gamma$ BCPs, whereas processing of I3 yields mature ELH, which is amidated.

PROHORMONE STRUCTURE EFFECTS 423

information on the kinetics, order of processing, and sorting of the peptides (Newcomb and Scheller, 1987; Fisher *et al.*, 1988; Sossin *et al.*, 1990a). As with other prohormones, the ELH prohormone is inserted cotranslationally into the ER via its signal sequence, and vectorially transported from the ER to the Golgi stacks. The first internal cleavage occurs at a unique and conserved tetrabasic site (Newcomb and Scheller, 1987). This cleavage occurs with a half-life of 30 minutes (Fig. 3). Cleavage at the tetrabasic site still occurs in bag cells treated with monensin (Yates and Berry, 1984; Sossin *et al.*, 1990a). Because monensin prevents protein exit from the Golgi (Crine and Dufuor, 1982; Tartakoff, 1983), this initial cleavage must occur late in the Golgi or at the trans-Golgi network (TG/TGN). Cleavage at this unique site allows physical separation of the amino-terminal (F2) and carboxy-terminal (I3) intermediates of the ELH prohormone, and consequently segregates the BCPs from ELH. The tetrabasic sequence (Arg-Arg-Lys-Arg) corresponds to the proposed consensus sequence for furin cleavage (Arg-X-Arg/Lys-Arg), and localization of the cleavage site to the TG or TGN is consistent with the localization of furin.

Once separated, the two intermediates are sorted away from each other in the TGN. As a result, the BCPs derived from the amino-terminal intermediate are found in one class of DCVs, and ELH derived from the carboxy-terminal intermediate is present in a separate set of DCVs (Fisher *et al.*, 1988). This differential sorting was discovered by performing immunoelectron microscopy on bag cell sections using antibodies generated against peptide regions corresponding to amino-terminal and to carboxy-terminal intermediates. These and other observations clearly demonstrate that different peptides derived from the common ELH prohormone are not colocalized, but instead are packaged into distinct sets of vesicles. The cellular distribution of these BCP- or ELH-containing vesicles within the bag cell neurons is also different. BCP and ELH DCVs are found in an approximately 1 : 1 ratio in cuff processes, whereas ELH DCVs are the primary class of DCV in processes that ramify through the sheath surrounding the abdominal ganglion (Sossin *et al.*, 1990b).

The rates of processing for the two intermediates are also different (Newcomb and Scheller, 1987). The carboxy-terminal intermediate is processed to release mature ELH with a half-life of 4 hours. However, the amino-terminal intermediate is processed much slower and after 20 hours there is still a significant amount of unprocessed intermediate. Analysis of the steady-state amounts of the peptides from the two intermediates reveals that there is 5- to 8-fold more ELH than bag cell peptides. Experimental results further suggest that there is degradation of DCVs containing primarily amino-terminal intermediate, resulting in nonstoichiometric amounts of peptides derived from the ELH precursor (Sossin *et al.*, 1990a).

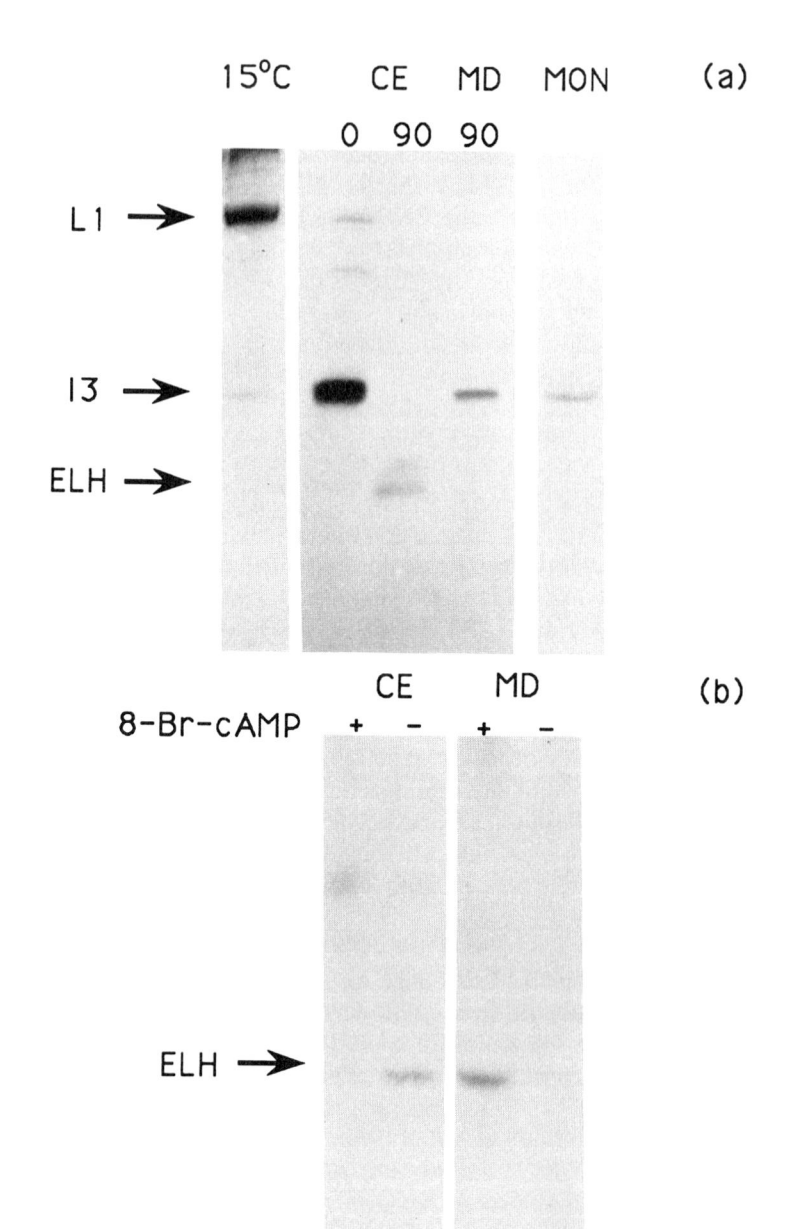

FIG. 4. Analysis of the ELH prohormone expressed in AtT-20 cells. (a) Processing and sorting of the carboxy-terminal intermediate. A product the size of the ELH prohormone is immunoprecipitated with antibody against ELH in cells pulsed with [$^{35}$S]methionine at 15°C. The first cleavage occurs at the tetrabasic site; I3 is immunoprecipitated after a 30-minute pulse (0 chase time, minutes) with 0.5 mCi [$^3$H]leucine at 37°C. After 90 minutes of chase, further processing occurs, and ELH is isolated from the cell extract (CE) and I3 is isolated from the media (MD). Cells were pretreated with 10 $\mu M$ monensin (MON) for 30 minutes and were labeled for 3 hours with 0.25 mCi [$^3$H]leucine in the presence of monensin; this results in I3 recovery from the cell extract but not from the media. (b) Release of ELH from the dense-core vesicles. Two identical plates were pulsed with 1 mCi [$^3$H]leucine for 16

PROHORMONE STRUCTURE EFFECTS 425

What is the physiological significance of the differential sorting, localization, and processing of the ELH prohormone? Differential sorting allows the neuron to control the local concentration of different peptides at separate release sites, as evidenced by the concentration of BCP-containing DCVs in cuff neurites. Release from different sets of peptides from two distinct sites allows interaction of the BCPs with autoreceptors and delivery of ELH to the circulation. Furthermore, by sorting BCP peptides from ELH, the need to coordinate the expression of a set of neuropeptides and the need to regulate the ratio of the peptides can be controlled by cellular degradation of a subset of vesicles. Future research on the physiology of the bag cell system will no doubt further enlighten us as to the need for this differential processing and sorting in regulating behavior.

The relationship between the order of processing and sorting of the ELH prohormone was explored by introducing an ELH prohormone encoding cDNA (Mahon et al., 1985) into a surrogate cell line (Jung and Scheller, 1991). The AtT-20 tumor cell line contains both constitutive and regulated secretory pathways (Gumbiner and Kelly, 1982) and is able to correctly process prohormones. AtT-20 cells normally express POMC, process the precursor, and package its derived peptides into DCVs. Furthermore, this cell line is capable of processing heterologous prohormones and sorting the derived peptides to the regulated pathway by packaging the products into DCVs.

Stably transfected AtT-20 clones expressing the ELH prohormone were isolated, and will be referred to as the ELH clones. Several biochemical studies were performed on one of the ELH clones to trace the biosynthesis of the ELH prohormone. The ELH clone was labeled with specific radioactive amino acids and then chased for various time points. Cell extracts and media were subjected to immunoprecipitation with the anti-ELH antibody and the products were analyzed by SDS–polyacrylamide gel electrophoresis followed by fluorography. Incubating cells at 15°C traps proteins in the ER (Saraste and Kuismanen, 1984), and under these conditions, a product of the expected size for the ELH prohormone was isolated (Fig. 4a). After a 30-minute pulse at 37°C, a 7-kDa band was immunoprecipitated, and after 90 minutes of chase a 3-kDa product was isolated (Fig. 4a) from the cell extracts. The 7- and 3-kDa products were subjected to radiosequencing and were shown to be the carboxy-terminal intermediate and mature ELH, respectively. A considerable amount of carboxy-

---

hours. After 5 hours of chase, fresh media was added and 5 m$M$ 8-Br-cAMP was added to one of the plates. After 3 hours of chase, ELH is immunoprecipitated from cell extract from untreated cells, and is absent in 8-Br-cAMP-treated cells. In 8-Br-cAMP-treated cells, but no untreated cells, ELH is found in the media, demonstrating regulated release.

terminal intermediate is also secreted constitutively into the media during the 90-minute chase; however, it is not uncommon for AtT-20 cells and other tumor cell lines to constitutively secrete unprocessed or intermediate forms of prohormones (Moore et al., 1983b). When protein transport out of the Golgi was blocked with monensin, the tetrabasic site was cleaved but further cleavages were hindered. Hence, the order and location of ELH prohormone cleavage in AtT-20 cells are very similar to the order and location of cleavage in bag cells.

AtT-20 cells treated with 8-Br-cAMP release the contents of their DCVs into the media (Fig. 4b) (Moore and Kelly, 1985). Treatment of the ELH clone with this substance caused release of mature ELH into the media, indicating that mature ELH is stored in the DCVs of the regulated pathway. Immunoelectron microscopy using antibodies against ELH and ACTH showed ELH immunoreactivity predominantly in DCVs, colocalized with the endogenous ACTH immunoreactivity (Fig. 5, A and B).

To follow the processing and sorting of the amino-terminal intermediate, we used an antibody directed against a region in the amino-terminal intermediate (Fig. 6a). As expected, an 18-kDa product, representing the amino-terminal intermediate, was isolated after the initial pulse. After 90 minutes of chase, the 18-kDa band was still present but at lower amounts, and no further processed products were detected. Eventually, after 2.5 hours of chase, there was no detectable amino-terminal intermediate and no products in the cells or media. Thus, although the carboxy-terminal intermediate is processed further to yield mature ELH, the amino-terminal intermediate is degraded. Light-level immunocytochemistry and immunoelectron microscopy using a variety of antibodies to the amino-terminal intermediate showed no apparent staining in the cells, lending support to observations that the amino-terminal intermediate is degraded (Sossin et al., 1990a).

Degradation could occur by sorting of the amino-terminal intermediate to lysosomes. In bag cells, studies indicated that the DCVs containing the amino-terminal intermediate turned over, possibly by fusion with lysosomes. To test this hypothesis, the ELH clone was treated with 200 $\mu M$ chloroquine throughout the labeling and chase (Fig. 6b). Chloroquine is weak base and is known to disrupt the pH gradient of the Golgi. In other tissues, lysosomal enzymes are also missorted to the constitutive secretory pathway, with the result that the enzymes are secreted (Gonzalez-Noriega et al., 1980; Hasilik and Neufeld, 1980). AtT-20 cells treated with chloroquine also missort POMC to the constitutive pathway (Moore et al., 1983a). After 2.5 hours of chase, no amino-terminal intermediate is found in media from untreated cells; in contrast, the intermediate is recovered from the media of chloroquine-treated cells. Hence, the intermediate is being sorted to a degradative pathway, perhaps lysosomal.

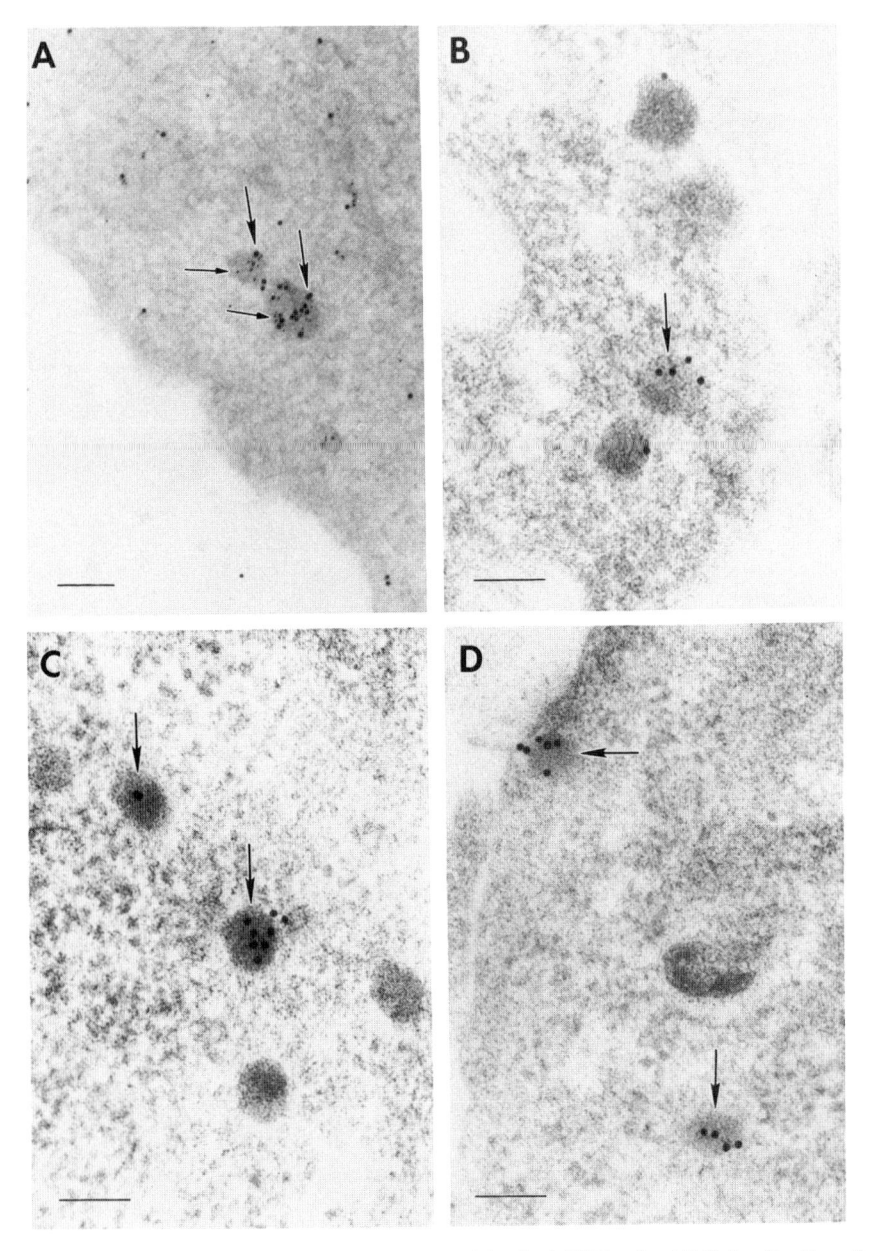

FIG. 5. ELH is sorted to the dense-core vesicles in AtT-20 cells. (A) Colocalization of ELH and ACTH to DCVs in the ELH AtT-20 clone. Small gold particles (5 nm, small arrows) indicate ELH immunoreactivity and large gold particles (10 nm, large arrows) indicate ACTH immunoreactivity. Bar, 100 nm, (B) Localization of ELH immunoreactivity to DCVs in the ELH AtT-20 clone. Bar, 100 nm. (C) Localization of ELH immunoreactivity to DCVs in the dibasic ELH mutant AtT-20 clone. Bar, 100 nm. (D) Localization of ELH immunoreactivity to DCVs in the tetrabasic ELH mutant AtT-20 clone. Bar, 100 nm.

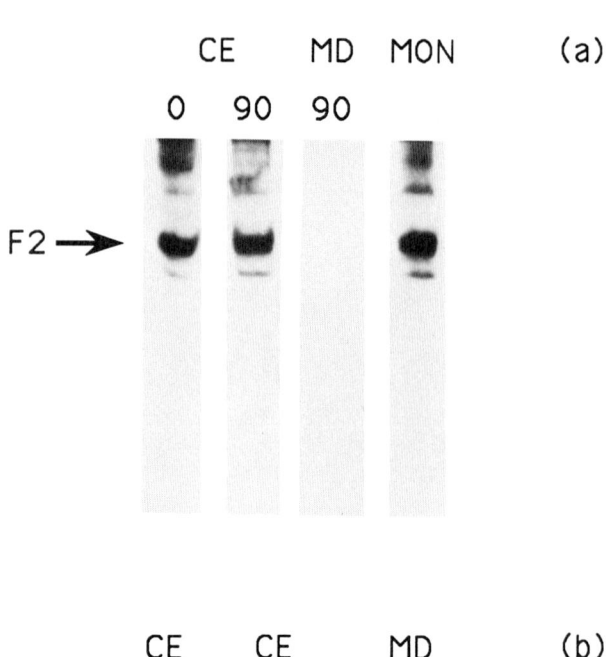

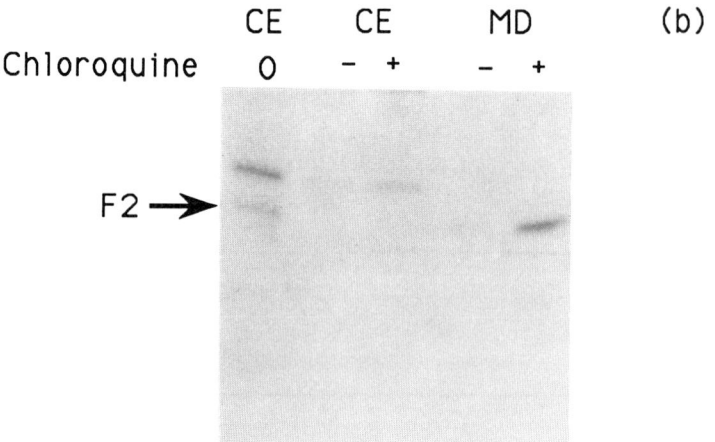

FIG. 6. Analysis of the amino-terminal intermediate in AtT-20 cells. (a) Processing and sorting of the amino-terminal intermediate. The amino-terminal intermediate (F2) is immuno-precipitated from the cell extract after the initial 30-minute pulse (0 chase time, minutes) with 0.5 mCi [³H]leucine. After 90 minutes of chase, F2 is present in the cell extract (E) at a lower level and is not found in the media (MD). F2 is also present in cells treated with monensin (MON) as described in Fig. 4a. (b) Chloroquine-treated ELH AtT-20 clones secreted amino-terminal intermediate into the media. Three 10-cm plates were pulsed for 30 minutes with 0.5 mCi [³H]leucine. One plate was treated with 200 $\mu M$ chloroquine 1 hour before the chase and throughout the pulse and chase periods. After the initial pulse (0 chase time), F2 is isolated from the CE. F2 is absent from the CE in treated and untreated cells after 2.5 hours of chase. However, F2 is isolated from the media of the treated cells, but not from untreated cells.

PROHORMONE STRUCTURE EFFECTS    429

In summary, this heterologous cell line is able to process and sort the ELH prohormone in a manner similar to that of the bag cells. In AtT-20 cells, the ELH prohormone is cleaved internally at the unique tetrabasic site in the late Golgi, producing two intermediates. This cleavage is probably performed by the Golgi-localized furin, which has been detected in AtT-20 cells. The intermediates are then differentially sorted in the TGN; the carboxy-terminal intermediate is sorted to the regulated pathway for further processing and storage of derived peptides in DCVs, while the amino-terminal intermediate is sorted to a degradative pathway. Clearly, the sorting and processing machinery are similar in mammalian AtT-20 cells and invertebrate bag cell neurons. More striking is that structural features in the ELH prohormone are able to direct its correct processing and sorting in a heterologous cell line.

From these analyses, it is reasonable to hypothesize that disrupting the order and localization of processing will cause incorrect processing and thus missorting of the ELH prohormone products. To test this hypothesis, mutations were made at the unique tetrabasic site of the ELH prohormone, which is critical for the correct sorting of the peptides. Two constructs were made: (1) a tetrabasic deleted mutant in which the four basic residues were deleted and (2) a dibasic mutant that was generated by deleting the first two basic residues of the four, leaving a Lys-Arg pair. Deleting the tetrabasic site should abolish the initial cleavage, which is crucial to the separation of the two intermediates and thus their differential processing and sorting. Replacing the tetrabasic group with a commonly used dibasic site could also cause missorting if other dibasic sites are used cotemporally or if processing is delayed. Of course, other structural features may also dictate the order and sites of cleavage. In the following section, we discuss our recent results on the processing and sorting of these two mutant ELH prohormone forms, and propose a model to explain the observations.

Analyses of the two mutants' prohormones were performed as described above for the ELH clone. After an initial pulse of radioactive [³H]leucine, the carboxy-terminal intermediate was isolated from the cell extract (Fig. 7a). This indicates that in the dibasic mutant, the first cleavage still occurs at the same location and at a similar rate, even though there is only a pair of basic residues. Ninety minutes after the pulse, mature ELH was isolated from the cell extract and some carboxy-terminal intermediate was isolated from the media. Mature ELH was found in DCVs as determined by secretion analysis and immunoelectron microscopy (Fig. 5C). Thus, with respect to the carboxy-terminal intermediate, processing of the dibasic mutant is similar to the wild-type ELH prohormone. In contrast, the amino-terminal intermediate resulting from cleavage of the dibasic mutant prohormone is treated differently than the wild-type amino-terminal intermediate (Fig. 7b). Amino-terminal intermediate is absent from the cells

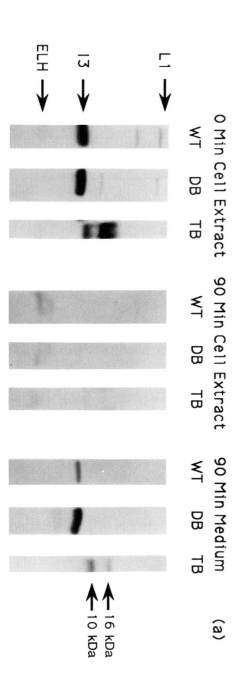

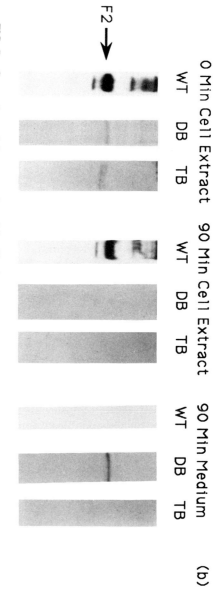

FIG. 7. Comparison of the processing of the dibasic and tetrabasic ELH prohormones. (a) Analysis of the processing and sorting of the carboxy-terminal intermediate. After a 30-minute pulse with 0.5 mCi [$^3$H]leucine, I3 (7 kDa) is immunoprecipitated with antibody against ELH, in the wild-type (WT) and dibasic (DB) clone cell extracts; however, in the tetrabasic (TB) clone, two proteins of 10 and 16 kDa are immunoprecipitated. ELH is isolated from all of the clones after a 90-minute chase time. I3 is found in the media of the wild-type and dibasic clones after the 90-minute chase period, but the 10- and 16-kDa proteins are found in the media of the tetrabasic clone. (b) Analysis of the processing of the amino-terminal intermediate. After the 30-minute pulse (0 chase time) with 0.5 mCi [$^3$H]leucine, F2 is isolated from the cell extract of the wild-type and dibasic clones, and a slightly smaller form from the tetrabasic clone. F2 is still present in the wild-type cell extract after a 90-minute chase period, but is absent from the dibasic and tetrabasic clone cell extracts. However, at this time point, F2 is isolated from the dibasic clone media and is absent from the media of the wild-type and tetrabasic clones.

but is detected in the media after 90 minutes of chase. Thus, although the order and site of cleavage are maintained, some of the amino-terminal intermediate is missorted to the constitutive pathway, whereas the wild-type amino-terminal intermediate is degraded.

Processing of the tetrabasic deleted ELH prohormone also yielded surprising results. After the initial pulse, two peptides along with some prohormone were observed (Fig. 7a). These two products were 16 and 10 kDa as opposed to the 7-kDa carboxy-terminal intermediate observed in wild type. Amino-terminal sequencing of these two products revealed that the initial cleavage occurs at two places upstream from the tetrabasic site: one at a tribasic site, and the other likely at the arginine immediately carboxy terminal to $\gamma$-BCP, which defines a potential furin cleavage site. Hence, in the absence of the tetrabasic site, upstream basic residue sites are chosen, and the rate of prohormone processing is markedly slower than processing of the wild-type or dibasic mutant prohormones. The two new carboxy-terminal intermediates were further processed to produce mature ELH. Again, as with wild-type and the dibasic mutant, the mature ELH was sorted and stored into DCVs (Fig. 5D). The path of the truncated amino-terminal intermediate is still in question because the antibody used in these experiments could only immunoprecipitate one of the two new amino-terminal intermediates generated. However, our initial observations indicate that the amino-terminal intermediate is intracellularly degraded (Fig. 7b).

## V. Model for Processing and Sorting of the ELH Prohormone in AtT-20 Cells

Analysis of the processing and sorting of the dibasic and tetrabasic deleted ELH prohormone mutants provided some rather interesting and surprising results. When the initial tetrabasic cleavage site is mutated to a dibasic site, the order and sites of cleavages are maintained; however, the amino-terminal intermediate is secreted into the media, rather than being degraded, as in wild type. In the tetrabasic deleted mutant, the order and sites of cleavages are, of course, not maintained, and two sites amino-terminal to the former tetrabasic site are used. The rate of cleavage in the tetrabasic mutant clone is also slower than in the wild-type ELH clone, indicating that furin may cleave some of the prohormone at these sites, but not efficiently. Curiously, the other dibasic sites in the carboxy-terminal intermediate are not chosen, and it will be interesting to know how this selection of processing site is made.

To explain these observations we propose the following model (Fig. 8). When the ELH prohormone reaches the area of the Golgi where furin resides, perhaps the TG, it is cleaved. The two intermediates travel to a

PROHORMONE STRUCTURE EFFECTS 433

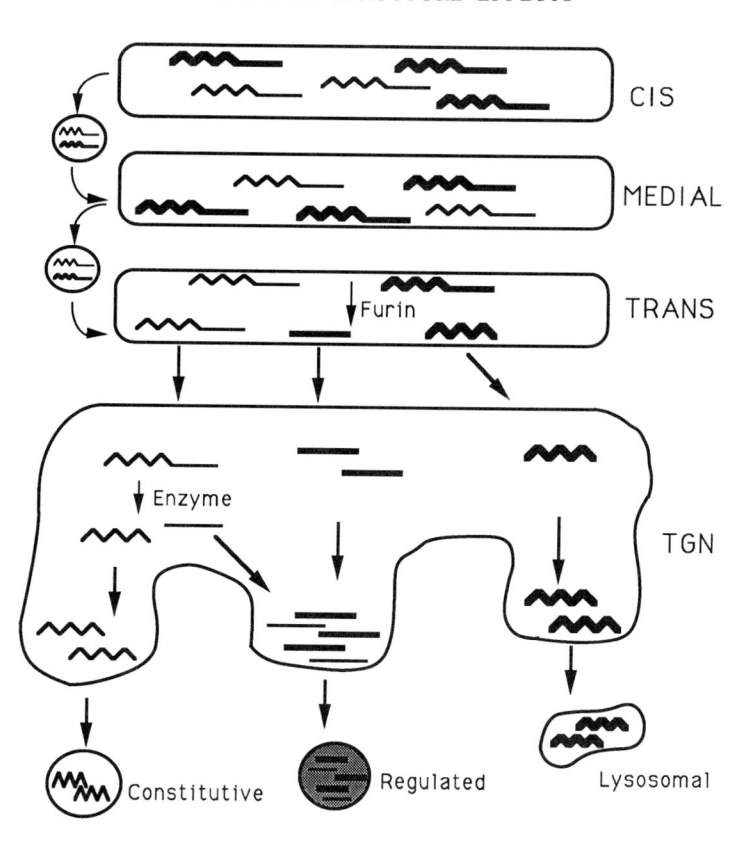

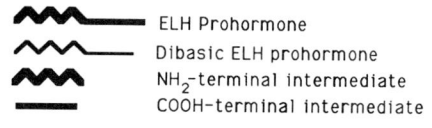
ELH Prohormone
Dibasic ELH prohormone
NH$_2$-terminal intermediate
COOH-terminal intermediate

FIG. 8. Model for processing and sorting the ELH and dibasic mutant ELH prohormones in the secretory pathway of AtT-20-cells. The ELH and dibasic mutant ELH prohormones travel vectorially through the Golgi stacks. The first endoproteolytic cleavage at the tetrabasic site is proposed to be performed by furin in the trans-Golgi. At this point, the amino-terminal intermediates from both prohormones are sorted in the TGN to a degradative pathway, perhaps lysosomal. The carboxy-terminal intermediates from both prohormones are sorted in the TGN to the regulated secretory pathway and packaged into DCVs. However, not all of the dibasic mutant ELH prohormone is cleaved by furin and the prohormone passes into the TGN, where it is cleaved at this site by a different endoprotcolytic enzyme specific for a pair of basic residues. Here, the carboxy-terminal intermediate is still sorted into the regulated pathway and packaged into DCVs. In contrast, the amino-terminal intermediate is now sorted to the constitutive secretory pathway instead of the degradative pathway.

434                    LINDA J. JUNG ET AL.

sorting area of the TGN: the amino-terminal intermediate is directed to
the degradative pathway and the carboxy-terminal intermediate is sorted
to the regulated pathway. In contrast, only a portion of the dibasic mutant
prohormones are similarly processed. The portion of the dibasic mutant
prohormone that was not cleaved by furin passes into the TGN and is then
cleaved at the initial site by a different endoproteolytic enzyme. Cleavage
at this spatial location of the TGN is past the sorting site to the degradative
pathway and thus the dibasic mutant amino-terminal intermediate is directed to a constitutive route of secretion, while the carboxy-terminal
intermediate is still correctly sorted to the regulated pathway. The tetrabasic deleted ELH prohormone should behave similarly, with some initial
cleavages at the tribasic and Arg-Arg sites carboxy terminal to the $\gamma$-BCP
by furin; the rest of the prohormone is cleaved at a later time point in a
different area of the TGN.

## VI.   Concluding Remarks

Using the ELH prohormone to study the relationship between processing and sorting has opened up new avenues of thought. In the past,
we have learned from these studies that cellular mechanisms can control
the relative levels and spatial localization of different neuropeptides derived from a common precursor. Our recent results indicate that perturba-
tion of a processing site can cause processing in a different locale in the
cell and result in altered pathways of sorting. Defining the hierarchy of
processing enzymes and their localization will be crucial to future work in
this area, and to understanding how the design of prohormones complements the repertoire of processing enzymes.

### REFERENCES

Barr, P. J. (1991). *Cell (Cambridge, Mass.)* **66,** 1–3.
Barr, P. J., Mason, O. B., Landsberg, K. E., Wong, P. A., Kiefer, M. C., and Brake, A. J.
      (1991). *DNA Cell Biol.* **10,** 319–328.
Benjannet, S., Rondeau, N., Day, R., Chrétien, M., and Seidah, N. G. (1991). *Proc. Natl.
      Acad. Sci. U.S.A.* **88,** 3564–3568.
Bradbury, A. F., and Smyth, D. G. (1991). *TIBS* **16,** 112–115.
Breshnahan, P. A., Leduc, R., Thomas, J., Gibson, H. L., Brake, A. J., Barr, P. J., and
      Thomas, G. (1990). *J. Cell Biol.* **111,** 2851–2859.
Brown, R. O., and Mayeri, E. (1989). *J. Neurosci.* **9,** 1443–1451.
Carroll, R. J., Hammer, R. E., Chan, S. J., Swift, H. H., Rubenstein, A. H., and Steiner,
      D. F. (1988). *Proc. Natl. Acad. Sci. U.S.A.* **85,** 8943–8947.
Crine, P., and Dufuor, L. (1982). *Biochem. Biophys. Res. Commun. (London)* **109,** 500–506.
Davidson, H. W., Rhodes, C. J., and Hutton, J. C. (1988). *Nature (London)* **333,** 93–96.
Docherty, K., and Steiner, D. F. (1982). *Annu. Rev. Physiol.* **44,** 625–638.
Douglass, J., Civelli, O., and Herbert, E. (1984). *Annu. Rev. Biochem.* **52,** 665–715.

## PROHORMONE STRUCTURE EFFECTS 435

Eipper, B. A., Mains, R. E., and Glembotski, C. C. (1983). *Proc. Natl. Acad. Sci. U.S.A.* **80**, 5144–5148.

Eipper B. A., Park, L. P., Dickerson, I. M., Keutmann, H. T., Thiele, E. A., Rodriguez, H., Schofield, P. R., and Mains, R. E. (1987). *Mol. Endocrinol.* **1**, 777–790.

Fisher, J. M., Sossin, W., Newcomb, R., and Scheller, R. H. (1988). *Cell (Cambridge, Mass.)* **54**, 813–822.

Frazier, W. T., Kandel, E. R., Kupfermann, I., Waziri, R., and Coggeshall, R. E. (1967). *J. Neurophysiol.* **30**, 1288–1351.

Fricker, L. D., Evans, C. J., Esch, F. S., and Herbert, E. (1986). *Nature (London)* **323**, 461–464.

Fricker, L. D., and Snyder, S. H. (1983). *J. Biol. Chem.* **258**, 10950–10955.

Fuller, R. S., Brake, A., and Thorner, J. (1989a). *Proc. Natl. Acad. Sci. U.S.A.* **86**, 1434–1438.

Fuller, R. S., Brake, A., and Thorner, J. (1989b). *Science* **246**, 482–486.

Fuller, R. S., Sterne, R. S., and Thorner, J. (1988). *Annu. Rev. Physiol.* **50**, 3545–3562.

Gonzalez-Noriega, A., Grubb, J. H., Talkad, V., and Sly, W. S. J. (1980). *J. Cell Biol.* **85**, 839–852.

Griffiths, G., and Simons, K. (1986). *Science* **234**, 438–443.

Gross, D. J., Villa-Komaroff, L., Kahn, C. R., Weir, G. C., and Halban, P. A. (1989). *J. Biol. Chem.* **264**, 21486–21490.

Gumbiner, B., and Kelly, R. B. (1982). *Cell (Cambridge, Mass.)* **28**, 51–59.

Hasilik, A., and Neufeld, E. F. (1980). *J. Biol. Chem.* **255**, 4937–4945.

Julius, D., Blake, A., Blair, L., Kunisawa, R., and Thorner, J. (1984). *Cell (Cambridge, Mass.)* **37**, 1075–1089.

Jung, L. J., and Scheller, R. H. (1991). *Science* **251**, 1330–1335.

Kaczmarek, L. K., Finbow, M., Revel, J. P., and Strumwasser, F. (1979). *J. Neurobiol.* **10**, 535–550.

Kandel, E. R. (1976). "The Cellular Basis of Behavior," Freeman, San Francisco.

Katapodis, A. G., Ping, D., and May, S. W. (1990). *Biochemistry* **29**, 6115–6120.

Kauer, J. A., Fisher, T. E., and Kaczmarek, L. K. (1987). *J. Neurosci.* **7**, 3623–3633.

Loh, Y. P., Brownstein, M. J., and Gainer, H. (1984). *Annu. Rev. Neurosci.* **7**, 189–222.

Lepage-Lezin, A., Joseph-Bravo, P., Devilliers, G., Benedetti, L., Launay, J.-M., Gomez, S., and Cohen, P. (1991). *J. Biol. Chem.* **266**, 1679–1688.

Mackin, R. B., and Noe, B. D. (1987). *J. Biol. Chem.* **262**, 6453–6456.

MacCumber, M. W., Snyder, S. H., and Ross, C. A. (1990). *J. Neurosci.* **10**, 2850–2860.

Mahon, A. C., Nambu, J. R., Taussig, R., Malladi, S., Roach, A., and Scheller, R. H. (1985). *J. Neurosci.* **5**, 707–719.

Mains, R. E., and Eipper, B. A. (1984). Endocrinology (*Baltimore*) **115**, 1683–1690.

Mayeri, E., Brownell, W. D., Branton, J., and Simons, S. B. (1979a). *J. Neurophys.* **42**, 1165–1184.

Mayeri, E., Brownell, W. D., and Branton, J. (1979b). *J. Neurophysiol.* **42**, 1185–1197.

Moore, H.-P. H., Gumbiner, B., and Kelly, R. B. (1983a). *Nature (London)* **302**, 434–436.

Moore, H.-P. H., Walker, M. D., Lee, F., and Kelly, R. B. (1983b). *Cell (Cambridge, Mass.)* **35**, 531–538.

Moore, H.-P. H., and Kelly, R. B. (1985). *J. Cell Biol.* **101**, 1773–1781.

Murthy, A. S. N., Mains, R. E., and Eipper, B. A. (1986). *J. Biol. Chem.* **261**, 1815–1822.

Newcomb, R., and Scheller, R. H. (1987). *J. Biol. Chem.* **263**, 12514–12521.

Orci, L., Ravazzola, M., Amherdt, M., Madsen, O., Vassalli, J.-D., and Perrelet, A. (1985). *Cell (Cambridge, Mass.)* **42**, 671–681.

Orci, L., Ravazzola, M., Amherdt, M., Perrelet, A., Powell, S. K., Quinn, D. L., and Moore, H.-P. H. (1988). *Cell (Cambridge, Mass.)* **51**, 1039–1051.

Perkins, S. N., Husten, E. J., and Eipper, B. A. (1990). *Biochem. Biophys. Res. Commun.* **171,** 926–932.

Pinkser, H. M., and Dudek, F. E. (1977). *Science* **197,** 490–493.

Rholam, M., Nicolas, P., and Cohen, P. (1986). *FEBS Lett.* **207,** 1–5.

Roebroek, A. J. M., Schalken, J. A., Bussemakers, M. J. G., Van Heerikuizen, H., Onnekink, C., Debruyne, F. M. J., Biomers, H. P. J., and van de Ven, W. J. M. (1986). *Mol. Biol. Rep.* **11,** 117–125.

Rothman, B. S., Mayeri, E., Brown, R. O., Yuan, P.-M., and Shively, J. E. (1983). *Proc. Natl. Sci. U.S.A.* **80,** 5753–5757.

Saraste, J., and Kuismanen, E. (1984). *Cell (Cambridge, Mass.)* **38,** 535–549.

Scheller, R. H., Jackson, J. F., McAllister, L. B., Schwartz, J. H., Kandel, E. R., and Axel, R. (1982). *Cell (Cambridge, Mass.)* **28,** 707–719.

Seidah, N. G., Gaspar, L., Mion, P., Marcinkiewicz, M., Mbikay, M., and Chrétien, M. (1990). *DNA Cell Biol.* **9,** 415–424.

Smeekens, S. P., Avruch, A. S., LaMendola, J., Chan, S. J., and Steiner, D. J. (1991). *Proc. Natl. Acad. Sci. U.S.A.* **88,** 5297–5301.

Smeekens, S. P., and Steiner, D. F. (1990). *J. Biol. Chem.* **265,** 2997–3000.

Smith, A. I., and Funder, J. W. (1988). *Endocr. Rev.* **9,** 159–179.

Sossin, W. S., Fisher, J. M., and Scheller, R. H. (1990a). *J. Cell Biol.* **110,** 1–12.

Sossin, W. S., Sweet-Cordero, A., and Scheller, R. H. (1990b). *Proc. Natl. Acad. Sci. U.S.A.* **87,** 4845–4848.

Suter, U., Heymach, J. V., Jr., and Shooter, E. M. (1991). *EMBO J.* **10,** 2395–2400.

Takahashi, K., Okamoto, H., Seino, H., and Noguchi, M. (1990). *U. Com.* **169,** 524–530.

Tartakoff, A. M. (1983). *Cell (Cambridge, Mass.)* **32,** 1026–1028.

Thomas, G., Thorne, B. A., Thomas, L., Allen, R. G., Hruby, D. E., Fuller, R., and Thorner, J. (1988). *Science* **241,** 226–230.

Thomas, L., Richard, L., Thorne, B. A., Smeekens, S. P., Steiner, D. F., and Thomas, G. (1991). *Proc. Natl. Acad. U.S.A.* **88,** 5297–5301.

Tooze, J., Hollinshead, M., Frank, R., and Burke, B. (1987). *J. Cell Biol.* **105,** 1551–1562.

Ullrich, A., Shine, J., Chirgwin, J., Pictet, R., Tischer, E., Rutter, W. J., and Goodman, H. M. (1977). *Science* **196,** 1313–1319.

van de Ven, W. J. M., Voorberg, J., Fontijn, R., Pannekoek, H., van den Ouweland, A. M. W., van Duijnhoven, H. L. P., Roebroek, A. J. M., and Siezen, R. J. (1990). *Mol. Biol. Rep.* **14,** 265–275.

Walter, P., and Blobel, G. (1981). *J. Cell Biol.* **91,** 551–556.

Wise, R. J., Barr, P. J., Wong, P. A., Kiefer, M. C., Brake, A. J., and Kaufman, R. J. (1990). *Proc. Natl. Acad. Sci. U.S.A.* **87,** 9378–9382.

Yates, M. F., and Berry R. W. J. (1984). *J. Neurobiol.* **15,** 141–155.

RECENT PROGRESS IN HORMONE RESEARCH, VOL. 48

# Endocrinology Alfresco: Psychoendocrine Studies of Wild Baboons

## Robert M. Sapolsky

Department of Biological Sciences, Stanford University, Stanford, California 94305 and
Institute of Primate Research, National Museums of Kenya, Karen, Nairobi, Kenya

## I. Introduction

As scientists, we collectively shudder at large error bars. They can mean that our techniques are not as replicable as we might want, that our measuring devices are noisy, that our ostensibly homogeneous study populations are anything but homogeneous. It is the latter case that is particularly interesting, because we so often yearn for that perfectly homogeneous population. Yet there is a tremendous potential to exploit heterogeneity in populations as a wedge for greater understanding.

As a stress physiologist, that notion is constantly brought home to me by a seeming paradox. We are learning of ever-increasing numbers of outposts of our bodies that are affected, disrupted, and made diseased by stress. Yet, given all the ways in which stress can make us sick, many of us are relatively resistant to stress-related disease, even in the face of severe stressors. I have tried to study such individual variability by asking, in effect, why are some bodies and some psyches more resistant to stress than are others?

Potentially, this could be studied with laboratory rodents, but the limits of such animals for extrapolating to our own psychoendocrinology are obvious. At the other extreme, one could study humans with the difficulties in uncovering underlying physiological mechanisms or in carrying out truly long-term studies. Alternatively, one can study these issues in captive primates, but one must wrestle then with the demographic and psychological confounds of the captivity.

Given these limitations, I have tried to study individual differences in stress physiology in a rather novel setting, namely with a population of wild baboons living in the Serengeti ecosystem of East Africa, returning to the same population of baboons annually for 14 years.

Olive baboons (*Papio anubus*) are among the most social of all primates, living for decades in stable troops of 50 to 200 individuals in the wood-

437

Copyright © 1993 by Academic Press, Inc.
All rights of reproduction in any form reserved.

438                    ROBERT M. SAPOLSKY

land–grassland mosaics of Africa. The animals are highly successful in this niche, being catholic in their omnivory, devoting relatively few hours each day to foraging, and suffering only low predation and infant mortality rates. Typically, much of each day is devoted to the social complexity at which they excell (cf. Altmann, 1980; Ransom, 1981; Smuts, 1986).

A major but not preeminent feature of this social complexity is social dominance. Fairly linear hierarchies of rank occur, with limited resources being divided unevenly according to rank. Thus, a high-ranking individual may have access to the most desirable food items, may be in the safest spot during a predator attack, may be groomed free of ticks most readily, and may have the easiest access to a sexual partner. Among females, rank tends to be stable and inherited over the lifetime, with an oldest daughter achieving a rank one below that of her mother (Altmann, 1980). Among males, rank shifts over the lifetime; typically, a high rank is initially achieved through overt aggression. Thereafter, it is typically maintained through a combination of occasional overt aggression, but more frequently through threats of aggression, psychological harassment, and bluff. And, inevitably, at some point, a lower ranking male successfully overcomes the harassment and bluff to overthrow the status quo and force a change in ranking. Thus, their lives are filled with social competition between rivals that can be overt and highly aggressive, or subtly built purely around psychological stress.

In addition to this landscape of competition and individualism, there are just as many instances of altruism and cooperation. Because animals will stay in the same troop for anywhere from years to their entire lives (20–30 years), there are often many relatives in the same troop and there is the potential for complex patterns of coalitional behavior among members of the same lineage. In addition, such cooperation can occur between nonrelatives with stable social affiliations (who, I think, can be termed "friends" without being anthropomorphic) (Strum, 1982; Smuts, 1986).

The similarities to our own society are striking. For both olive baboons in this habitat and ourselves, life has been freed from most of the exigencies of mere survival. Few of the stressors in the life of an olive baboon are of that sort; similarly, few of us are hypertensive because we must forage miles each day amid drought, or because we must physically wrestle a competitor for a canned food item in the supermarket. In the wake of that freedom, both they and we have lives filled with self-generated social stressors and with communities of relatives and friends that can shelter us from such stressors. Over the years, I have sought to determine how the bodies of these animals respond to stress, and whether individual differences in the endocrine stress response reflect differences in social rank, patterns of affiliation, coping, and personality.

To do so, one must first observe these animals at length, and the techniques of field primatology allow for observations of subtle behaviors in a way that is quantitative and unbiased (e.g., observation methods that avoid merely looking at whoever is doing something interesting, that avoid seasonal, circadian, or cohort confounds, and so on) (Altmann, 1971). In addition, I have had to obtain blood and tissue samples from these wild animals. For that, one must anesthetize the subjects (in my case, using an anesthetic blowgun, which has the advantage of being silent, thus reducing the stressfulness of a darting for both the subject and neighboring animals). One must use an anesthetic that does not affect the hormones under study (in this case, phencyclidine) (Sapolsky, 1982) and dart everyone at the same time of day and season to avoid circadian and circannual rhythms. Moreover, if one hopes to obtain basal concentrations of hormones, animals cannot be darted if they are sick or injured or have just mated or had a fight (or in the case of some metabolic studies, to be discussed, when they have already eaten that day). Finally, one must obtain an initial blood sample as rapidly as possible, and only consider to be "basal" the concentrations in that first sample of hormones, whose basal values are known to be deflected by stress more slowly than the lag time between darting and that first sample.

In addition to obtaining basal samples, I can anesthetize an animal and study it over the course of the day, conducting some of the experiments to be described. After recovering overnight in a holding cage, baboons are then released in the morning. Using this approach, I have been able to dart my study subjects annually over the years without any loss of habituation. The following sections review the workings of their various endocrine axes, the behavioral correlates of individual differences in their workings, and some of the underlying neuroendocrine mechanisms accounting for such differences. Because most mature females are either pregnant or nursing (more than 80% of the time in this population) and thus a darting might endanger either the fetus or the dependent infant, much of the work to be described concentrates on the males in this troop. The final section reviews some of my more recent findings with females.

## II. The Adrenocortical Axis

### A. BASIC FEATURES OF THE BABOON ADRENOCORTICAL AXIS

Both quantitatively and qualitatively, adrenocortical function in baboons and other Old World primates is quite similar to that in humans [this contrasts with the unique picture of New World primates, who have an order of magnitude higher basal cortisol concentrations and decreased corticosteroid receptor affinity (Chrousos *et al.*, 1982)]. Among the olive baboons [and the closely related yellow baboons (*Papio cynocephalus*) that I studied on one occasion], basal cortisol concentrations in the morning (i.e., the circadian peak) average approximately 15 $\mu$g/100 ml in both sexes. Because of the constraints of darting wild animals described in the previous section, it has not been possible to dart animals in the evening in order to obtain basal values during the circadian trough or, because of the speed with which its secretion is altered by the stress of darting, to obtain basal ACTH values.

Recent literature demonstrates that basal cortisol concentrations are elevated in extreme old age in humans (cf. Sapolsky, 1990a), and we have recently observed the same among these wild baboons (as well as a relative dexamethasone resistance among aged baboons) (Sapolsky and Altmann, 1991).

As would be expected, cortisol concentrations in these animals are enhanced by various stressors. We have documented this in response to a number of social stressors. For example, elevated basal cortisol concentrations were observed among males during a period of social instability in which the dominance hierarchy underwent a reorganization associated with high rates of aggression and dominance reversals (Sapolsky, 1983a). As another example, the transfer of an extremely aggressive male into the troop was associated with an approximate 70% increase in basal cortisol concentrations in males and an approximate doubling of concentrations in females (Alberts *et al.*, 1992). This appears to reflect both psychological and physical stressors connected with the male's transfer: although the rise in cortisol concentrations was most dramatic in animals who were direct victims of his physical attacks, there was still a significant rise even among animals who were never his overt targets.

More acute stressors also elevate cortisol concentrations. As one intrinsic to the obtaining of blood samples, the process of darting the baboons represents a stressor, and cortisol concentrations approximately double in both sexes during the first postdarting hour (Sapolsky, 1982).

Finally, in an interesting example of a stressor that did *not* elevate cortisol concentrations, baboons were studied during the East African drought of 1984, in which subjects spent significantly more time foraging and walking to find food. Although this period was associated with a significant suppression of testosterone concentrations in males (Sapolsky, 1986a) and of fertility in females (unpublished) cortisol concentrations were unaffected.

The acute rise in cortisol concentrations following a darting appears to play a role in some of the physiological sequelae of this stressor. The

BABOON PSYCHOENDOCRINOLOGY 441

cortisol rise contributes to the rapid decline in testosterone concentrations in these males, and of circulating lymphocyte counts. As evidence, these declines can be blunted with metyrapone treatment and mimicked with exogenous glucocorticoids (Sapolsky, 1985, and unpublished). In contrast, this acute rise in cortisol concentration appears to have little effect on cholesterol metabolism (Sapolsky and Mott, 1987).

## B. RANK-RELATED DIFFERENCES IN ADRENOCORTICAL FUNCTION

Does adrenocortical function differ by rank? Consistently, two such differences have emerged: (a) high-ranking individuals have low basal cortisol concentrations, relative to subordinate individuals; (b) despite the lower basal cortisol concentrations, dominant males are able to achieve levels equivalent to those of subordinate individuals in response to a stressor (i.e., darting), because of a faster initial rise in cortisol concentrations.

The elevated basal cortisol concentrations among low-ranking males over the years are demonstrated in Fig. 1 (from Sapolsky, 1989). *A priori,* this rank difference appears logical, given that in a stable dominance hierarchy it is subordinate animals who are subject to the highest rates of physical and psychological stressors: they are the most likely to have to relinquish a food item or a resting spot, are the most likely to have their

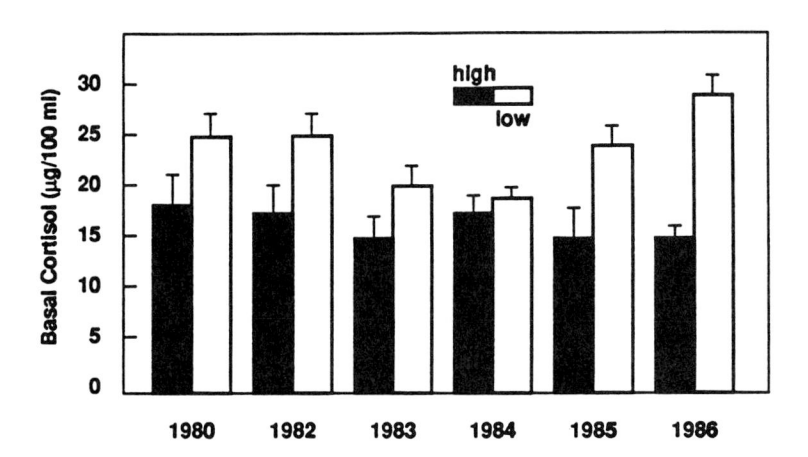

FIG. 1. Basal cortisol concentrations of the six highest ranking and six lowest ranking males in each of the 6 years of study. Total number of males under study/year ranged from 12 to 16. Values are derived from a single anesthetization/animal/season. In this and in Figs. 2–13, rankings were derived by approach–avoidance criteria. From Sapolsky (1989).

442                           ROBERT M. SAPOLSKY

grooming bouts or sexual consortships disrupted and terminated, and are most subject to unpredictable displacement aggression. In support of this, socially subordinate animals in stable social groups tend to have higher glucocorticoid concentrations in a wide range of species, including mice (Davis and Christian, 1957; Southwick and Bland, 1959; Bronson and Eleftheriou, 1964; Louch and Higginbotham, 1967; Archer, 1970; Schuhr, 1987), rats (Barnett, 1955; Popova and Naumenko, 1972; Sakai *et al.*, 1991), wolves (Fox and Andrews, 1973), birds (Wingfield and Moore, 1987), and fish (Ejike and Schreck, 1980).

My work has explored the neuroendocrine bases of the rank difference in basal cortisol concentrations. If the elevation in subordinate animals is a consequence of their being exposed most frequently to stressors, there should be hypersecretion of cortisol as a final step subsequent to hypersecretion of hypothalamic secretagogues. This appears to be the case:

1. Potentially, the elevated cortisol concentrations of low-ranking males might not reflect rank differences in secretion, but rather depressed clearance. However, cortisol clearance rate does not differ by rank (Sapolsky, 1983b).

2. Potentially, the elevated cortisol concentrations of low-ranking males might be due to enhanced adrenal sensitivity to ACTH and/or to elevated ACTH concentrations. As noted, it has not been possible to determine basal ACTH concentrations in these animals. However, adrenal sensitivity to a wide range of ACTH concentrations does not differ by rank in these animals (Fig. 2) (Sapolsky, 1990b). Thus, low-ranking males appear to secrete more ACTH basally than do dominant individuals.

3. Moving higher in the axis, the inferred elevated ACTH concentrations of low-ranking males might be due to enhanced pituitary sensitivity to a hypothalamic secretagogue(s) and/or to enhanced secretion of such secretagogues. Testing this possibility, I observed that not only were pituitaries of subordinate males not more sensitive to CRF than those of dominant individuals, they were *less* so (Fig. 3) (Sapolsky, 1989). This blunted sensitivity could be due to enhanced feedback inhibition at the pituitary level by the elevated circulating cortisol concentrations, and/or to intrinsic loss of pituitary sensitivity to CRF. The latter appears to be the case, as the experiment in Fig. 3 was conducted in animals treated with metyrapone, which suppressed cortisol concentrations to equivalently low levels in all subjects (although, as a confound, the methodological constraints intrinsic with working with these animals prevented treatment with mytyrapone over a sufficiently long period to be certain that delayed glucocorticoid feedback had been relieved). Thus, if the pituitary of a low-ranking baboon is less sensitive to the predominant releaser of ACTH, yet

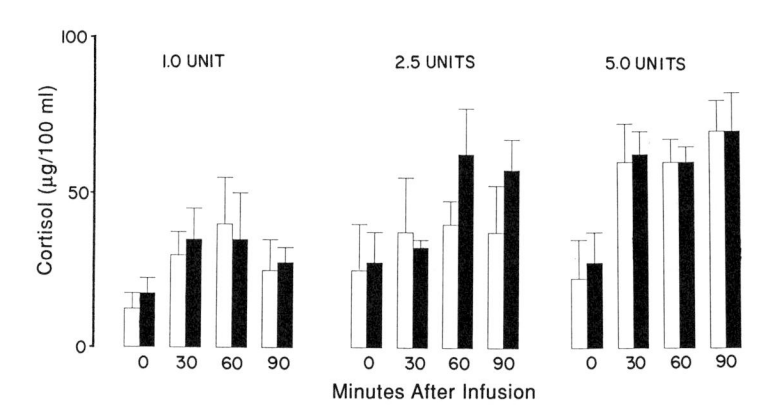

FIG. 2.  Adrenocortical responses to ACTH challenges, according to rank. High-ranking (black bars) and low-ranking (white bars) adult males ($n = 6$/group) were infused i.v. with indicated quantities of ACTH (ACTHAR, porcine ACTH); cortisol concentrations were determined at indicated times after the challenge. The study was initiated 1 hour after anesthetization; thus, the initial sample does not represent basal values, but rather those in response to the stressor of anesthetization. The two groups did not differ, as assessed by ANOVA. From Sapolsky (1990b).

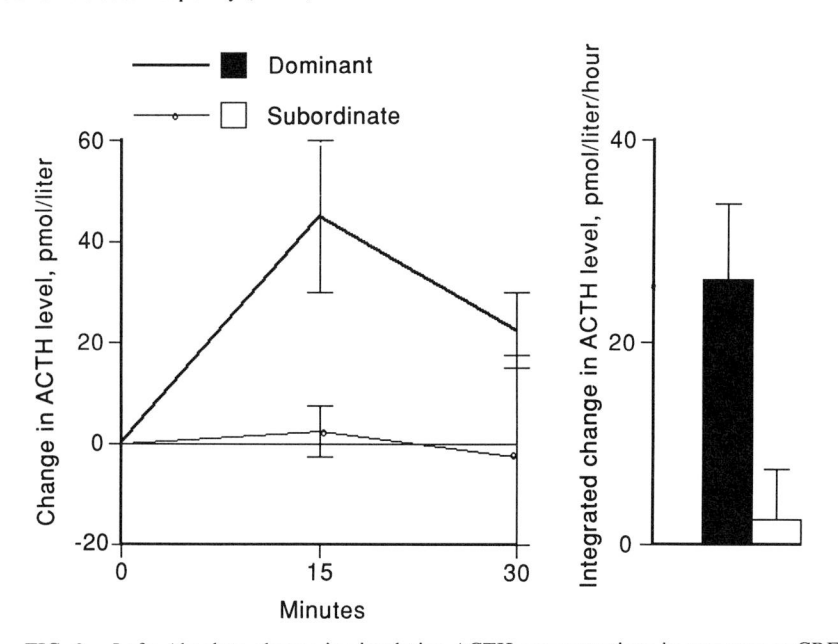

FIG. 3.  Left: Absolute change in circulating ACTH concentrations in response to CRF challenge among high-ranking and low-ranking males ($n = 6$/group) in the absence of cortisol feedback (following metyrapone administration). Dominant males were significantly more responsive to CRF. Data points were taken at 0, 15, and 30 minutes after CRF challenge. Right: This rank difference was also apparent when the areas under the curve of the data from the left were generated. Dominant animals had significantly greater integrated ACTH values than did subordinates. From Sapolsky (1989).

444                     ROBERT M. SAPOLSKY

still secretes more ACTH, this implies that there is enhanced hypothalamic secretagogue drive in subordinate animals.

Collectively, these experiments suggest that the enhanced cortisol secretion in low-ranking males is neurally driven, and this is certainly commensurate with the fact that low-ranking animals are subject to the highest rate of stressors (which are, of course, sensed in the brain). In addition, the cortisol hypersecretion can be due to a damping of feedback regulation, in that animals with elevated basal cortisol concentrations are also relatively dexamethasone resistant (Fig. 4) (Sapolsky, 1983b).

The "psychoendocrine" interpretation of the data (cortisol hypersecretion in subordinates occurs because of frequent stressors) need not conflict with the "neuroendocrine" interpretation (the hypersecretion occurs because of blunting of feedback sensitivity). Prolonged or repeated stressors will elevate basal glucocorticoid concentrations and cause feedback resis-

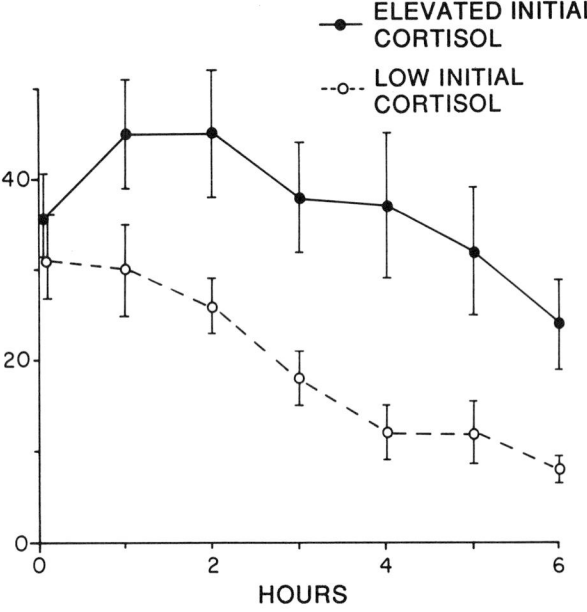

FIG. 4. Cortisol concentrations after 5 mg of dexamethasone administration. Subjects were divided into the 50% group with below-average basal cortisol concentrations (solid line, $n = 5$, consisting of four high-ranking males and one low-ranking male, with mean basal cortisol concentrations of $13 + 2 \mu g/100$ ml) and the 50% group with above-average concentrations (broken line, $n = 6$, consisting of three low-ranking and three middle-ranking males, with a mean basal cortisol concentration of $36 \pm 4 \mu g/100$ ml). The two groups did not differ in body weight. From Sapolsky (1983b).

tance in rodents, primates, and humans (Kalin *et al.*, 1981; Vernikos *et al.*, 1983; Baumgartner *et al.*, 1985; Ceulemans *et al.*, 1985; Young *et al.*, 1990). Such stressors can potentially down-regulate corticosteroid receptor numbers in various neural sites relevant to glucocorticoid feedback inhibition (Sapolsky *et al.*, 1983), blunting the efficacy with which those sites work and producing the subsequent hypersecretion and feedback resistance (Sapolsky *et al.*, 1984). Whether the same autoregulatory receptor changes occur in these subordinate baboons is, of course, not known, but the following scenario can tentatively be offered, given these laboratory findings: Shortly into the onset of social subordinance, low-ranking animals probably have relatively normal basal cortisol concentrations, but frequent stress-responses, due to the stressfulness of their social status. With time, the repeated cortisol excursions cause a down-regulatory blunting of feedback sensitivity, resulting in elevated basal cortisol concentrations and relative feedback resistance.

## C. SOCIAL SUBORDINANCE AND DEPRESSION

The elevated basal cortisol concentrations, the feedback resistance, and the underlying neuroendocrine mechanisms in low-ranking baboons echo some of the findings in biological psychiatry concerning neuroendocrine abnormalities in depression. Approximately half of depressives show some manifestation of hyperactivity of the adrenocortical axis, including elevated basal cortisol concentrations (either during the circadian trough, or throughout the circadian cycle) and/or dexamethasone resistance. There has been relatively little success in linking hypercortisolism with depression subtype, prognosis, or which drugs the individual is most likely to respond to. As exceptions to these negative findings, there is some consistency to hypercortisolism being more common among psychotic depressives, and among older depressives (reviewed in APA Taskforce, 1987).

The cases of basal cortisol hypersecretion and feedback resistance of depressives and of low-ranking baboons share some similarities. Most strikingly, in both cases, the hypersecretion appears to be driven at the level of the brain, with the pituitary showing blunted sensitivity to CRF [see Holsboer *et al.* (1984) and Gold *et al.* (1986) for demonstrations of this phenomenon in depressives]. Moreover, in both cases, stress may play a predisposing role: low-ranking males are subject to particularly high rates of social stressors, as discussed, whereas stress has often been posited to precipitate certain depression (reviewed in Anisman and Zacharko, 1982; Gold et al., 1988 [and in that vein, we have presented a model by which stress-induced corticosteroid receptor down-regulation might be

446      ROBERT M. SAPOLSKY

involved in the hypercortisolism of depressives (Sapolsky and Plotsky, 1990)].

Despite these similarities, the two models differ in a number of ways, or at least it is not yet clear if they are similar.

1. The basal hypercortisolism of depressives is typically most pronounced during the circadian trough (i.e., the evening) (APA Taskforce, 1987). It is not known if the same is the case for low-ranking baboons.

2. In neither case is it clear *which* hypothalamic secretagogue is hypersecreted (despite the blunted sensitivity to CRF) (reviewed in Sapolsky and Plotsky, 1990).

3. As noted, adrenal sensitivity to ACTH does not change with rank. However, adrenal hyperplasia and enhanced sensitivity to ACTH seem to be features of the adrenal of long-term depressives (Nasr *et al.*, 1982; Holsboer *et al.*, 1984; Gold *et al.*, 1986; Amsterdam *et al.*, 1987; Dorovini-Zis and Zis, 1987), which is thought to be due to the trophic effects of chronic overexposure to ACTH (Gold *et al.*, 1988).

4. Finally, and most importantly, despite the rigors of social subordinance, there is little reason to regard low-ranking baboons as being clinically depressed.

Thus, although social subordinance and depression bear some interesting similarities, they still represent two distinct routes toward altered adrenocortical function.


## D. WHAT ARE THE CONSEQUENCES OF ELEVATED BASAL CORTISOL CONCENTRATIONS IN LOW-RANKING BABOONS?

Naturally, rank differences such as those discussed are interesting only if they have some physiological consequence. The vast literature concerning glucocorticoid pathophysiology teaches two important lessons. First, excessive glucocorticoid exposure can be deleterious in an enormous number of ways (Munck *et al.*, 1984). Second, it is by no means the case that every increment of overexposure exacts a pathogenic price.

Thus, one cannot assume that the relatively mild hypercortisolism of the low-ranking animals is pathogenic. Naturally, the best way to determine that would be to manipulate cortisol profiles, but such chronic manipulations are anathema to the goals of research in this feral setting. Instead, one must rely on some rather correlative evidence suggesting that the elevated concentrations do have some consequences.

1. In studies of five different troops of olive and yellow baboons in two different national parks, elevated basal cortisol concentrations correlated

BABOON PSYCHOENDOCRINOLOGY 447

with depressed circulating lymphocyte counts (unpublished). This reflects the well-known suppressive effects of glucocorticoids on this immune parameter.

2. In three troops of olive baboons, social subordinance was associated with suppression of concentrations of high-density lipoprotein (HDL) cholesterol and of its associated apoprotein A (Sapolsky and Mott, 1987), and the elevated basal cortisol concentrations best explain such profiles (discussed below).

Thus, although tentative and correlative in nature, these data suggest that these rank-related adrenocorticol differences may be of some physiologic consequence.

### III. The Testicular Axis

#### A. MECHANISMS MEDIATING STRESS-INDUCED SUPPRESSION OF TESTOSTERONE CONCENTRATIONS

As with the vast majority of male mammals studied, physical and psychological stressors suppress testosterone concentrations in these males. Such suppression occurs in a number of situations among the baboons. Figure 5 (Sapolsky, 1987) demonstrates, on the left, the remarkable consistency of basal testosterone concentrations in olive baboons from stable populations in two different Kenyan study sites. This is contrasted with the suppressed values (Fig. 5, right) observed during a period of hierarchical reorganization and group instability, and during the East African drought of 1984.

It has not been possible to study the neuroendocrine mechanisms underlying such naturalistic occurrences of testosterone suppression. However, the stressor of darting for blood sampling also causes a prompt and dramatic suppression of both luteinizing hormone (LH) and testosterone concentrations (Fig. 6) (Sapolsky, 1986b), and, over the years, I have been able to uncover some of the mechanisms underlying this phenomenon.

The decline in LH concentrations is not secondary to stress-induced secretion of glucocorticoids or sympathetic activation. As evidence, blockade of each (with metyrapone and chlorisondamine, respectively) failed to prevent the decline, whereas exogenous glucocorticoids failed to inhibit pituitary responsiveness to gonadotropin-releasing hormone (GnRH). Moreover, prolactin concentrations are not elevated by the darting stressor in these animals, eliminating it as a possible mediator of the LH suppression.

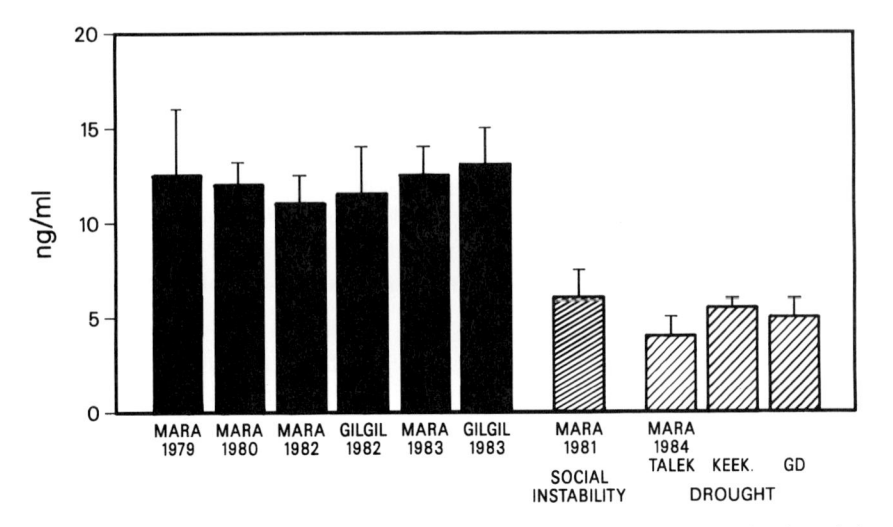

FIG. 5. Basal testosterone concentrations in baboons of different troops living in varied social and ecological settings. "Mara" refers to baboons living in the Masai Mara National Reserve in the Serengeti ecosystem of southwest Kenya. "Gilgil" refers to baboons living in a troop in the agricultural region of Rift Valley Province, Central Kenya, near the town of Gilgil. "Talek," "Keek," and "GD" refer to three different troops within the Masai Mara. Data from Mara baboons in 1979–1983 were derived from the Keek troop. From Sapolsky (1987).

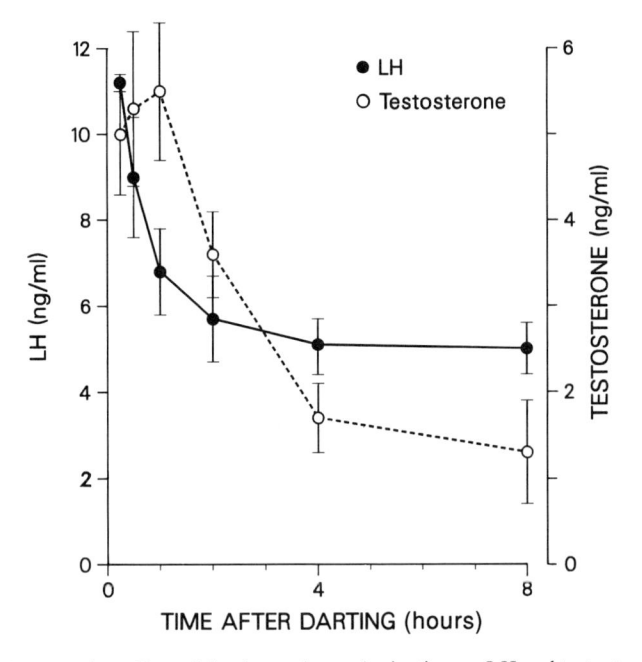

FIG. 6. Suppression effect of darting and anesthetization on LH and testosterone. From Sapolsky (1986b).

BABOON PSYCHOENDOCRINOLOGY 449

Instead, the decline appears to be due to stress-induced opiate secretion (Sapolsky and Krey, 1988), as the decline is prevented by opiate receptor antagonists. The suppression appears to involve both the $\mu$-opiate receptor, as evidenced by the protective effects of a dose of naloxone, which preferentially binds that receptor type, and the $\kappa$-opiate receptor, as evidenced by the efficacy of a $\kappa$ antagonist, MR 1452. These opioids not only blocked the LH decline but caused an increase in concentrations, suggesting a role for the endogenous opiates in tonic inhibition of basal LH secretion. Finally, the antireproductive actions of opiates during stress probably occur at the level of hypothalamic GnRH release, rather than directly at the pituitary (Rasmussen et al., 1983; Ching, 1983). This, of course, cannot be tested in these wild primates.

In addition to the decline in LH concentrations, the glucocorticoids secreted during stress blunt testicular responsiveness to LH (Fig. 7) (Sapolsky, 1985), a phenomenon noted in other species, where a likely mediating mechanism is glucocorticoid-induced decreases in LH receptor number (Bambino and Hsueh, 1981; Johnson et al., 1982; Cumming et al., 1983). This appears to be an important step, as the extent to which glucocorticoids inhibit testicular responsiveness to LH predicts the extent to which stress

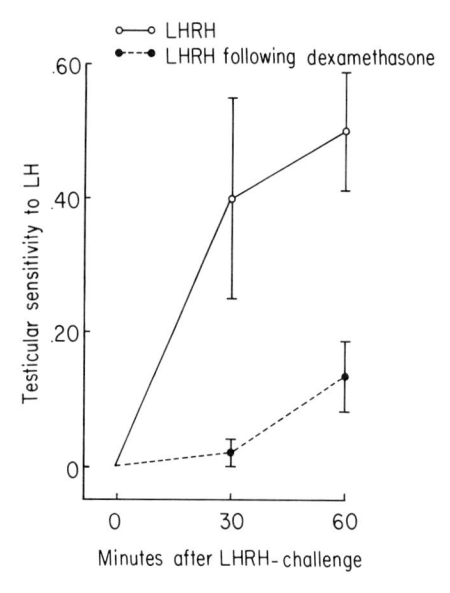

FIG. 7. Testicular responsiveness to GnRH-induced elevations of LH concentrations, with or without prior dexamethasone administration. Testicular sensitivity was determined by dividing the testosterone concentration at each time by the LH concentration at that time. From Sapolsky (1985). Note that in this and Fig. 8, GnRH is referred to as lutenizing hormone releasing hormone (LHRH).

450                    ROBERT M. SAPOLSKY

suppresses testosterone concentrations in these animals (Fig. 8) (Sapolsky, 1985).

Thus the (rather artificial) stressor of darting inhibits testosterone secretion through at least two distinct mechanisms. I have uncovered some striking rank differences in the workings of these mechanisms.

## B. DOMINANT MALES ARE RELATIVELY RESISTANT TO SUPPRESSIVE EFFECTS OF STRESS ON TESTOSTERONE CONCENTRATIONS

Before discussing rank-related differences in the testicular stress response, it should be noted that, over the years of this study, it has consistently been the case that dominance rank does *not* correlate with basal testosterone concentrations. This flies somewhat in the face of endocrine mythology, in which dominant male primates are often thought to have relatively elevated testosterone concentrations. This view was mostly

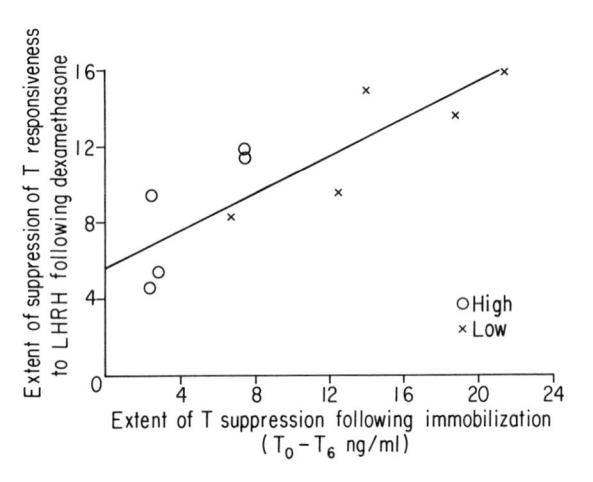

FIG. 8.   Relationship between sensitivity to immobilization/anesthetization-induced declines in testosterone (T) concentration and glucocorticoid-induced suppression of testicular responsiveness to LH. The $x$ axis indicates the absolute decline of T concentrations during the 6 hours after anesthetization ($T_0$–$T_6$). The $y$ axis indicates the extent to which dexamethasone suppressed the rise in T concentrations in GnRH administration (see Fig. 7) (absolute rise in T concentrations 1 hour after GnRH administration in the absence of dexamethasone minus the absolute rise at same time in the presence of dexamethasone). There was a highly significant correlation between the two indices, indicating that individuals most sensitive to the suppressive effects of stress on T concentrations were also most sensitive to the inhibitory effects of glucocorticoids on testicular responsivenss. Open circles indicate high-ranking males (upper 50th percentile). From Sapolsky (1991).

BABOON PSYCHOENDOCRINOLOGY 451

established by a rather influential *Nature* paper showing such a correlation (Rose *et al.*, 1971). In retrospect, subsequent studies have shown that elevated basal testosterone concentrations are a feature of dominance in unstable hierarchies, where such dominant animals typically are the most aggressive. In captive populations, this occurs after the initial formation of a social group, or after the membership of an established one is changed. In the wild, I have observed such an instability once over the years (discussed below), and at that time, dominant males did indeed have the highest basal testosterone concentrations and highest rates of aggression. In contrast, in stable hierarchies, dominant males are neither the most aggressive nor have the highest testosterone concentrations (Eaton and Resko, 1974; Gordon *et al.*, 1976). Such stability is rare in captive populations (requiring that they be undisturbed for months to years), but is the norm in ecologically stable settings in the wild. In my study population, it is low-ranking adolescent males who have both the highest basal testosterone concentrations and the highest rates of aggression (Sapolsky, 1983a).

In stable dominance hierarchies, however, dominant males show the distinctive feature of being able to *elevate* transiently testosterone concentrations immediately following the onset of darting stress, in contrast to the prompt decline seen in lower ranking males (Sapolsky, 1982). This appears not to be a pituitary phenomenon, as both rank cohorts show similar declines in LH (Sapolsky, 1985). Instead, a pair of rank-specific mechanisms at the testes appear to account for this effect.

First, the testes of dominant males are less vulnerable to the disruptive effects of glucocorticoids on testicular sensitivity to LH. This was shown in Fig. 8 (Sapolsky, 1991), where data points from high-ranking individuals are clustered on the left side of the regression line.

This effect would explain why testosterone concentrations decline with the onset of stress more slowly than do those of subordinates. However, it does not explain why, instead, concentrations are actually *elevated* at such times. This appears to involve sympathetic activation during stress because blockade of catecholamine release with the drug chlorisondamine blocks the effect (i.e., when chlorisondamine is administered immediately after darting, testosterone concentrations decline in dominant males just as in subordinates, whereas chlorisondamine has no effect on subordinate animals) (Sapolsky, 1986b). Sympathetic activation can increase testosterone concentrations in rodents (Frankel and Ryan, 1981). Possible mediating mechanisms might include direct autonomic innervation of the testes [which occurs in Old World monkeys (Hodson, 1970)]. In addition, catecholamines can vasodilate the testicular parenchyma, increasing testicular blood flow and absolute amounts of LH delivered. Whether either mechanism (or others) explains why sympathetic catecholamines can transiently

452                           ROBERT M. SAPOLSKY

elevate testosterone concentrations during stress, the effect is specific to
dominant males. Thus, they must either secrete more catecholamines
during stress and/or show greater tissue sensitivity to catecholamines. As
will be discussed below, the latter is the case.

## C.  CONSEQUENCES OF RANK DIFFERENCE IN TESTICULAR STRESS RESPONSE

The data just discussed are striking in that they demonstrate that baboons can have diametrically *opposite* endocrine responses to a stressor,
depending on their social rank. Is this difference of any consequence? As
has been discussed elsewhere, the time course and magnitude of the
differences are unlikely to affect testosterone-dependent aggression or
sexual behavior—very large, persistent, and rather unphysiological
changes in testosterone concentrations in such cases (Sapolsky, 1987).
Thus, one must probably rule out some attractive speculations about the
effects of social stress having rank-dependent and opposing effects on
aggression or sexual behavior. Instead, there may be consequences for
muscle metabolism. Androgens, of course, have anabolic effects on muscle, and they promote glucose uptake rather rapidly (Max and Toop, 1983).
Thus, it is conceivable that during a sustained and stressful confrontation
between two males, should one show a transient rise in testosterone
concentrations for a few hours while the other has a prompt decline, the
former might gain some metabolic advantage at the muscles. This idea is
currently being treated.

## IV.  Cholesterol Metabolism

It is well established in humans and baboons that atherosclerosis and
coronary heart disease are directly associated with total serum cholesterol
and low-density lipoprotein cholesterol (LDL-C) concentrations and in-
versely related to high-density lipoprotein cholesterol (HDL-C) concentrations (Gofman et al., 1954; Miller and Miller, 1975; McGill et al., 1981).
The atherogenic nature of LDL probably arises from the capacity of
cholesterol, transported in association with LDL, to accumulate in arterial
walls along with foam cells. In contrast, HDL appears to be antiathero-
genic because of its ability to transport cholesterol from peripheral tissues
and from arterial walls to the liver for catabolism.

Individual variability in levels of LDL-C, of HDL-C, or of their associated apolipoproteins (Apo-B and Apo-A-I, respectively) can arise from
diet, body weight, age, genetics, gender, and sex hormone concentrations
(McGill, 1979; Goldberg et al., 1985; Goldbourty and Neufeld, 1986).

Moreover, psychological stressors can decrease the HDL:LDL ratio, whereas psychotherapy to reduce such stressors raises the ratio (Grundy and Griffin, 1959; Avogaro, 1978; Martinsen et al., 1981; Gill et al., 1985).

Given these findings, it is quite interesting that we have observed that HDL-C and its associated apoliprotein A-I are suppressed in low-ranking males (Fig. 9) (Sapolsky and Mott, 1987). This rank difference was not attributable to differences in testosterone concentrations, body weight, age, diet, level of activity, or genetics. Instead, there was a significant association between elevated basal cortisol concentrations and suppressed concentrations of both HDL-C and apolipoprotein A-I. Sustained glucocorticoid overexposure can suppress HDL:LDL ratios (Adlersberg et al., 1950; Stern et al., 1973; Cavanee and Melnykovych, 1979).

The rank differences and their metabolic correlates might have some cardiovascular consequences, in that sustained social subordinance in stable dominance hierarchies in captive primates is associated with a depressed HDL:LDL ratio and increased stenosis of the coronary artery and aorta (Kaplan et al., 1982a,b; Hamm et al., 1983). Moreover, wild baboons have been shown to have some degree of atherosclerosis and fatty streaks. However, at present it is pure speculation to suggest that low-ranking animals are more at risk for such cardiovascular disease.

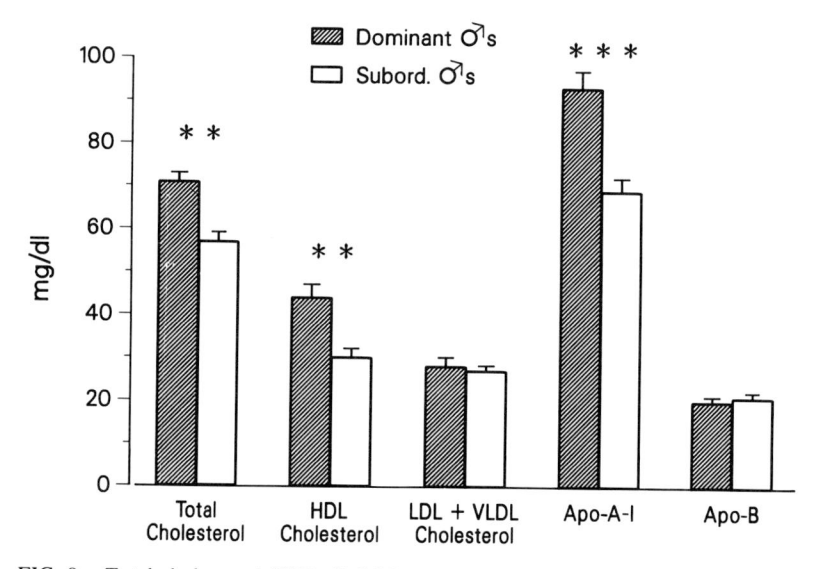

FIG. 9. Total cholesterol, HDL-C, LDL + VLDL-C, and apolipoprotein (Apo-A-I and Apo-B) concentrations as a function of social rank. From Sapolsky and Mott (1987). **, *** indicate significant differences by rank at the $p < 0.01$ and 0.001 levels, respectively.

## V.  Immunologic Profiles

One of the most exciting topics in current biology is the study of the interactions among the nervous, endocrine, and immune systems. Some of the most interesting findings in psychoneuroimmunology, as such studies are often termed, explore how individual differences in rate of exposure to stressors and in endocrine responses to stressors can alter immune profiles and vulnerabilities to diseases. One critical link between such stressors and immune profiles is glucocorticoid secretion; the hormone has long been recognized for its capacity to suppress a variety of immunologic measures (cf. Munck *et al.,* 1984).

Naturally, I have thus been interested in whether subordinate baboons with, among other things, their elevated basal cortisol concentrations show any indices of suppressed immune function. This has proved to be a particularly difficult question to answer regarding animals living in the wild. For example, Mark Laudenslager, of the University of Colorado, and I have attempted to immunize baboons against a neutral antigen (keyhole limpet hemocyanin) and to then redart the animals to measure antibody titers; however, to dart selected individuals on specific (postimmunization) days has proved dauntingly difficult. Furthermore, baboons consume a fair number of arthropods in their diet, which are phylogenetically similar to the keyhole limpet, producing high and variable preimmune antibody titers.

Amid these complications, it has been possible to at least obtain one rather crude index of immune function, namely to measure numbers of circulating lymphocytes. These studies show that there are indeed fewer circulating lymphocytes in subordinate males (Table I). Naturally, the vast complexity of immune function makes it presumptious to extrapolate from

TABLE I

*Relationship between Social Rank
and Lymphocyte Counts*[a]

| Rank | Lymphocyte count |
|---|---|
| **Talek troop** | |
| 1–6 | $1610 \pm 516$ |
| 7–12 | $980 \pm 310$ |
| 13–19 | $640 \pm 225$ |
| **Keekorok troop** | |
| 1–6 | $1060 \pm 360$ |
| 7–11 | $490 \pm 110$ |

[a]Number of lymphocytes/$10^6$ red blood cells.

numbers of lymphocytes to inferences about overall immune competence. More informative studies remain to be done in this area.

## VI. Autonomic Nervous System and Cardiovascular Function

As noted in the discussion on testicular function (Section III), the sympathetic nervous system appears to be involved in the ability of dominant males to elevate transiently testosterone concentrations during stress. As evidence, the elevation can be eliminated by sympathetic blockade, and a number of features of sympathetic innervation of the testes and neighboring blood vessels might mediate the effect. The fact that the sympathetically mediated elevation occurs only in dominant males suggests that they secrete more catecholamines during stress and/or have greater target tissue sensitivity to catecholamines.

The first possibility is essentially impossible to test, because of the labile nature of catecholamine secretion and the primitive means of preserving samples in the field. However, there appears to be a robust relationship between rank and catecholamine responsiveness. Figure 10 (Sapolsky, 1991) demonstrates that over a wide range of epinephrine concentrations, dominant males have a greater absolute rise in systolic pressure (as well as in diastolic pressure and heart rate; data not shown). This turned out to be but one of a number of features of autonomic function that are distinctly rank-related:

1. Dominant animals not only have a larger response to epinephrine, but a faster recovery from it as well (Table II, A).
2. This enhanced sympathetic responsiveness of dominant males seems to involve enhanced sensitivity to both alpha and beta receptor stimulation. Relative to low-ranking animals, dominant males have a faster rise in systolic pressure in response to the $\alpha$ agonist phenylephrine (0.2 mg) (Table II, B) and a faster recovery from that rise (Table II, C); similarly, they have a faster rise in systolic pressure in response to the $\beta$ agonist isoproterenol (0.5 mg) and a faster recovery (Table II, D and E).
3. Dominant males also appear to have enhanced parasympathetic tone. As evidence, they show a significantly greater disinhibition of heart rate following atropine administration than do subordinate animals (Table II, F).

These findings suggest marked differences in cardiovascular function according to rank, with a relatively more adaptive profile in dominant animals. They are likely to have lower resting systolic blood pressure (because of the enhanced vagal tone) and, in the face of a stressor that

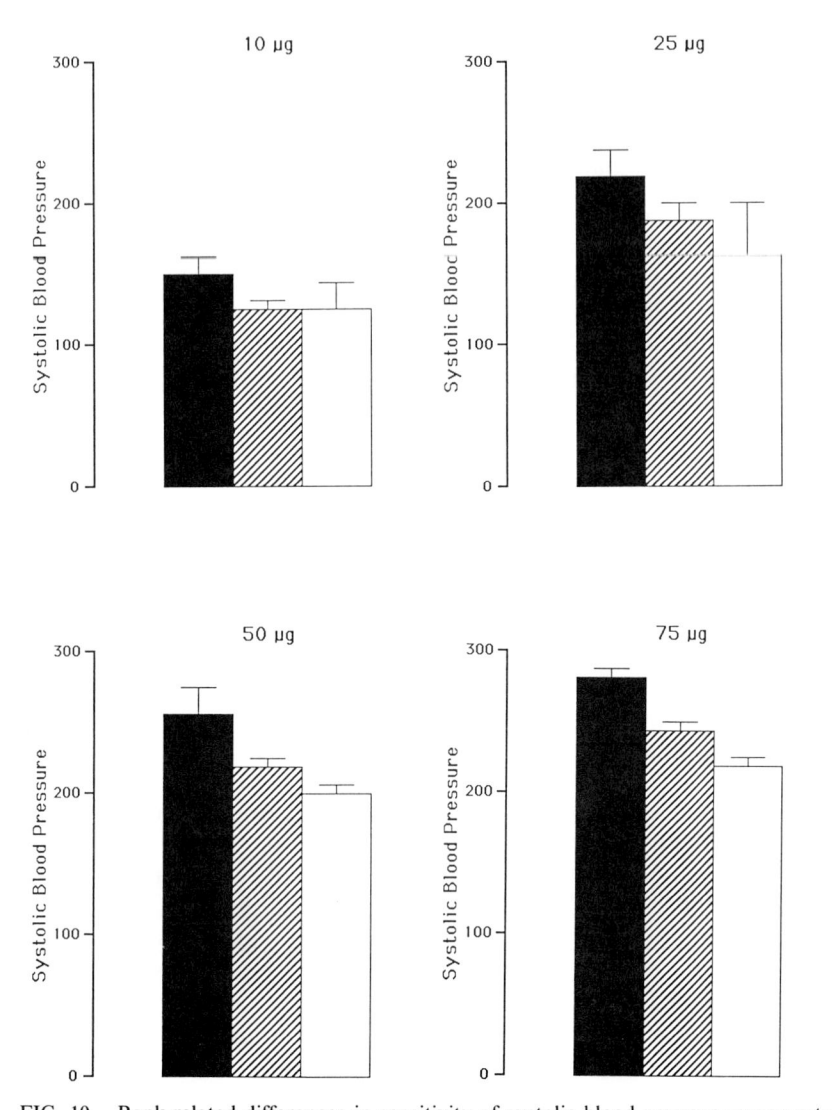

FIG. 10. Rank-related differences in sensitivity of systolic blood pressure response to catecholamines. Baboons were categorized as being in either the highest ranking third (black), middle-ranking third (striped), or lowest ranking third (white); animals were weight matched. Subjects were injected with chlorisondamine immediately after anesthetization in order to inhibit endogenous catecholamine release; there was no rank difference in postchlorisondamine systolic pressure. One hour later, baboons were injected i.v. with 10 μg epinephrine and the systolic response was monitored 1 minute later. Animals were then challenged with the successively higher epinephrine doses at 1-hour intervals. At all but the lowest epinephrine concentration, there was a stepwise relationship such that the higher the rank, the greater the rise in response to epinephrine. From Sapolsky (1991).

BABOON PSYCHOENDOCRINOLOGY 457

### TABLE II
*Effects of Social Rank on Cardiovascular Parameters*

| Rank | Change in Cardiovascular Parameter |
|------|-----------------------------------|
| A. Speed of recovery from epinephrine challenge (decline in systolic blood pressure 1 minute after peak blood pressure postepinephrine) | |
| 1–5 | −103 ± 6 mm Hg |
| 6–17 | −21 ± 20 mm Hg |
| B. Sensitivity to α stimulation with the α agonist phenylephrine (rise in systolic blood pressure 1 minute after infusion) | |
| 1–5 | 97 ± 2 mm Hg |
| 6–10 | 75 ± 9 mm Hg |
| 11–16 | 63 ± 5 mm Hg |
| C. Speed of recovery from phenylephrine challenge (decline in systolic blood pressure one minute after the post-phenylephrine peak) | |
| 1–5 | −63 ± 2 mm Hg |
| 6–10 | −47 ± 13 mm Hg |
| 11–16 | −39 ± 5 mm Hg |

| Rank | Change in Cardiovascular Parameter |
|------|-----------------------------------|
| D. Sensitivity to β stimulation with the β agonist isoproteronol (rise in heart rate at 1 minute after infusion) | |
| 1–5 | 75 ± 11 beats/minute |
| 6–17 | 40 ± 15 beats/minute |
| E. Speed of recovery from isoproteronol (decline in heart rate 1 minute after peak postisoproteronol response) | |
| 1–5 | −20 ± 7 beats/minute |
| 6–17 | +16 ± 5 beats/minute; i.e., heart rate was still rising |
| F. Effects of atropine on resting heart rate (increase in heart rate 1 minute after atropine injection) | |
| 1–5 | 30 ± 2 beats/minute |
| 6–10 | 17 ± 5 beats/minute |
| 11–17 | 12 ± 3 beats/minute |

activates the sympathetic nervous system, a larger response. On cessation of the stressor, they are likely to have the most rapid of recoveries. The contribution of both α and β receptors to this phenomenon suggests both cardiac and vascular involvement.

## VII. Some Complications of the Association between Particular Social Ranks and Particular Endocrine Profiles

Data discussed in the preceding sections suggest that among wild olive baboons, social rank is a strong predictor of certain endocrine and physiologic traits. However, it would be simplistic, knowing, for example, about the observed differences in cholesterol concentrations according to rank,

458                    ROBERT M. SAPOLSKY

to observe two males of differing rank and to make predictions about their lifelong vulnerabilities to cardiovascular disease. This is not only because of the relative weakness of the link between cholesterol profiles and such pathology, but also because among males, ranks change over time and in idiosyncratic ways. Thus, any consequences that these observations might have for the relationship between rank and disease must reflect lifelong patterns of rank.

But the significance of these observations is also limited by the fact that mere rank is not a particularly meaningful variable for a primate. Amid the complexities of primate social behavior, a dominance rank is not a monolithic state nor, in some social settings, a particular important variable. Primates are far too sophisticated behaviorally to be merely divided, in effect, into two flavors—dominant or subordinate—or to be characterized by a particular number in a dominance hierarchy that may be more a theoretical construct of the human observer than of the primate participant.

In this vein, some of the most interesting current trends in primatology explore the various forms that dominance hierarchies can take, the extent to which such hierarchical constructs do in fact predict anything about resource acquisition or reproductive success on the part of an individual, and the numerous behavioral strategies available to a primate that are alternatives to achieving high rank in a dominance hierarchy (cf. Goodall, 1986; Smuts, 1986; de Waal, 1982). Thus, it is not surprising that the seemingly straightforward associations between rank and endocrine parameters are, in fact, far from straightforward. Therefore, it would be simplistic to conclude that a certain dominance rank always equals a certain physiological profile, or that rank is the sole factor that correlates with physiological variability in an interesting way. Next we review some of these complications, drawing on data from a number of different physiological systems.

## A. ENDOCRINE CORRELATES OF RANK ARE INFLUENCED BY THE NATURE OF THE PRIMATE SOCIETY IN WHICH THE PARTICULAR RANK OCCURS

This idea is best illustrated when contrasting stable versus unstable dominance hierarchies. In a particular dyadic relationship in which one individual is dominant, that individual could win 51% of interactions and still be the "dominant" animal, or could win 100% of them. The former case, in which 49% of the interactions represent reversals of the direction of dominance, implies a shifting, unstable dyadic relationship. When there are many such unstable relationships within an overall hierarchy, the

## TABLE III
### Effects of Social Instability on Basal Cortisol Concentrations among High-Ranking Males[a]

| Season | Basal cortisol ($\mu$g/100 ml) |
| --- | --- |
| Stable ($n = 7$) | 15.5 ± 0.4 |
| Unstable ($n = 1$) | 24.2 ± 0.9 |

[a] Instability was defined as observational seasons in which the overall rate of reversals of approach–avoidance dominance interactions was greater than 10%.

hierarchy itself can be characterized as being unstable. In my own studies, I have defined this as being when more than 10% of the approach–avoidance interactions represented reversals of the direction of dominance (Sapolsky, 1983a).

In the wild, this is a rare event, following a rather major demographic shift within a troop; in the case of this that I have studied, this followed the abrupt crippling of the alpha male by a cooperative coalition that disintegrated the same day. In the aftermath of that collapse, 3 months of relative instability ensued (Sapolsky, 1983a). In settings with captive primates, such instability typically occurs when social groups are first formed, or when membership has been changed.

Such unstable periods involve high rates of aggression, shifting alliances, disruption of feeding, and affiliative social behavior—all hallmarks of psychological stress (Levine et al., 1989). Moreover, this picture is most pronounced among the males seeking to attain and maintain high rank in this shifting setting. Thus, it is not surprising that during unstable periods, highest ranking males no longer have the low basal cortisol concentrations typically observed at other times (Table III). The same has been observed among dominant males in unstable captive social groups (where such instability occurs following the initial group formation, or with a change of individuals in the group) (Keverne et al., 1982; Mendoza et al., 1979; Coe et al., 1979; Chamove and Bowman, 1976).

Similarly, the physiological correlates of low rank can reflect the quality of such subordinance. Figure 11 shows an example of this. As discussed, being a subordinate male typically involves being subject to high rates of displaced aggression, most often when a higher ranking male has lost a tense approach–avoidance or dominance interaction with another high-ranking male. Yet primate social groups will differ as to how common an occurrence this is; in effect, although it is probably never particularly psychologically restful to be a subordinate male baboon, it is definitely

460 ROBERT M. SAPOLSKY

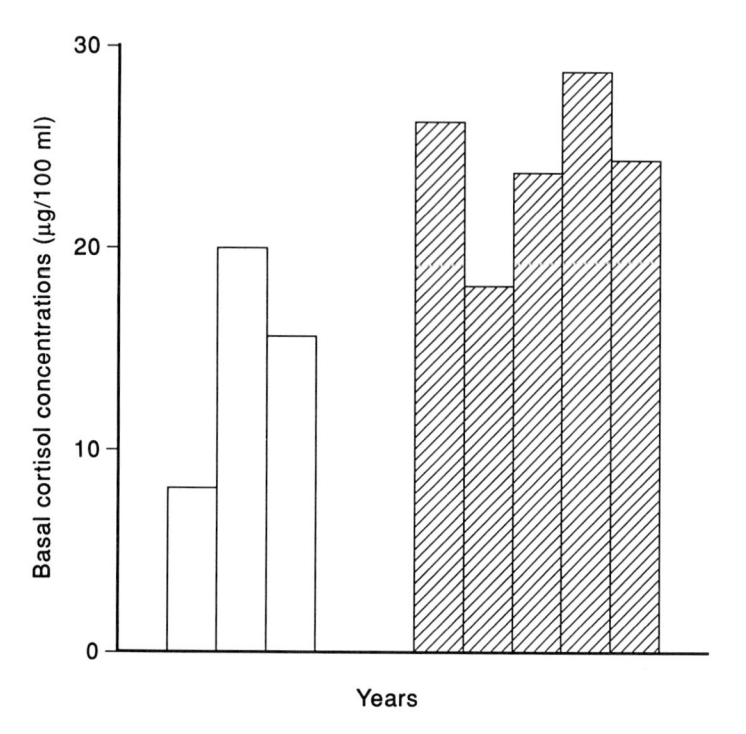

FIG. 11. Basal cortisol concentrations in low-ranking males during years in which there were low rates of displaced aggression by high-ranking males onto subordinates (open bars) and years in which there were high rates (striped). Basal cortisol concentrations are elevated among subordinate animals subjected to the highest rates of displaced aggression by dominant animals. Unpublished data.

worse in some social groups than others. During years in which there was a low rate of displacement attacks by high-ranking males on low-ranking males (3 years with a mean of $0.4 \pm 0.08$ attacks/10 hours; Fig. 11, open bars), basal cortisol concentrations in such subordinates were far lower than during years where there were more frequent attacks (5 years with a mean of $1.1 \pm 0.1$ attacks/10 hours).

Thus, low basal cortisol concentrations are not automatic markers of high rank. Instead, the trait appears to mark the circumstances wherein dominance is psychologically advantageous. Similarly, the extent to which elevated basal cortisol concentrations are markers of low social rank seems to reflect how aversive such subordinance is.

## B. ENDOCRINE CORRELATES OF RANK ARE INFLUENCED BY THE INDIVIDUAL EXPERIENCES OF AN ANIMAL

For a low-ranking animal, it does not merely matter how benevolent high-ranking males are to subordinate animals in general. Of even greater importance, obviously, is how that individual animal fares within the troop. We observed a striking example of this recently.

As is the case for most Old World monkeys, male baboons transfer from their natal troop around the time of puberty into their adult troop. Thereafter, they might transfer again at later opportune times (for example, after losing a major fight in their resident troop). Typically, the period following a transfer is a rather harrowing one for the individual, particularly when it is his first, adolescent transfer. He is subject to extremely high rates of aggression and approach–avoidance interactions, has no friends or coalitional partners, and often has not been groomed for months. Not surprisingly, most recent transfer males are trepidous, peripheral, and subordinate.

However, in rare circumstances, a transfering male will take a highly aggressive strategy; typically this is a physically imposing adult for whom this is not a first transfer. A sufficiently aggressive and intimidating individual may, in effect, be given a grace period before the males in his new troop dare challenge him, and in this interim, he can often form a number of sexual consortships.

We were fortunate to have such a rare transfer occur in the middle of one study season, affording the possibility to compare physiologic values before and after the baboon made the transfer. We observed that his entry into the troop and onset of marked aggressiveness were associated with elevations of cortisol concentrations in both sexes, declines in circulating lymphocyte counts, a mild suppression of testosterone concentrations in males, and dramatic disruptions of reproductive physiology in females (i.e., a number of spontaneous abortions during the first few weeks of his residency) (Alberts *et al.*, 1992). Strikingly, these changes were sensitive to the quality of the interactions with this highly aggressive individual. This is shown in Fig. 12, wherein circulating lymphocyte counts reflected, in a remarkable dose–response manner, the number of times that an individual had been attacked by the transfer male during the preceding weeks.

This study offered an additional and ironic demonstration of how individuals differ in their sensitivity to generalized social phenomena. As noted, basal cortisol concentrations were elevated and lymphocyte counts suppressed following the transfer of the male. Of note, the male himself

462 ROBERT M. SAPOLSKY

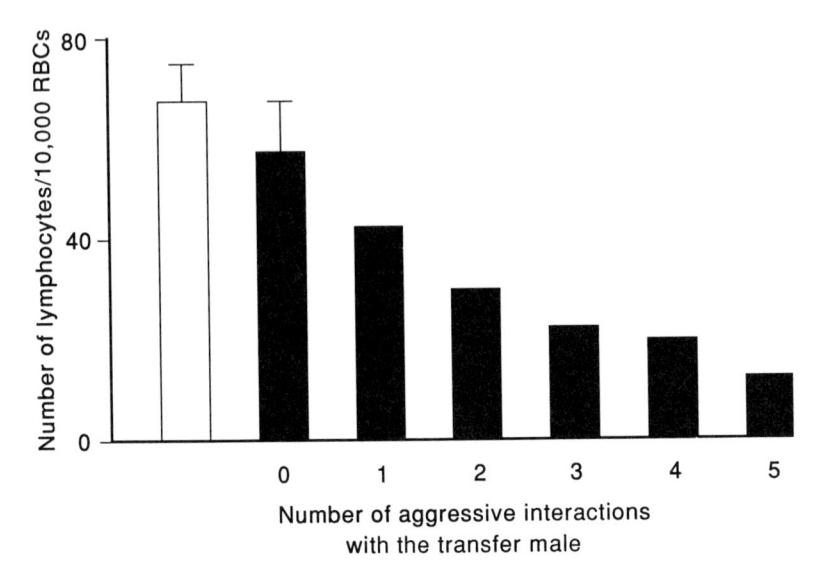

FIG. 12. Relationship between numbers of circulating lymphocytes and frequency with which individuals were attacked by the aggressive transfer male. The open bar indicates average lymphocyte counts prior to the transfer of that individual into the troop. Unpublished data.

had the highest cortisol concentrations and lowest lymphocyte counts of anyone in the troop—a measure, perhaps, of the costs of this potentially very succesful behavioral strategy. Physiologically, a highly aggressive transfer into a new troop does not appear to constitute a free lunch.

Another example concerns an individual's experiences within the larger context of rank and social setting. As was shown in Table III, when the overall dominance hierarchy is unstable, being dominant is associated with relatively elevated mean basal cortisol concentrations. However, even within a stable dominance hierarchy, basal cortisol concentrations are elevated in males whose *individual* dyadic relationships are unstable. This is shown in Fig. 13; cortisol concentrations are elevated in males in whom more than 20% of their approach–avoid dominance interactions represented reversals of the direction of dominance (remember that 49% represents the maximum percentage of reversals possible). These data are particularly subtle and informative, I believe. They present the relationship between cortisol concentrations and the degree of instability in dyadic interactions *only between the male and the three males ranking immediately below him in the hierarchy*. If the male under study is having numerous interactions with these other three, interactions that represent reversals of the direction of dominance, this implies that those three lower

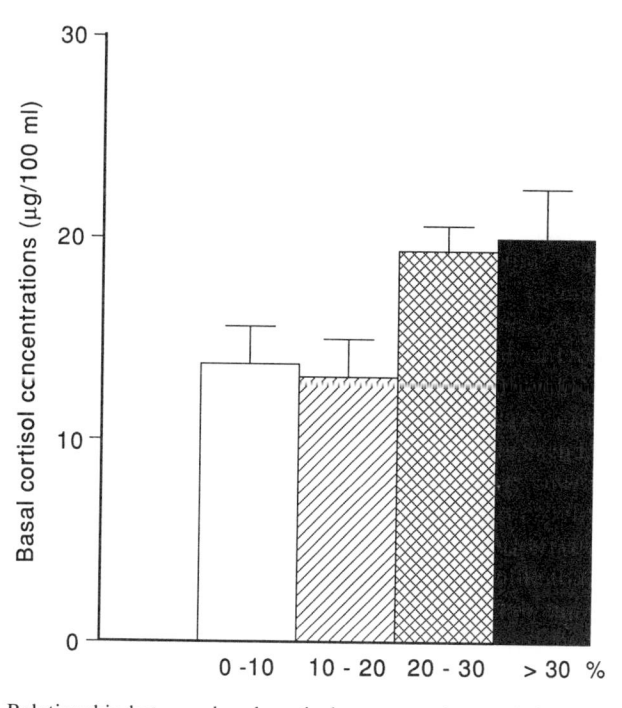

FIG. 13. Relationship between basal cortisol concentrations and the percentage of approach–avoid interactions that represented reversals of the direction of dominance with the three lower ranking males in the hierarchy. Unpublished data.

ranking males are effectively pressuring him for his rank. In contrast, there was no significant relationship between cortisol concentrations in a male and the extent of instability in his interactions with the three males ranking immediately *above* him in the hierarchy. In other words, "social instability" is not necessarily always a stressor associated with cortisol elevations; it appears to be when the instability implies that someone behind you in the hierarchy is gaining on you, but it is not associated with hypercortisolism when the instability implies that you are gaining ground on someone ahead of you.

## C. PERSONALITY MAY BE AN ADDITIONAL CRITICAL VARIABLE RELATED TO INDIVIDUAL DIFFERENCES IN ENDOCRINE PROFILES

Clearly, the style with which two baboons of the same rank behave may differ, and the preceding discussions review how this may reflect the nature of the troop's particular dominance hierarchy, and how that individual is treated within that hierarchy. In addition, differences in behavior may arise from features of the individuals. For example, this can reflect life-history factors—a low-ranking baboon is typically either a subadult rising in the hierarchy or an aged animal on the way down, and age has an important influence on certain behaviors. These differences may reflect patterns of relatedness within the troop—two males of similar ranks may differ behaviorally because only one of them has a brother residing in the same troop with him. Some of my recent work suggests that "personality" is also an important factor in linking behavior to physiology.

What do I mean by this term? Two baboons of similar rank may differ in their tendency to form cooperative coalitions with other males or to form affiliative "friendships" with females. They may differ in how readily they initiate fights, if they respond differently to losing minor and major dominance interactions, if they play with infants. Furthermore, there are considerable data indicating that these individual differences in "styles" of behavior are stable over years, justifying the tentative use of the term "personality" in these animals.

We have reviewed the many years of behavioral data obtained from these animals and formalized numerous stylistic features of behavior among them. We have then examined whether these stylistic features of individual behavior are related to any endocrine measures, after eliminating rank as a variable.

We have observed that low basal cortisol concentrations are not, in fact, an automatic correlate of high rank in stable hierarchies. Instead, the marker is observed only among dominant males with any of a number of behavioral traits (and strikingly, males of equivalently high rank without those traits can have basal cortisol concentrations as high as those found in subordinate animals). One set of "low basal cortisol" traits is related to how males compete with each other (Sapolsky and Ray, 1989; Ray and Sapolsky, 1992):

1. Can a male differentiate between truly threatening interactions with rival males versus interactions of lesser importance? This is determined by examining the effect of the presence of a rival on the male's behavior: the rival might threaten with high intensity (e.g., giving a threat yawn at close quarters), with low intensity (e.g., sitting a few yards away and giving a directed stare), or not at all (e.g., going to sleep 20 yards away). What does the male under study do in each case? We have found that low basal cortisol concentrations are found in the males who behaviorally distinguish high-intensity threats from the other two categories.

2. Does a male tend to initiate fights once a high-intensity threat is given? When a rival threat yawns from a yard away (a fairly reliable indicator

that the rival is about to initiate a fight), does the male under study passively wait for the fight to occur or does he initiate it himself? We observe significantly lower cortisol concentrations in males who initiate the fights in those circumstances.

3. Does the male differentiate behaviorally between winning and losing the fight? A male adept at this may, perhaps, go and groom a female after winning a fight yet may displace aggression onto a third party after losing a fight. In contrast, males who do differentiate between winning and losing might typically displace aggression in either case. We observe significantly lower cortisol concentrations in those who distinguish between winning and losing.

4. When losing, does the male displace aggression onto a third party? Those with the highest tendency to do so have significantly lower basal cortisol concentrations. In contrast, displacing aggression at high rates after *winning* a fight is unrelated to cortisol profiles.

In addition, we have observed certain behavioral traits related to affiliative behavior that mark low basal cortisol concentrations. These are the males who have the highest rates of grooming nonestrus females, being groomed by nonestrus females, and of interacting with infants. In contrast, rates of interactions with estrus females, and other markers of sexual rather than affiliative behavior (such as frequency of copulation), are unrelated to cortisol profiles.

As a final interesting observation, we have observed that such "low-cortisol" dominant males remain in the dominant cohort of the troop for a significantly longer time than do "high-cortisol" males of equivalent rank. Potentially, this could imply that as a male is dominant for a longer period (for some other, unrelated reason), he becomes more psychologically comfortable with such a position, and his basal cortisol concentrations decline progressively. Instead, the low basal cortisol trait is present in the very first season of such males' long tenures; it is a predictor, rather than a consequence, of such social success.

These findings regarding individual stylistic differences reinforce themes that resonate through the psychoendocrine literature and in strategies of human stress management (cf. Wolff *et al.*, 1964; Weiss, 1984; Levine *et al.*, 1989; Weiner, 1991). Such research shows that in the face of overt and undeniable external stressors, the magnitude of the physiological stress response can be modulated enormously by psychological factors. Protection is afforded by increasing the individual's sense of control, of predictability, by providing outlets for frustration, and by strengthening social support networks.

Extending this approach, we are now finishing an analysis of the subordi-

466                                  ROBERT M. SAPOLSKY

nate half of the population over the years, once again trying to formalize different styles of behavior. Although preliminary, we observe that some of the relatively deleterious physiologic markers of subordinance are blunted in subordinate animals with particularly strong patterns of social affiliation (Virgin and Sapolsky, in preparation).

The studies reviewed in earlier sections suggest that although social rank is an important predictor of some physiologic parameters, just as important can be the type of society in which that rank occurs, and the way in which one experiences such a rank. The data presented in the section add additional factors to this picture: the filters of personality with which an individual views these events and the varying strategies available for coping with them are probably immensely important variables as well. For primates, both human and otherwise, living in a world rife with unavoidable stressors, the importance of such filters seems encouraging news.

## ACKNOWLEDGMENTS

These studies were made possible by the long-standing generosity of the Harry Frank Guggenheim Foundation and the John D. and Catherine T. MacArthur Foundation. Permission for these studies was granted by the Office of the President, Republic of Kenya. Field assistance was provided by Richard Kones, Francis Onchiri, Hudson Oyaro, Diane Rich, Lisa Share, and Reed Sutherland.

## REFERENCES

Adlersberg, D., Schaefer, L., and Drachman, S. (1950). *J. Am. Med. Assoc.* **144**, 909–914.
Alberts, S., Altmann, J., and Sapolsky, R. (1992). *Horm. Behav.* **26**, 163–173.
Altmann, J. (1971). *Behaviour* **39**, 73–90.
Altmann, J. (1980). "Baboon Mothers and Infants." Harvard Univ. Press, Cambridge, MA.
Amsterdam, J., Marinelli, D., Arger, P., and Winokur, A. (1987). *Psychiatry Res.* **21**, 189–197.
Anisman, H., and Zacharko, R. (1982). *Behav Brain Sci.* **5**, 89–106.
APA Taskforce on Laboratory Tests in Psychiatry. (1987). *Am. J. Psychiatry* **144**, 1253–1268.
Archer, J. (1970). *J. Mammol.* **51**, 327, 335.
Avogaro, P. (1978). *Eur. J. Clin. Invest.* **8**, 121–129.
Bambino, T., and Hsueh, A. (1981). *Endocrinology (Baltimore)* **108**, 2142–2148.
Barnett, S. (1955). *Nature (London)* **175**, 126–128.
Baumgartner, A., Graf, K., and Kurten, I. (1985). *Biol. Psychiatry* **29**, 675–681.
Bronson, F., and Eleftheriou, B. (1964). *Gen. Comp. Endocrinol.* **4**, 9–5.
Cavanee, W., and Melnykovych, G. (1979). *J. Cell Physiol.* **98**, 199–206.
Ceulemans, D., Westenberg, H., and van Praag, H. (1985). *Psychiatry Res.* **14**, 189–196.
Chamove, A., and Bowman, R. (1976). *Folia Primatol.* **26**, 57–68.
Ching, M. (1983). *Endocrinology (Baltimore)* **112**, 2209–2214.
Chrousos, G., Renquist, D., and Brandon, D. (1982). *Proc. Natl. Acad. Sci. U.S.A.* **79**, 2036–2039.

BABOON PSYCHOENDOCRINOLOGY 467

Coe, C., Mendoza, S., and Levine, S. (1979). *Physiol. Behav.* **26**, 633–641.
Cumming, D., Quigley, M., and Yen, S. (1983). *J. Clin. Endocrinol. Metab.* **57**, 671–678.
Davis, D., and Christian, J. (1957). *Proc. Soc. Exp. Biol. Med.* **94**, 728–735.
de Waal, F. (1982). "Chimpanzee Politics." Harper Colophon, New York.
Dorovini-Zis, K., and Zis, A. (1987). *Am. J. Psychiatry* **144**, 1214–1215.
Eaton, G., and Resko, J. (1974). *Horm. Behav.* **5**, 251–263.
Ejike, C., and Schreck, C. (1980). *Trans. Am. Fisheries Soc.* **109**, 423–429.
Fox, M., and Andrews R. (1973). *Behavior* **46**, 129–136.
Frankel, A., and Ryan, E. (1981). *Biol. Reprod.* **24**, 491–496.
Gill, J., Price, V., and Friedman, M. (1985). *Am. Heart J.* **110**, 503–512.
Gofman, J., Glazier, F., Tamplin, A., Strisower, B., and De Ialla, O. (1954). *Physiol. Rev.* **34**, 589–611.
Gold, P., Goodwin, F., and Chrousos, G. (1988). *N. Engl. J. Med.* **319**, 348–353.
Gold, P., Loriaux, L., and Roy, A. (1986). *N. Engl. J. Med.* **314**, 1329–1335.
Goldberg, R., Rabin, D., Alexander, A., Doelle, G., and Goetz, G. (1985). *J. Clin. Endocrinol. Metab.* **60**, 203–211.
Goldbourty, U., and Neurfeld, H. (1986). *Arteriosclerosis* **6**, 357–363.
Goodall, J. (1986). "The Chimpanzees of Gombe" Belknap Press, Cambridge, MA.
Gordon, T., Rose, R., and Bernstein, I. (1976). *Horm. Behav.* **7**, 229–236.
Grundy, S., and Griffin, A. (1959). *J. Am. Med. Assoc.* **171**, 1794–1798.
Hamm, T., Kaplan, J., Clarkson, T., and Bullock, B. (1983). *Atherosclerosis* **48**, 221–229.
Hodson, N. (1970). *In* "The Testis," (D. Johnson, W. Gome, and N. Vandermark, eds.) Vol. 1, pp. 47–68. Academic Press, New York.
Holsboer, F., Bardeleben, U. von, and Gerken, A. (1984). *N. Engl. J. Med.* **311**, 1127–1131.
Johnson, B., Welsh, T., and Juniewicz, P. (1982). *Biol. Reprod.* **26**, 305–314.
Kalin, N., Cohen, R., and Kraemer, G. (1981). *Neuroendocrinology* **32**, 92–95.
Kaplan, J., Adams, M., Clarkson, T., and Koritnik, D. (1982a). *Arteriosclerosis* **2**, 359, 368.
Kaplan, J., Manuck, S., Clarkson, T., Lusso, F., and Taub, D. (1982b). *Arteriosclerosis* **2**, 359–368.
Keverne, E., Meller, R., and Eberhart, J. (1982) *In* "Advanced Views in Primate Biology" (V. Chiarelli and J. Corruccini, eds.), pp. 213–247. Springer-Verlag, Berlin.
Levine, S., Coe, C., and Wiener, S. (1989). *In* "Psychoendocrinology" (S. Levine and R. Brushk, eds.). pp. 181–207. Academic Press, New York.
Louch, C., and Higginbotham, M. (1967). *Gen. Comp. Endocrinol.* **8**, 441–448.
Martinsen, K., Ehnholm, C., Huttunen, J., Teruila, L., and Kistiainen, E. (1981). *Eur. J. Clin. Invest.* **11**, 351, 358.
Max, S., and Toop, F. (1983). *Endocrinology (Baltimore)* **113**, 119–126.
McGill, H. (1979). *Am. J. Clin. Nutr.* **32**, 2664–2671.
McGill, H., McMahan, C., Kruski, A., and Mott, G. (1981). *Arteriosclerosis* **1**, 3–15.
Mendoza, S., Coe, C., Lowe, D., and Levine, S. (1979). *Psychoneuroendocrinology* **3**, 221–230.
Miller, G., and Miller, N. (1975). *Lancet* **1**, 16–18.
Munck, A., Guyre, P., and Holbrook, N. (1984). *Endocr. Rev* **5**, 25–47.
Nasr, S., Rodgers, G., and Pandey, E. (1982). *Proc. Soc. Biol. Psychiatry* **37**, 68–75.
Popova, N., and Naumenko, E. (1972). *Anim. Behav.* **20**, 108–115.
Ransom, T. (1981). "Beach Troop of the Gombe." Bucknell Press, Lewisburg.
Rasmussen, D., Liu, J, Wolf, P., and Yen, S. (1983). *J. Clin. Endocrinol. Metab.* **57**, 881, 887.
Ray, J., and Sapolsky, R. (1992). *Am. J. Primatol.*, in press.
Rose, R., Holaday, J., and Bernstein, I. (1971). *Nature (London)* **231**, 366–369.

Sakai, R., Weiss, S., Blanchard, C., Blanchard, R., Spencer, R., and McEwen, B. (1991). *Soc. Neurosci. Abstr.* **17,** 621.

Sapolsky, R. (1982). *Horm. Behav.* **16,** 279–292.

Sapolsky, R. (1983a). *Am. J. Primatol.* **5,** 365–379.

Sapolsky, R. (1983b). *Endocrinology (Baltimore)* **113,** 2263–2268.

Sapolsky, R. (1985). *Endocrinology (Baltimore)* **116,** 2273–2279.

Sapolsky, R. (1986a). *Am. J. Primatol.* **11,** 217–222.

Sapolsky, R. (1986b). *Endocrinology (Baltimore)* **118,** 1630–1636.

Sapolsky, R. (1987). *In* "Psychobiology of Reproductive Behavior: An Evolutionary Perspective" (D. Crews, ed.), pp. 149–175. Prentice-Hall, Englewood Cliffs, NJ.

Sapolsky, R. (1989). *Arch. Gen. Psychiatry* **46,** 1407–1452.

Sapolsky, R. (1990a). *In* "Handbook of the Biology of Aging (E. Schneider and J. Rowe, eds.), 3rd ed., pp. 330–354. Academic Press, New York.

Sapolsky, R. (1990b). *Biol. Psychiatry* **28,** 862–881.

Sapolsky, R. (1991). *Psychoneuroendocrinology.* **16,** 281–297.

Sapolsky, R., and Altmann, J. (1991). *Biol. Psychiatry,* in press.

Sapolsky, R., and Krey, L. (1988). *J. Clin. Endocrinol. Metab.* **66,** 722–726.

Sapolsky, R., Krey, L., and McEwen, B. (1983). *Endocrinology (Baltimore)* **114,** 287–294.

Sapolsky, R., Krey, L., and McEwen, B. (1984). *Proc. Natl. Acad. Sci. U.S.A.* **81,** 6174–6178.

Sapolsky, R., and Mott, G. (1987). *Endocrinology (Baltimore)* **121,** 1605–1610.

Sapolsky, R., and Plotsky, P. (1990). *Biol. Psychiatry* **27,** 937–952.

Sapolsky, R., and Ray, J. (1989). *Am. J. Primatol.* **18,** 1–9.

Schuhr, B. (1987). *Physiol. Behav.* **40,** 689–694.

Smuts, B. (1986). "Sex and Friendship in Baboons." Aldine Press, Hawthorne.

Southwick, C., and Bland, V. (1959). *Am. J. Physiol.* **197,** 111–119.

Stern, M., Kilterman, O., Fries, J., McDevitt, H., and Reaven, G. (1973). *Arch. Intern. Med.* **132,** 97–103.

Strum, S. (1982). *Int. J. Primatol.* **3,** 175–189.

Vernikos, J., Dallman, M., Bonner, C., Katzen, A., and Shinsako, J. (1983). *Endocrinology (Baltimore)* **110,** 413–419.

Wingfield, J., and Moore, M. (1987). *In* "Psychobiology of Reproductive Behavior: An Evolutionary Perspective" (D. Crews, ed.), pp. 149–175. Prentice–Hall, Englewood Cliffs, NJ.

Weiner, H. (1991). *In* "Stress: Neurobiology and Neuroendocrinology" (M. Brown, G. Koob, and C. Rivier, eds.), pp. 23–54. Dekker, New York.

Weiss, J. (1984). *In* "Handbook of Behavioral Medicine" (W. Gentry, ed.), pp. 321–331. Guilford, New York.

Wolff, C., Friedman, S., Hofer, M., and Mason, J. (1964). *Psychosom. Med.* **26,** 576–588.

Young, E., Akana, S., and Dallman, M. (1990). *Neuroendocrinology,* **51,** 536–541.

# Short Communications

# Heterogeneous Secretory Response of Parathyroid Cells

LORRAINE A. FITZPATRICK

*Division of Endocrinology, Department of Medicine, Mayo Clinic and Mayo Foundation, Rochester, Minnesota 55905*

## I. Introduction

The parathyroid cell is unusual in that decreasing concentrations of calcium inhibit the secretion of parathyroid hormone (PTH). In turn, PTH maintains calcium homeostasis through its effects on target tissues, bone and kidney. The inverse relationship between calcium and PTH secretion is well established. Early studies in animals and in isolated cells demonstrated that hypocalcemia increased PTH secretion and hypercalcemia decreased PTH release (Pocotte *et al.*, 1991). In these studies, the observations have been made with respect to intact animals or to groups of dispersed cells. Recent work has demonstrated the heterogeneity of cell populations in many endocrine and nonendocrine cells. In order to determine the mechanism responsible for the secretion of PTH in response to differing concentrations of calcium, we established a reverse hemolytic plaque assay for quantitative measurement of parathyroid hormone from single, isolated parathyroid cells.

## II. Reverse Hemolytic Plaque Assay for PTH

The reverse hemolytic plaque assay (RHPA) was originally designed to measure release of immunoglobulins from immune cells (Jerne *et al.*, 1974). The RHPA utilizes complement-mediated cell lysis to detect antigen release from individual cells. There are many advantages to the system: secretion can be measured from a single, viable cell in a quantitative fashion, and other parameters of cellular activity such as receptor number, mRNA expression, or intracellular signals can be temporally and spatially correlated with the secretory response. The RHPA has been adapted to allow measurement of hormone release from a variety of endocrine cells. With the RHPA, there can be correlation of membrane-bound events with cell secretory activity. The assay also allows examination of secretory heterogeneity that occurs in a large number of tissues and is underesti-

472                    LORRAINE A. FITZPATRICK

mated by the traditional method of evaluating secretory activity by measurement of hormone release in cell supernatants.

We have previously validated an RHPA for PTH, established criteria for specificity and sensitivity, and tested the cell response to changes in extracellular calcium (Fitzpatrick and Leong, 1990). In our bovine RHPA, glands obtained from a local abattoir were isolated with trypsin, allowing monodispersion of parathyroid cells. This dispersion technique does not interfere with the parathyroid cell response to calcium, possibly due to the low concentrations of trypsin that are necessary for the procedure. Ovine erythrocytes were conjugated to protein A with chromium chloride. After extensive washing, the erythrocytes and parathyroid cells were mixed together and plated into a Cunningham chamber. A Cunningham chamber is a poly-L-lysine-coated microscope slide with a cover slip attached by double-sided tape. After addition of a stimulatory (or inhibitory) agent and appropriate incubation at 37°C, complement was added. The erythrocytes lyse around a secreting cell, forming a "halo" or "plaque." The size of the plaque can be determined by a segmentation routine on an image analysis system and expressed in square micrometers (Fig. 1).

## III.  Validation of the RHPA

Specificity of the RHPA for parathyroid hormone was validated by several procedures. The antibody was absorbed for 16 hours at 4°C with PTH and inhibition of plaque formation was demonstrated. Preabsorption with 50- to 100-fold excess of another peptide such as calcitonin did not prevent the formation of plaques. Omission of any of the plaque assay reagents prevented plaque formation. In addition, serial dilutions of the anti-PTH antibody reduced plaque size (Fitzpatrick and Leong, 1990).

To use the RHPA as a quantitative measure of PTH, bovine parathyroid cells were placed in the RHPA at varying calcium concentrations (0.1–2.0 m$M$). A cohort cell suspension was incubated in borosilicate glass tubes and the cell supernatant was assayed for PTH by radioimmunoassay. A standard curve comparing plaque area (in square micrometers) and PTH release (in moles per cell) was generated, suggesting that a linear correlation existed ($r = +0.91$).

## IV.  Parathyroid Cell Heterogeneity

The development of the RHPA for PTH allowed further investigation of the differences in cell responsiveness to changes in extracellular calcium. Although the classical stimulus–secretion coupling theory suggests that

PARATHYROID CELL SECRETORY RESPONSE 473

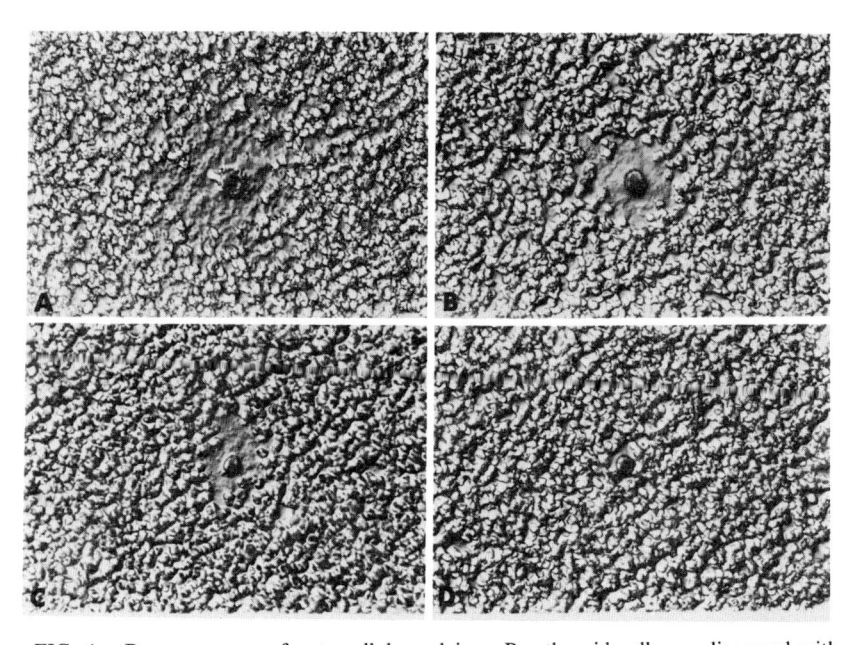

FIG. 1. Dose response of extracellular calcium. Parathyroid cells are dispersed with trypsin in the RHPA at the following extracellular calcium concentrations: (A) 0.25 m$M$; (B) 0.5 m$M$; (C) 1.0 m$M$; and (D) 2.0 m$M$. Plaque size decreases with increasing concentrations of extracellular calcium. This photomicrograph of cells fixed with formaldehyde was taken with differential interference contrast microscopy.

each cell responds to a given stimulus with the release of secretory product, a number of investigators are recognizing that a heterogeneous secretory response to a given agonist occurs. For example, Kineman and colleagues (1990) have identified a unique subpopulation of pituitary cells that does not release growth hormone under basal conditions, but does release GH after stimulation. We have found further evidence for the heterogeneity of secretory responses in isolated parathyroid cells. In response to low extracellular calcium, approximately 48% of the parathyroid cells form plaques (Table I). As the calcium concentration rises to 2.0 m$M$, only 23% of the parathyroid cells respond to the extracellular stimulus. Several technical questions arose, including cell viability. At each step during the assay, cell viability exceeded 90%. Most cells (>95%) contained stores of PTH as detected by immunocytochemistry and PTH mRNA as detected by *in situ* hybridization.

The "secretory reserve" that appears to be present in parathyroid cells may provide the parathyroid cell with the ability to respond to a second

## TABLE I
### Calcium Concentration and Plaque Formation[a]

| Calcium concentration (mM) | Cells forming plaques (%) |
|---|---|
| 0.1 | 48.5 ± 1.7 |
| 0.2 | 45.0 ± 0.9 |
| 0.4 | 35.2 ± 1.0 |
| 1.0 | 28.0 ± 0.7 |
| 1.5 | 29.1 ± 1.3 |
| 2.0 | 22.9 ± 1.8 |

[a] Bovine parathyroid cells were placed in the RHPA at the calcium concentrations indicated. The percentage of plaque-forming cells was calculated from 100 cells per slide, performed in duplicate. Data represent the mean of 5–13 experiments. As calcium concentration increased, fewer cells formed plaques.

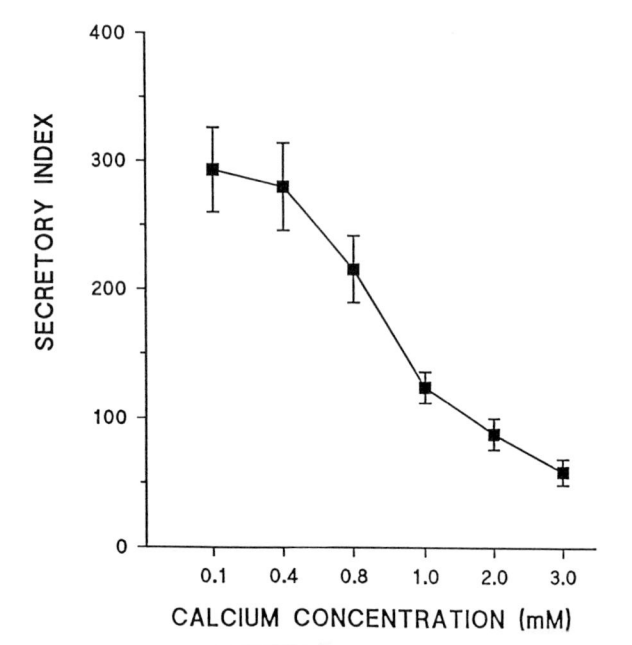

FIG. 2. Secretory index of normal human parathyroid cells. The secretory index is calculated as the product of percentage of cells forming plaques and plaque area. This reflects total secretion from a particular tissue sample. Maximum inhibition of PTH release was 80%.

or third sequential stimulus. Shannon and Roth (1974) proposed a secretory cell cycle in the parathyroid cell based on morphological findings on electron micrographs. The data presented here support the premise that not all parathyroid cells are capable of responding to a stimulus at a particular time. This secretory cycle demonstrated *in vitro* may reflect *in vivo* parathyroid responses to repeated alterations in serum calcium concentrations.

The characteristics of secretory control in abnormal human parathyroid tissue remain unknown. We have adapted the RHPA to allow measurement of parathyroid hormone release from normal human tissue (Herrera *et al.*, 1993). In normal human parathyroid tissue, the set point (point where there is half-maximal suppression of PTH) is 0.9 m$M$ extracellular calcium, which is consistent with previously published data. In normal tissue, however, secretory heterogeneity was also present, suggesting that human tissue may also behave in a "secretory cycle" (Fig. 2). Further evaluation of parathyroid adenomas and hyperplastic tissue with this assay may lend insight into the mechanisms responsible for the abnormal secretion that occurs in these disorders.

## REFERENCES

Fitzpatrick, L. A., and Leong, D. A. (1990). *Endocrinology (Baltimore)* **126,** 1720–1727.
Herrera, M. F., Grant, C. S., Maercklein, P., van Heerden, J. A., and Fitzpatrick, L. A. (1993). *Surgery,* in press.
Jerne, N. K., Henry, C., Nordin, A. A., Fuji, H., Koros, A. M. C., and Lefkovitz, I. (1974). *Transplant. Rev.* **18,** 130–191.
Kineman, R. D., Faught, W. J., and Frawley, L. S. (1990). *Endocrinology (Baltimore)* **127,** 2229–2235.
Pocotte, S. L., Ehrenstein, G., and Fitzpatrick, L. A. (1991). *Endocr. Rev.* **12,** 291–301.
Shannon, W. A., and Roth, S. I. (1974). *Am. J. Pathol.* **77,** 493–506.

# Progesterone Inhibits Estrogen-Induced Increases in c-*fos* mRNA Levels in the Uterus

JOHN L. KIRKLAND,* LATA MURTHY,* AND GEORGE M. STANCEL[†]

* Department of Pediatrics, Baylor College of Medicine, Houston, Texas 77030 and
[†] Department of Pharmacology, The University of Texas Medical School of Houston,
Houston, Texas 77225

## I. Introduction

Uterine growth and function are regulated by the interplay of estrogens and progestins. Estrogens stimulate a number of metabolic events and have a major role in regulating proliferation of the endometrium. Progestins inhibit continued proliferation stimulated by estrogen and cause differentiation of endometrium to the secretory type. This antagonism of estrogen-induced uterine growth by progesterone has been recognized for many years. The specific mechanisms responsible for this inhibition are unknown.

Several labs have identified recently a number of protooncogenes and growth factors/receptors that are stimulated in the uterus by estrogen. These include c-*fos*, c-*jun*, c-*myc*, N-*myc*, c-*ras*[Ha], EGF, the EGF receptor, IGF-1, and the IGF-1 receptor (1). These cellular oncogenes and growth factors are attractive candidates for mediators of estrogen-induced growth because they are all capable of amplifying the signal emanating from the interaction of estrogen with its receptor. Furthermore, some endometrial cancers and endometrial adenocarcinoma cell lines show amplified expression of various protooncogenes, growth factors, and growth factor receptors. We now have initiated a series of studies to determine if progesterone treatment alters the induction of these factors by estradiol, because such an effect could play a role in antagonism of estrogen-induced uterine growth. In this abstract, we demonstrate that progesterone decreases the induction of c-*fos* mRNA in the rodent uterus by estrogen.

## II. Materials and Methods

Immature female rats or mice were ovariectomized at 22 days of age. The animals were allowed to recover for 2 days, and then were administered priming doses of estradiol (40 $\mu$g/kg) on each of the next 2 days. The

477

478                    JOHN L. KIRKLAND ET AL.

animals remained untreated for the next 2 days, and then were treated
with estradiol alone (4 μg/kg) or a combination of estradiol plus progesterone. The dosage of progesterone was 2.5 mg for rats and 0.5 mg for mice.
The dosages of dexamethasone and aldosterone were 600 μg/kg, and the
dosage of dihydrotestosterone was 400 μg/kg. In all experiments, animals
were sacrificed 3 hours after estradiol administration, regardless of the
time of progesterone treatment. RNA then was prepared and analyzed for
c-*fos* mRNA levels by blot analysis using $^{32}$P-labeled antisense RNA
probes (2). The resultant films were scanned with a laser densitometer to
obtain quantitative estimates of the transcript levels. The isolation of
uterine nuclei and the measurement of occupied nuclear estrogen utilized
the basic procedures previously described (3).

## III.  Results

The time course for induction of c-*fos* mRNA in the rat uterus following
estrogen administration is very rapid, and the protooncogene transcript
levels are maximum 3 hours after treatment. Previous studies have established that this induction is dose dependent, exhibits hormonal specificity
for estrogens, and is blocked by inhibitors of RNA synthesis (2). We
first determined the effects of progesterone administration on estrogen
induction of c-*fos*. Figure 1 is a graphical representation obtained by
densitometric analysis of films from several experiments. These results
indicate that progesterone markedly decreases the estrogen-induced levels
of the c-*fos* protooncogene transcript. This effect is very rapid and can be
observed as early as 1 hour after progesterone treatment. The effect of
progesterone persists for up to 9 hours and is then reversible at 15 hours.

Other experiments indicated that progesterone inhibits estradiol-induced uterine growth (reflected by a 50% decrease in net weight of the
organ) under the dosing regimen used to obtain the *fos* mRNA data.

Our next experiments were to determine the dose–response curve for
the progesterone inhibition of c-*fos* induction. The results indicated
a typical dose–response curve, with a dose of 2.5 mg of progesterone
producing a maximum decrease of approximately 60% in the levels of
estradiol-induced c-*fos* mRNA.

We next determined the hormonal specificity of progesterone inhibition
by utilizing androgens, glucocorticoids, and mineralocorticoids. Neither
dihydrotestosterone nor aldosterone decreased the levels of c-*fos* mRNA
following estrogen treatment. However, the glucocorticoid dexamatha-
sone has an inhibitory effect that is comparable to that produced by
progesterone.

# UTERINE c-*fos* mRNA INHIBITION

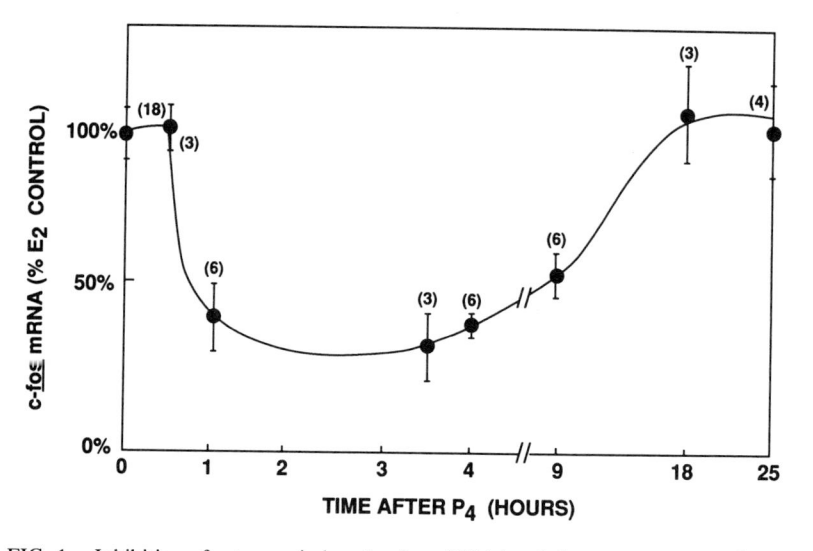

FIG. 1. Inhibition of estrogen-induced c-*fos* mRNA levels by progesterone. All animals received estradiol ($E_2$) 3 hours prior to sacrifice. Each group was treated also with progesterone ($P_4$) for the times indicated. Total RNA was prepared and subject to blot analysis to detect c-*fos* mRNA. Densitometric analysis was performed, with the maximum level of 100% representing the c-*fos* mRNA levels observed in control animals treated only with estradiol and sacrificed at 3 hours. Values represent means, with the indicated SEM and number of determinations. Each determination was performed with an RNA sample prepared from three pooled uteri.

The results in all previous experiments were seen in the rat. In order to determine if the effect was observed in other species, we utilized the same protocol in the mouse. The induction of c-*fos* mRNA in the mouse is similar to that in the rat. The inhibitory effects of progesterone on c-*fos* were present also in the mouse.

To investigate the mechanism of this effect we determined if progesterone decreased nuclear levels of occupied estrogen receptors and/or if the induction of other estrogen-responsive genes was decreased. The result showed that progesterone did not decrease occupancy of nuclear estrogen receptors nor did it block the induction of c-*myc* or c-*jun* following estradiol administration.

In a final set of experiments we investigated the effect of progesterone on the induction of c-*fos* by a nonestrogenic agent, the phorbol ester TPA. *In vivo* TPA treatment clearly induces c-*fos* mRNA levels in the uterus, and this induction is not blocked by progesterone. This result argues against a progesterone effect on the basal *fos* promoter or a steroid effect to increase the degradation of *fos* transcripts.

480                     JOHN L. KIRKLAND ET AL.

## IV. Discussion

The information presented in this abstract reveals that progesterone antagonizes the estrogen-induced increase in uterine c-*fos* mRNA. This inhibition raises the possibility that progesterone antagonism of estrogen-induced uterine growth is mediated in part by suppression of *fos* transcript levels. This effect of progesterone occurs in both the rat and mouse, and it will be interesting to determine in future work if it occurs in other species. The dosage of progesterone that elicits a decrease in estrogen-induced *fos* mRNA is in the range that produces other tissue responses to progesterone. These studies with other classes of steroids indicate that the effect of progesterone exhibits some specificity, because neither an androgen nor a mineralocorticoid inhibits estrogen induction of c-*fos* mRNA.

Our results suggest that the most likely mechanism of the observed effect is a direct inhibition of transcription at the level of the c-*fos* gene. This inhibition could occur by one of several mechanisms. First, the progesterone receptor could bind to a regulatory region of the *fos* gene and prevent or alter the binding of estrogen receptor and or other transcription factors. Alternatively, the antagonistic effect of progesterone could be mediated by way of protein–protein interactions between its receptor and transcription factors required for estrogen induction of *fos* transcription.

### ACKNOWLEDGMENTS

The authors would like to thank Dr. Salman Hyder and Dr. David Loose-Mitchell for helpful discussions and suggestions, Ms. Constance Chiappetta for technical assistance, Dr. Michael Crow for the murine c-*myc* probe, and Dr. Daniel Nathans for the murine c-*jun* probe. This work was supported by NIH Grant HD-08615, and a grant from The John P. McGovern Foundation.

### REFERENCES

1. Stancel, G. M., Chiapetta C., Gardner, R. M., Hyder, S. M., Kirkland, J. L., Lin, T. H., Lingham, R. B., Loose-Mitchell, D. S., Mukku, V. R., and Orengo, C. A. (1991). *In* "Cellular Signals Controlling Uterine Function" (L. A. Lavia, ed.), pp 49–91. Plenum, New York.
2. Loose-Mitchell, D. S., Chiappetta, C., and Stancel, G. M. (1988). *Mol. Endocrinol.* **2,** 946–951.
3. Mukku, V. R., Kirkland, J. L., Hardy, M., and Stancel, G. M. (1981). *Endocrinology (Baltimore)* **109,** 1005–1010.

# Genotoxic Damage and Aberrant Proliferation in Mouse Mammary Epithelial Cells

N. T. TELANG, A. SUTO, H. LEON BRADLOW, G. Y. WONG, AND MICHAEL P. OSBORNE

*Strang Cancer Prevention Center, New York, New York 10021*

The role of ovarian secretions in the induction of mammary and endometrial tumors has been clear for a century. Two secretory products of the ovary, estradiol and progesterone, profoundly influence hormone target tissue epithelial cell proliferation, morphogenesis, cytodifferentiated function, and neoplastic transformation (Bannerjee, 1986; Mauvais-Jarvis *et al.*, 1986; Siiteri *et al.*, 1986). The responses, mediated by intracellular receptors acting together with induced enhancer factors, modulate gene transcription, including specific oncogenes (O'Malley, 1990; Sekeris, 1991).

Because of the well-known mitogenic effects of estradiol, much attention has been focused on the metabolic pattern of estradiol metabolism with a view toward determining the specific role of these metabolites on the tumorigenic process. Prior studies have established that $16\alpha$-hydroxylation, the reaction leading to $16\alpha$-OHE$_1$, is activated in women with breast cancer (Schneider *et al.*, 1982), in women at increased risk for breast cancer (Osborne *et al.*, 1988), and in mice at high risk for mammary tumors (Bradlow *et al.*, 1985). This reaction is also elevated in murine and human mammary tissue grown in organ culture (Telang, 1986, Telang *et al.*, 1991a) and is further increased by treatment with various initiators (Telang *et al.*, 1991b). These results suggest that because $16\alpha$-hydroxylation alone is altered well before the onset of hormone-sensitive target tissues, perturbations in this pathway represent an endocrine biomarker for breast cancer risk. The product of this reaction, $16\alpha$-OHE$_1$, exhibits persistant covalent binding to the nuclear estrogen receptors (Fishman and Martucci, 1984; Swaneck and Fishman, 1988; Miyairi *et al.*, 1991), little binding to sex hormone binding globulin (SHBG), and interaction with nuclear histones and DNA (Yu and Fishman, 1985). Because of these properties, $16\alpha$-OHE$_1$ can function as a potent estrogen agonist (Lustig *et al.*, 1988, 1989). The findings in this paper demonstrate that expression of conventional biomarkers of tumorigenic transformation can be induced

481

Copyright © 1993 by Academic Press, Inc.
All rights of reproduction in any form reserved.

482    H. LEON BRADLOW ET AL.

and/or promoted in a cell culture model by 16α-OHE$_1$ but not by estradiol (E$_2$) or estriol (E$_3$).

## I.   Material and Methods

Mammary epithelial cells were obtained from 6- to 8-week-old virgin female C57BL mice. This is a mouse mammary tumor virus-negative (MMTV$^-$) strain with a <1% tumor incidence. The mammary tissue was finely minced and enzymatically digested by standard procedures (Telang et al., 1990b). The cells were cultured in DME/F12 containing fetal calf serum and other cofactors. The epithelioid colonies from the primary cultures were trypsinized in situ and expanded up to passage 10 by 1 : 10 splits. The final cell chosen was termed C57/MG CI$_5$. It was used for this study at passage 14.

### A.   ESTRADIOL METABOLISM

The effect of dimethylbenzanthracene acid (DMBA) on the relative extent of biotransformation of E$_2$ by the C-2 and C-16α pathways was assessed with a radiometric method previously used in the authors' laboratory (see Telang, 1986, Telang et al., 1991a).

### B.   EXPOSURE TO STEROIDS AND DMBA

Stock solutions of E$_2$, E$_3$, and 16α-OHE$_1$ in alcohol were diluted in culture medium to obtain a final concentration of 200 ng/ml. DMBA in dimethyl sulfoxide (DMSO) was also diluted in medium to give the same final concentration. In both cases the solvent concentration was less than 0.2% and was not cytotoxic.

### C.   DNA REPAIR ASSAY

Genotoxicity of the compounds was assayed by measuring unscheduled DNA synthesis in cultures treated with the test compounds. One-day-old cultures were treated with the compounds in a medium containing hydroxyurea and labeled thymidine. After a 24-hour incubation, the cultures were washed to remove unincorporated thymidine, lysed, and the extent of uptake determined after precipitation with trichloroacetic acid (TCA). The precipitated macromolecules were counted (Telang et al., 1984; Williams, 1990).

MOUSE MAMMARY EPITHELIA                    483

## D. CELL PROLIFERATION AND ANCHORAGE-INDEPENDENT ASSAY

The effect of the estrogens on cell proliferation was determined during a 4-day exposure of the cells to 200 ng/ml of the compounds. The cells were then trypsinized and counted. Viability was established by trypan blue exclusion studies. The ability of the various compounds to induce anchorage-independent growth in C57/MG cells was examined by measuring the formation of tridimensional colonies in 0.33% agar made up in DME/F12 medium of cells pretreated with the various compounds. After growing for 14 days the cells were fixed with formalin, stained with Giemsa stain, and the colonies were counted.

## II. Results

As shown in Fig. 1, 16$\alpha$-OHE$_1$ caused a 58% increase in unscheduled DNA repair, substantially greater than that induced by E$_2$ or E$_3$ and comparable to that observed with DMBA. Increased [$^3$H]thymidine uptake in the presence of hydroxyurea in treated cells is indicative of increased unscheduled DNA synthesis.

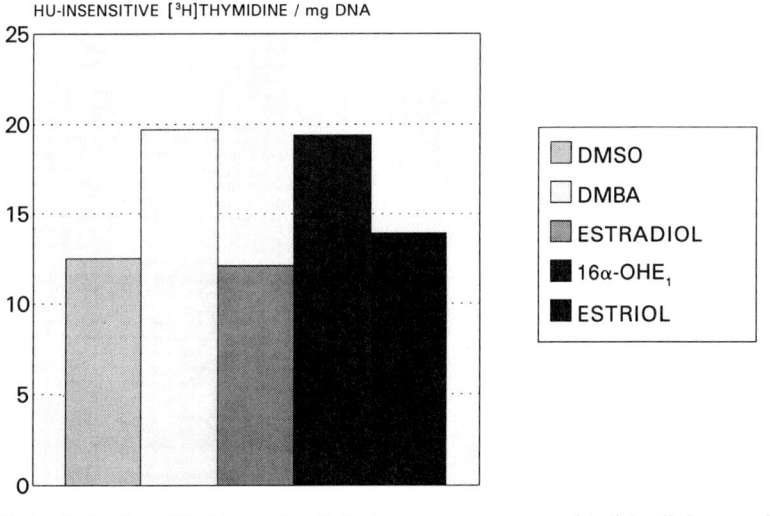

FIG. 1. Induction of DNA repair activity in mouse mammary epithelial cells by estradiol metabolites. One-day-old cultures were treated with the test compounds plus 5$\mu M$ hydroxyurea (HU) and [$^3$H]thyrmidine for 24 hours. Bound thymidine in the cellular lysates was then determined by counting the TCA precipitate fraction. DMSO vs. DMBA, DMSO vs. 16$\alpha$-OHE$_1$, $p < 0.0001$; E$_2$ or E$_3$ vs. DMSO not significant.

## A. ESTRADIOL METABOLISM

As previously observed in other systems, DMBA simultaneously decreases C-2 hydroxylation and increases 16α-hydroxylation. This shows that C57/MG cells retain their responsiveness to $E_2$ and that the metabolite ratio is altered by exposure to DMBA (Fig. 2)

## HYPERPROLIFERATION

As shown in Fig. 3, hyperproliferation was increased in DMBA (positive control) and 16α-OHE$_1$-treated cultures relative to control cultures. The increase was 68 and 24%, respectively, for the two compounds. $E_2$ was totally unresponsive and $E_3$ caused only a minimal response.

## C. ANCHORAGE-INDEPENDENT GROWTH

Exposure of mammary epithelial cells to DMBA or 16α-OHE$_1$ also induced increased formation of tridimensional colonies under anchorage-independent conditions. Figure 4 shows the relative frequency of colonies arising from cells treated with DMSO, DMBA, $E_2$, 16α-OHE$_1$, and $E_3$. DMBA caused a 117-fold increase and 16α-OHE$_1$ caused an 18-fold in-

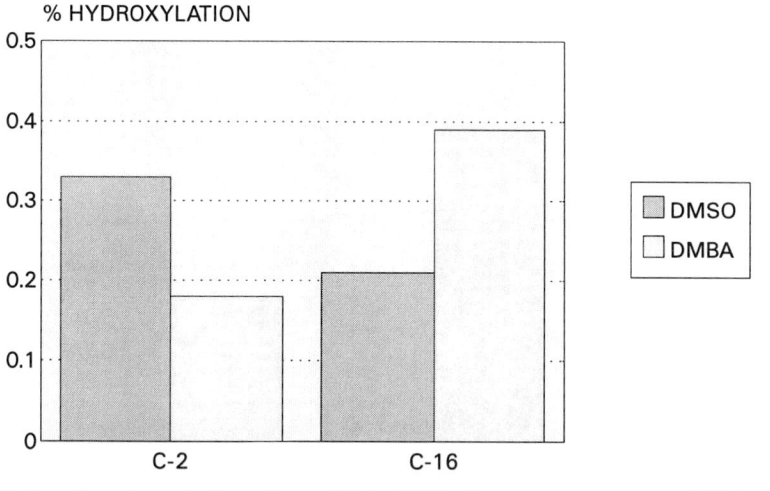

FIG. 2. Altered metabolism of estradiol by DMBA in mouse mammary epithelial cells. C57/MG cells were exposed to DMBA for 24 hours and 2-OH- or 16α-HE$_2$ was then added for 48 hours and the percent hydroxylation reaction determined radiometrically. Results were normalized per $10^4$ cells for 2-hydroxylation (DMBA vs. DMSO, $p < 0.001$) and for 16α-hydroxylation (DMBA vs. DMSO, $p = 0.0001$).

MOUSE MAMMARY EPITHELIA 485

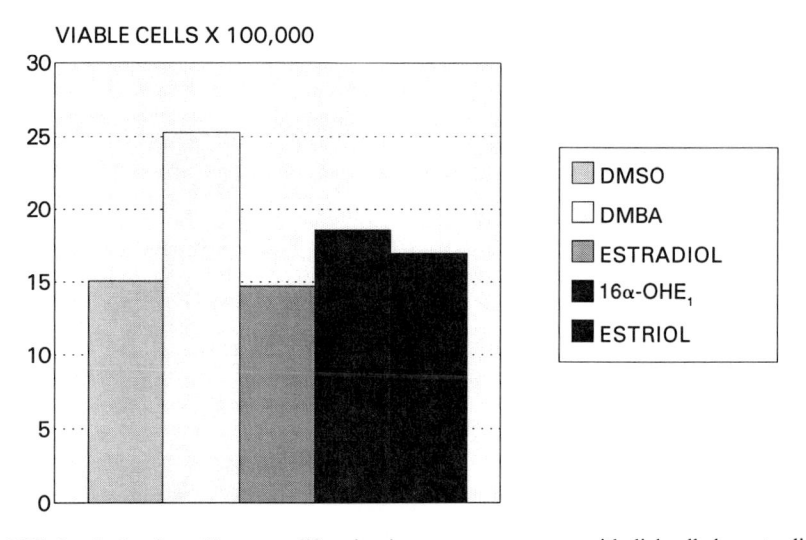

FIG. 3. Induction of hyperproliferation in mouse mammary epithelial cells by estradiol metabolites. $1.0 \times 10^4$ C57/MG cells were plated in individual wells and exposed to DMSO (solvent control), DMBA (positive control), $E_2$, $16\alpha$-$OHE_1$, or $E_3$ for 4 days. The cells were then trypsinized and counted to determine the number of trypan blue-negative cells. DMBA, $16\alpha$-$OHE_1$ vs. control, $p = 0.0001$; $E_2$ or $E_3$ vs. control, not significant.

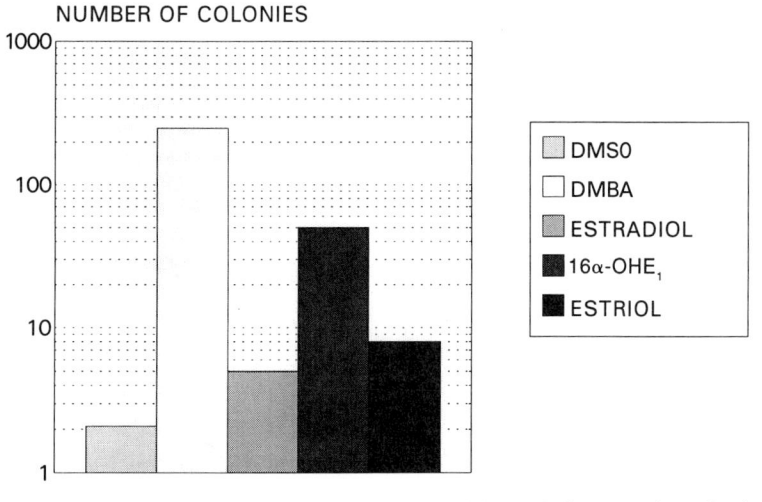

FIG. 4. Anchorage-independent growth by estradiol metabolites: number of colonies formed from C57/MG cells treated with DMSO (solvent control), DMBA (positive control), $E_2$, $16\alpha$-$OHE_1$, or $E_3$. $1.0 \times 10_4$ treated cells were then suspended in 2 ml of 0.33% agar. Colonies were counted 14 days postseeding. DMSO vs. DMBA, $p = 0.0001$; DMSO vs. $16\alpha$-$OHE_1$, $p < 0.0001$.

486                    H. LEON BRADLOW ET AL.

crease relative to that observed in control (DMSO) cells. In contrast, colony-forming ability in $E_2$- and $E_3$-treated cells was not different from control cells.

## D.  PROMOTIONAL EFFECTS OF ESTROGEN METABOLITES ON DMBA-PRETREATED C57/MG CELLS

Cells were pretreated with 200 ng/ml of DMBA and the treated cells in their second passage were challenged with estrogens for 4 days and then transferred to agar for the colony-forming assay (Fig. 5). Preinitiated cells formed $254 \pm 24$ colonies in 14 days. $16\alpha$-OHE$_1$ caused an increase to $298 \pm 46$ colonies whereas no response was observed with $E_2$ or $E_3$. No response in any of the assays was observed with the catechol estrogens.

## IV.  Discussion

Conversion to the fully tranformed tumorigenic phenotype is the end result of a series of molecular, metabolic, and cellular events leading to

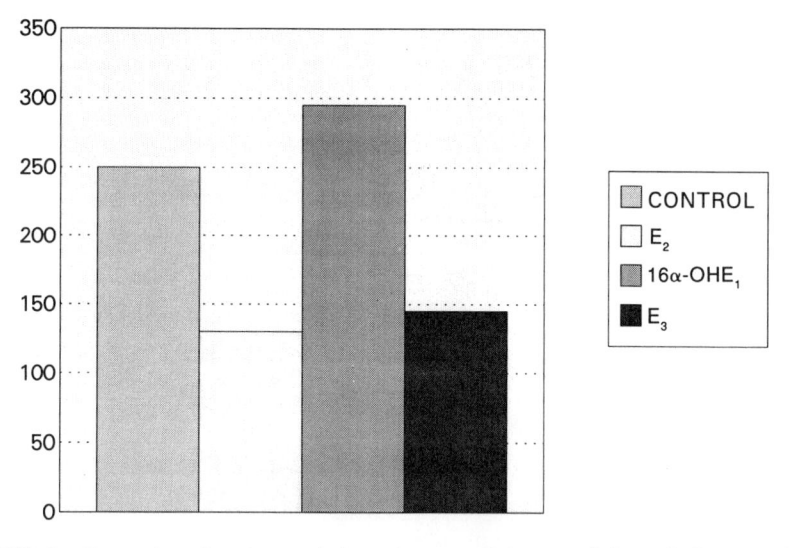

FIG. 5.  Promotion of anchorage-independent growth by estradiol metabolites: number of colonies found in C57/MG cells preinitiated with low-dose DMBA (24 hours) and exposed to $E_2$, $16\alpha$-OHE$_1$, or $E_3$, for 96 hours. $1 \times 10^4$ treated cells were suspended in 20 ml of 0.33% agar. Colonies were counted 14 days postseeding. $16\alpha$-OHE$_1$ vs. control, $p < 0.05$; $E_2$ or $E_3$ vs. control, $p < 0.001$ for a negative response.

the neoplastic state. This series of events has been reported *in vivo* as well as *in vitro* during mammary carcinogenesis (Mauvais-Jarvis *et al.*, 1986; Bannerjee *et al.*, 1986). We have previously reported *in vitro* studies demonstrating that the C-2 : C-16$\alpha$ estrogen metabolite ratio occurred in normal mammary tissue in response to treatment with chemical carcinogens or transfection with oncogenes (Telang *et al.*, 1991b). Thus, in addition to intermediate biomarkers such as deregulated expression of oncogenes, mutagenic changes, and altered proliferation (Telang *et al.*, 1991b,c), changes in estrogen metabolism may be a biomarker for target tissue susceptibility to tumorigenic changes. In this study we have utilized induction of DNA repair (the flip side of DNA damage) and persistance of aberrant proliferation as measures of whether estrogens can function as initiators or promoters of carcinogenesis in mammary epithelial cells.

Increases in DNA repair have been extensively validated as a marker for genotoxic DNA damage (Williams, 1980). Using hydroxyurea to suppress replicative DNA synthesis, we can readily measure DNA repair activity. Only 16$\alpha$-OHE$_1$ among the estrogens increased this response to levels comparable to that of DMBA. E$_2$ and E$_3$ were totally inactive. 16$\alpha$-OHE$_1$ can directly interact with cellular DNA and thereby function as an initiator. Direct evidence for this binding has been reported by J. Liehr (private communication).

Hyperproliferation of the epithelial component has been reported to occur in tissues at increased risk for tumors or following exposure to carcinogens (Lin *et al.*, 1976; Kundu *et al.*, 1978; Telang, 1986), and is implicated as an intermediate biomarker preceding transformation. Only 16$\alpha$-OHE$_1$ exhibited this response; the other estrogens were inactive. A similar correlation with the ability to induce anchorage-independent growth was also observed as well as an increase in promotional activity. A similar differential was observed in the proliferative response of MCF-7 cells to 16$\alpha$-OHE$_1$ and E$_3$, respectively (Schneider *et al.*, 1984).

The ability to acquire anchorage-independent growth has been correlated with treatment with various initiators of tumorigenic transformation (Suto *et al.*, 1991) and may be a cellular marker for tumorigenesis. Our studies clearly showed that 16$\alpha$-OHE$_1$ was much more potent than E$_2$ or E$_3$ in inducing a greater number of tridimensional soft agar colonies. Following preincubation with DMBA, 16$\alpha$-OHE$_1$ was also more potent at promotion than were the other estrogens tested.

As an endogenous estrogen metabolite, 16$\alpha$-OHE$_1$ represents the first case of a substance normally present in the body acting as an initiator and promoter. Preliminary studies have shown that inhibiting the formation of 16$\alpha$-OHE$_1$ by treatment with indole-3-carbinol decreases the incidence of tumors in susceptible mice.

## REFERENCES

Bannerjee, M. R. (1986). *In* "In Vitro Models for Cancer Research" (M. Weber and L. I. Sekely, eds.), pp. 68–114. CRC Press, Boca Raton, FL.

Bradlow, H. L., Herschcopf, R. J., Martucci, C. P., and Fishman, J. (1985). *Proc. Natl. Acad. Sci. U.S.A.* **82,** 6295–6299.

Fishman, J., and Martucci, C. P. (1980). *J. Clin. Endocrinol. Metab.* **51,** 711–15.

Kundu, A. B., Telang, N. T., and Banerjee, M. R. (1978). *J. Natl. Cancer Inst.* **61,** 465–469.

Lin, F. K., Banerjee, M. R., and Crump, L. R. (1976). *Cancer Res.* **36,** 1607–1614.

Lustig, R., Mobbs, C., Pfaff, D., and Fishman, J. (1988). *J. Ster. Biochem.* **33,** 417–425.

Lustig, R., Bradlow, H. L., McEwen, B., and Pfaff, D. (1989). *Endocrinology (Baltimore)* **125,** 2701–2709.

Mauvais-Jarvis, P., Kutten, F., and Gompel, A. (1986). *Ann. N.Y. Acad. Sci.* **464,** 152–167.

Miyairi, S., Ichikawa, T., and Nambara, T. (1991). *Steroids* **56,** 361–366.

O'Malley, B. W. (1990). *Mol. Endocrinol.* **4,** 363–369.

Osborne, M. P., Karmali, R. A., Herschkopf, H., Bradlow, H. L., Kourides, I. A., Williams, W. R., Rosen, P. P., and Fishman, J. (1988). *Cancer Invest.* **6,** 629–631.

Schneider, J., Huh, M. M., Bradlow, H. L., and Fishman, J. J. (1984). *J. Biol. Chem.* **259,** 4840–4845.

Schneider, J., Kinne, D., and Fracchia, A. (1982). *Proc. Natl. Acad. Sci. U.S.A.* **79,** 3047–3051.

Sekeris, C. E. (1991). *J. Cancer Res. Clin. Oncol.* **117,** 96–101.

Siiteri, P. K., Simberg, N., and Murai, J. (1986). *Ann. N.Y. Acad. Sci.* **464,** 100–105.

Suto, A., Bradlow, H. L., Wong, G. Y., Telang, N. T., and Osborne, M. P. (1992). *Steroids* **57,** 262–268.

Swaneck, G., and Fishman, J. (1988). *J. Proc. Natl. Acad. Sci. U.S.A.* **85,** 7831–7835.

Telang, N. T., Bockman, R. S., and Sarkar, N. H. (1984). *Carcinogenesis* **5,** 1123–1127.

Telang, N. T. (1986). "Dietary Fat and Cancer," pp. 707–728.

Telang, N. T., Bradlow, H. L., Kurihara, H., and Osborne, M. P. (1990a). *Breast Cancer Res. Treat.* **13,** 173–181.

Telang, N. T., Osborne, M. P., Swterlitsch, L., and Narayanan, R. (1990b). *Cell Regul.* **1,** 863–872.

Telang, N. T., Axelrod, M., and Wong, G. Y. (1991a). *Steroids* **56,** 37–43.

Telang, N. T., Kurihara, H. M., Wong, G. Y. (1991b). *Anticancer Res.* **11,** 1021–1028.

Telang, N. T., Narayanan, R., Bradlow, H. L., and Osborne, M. P. (1991c). *Breast Cancer Res. Treat.* **18,** 155–165.

Yu, S. C., and Fishman, J. (1985). *Biochemistry* **29,** 8017–8021.

Williams, G. M. (1980). *Ann. N.Y. Acad. Sci.* **349,** 273–282.

# Development of Hypophysiotropic Neuron Abnormalities in GH- and PRL-Deficient Dwarf Mice

CAROL J. PHELPS* AND DAVID L. HURLEY[†]

*Department of Anatomy, Tulane University School of Medicine, New Orleans, Louisiana 70112 and [†] Department of Cell and Molecular Biology, Tulane University, New Orleans, Louisiana 70118

## I. Introduction

Two independent, spontaneous mutations in mice, "Snell" (dw) (Snell, 1929) and "Ames" (df) (Schaible and Gowen, 1961), result in hypopituitary dwarfism. In the homozygous condition (dw/dw or df/df), the most notable deficiency is an absence of growth hormone (GH) and prolactin (PRL) expression. Although both GH and PRL genes are grossly intact (Phillips *et al.*, 1982; Slabaugh *et al.*, 1982), neither the proteins nor the transcripts are detectable (Cheng *et al.*, 1983; Sinha *et al.*, 1972; Slabaugh *et al.*, 1982). In the Snell dwarf, a point mutation in the pituitary transcription factor Pit-1 has been discovered (Li *et al.*, 1990).

Accompanying these pituitary deficits are abnormalities in the expression of hypothalamic hypophysiotropic factors that regulate GH and PRL. Somatostatin (SRIH) levels (Fuhrmann *et al.*, 1985; Webb *et al.*, 1985) are severely depressed in dwarf hypothalamus; the deficit is limited morphologically to neurons of the anterior periventricular hypothalamus that project to the external zone of the hypothalamic median eminence and thus to pituitary portal vasculature (Phelps and Hoffman, 1987). Similarly, PRL-inhibiting hypothalamic dopamine (DA) in tuberoinfundibular neurons of the arcuate nucleus is markedly decreased in dwarfs, as assessed by HPLC levels (Morgan *et al.*, 1981) and by direct visualization using histofluorescence (Phelps *et al.*, 1985); the DA deficit is reflected in decreased numbers of tyrosine hydroxylase (TH)-expressing tuberoinfundibular neurons (Phelps, 1987). As was found for SRIH, expression of DA and TH in CNS areas not afferent to the median eminence is unaffected in dwarfs. Conversely, expression of growth hormone-releasing hormone (GHRH), which is restricted in brain to the hypothalamic arcuate nucleus, is increased in adult dwarf mouse hypothalamus (Hurley and Phelps, 1991).

In order to assess whether these hypophysiotropic abnormalities in adult

490 CAROL J. PHELPS AND DAVID L. HURLEY

dwarfs represent a primary pleiotropic phenotype of the mutation, or, rather, a regressive condition secondary to pituitary deficits, the developmental patterns of DA/TH, SRIH, and GHRH gene expression were examined in Ames dwarf (df/df) and phenotypically normal (DF/?) littermate mice using catecholamine (CA) histofluorescence, peptide immunocytochemistry, and mRNA *in situ* hybridization.

## II. Method

Mice were selected from a colony whose original heterozygous (DF/df) breeding pairs were provided by Dr. A. Bartke (Southern Illinois University). On days 1 (day of birth), 4, 7, 14, 21, 60, and >120 (adult), deeply anesthetized (pentobarbital, 65 mg/kg body weight) animals were vascularly perfused with saline followed by formaldehyde/glutaraldehyde (FAGLU) for CA fluorescence induction (Furness *et al.*, 1978) or by 4% paraformaldehyde for fixation. Brains and pituitaries were removed and postfixed overnight. A minimum of six mice (three male, three female) of each genotype at each postnatal age was examined.

Pituitaries were paraffin embedded, sectioned at 5 $\mu$m, and stained immunocytochemically for GH (monkey anti-mouse GH, courtesy of A. F. Parlow), in order to verify the df/df or DF/? phenotype, because dwarfs and normals are not reliably distinguishable by size prior to 15 days of age. Brains were sectioned frozen in the coronal plane at 30 $\mu$m.

For CA histofluorescence, brain sections were mounted and examined using blue–violet epi-illumination (Phelps *et al.*, 1985). Immunocytochemical (ICC) staining for TH and SRIH was performed on free-floating sections incubated in primary antisera for 3 days, followed by exposure to biotinylated secondary antisera and avidin–biotin complex reagents from Vector Laboratories; anti-TH (ETI, Inc.) was used at 1 : 3000; sheep anti-SRIH$_{1-28}$ (1 : 5000) was a gift of T. Görcs (U.S.–Japan Biomedical Research Laboratories, Tulane University). *In situ* hybridization for pre-proSRIH and mGHRH was accomplished using $^{35}$S-labeled antisense RNA probes (Simmons *et al.*, 1989); SRIH cDNA was provided by R. Goodman (New England Medical Center), and mGHRH cDNA was a gift from K. Mayo (Northwestern University). Importantly, two or all three techniques were performed using alternate sections in single brains, allowing simultaneous evaluation of transmitter and message as evidence of hypophysiotropic neuron differentiation in each mouse (Phelps *et al.*, 1991).

## III. Results

### A. DA/TH

At postnatal day 7 (the youngest age examined), DA histofluorescence in PRL-inhibiting tuberoinfundibular neurons was more intense in df/df than in DF/? hypothalamus, even though median eminence (ME) terminal fluorescence was rudimentary in both normal and dwarf mice. Neuronal cell body and ME fluorescence developed further in both phenotypes through day 21 (df/df shown in Fig. 1A). Thereafter, DA fluorescence was progressively reduced in dwarfs, to the point of virtual absence at 6 months (Fig. 1B), compared with robust histofluorescence in DF/? hypothalamus. These results have been confirmed by HPLC (unpublished results).

Arcuate hypothalamic TH ICC was more robust in df/df mice, compared

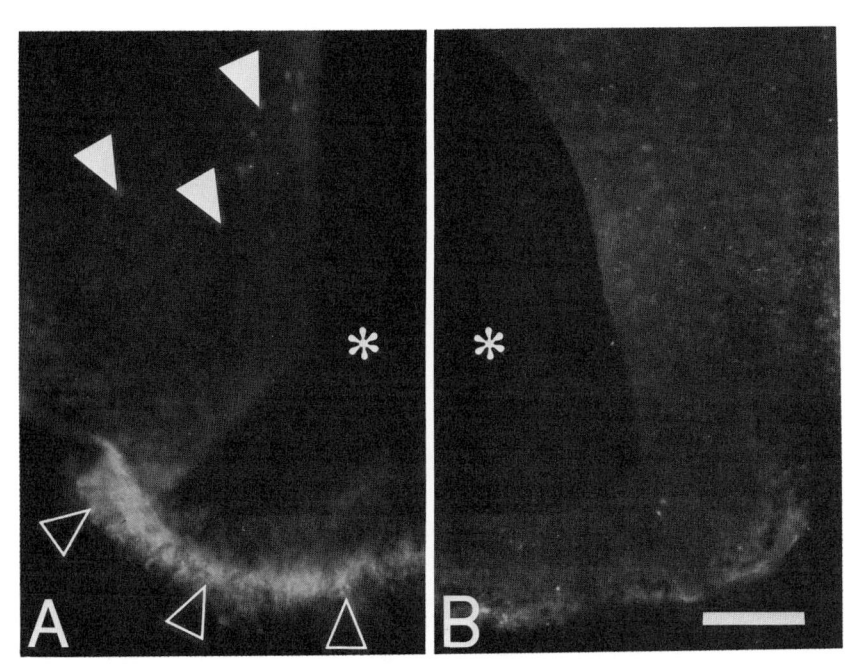

FIG. 1. DA histofluorescence in the arcuate nucleus of (A) a 21-day-old and (B) a 6-month-old Ames dwarf mouse (df/df). (A) Brightly fluorescent neuronal perikarya are indicated with solid arrowheads; fluorescence in the neuronal terminals of the external zone of the median eminence is indicated with open arrowheads. (B) Fluorescence in both areas is lacking. The third ventricle is marked with an asterisk. Bar = 100 μm; coronal sections; original magnification ×20.

492                CAROL J. PHELPS AND DAVID L. HURLEY

with TH in DF/? mice, as early as day 7, and continued to increase in
intensity through day 21 (Fig. 2A), when ME TH terminal ICC as well as
perikaryal staining was distinct. Thereafter, hypothalamic TH ICC staining decreased in intensity in df/df mice and DF/? adults showed no
further TH development (Fig. 2B), such that arcuate nucleus TH-positive
cell numbers were reduced in dwarfs to 40% of numbers in age-matched
normals and ME TH was negligible. DA and TH in brain areas not afferent
to ME (zona incerta, substantia nigra) were comparable in perikaryal
intensity and numbers for DF/? and df/df.

## B.   SRIH

At 4 days of age, the SRIH mRNA signal in ME-afferent neurons of the
anterior periventricular hypothalamus (periventricular nucleus; PeN) was
strong in dwarfs (Fig. 3A) as well as in normals. SRIH-immunostained

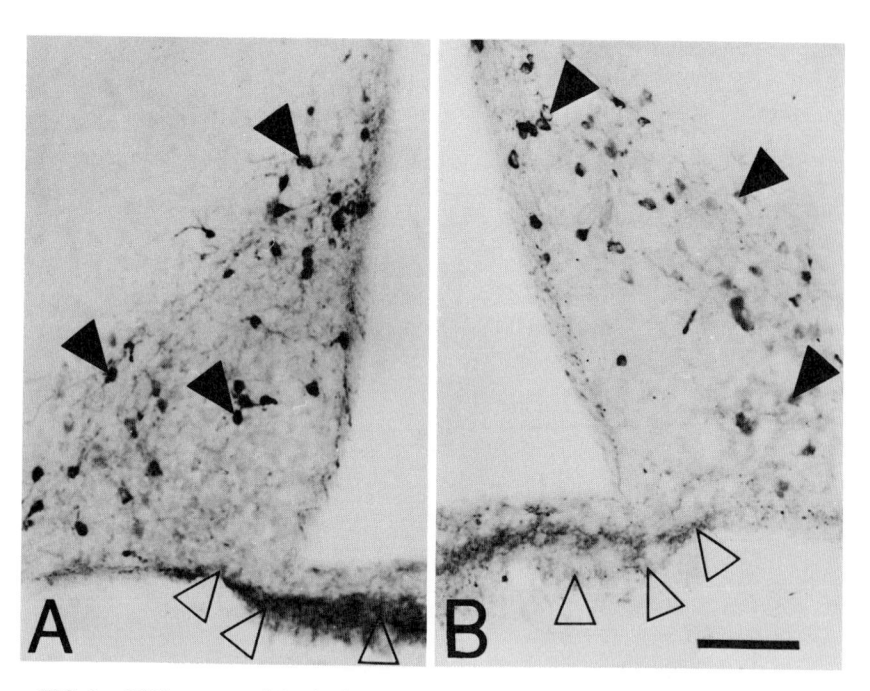

FIG. 2.   TH immunostaining in the arcuate nucleus of (A) a 21-day-old and (B) a 2-month-
old Ames dwarf mouse (df/df). Robust cellular (solid arrowheads) and terminal staining
(open arrowheads) are present at 21 days (A). At 2 months of age, the intensity of staining
is reduced, especially in the median eminence (B). Bar = 100 μm; coronal sections; original
magnification ×20.

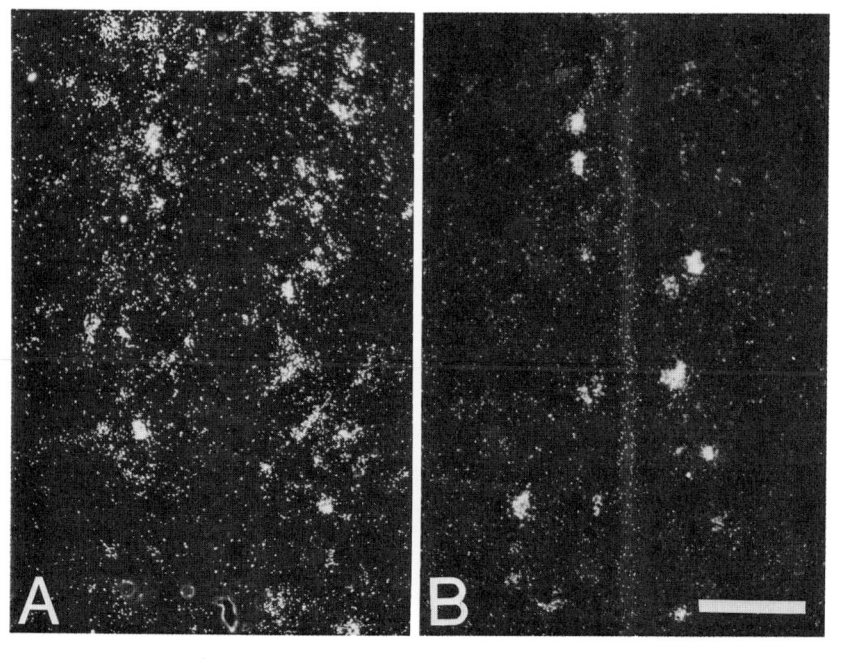

FIG. 3. SRIH mRNA expression detected by *in situ* hybridization in the anterior periventricular hypothalamus of Ames dwarf mice (df/df) at (A) 4 days and (B) 21 days of age. Reduced silver grains over cells expressing SRIH mRNA are visible as bright dots on these darkfield micrographs. There is a noticeable reduction in the number of SRIH mRNA-expressing cells at 21 days relative to that at 4 days. Bar = 100 μm; coronal sections; original magnification ×20.

neurons at this age were fewer in number than indicated by mRNA expression in both df/df and DF/? mice. At day 14, the SRIH neuron number in the PeN was reduced in dwarfs relative to normals, as assessed by ICC staining. By day 21, reduction in both SRIH mRNA and SRIH immunostaining was marked in df/df brain (Fig. 3B).

## C. GHRH

In 7-day-old mouse basal hypothalamus, the GHRH *in situ* mRNA signal was strong. A similar intensity for df/df versus DF/? continued through day 21. Thereafter, the GHRH message signal decreased in normal mouse brain, but was maintained at a level comparable to the early postnatal pattern in dwarfs. Adult normal and dwarf hypothalamic GHRH mRNA *in situ* hybridization patterns are illustrated in Fig. 4.

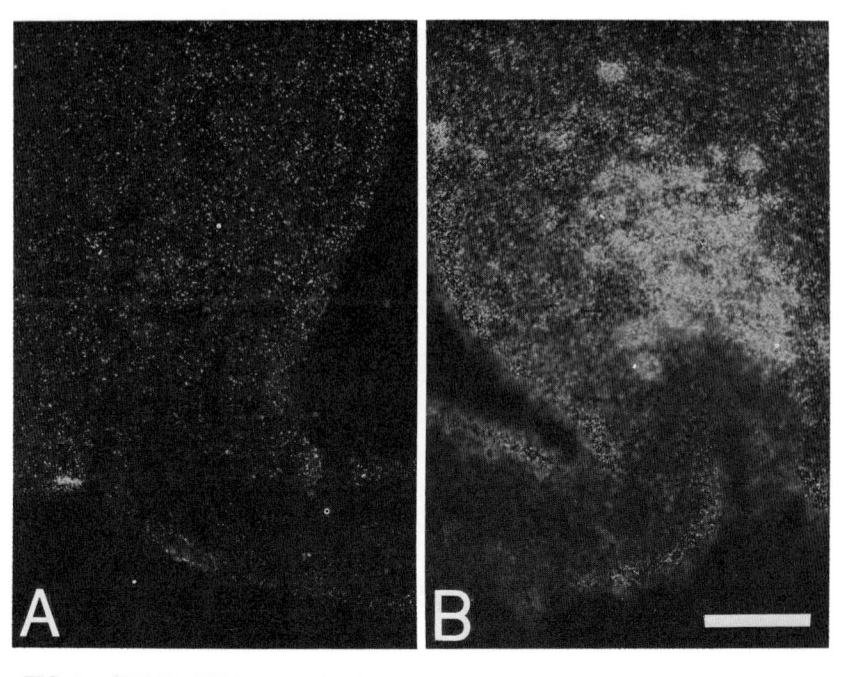

FIG. 4.   GHRH mRNA expression detected by *in situ* hybridization in the arcuate nucleus of adult (A) normal (DF/?) and (B) Ames dwarf (df/df) mice. Reduction in signal from that observed in young animals is seen in the normal. Bar = 100 $\mu$m; coronal sections; original magnification ×20.

## IV.  Discussion

These observations indicate that alterations in both mRNA and peptide expression of hypothalamic hypophysiotropic factors in dwarf mice are a gradual developmental response to the primary absence of pituitary PRL and GH.

Although numerous previous studies on the pituitary and peripheral endocrinology of dwarf mutant mice have concluded that the essential defect is that somatotrophs and lactotrophs fail to differentiate, consideration of hypothalamic abnormalities as a possible primary cause has been unexplored. An exception to this view was a study by Carsner and Rennels (1960), wherein pituitaries from Snell dwarfs, when transplanted to the sella turcica of young normal mice, failed to promote growth, whereas normal donor pituitaries promoted growth when transplanted into dwarfs. Fuhrmann and colleagues (1985) proposed that early developmental excess of SRIH might account for GH deficit in Snell dwarfs, but treatment of normal mice with SRIH (O'Hara *et al.*, 1988) failed to support the hypothesis. The elucidation of a pituitary-specific transcription factor mutation in the Snell dwarf (Li *et al.*, 1990) confirms the primary pituitary mutation site in that strain.

The initial normal differential progress of hypophysiotropic DA, SRIH, and GHRH neurons, followed by eventual regression, suggests that the hypothalamus of dwarf mice is genetically normal. The present studies thus may be viewed as a further confirmation of the positive feedback of pituitary hormones on respective hypothalamic inhibitory factors (SRIH and DA), and extend the findings of others on the results of acute hypophysectomy on SRIH reduction (Rogers *et al.*, 1988) and GHRH increase (Merchenthaler and Arimura, 1985; Katakami *et al.*, 1987): chronic developmental absence of PRL and GII results in even more severe changes in DA, SRIH, and GHRH neuronal gene expression.

The effect of GH and/or PRL replacement treatment is an obvious next question. Attempts to stimulate hypothalamic DA neurons by treatment with PRL in adult dwarf mice has led to only meager restoration in DA (Morgan *et al.*, 1981) and TH (Morgan and Besch, 1990). Morgan's results suggest that the adult dwarf hypothalamus has become refractory to pituitary hormone feedback. A purpose of the present studies was identification of critical developmental periods at which target hormone feedback might be effective; these studies are underway.

## ACKNOWLEDGMENTS

The studies were supported financially by NIH Grant NS25987. Gifts of antisera from Dr. A. F. Parlow and Dr. T. Görcs, and of cDNAs from Dr. R. Goodman and Dr. K. Mayo, are gratefully acknowledged. Invaluable technical assistance was contributed by Susan Carlson, Melanie Kratzman, and Myra Vaccarella. Dr. Andrzej Bartke, who contributed the original Ames mice, continues to be a valuable resource.

## REFERENCES

Carsner, R. L., and Rennels, E. G. (1960). *Science* **131,** 829.

Cheng, T. C., Beamer, W. G., Phillips, J. A., III, Bartke, A., Mallonee, R. L., and Dowling, C. (1983). *Endocrinology (Baltimore)* **113,** 1669–1678.

Fuhrmann, G., Scala-Guenot, D. D., and Ebel, A. (1985). *Brain Res.* **328,** 161–164.

Furness, J. B., Heath, J. W., and Costa, M. (1978). *Histochemistry* **57,** 285–295.

Hurley, D. L., and Phelps, C. J. (1991). *Endocrinology (Baltimore)* **128**(Suppl.), 1271. [abstract]

Katakami, H., Downs, T. R., and Frohman, L. A. (1987). *Endocrinology (Baltimore)* **120,** 1079–1082.

Li, S., Crenshaw, E. B., III, Simmons, D. M., Swanson, L. A., and Rosenfeld, M. G. (1990). *Nature (London)* **347,** 528–533.

Merchenthaler, I., and Arimura, A. (1985). *Peptides* **6,** 865–867.

Morgan, W. W., Bartke, A., and Pfiel, K. (1981). *Endocrinology (Baltimore)* **109,** 2069–2075.

Morgan, W. W., and Besch, K. (1990). *Neuroendocrinology* **52,** 70–74.

O'Hara, B. F., Bendotti, C., Reeves, R. H., Oster-Granite, M. L., and Gearhart, J. D. (1988). *Mol. Brain Res.* **4,** 283–292.

Phelps, C. J. (1987). *Brain Res.* **416,** 354–358.

Phelps, C. J., Carlson, S. W., and Hurley, D. L. (1991). *Anat. Rec.* **231,** 446–456.

Phelps, C. J., and Hoffman, G. E. (1987). *Peptides* **8,** 1127–1133.

Phelps, C. J., Sladek, J. R., Jr., Morgan, W. W., and Bartke, A. (1985). *Cell. Tissue Res.* **240,** 19–25.

Phillips, J. A., III, Beamer, W. G., and Bartke, A. (1982). *J. Endocrinol.* **92,** 405–407.

Rogers, K. V., Vician, L., Steiner, R. A., and Clifton, D. K. (1988). *Endocrinology (Baltimore)* **122,** 586–591.

Schaible, R., and Gowen, J. W. (1961). *Genetics* **46,** 896–899.

Simmons, D. M., Arriza, J. L., and Swanson, L. W. (1989). *J. Histotechnol.* **12,** 169–181.

Sinha, Y. N., Selby, F. W., Lewis, D. J., and VanderLaan, W. P. (1972). *Endocrinology (Baltimore)* **91,** 1045–1053.

Slabaugh, M. B., Hoffman, L. M., Lieberman, M. E., Rutledge, J. J., and Gorski, J. (1982). *Mol. Cell. Endocrinol.* **28,** 289–297.

Snell, G. W. (1929). *Proc. Natl. Acad. Sci. U.S.A.* **15,** 733–734.

Webb, S. M., Lewinski, A. K., Steger, R. W., Reiter, R. J., and Bartke, A. (1985). *Life Sci.* **36,** 1239–1245.

# Germ Cell Factor(s) Regulates Opioid Gene Expression in Sertoli Cells

MASATO FUJISAWA, C. WAYNE BARDIN, AND PATRICIA L. MORRIS

*The Population Council, New York, New York 10021*

## I. Studies

Preproenkephalin (PPenk) is one of the three opioid precursors in the testis (Kilpatrick and Rosenthal, 1986). In the seminiferous tubule, PPenk mRNA (1450–1500 bases) and Met-enkephalin immunoreactive peptides have been shown to be present in Sertoli cells (Yoshikawa and Aizawa, 1988a; Kew and Kilpatrick, 1989). In both pachytene spermatocytes (PSs) and round spermatids (RSds), a larger mRNA (1750–1900 bases) has been observed (Kilpatrick and Millette, 1986; Yoshikawa and Aizawa, 1988b); polysome analysis indicates that this germ cell-specific mRNA is not efficiently translated (Garrett *et al.*, 1990). Although the function of PPenk gene expression in Sertoli cells is unknown, PPenk-derived peptides may play an important role in the testis as local paracrine regulators, similar to the role of the POMC-derived peptides (Boitani *et al.*, 1986; Gerendai *et al.*, 1986; Morris *et al.*, 1987). The expression of Sertoli cell PPenk mRNA is known to be regulated by follicle-stimulating hormone (FSH) and cAMP (Kilpatrick and Millette, 1986; Yoshikawa and Aizawa, 1988a). In this study, we sought to determine whether the cell-to-cell interactions within the tubule, as well as the secreted proteins, regulate PPenk gene expression in Sertoli cells.

Immunocytochemistry with an antiserum raised against Met-enkephalin was used to detect the presence of preproenkephalin-derived peptide in control Sertoli cells cultured in serum-free medium for 4 days (Fig. 1A). An increase in fluorescence was observed when the cAMP stimulator forskolin (50 $\mu M$) was added to the cells for 2 hours (Fig. 1B).

Next, we determined the expression of PPenk gene in Sertoli cells after coculture with elutriator-purified RSds and PSs. Sertoli cells were cocultured with either PSs or RSds for up to 24 hours. Following the removal of germ cells by hypotonic treatment and three rinses with phosphate-buffered saline, total RNA was isolated from Sertoli cells. Northern blotting with PPenk cRNA showed a single hybridization band

497

Copyright © 1993 by Academic Press, Inc.
All rights of reproduction in any form reserved.

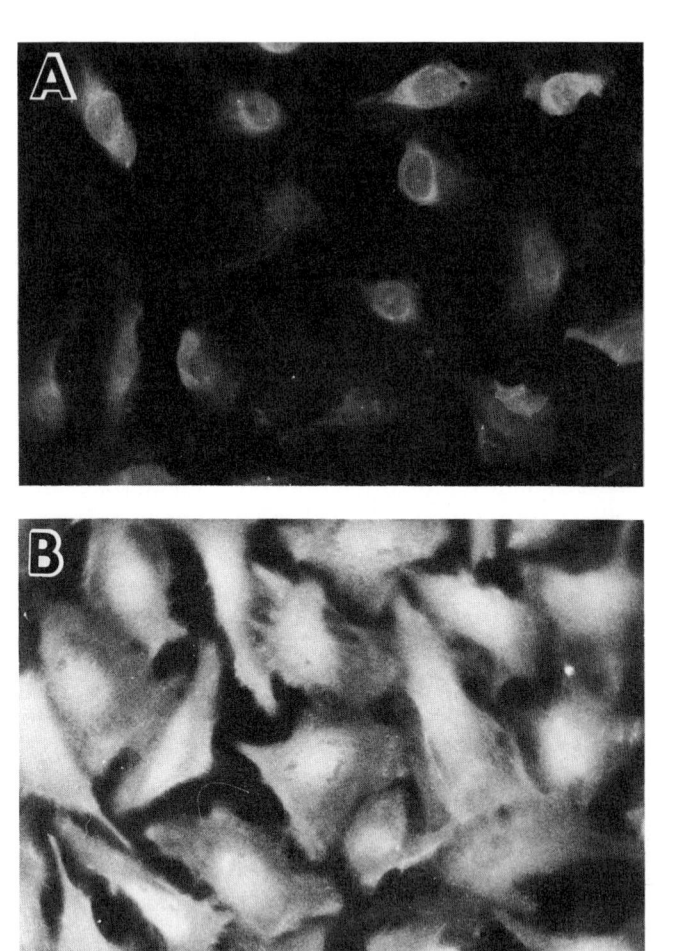

FIG. 1. Immunodetection of PPenk-derived peptides in Sertoli cells; basal (A) and forskolin-stimulated (B) cells. Sertoli cells were cultured for 4 days at a density of $2.25 \times 10^5$ cells per 1.78 cm$^2$ glass chamber area.

of 1450 bases corresponding to that previously reported for Sertoli cells (Yoshikawa and Aizawa, 1988a; Kew and Kilpatrick, 1989). No hybridization band was observed at 1750–1900 kb, attesting to the lack of any detectable contamination with germ cell mRNA. No differences were observed between the levels of PPenk mRNA from Sertoli cells only (control) and those cocultured with either PSs or RSds for 12 hours. However, after the Sertoli cells were cocultured with germ cells for 24 hours, Northern blot analysis showed increases in Sertoli cell PPenk mRNA accumulation. Similar results were found in both duplicates in the three Sertoli cell preparations so treated. Representative densitometric analyses showed 6.4- and 1.9-fold increases following PS and RSd coculture, respectively.

We then determined the effect of germ cell-conditioned media on PPenk gene expression. PPenk gene expression by Sertoli cells was examined at various points up to 24 hours after adding germ cell-conditioned media. Sertoli cell PPenk mRNA levels increased twofold at 2 hours after exposure to PS-conditioned media. Maximal stimulation of 3.5 times that of the control was observed at 12 hours; the PPenk mRNA levels gradually decreased with time but were still elevated above control at 24 hours. Similarly, media from RSds induced a twofold increase in PPenk gene expression by 2 hours and a 7.6-fold increase at 12 hours. Control media had no effect on Sertoli cell PPenk gene expression. A representative densitometric analysis is shown in Fig. 2.

Three fractions (<10, 10–30, and >30 kDa) of proteins in conditioned media were obtained by ultrafiltration and added to Sertoli cells to assess their effects on PPenk mRNA levels. In two of three experiments, the fraction containing PS proteins >30 kDa stimulated higher levels of

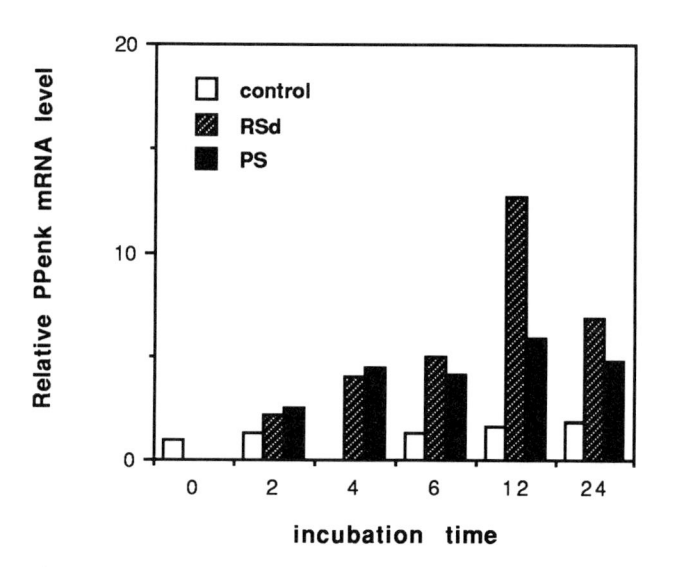

FIG. 2. Time course of PPenk gene expression following exposure to PS- or RSd-conditioned medium as indicated. A representative densitometric analysis is shown. The relative level of PPenk mRNA was obtained after dividing by the total amount of RNA present (using $\beta$-actin autoradiograms). The ratio PPenk mRNA:$\beta$-actin mRNA at time = 0 is arbitrarily assigned as 1.

## TABLE I
### Effect of Germ Cell-Conditioned Media on Sertoli Cell cAMP Levels[a]

| Time (hours) | Treatment | | |
|:---:|:---:|:---:|:---:|
| | Control | PS | RSd |
| 2 | 16.7 ± 1.7 | 17.8 ± 4.2 | 30.7 ± 4.4[b] |
| 6 | 18.1 ± 1.2 | 20.5 ± 1.5 | 35.2 ± 1.6[c] |
| 12 | 18.8 ± 1.1 | 16.9 ± 7.0 | 29.0 ± 1.6[d] |

[a] pmol/ml ± SD.
[b] $p < 0.05$.
[c] $p < 0.001$.
[d] $p < 0.005$.

PPenk mRNA compared with the other two fractions obtained from PS-conditioned media. In contrast, in all experiments RSd-conditioned media containing proteins >30 kDa increased PPenk expression compared with the other fractions.

We next determined the effect of germ cell-conditioned media on levels of cAMP in Sertoli cell cultures. Lactate- and pyruvate-supplemented media-blank-treated Sertoli cells were used as controls for the germ cell-conditioned media-treated Sertoli cell cultures. At 2, 6, and 12 hours after the addition of nonconcentrated, nonfractionated RSd-conditioned media, small but significant increases (1.8- to 2.0-fold; $p < 0.05-0.001$) in non-FSH-stimulated extracellular cAMP levels were observed (Table I).

Because the increases in cAMP were one-tenth that observed following stimulation with FSH, and the stimulation of PPenk was greater than that observed with FSH, we postulated that a second transduction system may be involved. Therefore, the expression of c-*fos* mRNA by Sertoli cells following treatment with germ cell-conditioned media was determined.

Sertoli cells were cultured for 3 days. Conditioned medium from PS ($1 \times 10^7$ PS cells, 20- to 24-hour germ cell only culture, 1 dose-equivalent) was added to the Sertoli cell monolayers for 24 hours. On day 4, the Sertoli cells were treated with 100 ng/ml FSH for 2 hours. Total RNA was prepared from Sertoli cells. Northern blots were performed using 30 $\mu$g/lane. Results shown in Fig. 3 are as follows: *No FSH treatment:* Lane 1, control medium; lane 2, PS-conditioned medium, 1 dose-equivalent; lane 3, PS-conditioned medium, 0.5 dose-equivalent. *FSH-stimulated:* Lane 4, control medium; lane 5, PS-conditioned medium, 1 dose-equivalent; lane 6, PS-conditioned medium, 0.5 dose-equivalent.

Treatment with PS-conditioned media had no effect on c-*fos* mRNA levels in Sertoli cells, nor did its addition change the response to FSH stimulation (Fig. 3). No stimulation of c-*fos* mRNA was observed follow-

# GERM CELL FACTOR GENE REGULATION 501

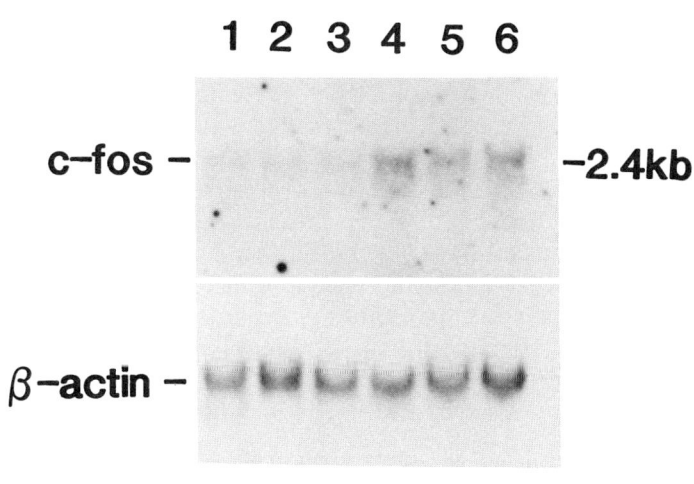

FIG. 3. Expression of c-*fos* mRNA in Sertoli cells following a 2-hour incubation without 100 ng/ml FSH (lanes 1, 2, and 3) or with concomitant FSH and PS-conditioned medium (lanes 4, 5, and 6). Total RNA, 30 μg/lane.

ing the exposure of Sertoli cells to either germ cells or their conditioned media for intervals from 1 to 12 hours. Consistent with other studies, the Sertoli cells did show twofold increases in c-*fos* mRNA levels when stimulated by FSH (1 μg/ml) for 1 hour.

## II. Discussion

There are many potential paracrine and autocrine factors mediating interactions between different testicular cells (Skinner and Griswold, 1980; Jutte *et al.*, 1982; Holmes *et al.*, 1983; Galdieri *et al.*, 1984; Morris *et al.*, 1987; Djakiew and Dym, 1988). Several investigations of cell-to-cell interactions within the tubule indicate that inhibin and transferrin gene expression in Sertoli cells are under the effect of paracrine factors potentially of germ cell origin (Pineau *et al.*, 1990; Stallard and Griswold, 1990). Here, we demonstrate that both PSs and RSds stimulate PPenk mRNA accumulation in Sertoli cells.

Germ cells are likely to produce factors that affect the maturation of Sertoli cells or the acquisition of specific functions (Galdieri *et al.*, 1984; Le Magueresse *et al.*, 1986; Le Magueresse and Jegou, 1988). After the appearance of PSs and RSds in the seminiferous tubules, cell-to-cell contacts and germ cell secretory products may affect the onset and increase of PPenk gene expression in Sertoli cells. In turn, Sertoli cells may regulate spermatogenesis through paracrine signaling factors, many of

502                    MASATO FUJISAWA ET AL.

which may be responsive to gonadotropin. Because Sertoli cells exhibit their greatest sensitivity to FSH during testicular maturation (Steinberger *et al.*, 1978), it may be that major regulatory changes in Sertoli cell expression of the PPenk gene occur in accordance with both FSH and specific germ cells during this period.

An initial estimation of the molecular weight of the factor(s) that regulates Sertoli PPenk gene expression is presented. For both PSs and RSDs, the factor(s) was greater than 30 kDa. Further studies will be required to establish the identity of this opioid-regulating factor(s). Studies from other laboratories indicated that the factor in germ cell-conditioned media that stimulated androgen-binding protein secretion and transferrin gene expression were >10 and 10–30 kDa, respectively (Le Magueresse and Jegou, 1988; Stallard and Griswold, 1990). The identity of this factor(s) is unknown.

In conclusion, we have demonstrated that either germ cells or a factor(s) in the germ cell-conditioned media stimulates the expression of PPenk gene in Sertoli cells, although the mechanism of this regulatory effect is yet to be determined. These data are consistent with a role for a paracrine factor(s) within the seminiferous tubule that regulates expression of this testicular opioid gene. In addition, such regulation appears to be dependent on specific germ cells.

### ACKNOWLEDGMENTS

These studies were supported in part by NIH Grant HD-13541. Fellowship support for Dr. Fujisawa was generously provided by The Andrew W. Mellon Foundation. The authors express their appreciation for excellent technical assistance by Lyann R. Hodgskin and Florence J.Kaczorowski.

### REFERENCES

Boitani, C., Mather, J. P., and Bardin, C. W. (1986). *Endocrinology (Baltimore)* **118**, 1513–1518.
Djakiew, D., and Dym, M. (1988). *Biol. Reprod.* **39**, 1193–1205.
Galdieri, M., Monaco, L., and Stefanini, M. (1984). *J. Androl.* **5**, 409–415.
Garrett, J. E., Collard, M. W., and Douglass, J. O. (1990). *Mol. Cell Biol.* **9**, 4381–4389.
Gerendai, I., Shaha, C., Gunsalus, G. L., and Bardin, C. W. (1986). *Endocrinology (Baltimore)* **118**, 2039–2044.
Holmes, S. D., Bucci, L. R., Lipshultz, L. I., and Smith, R. G. (1983). *Endocrinology (Baltimore)* **113**, 1916–1918.
Jutte, N. H. P. M., Jansen, R., Grootegoed, J. A., Rommerts, F. F. G., Clausen, O. P. F., and Van Der Molen, H. J. (1982). *J. Reprod. Fertil.* **65**, 431–438.
Kew, D., and Kilpatrick, D. L. (1989). *Mol. Endocrinol.* **3**, 179–184.
Kilpatrick, D. L., and Millette, C. F. (1986). *Proc. Natl. Acad. Sci. U.S.A.* **83**, 5015–5018.
Kilpatrick, D. L., and Rosenthal, J. L. (1986). *Endocrinology (Baltimore)* **119**, 370–374.

Le Magueresse, B., Le Gac, F., Loir, M., and Jegou, B. (1986). *J. Reprod. Fertil.* **77,** 489–498.

Le Magueresse, B., and Jegou, B. (1988). *Mol. Cell. Endocrinol.* **58,** 65–72.

Morris, P. L., Vale, W. W., and Bardin, C. W. (1987). *Biochem. Biophys. Res. Commun.* **148,** 1513–1519.

Pineau, C., Sharpe, R. M., Saunders, P. T. K., Gerard, N., and Jegou, B. (1990). *Mol. Cell. Endocrinol.* **72,** 13–22.

Skinner, M. K., and Griswold, M. D. (1980). *J. Biol. Chem.* **225,** 9923–9925.

Stallard, B. J., and Griswold, M. D. (1990). *Mol. Endocrinol.* **4,** 393–401.

Steinberger, A., Hintz, M., and Heidel, J. J. (1978). *Biol. Reprod.* **19,** 566–572.

Yoshikawa, K., and Aizawa, T. (1988a). *Fed. Eur. Biochem. Soc. Lett.* **237,** 183–186.

Yoshikawa, K., and Aizawa, T. (1988b). *Biochem. Biophys. Res. Commun.* **151,** 664–671.

# LHRH- and (Hydroxyproline⁹) LHRH-Stimulated hCG Secretion from Perifused First-Trimester Placental Cells

W. David Currie,* Gillian L. Steele*, Basil Ho Yuen*, Claude Kordon,† J.-Pierre Gautron,† and Peter C. K. Leung*

*Obstetrics and Gynecology, Grace Hospital, University of British Columbia, Vancouver, British Columbia V6H 3V5, Canada and †Unité de Dynamique des Systèmes Neuroendocrines (U159) de l'INSERM, Centre Paul Broca, 75014 Paris, France*

## I. Introduction

Synthetic mammalian luteinizing hormone-releasing hormone (LHRH) stimulates human chorionic gonadotropin (hCG) secretion from the placenta in a dose-dependent manner (Khodr and Siler-Khodr, 1978; Siler-Khodr and Khodr, 1981). Most LHRH-like material in the maternal circulation may be synthesized and secreted from cytotrophoblasts to stimulate hCG synthesis and secretion from syncytiotrophoblasts (Petraglia *et al.*, 1990). However, the fetal hypothalamus is also a potential source of LHRH or LHRH-like material. LHRH-like immunoreactivity is detectable in human fetal hypothalamic extracts by $5\frac{1}{2}$ weeks of gestation (Jewelewicz and Shortle, 1987).

Multiple forms of LHRH have been isolated from the human placenta and hypothalamus. (Hydroxyproline⁹)LHRH [(Hyp⁹)LHRH] is present in human, sheep, rat, and frog hypothalamus and is the major LHRH moiety in fetal rat hypothalamus (Gautron *et al.*, 1991). (Hyp⁹)LHRH stimulates LH secretion from rat pituitary (Gautron *et al.*, 1992). This study tested (Hyp⁹)LHRH as a potential hCG secretagogue. The ability and specificity with which LHRH and (Hyp⁹)LHRH stimulated hCG secretion were examined using physically dissociated and perifused first-trimester human placental cells.

## II. Materials and Methods

The protocol was approved by the Clinical Screening Committee for Research (University of British Columbia). Tissues from therapeutic procedures at 8–12 weeks of gestation were physically dissociated by passage through a 150-$\mu$m pore size screen (Sigma) and were filtered with a 48-$\mu$m pore-size cloth (Nitex) in M199 (1% FCS, 100 U penicillin G and 100 $\mu$g streptomycin/ml; 25 m$M$ NaHCO$_3$; 15 m$M$ HEPES, Gibco). Cell pellets (1000$g$/6 minutes) were resuspended and RBCs were removed (1700$g$/20 minutes) on Percoll in HBSS (40%, Gibco). Viability was >85%. Eight cell preparations from three to four placentas each were plated on beads (Cytodex-3, Sigma) for 4 days (37°C, humidified air, 5% CO$_2$). Cells were loaded into chambers (1.5 × 10$^6$ cells/1.5-ml chamber, Endotronics, Minneapolis, Minnesota) in a 36°C bath 24 hours before trials.

Perifusion was with supplemented M199 (15 ml/hour) over 2 hours with treatments over the first 5 minutes of the second hour. Effluent was collected in 5-minute fractions. Treatments were (1) LHRH (10$^{-9}$ $M$, Sigma, $n$ = 15 chambers), (2) LHRH + (Nal-Glu)LHRH (equimolar, 10$^{-9}$ $M$, $n$ = 5), (3) synthetic (Hyp$^9$)LHRH (10$^{-9}$ $M$, 99% pure, Neosystem Laboratories, Strasbourg, $n$ = 8), (4) (Hyp$^9$)LHRH + (Nal-Glu)LHRH (equimolar, 10$^{-9}$ $M$, $n$ = 5), and (5) (Nal-Glu)LHRH (10$^{-9}$ $M$, Salk Institute, San Diego, California, $n$ = 5). Perifusion with LHRH + (Nal-Glu)LHRH or (Hyp$^9$)LHRH + (Nal-Glu)LHRH determined specificity of effect. Perifusion with (Nal-Glu)LHRH alone tested the antagonist for agonist activity and determined dependence of basal hCG secretion on endogenous LHRH.

Samples were assayed (Serono) in duplicate and hCG concentrations expressed in terms of the first IRP/third IS 75/537. Assay sensitivity was 0.5 mIU hCG/ml media ($p$ < 0.05). Intra- and interassay CVs in the sample range were ≈8 and 11%, respectively. Control and treatment means were compared by a paired $t$-test and the potency of effects were examined by ANOVA of percent differences between control and treatment means. Means were separated by Scheffe's test.

## III. Results

LHRH and (Hyp$^9$)LHRH (10$^{-9}$ $M$) stimulated hCG secretion ($p$ < 0.01; Fig. 1) equipotently (72.1 vs. 60.7% increase over control, respectively; $p$ > 0.05). However, final determination of (Hyp$^9$)LHRH versus LHRH potency may not be evaluated by a single-dose comparison. LHRH- and (Hyp$^9$)LHRH-stimulated hCG secretion from individual chambers peaked within 5–40 minutes and returned to basal levels within 30–50 minutes. LHRH or (Hyp$^9$)LHRH perifused simultaneously with (Nal-Glu)LHRH (equimolar, 10$^{-9}$ $M$) did not affect hCG secretion ($p$ > 0.05, Fig. 1). (Nal-Glu)LHRH alone (10$^{-9}$ $M$) did not affect hCG secretion ($p$ > 0.05, Fig. 1).

PLACENTAL CELL hCG SECRETION 507

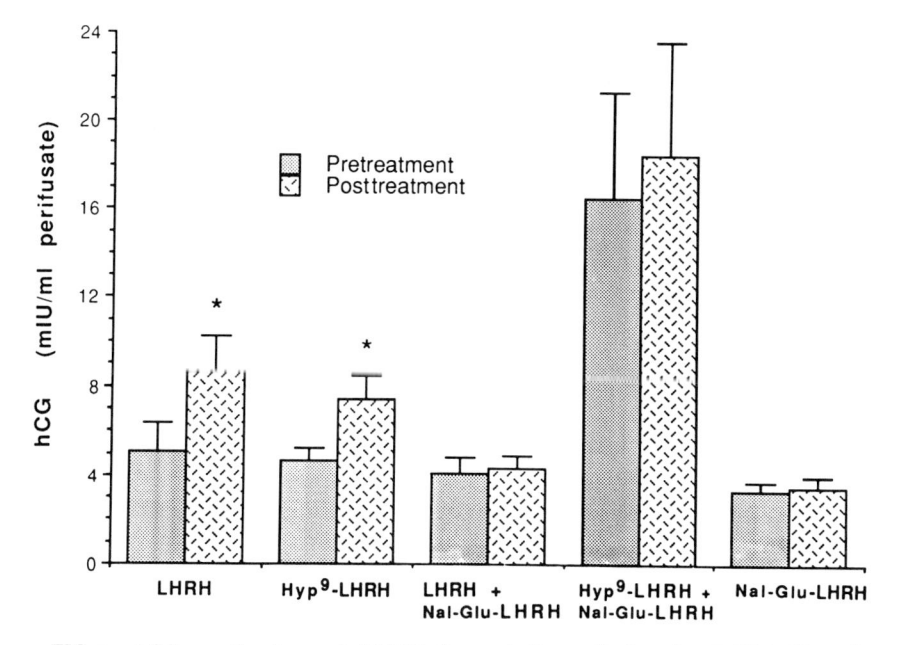

FIG. 1. hCG secretion (mean ± SEM) before and after perifusion of pooled 8- to 12-week placental cells with LHRH ($10^{-9}\,M$, $n = 15$), (Hyp⁹)LHRH ($10^{-9}\,M$, $n = 9$), LHRH + (Nal-Glu)LHRH (equimolar, $10^{-9}\,M$, $n = 5$),(Hyp⁹)LHRH + (Nal-Glu)LHRH (equimolar, $10^{-9}\,M$, $n = 5$), or (Nal-Glu)LHRH ($10^{-9}\,M$, $n = 5$). Asterisks indicate that hCG was increased by the hormone treatment ($p < 0.01$).

## IV. Discussion

Filtration of cell preparations prior to culturing on carrier beads reduced syncytiotrophoblast numbers. In keeping with the 2-day period required for transformation of cultured cytotrophoblasts to syncytiotrophoblasts (Kliman *et al.*, 1986), basal and LHRH-stimulated hCG secretion were detectable within 4 days of initial culture. Dependence of basal hCG secretion on LHRH is equivocal.

Evidence suggests that hCG secretion from placental tissue obtained prior to 13 weeks of gestation may not be suppressed by LHRH antagonists (Siler-Khodr *et al.*, 1987). Results of this study indicate that basal hCG secretion from first-trimester placental cells does not depend on exogenous LHRH. However, cytotrophoblast cells synthesize and secrete LHRH (Petraglia *et al.*, 1990), which could be responsible for hCG secretion in the absence of exogenous LHRH treatment. The failure of an LHRH antagonist to decrease basal hCG secretion suggested that basal hCG secretion is independent of endogenous LHRH activity. However (Nal-Glu)LHRH may not have been administered in sufficient concentration or for sufficient time to antagonize endogenous LHRH activity. In general, basal hCG concentrations marginally exceeded assay sensitivity. Longer term culture may be required to establish sufficiently high baselines to demonstrate potential suppression of hCG by LHRH antagonists. Basal hCG secretion increases with time in culture. The cells used may not yet have depended on LHRH for basal hCG secretion. The responsiveness and specificity of hCG secretory response to exogenous LHRH and (Hyp$^9$)LHRH suggested that LHRH receptors were present on the cells used.

In this study, (Hyp$^9$)LHRH stimulated hCG secretion specifically and equipotently with LHRH at the single concentration tested. Gonadotropes respond to (Hyp$^9$)LHRH at concentrations about 10-fold higher than used here (Gautron et al., 1992). Placental receptors may be more sensitive to (Hyp$^9$)LHRH than pituitary receptors. Pituitary binding studies and in vitro LH and FSH release from dispersed pituitary cells indicated a 22-fold higher potency of LHRH. Different potency ratios of (Hyp$^9$)LHRH and LHRH suggest that placental and pituitary LHRH receptors may differ. Placental LHRH receptors may be less selective to the sequence or conformation of LHRH-like molecules, as suggested by previous studies (Currie et al., 1981; Bélisle et al., 1984, 1987).

LHRH is detectable in human fetal hypothalamus by $5\frac{1}{2}$ weeks of gestation (Jewelewicz and Shortle, 1987). (Hyp$^9$)LHRH is present in human hypothalamus and is the major LHRH moiety in fetal rat hypothalamus (Gautron et al., 1991). Though most LHRH in maternal circulation may be of placental origin (Petraglia et al., 1990), the fetal hypothalamus is a potential source of (Hyp$^9$)LHRH and LHRH-like activity stimulating hCG secretion.

In summary, perifusion confirmed LHRH stimulation of hCG secretion by first-trimester placental cells. The results suggested that (Hyp$^9$)LHRH is a potential physiological hCG secretagogue. The study adds to the notion that fetal LHRH moieties are a component of the total LHRH stimulus for hCG secretion from the placenta.

### ACKNOWLEDGMENTS

(Nal-Glu)LHRH synthesized at the Salk Institute (NIH NO1-HD-0-2906) was from the Contraceptive Development Branch, Center for Population Research, NICHD. Funding was by the Medical Research Council of Canada and B.C. Health Care Research Foundation.

## REFERENCES

Bélisle, S., Guévin, J. F., Bellabarba, D., and Lehoux, J. G. (1984). *J. Clin. Endocrinol. Metab.* **59,** 119–126.

Bélisle, S., Lehoux, J. G., Bellabarba, D., Gallo-Payet, N., and Guévin, J. F. (1987). *Mol. Cell. Endocrinol.* **49,** 195–202.

Currie, A. J., Fraser, H. M., and Sharpe, R. M. (1981). *Biochem. Biophys. Res. Commun.* **99,** 332–338.

Gautron, J. P., Leblanc, P., Bluet-Pajot, M. T., Pattou, E., L'Héritier, A., Mounier, F., Ponce, G., Audinot, V., Rasolonjanahary, R., and Kordon, C. (1992). *Mol. Cell. Endocrinol.* **85,** 99–107.

Gautron, J. P., Pattou, E., Bauer, K., and Kordon, C. (1991). *Neurochem. Int.* **18,** 221–235.

Jewelewicz, R., and Shortle, B. (1987). *In* "Gynecologic Endocrinology" (J. J. Gold and J. B. Josimovich, eds.), Vol. IV, pp. 89–107. Plenum, New York.

Khodr, G. S., and Siler-Khodr, T. M. (1978). *Fert. Steril.* **30,** 301–304.

Kliman, H. J., Nestler, J. E., Sermasi, E., Sanger, J. M., and Strauss, J. F. (1986). *Endocrinology (Baltimore)* **118,** 1567–1582.

Petraglia, F., Volpe, A., Genazzani, A. R., Rivier, J., Sawchenko, P. E., and Vale, W. (1990). *Front. Neuroendocr.* **11,** 6–37.

Siler-Khodr, T. M., and Khodr, G. S. (1981). *Biol. Reprod.* **25,** 353–358.

Siler-Khodr, T. M., Khodr, G. S., Rhode, J., Vickery, B. H., and Nestor, J. J. (1987). *Placenta* **8,** 1–14.

# Tyrphostins Inhibit Sertoli Cell-Secreted Growth Factor Stimulation of A431 Cell Growth

DOLORES J. LAMB,[*,†] AND SANKARARAMAN SHUBHADA[*]

*The Scott Department of Urology and †The Department of Cell Biology, Baylor College of Medicine, Houston, Texas 77030*

Spermatogenesis is regulated by hormones, for example, the gonadotropins and testosterone, as well as by locally active autocrine and paracrine factors. Growth factors such as fibroblast growth factor (FGF)-like protein, transforming growth factor-$\beta$ (TGF$\beta$)-like protein, TGF$\alpha$, insulin-like growth factor I, nerve growth factor, and epidermal growth factor (EGF)-like protein have been reported in testicular homogenates or secretions (reviewed in Buch *et al.*, 1991). Sertoli cell-secreted growth factor (SCSGF) is secreted by rat and human Sertoli cells *in vitro*. The exact role of most of these growth factors in spermatogenesis is unknown, but it is likely that they regulate mitosis, meiosis, and differentiation in the testis. The goal of the current study was to determine the signal transduction pathway activated by SCSGF using a model cell line that exhibits growth stimulation in response to this growth factor.

Only now are we beginning to understand the diversity of the protein kinase family. Protein phosphorylation is an important modification that regulates receptor function and/or cellular trafficking (reviewed in Sibley *et al.*, 1988). A number of growth factor and hormone receptors exhibit enhanced tyrosine kinase activity after ligand interaction and receptor autophosphorylation. Activation of receptor tyrosine kinase by specific growth factors is a crucial signal that initiates multiple cellular responses, resulting in DNA synthesis and cellular proliferation.

Inhibitors of specific tyrosine kinases have been designed to bind to the substrate subsite of the tyrosine kinase domain. Erbstatin, purified from actinomycete medium, inhibits EGF receptor phosphorylation (Fig. 1) (Umezawa *et al.*, 1986). Compounds derived from a benzylidenemalononitrile (BMN) nucleus have been synthesized (Gazit *et al.*, 1989; Yaish *et al.*, 1988). These compounds resemble the moieties of tyrosine and erbstatin and demonstrate a progressive increase in affinity for the substrate site of the EGF receptor kinase domain. Some exhibit receptor tyrosine kinase specificity and block EGF-dependent autophosphorylation

511

Copyright © 1993 by Academic Press, Inc.
All rights of reproduction in any form reserved.

512 DOLORES J. LAMB AND SANKARARAMAN SHUBHADA

FIG. 1. Tyrosine and selected tyrphostins.

of the receptor, inhibit EGF receptor tyrosine kinase activity *in vitro* in living cells, and suppress cell proliferation (Gazit *et al.,* 1989; Lyall *et al.,* 1989; Posner *et al.,* 1989; Yaish *et al.,* 1989). Higher concentrations (1000-fold) of inhibitor were required to inhibit insulin receptor kinase activity (Gazit *et al.,* 1989). The compounds are not cytotoxic.

Saperstein *et al.* (1989) designed insulin receptor tyrosine kinase inhibitors based on the structure of (hydroxy-2-naphthalenylmethyl)phosphoric acid (Fig. 1). The resemblance of the naphthalene region of the inhibitor to the tyrosine substrate is presumed to account for the specificity of the insulin receptor tyrosine kinase inhibitor. These insulin receptor tyrosine kinase inhibitors exhibit antiproliferative activity that relates to their ability to penetrate cells.

TYRPHOSTIN INHIBITION OF SCSGF 513

The signal transduction mechanism for SCSGF is unknown. To test whether SCSGF acts through a tyrosine kinase pathway, we synthesized a series of "tyrphostins" (tyrosine-like inhibitors). Highly purified SCSGF was not required in this approach for *in vitro* phosphorylation studies. We assessed the ability of several tyrphostins to antagonize SCSGF-stimulated cell proliferation *in vitro*.

Tyrphostin inhibitors I and II are shown in Fig. 1. Inhibitor I was reported to preferentially suppress EGF-stimulated growth, whereas inhibitor II exclusively blocked EGF-stimulated growth (Yaish *et al.*, 1989). To synthesize the inhibitors, 2.75 g of 3,4-dihydroxybenzaldehyde (0.002 mol) was dissolved in 10 ml of methanol and 0.02 mol of either cyanocetamide (3.00 g; inhibitor I) or malononitrile (2.36 g; inhibitor II). Fifteen milliliters of methanol was added to it with 0.2 m of piperidine. The reaction mixture was stirred for 30 minutes at room temperature. A yellow solid was filtered and recrystallized from ethanol twice. $^1$H-NMR (proton magnetic resonance) spectra are as follows.

Inhibitor I: $\alpha,\alpha$-dicyano-3,4-dihydroxystyrene: $\delta$ 7.76 (1H, s, vinylic proton), 7.62 (1 H, d, $J_{2,6}$ = 2.5 Hz, H2), 7.43 (1H, dd, $J_{5,6}$ = 8.3 Hz, $J_{2,6}$ = 2.4 Hz, H6), 6.43 (1H, d, $J_{5,6}$ = 8.3 Hz, H5).

Inhibitor II: $\alpha$-cyano,$\alpha$-amido-3,4-dihydroxystyrene: $\delta$ 8.05 (1H, s, vinylic), 7.67 (1H, d, $J_{2,6}$ = 2.2 Hz, H2), 7.39 (2H, dd, $J_{5,6}$ = 8.3 Hz, $J_{2,6}$ = 2.2 Hz, H6), 6.97 (1H, d, $J_{5,6}$ = 8.3 Hz, H5).

Using an *in vitro* assay to measure SCSGF (Lamb *et al.*, 1991), we determined if the inhibitors would block the stimulation of A431 cell growth by SCSGF using Sertoli cell-conditioned medium (SCCM) as the source of SCSGF. The inhibitors were tested over a dose range of 0–50 $\mu M$. For the initial studies, the A431 cells were cultured in the presence of 10% fetal bovine serum (FBS) in Dulbecco's modified Eagle's medium (DME). Kinase inhibitors I and II inhibited both SCCM and FBS-stimulated growth in a dose-dependent manner. The cells remained healthy and viable. The inhibitors antagonized SCCM stimulation of A431 cell growth in serum-free medium with an $EC_{50}$ of 10 $\mu M$ (Fig. 2). The inhibitors were not toxic because, after removal, with a medium change, the cells proliferated (Fig. 3).

Receptor polypeptides consist of families of structurally related proteins with intrinsic tyrosine kinase activity. Inhibitors I and II both inhibited SCSGF stimulation of cell proliferation. These two inhibitors were originally designed to block EGF receptor-dependent tyrosine kinase activity and exhibit different potencies for the EGF receptor (Yaish *et al.*, 1989), suggesting that SCSGF may stimulate growth through the EGF receptor. The receptor family for epidermal growth factor includes v-*erbB*, EGF/

514    DOLORES J. LAMB AND SANKARARAMAN SHUBHADA

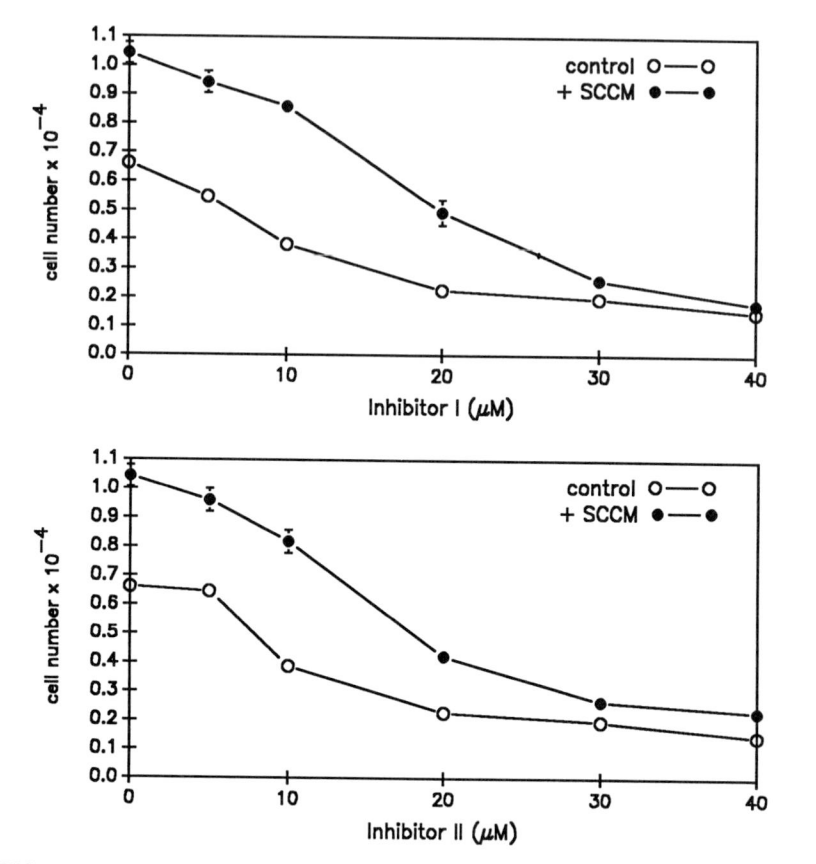

FIG. 2. Inhibition of basal and SCSGF-stimulated A431 cell growth by tyrphostins. An A431 cell growth assay in serum-free medium (Lamb *et al.*, 1991) was used to determine the effect of inhibitor I and II on both basal and SCSGF-stimulated growth. A431 cells were incubated with DME or 50% SCCM for 3 days in the presence or absence of increasing concentrations of inhibitor I (upper panel) and II (lower panel) prior to counting the cells in a Coulter counter. Results are expressed as mean ± SD of triplicate determinations. Where not visible, error bars are smaller than the symbols.

transforming growth factor$_\alpha$, and HER2/*neu*. The preliminary studies on tyrphostin inhibition of SCSGF-stimulated growth provide evidence that the SCSGF stimulates cell growth through a tyrosine kinase mechanism.

It is feasible to prepare selective and nontoxic protein tyrosine kinase inhibitors. Theoretically it should be possible to synthesize inhibitors for each of the known protein tyrosine kinases involved in cell proliferation and transformation (Gazit *et al.*, 1989). These compounds could serve as antiproliferative agents that inhibit tumor growth.

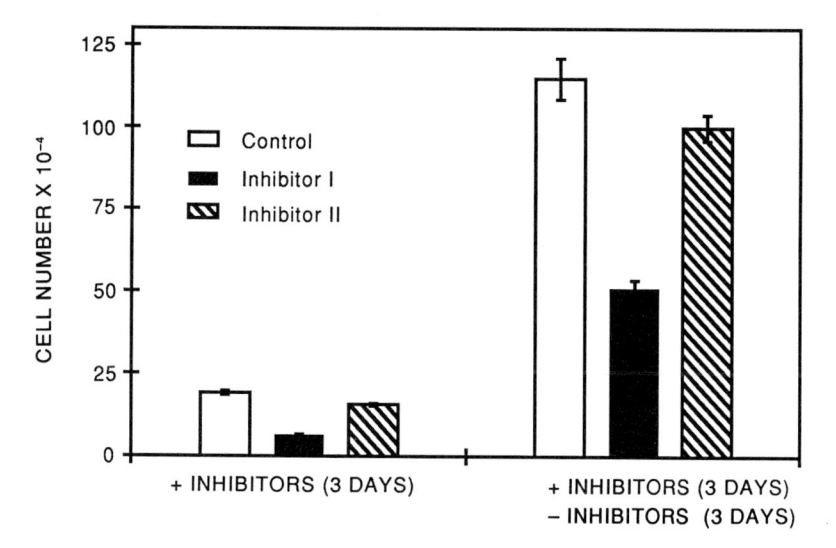

FIG. 3.   The tyrphostins are not cytotoxic *in vitro*. A431 cells were incubated with 50% SCCM (control) and 20 $\mu M$ inhibitor I or 20 $\mu M$ inhibitor II. After a 3-day culture period [as described in Lamb *et al.* (1991)], the cells were counted. Following a medium change to 50% SCCM after removal of the inhibitors, replicate dishes were incubated for an additional 3 days prior to cell count. Results are expressed as the mean ± 2 SD.

To date, most experimental approaches to reversible male contraception have involved hormonal manipulations of the pituitary–gonadal axis (which may have adverse effects on libido or sexual potency), drugs (such as gossypol, α-chlorohydrin, and hexose derivatives), as well as immunological strategies that are frequently irreversible. Because tyrphostins that demonstrate kinase specificity have been synthesized, inhibition of the specific receptor tyrosine kinase should therefore result in nullifying the effect of the growth factor and thus inhibit spermatogenesis. Tyrphostins may provide a new approach to male contraception.

## REFERENCES

Buch, J. P., Tindall, D. J., Rowley, D. W., and Lamb, D. J. (1991). *In* "Infertility in the Male" (S. Howards and L. I. Lipshultz, eds.), 2nd ed., pp. 54–83, Mosby-Yearbook Publishers, St. Louis.

Gazit, A., Pnina, Y., Gilon, C., and Levitzki, A. (1989). *J. Med. Chem.* **32**, 2352–2357.

Lamb, D. J., Spotts, G. S., Shubhada, S., and Baker, K. R. (1991). *Mol. Cell. Endocrinol.* **79**, 1–12.

Lyall, R. M., Zilberstein, A., Gazit, A., Gilon, C., Levitzki, and Schlessinger, J. (1989). *J. Biol. Chem.* **264**, 14503–14509.

Posner, I., Gazit, A., Gilon, C., and Levitzki, A. (1989). *Eur. J. Biochem.* **257**, 287–291.

516     DOLORES J. LAMB AND SANKARARAMAN SHUBHADA

Saperstein, R., Vicario, P. P., Strout, H. V., Brady, E., Slater, E. E., Greenlee, W. J., Ondeyka, D. L., Patchett, A. A., and Hangauer, D. G. (1989). *Biochemistry* **28**, 5694–5701.

Sibley, D. R., Benovic, J. L., Caron, M. G., and Lefkowitz, R. J. (1988). *Endocr. Rev.* **9**, 38–56.

Umezawa, M., Imoto, M., Sawar, T., Isshiki, K., Matsuda, N., Uchida, T., Iinuma, H., Hamada, M., and Takeuchi, T. (1986). *J. Antibiot.* **39**, 170.

Yaish, P., Gazit, A., Gilon, C., and Levitzki, A. (1988). *Science* **242**, 933–935.

# Induction of Calcium Transport into Cultured Rat Sertoli Cells and Liposomes by Follicle-Stimulating Hormone

PATRICIA GRASSO AND LEO E. REICHERT, JR.

*Department of Biochemistry and Molecular Biology, Albany Medical College, Albany, New York 12208*

## I. Introduction

The recently identified ability of follicle-stimulating hormone (FSH) to bind calcium (Santa-Coloma *et al.*, 1992), activate calcium channels (Grasso and Reichert, 1989, 1990; Grasso *et al.*, 1991b), and influence $Na^+/Ca^{2+}$ exchange (Grasso *et al.*, 1991a) in cultured rat Sertoli cells and FSH receptor-containing proteoliposomes suggests that the mechanism of FSH signal transduction in the testis is a complex process that may involve second messengers other than those associated with hormone-stimulated elevation of cAMP (Means *et al.*, 1980). Although the latter is a characteristic of the pituitary glycoprotein hormones, modulation of intracellular free $Ca^{2+}$ has also been proposed as a mechanism associated with cellular response(s) to luteinizing hormone (LH) in rat Leydig cells (Cooke, 1990) and granulosa cells (Asem *et al.*, 1987), and adrenocorticotropic hormone (ACTH) in bovine adrenocortical cells (Li *et al.*, 1989). The involvement of FSH in regulation of $Ca^{2+}$ flux into Sertoli cells suggests that cyosolic free $Ca^{2+}$ may be an important second messenger in FSH signal transduction. Evidence recently developed in this laboratory supporting this notion is summarized below.

## II. FSH-Stimulated Calcium Channel Activity

The ability of FSH to induce changes in intracellular free calcium in Sertoli cells is well documented (Means *et al.*, 1980), although the mechanisms involved in this process are not fully understood. We (Quirk and Reichert, 1988) and others (Monaco *et al.*, 1988) have shown that although the phosphatidylinosotol pathway, associated with mobilization of calcium from intracellular stores (Carafoli, 1987), is present in cultured Sertoli cells from immature (15- to 18-day-old) rats, it does not appear to be activated by FSH. Thus, the effect of FSH on cytosolic free calcium levels presumably is more directly related to internalization of extracellular calcium rather than mobilization from intracellular stores.

Successful incorporation of hormone-responsive Triton X-100-solubilized FSH receptors from bovine calf testes into lipid vesicles (Grasso et al., 1988) provided a model with which to examine the effects of FSH on calcium flux. Our studies indicated that FSH-induced hormone-specific and concentration-dependent uptake of calcium (as $^{45}Ca^{2+}$) could be inhibited by antagonists specific for either voltage-sensitive (methoxy-verapamil and nifedipine) or voltage-independent (ruthenium red and gadolinium chloride) calcium channels (Grasso and Reichert, 1989). Similar results were obtained with cultured rat Sertoli cells. Thus, it appears that FSH-stimulated uptake of extracellular calcium may be facilitated by channels gated by hormonally induced changes in membrane potential, as well as by channels that are insensitive to such changes.

Recently, a class of ion channels sensitive to specific signal-transducing membrane-associated G proteins has been identified (Brown and Birnbaumer, 1988). These channels, when activated by nonhydrolyzable analogs of GTP independent of receptor stimulation, mimic the response produced by an occupied receptor. Several lines of evidence suggest that the calcium channel regulatory protein is $G_s$, the activator of adenylate cyclase (Brown and Birnbaumer, 1988). The Triton X-100-solubilized FSH receptor incorporated into liposomes was found to be complexed to cholera toxin-sensitive G protein, presumably $G_s$ (Dattatreyamurty et al., 1987). If the calcium-conducting channels activated by FSH binding to its receptor were gated by $G_s$, it would be expected that GTP, GTPγS, or Gpp(NH)p, in the absence or presence of fluoride, might stimulate $^{45}Ca^{2+}$ uptake by the FSH receptor-containing proteoliposomes. Such stimulation of $^{45}Ca^{2+}$ uptake, however, was not observed (Grasso and Reichert, 1989), nor was cholera toxin able to induce uptake of extracellular calcium (as $^{45}Ca^{2+}$) by cultured rat Sertoli cells (Grasso and Reichert, 1990). Receptor occupancy by FSH appears to exert a direct effect on calcium influx that is not mediated by activation of receptor-associated $G_s$ protein.

Dibutyryl-cAMP, a membrane-permeable analog of cAMP, did not enhance $^{45}Ca^{2+}$ uptake over basal levels in Sertoli cells, although a concentration-related increase was evident in conversion of androstenedione to estradiol, a Sertoli cell response known to be mediated, at least in part, by activation of adenylate cyclase (Grasso and Reichert, 1990). The influx of calcium observed on interaction of FSH with its receptor apparently does not require accumulation of cAMP. The data suggest that FSH stimulation of calcium channels does not involve gating by activated $G_s$ protein either directly or indirectly through generation of second messengers. The possibility that the FSH receptor may function as a calcium channel is under investigation.

## III. FSH Inhibition of Sodium-Dependent Calcium Uptake

Incubation of cultured Sertoli cells from immature rat testes and FSH receptor-containing proteoliposomes in $Na^+$-free buffer enhanced calcium (as $^{45}Ca^{2+}$) uptake when intracellular (or liposome-encapsulated) $Na^+$ levels were elevated (Grasso et al., 1991a). FSH attenuated this effect by approximately 30% in each system. Removal of the outwardly directed $Na^+$ gradient, achieved by increasing external $Na^+$, reduced $Na^+$-dependent $Ca^{2+}$ influx and eliminated the inhibitory effect of FSH.

Treatment of Sertoli cells with ouabain also resulted in significantly greater uptake of $^{45}Ca^{2+}$ compared to untreated cells, possibly by preventing extrusion of intracellular $Na^+$ via the $Na^+/K^+$-ATPase known to be present in Sertoli cell membranes (Joffre and Roche, 1988). The inhibitory effect of FSH was evident in the presence of ouabain. Thus, pharmacological manipulation of the electrochemical gradient for $Na^+$ resulted in increased uptake of calcium that was sensitive to FSH inhibition. This observation provides additional evidence supporting FSH-mediated $Na^+/Ca^{2+}$ exchange activity in the testis.

In conjunction with its influence on calcium channel activation, the inhibitory effect of FSH on $Na^+$-dependent $Ca^{2+}$ uptake by cultured Sertoli cells and receptor-containing proteoliposomes suggests that FSH-regulated influx of extracellular $Ca^{2+}$ involves more than one mechanism. Inhibition of the $Na^+$-dependent component of $Ca^{2+}$ influx by FSH, in association with its stimulatory effect on $Ca^{2+}$ channel activity, apparently helps to maintain the appropriate level of intracellular $Ca^{2+}$ required for modulation of FSH receptor-mediated postbinding events in the Sertoli cell.


## IV. A New Role for FSH: Calcium Binding and Transport

Previous studies from this laboratory demonstrated that calcium is required for optimal binding of FSH to receptor (Andersen and Reichert, 1982). To determine the basis for this requirement, 11 overlapping peptide amides representing the entire primary structure of hFSH $\beta$ subunit were screened for their ability to bind calcium (as $^{45}Ca^{2+}$) (Santa-Coloma et al., 1992). These studies indicated that hFSH-$\beta$-(1–15), which contains an amino acid sequence similar to that found in the loop structures of the calcium-binding domains of calmodulin, bound significant amounts of $^{45}Ca^{2+}$ with a $K_d$ of $1.2 \pm 0.3$ m$M$, an affinity similar to that reported for a peptide corresponding to calmodulin binding site III (Reid, 1990).

The ability of hFSH-$\beta$-(1–15) to bind calcium correlates well with its ability to induce uptake of $^{45}Ca^{2+}$ by liposomes (Grasso et al., 1991b). The physiological significance of these heretofore unrecognized properties of this peptide is unknown. We have suggested that, in addition to its

520     PATRICIA GRASSO AND LEO E. REICHERT, JR.

effect on voltage-sensitive calcium channel activity, interaction of FSH with its receptor may induce a conformational change in the hormone that results in the formation of calcium-conducting transmembrane channels (Grasso et al., 1991b). Alternatively, fragments of FSH-$\beta$-containing calcium-binding regions may be generated after FSH–receptor complex internalization and FSH degradation (Fletcher and Reichert, 1984). Because internalized vesicles contain higher calcium concentrations (similar to extracellular calcium) than cytosol, calcium-binding fragments of FSH-$\beta$ resulting from proteolytic cleavage may be able to induce calcium entry into the cytosol via a process thermodynamically favored by these high levels of intravesicular calcium.

## V.  Concluding Remarks and Future Directions

A substantial body of evidence indicates that precise control of cytosolic free $Ca^{2+}$ is a critical requirement for Sertoli cell function. The recently identified ability of FSH to bind calcium and to form transmembrane calcium-conducting channels, the presence of multiple mechanims for regulating $Ca^{2+}$ influx, i.e., FSH-sensitive calcium channel activity and $Na^+/Ca^{2+}$ exchange, all attest to the importance of this ion in FSH action. Although we do not as yet fully understand the role of calcium in Sertoli cell function, initial data indicate that it has a profound effect on FSH-stimulated conversion of androstenedione to estradiol (Grasso and Reichert, 1989). It remains to be determined what other influences calcium may exert on Sertoli cell responsivenss to FSH stimulation.

### ACKNOWLEDGMENTS

This work was supported by NIH Grant HD-13938.

### REFERENCES

Andersen, T. T., and Reichert, L. E., Jr. (1982). *J. Biol. Chem.* **257**, 11551–11557.
Asem, E. K., Molnar, M., and Hertelendy, F. (1987). *Endocrinology (Baltimore)* **120**, 853–859.
Brown, A. M., and Birnbaumer, L. (1988). *Am. J. Physiol.* **254**, H401–H410.
Carafoli, E. (1987). *Annu. Rev. Biochem.* **56**, 395–433.
Cooke, B. A. (1990). *Mol. Cell. Endocrinol.* **69**, C11–C15.
Dattatreyamurty, B., Figgs, L. W., and Reichert, L. E., Jr. (1987). *J. Biol. Chem.* **262**, 11737–11745.
Fletcher, P. W., and Reichert, L. E., Jr. (1984). *Mol. Cell. Endocrinol.* **34**, 39–49.
Grasso, P., Dattatreyamurty, B., and Reichert, L. E., Jr. (1988). *Mol. Endocrinol.* **2**, 420–430.
Grasso, P., and Reichert L. E., Jr. (1989). *Endocrinology* (Baltimore) **125**, 3029–3036.

Grasso, P., and Reichert, L. E., Jr. (1990). *Endocrinology* (Baltimore) **127**, 949–956.

Grasso, P., Joseph, M. P., and Reichert, L. E., Jr. (1991a). *Endocrinology (Baltimore)* **128**, 158–164.

Grasso, P., Santa-Coloma, T. A., and Reichert, L. E., Jr. (1991b). *Endocrinology (Baltimore)* **128**, 2745–2671.

Joffre, M., and Roche, A. (1988). *J. Physiol. (London)* **400**, 481–499.

Li, Z-G., Park, D., and LaBella, F. S. (1989). *Endocrinology (Baltimore)* **125**, 592–596.

Means, A. R., Dedman, J. R., Tash, J. S., Tindall, D. J., van Sickle, M., and Welsh, M. J. (1980). *Annu. Rev. Physiol.* **42**, 59–70.

Monaco, L., Adamo, S., and Conti, M. (1988). *Endocrinology (Baltimore)* **123**, 2032–2039.

Quirk, S. M., and Reichert, L. E., Jr. (1988). *Endocrinology (Baltimore)* **123**, 230–237.

Reid, R. E. (1990). *J. Biol. Chem.* **265**, 5971–5976.

Santa-Coloma, T. A., Grasso, P., and Reichert, L. E., Jr. (1992). *Endocrinology (Baltimore)* **130**, 1103–1107.

# Pituitary *in Vitro* LH and FSH Secretion after Administration of the Antiprogesterone RU486 *in Vivo*

K. L. KNOX AND N. B. SCHWARTZ

*Department of Neurobiology and Physiology, Northwestern University, Evanston, Illinois 60208*

## I. Introduction

The secretion of LH and FSH during the 4-day rat estrous cycle is remarkably similar except following the preovulatory gonadotropin surges on the afternoon of proestrus (Nequin *et al.*, 1979). Although levels of serum LH drop immediately following the proestrous surge, serum FSH remains elevated throughout the evening of proestrus until noon on the day of estrus, a phenomenon known as the secondary FSH surge. The primary surges of LH and FSH and the secondary FSH surge can be completely abolished by administration of a GnRH antagonist before the primary surges have taken place (Schwartz *et al.*, 1985). If a GnRH antagonist is given after the primary surge of LH, the secondary FSH surge cannot be blocked by a GnRH antagonist (Schwartz *et al.*, 1985). This suggests that an increase in GnRH secretion is not required for the secondary FSH surge to occur once the primary surges have taken place. We have proposed that the secondary FSH surge results from a drop in inhibin after the primary surges (Schwartz and Channing, 1977). Serum inhibin and mRNA levels of inhibin subunits in the ovary drop following the primary gonadotropin surges, and when the primary surges are blocked using a GnRH antagonist, levels of inhibin mRNA and serum inhibin remain elevated throughout the night of proestrus and the secondary surge of FSH is abolished (Woodruff *et al.*, 1989).

In response to rising levels of LH on the afternoon of proestrus, progesterone secretion also increases and elevated levels of progesterone facilitate the full expression of the preovulatory gonadotropin surges (Banks and Freeman, 1978). We have used the antiprogesterone RU486 to block the actions of progesterone on proestrus, which results in a suppression of the primary LH and FSH surges. Most important, despite a normal decline in levels of serum inhibin, RU486-treated animals do not experience a secondary FSH surge (Knox and Schwartz, 1992). On the basis of

524 K. L. KNOX AND N. B. SCHWARTZ

these findings, we propose that the secondary FSH surge occurs because of both the drop in serum inhibin and the elevation of serum progesterone prior to or following the gonadotropin surges on proestrus afternoon. The action of progesterone that is important for the secondary FSH surge could be envisioned at the level of the hypothalamus or the pituitary, because receptors for progesterone have been localized in both (Attardi, 1984). In the studies we report here, we administered RU486 *in vivo* and studied the effects at the pituitary level *in vitro* using a dynamic perifusion system.

## II.  Materials and Methods

### A.  ANIMALS

Female Sprague–Dawley CD rats (Charles River, Portage, Michigan) were obtained at 60 days of age and maintained in controlled temperature (20°C) and lighting conditions (14 hours of light, 10 hours of dark; lights on at 0500 hours), with water and food provided *ad libitum*. Proestrus females were selected after exhibiting at least two consecutive 4-day estrous cycles, determined by daily vaginal smears.

### B.  RU486

RU486 [17$\beta$-hydroxy-11$\beta$-(4-dimethylaminophenyl-1)-17$\delta$-(prop-1-ynyl) -estra-4,9-dien-3-one], the gift of Roussel-UCLAF (Romainville, France), was dissolved in ethanol, suspended in oil, and given as a subcutaneous injection. The dose of RU486 was 6 mg/kg body weight. Rats were injected at 1200 hours proestrus with corn oil ($N = 8$) or RU486 ($N = 8$), decapitated at 0600 hours estrus, and pituitaries were collected.

### C.  PROTOCOL FOR PERIFUSION RUNS

Anterior pituitaries (APs) were cut into eighths and placed into a medium previously warmed and oxygenated, then were rinsed twice with fresh medium. Fragments from an individual pituitary were placed into a 100-$\mu$l microchamber and perifused for 6 hours in the ACUSYST (Model 1501, Endotronics Inc., Coon Rapids, Minnesota). Four APs were used in each perifusion. Details of the perifusion protocol can be found in Fallest and Schwartz (1990). Fractions were collected every 5 minutes and frozen for RIA of LH and FSH. AP membrane viability was verified using a medium containing 60 m$M$ K$^+$ as a depolarizing stimulus for the last 30 minutes of the perifusion. In some runs, pulsatile GnRH (LRF; U.S. Biochemical Corp., Cleveland, Ohio) was administered as five brief pulses delivered one per hour at 30, 90, 150, 210, and 270 minutes. The pulse parameters programmed into the ACUSYST computer consisted of a rapid 30-second rising phase, reaching a peak concentration of 50 ng/ml, and a 30-second falling phase. The half-maximal concentration of GnRH was attained 1.5 minutes after the peak of the pulse.

## D.  RADIOIMMUNOASSAYS

### 1.  LH

Measurements of medium LH were made using the ovine–rat RIA system (NIH LH S-25 as standard and antirat antibody S-10 from the NIDDK kit). Mean intra- and interassay coefficients of variation were 2 and 4%, respectively.

### 2.  FSH

Measurements of medium FSH were made using materials in the rat–rat RIA system (FSH RP-2 as standard and antibody S-11 from the NIDDK kit). Mean intra- and interassay coefficients of variation were 5 and 9%, respectively.

## E.  DATA ANALYSIS AND STATISTICS

The group mean ± the SEM ($N = 4$) was calculated in nanograms/ milliliter for each 5-minute fraction of LH and FSH. LH and FSH mean hourly secretion rates were calculated by summing the 12 fractions for each 1-hour interval. Two-way analysis of variance (ANOVA) was used to analyze the data using treatment as the between-subjects factor. All statistics were computed using the Systat Statistical Package (Evanston, Illinois). Levels of significance attained are indicated in the legends to Figs. 1–3.

## III.  Results

## A.  BASAL GONADOTROPIN SECRETION FROM RATS TREATED *in Vivo* WITH RU486

The mean ± the SEM for hourly FSH and LH secretion (ng/hour) is shown in Fig. 1. Secretion rates of FSH were significantly suppressed in rats treated the previous day with RU486. Secretion rates of LH were significantly higher from pituitaries of rats treated *in vivo* with RU486.

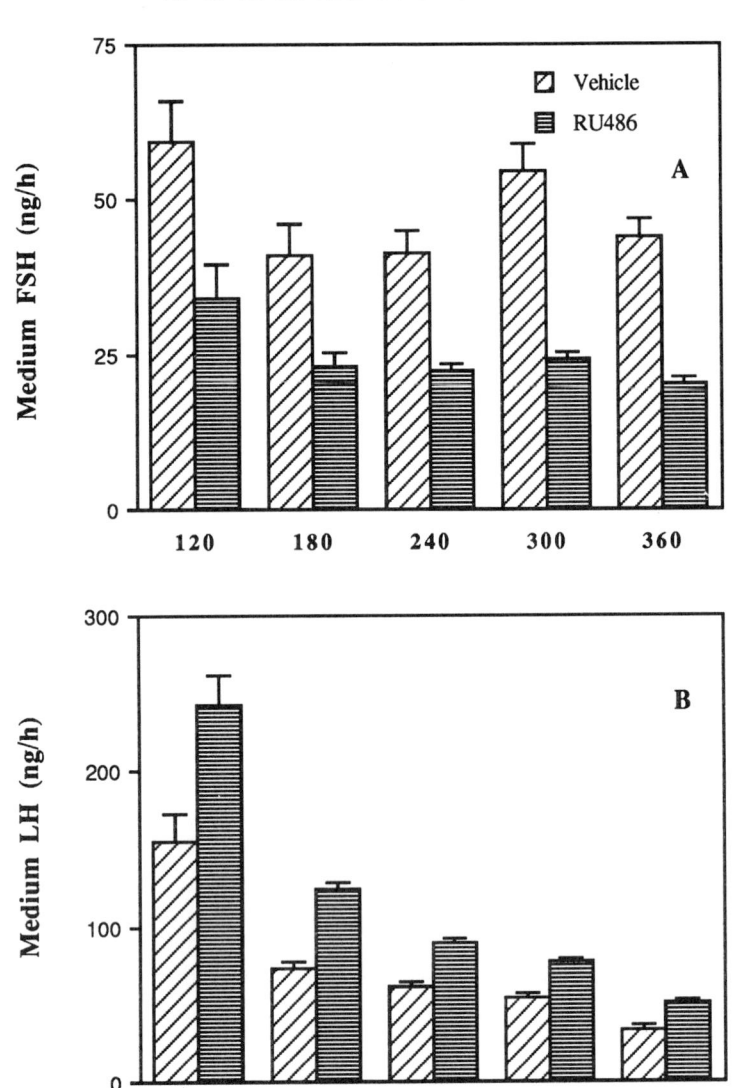

FIG. 1. Basal gonadotropin secretion from rats treated *in vivo* with RU486. Mean ± the SEM for hourly FSH (A) and LH (B) secretion rates from pituitaries of rats treated *in vivo* on the day of proestrus. Levels of FSH were significantly suppressed in RU486-treated rats compared to controls ($p < 0.0001$). Levels of LH were significantly elevated in RU486-treated rats compared to controls ($p < 0.0001$).

RU486 EFFECTS ON FSH AND LSH                527

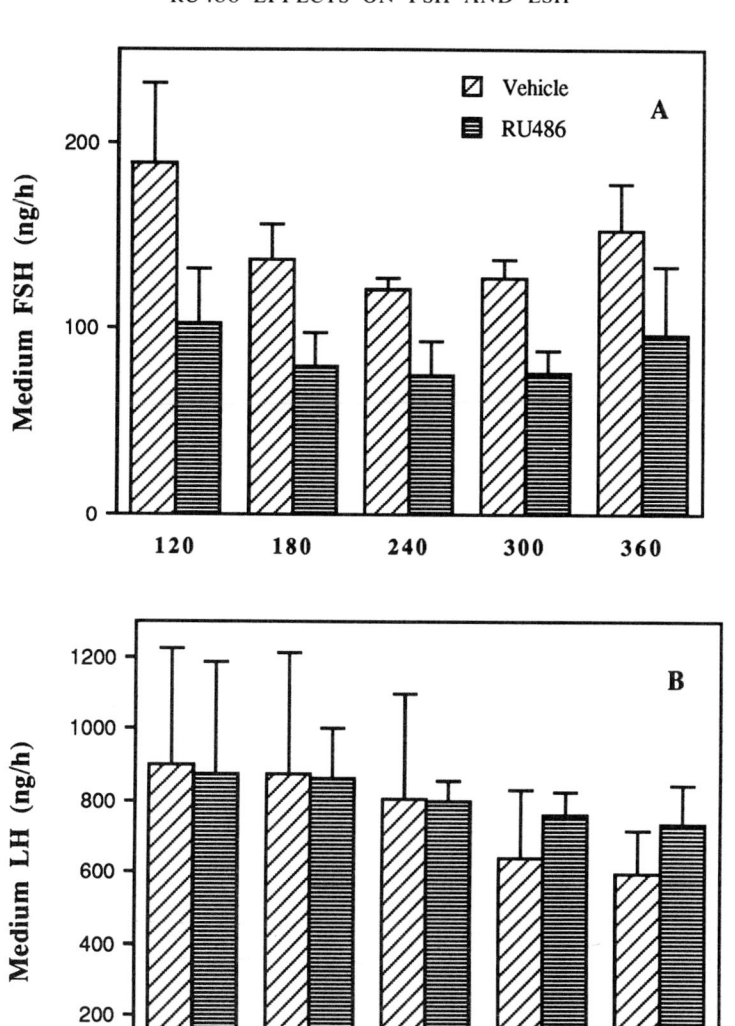

FIG. 2.  GnRH-stimulated gonadotropin secretion from rats treated *in vivo* with RU486. Mean ± the SEM for hourly FSH (A) and LH (B) GnRH-stimulated secretion rates from pituitaries of rats treated *in vivo* on the day of proestrus. FSH secretion was significantly suppressed in rats treated with RU486 compared to controls ($p < 0.0001$). No significant differences for treatment were detected for mean LH secretion between controls and rats treated with RU486.

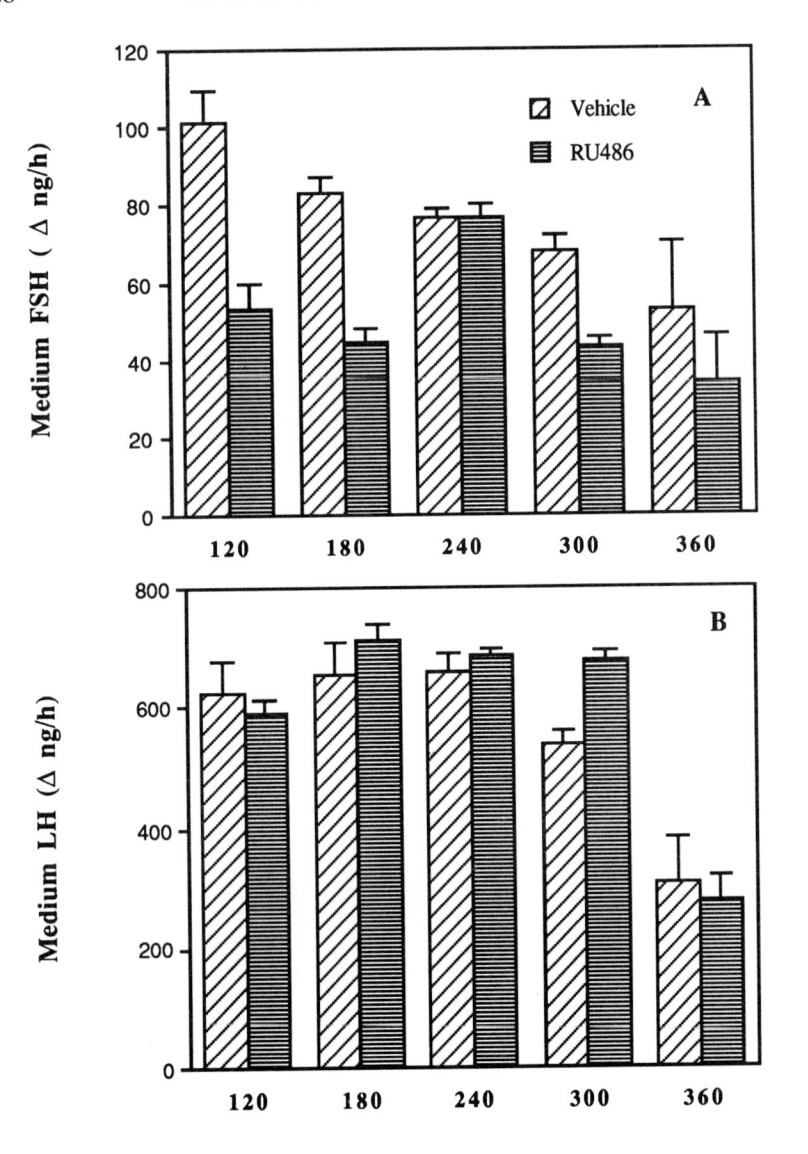

**Minutes of Perifusion**

FIG. 3. Calculated changes from baseline (△) for GnRH-stimulated gondatropin mean secretion of FSH (A) and LH (B). Net secretion of FSH due to GnRH stimulation was significantly suppressed in RU486-treated rats compared to controls ($p > 0.0001$). No significant differences for treatment were detected for net secretion of LH due to GnRH stimulation between controls and rats treated with RU486.

## B. TOTAL GnRH-STIMULATED GONADOTROPIN SECRETION FROM RATS TREATED in Vivo WITH RU486

The mean ± the SEM for hourly FSH and LH GnRH-stimulated secretion rates from pituitaries of rats treated *in vivo* (ng/hour) is shown in Fig. 2. GnRH-stimulated secretion was significantly suppressed in rats treated *in vivo* with RU486 compared to control rats. There were no significant differences in GnRH-stimulated secretion of pituitary LH between control rats and rats treated *in vivo* with RU486.

## C. GnRH-STIMULATED GONADOTROPIN SECRETION MINUS BASAL SECRETION

In Fig. 3 we report GnRH-stimulated pituitary FSH and LH secretion after basal hourly group means are subtracted from hourly secretion rates of each AP that received pulsatile GnRH. This provides a clearer profile of the significant suppression of FSH pituitary secretion, but not of LH pituitary secretion, in rats treated *in vivo* with RU486 and stimulated with GnRH pulses.

## IV. Discussion

We have shown that when RU486 is administered on the afternoon of proestrus, the pattern of basal LH and FSH pituitary secretion is remarkably different the next morning. Basal secretion rates of FSH from pituitaries were significantly suppressed in RU486-treated rats. This observation concurs with our previous *in vivo* observations that the secondary FSH surge is suppressed following treatment with RU486, despite a normal drop in inhibin. In contrast to our observation *in vivo*, basal secretion rates of LH from pituitaries of rats treated with RU486 were significantly higher than basal secretion rates from pituitaries of control rats. Because previous work in our laboratory has demonstrated that there is a normal drop in pituitary LH response to GnRH stimulation on the morning of estrus in comparison to proestrus (Fallest and Schwartz, 1990), we interpret the present data to mean that progesterone on proestrus normally inhibits pituitary LH secretion on the morning of estrus, and treatment with RU486 has released this suppression.

When pulsatile GnRH was delivered to pituitaries of rats treated *in vivo*, we again found there was a significant suppression of secretion of FSH. No differences were found between LH response to GnRH stimulation in controls compared to rats treated with RU486. The data on pituitary FSH secretion, either basal or GnRH-stimulated secretion, suggest a role for progesterone, at the pituitary level, in maintaining FSH secretion, which may have implications for the producing the secondary FSH surge. In contrast, these data suggest that progesterone inhibits basal pituitary secretion of LH, and that the effects of GnRH stimulation *in vitro* can override this inhibition.

## ACKNOWLEDGMENTS

We thank Brigitte R. Mann, Robert J. Valadka, and William Talley for excellent technical assistance; Dr. Jacqueline Ackland for helpful comments on the manuscript; and NIDDK for provision of reagents for the LH and FSH assays. RU486 was the gift of Roussell UCLAF (Romainville, France).

## REFERENCES

Attardi, B. (1984). Endocrinology (*Baltimore*) **115**, 2113–2122.
Banks, J. A., and Freeman, M. E. (1978). *Endocrinology* (*Baltimore*) **102**, 426–432.
Fallest, P., and Schwartz, N. B. (1990). *Biol. Reprod.* **43**, 977–985.
Knox, K. L., and Schwartz, N. B. (1992). *Biol. Reprod.*, **46**, 220–225.
Nequin, L. G., Alvarez, J., and Schwartz, N. B. (1979). *Biol. Reprod.* **20**, 659–670.
Schwartz, N. B., and Channing, C. P. (1977). *Proc. Natl. Acad. Sci. U.S.A.* **74**, 5721–5724.
Schwartz, N. B., Rivier, C., Rivier, J., and Vale, W. (1985). *Biol. Reprod.* **32**, 391–398.
Woodruff, T. K., D'Agostino, J. B., Schwartz, N. B., and Mayo, K. (1989). *Endocrinology* (*Baltimore*) **124**, 2193–2199.

# Release of Immunoreactive Inhibin from Perifused Rat Ovaries: Effects of Forskolin and Gonadotropins during the Estrous Cycle

JACQUELINE F. ACKLAND, BRIGITTE G. MANN, AND
NEENA B. SCHWARTZ

*Department of Neurobiology and Physiology, Northwestern University, Evanston,
Illinois 60208*

## I. Introduction

Inhibin, a glycoprotein hormone produced by the gonads, specifically suppresses the release of follicle-stimulating hormone (FSH) from the pituitary (Ling *et al.*, 1985). It is comprised of two disulfide-linked subunits: an $\alpha$ and either $\beta_A$ or $\beta_B$ (Esch *et al.*, 1987). *In vivo* control of FSH by inhibin has been demonstrated with recombinant inhibins (Rivier *et al.*, 1991) and by administering inhibin antiserum to intact rats (Rivier and Vale, 1989). *In vitro* inhibin inhibits the basal release of FSH and the gonadotropin-releasing hormone (GnRH)-stimulated release of luteinizing hormone (LH) and FSH from rat pituitary cells (Campen and Vale, 1988). In the female rat the main source of inhibin is the granulosa cells of the ovary. Levels of both serum inhibin (Hasegawa *et al.*, 1988) and ovarian inhibin mRNA (Woodruff *et al.*, 1988) change during the estrous cycle. Previous studies of factors affecting the synthesis and release of inhibin *in vitro* have used granulosa cell cultures, with cells obtained from pregnant mare's serum gonadotropin (PMSG)-stimulated immature rats (Bicsak *et al.*, 1986). Using these cultures a number of factors have been shown to regulate the release of inhibin *in vitro*, including gonadotropins, forskolin, and growth factors (Bicsak *et al.*, 1986). We have developed an *in vitro* dynamic perifusion system using quartered ovaries from adult rats to study the release of immunoreactive inhibin-$\alpha$ (irl$\alpha$) at different stages of the estrous cycle. Ovaries were perifused in an ACUSYST perifusion system (Endotronics Inc., Coon Rapids, Minnesota) with and without forskolin and gonadotropins. For ovaries removed on the day of metestrus, forskolin alone significantly stimulated irl$\alpha$ release by 45% over basal levels; gonadotropins stimulated irl$\alpha$ release by 42% over basal levels. For ovaries removed during proestrus, neither forskolin nor gonadotropins

531

Copyright © 1993 by Academic Press, Inc.
All rights of reproduction in any form reserved.

532                        JACQUELINE F. ACKLAND ET AL.

significantly stimulated irlα release over basal levels. For ovaries removed during estrus, forskolin alone did not significantly alter Irlα release over basal levels, but LH and FSH stimulated irlα release by 28% over basal levels. There was a significant effect of cycle stage on irlα release ($p < 0.0001$). Total irlα release was highest for ovaries removed during proestrus and lowest for ovaries removed during estrus for all treatment groups. These results indicate that this perfusion system is suitable for studying factors controlling the release of irlα *in vitro*, and that the effects of forskolin and gonadotropins vary depending on the cycle stage of the donor rat and presumably the microanatomy of the follicular pool.

## II.  Methods

### A.  ANIMALS AND TISSUE

Female Sprague–Dawley rats were obtained from Charles River (Portage, Michigan). They were housed four to a cage and maintained on a 14:10 light:dark (hours) cycle, with lights on at 0500 hours. They were provided with food and water *ad libitum*. Estrous cyclicity was monitored by daily vaginal smears and only rats that displayed at least two consecutive 4-day estrous cycles were used. Animals (three/experiment) were killed by decapitation at 0900 hours. Ovaries were collected, quartered, and washed in warmed, gassed basal medium containing 0.33 U/liter heparin (Elkins-Sinn, Cherry Hill, North Carolina) before being placed in the ACUSYST chambers.

### B.  PERIFUSION MEDIA AND METHODS

Ovaries were perifused in an ACUSYST six-channel perifusion system using one quartered ovary per 500-µl chamber. Basal perifusion medium was Medium 199 (Gibco) supplemented with 0.5% bovine serum albumin (BSA). 0.2 U/liter insulin (Lente Ilentin I, Eli Lilly), 25 µg/ml gentamicin (Sigma), and 1.25 mg/ml bacitracin (Sigma). Medium was warmed to 37°C and gassed with 95% $O_2$/5% $CO_2$ before delivery to the tissue. Ovaries were perifused at a flow rate of 6ml/hour for a total of 5.5 hours. Fractions were collected every 10 minutes and stored at −20°C until assay for irlα. Three different treatments were used at each cycle stage: (1) basal medium only, (2) basal medium plus $10^{-7}$ M forskolin, and (3) basal medium plus 10 ng/ml LH and 40 ng/ml FSH. All treatments were included continuously throughout each experiment. Ovine LH (oLH) (G3-330BR) and oFSH (G42 211B) were provided by Dr. H. Papkoff (University of California,

INHIBIN RELEASE FROM PERIFUSED OVARIES          533

San Francisco). In each experiment one ovary from each rat was perifused with basal medium and one overy was perifused with forskolin or LH and FSH.

## C. RADIOIMMUNOASSAYS

Irl$\alpha$ was measured using an antiserum raised against synthetic porcine inhibin $\alpha$ (1–26)-Gly-Tyr. The same synthetic peptide was used for standards and for the preparation of tracer. This assay has been described in detail elsewhere (Ackland *et al.*, 1990). The intra- and interassay CVs were 9.7 and 16.3%, respectively. Results are expressed as picograms/milliliter of synthetic peptide standard.

## D. STATISTICS

All results are expressed as mean $\pm$ SEM. Differences between treatments and cycle stages were calculated from the hourly secretion rates of irl$\alpha$ for each ovary by analysis of variance (ANOVA). All statistics were computed using the SYSTAT computer program (Systat, Inc., Evanston, Illinois).

## III. Results

There was a significant effect of cycle stage on irl$\alpha$ released from ovaries perifused *in vitro* ($p < 0.001$). For all treatments, irl$\alpha$ release was highest for ovaries removed on the morning of proestrus and lowest for ovaries removed on the morning of estrus. These secretion rates reflect the serum irl$\alpha$ levels measured in donor rats, which are also highest during proestrus and lowest during estrus (Hasegawa *et al.*, 1988). Additionally, for ovaries removed on the morning of estrus, the secretion rate of irl$\alpha$ declined more rapidly toward the end of the experiment than was the case for ovaries removed on the morning of metestrus or proestrus.

The effects of forskolin on irl$\alpha$ release from perifused ovaries varied with cycle stage of donor rats (Fig. 1). For ovaries removed on the morning of metestrus, forskolin significantly increased the release of irl$\alpha$ from these ovaries by an average of 45% over basal levels ($p < 0.001$). For ovaries removed from rats on the morning of proestrus, forskolin did not have a significant stimulatory effect on irl$\alpha$ release when administered alone.

The effects of LH and FSH on irl$\alpha$ release also depended on cycle stage (Fig. 2). For ovaries removed on the morning of metestrus, LH stimulated irl$\alpha$ release by an average of 42% ($p < 0.001$). For ovaries removed on the morning of proestrus, LH and FSH did not have any significant

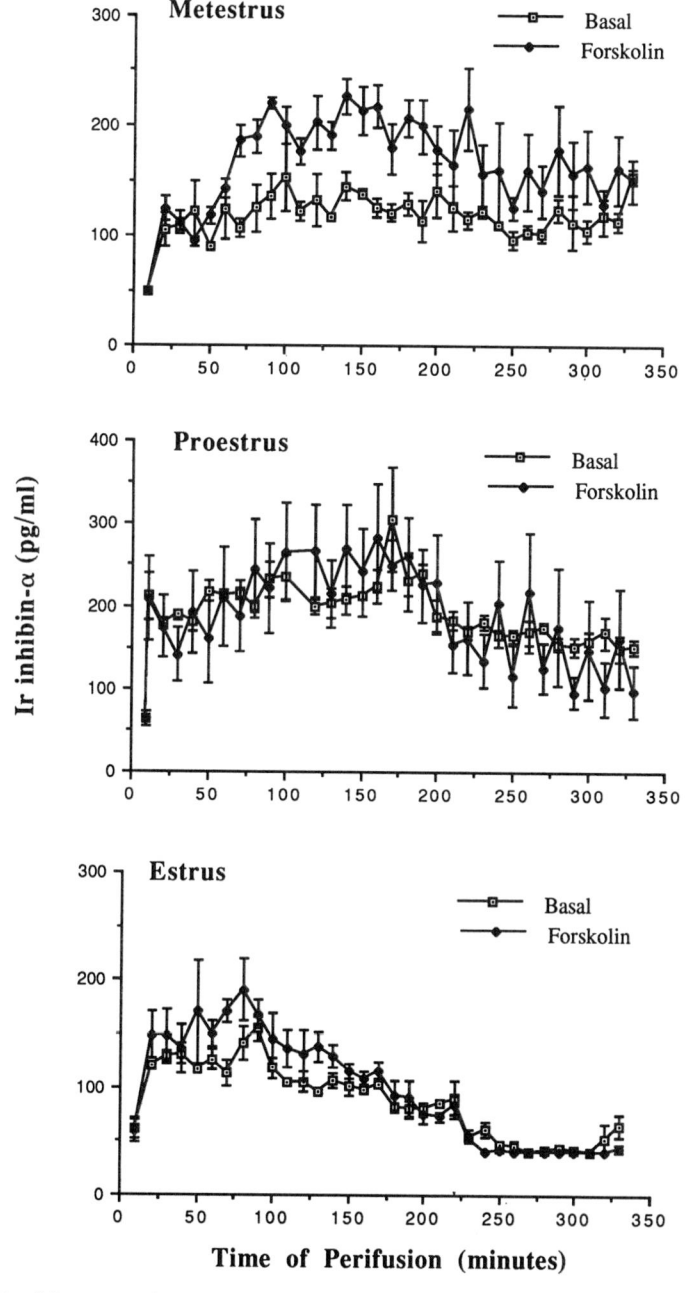

FIG. 1. Mean secretion rates of immunoreactive (Ir) inhibin-$\alpha$ (irI$\alpha$) from ovaries re-moved on the morning of metestrus (upper panel), proestrus (center panel), and estrus (bottom panel), and perifused with either basal medium (open symbols) or basal medium plus $10^{-7}$ $M$ forskolin (solid symbols). All values are the mean $\pm$ SEM, $n = 3$.

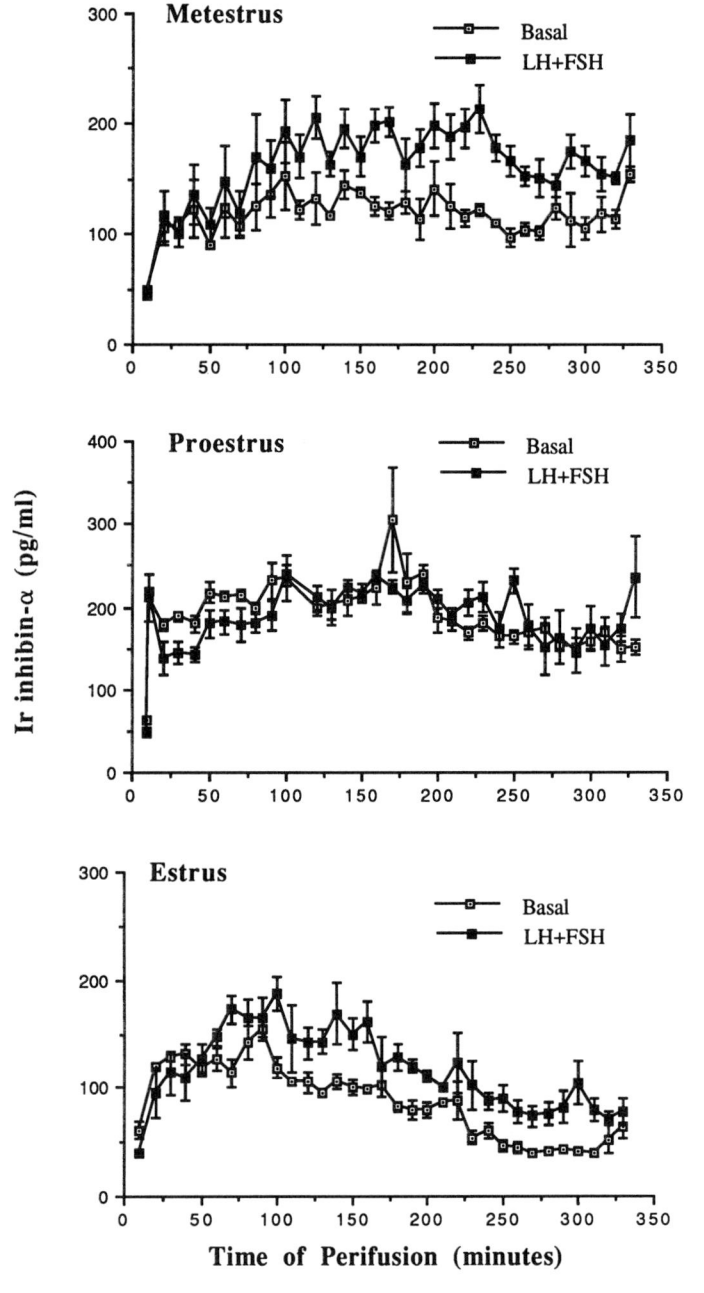

FIG. 2. Mean secretion rates of immunoreactive (Ir) inhibin-$\alpha$ (irI$\alpha$) from ovaries removed from rats on the morning of metestrus (upper panel), proestrus (center panel), and estrus (bottom panel), and perifused with basal medium (open symbols) or basal medium plus LH and FSH (solid symbols). All values are the mean $\pm$ SEM, $n$ = 3.

536                    JACQUELINE F. ACKLAND ET AL.

stimulatory effect, whereas for ovaries removed on the morning of estrus, LH and FSH significantly stimulated irIα release by 28% over basal levels ($p < 0.005$).

## IV. Discussion

For ovaries removed on the morning of metestrus, either forskolin and gonadotropins significantly stimulated irIα release. The doses of forskolin and gonadotropins used have been shown to be give maximum stimulation of inhibin release from cultured granulosa cells (Bicsak et al., 1986). However the stimulation observed in this study is less than that observed for granulosa cells. This could be because the length of incubation time with the factors is considerably shorter or because there is less penetration of factors into the cells due to the use of tissue fragments rather than dispersed cells. For ovaries removed on proestrus morning, neither forskolin nor gonadotropins had a significant effect on irIα release, probably because inhibin secretion is already maximal at this time. For ovaries removed on the morning of estrus, only gonadotropins, and not forskolin, significantly stimulated irIα release. The differences in the actions of forskolin and gonadotropins are presumably due to their different sites of action on the cells and the differences in the follicles at the various cycle stages.

The stimulatory effects of gonadotropins and forskolin on inhibin release from cultured granulosa cells are well documented (Bicsak et al., 1986). However, serum levels of gonadotropins and inhibin and ovarian levels of mRNA for inhibin subunits change frequently during the 4-day rat estrous cycle (Hasegawa et al., 1988; Woodruff et al., 1988). Thus, incubating granulosa cells with potential stimulants for 48 hours or longer does not necessarily reflect true physiological circumstances. In this study we have attempted to examine the effects of gonadotropins and forskolin in a more dynamic system, monitoring minute-to-minute changes in inhibin release. Additionally, the system permits the study of effects of potential regulators on ovaries containing intact follicles, including thecal cells and granulosa cells, and allows us to study the effects of potential regulators on ovaries removed under varying physiological conditions.

### REFERENCES

Ackland, J. F., D'Agostino, J. B., Ringstrom, S. J., Hostetler, J. P., Mann, B. G., and Schwartz, N. B. (1990). *Biol. Reprod.* **43**, 347–352.
Bicsak, T. A., Tucker, E. M., Cappel, S., Vaughan, J., Rivier, J., Vale, W., and Hsueh, A. J. W. (1986). *Endocrinology (Baltimore)* **119**, 2711–2719.

Campen, C. A., and Vale, W. (1988). *Endocrinology (Baltimore)* **123**, 1320–1327.
Esch, F. S., Shumasaki, S., Cooksey, K., Mercado, M., Mason, A. J., Ying, S.-Y., Ueno, N., and Ling, N. (1987). *Mol. Endocrinol* **1**, 388–396.
Hasegawa, Y., Miyamoto, K., and Igarashi, M. (1988). *J. Endocrinol.* **121**, 91–100.
Ling, N., Ying, S.-Y., Ueno, N., Esch, F., Denoroy, L., and Guillemin, R. (1985). *Proc. Natl. Acad. Sci. U.S.A.* **82**, 7217–7221.
Rivier, C., Schwall, R., Mason, A., Burton, L., Vaughan, J., and Vale, W. (1991). *Endocrinology (Baltimore)* **128**, 1548–1554.
Rivier, C., and Vale, W. (1989). *Endocrinology (Baltimore)* **125**, 152–157.
Woodruff, T. K., D'Agostino, J. B., Schwartz, N. B., and Mayo, K. E. (1988). *Science* **239**, 1296–1299.

RECENT PROGRESS IN HORMONE RESEARCH, VOL. 48

# Identification and Partial Purification of a Germ Cell Factor That Stimulates Transferrin Secretion by Sertoli Cells

C. PINEAU,* V. SYED,[†] C. W. BARDIN,* B. JÉGOU,[†] AND C. Y. CHENG*

*The Population Council, New York, New York, 10021 and [†] Groupe d'Etude de la Reproduction chez le Mâle, INSERM, Contrat Jeune Formation 91-04, Université de Rennes I, 35042 Rennes Cedex, France*

The Sertoli cell is the major secretory component in the seminiferous tubule; it secretes many serum and testis-specific proteins and determines the fluid composition in the seminiferous tubular lumen. Morphological studies have revealed that the cytoplasm of Sertoli cells is extensively interdigitated into the developing germ cells. Based on this morphological intimacy between Sertoli cells and germ cells and the fact that all of the processes of spermatogenesis and spermiogenesis are segregated from the systemic circulation by the blood–testis barrier, it was postulated that the Sertoli cell plays a major role in germ cell development. It has been shown that germ cell-conditioned media can affect Sertoli cell secretory functions, including the secretions of androgen binding protein (Le Magueresse and Jégou, 1986), transferrin (Ireland and Welsh, 1987; Le Magueresse *et al.*, 1988; Djakiew and Dym, 1988), and inhibin (Pineau *et al.*, 1990). Recent studies from our laboratory have shown that germ cell-conditioned media affected the secretions of clusterin (sulfated glycoprotein-2), $\alpha_2$-macroglobulin, and testins by Sertoli cells. However, the molecular identity of these biological factors is not yet known. It is the aim of this study to identify and purify these biological factors. We have now described a purification scheme for the isolation of a biological factor that stimulates transferrin secretion by Sertoli cells.

Germ cells were isolated from adult rat testes by trypsinization. Pachytene spermatocytes (SPC) and early spermatids (SPT) were enriched to a purity of greater than 85% by centrifugal elutriation judged by PAS–hematoxylin staining. Total germ cells (GC), SPC, and SPT at a concentration of $2.5 \times 10^6$ cells/ml were maintained in serum-free Ham's F12/DME medium supplemented with 6 m$M$ sodium lactate and 2 m$M$ sodium pyruvate; cells were cultured for 20 hours at 32°C in a humidifed atmosphere of 5% $CO_2$/95% air (v/v).

GC-, SPC-, and SPT-conditioned media were equilibrated and concentrated to about 10 ml using a Millipore Minitan tangential ultrafiltration

539

Copyright © 1993 by Academic Press, Inc.
All rights of reproduction in any form reserved.

540                                    C. PINEAU ET AL.

unit equipped with eight Minitan plates with a $M_r$ cutoff at 10,000. The concentrates were then fractionated individually by anion-exchange HPLC as previously described (Cheng and Bardin, 1986; Cheng et al., 1989, 1990). When an aliquot from each of these HPLC fractions was resolved by SDS–PAGE onto 10% T polyacrylamide gels, and proteins were visualized by silver staining, it was noted that media derived from GC, SPC, and SPT displayed very similar patterns; however, defined stages of germ cells contained specific proteins.

An aliquot from each of these fractions was also bioassayed for possible effects on Sertoli cell secretory functions; each aliquot was incubated with primary cultures of Sertoli cells ($2 \times 10^5$ cells/500 $\mu$l/well) for 24 hours. Thereafter, spent media were collected and the concentrations of transferrin and clusterin were quantified by specific radioimmunoassays (Rossi et al., 1989; Grima et al., 1990). A laboratory standard was established using a pool of conditioned medium derived from total germ cells, which was run in every bioassay at multiple doses. Negative controls were prepared using HPLC buffers. One unit of biological activity was defined as 50% stimulation (or inhibition) of transferrin (or clusterin) secretion by Sertoli cells compared to control incubations in which no germ cell-conditioned medium was present.

Using a pool of 10 liters of germ cell-conditioned media (total protein content was about 50 mg) fractionated by preparative anion-exchange

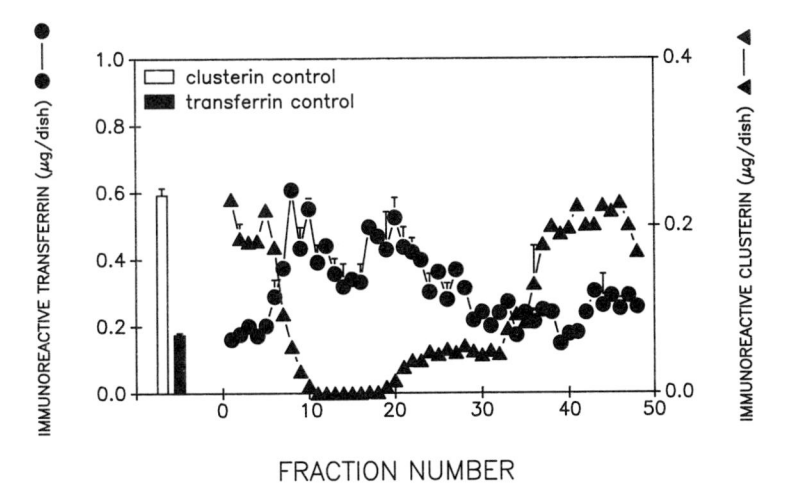

FIG. 1.   Bioassays on the effects of the anion-exchange column fractions on Sertoli cell transferrin and clusterin secretion following the preparative fractionation of total germ cell-conditioned media onto a Mono Q HR 10/10 (10 × 100 mm inner diameter) column. Aliquots of 20 and 2 $\mu$l from each of these fractions were bioassayed for their effects on Sertoli cell transferrin and clusterin secretion, respectively. Mean ± SEM ($n = 3$).

SERTOLI CELL TRANSFERRIN SECRETION 541

HPLC, two distinctive peaks of transferrin stimulatory activity were noted (Fig. 1), whereas the clusterin inhibitory activity was eluted as a broad peak. These observations were also noted using SPT-conditioned media (data not shown). We have partially purified the germ cell factor that stimulates transferrin secretion by Sertoli cells using sequential HPLC on anion-exchange, gel permeation, $C_8$, and $C_{18}$ reversed-phase columns. This biological factor displayed an apparent $M_r$ of 28,000 on a 14% T SDS–polyacrylamide gel under reducing conditions. However, an apparent $M_r$ of about 22,000 was noted when the purified protein was resolved on a 14% T SDS–polyacrylamide gel under nonreducing conditions, suggesting that this is a globular polypeptide chain (Fig. 2).

In summary, proteins contained in GC-, SPC-, and SPT-conditioned media displayed similar patterns following anion-exchange HPLC and SDS–PAGE; however, defined stages of germ cells contain specific proteins. GC-conditioned media also contain multiple biological factors that modulate Sertoli cell secretory functions. A factor that stimulates Sertoli cell transferrin secretion has been purified to apparent homogeneity.

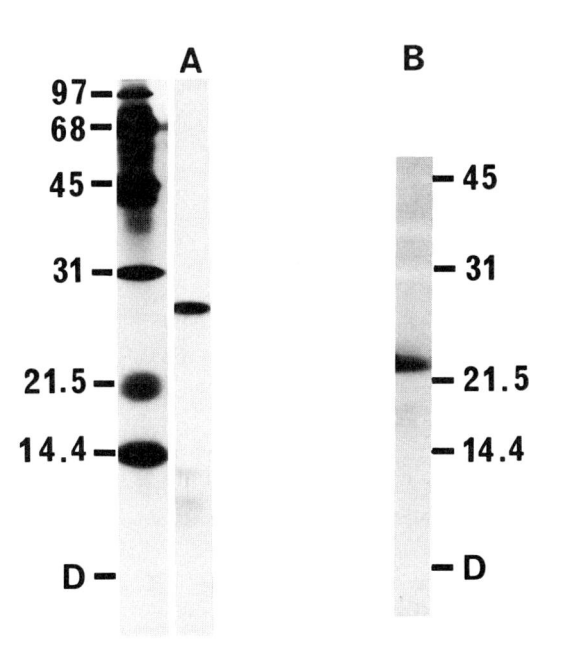

FIG. 2. Purification of the biological factor in germ cell-conditioned media that stimulates Sertoli cell transferrin secretion. Following four sequential HPLC steps, the purity of the biological factor that stimulated Sertoli cell transferrin secretion was confirmed by SDS–PAGE on silver-stained SDS–polyacrylamide gels (14% T) under reducing (A) and nonreducing (B) conditions.

542                    C. PINEAU ET AL.

The identification and purification of the biological factors in germ cell-conditioned media that affect Sertoli cell functions will enhance our understanding of germ cell and Sertoli cell interactions in the seminiferous epithelium.

REFERENCES

Cheng, C. Y., and Bardin, C. W. (1986). *Biochemistry* **25,** 5276–5288.
Cheng, C. Y., Grima, J., Stahler, M., Guglielmotti, A., Silvestrini, B., and Bardin, C. W. (1989). *J. Biol. Chem.* **264,** 21386–21393.
Cheng, C. Y., Grima, J., Stahler, M., Guglielmotti, A., Silvestrini, B., and Bardin, C. W. (1990). *Biochemistry* **29,** 1063–1068.
Djakiew, D., and Dym, M. (1988). *Biol. Reprod.* **39,** 1193–1205.
Grima, J., Zwain, I., Lockshin, R. A., Bardin, C. W., and Cheng, C. Y. (1990). *Endocrinology (Baltimore)* **126,** 2989–2997.
Ireland, M. E., and Welsh, M. J. (1987). *Endocrinology (Baltimore)* **120,** 1317–1326.
Le Magueresse, B., and Jégou, B. (1986). *Biochem. Biophys. Res. Commun.* **141,** 861–869.
Le Magueresse, B., Pineau, C., Guillou, F., and Jégou, B. (1988). *J. Endocrinol.* **118,** R13–R16.
Pineau, C., Sharpe, R. M., Saunders, P. T. K., Gérard, N., and Jégou, B. (1990). *Mol. Cell. Endocrinol.* **72,** 13–22.
Rossi, V., Cheng, C. Y., Gunsalus, G. L., Bardin, C. W., and Spitz, I. M. (1989). *J. Androl.* **10,** 466–471.

# INDEX

## A

A431 cell growth, stimulation by Sertoli cell-secreted growth factor, 511–515

ACTH, concentrations, rank-related differences in baboons, 442–444

Acute promyelocytic leukemia
chromosomal translocation, 112–113
fusion retinoic acid receptor $\alpha$, 111–119
proposed model, 118

Adrenergic receptor, subtypes, properties, 278–279

$\alpha_1$-Adrenergic receptor, as protooncogene, 285

$\alpha_{1B}$-Adrenergic receptor
constitutive activation, 283–284
membrane topography model, 282–283

Adrenocortical axis, baboon, 439–441

Adrenocortical function, baboons, rank-related differences, 441–445

Aging subjects, growth hormone therapy, 206–207

Alanine scanning, growth hormone binding protein, 265–266

Allosteric enzyme, insulin receptor as, 308–309

$\alpha$ subunit, cDNA and gene, 394–396
putative nTREs, 400–401

AMH$_{Bruxelles}$, 45–48

AMH$_{Chicago}$, 48–52

AMH$_{New\ York}$, 51–53

Amino acid sequence
anti-Müllerian hormone, 4–5
FSH receptor, 65

Antibody formation, growth hormone side effect, 220

Anticancer effects, anti-Müllerian hormone, 32–33

Antiidiotypic antibodies, AMH effects, 22–23

Anti-Müllerian hormone, 1–54, *see also* Persistent Müllerian duct syndrome
amino acid sequence, 4–5
anticancer effects, 32–33
bilateral cryptorchidism, 37, 40
bioactive site mapping, 33–35
biochemical structure, 2–7
bovine, 2–3
human recombinant, 2–4
mature proteins and structure predictions, 4–7
N- and C-terminal proteolytic fragment homology, 4, 6
precursors, 4–5
calspermin, 81–83
calmodulin kinase IV and, 83–88
C terminus, 6, 8–9
gene
cloning, 9–11
mapping, 15
mutations, *see* Persistent Müllerian duct syndrome
regulation, 19–21
structure, 11–14
hydrophobic cluster analysis, 6–7
immunoenzyme-linked assay, 35–39
intersex states, 37, 40
lung maturation and immune system effects, 33
morphological effects on female genital primordia, 21–28
antiidiotypic antibodies, 22–23
freemartin model, 23–26
germ cell deprivation role in AMH-induced ovarian lesions, 25, 27–28
Müllerian duct effect, 21–22
transgenic mice with chronic, unregulated AMH production, 25, 27
ontogeny, 15–21
gene regulation, 19–21

543

ovarian production, 18–19
testicular production, 16–18
oocyte meiosis-retarding effect, 32
ovarian steroidogenesis pathway effects,
28–31
preparation proteolytic fragments, 9
processing, 9
protein phosphorylation effect, 33
testes
descent effect, 32
differentiation effect, 31–32
Antiprogesterones, RU486, 523–530
ATP/GTP binding sites, prolactin receptors,
138
AtT-20 cells, see ELH prohormone
Autonomic nervous system, social rank in
baboons and, 455–456
Autophosphorylation, insulin receptor,
305–306, 309
adrenocortical axis, 439–441
in vitro, 303
social dominance, 438

**B**

Baboons, psychoendocrine studies, 437–466
ACTH concentrations, 442–444
adrenocortical axis, 439–441
autonomic nervous system and cardiovascular function, 455–457
cholesterol metabolism, 452–453
complications of association between social ranks and endocrine profiles,
457–466
effect of individual experiences of animal, 461–463
effect of primate society nature,
458–460
personality effect, 463–466
cortisol concentrations, 440–441
rank-related differences, 441–442
elevated basal cortisol concentrations in
low-ranking baboons, 446–447
immunologic profiles, 454–455
LH suppression, 447–450
rank-related differences in adrenocortical
function, 441–445
social dominance, 438
depression and, 445–446

testosterone concentrations
consequences of rank difference, 452
dominant males, 450–452
mechanisms mediating stress-induced
suppression, 447–450
βARK, mRNA, 286–288
Biosynthetic growth hormone, 181–182
Body composition, modification, growth
hormone therapy, 205–207
Bombesin, 366–369
Bombesin-like peptides, 365–386
bombesin, 366–369
classes, 365–366
expression in SCLC cells, 384, 386
gastrin-releasing peptide, 366–369
neuromedin B, 369–371
perspectives, 384, 386
phyllolitorins, 371–372
ranatensin, 369–371
receptors for the bombesin-like, 372–385
GRP receptor, 373–383
NMB receptor, 382, 384–385
Bone physiology, growth hormone therapy,
216–217
Burns, growth hormone therapy, 213

**C**

Calcium
channel activity, FSH-stimulated,
517–518
concentration and parathyroid cells plaque
formation, 473–474
transport in Sertoli cells and liposomes,
induction by FSH, 517–520
Calmodulin-binding region, 82–83
Calmodulin kinase II, 87–88
Calmodulin kinase IV
amino acid sequence, 83
baculovirus-expressed, activity, 86–87
calspermin and, 83–88
catalytic domains, 85
changes during germ cell development,
92–93
functions, 93–95
gene structure, 86–87
mRNA from contiguous DNA, 89–92
transfer vector, 86
Calspermin, 81–83
calmodulin kinase IV and, 83–88

changes during germ cell development, 92–93

functions, 93–95

mRNA, 84

transcriptional start site, 89–90

from contiguous DNA, 89–92

polymerase chain reaction, 84

cAMP levels, Sertoli cells, germ cell-conditioned media effect, 500

Carbohydrate metabolism, growth hormone effects, 203

Carboxypeptidase E, 420

Carboxypeptidase-like enzyme, 420

Cardiovascular function, social rank in baboons and, 455, 457

Catabolism, growth hormone as defense against

critical illness, 212–215

from glucocorticoid therapy, 215–216

hypocaloric nutrition, 211

nitrogen retention, 215

in trauma, 211

cDNA

bovine anti-Müllerian hormone, 9–10

encoding NMB, 370

TRH-R, cloning, 344–348

C-erbA, 405–406

C-*fos* mRNA

expression in Sertoli cells, 500–501

uterine, 477–480

Chimeras, using embryonic stem cell clones, 240–241

Chloramphenicol acetyltransferase expression, estradiol stimulation of, 13–14

Cholesterol metabolism, social ranking in baboons and, 452–453

Chronic renal failure, growth hormone therapy, 196–197

Coelectroporation, 243–245

Congenital anomalies, growth hormone therapy, 200–201

Corticosteroid-induced growth failure, growth hormone therapy, 201–202

Cortisol concentration

elevated, consequences in low-ranking baboons, 446–447

low-ranking males, displacement and, 459–460

percentage of approach–avoid interactions and, 462–463

social instability and high-ranking males, 459

COUP-TF

bacterially expressed, 106–108

interactions with retinoid X receptors, 107–109

CRBPII, 103

Critical illness, growth hormone therapy, 212–215

Cryptorchidism, bilateral, AMH, 37, 40

Cytokine receptors, expanded family, 140–141

## D

Depression, neuroendocrine abnormalities, 445

Diabetes, regulation of IRS-1, 331

Diacylglycerol, 329

1,25-Dihydroxyvitamin $D_3$, 102

$\alpha-\beta$ Disulfide bonding, 298

DNA repair assay, mouse mammary epithelia, 482

Dopamine, hypophysiotropic neuron abnormalities, 491–492

Dopamine receptor, subtypes, properties, 279–281

Double-hit gene replacement, site-directed mutagenesis, 247–248

Down-regulation, receptors, 341

Down syndrome, growth hormone therapy, 199

## E

EGF receptor, 315

Egg-laying prohormone, *see* ELH prohormone

ELH prohormone

model for processing and sorting, 432–434

structure and processing, 422–432

amino-terminal intermediate, 423, 426, 428

biosynthesis, 425

carboxy-terminal intermediate, 423–426

cleavage, 429

degradation, 426

dense-core vesicles, 426–427

differential sorting, 423, 425
mutations, 429–432
Embryonic stem cells
clonal heterogeneity and maintenance of
totipotency, 239–241
modifying mice genetically, 237–238
Enzymes, prohormone-processing, 418–421
Erbstatin, 511–512
Estradiol
metabolism, mouse mammary epithelia,
482, 484
stimulation of chloramphenicol acetyltransferase expression, 13–14
Estrogen
induced c-*fos* mRNA levels in uterus, progesterone inhibition, 477–480
metabolites, effects on DMBA-pretreated
C57/MG cells, 486
response element, 13–14

## F

Fetal ovary/aromatase assay, 29–31
Follicle-stimulating hormone
actions in spermatogenesis, 64
calcium transport induction in Sertoli cells
and liposomes, 517–520
physiological response to testis, 61–64
pituitary secretion after RU486 administration, 523–530
spermatogenesis, 80
Follicle-stimulating hormone receptor, *see*
FSH receptor
Forskolin
effect on inhibin release from rat ovaries
during estrous cycle, 531–536
TRH-R mRNA levels, 355–356
Freemartin model, 23–26
FSH receptor, 61–75
amino acid sequence, 65
expression
control, 72–73
in testis, 66, 68–71
gene, 65–68
mRNA, 70–75
promoter activity in transfected Sertoli
cells, 66, 68
protein structure, 66–67
Sertoli cells, 61
spermatogenesis, 70

Functional assay, growth hormone and prolactin receptors, 150–152

## G

Gastrin-releasing peptide, 365, 366–369, *see
also* GRP receptor
Gastrointestinal tract, GRP, 368–369
Gastrointestinal surgery, growth hormone
therapy, 212–213
Gene expression, *see also* Spermatogenesis;
Thyrotropin gene expression
exclusively in male germ cells, 80–81
growth hormone receptors, 146–147
opioid genes, in Sertoli cells, 497–502
Gene mapping, anti-Müllerian hormone, 15
Genetic diseases
growth hormone receptors, 156–157
prolactin receptors, 157–158
Germ cell factor, stimulation of transferrin
secretion by Sertoli cells, 539–542
Germ cells, development, calspermin and
calmodulin kinase IV changes, 92–93
Glucocorticoid therapy, defense against catabolism, growth hormone therapy,
215–216
Glycogen synthase, insulin stimulation,
317–318
Glycoprotein hormones, gene structures,
66, 68
Golgi complex, *see* Prohormone
Gonadotropin
AMH gene regulation, 20
basal secretion after RU486 administration, 525
effect on inhibin release from rat ovaries
during estrous cycle, 531–536
GnRH-stimulated secretion, 529
G protein-coupled receptors, 277–288
activated conformation, 284–285
agonist-mediated desensitization, 285–286
regulation of function by phosphorylation
mechanisms, 285–288
seven-transmembrane domain, 278–279
structure
function relationships, 281–285
heterogeneity and, 277–281
Growth factors, bombesin-like peptides, role
of, 372

# INDEX

Growth hormone, 179–224, 253–274
  action, 165
  alanine-scanning mutagenesis, 262–263
  binding affinity optimization for GH binding protein by phage display, 259–261
  biosynthetic, 181–182
  deficiency, *see also* Laron-type GH insensitivity syndrome
    adults, 204–205
    diagnosis, 181
    etiology, 183
    growth hormone efficacy, 183–185
    somatropin preparations, 185
    use of growth hormone to increase stature, 182–186
  high-resolution functional data, 257–264
  immune system and, 126–127
  induced GH binding protein dimerization, 267–272
  metabolic effects, 202–203
  molecular endocrinology, 272–273
  pituitary, 179–181
  potential adverse effects, 219–223
    growing child, 220–221
    oncogenesis, 221–223
  receptor-selective variant design, 261–264
  scanning mutational analysis, 254–257
  signal transduction pathways, 152–153
  site 2 functional characterization, 268–269
  storage as $(Zn^{2+} \cdot hGH)_2$, 264–265
  therapeutic failure, 221
  use in improve metabolic status
    aging subjects, 206–207
    bone physiology and osteoporosis, 216–217
    clinical trials, 204–205
    defense against catabolism, 210–216
    induction of ovulation, 208–210
    lactation failure, 210
    obesity, 205–206
    rationale for, 204
  uses, 180
    metabolic status improvement, immunologic status modulation, 217–219
  use to increase stature
    chronic renal failure, 196–197
    classical growth hormone deficiency, 182–186
    corticosteroid-induced growth failure, 201–202
    Down syndrome, 199
    idiopathic short stature, 187–193
    intrauterine growth retardation, 198
    Noonan syndrome, 199
    Prader–Willi syndrome, 199–200
    radiation-induced GH secretory dysfunction, 186–187
    skeletal abnormalities and other congenital anomalies, 200–201
    Turner syndrome, 194–196
Growth hormone binding protein, 133
  alanine scanning, 265–266
  biological relevance of dimerization, 272
  expression and biochemical characterization, 253–254
  GH binding affinity optionization by phage display, 259–261
  induced GH dimerization, 267–272
  recruitment of prolactin and placental lactogen for binding, 257–259
  sequential binding model, 268–269
Growth hormone receptor, 123–158
  amino acid sequences, 134–136
  biological and biochemical actions, 124–125
  cDNA
    clone expression, 173, 175
    sequence, 167–168
  conserved sequences, 173–174
  decoding structure–function relationships, 173–176
  distribution and regulation, 127–129
  functional domains, 150–152
  gene expression, 146–147
  genetic diseases and models of receptor defects, 156–157
  homology, 124
  immunolocalization, 175–176
  ligand-binding determinant identification, 142–143
  structure, 131–136
Growth hormone receptor gene
  haplotypes, 168–169
  linkage analysis with Laron phenotype, 166–170
  stop mutations and Laron's syndrome, 170–172
Growth hormone-releasing hormone, hypophysiotropic neuron abnormalities, 493–495

GRP receptor, molecular characterization,
  373–383
  DNA sequences, 379–381
  luminometric assay, 374–376
  oncogenic potential of mutations, 379, 382
  structure, 374, 377–378, 380–381
Guanine nucleotide regulatory proteins, *see*
  G proteins

# H

hGH·(hGHbp)$_2$ complex, 270–272
Hit and run, site-directed mutagenesis,
  245–247
Hormonal signaling, pathways, 99–100
16$\alpha$-Hydroxylation, 481
(Hydroxyproline[9])LHRH, stimulated hCG
  secretion in placental cells, 505–508
Hypermetabolism, growth hormone side effect, 220–221
Hyperproliferation, mouse mammary epithelia, 484–485, 487
Hypocaloric nutrition, growth hormone and,
  211
Hypoglycemia, growth hormone side effect,
  220–221
Hypophysiotropic neuron abnormalities,
  GH- and PRL-deficient dwarf mice,
  489–495
  dopamine/tyrosine hydroxylase, 491–492
  growth hormone-releasing hormone,
    493–495
  method, 490
  somatostatin, 492–494

# I

$\lambda$-ICM-1, 84
Idiopathic short stature, growth hormone
  treatment, 187–193
  clinical trials, 188–193
  rationale for, 187–188
IGF-I, promotion of nitrogen retention, 215
Immune function, social ranking in baboons
  and, 454–455
Immune system
  AMH effect, 33
  growth hormone
    prolactin and, 126–127
    therapy, 217–219

Immunoenzyme-linked assay, anti-
  Müllerian hormone, 35–39
Inhibin, release from rat ovaries, forskolin
  and gonadotropin effects during estrous
  cycle, 531–536
Insulin, action, 291–292
  cascade, 325–326
  phosphorylations, 310–311
  regulation in physiologic and pathologic
    states, 329–332
Insulin receptor, 291–332
  as allosteric enzyme, 308–309
  $\alpha$ subunit, 292–293
    cysteine residues, 298–300
    domain structure, 296–300
    function, 309
    negative cooperative properties,
      297–298
  autophosphorylation, 305–306, 309
  behavior in artificial lipid membranes,
    301–302
  $\beta$ subunit, 292–293
    domain structure, 300–301
    intracellular region, 304–308
  $\beta$ unit
    catalytic domain, 304–305
    C-terminal domain, 306
  biosynthesis, 294–296
  *in vitro* autophosphorylation, 303
  *in vitro* mutants, transmembrane domain,
    301–304
  mRNAs, 295
  multisite phosphorylation regulation,
    310–313
  mutant function, 310–312
  structure, 292–293
  substrates, 313–329
    alternative signaling mechanisms,
      328–329
    model, 313
    pp185/IRS-1, 315–327
    tyrosine phosphorylation, 327–328
    tyrosine kinase activation, 306–308
Insulin receptor gene, 294
Insulin receptor tyrosine kinase inhibitors,
  512
Intersex states, anti-Müllerian hormone, 37,
  40
Intracellular receptor, 99–100
Intrauterine growth retardation, growth hormone therapy, 198

# INDEX

IRS-1, 315–317, 323
  abundance, 319
  characteristics, 316
  phosphorylation, 316, 317
    potential sites, 320–321
  regulation, 330
  sequencing, 319
  SH2 domains, 323–324
  structural features, 320–321
  YMXM motifs, 322–323
1-(5-Isoquinolinesulfonyl)-2-
    methylpiperazine dihydrochloride, 353

## J

Jost factor, see Anti-Müllerian hormone

## K

KEX2 gene, 418–419

## L

Lactation failure, growth hormone use, 210
Laron phenotype, linkage analysis with
    growth hormone receptor gene, 166–170
Laron-type GH insensitivity syndrome,
    165–176
  growth hormone receptor
    linkage analysis of receptor gene, 166–170
    stop mutations, 170–172
    structure–function relationships,
      173–176
  molecular heterogeneity associated with,
    172
Ligand-binding determinants, identification
  growth hormone receptors, 142–143
  prolactin receptors, 143–146
Lipid membranes, artificial, insulin receptor
    behavior, 301–302
Lipid metabolism, GH effects, 203
Liposomes, calcium transport induction by
    FSH, 517–520
Luminometric assay, for bombesin receptor
    expression, 374–376
Lung maturation, AMH effect, 33
Luteinizing hormone
  pituitary secretion after RU486 adminis-
    tration, 523–530
  suppression, in baboons, 447–450

Luteinizing hormone-releasing hormone,
    stimulated hCG secretion in placental
    cells, 505–508

## M

Malignancies, growth hormone effects,
    221–223
Mammary epithelial cells, mouse, 481–487
  anchorage-independent growth, 484–486
  cell proliferation and anchorage-indepen-
    dent assay, 483
  DNA repair assay, 482
  estradiol metabolism, 482, 484
  estrogen metabolites, 486
  exposure to steroids and DMBA, 482
  hyperproliferation, 484–485, 487
Membrane topography model, $\alpha_{1B}$-adrener-
    gic receptor, 282–283
Milk protein genes, stimulation by prolactin,
    154–155
Mitogen-activated protein kinase, 325–326
Mitogenic activity, prolactin, 154
Molecular defects, growth hormone receptor
    gene, patients with Laron's syndrome,
    172
Mouse, see Hypophysiotropic neuron abnor-
    malities; Mammary epithelial cells; Site-
    directed mutagenesis
mRNA
  $\beta$ARK, 286–288
  calmodulin kinase IV and calspermin,
    from contiguous DNA, 89–92
  calspermin, 84
    encoding, 83
    transcriptional start site, 89–90
  FSH receptor, 70–75
  insulin receptor, 295
  prolactin receptors levels, 148–149
Müllerian duct, AMH effect, 21–22
Müllerian inhibiting substance, see Anti-
    Müllerian hormone
Mutagenesis, see Site-directed mutagenesis

## N

Negative thyroid hormone-response ele-
    ments
  rat $\alpha$ subunit gene, 400–401
  TSB$\beta$ subunit gene, 401–404

Neuromedin B, 365, 369–371
Neuromedin B receptor, 382, 384–385
Neuropeptides, 415–416, *see also* Pro-
    hormone
    structure, 416–418
Nitrogen retention, IGF-I versus GH in pro-
    motion of, 215
Noonan syndrome, growth hormone ther-
    apy, 199
Nuclear receptors, 3–4–5 rule, 106–107
Null alleles, generation by gene targeting,
    241–248
    coelectroporation, 243–245
    double-hit gene replacement, 247–248
    hit and run, 245–247

## O

Obesity, growth hormone therapy, 205–206
Oncogenesis, growth hormone effects,
    221–223
Ontogeny, anti-Müllerian hormone, 15–21
Oocyte meiosis, retardation, 32
Opioid genes, expression in Sertoli cells,
    497–502
Osteoporosis, growth hormone therapy,
    216–217
Ovary
    anti-Müllerian hormone production, 18–19
    fetal, morphological virilization and
        AMH, 23–26
    lesions, AMH-induced, germ cell depriva-
        tion role, 25, 27–28
    steroidogenesis pathway, AMH effects,
        28–31
Ovulation, induction, growth hormone use,
    208–210

## P

Parathyroid cells, 471–475
    heterogeneity, 472–475
Parathyroid hormone, reverse hemolytic
    plaque, 471–472
Pathophysiologic states, 330–332
PC1/PC3, 419–420
PC2, 419–420
Peptides, see Bombesin-like peptides
Peptidylglycine-α-amidating-
    monooxygenase, 421

Persistent Müllerian duct syndrome
    AMH gene mutations, 45–54
        AMH$_{Bruxelles}$, 45–48
        AMH$_{Chicago}$, 48–52
        AMH$_{New York}$, 51–53
        heterogeneity of mutations, 52, 54
    AMH-positive and -negative forms, 42–45
    anatomy, 39, 41–42
    genetic transmission, 45
    heterogeneity, 42–43
    preservation of fertility, 42
Phage display, GH binding affinity optimiza-
    tion for GH binding protein, 259–
    261
Phorbol-12-myristate-13-acetate, TRH-R
    mRNA levels and, 353–354, 356
Phosphatidylinositol 3-kinase (PI 3-kinase)
    insulin stimulation, 324–325
    role in insulin action, 323
Phospholipase C, phosphatidylinositol
    glycan-specific, 328–329
Phosphorylation
    G protein-coupled receptors function reg-
        ulation, 285–288
    multisite, insulin receptor, 310–313
Phyllolitorins, 371–372
Pituitary
    dwarfism, treatment with growth hor-
        mone, 179–180, 182–186
    LH and FSH secretion after RU486 admin-
        istration, 523–530
Pituitary growth hormone, 179–181
Placental cells, hCG secretion, LHRH- and
    (hydroxyproline⁹)LHRH-stimulated,
    505–508
Placental lactogen, recruitment to bind
    growth hormone binding protein,
    257–259
PML protein family, 115–116
PML-1 protein, 114–115
Posttranscriptional regulation, TSH gene ex-
    pression, 404
pp185, see Insulin receptor substrate
Prader–Willi syndrome, growth hormone
    therapy, 199–200
Preproenkephalin, expression in Sertoli
    cells, 497–502
    time course, 499
Progesterone, inhibition of estrogen-induced
    uterine c-*fos* mRNA levels, 477–480

INDEX 551

Prohormone, 415–434
  ELH prohormone, 422–432
  processing enzymes, 418–421
Prolactin
  biological and biochemical actions,
    125–126
  conserved sequences, 173–174
  immune system and, 126–127
  mitogenic activity in Nb2 cells, 154
  recruitment to bind growth hormone bind-
    ing protein, 257–259
  signal transduction pathways, 153–155
  stimulation of milk protein genes, 154–
    155
Prolactin receptors, 123–158
  amino acid sequences, 134–136
  ATP/GTP binding sites, 138
  chromosomal localization of genes, 139
  cysteine-rich domain residues, 145–146
  distribution and regulation, 129–131
  expanded family, 140–141
  functional domains, 150–152
  gene expression, 147–150
  genetic diseases and models of receptor
    defects, 157–158
  ligand-binding determinant identification,
    143–146
  mRNA levels, 148–149
  multiple forms, 137–139
  secondary structure, 143–144
  structure, 133–139
  transcript sizes, 149
Protein kinases, catalytic domains, 85
Proteins
  metabolism, growth hormone effects,
    202–203
  phosphorylation, AMH effect, 33
  YMXM motifs, 322–323
Proteolytic processing, *see* Prohormone
PtdIns(4,5)$P_2$-1,2-diacylglycerol-$Ca^{2+}$, 343
Puberty, growth hormone therapy effects,
  184, 192–193

**R**

Radiation-induced GH secretory dysfunction, 186–187
raf-1 kinase, 326
Ranatensin, 369–371

Receptors
  bombesin-like peptides, 372–385
    GRP receptor, 373–383
    NMB receptor, 382, 384–385
  down-regulation, 341
Recombinant DNA techniques, anti-Müllerian hormone production, 2–4
9-*cis*-Reninoic acid, 102–104
Retinoic acid receptor $\alpha$, fusion, acute promyelocytic leukemia, 111–119
Retinoic acid receptors, direct interaction
  with retinoid X receptors, 110–112
Retinoid transduction pathway, 101
Retinoid X receptors, 99–119
  amino acid sequence, 101
  chromosomal localization of genes,
    139
  direct interaction with VDR, TR, and
    RAR, 110–112
  expanded family, 140–141
  expression patterns, 104–106
  half-site spacing, 109–110
  interactions with COUP-TF, 107–109
  9-*cis*-retinoic acid, 102–104
  retinoid transduction pathway, 101
  3–4–5 rule, 106–110
  signaling pathway, 100–101
Reverse hemolytic plaque, parathyroid hormone, 471–472
RU486, pituitary LH and FSH secretion
  after administration, 523–530

**S**

Scanning mutational analysis, growth hormone, 254–257
Sepsis, growth hormone therapy, 213–214
Sertoli cells
  calcium transport induction by
    follicle-stimulating hormone, 517–520
  cAMP levels, germ cell-conditioned media
    effect, 500
  FSH receptor, 61
  preproenkephalin expression, 497–502
  transferrin secretion stimulation by germ
    cell factor, 539–542
Sertoli cell-secreted growth factor, tyrphostin inhibition, 511–515
Short stature, *see* Growth hormone, use to
  increase stature

## INDEX

Signal transduction
  pathways
    growth hormone, 152–153
    prolactin, 153–155
    transmembrane domain role, 301–304
Silver–Russell syndrome, growth hormone
    therapy, 198
Single-strand conformation polymorphism,
    49
Site-directed mutagenesis, mouse, 237–249
  clonal heterogeneity and totipotency
    maintenance, 239–241
  null alleles generated by gene targeting,
    241–248
    coelectroporation, 243–245
    double-hit gene replacement, 247–248
    hit and run, 245–247
Skeletal abnormalities, growth hormone
    therapy, 200–201
Slipped capital femoral epiphysis, growth
    hormone side effect, 220
Somatostatin, hypophysiotropic neuron ab-
    normalities, 492–494
Spermatogenesis
  differential gene expression, 79–95
    calmodulin kinase IV and calspermin
      mRNAs from contiguous DNA,
      89–92
    changes in calspermin and CaM kinase
      IV during germ cell development,
      92–93
    functions of calspermin and CaM kinase
      IV, 93–95
    testis-specific promoter, 91–92
  FSH, 64, 80
  FSH receptor, 70
Stop mutations, growth hormone receptors,
    patients with Laron's syndrome,
    170–172
Stress, *see* Baboons, psychoendocrine
    studies

### T

Testis
  anti-Müllerian hormone production, 16–18
  descent, AMH effect, 32
  differentiation, AMH effect, 31–32
  FSH receptor expression, 66, 68–71

  physiological response to FSH, 61–64
Testis-determining factor, AMH gene regu-
    lation, 19–21
Testosterone concentrations
  consequences of rank difference, 452
  dominant males, 450–452
  mechanisms mediating stress-induced
    suppression in baboons, 447–450
3–4–5 rule, 106–110
Thyroid hormone
  action, role of TR heterodimers, 408–409
  mechanisms of action, 404–408
  thyrotropin gene expression, *see* Thyro-
    tropin gene expression
Thyroid hormone receptor
  direct interaction with retinoid X recep-
    tors, 110–112
  heterodimers, thyroid hormone action
    role, 408–409
  homodimers, binding to TREs, 409
Thyroid hormone-response elements
  regulatory actions, 407
  TR homodimer binding, 409
Thyroid-stimulating hormone, 393
Thyrotropin
  function, 393
  gene expression, posttranscriptional regu-
    lation, 404
  regulation of synthesis and release by thy-
    roid hormone, 394
  subunit gene expression, models of nega-
    tive thyroid hormone regulation,
    409–411
Thyrotropin gene expression, thyroid hor-
    mone regulation, 393–411, *see also* Neg-
    ative thyroid hormone-response ele-
    ments
  $\alpha$ subunit, cDNA and gene, 394–396
  posttranscriptional TSH gene expression
    regulation, 404
  TSH subunit gene expression, negative
    thyroid hormone regulation, 409–411
Thyrotropin-releasing hormone
  electrophysiological responses, 346–347
  TRH-R mRNA levels and, 353–354, 356
Thyrotropin-releasing hormone receptor,
    341–361
  cDNA cloning, 344–348
  down-regulation, 341–342

# INDEX

RNA
  antisense, 347–348
  number, down-regulation, 350–352
  mRNA
    cAMP, 355
    endogenous, down-regulation, 351,
      353–356
    mechanism of down-regulation,
      356–360
    TRH effects on mouse, 358–360
    speculations and future directions,
      360–361
    structure, 348–350
    studies in *Xenopus* oocytes, 342–343
Totipotency, maintenance, 239–241
Transcription initiation sites, AMH promoter, 11–12
Transferrin, secretion by Sertoli cells, stimulation by germ cell factor, 539–542
Transforming growth factor-$\beta$ family, rela-
  tionship to anti-Müllerian hormone C
  terminus, 6, 8
Transgenic mice, chronic, unregulated pro-
  duction of AMH, 25, 27
Transmembrane domain, signal transduction
  role, 301–304
Trauma, growth hormone therapy, 211
L-Triiodothyronine, TRH-R number and,
  350–351
TSH, *see* Thyrotropin
TSH$\alpha$, subunit gene expression, regulation,
  397–400
TSH$\beta$
  cDNA and gene, 396–397

subunit gene
  expression, regulation, 397–400
  putative nTREs, 401–404
TTNPB, 102
Turner syndrome, growth hormone treatment, 194–196
Tyrosine hydroxylase, hypophysiotropic
  neuron abnormalities, 491–492
Tyrosine phosphorylation
  insulin-induced, 314–315
  involvement in insulin action, 327–328
  IRS-1, 324–326
  stimulation, 316
Tyrphostine, inhibition of Sertoli cell
  secreted growth factor stimulation of
  A431 cell growth, 511–515

## U

Uterus, c-*fos* mRNA, 477–480

## V

Vitamin D receptor, direct interaction with
  retinoid X receptors, 110–112

## X

*Xenopus* oocytes, TRH-R studies, 342–343

## Z

$(Zn^{2+} \cdot hGH)_2$ complex, 264–265

ISBN 0-12-571148-4

90065

9 780125 711487